MW00426664

PHYSICS FOR RADIATION PROTECTION

PHYSICS FOR RADIATION PROTECTION

JAMES E. MARTIN
School of Public Health
The University of Michigan

WILEY-VCH Verlag GmbH Co. KGaA

Library of Congress Card No.: Applied for
British Library Cataloging-in-Publication Data: A catalogue record for this book is available from the British Library

Bibliographic information published by
Die Deutsche Bibliothek
Die Deutsche Bibliothek lists this publication in the Deutsche Nationalbibliografie;
detailed bibliographic data is available in the Internet at <http://dnb.ddb.de>.

Printed in the Federal Republic of Germany
Printed on acid-free paper

Printing betz-druck GmbH, Darmstadt
Bookbinding Litges & Dopf Buchbinderei GmbH, Heppenheim

ISBN 0-471-35373-6

To

Barbara W. and Jenifer J. Martin

the memory of
Frank A. and Virginia E. Martin and JoAnne Martin Burkhart

and Graduates
of The University of Michigan
Radiological Health Program

You are in God's service.
—Mahalia Jackson

CONTENTS

PREFACE

This book is the outcome of teaching radiation physics to students beginning a course of study in radiation protection, or health physics. It is intended to be a text of basic physics concepts that health physicists and other radiation protection specialists need, presented at a level that can be understood by people with limited science background. In this context, I have resisted the more theoretical approach that one finds in many books on nuclear or modern physics as well as the temptation to try to cover everything in the field of health physics. Instead, I have attempted to provide in one place a comprehensive treatise of the major physics concepts required of practicing professionals in this broad interdisciplinary field of radiation protection and to present these concepts in an applied manner. Numerous real-world examples and practice problems are provided to demonstrate concepts and hone skills, and even though its limited uses are thoroughly developed and explained, some familiarity with calculus would help the reader to grasp some of the subjects more quickly. Armed with such knowledge, the reader can go on to study the specialized areas of radiation dosimetry, radiation measurement, and other related subjects.

Health physics problems require resource data. To this end, decay schemes and associated radiation emissions are included for about 100 of the most common radionuclides encountered in radiation protection. These are developed in the detail needed for health physics uses, such as measurement and dosimetry, and are cross-referenced to standard compendiums for straightforward use when these more-in-depth listings need to be consulted. Resources are also provided on activation cross sections, fission yields, fission-product chains, photon absorption coefficients, nuclear masses, and abbreviated excerpts of the *chart of the nuclides*. These are current from the National Nuclear Data Center at Brookhaven National Laboratory; the Center and its staff are truly a national resource and were indispensable to my goals.

The fifteen chapters in this compendium can be used in a variety of ways, for both instruction and reference. A selective use may be chosen for a one-semester course in radiation physics. The first two chapters constitute a review for those with a background in elementary physics but are designed to tie down basic elements of the atom as an energy system and the terminology commonly used in radiation protection. As such, they may be of most use for those with a minimal science background. The major discoveries in nuclear physics are revisited in Chapter 3 to provide additional insight and understanding of the fundamental structure and energy makeup of the atom. In combination, the first three chapters can be used to lay the basic groundwork for the atom as a source of the radiant energy addressed by the health physics profession.

Chapters 4 and 5 deal with nuclear interactions and the special condition of radioactive transformation (or disintegration) of atoms with excess energy, regardless of how acquired. These two chapters emphasize the processes by which changes occur in atoms and the amount of energy gained or lost; these define many of the sources that are addressed in radiation protection. Since radioactivity is a major part of radiation protection, it has been developed extensively in Chapter 5, and any course in radiation physics would be expected to include thorough coverage of the material. The natural sources of radiation and radioactive materials are developed in Chapter 6 primarily as reference material; however, the sections on radioactive dating and radon, although specialized, could be used to supplement the material in Chapter 5.

The interaction of radiation with matter and the resulting deposition of energy together constitute some of the more basic concepts of radiation protection and are covered in Chapter 7 along with the corollary subjects of radiation exposure and dose. Radiation shielding, also related to interaction processes, is described in Chapter 8 with details on various geometries encountered in radiation protection. Chapters 9 and 10, on activation products and fission, are also fundamental for understanding the unique sources of radiation and radioactivity and how these systems present radiation issues. These are followed by specialty chapters on nuclear criticality (Chapter 11); radiation detection and measurement (Chapter 12); applied statistics (Chapter 13); sources, interactions, shielding, and detection of neutrons (Chapter 14); and finally, x rays (Chapter 15).

A course in radiation physics that is based on this book could be expected to include substantial treatment of the material in Chapters 4, 5, 7, 8, 9, 10, and 12, with selections from the other chapters to develop needed background and to address specialty areas of interest to instructor and student. In anticipation of such uses, an attempt has been made to provide comprehensive and current coverage of the material in each chapter and accompanying data sets. Some repetition may thus occur, but only, it is hoped, in the context required for the subject under discussion.

Units

The units used in the physics of radiation protection have evolved over the hundred years or so encompassing the basic discoveries and applications of this science to radiation safety for workers and the public. They continue to do so with a fairly recent, but not entirely accepted, emphasis on Systeme International (SI) units. This trend is somewhat vexing because the U.S. standards and regulations for control of radiation and radioactivity have continued to use conventional units.

To the degree possible, this book uses fundamental quantities such as eV, transformations, time, distance, and the number of atoms or emitted particles and radiation to describe nuclear processes, primarily because they are basic to the concepts being described but partially to avoid conflict between SI and conventional units. Both sets of units are defined as they apply to radiation protection, but in general the more fundamental parameters are used. For the specific units of radiation protection such as exposure, absorbed dose, dose equivalent, and activity, text material and examples are generally presented in conventional units because the field is very much an applied one and is governed by regulations that are stated in conventional units; however, the respective SI units are also included where feasible. By doing so, I believe the presentation is clearer and relevant to the current conditions, but I recognize that sorting through this quandary is not easy and will surely continue for some time. In the meantime, students and practitioners need information that reflects current practice.

Errors

In an undertaking of this type, it is inevitable that undetected mistakes creep in and remain despite the best efforts of preparers and editors. There are at least two that remain to encourage the reader to be watchful and with the hope that they will be reported to the author (email: jemartin@umich.edu). Those who communicate errors found will become eligible for a prize, either a current edition, or if it should occur, a second edition that will contain the noted correction and an acknowledgment of the discoverer.

Acknowledgments

This endeavor has been possible because of the many contributions of my research associates Chul Lee, Arthur Ray Morton, Suellen Cook, and Ihab Kamel, who compiled and checked materials and did the expert computer work required, and my longtime secretarial associate Rebecca Pintar, who typed many of the sections several times. I am particularly indebted to Chul Lee, who hung with me through the final stages and contributed expertise, skill, and attention to detail with patience, persistence, and understanding. I am also indebted to my students, whose feedback shaped the teacher on the extent and depth of the physics materials necessary to function as a professional health physicist. My greatest satisfaction will occur if it helps you, the reader, understand and appreciate the basic physics of the exciting and rewarding field of radiation protection.

JAMES E. MARTIN, PHD, CHP

The University of Michigan, 1999

1

INTRODUCTORY CONCEPTS

Nothing in life is to be feared. It is only to be understood.
—Marie Curie

Radiation need not be feared, but it certainly deserves respect. Each perspective requires an understanding of the origins and effects of radiation in all forms of human endeavor. The profession that seeks such understanding and a proper degree of respect is known as *health physics*; it brings together science, technology, human values, and public policy to provide safe levels of protection for workers and the public from radiation, and once these safe levels have been met, to further optimize protection below them. This book seeks, in the spirit of Madame Curie's insights, to foster understanding of those principles of physics that are fundamental to radiation protection.

Determining or controlling radiation and any associated effects requires knowledge of:

- The sources, behavior, and energy of emitted radiations
- Deposition of radiant energy in various media to produce dose
- Biological effects that occur due to a given radiation dose and any associated risks, and
- Application of controls to change the relations

Schematically, this may appear as follows:

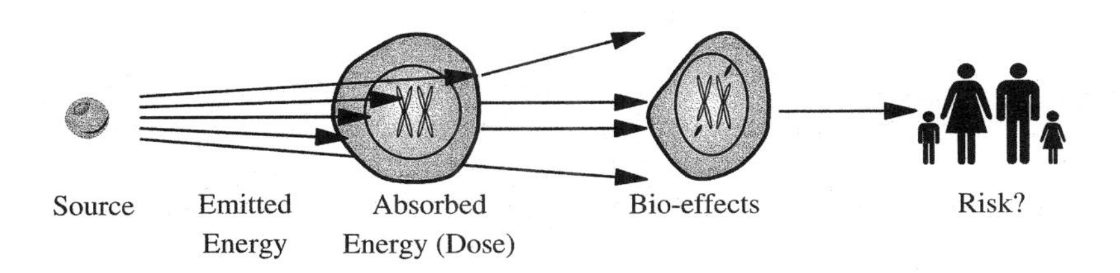

To formulate radiation protection policies to limit risks to persons, all of the parts of this interconnected system must be fully understood, but many aspects of radiation control begin with the physics of radiation emission and its absorption in matter, whether for determining the energy deposited to produce a dose or modifying it to change the dose. Since radiant energy, or radiation, is emitted from various devices and radioactive materials, its effective control is directly related to the unique properties of the radiation source, especially the energy emitted, why such energy is released, and the types and properties of the emitted radiation(s).

Radiation sources include x-ray machines and accelerators that emit radiation directly and radioactive materials that emit radiation of various types and vary over time. Electromagnetic radiation is produced by all of these sources, and it is important to examine its characteristics of production and the parameters that determine its energy and behavior. We look first at the structure of atoms and the concepts of energy followed by basic concepts that allow a reasonably good description of the atom, its dynamics, and its existence as an energy system. The reader who is familiar with these basic concepts can skip to a review of the checkpoints at the end of the chapter and proceed to Chapter 2, which lays the groundwork for energy concepts used in radiation protection.

STRUCTURE OF ATOMS

Radioactive materials are a common source of radiation. They are first of all, materials (i.e., they have volume, mass, etc.), but they have the additional property of emitting radiant energy as particles or electromagnetic energy. Atoms that are radioactive contain excess energy and therefore are unstable. This simple statement contains several important concepts:

- The structure of matter is determined by the atoms it contains
- Many atoms are stable (i.e., they neither can nor do they need to emit energy)
- Radioactive atoms are unstable and need to emit energy
- Energy is produced by atoms when they undergo change

Atoms contain three types of building blocks: protons, neutrons, and electrons. Modern theory has shown that protons and neutrons are made of more fundamental particles, or quarks, but it is not necessary to go into such depth to understand the fundamental makeup of atoms and how they behave to produce radiant energy. The number and array of neutrons and protons in atoms establish:

- What the element is
- Whether the atoms in the element are stable or unstable
- If unstable, how the atoms will emit energy (we deal with energy later)

The *proton* has a reference mass of about 1.0. It also has a positive electrical charge.

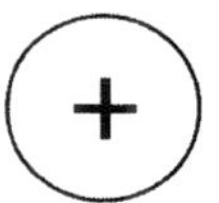

mass = 1 charge = +1

The *electron* is much lighter than the proton. Its mass is about 1/1800 of the proton and it has an electrical charge of -1.

Electron

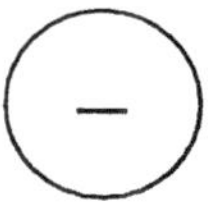

mass = 1/1800 charge = −1

The *neutron* is almost the same size as the proton, but slightly heavier. It has no electrical charge.

Neutron

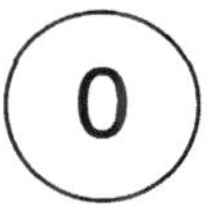

mass = 1 charge = 0

By themselves these basic building blocks are a bit boring. But if one started to put them together, which is what happened at the beginning of time, very important things become evident. First, if a proton is placed in the room, a free electron will join with it, and we get an atom:

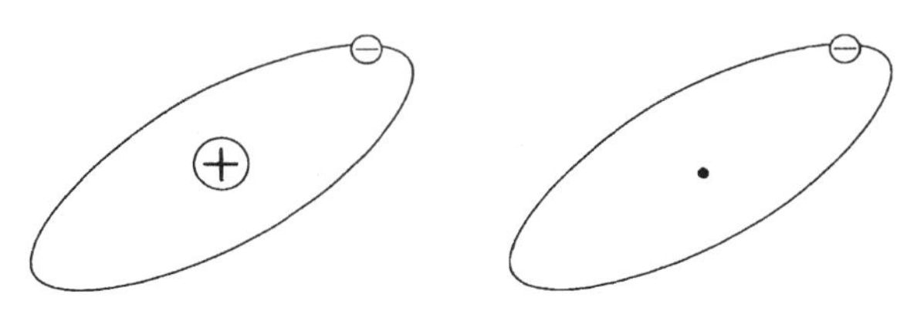

Schematic Closer to "actual"

The total atom (proton plus an electron) has a diameter of about 10^{-8} cm and is much bigger than the central nucleus, with a radius of about 10^{-13} cm. In fact, the "actual" representation must be imagined because of the very small sizes. Consequently, the atom is mostly empty space.

The resulting atom is electrically neutral. That is, the −1 charge on the electron that joins with the proton matches the +1 positive charge on the proton in the nucleus. This is one of the first principles of atoms: Nature calls for them to be electrically neutral (i.e., each proton in an atom is offset by an electron, and in any atom the number of protons and electrons will be the same). Electrons are common. There are zillions of them floating about, as we learn from crossing a carpeted room on a dry day and getting a shock when we touch the doorknob.

If a neutron is thrown out into the room, it continues to float around because it is electrically neutral, and in contrast to a proton, an atom does not form (i.e., it is just a free neutron subject to thermal forces of motion):

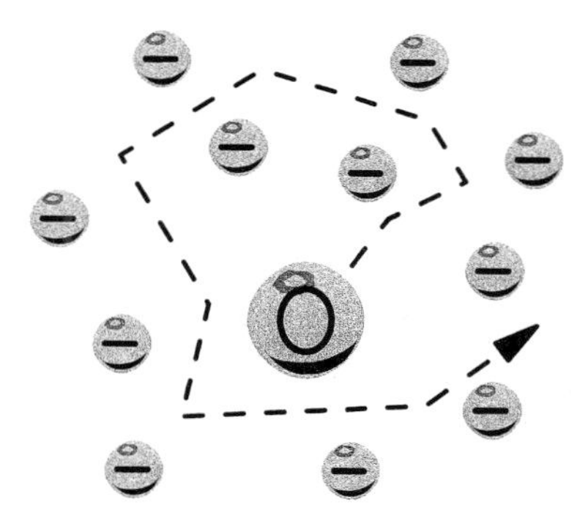

Similarly, if two neutrons are put into the room, they too just float around; they don't join together. They also ignore any electrons present. A neutron will move around in the air of the room and its motion will be governed by the room temperature. If left alone for awhile, a free neutron will undergo transformation (commonly referred to as *decay*) into a proton and an electron. In an unbound "free" state, the neutron, although not an atom (no orbiting electrons), will behave like one and emit a negatively charged electron:

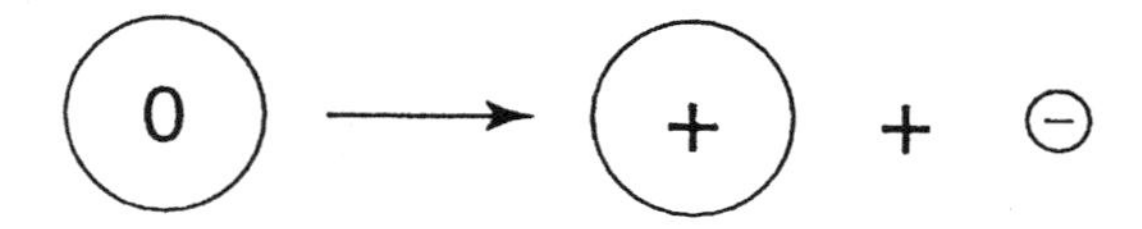

A neutron in the nucleus of an atom is, however, quite stable. It can combine with a proton to produce a stable atom:

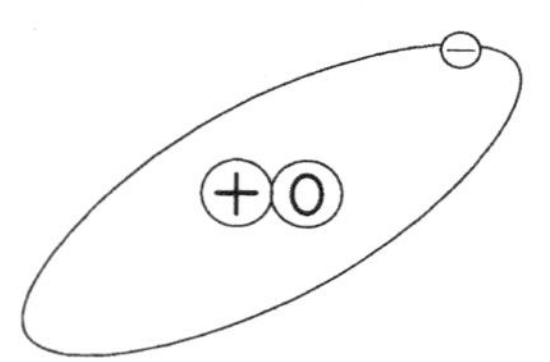

Note that we have the same electrically neutral atom we had before, but it now weighs about twice as much as the other one because of the added neutron mass;

and if we add another (a second) neutron, we get

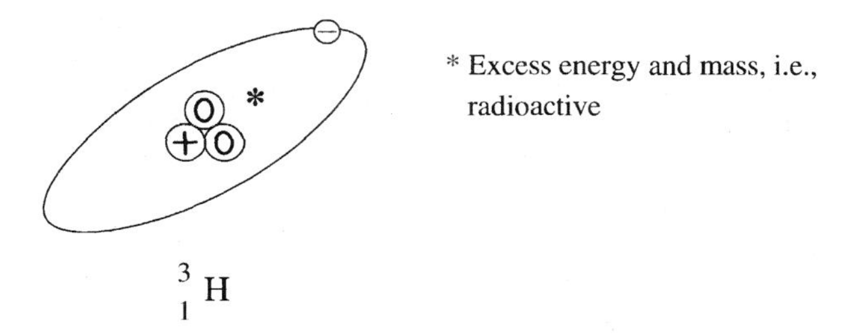

This atom is the same electrically neutral atom (one proton balanced by one electron) that we started with, but it weighs about three times as much, due to the two extra neutrons, and it has excess energy (i.e., it is unstable or radioactive). Each atom is different, but each has only one proton, which is balanced by one electron. Each atom is an atom of hydrogen because hydrogen is defined as any atom containing one proton. But because each atom has a different weight, we call them *isotopes* (Greek: *iso* = same; *tope* = place) of hydrogen to recognize their particular features. We give each of these three isotopes of hydrogen the following symbols:

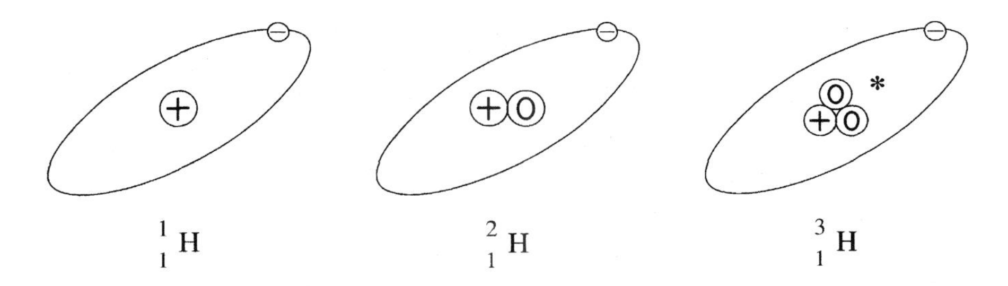

These symbols establish the nomenclature used to identify atoms: The subscript on the lower left denotes the number of protons in the atom; the superscript on the upper left refers to the mass number, an integer that is the sum of the number of protons and neutrons in the nucleus. It is common practice to leave off the subscript for the number of protons because the elemental symbol, H, defines the substance as hydrogen and by definition hydrogen has only one proton. The isotopes of hydrogen are identified as protium (or hydrogen), deuterium, and tritium; the first two are stable and exist in nature, but tritium is radioactive and will disappear through radioactive transformation unless replenished. Almost all elements exist, or can be produced, in several isotopes. A particular substance is often identified by its element and the mass number of the isotope present [e.g., carbon-14 (^{14}C), hydrogen-3 (^{3}H, or tritium)].

Two-Proton Atoms

If we try to put two protons together, they repel each other with such force at this very close distance that they won't stay together; thus an atom (actually, a nucleus) cannot be assembled from just these two particles.

But if a neutron is added, everything is fine and two electrons join up to balance the two plus (+) charges of the protons:

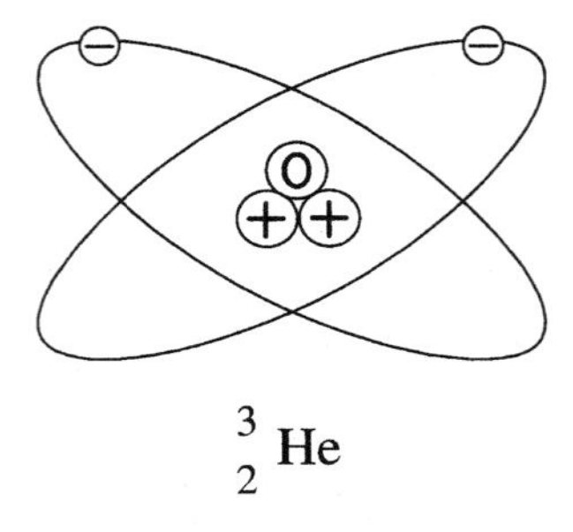

$^{3}_{2}\text{He}$

This atom is helium because it has two protons. It has a mass of 3 (two protons plus one neutron) and is written as helium-3 (^{3}He). Helium-2 (^{2}He) cannot be made because the repulsive force of the protons is so great at this close distance that it blows the nucleus apart. But the neutron adds a little more "glue" and redistributes the force of repulsion. Because neutrons provide a cozy effect, yet another neutron can be added to obtain ^{4}He:

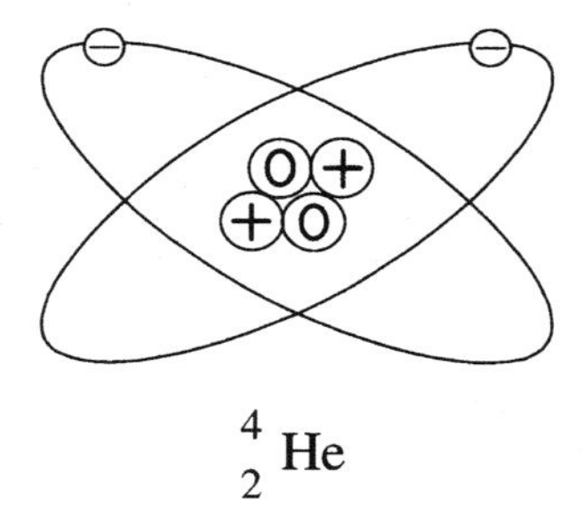

$^{4}_{2}\text{He}$

Although extra mass was added in forming ^{4}He, only two electrons are needed to balance the two positive charges. This atom is the predominant form of helium (isotope, if you will) on earth, and it is very stable (we will see later that this same atom, minus the two orbital electrons, is ejected from some radioactive atoms as an alpha particle, i.e., a charged helium nucleus).

If yet another neutron is stuffed into helium to form helium-5 (^{5}He), we have added too much mass for the atom to handle and it breaks apart very fast (in 10^{-21} s or so); it literally spits the neutron back out. There is just not enough room for the third neutron, *and* by putting it in we create a highly unstable atom. But as we observed for hydrogen and as we will see for other atoms, an added neutron (or proton) will often stay for quite awhile, and we have an unstable, or radioactive, atom due to the "extra" particle mass. This is why it is important to know the isotope of a given element.

Three-Proton Atoms

Following the foregoing principles, we can assemble atoms with three protons and three neutrons to form lithium-6 (^{6}Li), or with four neutrons, lithium-7 (^{7}Li). Note that we put in another orbit for the third electron. We had to add this orbit farther away because the first orbit can only hold two electrons. (There is an important reason for this, which is explained by quantum theory.)

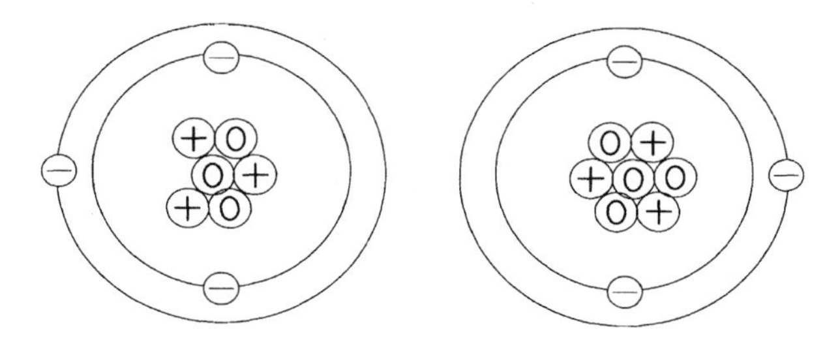

Atom Dimensions

Complete atoms with orbiting electrons have a radius of about 10^{-10} m. The nucleus alone has a radius proportional to $A^{1/3}$, where A is the atomic mass number of the atom in question, or

$$r = r_o A^{1/3}$$

The constant r_o varies according to the element but has an average value of about 1.3×10^{-15} m, or femtometers. The femtometer (10^{-15} m) is commonly referred to as a *fermi*, in honor of the great Italian physicist and nuclear navigator, Enrico Fermi. It is clear from these dimensions that the atom is mostly empty space.

NUCLIDE CHART

This logical pattern of atom building can be plotted in terms of the number of protons and neutrons in each, as shown in Figure 1-1. The plot has been extended to include all known atoms, or nuclides, to create a *chart of the nuclides*, a very useful compilation of data for radiation protection. The chart is also a useful way to describe radioactive transformation and nuclear interactions; therefore, this convention will be maintained throughout the book, although many authors plot them reversed, perhaps due to precedent.

The chart of the nuclides contains a lot of basic information on each element, how many isotopes it has (atoms on the horizontal lines), and which are stable (shaded) or unstable (unshaded). We will refer to this chart often as we develop the basic concepts of radiation physics because it contains valuable information for understanding each element and its various isotopes, including parameters needed in various calculations.

A good example of the information available from the chart of the nuclides is the line on the chart for six protons (carbon), and the lines for five protons (boron)

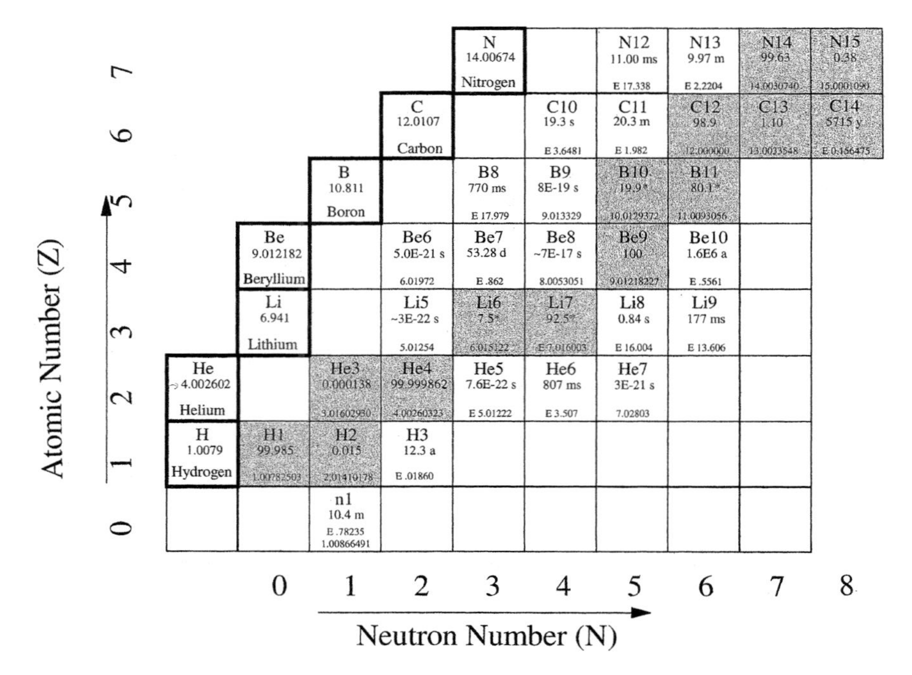

FIGURE 1-1. Part of the chart of the nuclides. [From General Electric (GE), 1996.]

and seven protons (nitrogen), as shown in Figure 1-2. Actually there are eight measured isotopes of carbon but ^{11}C, ^{12}C, ^{13}C, and ^{14}C are the most important. They are all carbon because each contains six protons, but each of the four has a different number of neutrons; hence they are isotopes and have different weights. Note that the two blocks in the middle for ^{12}C and ^{13}C are shaded, which indicates that these isotopes of carbon are stable, as are the shaded blocks for boron and nitrogen. The nuclides in the unshaded blocks (e.g., ^{11}C and ^{14}C) on each side of the shaded (i.e., stable) blocks are unstable simply because they don't have the right array of protons and neutrons to be stable (we use these properties later to discuss radioactive transformation). The dark band at the top of the block for ^{14}C denotes that it is a naturally occurring radioactive isotope, a convention used for other such radionuclides. The block to the far left contains information on naturally abundant carbon: It contains the chemical symbol, C, the name of the element, and the atomic weight of natural carbon, or 12.0107 g/mol, weighted according to the percent abundance of the two naturally occurring stable isotopes. The shaded blocks contain the atom percent (at %) abundance of ^{12}C and ^{13}C in natural carbon at 98.90 and 1.10 at %, respectively; these are listed just below the chemical symbol. Similar information is provided at the far left of the chart of the nuclides for all the elements (e.g., as shown in Figure 1-1).

The atomic weight of an atom is numerically equal to the mass, in unified mass units, or u, of the atom in question. The atomic mass is listed for all the stable isotopes (e.g., ^{12}C and ^{13}C) at the bottom of their respective blocks in the chart of the nuclides. The atomic mass of most of the stable and unstable isotopes of each of the elements can be obtained from the list in Appendix B.

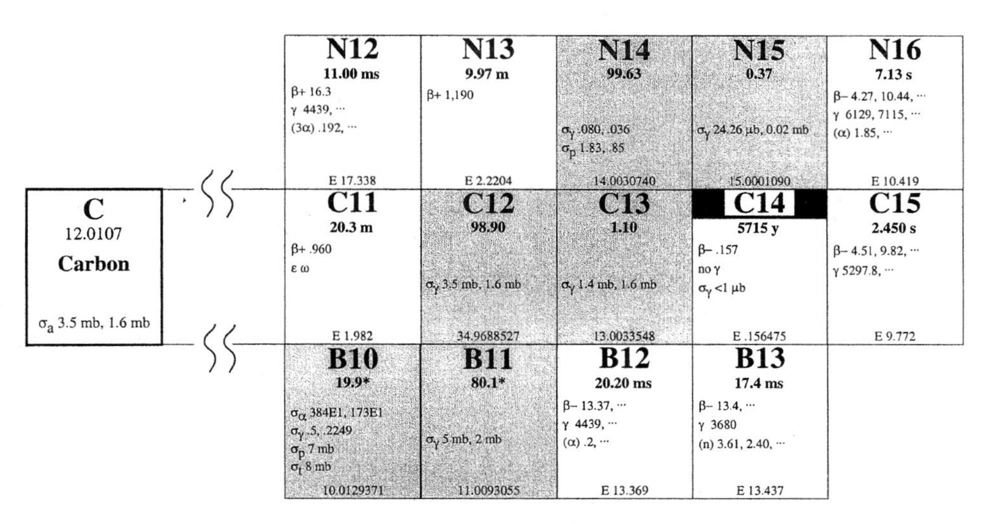

FIGURE 1-2. Excerpt from the chart of the nuclides for the two stable isotopes of carbon ($Z = 6$) and its two primary radioactive isotopes, in relation to primary isotopes of nitrogen ($Z = 7$) and boron ($Z = 5$). (Adapted from GE, 1996.)

If we keep combining protons and neutrons, we get heavier and heavier atoms but they obey the same general rules. The heaviest element in nature is ^{238}U, with 92 protons and 146 neutrons; it is radioactive. The heaviest stable element in nature is ^{209}Bi, with 83 protons and 126 neutrons (earlier charts show ^{209}Bi to be radioactive but with a very long half-life). Lead with 82 protons is much more common in nature than bismuth and for a long time was thought to be the heaviest of the stable elements; it is also the stable endpoint of the radioactive transformation of uranium and thorium, two primordial naturally occurring radioactive elements (see Chapter 6). The ratio of neutrons to protons is fairly high in heavy atoms because the extra neutrons are necessary to distribute the nuclear force and moderate the repulsive electrostatic force between protons in such a way that the atoms stay together.

There are other atoms heavier than and beyond uranium on the chart of the nuclides. These are referred to as *transuranic* (TRU) *elements* and can be produced by various nuclear interactions. All such atoms are unstable. As a group, TRUs warrant special consideration in radiation protection.

ATOM MEASURES: NUMBER, MASS, AND ENERGY

Many health physics problems require knowledge of how many atoms there are in common types of matter, the total energy represented by each atom, and the energy of its individual components, or particles that can be derived from the masses of the atoms and its particles. Avogadro's number and the atomic mass unit are basic to these concepts, especially the energy associated with mass changes that occur in and between atoms. The electron volt is another basic atom measure that is used to express energy of particles and emitted radiation.

Avogadro's Number

In 1811, an Italian physicist, Amedeo Avogadro, proposed three remarkable hypotheses which, although not accepted for some time, eventually provided the foundation for the atomic theory of chemistry. Avogadro assumed that:

1. Particles of a gas are small compared with the distances between them.
2. The particles of elements sometimes consist of two or more atoms stuck together to form *molecules*, which are distinguishable from atoms.
3. Equal volumes of gases at constant temperature and pressure contain equal numbers of atoms or molecules, whichever makes up the basic components of the gas.

The third assumption established the concept of atoms as basic units and that combining weights of substances contained a fixed number of these basic units. Avogadro had no knowledge of the magnitude of the number of molecules (atoms) in a mole volume of gas, only that the number was very large. Avogadro reasoned that the number of atoms or molecules in a mole of any substance is a constant, independent of the nature of the substance, and because of this insight, the number of atoms or molecules in a mole is called *Avogadro's number*, N_A. It has the following value:

$$N_A = 6.0221367 \times 10^{23} \text{ atoms/mol}$$

Example 1-1: Calculate the number of atoms of ^{13}C in 0.1 g of natural carbon.

SOLUTION: From Figure 1-2, the atomic weight of carbon is 12.0107 g and the atom percent abundance of ^{13}C is 1.10%. Thus

$$\text{number of atoms of } ^{13}\text{C} = \frac{0.1 \text{ g} \times N_A \text{ atoms/mol}}{12.0107 \text{ g/mol}} \times 0.011$$

$$= 5.515 \times 10^{19} \text{ atoms}$$

Atomic Mass Unit

Actual weights of atoms and constituent particles are extremely small and difficult to relate to, except that they are small. Elements are identified by name, and isotopes of an element are identified by name and mass number (e.g., ^{12}C). Thus a natural unit for expressing the masses of particles in individual atoms would be a mass unit that approximates the weight of a proton or neutron (since the sum of these is the mass number, one of these must be a mass unit). The *atomic mass unit* (amu), has been defined for this purpose. To be precise, the unit is the unified mass unit, denoted by the symbol u, but by precedent most refer to it as the atomic mass unit, or amu.

One amu, or u, is defined as $\frac{1}{12}$ the mass of the neutral ^{12}C atom. One mole of ^{12}C by convention is defined as weighing exactly 12.000000... g. All other elements and their isotopes are assigned weights relative to ^{12}C. The amu was originally defined relative to ^{16}O at 16.000000 g/mol, but ^{12}C has proved to be a better reference nuclide, and since 1962, atomic masses have been based on the unified mass scale, referenced to the mass of ^{12}C at exactly 12.000000... g.

Avogadro's number can be used to compute the mass of a single atom of ^{12}C. Since 1 mol of ^{12}C has a mass of exactly 12.000000 g and contains N_A atoms, it follows that the mass of one atom of ^{12}C is

$$m(^{12}\mathrm{C}) = \frac{12.000000 \text{ g/mol}}{6.0221367 \times 10^{23} \text{ atoms/mol}} = 1.9926482 \times 10^{-23} \text{ g/atom}$$

This mass is shared by six protons and six neutrons; thus the average mass of each of the 12 building blocks of the ^{12}C nucleus can be calculated simply by dividing the mass of one atom by 12. This quantity is defined as 1 amu, or u, and has the value

$$1 \text{ u} = \frac{1.9926482 \times 10^{-23} \text{ g}}{12} = 1.66054 \times 10^{-24} \text{ g}$$

which is close to the actual mass of the proton (actually, $1.6726231 \times 10^{-24}$ g) or the neutron (actually, $1.6749286 \times 10^{-24}$ g). In unified mass units, the mass of the proton is 1.00727647 u and that of the neutron is 1.008664923 u. Each of these values is so close to unity that the mass number of an isotope is thus a close approximation of its atomic weight. These useful parameters and other constants of nature and physics are included in Appendix A.

Measured masses of elements are some of the most precise measurements in physics and are accurate to six or more decimal places. As we will see in Chapter 2, mass changes in nuclear processes represent energy changes; thus these very accurate mass values, which are very useful for calculations of energy changes in atoms, are included in Appendix B.

Electron Volt

A practical unit is also necessary to characterize the energy in atoms, groups of atoms, or their constituent particles. The unit *electron volt* (eV) is used for this purpose. The eV is defined as the increase in kinetic energy of a particle with 1 unit of electric charge (e.g., an electron) when it is accelerated through a potential difference of 1 V. This can be represented schematically as

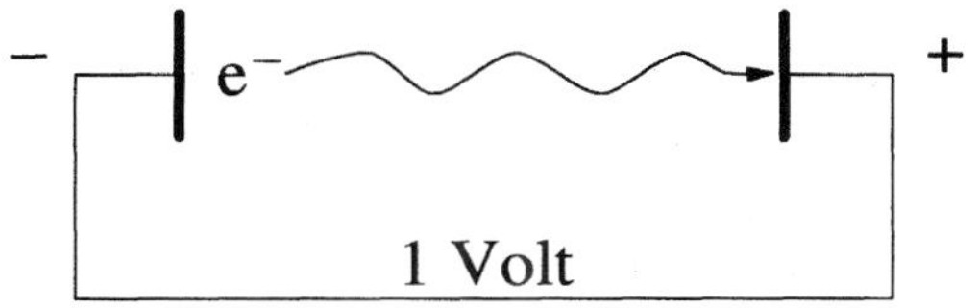

− e^- +

1 Volt

The electron in the figure will be repelled by the negatively charged electrode and attracted to the positively charged one. When it slams into the positive electrode it will have kinetic energy equal to 1 eV. This energy is calculated by multiplying the unit charge of the electron by the potential drop:

$$1 \text{ eV} = (q)(\Delta V) = (1.602177 \times 10^{-19} \text{ C})(1 \text{ V})$$
$$= 1.602177 \times 10^{-19} \text{ V} \cdot \text{C}$$
$$= 1.602177 \times 10^{-19} \text{ J}$$
$$= 1.602177 \times 10^{-12} \text{ erg}$$

This relationship, $1 \text{ eV} = 1.602177 \times 10^{-19} \text{ J} = 1.602177 \times 10^{-12}$ erg, is used frequently in health physics calculations of deposited energy in a medium. In absolute terms, 1 eV is not very much energy. In fact, the energy of atomic changes is commonly expressed in keV (10^3 eV) and MeV (10^6 eV). The concept of representing the energy of small particles by the energy they possess in motion is, however, very useful in describing how they interact. For example, one can think of the energy of a 1 MeV beta particle or proton as a unit-charged particle that gained an acceleration equal to being subjected to a jolt of 1 million volts of electrical energy.

Example 1-2: An x-ray tube accelerates electrons from a cathode into a tungsten target anode to produce x rays. If the electrical potential across the tube is 90 kV, what will be the energy of the electrons when they hit the target, in eV, joules, and ergs?

SOLUTION:

$$E = \text{eV} = 90{,}000 \text{ eV}$$
$$= 90{,}000 \text{ eV} \times 1.6022 \times 10^{-19} \text{ J/eV} = 1.442 \times 10^{-14} \text{ J}$$
$$= 1.44 \times 10^{-14} \text{ J} \times 10^{7} \text{ ergs/J} = 1.442 \times 10^{-7} \text{ erg}$$

CHECKPOINTS

These introductory concepts represent several important things that took physicists and chemists a long time to figure out. These are:

- An element is determined by the number of protons in each of its atoms. Hydrogen, by definition, has one proton. Helium has two protons, lithium has three, and so on.
- Isotopes are atoms of the same element but with different weights. Hydrogen (^{1}H), with a weight of 1, deuterium (^{2}H), with a weight of 2, and tritium (^{3}H), with a weight of 3, are isotopes of hydrogen. Similarly, ^{3}He and ^{4}He are isotopes of helium.
- Atoms left to themselves are electrically neutral. Electrons are attracted toward positively charged nuclei, where they orbit the nucleus in a bound energy state

with negative potential energy. The number of orbital electrons just balances the positive charge provided by the protons in the nucleus.

- Each atom has a nucleus that contains one or more protons (to establish the identity of the atom) and, except for hydrogen-1, one or more neutrons.
- The nucleus is very small compared with the electron orbits (or with anything else). The atom itself is about 10^{-8} cm in size; the nucleus, which is at the center of the atom, is less than 10^{-12} cm in diameter, so the atom is mostly empty space.
- The nucleus is very small and therefore very dense; it contains essentially all the mass of the atom because the protons and neutrons in the nucleus each weigh about 1800 times more than the electrons orbiting about it.
- According to Avogadro, the number of atoms (or molecules) in a mole of a substance is the same, given by Avogadro's number, $N_A = 6.0221376 \times 10^{23}$ atoms/mol.
- The unified mass unit, u, has a mass of 1.66054×10^{-24} g. The u, often referred to as the atomic mass unit, has a mass on the order of a neutron or proton (called nucleons because they are constituent parts of the nucleus of atoms); thus the mass number, A, of the isotope of an element is close to, but not exactly equal to, the atomic mass of the individual atoms.
- The masses of the elements are given in atomic mass units, u, which are close to the sum of Z protons and $(A - Z)$ neutrons in the nucleus. The mass of each proton is 1.007265 u and of each neutron, 1.008665 u; however, the total mass of each atom in an element is less than the sums of these individual masses, due to the binding forces, which reduce the potential energy of each. This net deficit of mass corresponds to the binding energy of the atoms of each element. This binding energy is not necessarily uniformly distributed between protons and neutrons, one of which is usually bound more or less tightly than the other, depending on the element.
- The electron volt is a reference amount of energy used to describe nuclear and atomic events, defined as the energy that would be gained by a particle with a unit charge when it is accelerated through a potential difference of 1 V. It is equivalent to 1.602177×10^{-19} J, or 1.602177×10^{-12} erg.

ACKNOWLEDGMENTS

This chapter was compiled with the substantial help of Chul Lee and Suellen K. Cook, both graduates of the University of Michigan Radiological Health Program.

REFERENCES AND ADDITIONAL RESOURCES

Chart of the nuclides, *Nuclides and Isotopes*, 15th ed., General Electric Company, San Jose, CA, 1996.

National Nuclear Data Center, Brookhaven National Laboratory, Upton, NY. Data resources are accessible through the Internet at *www.nndc.bnl.gov.*

PROBLEMS

1-1. How many neutrons and how many protons are there in (**a**) ^{14}C, (**b**) ^{27}Al, (**c**) ^{133}Xe, and (**d**) ^{209}Bi?

1-2. If one were to base the atomic mass scale on ^{16}O at 16.000000 atomic mass units (amu), calculate the mass of the ^{16}O atom and the mass of one amu. Why is its mass on the ^{12}C scale different from 16.000000?

1-3. Calculate the number of atoms in 1 g of pure hydrogen.

1-4. Hydrogen is a diatomic molecule or H_2. Calculate the mass of one molecule of H_2.

1-5. Calculate the radius of the nucleus of ^{27}Al in (**a**) meters and (**b**) fermi.

1-6. A 1 g target of natural lithium is to be put into an accelerator for bombardment of ^6Li to produce ^3H. Calculate the number of atoms of ^6Li in the target.

1-7. A linear accelerator is operated at 700 kV to accelerate protons. What energy will the protons have when they exit the accelerator in (**a**) eV and (**b**) joules? (**c**) If deuterium ions are accelerated, what will be the corresponding energy?

2

FORCES AND ENERGY IN ATOMS

I want to know the thoughts of God. The rest is just details.
—Albert Einstein

When we consider the energy of the atom, we usually focus on the energy states of the orbital electrons (referred to by many as *atomic physics*) or the arrangement of neutrons and protons in the nucleus (or *nuclear physics*). The energy states of electrons, protons, and neutrons in an atom constitute a bound system because nature forces the constituents of atoms toward the lowest potential energy possible; when they attain it they are stable, and until they do they have excess energy and are thus unstable, or *radioactive*.

There are four basic forces of nature that control the dynamics (i.e., position, energy, work, etc.) of all matter. The forces of nature range over some 40 orders of magniude that can be related to gravity, the weakest of the four, as

- Gravitational force between masses = G
- Weak force of radioactive transformation = $10^{24}G$
- Electromagnetic force between electric charges = $10^{37}G$
- Nuclear force between nucleons = $10^{39}G$

The energy states of particles in the atom are determined by two of the forces of nature:

1. The force of electrical attraction between the positively charged nucleus and the orbital electrons, which not only holds the electrons within the atom, but influences where they orbit
2. The nuclear force, which is so strong that it overcomes the electrical repulsion of the protons and holds the protons and the neutrons together to form the nucleus

Gravitational forces are insignificant for the masses of atom constituents, and the weak force is a special force associated with the process of radioactive transformation of unstable atoms.

The *nuclear force*, or *strong force*, is amazing and a bit strange. It exists only between protons and neutrons or *any* combination of them; consequently, it exists *only* in the nucleus of atoms. The nuclear force is not affected by the charge on neutrons and protons or by the distance between them. It is strongly attractive, so much so that it overcomes the natural repulsion between protons at the very short distances in the nucleus since it is about 100 times stronger than the electromagnetic force.

Nuclear Attraction

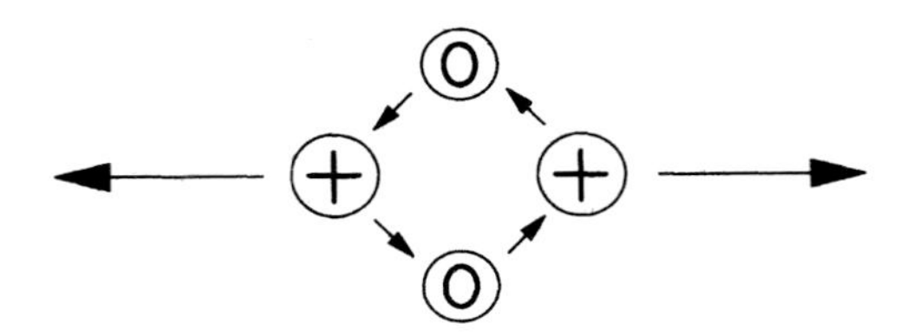

The nuclear force exists only in the nucleus—particles outside the nucleus are unaffected by it.

The *electromagnetic force*, on the other hand, exists between charged particles no matter where they are (a nucleus can also be thought of as a large charged particle, although it contains several protons, each of which has a unit positive charge). The electromagnetic force is inversely proportional to the square of the distance r between two particles with a charge of q_1 and q_2:

$$F_{em} = \frac{1}{4\pi\varepsilon_0}\frac{q_1 q_2}{r^2} = k_0 \frac{q_1 q_2}{r^2}$$

The constant k_0 is for two charges in a vacuum and has the value $k_0 = 8.9876 \times 10^9 \mathrm{N \cdot m^2/C^2}$. The charges are expressed in coulombs and r is in meters. This fundamental relationship, described by Coulomb's law, named after its developer, is called the *Coulomb force*. If q_1 and q_2 are of the same sign (i.e., positive or negative), F will be a repulsive force; if they are of opposite signs, F will be attractive. In any given atom, the electromagnetic force between the positive nucleus and the negative electrons largely establishes where the electrons will orbit.

Example 2-1: Calculate the electromagnetic force of repulsion between two protons, each of charge 1.6022×10^{-19} C, separated by a distance of 5×10^{-14} m.

SOLUTION:

$$F = 8.9876 \times 10^9 \mathrm{N \cdot m^2/C^2} \frac{(1.6022 \times 10^{-19}\ \mathrm{C})^2}{(5 \times 10^{-14}\ \mathrm{m})^2} = 9.24 \times 10^{-2}\ \mathrm{N}$$

An electron orbiting a nucleus experiences two forces: the electromagnetic force between the electron and the positively charged nucleus and the centrifugal force due to its angular rotation. The centrifugal force can be thought of as a force in which the electron "falls away" from the nucleus, whereas the centripetal force F_c is one that "pulls back" on the electron. It has a magnitude of

$$F_c = \frac{mv^2}{r}$$

where v is the velocity of the particle traveling in a radius r. The electrons travel in an orbit where the centrifugal force due to orbital rotation is just balanced by the attractive electromagnetic force exerted on the electrons by the positively charged nucleus.

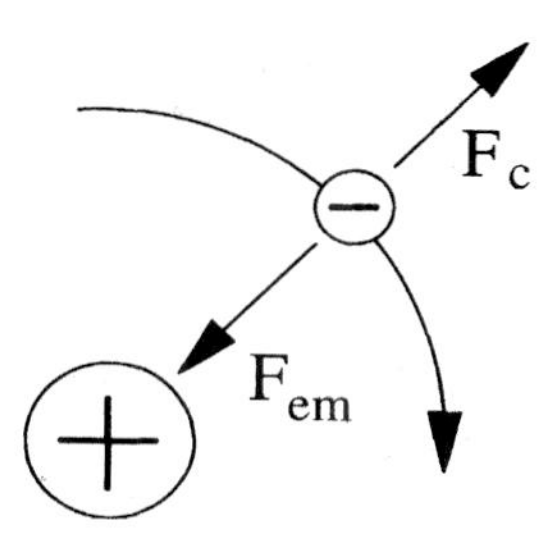

Actually, the position of an electron is dictated by other major factors determined by the quantum theory, which is discussed in Chapter 3. One major consequence of the quantum theory that we do need to mention here is that the electrons in their orbits have discrete, specific energy states for any given atom (we'll return to this later).

BASIC ENERGY CONCEPTS

Since atoms are arrays of particles bound together under the influence of the nuclear force and the electromagnetic force, the particles in atoms (or anywhere else for that matter) have energy states that are directly related to how the force fields act upon them. The concepts of energy and bound energy states of atoms lead naturally to the question: What is energy?

Energy is the ability to do work. Okay, what is work? *Work* is done when a force acts on a body through some distance. When a force of 1 newton ($kg \cdot m/s^2$) is applied to a 1 kg mass sufficient to move it a distance of 1 m, the amount of energy consumed (i.e., work done) is equal to 1 newton-meter or 1 joule, the SI unit of energy. The relation between energy and work, which appears to be a somewhat circular argument, can perhaps be better illustrated with a macroworld example of lifting a rock that weighs 1 kg (2.2 lb) from ground level onto a perch 1 m (about 3.1 ft) high.

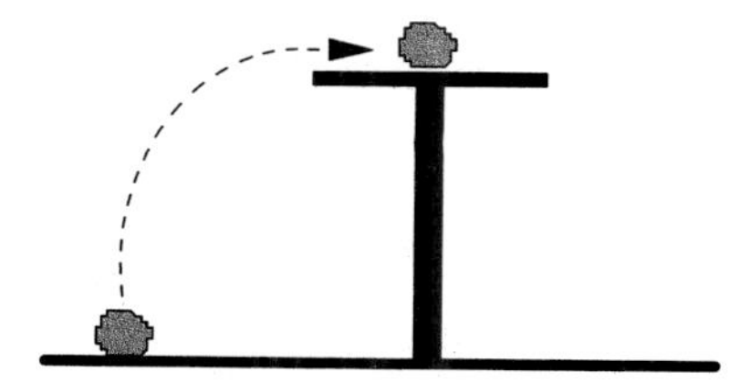

The work done by the lifter is obviously related to the amount of effort to lift the weight of the rock against the force due to gravity and how high it is to be lifted. The physicist would characterize this as

$$\text{work} = \text{force} \times \text{height of perch}$$

or

$$W = mgh$$

where m is the mass of the rock in kilograms (2.2 lb), g is the acceleration due to gravity (9.81 m per second per second), and h is the height in meters. In this example, the amount of work done would be

$$\text{work} = 1 \text{ kg} \times 9.81 \text{ m/s}^2 \times 1 \text{ m}$$
$$= 9.81 \text{ N} \cdot \text{m} = 9.81 \text{ J}$$

In this example, work is expressed in the energy unit of joules (i.e., energy and work are interrelated). The joule is admittedly a hard unit to think in, but the concept of lifting a mass of about 2.2 lb (1 kg) to about 3.1 ft (1 m) should be relatively familiar. The key concept is that an effort was expended against gravity and over a distance (i.e., work was done or energy was expended to get it there).

The basic concepts of energy, work, and position associated with this rock can be extrapolated to the atom, which contains particles in various energy states (i.e., with position influenced by a force). The rock on the perch can be thought of as "bound" with the ground. Since work was done to get it on the perch, it has positive energy relative to the ground by virtue of the work done on it to get it up there. If it were raised to another perch, 1 m higher, the same amount of work would need to be done again.

The rock on the perch contains energy in the form of stored work, or ability to do work. If the rock is pushed off the perch and allowed to fall under the force of gravity, the stored work would be recovered when it hit the ground.

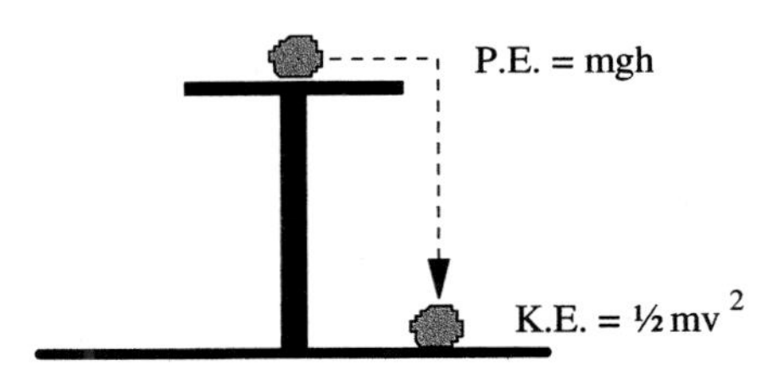

The stored energy (work = mgh) in the rock is called *potential energy*—it is setting there ready to do work. As the rock falls back to the ground it has *kinetic energy*, energy due to its motion. Potential energy is converted to kinetic energy as the rock falls under the influence of the gravitational force. The kinetic energy is delivered to the ground when it hits it. If you put your big toe in its path, you would easily experience the work done on your toenail. Kinetic energy is simply

$$\text{KE} = \tfrac{1}{2}mv^2$$

At any point above the ground the rock has both potential energy due to its height and kinetic energy due to its motion. Its total energy at any point is the sum of the two, or

$$E = mgh + \tfrac{1}{2}mv^2$$

where h and v are both variables. When the rock is at rest on the perch, all the energy is potential energy; when it strikes the ground, h is zero and all the potential energy that the rock had on the perch is converted to kinetic energy, or a total conversion of potential energy, mgh, into kinetic energy, $\tfrac{1}{2}mv^2$, or

$$mgh = \tfrac{1}{2}mv^2$$

which can be used to determine the velocity of the rock when it strikes the ground, called the *terminal velocity*:

$$v = \sqrt{2gh}$$

This expression can also be used to determine the velocity at any intermediate height and because of the conservation of energy, the relative proportions of potential energy and kinetic energy (always equal to the total energy) along the path of the rock.

Another good example of potential (stored) energy and kinetic energy is a pebble in a slingshot. Work is done to stretch the elastic in the slingshot so that the pebble has potential energy (stored work). When let go, the pebble is accelerated by the elastic returning to its relaxed position and it gains kinetic energy due to its motion, or velocity, v. The main point of both these examples is that a body (particle) with potential energy has a stored potential to do work; when the potential energy is released, it shows up in the motion of the body (particle) as kinetic energy, or, if you will, "active" energy.

We characterize potential energy as positive or negative. A rock on the perch above the ground (or in a slingshot) has positive potential energy that can be recovered as kinetic energy by letting it go. But suppose that the rock were in a hole 1 m deep instead:

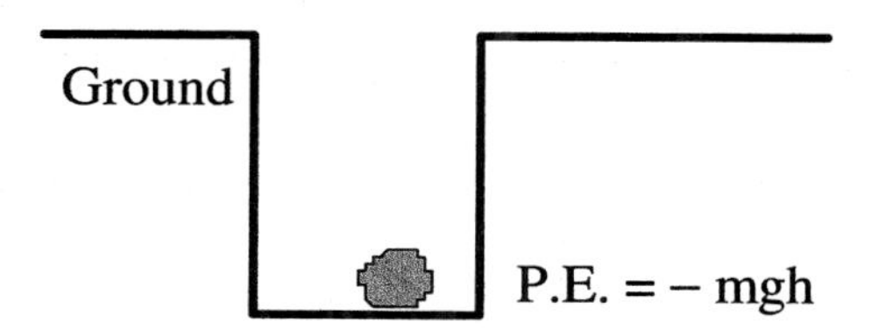

In this case the rock would have negative potential energy of $-mgh$ relative to the ground surface. Kinetic energy cannot be recovered from it, but instead, energy would need to be supplied (i.e., do work on it) to get it out of the hole and back to the ground surface. Of course, the deeper the hole, the more negative would be the potential energy and the more tightly bound it would be relative to the ground surface. If, however, it exists first on a ledge in the hole and then it fell even deeper into the hole, kinetic energy could be recovered from the distance it fell.

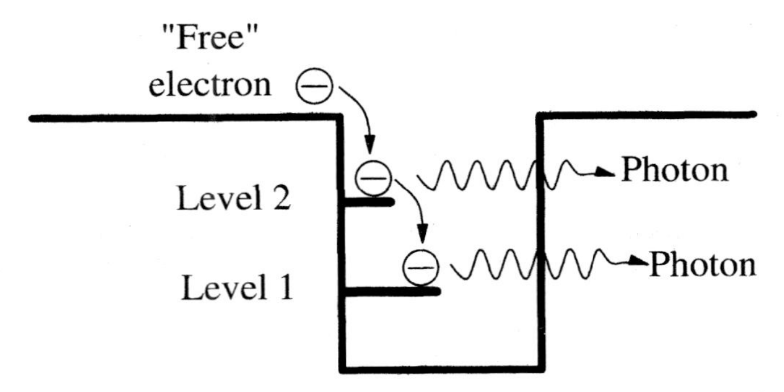

FIGURE 2-1. Conceptual representation of the potential energy states of electrons in an atom; when a free or less tightly bound electron becomes bound in a lower-energy orbit, the reduction in potential energy causes the emission of a photon to conserve energy.

Potential and Kinetic Energy in Atoms

It is relatively easy to relate to the potential energy of a raised rock and to the kinetic energy of a falling rock or a pebble ejected from a slingshot. But what do these have to do with atoms? Everything, it turns out—because the way atoms behave is determined by the potential energy states of its various particles and the way this potential energy can be converted to kinetic energy that appears as excited "radiation" either as a particle or as a wave.

Perhaps the most important concept for particles in atoms is that they will always have a total energy determined by their position and their motion, or

$$E_{tot} = \text{potential energy} + \text{kinetic energy}$$

An electron that exists "free" of the nucleus can be represented as having a neutral potential energy state. If, however, it becomes bound with a nucleus to form an atom, the various positions near the nucleus can be represented as perches in a hole as shown in Figure 2-1. The perches in the hole represent negative energy states that electrons would have in an atom because it would take work to get them out, that is, they possess negative potential energy relative to the region in which they would be considered "free" electrons. But if we have the electron go from the surface (the free state) to level 2, energy will have to be given up, and similarly on down to level 1.

CHECKPOINTS

Consideration of potential energy yields two very important concepts directly applicable to atoms:

- An electron bound to an atom has less potential energy than if it were floating around "free."
- The process of becoming bound causes the emission of energy as the particle goes to a lower potential energy state; the same amount of energy must be supplied to free a particle from its bound state.

We know this because we have to supply energy to get electrons out of atoms, and the amount of energy required to free a given electron is determined by the energy level it occupies. These energy changes have been observed by measuring the radiation energy emitted as photons (more about these later) when electrons are disturbed in atoms of an element such as hydrogen, helium, neon, nitrogen, and so on.

These concepts can be extended in a similar way to the nucleus; however, the energy released due to rearranging neutrons and protons to lower potential energy states can be in the form of an ejected particle (mass = energy) or pure electromagnetic radiation or both:

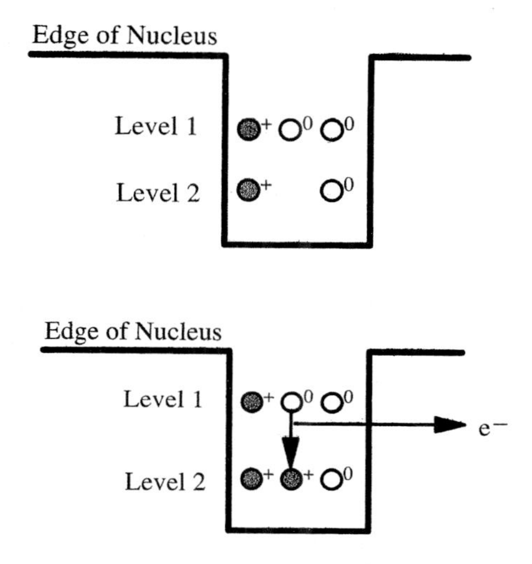

Processes that change the array of particles in the nucleus are fundamental to radioactive transformation, which is discussed in detail in Chapter 5.

RELATIVISTIC ENERGY

An atom's energy is governed by the concept that mass and energy are one and the same, as established by the special theory of relativity put forth by Einstein in 1905 to describe the role of mass and energy and how mass changes at high speeds. Einstein considered the relative dynamics of bodies in motion in different inertial systems and concluded that two key postulates govern their relative dynamics:

1. For two inertial systems moving at constant velocity relative to the other, the laws of physics are the same in each system.
2. The measured value for the velocity of light in a vacuum is the same for all observers and is independent of the relative velocity of the source of light and of the observer.

When Einstein deduced these postulates and applied them to the dynamics of moving bodies, he did so in search of fundamental laws of nature; he had no idea of atom systems, but his discoveries about mass and energy also govern the energetics of atoms and the radiation they emit. For example, we gave the rock potential energy by using our muscles to do the work necessary to lift it onto the perch. But where does a particle in an atom get its potential energy? Answer: from the mass that is available to be transformed into energy as an electron or a nucleon (proton or neutron) undergoes a change in its bound state (i.e., as it becomes more tightly bound it gives up energy). The atom with bound electrons has been determined to be slightly lighter than without; the loss of mass exactly matches the energy of the photon(s) emitted. This relationship between mass and energy is most valuable because energy changes in the atom can be determined by changes in mass, which can be measured with incredible accuracy (to six to eight decimal places).

Variation of Mass with Velocity

If the motion and dynamics of a particle are considered by an observer in an inertial system that is at rest and by another observer in motion relative to the stationary observer, each observer will obtain different values for length, velocity, and momentum. In his theory of special relativity, Einstein showed that all observers will find classical momentum principles to hold if the mass m of a body varies with its speed v according to

$$m = \frac{m_0}{\sqrt{1-(v^2/c^2)}}$$

where m_0, the rest mass, is the mass of the body measured when it is at rest with respect to the observer, and c is the speed of light in a vacuum. This relationship states that the mass of a body will increase with velocity, and it is necessary to treat momentum in a way that recognizes this mass change. In doing so, it is necessary to state *Newton's second law* precisely as he stated it: that the net force on a body is equal to the time rate of change of the momentum of the body, or

$$F = \frac{d(mv)}{dt}$$

If mass is treated as a variable quantity as Einstein's special theory of relativity requires, the relativistic specification of Newton's second law is

$$F = \frac{d(mv)}{dt} = \frac{d}{dt}\left(\frac{m_0 v}{\sqrt{1-v^2/c^2}}\right)$$

Since the basic concepts of force, work, and energy have been defined and derived from Newton's second law, the variation of mass with velocity is quite significant. At low speeds (less than 10% of the speed of light) the mass change of a body in motion can be assumed to remain constant, and Newton's second law reduces to the classical relationship that $F = ma$. But at higher speeds, many of which are associated with particles in atoms, allowance must be made for the fact that the mass of a body varies with its velocity.

RELATIVISTIC MECHANICS

In relativistic mechanics, as is classical mechanics, the kinetic energy (KE) of a body is equal to the work done by an external force acting through a distance to increase the speed of the body from zero to some value v. For a small change in distance ds, the kinetic energy is

$$\text{KE} = \int F\,ds$$

Using $ds = v\,dt$ and the relativistic expression of Newton's second law,

$$\text{KE} = \int_0^s \frac{d(mv)}{dt} v\,dt = \int_0^{mv} v d(mv) = \int_0^v vd\left(\frac{m_0 v}{\sqrt{1 - v^2/c^2}}\right)$$

If the term in parentheses is differentiated and the integration performed,

$$\text{KE} = \frac{m_0 c^2}{\sqrt{1 - v^2/c^2}} - m_0 c^2$$

which reduces to

$$\text{KE} = mc^2 - m_0 c^2 = (m - m_0)c^2$$

The kinetic energy (KE) thus represents the difference between the total energy mc^2 of the moving particle and the rest energy $m_0 c^2$ of the particle when it is at rest. This is the same logical relationship between potential energy and kinetic energy that we developed for a macroworld rock, but in this case the key variable is the change in mass of the moving body. Rearrangement of this equation yields Einstein's famous relation

$$E = mc^2 = \text{KE} + m_0 c^2$$

which shows that the total energy of a body is similar to the principles of classical physics, the sum of the kinetic and potential energy at any given point in time and space with the important distinction that its potential energy is a property of its rest mass; even when it is at rest, a body has an energy content of $E_0 = m_0 c^2$. In principle, the potential energy inherent in the mass of an object such as an atomic particle can be completely converted into another, more familiar form of kinetic energy, and vice versa; atomic and nuclear processes that occur in atoms do in fact convert mass into energy and pure energy into mass, and these mass changes are used in descriptions of nuclear processes that yield or consume energy.

Momentum and Energy

Since momentum is conserved, but not velocity, it is often useful to express the energy of a body in terms of its momentum rather than its velocity. To this end, if the expression

$$m = \frac{m_0}{\sqrt{1 - v^2/c^2}}$$

is squared, both sides are multiplied by c^4 and terms are collected,

$$m^2c^4 - m^2v^2c^2 = m_0^2c^4$$

Since $E = mc^2$, $E_0 = m_0c^2$, and $p = mv$, E is related to p as follows:

$$E^2 = (pc)^2 + E_0^2 = (pc)^2 + (m_0c^2)^2$$

Effects of Velocity

First, it can be shown that the following expression for kinetic energy of a particle is universally applicable for any velocity, including bodies that move at ordinary speeds as well as those at relativistic speeds:

$$\text{KE} = mc^2 - m_0c^2 = \frac{m_0c^2}{\sqrt{1 - v^2/c^2}} - m_0c^2$$

Expanding this expression by the binomial theorem yields

$$\text{KE} = m_0c^2\left[\left(1 + \frac{v^2}{2c^2} + \frac{3v^4}{8c^4} + \cdots\right) - 1\right]$$

If the velocity of a particle, v, is $0.1c$ or less, this expression reduces to $\text{KE} = \frac{1}{2}mv^2$, which means that the rest mass can be used to calculate the kinetic energy of a particle with an error of less than 1%. Also, when v is $0.1c$ or less,

$$m \simeq m_0 \qquad \text{and} \qquad p \simeq m_0v$$

When v approaches c, E (or mc^2) and KE are very large compared with m_0c^2, and the relativistic equations reduce to the following simplified form:

$$p \simeq \frac{E}{c} \qquad \text{and} \qquad \text{KE} \simeq E$$

If $v/c > 0.99$, $E = pc$ with an error of less than 1%; that is, at such velocities the total energy E is almost all kinetic energy since the rest-mass energy, although still present, is negligible by comparison.

Natural Limit

Einstein's discoveries also yield a natural limit for particle dynamics. For a particle that contains a positive rest mass, m_0, to achieve a velocity v equal to c, its rest mass would increase to

$$m = \frac{m_0v}{\sqrt{1 - v^2/c^2}} = \frac{m_0}{0} = \infty$$

which is not possible; therefore, the velocity of light is a natural limit that no particle with mass can reach. Since it is impossible to provide infinite mass for a material particle, it is impossible for it to move with a velocity equal to that of the speed of light (in a vacuum). If, however, it is assumed that the rest mass $m_0 = 0$, then $m = 0/0$ when $v = c$. This is an indeterminate quantity mathematically. Thus only those particles that have zero rest mass can travel at the speed of light. When such a "particle" exists,

$$m_0 = 0, \qquad E = pc, \qquad \text{KE} = E$$

That is, the particle has momentum and energy but no rest mass. According to classical mechanics, there can be no such particle. But according to Einstein's special theory of relativity, a particle with such characteristics is indeed a reality; it is called a *photon*. A few examples illustrate these important relationships for nuclear constituents.

Example 2-2: What is the maximum speed that a particle can have such that its kinetic energy can be written as $\frac{1}{2}mv^2$ with an error no greater than 0.5%?

SOLUTION: The kinetic energy (KE) when $m_0/m = 0.005$,

$$\frac{\text{KE} - \frac{1}{2}m_0v^2}{\text{KE}} = 0.005$$

but

$$\text{KE} = m_0c^2\left(\frac{1}{\sqrt{1 - v^2/c^2}} - 1\right) = m_0c^2\left[\left(1 + \frac{v^2}{2c^2} + \frac{3v^4}{8c^4} + \cdots\right) - 1\right] \approx \frac{1}{2}\left(\frac{m_0v^2}{0.995}\right)$$

if higher-order terms are neglected, thus

$$v \simeq 0.082c$$

Example 2-3: Calculate the velocity of an electron that has a kinetic energy of 2 MeV (rest-mass energy = 0.511 MeV).

SOLUTION:

$$\text{KE} = \frac{m_0c^2}{\sqrt{1 - v^2/c^2}} - m_0c^2$$

$$2\text{ MeV} = \frac{0.511\text{ MeV}}{\sqrt{1 - v^2/c^2}} - 0.511\text{ MeV}$$

$$v = 0.979c$$

Example 2-4: Compute the effective mass of a 2 MeV photon.

SOLUTION:

$$E_{\text{photon}} = m_{\text{eff}}c^2$$

For E in joules, mass in kilograms, and the speed of light in m/s:

$$2 \text{ MeV} \times 1.6022 \times 10^{-13} \frac{\text{J}}{\text{MeV}} = m_{\text{eff}}(3 \times 10^8 \text{ m/s})^2$$

$$m_{\text{eff}} = \frac{3.2044 \times 10^{-13} \text{ J}}{(3 \times 10^8 \text{ m/s})^2} = 3.56 \times 10^{-30} \text{ kg}$$

Mass–Energy

A most useful quantity is the energy equivalent of masses on the atomic and nuclear scales, as expressed in amu. The energy equivalent of the atomic mass unit, u, calculated above, is

$$E = m_0 c^2 = \frac{1.66054 \times 10^{-27} \text{ kg/u}(2.99792458 \times 10^8 \text{ m/s})^2}{1.6021892 \times 10^{-13} \text{ J/MeV}} = 931.502 \text{ MeV/u}$$

A similar calculation for the rest-mass energy of the electron yields

$$E = m_0 c^2 = 9.10953 \times 10^{-31} \text{ kg}(2.9979 \times 10^8 \text{ m/s})^2 = 8.187 \times 10^{-14} \text{ J}$$

$$= \frac{8.187 \times 10^{-14} \text{ J}}{1.6022 \times 10^{-13} \text{ J/MeV}}$$

$$= 0.511 \text{ MeV}$$

The energy equivalent of the electron mass is expressed in various nuclear processes in which electron masses are converted to 0.511 MeV photons; other processes occur in which photons of sufficient energy are converted to electron masses.

BINDING ENERGY OF NUCLEI

The equivalence of mass and energy, and the law of conservation of mass–energy, take on special significance for particles bound into an atom such that their potential energy states are less than they would be if they existed as separate particles (i.e, the rest mass of a combined system is less than the sum of the separate masses that comprise it). If two masses, m_1 and m_2, are brought together to form a larger mass M, it will hold together only if

$$M < m_1 + m_2$$

In this circumstance a quantity of energy E_b is released when the two masses become bound and the rest mass of the system is decreased, or

$$m_1 + m_2 \longrightarrow M + E_b \text{ (released)}$$

where E_b is the amount of energy released in binding the masses together; it is calculated (as shown in Examples 2-5 and 2-6) simply by subtracting the mass of the bound system from the sum of the masses of the individual particles in their unbound states, or

$$E_b = (m_1 + m_2 - M)c^2 = \Delta mc^2$$

E_b is called the *binding energy* because it is responsible for holding the parts of the system together. It is also the energy that must be supplied to break M into separate masses, m_1 and m_2:

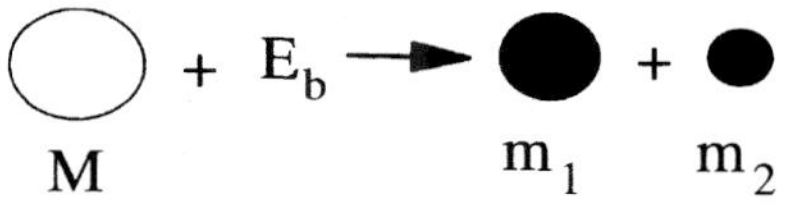

$$M + E_b \longrightarrow m_1 + m_2$$

The potential energy of the bound particles is negative due to the forces that hold the masses in the system together. These relationships are perhaps the most important consequences of Einstein's mass–energy relation.

Example 2-5: Find the binding energy of the deuterium atom.

SOLUTION: From Appendix B, the sum of the masses of the individual particles is

1.00727647 u	(proton)
1.00866491 u	(neutron)
0.00054858 u	(electron)
2.01648996 u	(total)

which differs from the measured mass of deuterium at 2.01410178 u by 0.00238818 u; therefore, the binding energy holding the deuterium atom together is

$$E_b = 0.00238818 \text{ u} \times 931.502 \text{ MeV/u} = 2.2246 \text{ MeV}$$

which is the amount of energy released (as a photon) when a proton and a neutron are combined to form a deuterium atom; it is also the amount of energy required to break deuterium into a proton and a neutron, which has been demonstrated experimentally. The binding energy of the orbital electron is ignored implicitly in this example; however, it is only about 13.6 eV, which is negligible compared to 2.225 MeV.

Example 2-6: Find the binding energy per nucleon for tritium, which contains one proton, one electron, and two neutrons.

SOLUTION: The constituent masses of the tritium atom are:

mass of proton	= 1.00727647 u
mass of neutrons	= 2 × 1.008664923 u
mass of electron	= 0.0005485799 u
Total	= 3.025155 u

which is larger than the measured mass of ^{3}H at 3.016049 u; thus the net mass difference is 0.0091061 u, and the total binding energy is

$$E_b = 0.009106 \text{ u} \times 931.502 \text{ MeV/u} = 8.48 \text{ MeV}$$

Since tritium (^{3}H) contains three nucleons, the average binding energy for each nucleon (E_b/A) is

$$\frac{E_b}{A} = \frac{8.48 \text{ MeV}}{3} = 2.83 \text{ MeV/nucleon}$$

Appendix B contains a listing of total binding energy for important isotopes of each of the elements calculated this same way. The binding energy per nucleon, E_b/A, which can also be calculated from the data in Appendix B, is shown in Figure 2-2 for each element; this plot provides interesting information on the structure of

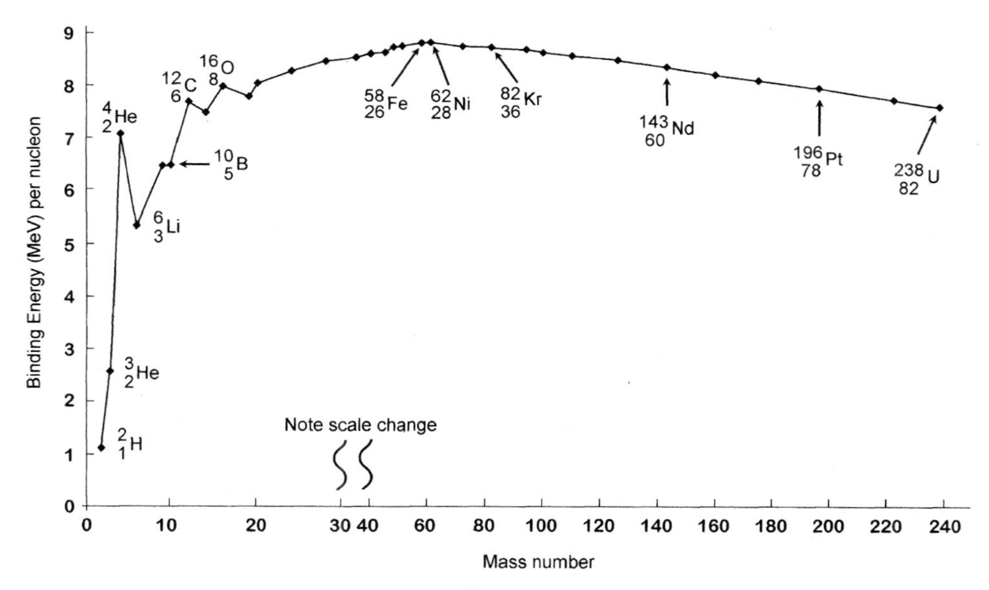

FIGURE 2-2. Curve of binding energy per nucleon versus atomic mass number (plotted from data in Appendix B).

matter. The curve in Figure 2-2 has several significant features:

- The elements in the middle part of the curve are the most tightly bound, the highest being ^{62}Ni at 8.7945 MeV/nucleon and ^{58}Fe at 8.7921 MeV/nucleon.
- Certain nuclei—^{4}He, ^{12}C, ^{16}O, ^{28}Si, and ^{32}S—are extra stable because they are multiples of helium atoms (mass number of 4), with one important exception, ^{8}Be, which breaks up very rapidly (10^{-16} s or so) into two atoms of ^{4}He.
- Fission of a heavy nucleus such as ^{235}U or ^{239}Pu produces two lighter nuclei (fragments), which according to Figure 2-2 are more tightly bound, resulting in a net energy release; fusion of light elements such as ^{2}H and ^{3}H produces larger atoms that are very tightly bound, which results in a greater energy release per unit mass.

Q-VALUE

Calculations of energy changes due to nuclear reactions which involve a net change in the nuclear mass are called *Q-values*. The *Q*-value represents the energy that is necessary to balance a given nuclear reaction; it is the amount of energy that is gained when atoms change their bound states or that must be supplied to break them apart in a particular way. For example, the calculation in Example 2-5 could be written

$$^{1}_{1}\text{H} + ^{1}_{0}\text{n} \rightarrow ^{2}_{1}\text{H} + Q$$

where Q represents the net energy change. In this case, Q is positive and 2.2246 MeV would be released in forming the deuterium atom. Q also represents the energy that must be added to the nucleus of ^{2}H to break it into a proton and a neutron, which in this case would be represented by the reaction

$$^{2}_{1}\text{H} + Q \rightarrow ^{1}_{1}\text{H} + ^{1}_{0}\text{n}$$

In this latter case Q would have a negative value, indicating that energy must be supplied to break the atom apart.

SUMMARY

The dynamics of all objects in the universe from subatomic particles to stars and planets are governed by four fundamental forces of nature: these are the gravitational, electromagnetic, weak, and nuclear (or strong) forces. These four forces have field strengths that vary by 40 orders of magnitude. The gravitational force field of object masses extends over all space, as do electromagnetic force fields of charged objects; and as far as is known, the weak force and nuclear force exist only between nuclear particles.

Einstein's theory of special relativity is applicable to atom systems where particles move at high speeds and hence undergo interchangeable mass–energy processes. Although Einstein's concepts are fundamental to atomic phenomena, they are even more remarkable because when he stated them in 1905 he had no idea of atom systems, and no model of the atom existed. He had deduced the theory in search of the basic laws of nature that govern the dynamics and motion of objects. Einstein's discoveries encompass Newton's laws for the dynamics of macroworld objects, but doesn't break down (as Newton's does) for microworld objects at velocities approaching that of the speed of light.

Einstein suggested that the laws of nature are discoverable but may be obscured by subtleties that require new ways of thinking if they are to be recognized as reflected in his statement that "subtle is the Lord, but vicious he is not." He believed that there was a more fundamental connection between the four forces of nature, and he sought, without success, a unified field theory to elicit an even more fundamental law of nature. Even though his genius was unable to find the key to interconnect the four forces of nature, or perhaps to describe a unified force that encompassed them all, his brilliant and straightforward mass–energy concepts provided the foundation for later descriptions and understanding of the origins of atomic and nuclear phenomena that are so basic to radiation protection.

ACKNOWLEDGMENTS

This chapter was compiled with the help of Suellen K. Cook, Arthur Ray Morton, and Chul Lee, all graduates of the University of Michigan Radiological Health Program.

PROBLEMS

2-1. A rowdy student in a food fight threw a fig (weight of 200 g) so hard that it had an acceleration of 10 m/s^2 when it hit the receptor. What "fig newton" force was applied?

2-2. Calculate the mass of an electron with a velocity that is (**a**) 10% of the speed of light and (**b**) 99% of the speed of light.

2-3. Calculate the kinetic energy of a proton that has a velocity of 0.8c.

2-4. Calculate the kinetic energy of a neutron that has momentum of 200 MeV/c.

2-5. What is the velocity of a proton with a kinetic energy of 200 MeV?

2-6. Calculate the velocity required to double the mass of a particle.

2-7. Calculate the multiple of the rest mass for an electron accelerated from rest through a potential difference of 15 million volts, and compare it with that for a proton under the same conditions.

2-8. Show that for practical purposes all electrons with kinetic energy of 5 MeV or more are traveling at essentially the speed of light.

2-9. Calculate from the masses in the mass table of Appendix B the total binding energy and the binding energy per nucleon of (**a**) beryllium-7, (**b**) iron-56, (**c**) nickel-62, and (**d**) uranium-238.

2-10. The binding energy of ^3H is 8.48 MeV, which is distributed between a proton and two neutrons. The binding energy of ^3He is 7.72 MeV, which is smaller due presumably to the coulombic repulsion between the two protons. If this is the reason, what would be the separation distance between the two protons in the ^3He nucleus?

3

MAJOR DISCOVERIES IN RADIATION PHYSICS

Chance only favors the prepared mind.
—Louis Pasteur

Radiation physics can be dealt with quite naturally in two parts: atomic physics and nuclear physics. The basic concepts upon which these rest were developed over a period of about 50 years that began just as the nineteenth century was ending. This short period saw a remarkable series of discoveries that quite literally changed the world forever.

Physical laws for ordinary phenomena were well understood by the year 1890. The dynamics of everyday objects were described precisely by Newton's laws of motion. The experimental facts of static electricity and magnetism were incorporated in Maxwell's famous equations which were well used and accepted. Electric fields were reasoned to have a magnetic field equal and opposite to each, and vice versa, and light had been shown to consist of electromagnetic waves. The first and second laws of thermodynamics were known in the form now used for heat engines of all kinds. Matter was thought to consist of atoms, but physicists believed that the recent kinetic theory of gases could be used to describe their motion similar to that of tiny billiard balls. Physics was thought to be intact, with the possible exception of more precise measurement of the basic constants.

The following discoveries, which occurred in a period of less than 20 years, shook the foundation of this confidence and set a new course for physics as well as humankind:

- Roentgen's discovery of x rays (1895)
- Becquerel's discovery of radioactivity shortly thereafter (1896)
- Thomson's discovery of the electron (1897)
- Planck's basic radiation law (1900)
- Einstein's theory of special relativity (1905)
- Rutherford's alpha-scattering experiments and the atomic nucleus (1911)
- Bohr's model of the atom (1913)

These phenomena and the information they provided made it clear that atoms were not solid little balls, but had structure and properties that required new and elaborate means to describe. We pick up three major threads in this tapestry of what is called modern physics: discoveries that determined the substructure of atoms and their energy states, the fundamentals of quantum emissions of radiation, and combinations of these to describe the structure and dynamics of the atom system. The insight required to formulate fundamental theories to resolve conflicts between observed phenomena and prevailing ideas in physics not only explained the phenomena but provided new discoveries of the laws of nature that led to even greater discoveries. These insights are presented as a means of understanding physics concepts that are basic to radiation science and their importance to physics as well.

GREAT DISCOVERIES FROM SIMPLE TOOLS

Einstein's *gedanken experiments* on the laws of physics in different inertial frames led to the monumental, but simple relation between mass and energy. His inspiration for the concepts of special relativity appears to have come from his daily observations of a large clock tower and surrounding buildings as he rode the train, a moving inertial system, to and from his work in the telegraph office. These principles, which were developed before any model of the atom was known, are directly applicable to describing the energetics of atoms as bound systems.

The discovery of x rays, radioactivity, and the electron were prompted by the study of gases. Since it was necessary to enclose a gas to study it, two simple tools were used: glass-blowing techniques and the vacuum pump, which had recently been developed. Such experiments eventually took the form of creating an electric potential across electrodes embedded in closed glass tubes; these studies led to Roentgen's discovery of x rays in 1895 and Thomson's discovery of the electron in 1897. The simplest form of this experimental tube, called a cathode-ray tube, is shown in Figure 3-1.

When a cathode-ray tube is filled with a gas, a discharge will occur, emitting light containing wavelengths characteristic of the gas in the tube. But when such a tube is evacuated to very low pressure (made possible with a good vacuum pump), the glow changes and diminishes, and at a pressure of about 10^{-3} atm, the tube produces a luminous glow filling the tube, as in a neon sign. Below 10^{-6} atm, the

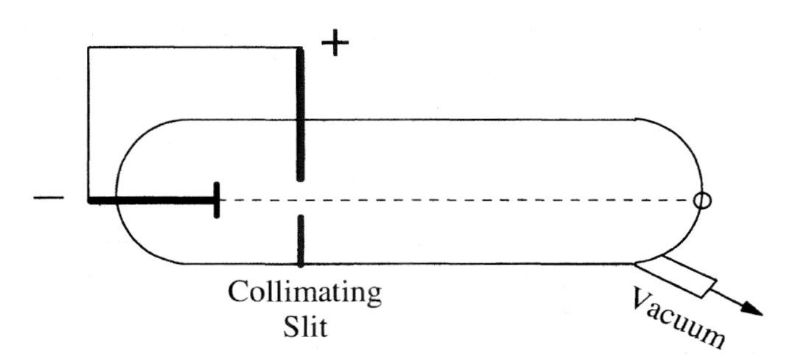

FIGURE 3-1. Cathode-ray tube.

negative electrode, or cathode, emits invisible rays that propagate through the nearly empty space in the tube. These emanations were quite logically called cathode rays. Although the rays by themselves are invisible, they make their presence known by producing a blue-green phosphorescence in the walls of the glass tube. The rays have notable characteristics:

- They travel in straight lines, projecting a glow on the end of the tube. Obstructions in their path cast a shadow.
- They can penetrate a small thickness of matter. If a "window" of thin aluminum or gold foil is built into the end of the tube, the rays produce luminous blue streamers in the air outside.
- They carry a negative electric charge, as evidenced by an electroscope placed in their path becoming negatively charged.
- They are deflected by both electrostatic and magnetic fields.
- They have considerable kinetic energy; metal obstacles put in the path of the rays glow brightly.

Discovery of X Rays

Two other simple tools—photographic plates and light-emitting phosphors such as zinc sulfide and barium-platinocyanide—contributed to the discovery of x rays. These tools, in combination with a highly evacuated cathode-ray tube (called a *Crooke's tube*) and the prepared mind of Wilhelm Conrad Roentgen, set the stage for this major discovery. One evening in December 1895, in an attempt to understand the glow produced in such a tube, he covered it with opaque paper. Then one of those events that triggers great minds to discovery happened: He had darkened his laboratory to better observe the glow produced in the tube, and in the dim light he noticed flashes of light on a barium-platinocyanide screen that happened to be near the apparatus. Because the tube (now covered with black paper) was obviously opaque to light emitted from the tube, he realized that the flashes on the screen must be due to emissions from the tube because they disappeared when the electric potential was disconnected.

Roentgen called the emissions x rays to denote their "unknown" nature, and in a matter of days went on to describe their major features. The most startling property was that the rays could penetrate dense objects and produce an image of the object on a photographic plate, as shown in Figure 3-2, the classic picture of the bones in his wife's hand. He also discovered that x-radiation could produce ionization in any gas through which it passes, a property that is used to measure x-ray intensity. The x rays could be reflected, refracted, and diffracted; they are a form of electromagnetic radiation like light, only of much shorter wavelength.

Very few discoveries have been as important to human existence as Roentgen's x rays. The rays were used almost immediately in diagnosis and treatment of disease, uses that are even more common a century later, and new applications continue. Roentgen never considered patenting this remarkable invention; his was truly a gift to science and humankind. X rays are discussed more completely in Chapter 15, but at this stage it is important to highlight their major role in discoveries so basic to radiation physics.

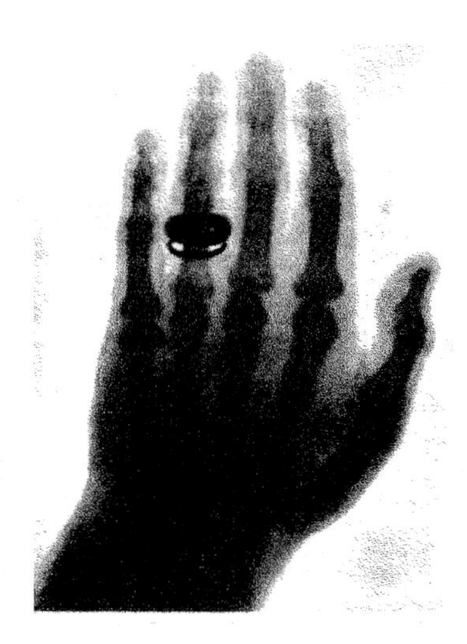

FIGURE 3-2. X-ray photograph of Frau Roentgen's hand.

Discovery of Radioactivity

The discovery of x rays appears to have triggered the discovery of radioactivity. Since x rays appeared to emanate from the portion of the Crooke's tube that glowed the brightest, Becquerel postulated that fluorescence or phosphorescence might be the source. To test the theory, he placed a salt of potassium uranyl sulfate on a tightly wrapped photographic plate and exposed it to sunlight to determine if the induced fluorescence perhaps contained Roentgen's x rays. When the plate was developed the outline of the crystal was clearly visible, which he assumed was due to fluorescence induced in the crystal. Apparently, he prepared several other photographic plates to repeat the experiment, but since the sky was cloudy for several days he placed the plates in the back of a drawer with the uranium crystals still attached. Even though the crystals had not been exposed to light, he developed the plates anyway (perhaps for use as a reference comparison for his earlier observation) and discovered that it had the same clear outline of the crystal on the plate as before. Aha! The images were much like those produced by Roentgen's rays, but since the crystals had not been exposed to sunlight, they could not be due to fluorescence or phosphorescence. Becquerel decided that the radiations affecting his plates originated from within the uranium salt itself and were spontaneous. Luckily, he had chosen a crystal that contained uranium; had he used a different fluorescent crystal among the many available, he would not have observed the image and probably would have just concluded that phosphorescence was not the source of Roentgen's rays.

Marie S. Curie and her husband, Pierre, studied this new property of materials and named it *radioactivity*. Her discovery of radium and polonium and her work on describing radioactivity far surpassed Becquerel's work. Nevertheless, Becquerel is credited with the discovery of radioactivity, an intrinsic property of certain substances. These events are discussed in further detail in Chapter 6.

Three major types of radioactive emissions were determined, as shown in Figure 3-3, by placing a well-collimated radium source in an evacuated chamber

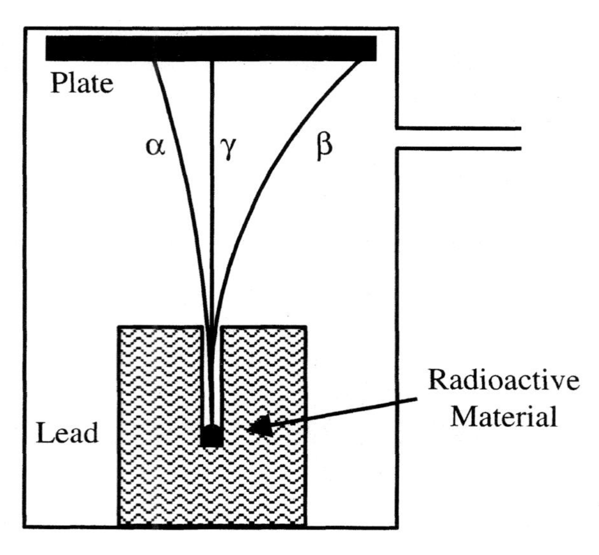

FIGURE 3-3. The three radiations from radioactive materials and their paths in a magnetic field perpendicular to the plane of the diagram.

and observing the patterns produced on a photographic plate by varying a magnetic field perpendicular to the path of the emitted "rays." If a weak magnetic field is applied, an undeviated spot is observed in the center of the phosphoric plate due to neutral rays (referred to as *gamma rays*), and a second spot to the right which exhibits the same deflection as that of negatively charged electrons (these were called *beta rays*). If the weak magnetic field is replaced by a strong field in the same direction, the undeviated spot appears again; however, no spot for the β rays is observed since the strong field deflects these rays completely off the plate. But the strong field produces a third spot slightly deflected to the left, indicating positively charged particles that were called *alpha rays*. The strong field is required because the mass of the positively charged particles is much greater than that of the negatively charged particles. Clearly, the α particles (later established to be helium ions) were much more massive than β particles, which were shown to be high-speed electrons. These radiations are not emitted simultaneously by all radioactive substances. Some elements emit α rays, others emit β rays, and γ rays may or may not be emitted.

In a series of comprehensive experiments on the nature of radioactivity, Madame Curie established that the activity of a given material is not affected by any physical or chemical process such as heat or chemical combination, and that the activity of any uranium salt is directly proportional to the quantity of uranium in the salt. Further work by Rutherford and Soddy clearly established that radioactivity is a subatomic phenomenon (see Chapter 6). Since helium nuclei and electrons were emanated from radioactive atoms, all atoms must be made up of smaller units. They could no longer be considered little round balls, and a theory of atomic structure that incorporated these discoveries was needed. Formulation of an adequate theory required two fundamental concepts not yet discovered: Planck's quantum theory

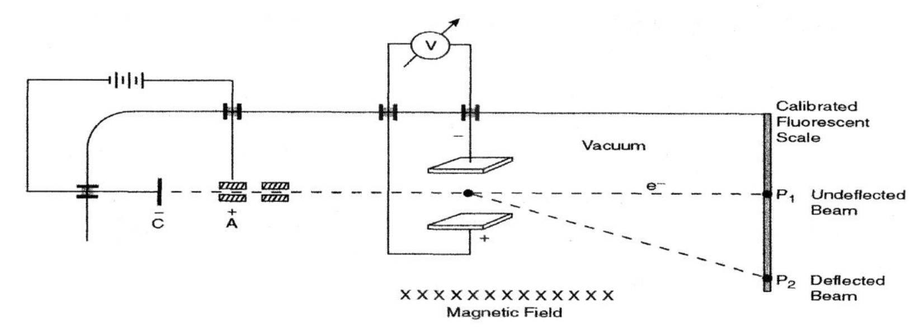

FIGURE 3-4. J. J. Thomson's apparatus for measuring the e/m value of cathode rays by establishing the electric field required to exactly cancel the deflection (shown by the scale glued onto the end of the tube) induced by the magnetic force field.

of electromagnetic radiation and Einstein's equivalence of mass and energy (as discussed in Chapter 2).

Discovery of the Electron

In other work with cathode rays, J. J. Thomson established that they are in fact small "corpuscles" with a negative electric charge. Thomson used a specially designed cathode-ray tube to investigate and quantitate the deflection ("deflexion" per Thomson) of the rays by electric and magnetic fields, as shown in Figure 3-4.

A narrow beam of cathode rays was directed between two plates which contained a uniform vertical electric field. A uniform magnetic field was established in the horizontal direction by an external electromagnet. The magnetic field deflected the rays upward; however, the electric field could be adjusted to cancel this "deflexion" exactly, yielding a net vertical force of zero. Thomson's work succeeded because he was able to establish a good vacuum in the tube. In 1883 (some 14 years earlier), Hertz had subjected the rays to an electric field but did not observe a deflection because the rays ionized gas still in the tube and the ions quickly neutralized the electric field between the plates.

The measured values of the field strengths allowed Thomson to calculate the ratio e/m_e in terms of known and measurable quantities, although his value was too small by a factor of about 2. The best modern value for the charge-to-mass ratio of an electron is

$$\frac{e}{m_e} = 1.758805 \times 10^{11} \text{ C/kg}$$

The same value of e/m was measured for cathode rays no matter what element was used for the cathode, suggesting that the rays were an intrinsic component of all elements. Thompson compared his measured e/m values to those of ionized hydrogen, the lightest element. The values suggested that the charge on cathode rays was either very large or that the mass of the hydrogen atom was about 1800 times larger than the mass of the cathode rays. Thomson chose the latter: that is, that the rays contained a unit charge and were individual "corpuscles" of matter that are approximately 1836 (the modern value) times smaller than hydrogen. He called these *electrons*.

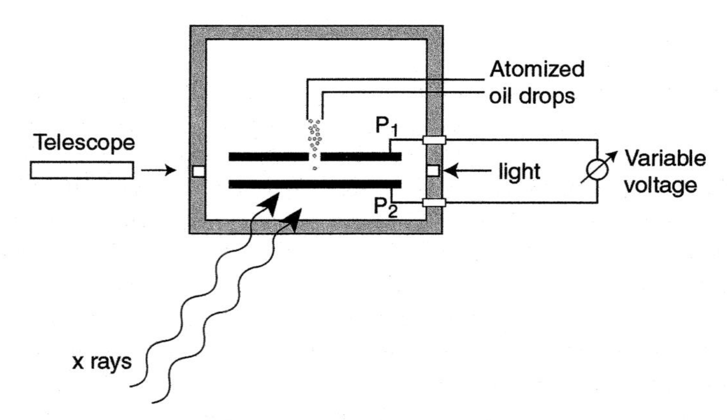

FIGURE 3-5. Millikan's oil-drop experiment. Illuminated drops could be held in space by adjusting the electric field between plates P_1 and P_2; the drops appeared as starlike dots when viewed by an external telescope.

Deflection experiments with electrons in electric or magnetic fields could determine the ratio e/m_e, but not e or m_e separately. If either is known, however, the other could be calculated from Thomson's measurements. Accurate determination of the unit charge of the electron is of most value because of its wide use in atomic phenomena.

Measuring an Electron's Charge

In 1909, R. A. Millikan performed a landmark experiment to measure the charge on the electron. He observed the motion of individual droplets of oil in an apparatus like that shown in Figure 3-5. Millikan used oil droplets rather than water to avoid uncertainties introduced by partial evaporation during the measurement. Oil droplets from an atomizer settle through a small hole in the plate into a uniform electric field, which can be adjusted to modify the fall of a given droplet. The oil drops, illuminated by a light beam, look like tiny bright stars when viewed through a telescope. The droplets move slowly under the combined influence of their weight, the buoyant force of the air, and the viscous force opposing their motion. The oil droplets usually have a negative charge from the atomizer, thus the electric field will affect their motion. Additional ions, both positive and negative, can be produced in the space with x rays (another value of Roentgen's discovery).

When a spherical body falls freely in a viscous medium, it attains a terminal velocity v_g such that the gravitational settling force mg is just equal to the viscous force kv_g

$$mg = kv_g$$

where k is a constant of proportionality from Stokes' law and the mass of the droplets was calculated using the measured free-fall velocities.

The experiment allowed Millikan to isolate an oil droplet in free fall. He would then switch the electric field on, which usually caused the droplet to undergo acceleration upward and attain a new terminal velocity, which was measured. By careful adjustment of the electric field strength, the droplet could be held in midair, which corresponded to the settling force and was a function of known and measured quantities. Millikan found that the charges of a large number of droplets could be expressed as integral multiples of a unit of charge, which he identified as the magnitude of the charge of a single electron. The best result obtained by Millikan was $e = 1.592 \times 10^{-19}$ C because he had used the wrong value for the viscosity of air (even the most meticulous researcher can make a mistake, but he still received a Nobel prize for this most important determination). The best modern result is

$$e = 1.60217733 \times 10^{-19}\ \mathrm{C}$$

This charge is the fundamental quantum of charge; electric charges of all particles found in nature are always 0, $\pm e$, $\pm 2e$, $\pm 3e$, and so on. Millikan could also now calculate the mass of the electron from measured values of e/m_e. The modern value for the mass of the electron is

$$m_e = 9.10953 \times 10^{-31}\ \mathrm{kg}$$

FIRST CONCEPT OF THE ATOM

Atoms were first proposed by philosophy, not physics, as indivisible units. Early Greek philosophers had reasoned that if a small piece of matter were continually subdivided, one would eventually obtain a particle so small that it could not be divided further and still have the same properties. They called these "pieces" atoms. Some 2000 years later, John Dalton proposed a similar atomic theory of matter in an attempt to consolidate observations that chemicals combined in multiples of whole numbers, or definite proportions. These observations suggested to Dalton that all matter is made up of elementary particles (atoms) which retain their identity, mass, and physical properties in chemical reactions. He visualized that these "atoms" linked up in chemical reactions in a way that resembled a hook-and-eye connection that could of course be "hooked" and "unhooked" when materials combined or disassociated, as shown in Figure 3-6. Dalton's formulation is an important landmark in the development of modern atomic physics.

J. L. Gay-Lussac added additional evidence of the discreteness or atomic nature of matter with his *law of combining volumes*: When two gases combine to form a

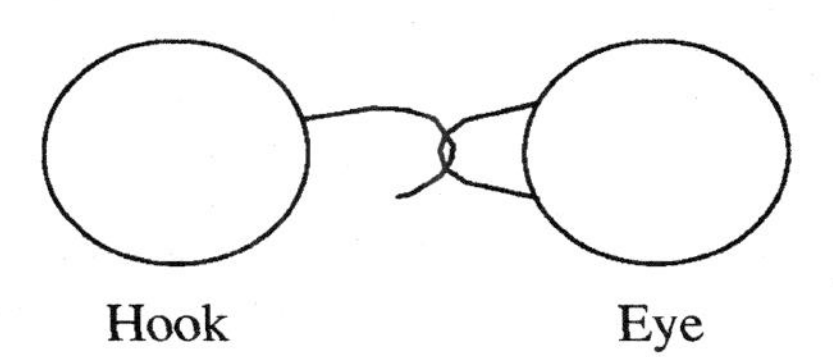

FIGURE 3-6. Dalton visualized that atoms linked up in ways that resembled the hooks and eyes used on clothing.

third, the ratios of the volumes are ratios of integers. For example, when hydrogen combines with oxygen to form water vapor, the ratio of the volume of hydrogen to that of oxygen is 2 : 1.

In 1811, an Italian physicist, Amedeo Avogadro, proposed three remarkable hypotheses which although not accepted for some time, eventually provided the foundation for the atomic theory of chemistry. Avogadro assumed that:

- Particles of a gas are small compared with the distances between them.
- The particles of elements sometimes consist of two or more atoms stuck together to form *molecules*, which are distinguishable from atoms.
- Equal volumes of gases at constant temperature and pressure contain equal numbers of atoms or molecules, whichever makes up the basic components of the gas.

The third assumption established the concept of atoms as basic units and that combining weights of substances contained a fixed number of these basic units. Avogadro had no knowledge of the magnitude of the number of molecules (atoms) in a mole volume of gas, only that the number was very large.

The first significant step in understanding and quantifying Avogadro's basic number came from Faraday's study of electrolysis. Faraday, an English physicist, established that a fundamental relationship exists between Avogadro's number, the unit of charge on the electron, and the amount of total charge necessary to deposit a gram equivalent weight of a metal on the cathode of an electrolysis experiment, as shown in Figure 3-7. Two electrodes, a positively charged anode and a nega-

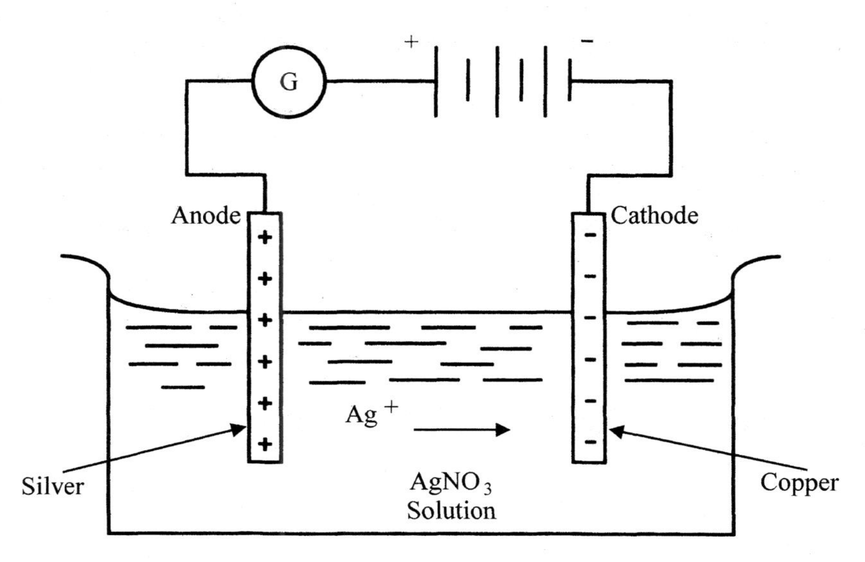

FIGURE 3-7. Typical electrolysis arrangement. Faraday showed that the charge (measured by the galvanometer, *G*) required to deposit one gram equivalent weight of any metal (in this case, silver) on the copper was always 96,485 C.

tively charged cathode, are immersed in a solution and connected electrically to a source of potential difference through a meter G, which measures current. If the anode is silver, the cathode copper, and the solution contains silver nitrate in water, the copper cathode will gradually become silver-plated and the silver anode will decrease in mass. Evidently, silver atoms are being transferred from the anode to the cathode. Under the influence of the electric current, a neutral silver atom, Ag, breaks away from the anode and goes into the solution as a positive ion, Ag^+, leaving behind a single electron that is free to flow through the connecting circuit. The electric field drives the positive ion to the cathode, where it becomes a neutral atom again by acquiring an electron, perhaps the same one freed up when the Ag^+ ion was released from the anode. The same phenomenon occurs for other metals, which means that atoms contain a basic unit that breaks off in electrolysis and is transmitted through the circuit to match up with each atom being deposited on the cathode. This discrete quantity, which is separable from an atom, is, of course, the same electron as that discovered by Thomson in his study of cathode rays.

Faraday showed in measurement after measurement that no matter what metal was used for the anode, it took exactly 96,485 C to deposit 1 gram equivalent weight (or *mole*) of it on the cathode. This quantity, called the *Faraday*, is related to the unit of electrical charge by Avogadro's number as follows:

$$F = N_A e$$

where F is 1 Faraday, N_A is Avogadro's number of atoms in 1 gram equivalent weight (or mole), and e is the charge on each electron liberated to flow through the circuit. Since the Faraday was well established by Faraday's work, it would be possible to determine Avogadro's number or the charge on the electron if either were known. Thus Millikan's determination of the electron's charge in his oil drop experiment produced the important constant along with Faraday's constant for determining Avogadro's number. Later work on the scattering of x rays from crystals would provide a more precise determination of Avogadro's number, but from Faraday's work, it was now clear that every atom contains electrons that are exactly alike, each carrying a definite amount of negative charge, and that each neutral atom contains just enough electrons to balance whatever positive charge it contains.

CHECKPOINTS

The Faraday, Avogadro's number, and the charge on the electron are three of those constants known as the fundamental constants of nature. Modern values of each are:

$$F = 96{,}485 \text{ C/mol}$$

$$N_A = 6.0221367 \times 10^{23} \text{ atoms/mol}$$

$$e = 1.60217733 \times 10^{-19} \text{ C}$$

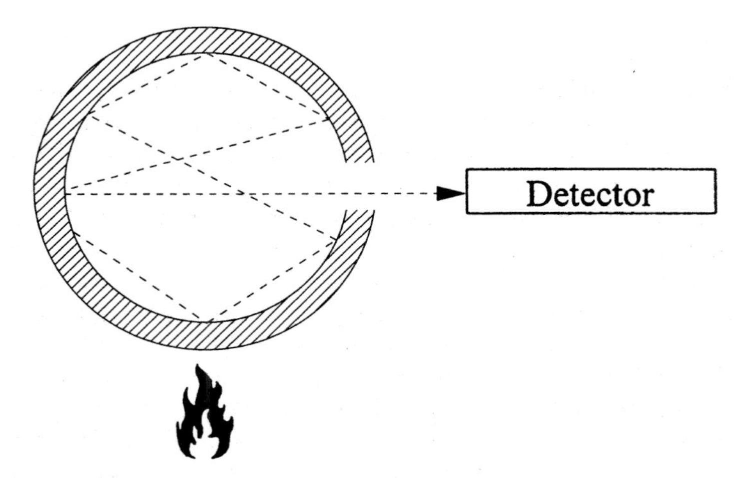

FIGURE 3-8. Blackbody cavity with an external heat source.

THEORY OF ELECTROMAGNETIC RADIATION: THE QUANTUM

A warm body emits heat, which is a form of radiation. Most heat radiation is felt but not seen, because the human eye is not sensitive to these wavelengths. Objects can be heated enough to cause them to glow, first with a dull red color and if heated still further, bright red, orange, and eventually "white hot." Intuitively, the amount of energy emitted increases with temperature and the frequency of the emitted radiation, but what is the basic law that describes the relationship exactly? Physicists responded to this challenge in the late nineteenth century with startling and far-reaching insights into the nature of electromagnetic radiation. Providing a fundamental theory for this relatively benign phenomenon gave birth to the quantum theory, the greatest revolution in physical thought during the twentieth century.

The relationship between temperature and wavelength is not as easy to explain as it is to measure. An empirical rule called the *Stefan–Boltzmann law* relates the intensity of radiant energy from a heated body and its temperature as

$$W = ekT^4$$

where W is the rate of emission of radiant energy per unit area, T the absolute temperature in kelvin, e the emissivity of the surface, and k the Stefan–Boltzmann constant, derived independently by both Stefan and Boltzmann.

A blackbody was used to study the relationship between the radiated thermal energy and wavelength. A blackbody does not really exist in nature, but the radiation from a small opening of a uniformly heated cavity is a good substitute, as shown in Figure 3-8. If a small hole is made through the wall, an optical spectrometer can be used to measure the emissive power as a function of wavelength. Such measurements made at fixed temperatures yield curves like those shown in Figure 3-9. It is clear from the figure that the monochromatic energy density from a blackbody cavity is very small at both short and long wavelengths, with a maximum value at

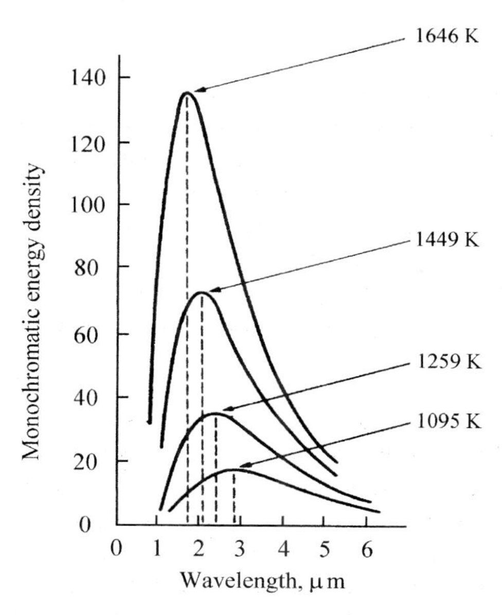

FIGURE 3-9. Distribution of energy in the spectrum of a blackbody at different temperatures.

a definite wavelength λ_m. The curves are similar in shape, but as the temperature increases, the height of the maximum increases and the position of the maximum is shifted in the direction of smaller wavelengths (or higher frequencies). The total radiant energy for a given temperature is simply the area under the curve, and according to the Stefan–Boltzmann law, it increases as the fourth power of the absolute temperature. Wein made the first significant attempt to explain these experimental results.

Wein's Law

According to Maxwell's electromagnetic theory of light, radiation in an enclosure exerts a pressure on the walls of the enclosure that is proportional to the energy density of the radiation. Wein treated blackbody radiation as though produced by a thermodynamic engine to which the first and second laws of thermodynamics could be applied, and developed several important relationships. First, the shift toward smaller wavelength when the temperature is increased is

$$\lambda T = \text{constant}$$

Second, the monochromatic energy density μ_λ and temperature T are connected by the relationship

$$\frac{\mu_\lambda}{T^5} = \mu_\lambda \lambda^5 = \text{constant}$$

Since the variations of both of these with respect to each other are a constant, but not necessarily the same one, $\mu_\lambda \lambda^5$ must be a function of λT such that

$$\mu_\lambda = C\frac{f(\lambda T)}{\lambda^5}$$

where C is a constant.

The challenge in explaining blackbody radiation was to determine the distribution function $f(\lambda T)$. Wein assumed that the radiation was produced by oscillators of molecular size, that the frequency of the radiation was proportional to the kinetic energy of the oscillator, and that the intensity in any particular wavelength range was proportional to the number of oscillators with the requisite energy. With these assumptions, Wein derived the following expression for the distribution:

$$\mu_\lambda = \frac{c_1}{\lambda^5} e^{-c_2/\lambda T}$$

where c_1 and c_2 are constants as yet undefined. This distribution has the required general form and it fits the experimental data quite well at short wavelengths, but at longer wavelengths it predicts values of the energy density that are too small (see Figure 3-10). The exponential function, however, is so satisfactory at short wavelengths that any successful radiation law must reduce approximately to such a function in this region of the spectrum.

Rayleigh–Jeans Law

Rayleigh and Jeans derived a distribution function by considering a hollow cavity that contains standing electromagnetic waves of energy produced by linear harmonic oscillators of thermal radiation. Each oscillator has an energy ε that may take on any value between zero and infinity, a crucial assumption. When the oscillators are at equilibrium, the value ε for any energy of the oscillator occurs with the relative probability $e^{-\varepsilon/kT}$ given by Boltzmann's distribution law, where k is Boltzmann's constant. The Rayleigh–Jeans theory requires two quantities: (1) the number of waves, which can be calculated as

$$N(\nu)d\nu = \frac{8\pi\nu^2 d\nu}{c^3}$$

where ν is the frequency; and (2) the average energy $\overline{\varepsilon}$ carried by each wave. The value of $\overline{\varepsilon}$ is obtained by averaging over all values of ε, using the Boltzmann distribution. If $\beta = 1/kT$, then

$$\overline{\varepsilon} = \frac{\int_0^\infty \varepsilon e^{-\beta\varepsilon} d\varepsilon}{\int_0^\infty e^{-\beta\varepsilon} d\varepsilon} = -\frac{d}{d\beta}\log\int_0^\infty e^{-\beta\varepsilon} d\varepsilon = -\frac{d}{d\beta}\log\frac{1}{\beta} = \frac{1}{\beta} = kT$$

If this average energy, kT, is multiplied by the number of standing waves in the cavity, the energy density is

$$\mu_\nu d\nu = N(\nu)\overline{\varepsilon} d\nu = \frac{8\pi\nu^2 kT}{c^3} d\nu$$

or

$$\mu_\lambda = \frac{8\pi kT}{\lambda^4}$$

This is the Rayleigh–Jeans formula, and it agrees well with the experimental data for long wavelengths (see Figure 3-10). At small values of λ, however, the formula fails because it predicts infinite energy, which is simply not observed; this failure has been dubbed the "ultraviolet catastrophe." We present it only because it provided the starting point for Planck's discovery of the quantum nature of electromagnetic radiation.

Planck's Quantum Theory of Radiation

The problem of the spectral distribution of thermal radiation was solved by Planck in 1900 by means of a revolutionary hypothesis. Planck assumed, as others had, that the emitted radiation was produced by linear harmonic oscillators; however, he made the ad hoc assumption that each oscillator does not have an energy that can take on any value from zero to infinity, but instead, only discrete values:

$$\varepsilon = n\varepsilon_0$$

where ε_0 is a discrete, finite amount, or *quantum*, of energy and n is an integer. By requiring the energy of an oscillator to be emitted only in discrete jumps, calculation of the average value of energy $\overline{\varepsilon}$ is based on a summation of discrete quantities rather than integrals:

$$\overline{\varepsilon} = \frac{\sum_{n=0}^{\infty} n\varepsilon_0 e^{-\beta n\varepsilon_0}}{\sum_{n=0}^{\infty} e^{-\beta n\varepsilon_0}}$$

This expression is the ratio of two infinite series, which reduces to

$$\overline{\varepsilon} = -\frac{d}{d\beta}\log\sum_{n=0}^{\infty} e^{-\beta n\varepsilon_0} = -\frac{d}{d\beta}\log\frac{1}{1-e^{-\beta\varepsilon_0}}$$

and

$$\overline{\varepsilon} = \frac{\varepsilon_0 e^{-\beta\varepsilon_0}}{1-e^{-\beta\varepsilon_0}} = \frac{\varepsilon_0}{e^{\varepsilon_0/kT}-1}$$

Note in particular that the value of n falls out in the solution; however, it must be remembered that it was crucial to obtaining the average value for $\overline{\varepsilon}$. The energy density is now obtained by multiplying the number of standing waves and the average energy $\overline{\varepsilon}$ contained in each wave:

$$\mu_\nu\, d\nu = N(\nu)\overline{\varepsilon}\, d\nu = \frac{8\pi\nu^2}{c^3}\frac{\varepsilon_0}{e^{\varepsilon_0/kT}-1}\, d\nu$$

or

$$\mu_\nu = \frac{8\pi\nu^2}{c^3}\frac{\varepsilon_0}{e^{\epsilon_0/kT}-1}$$

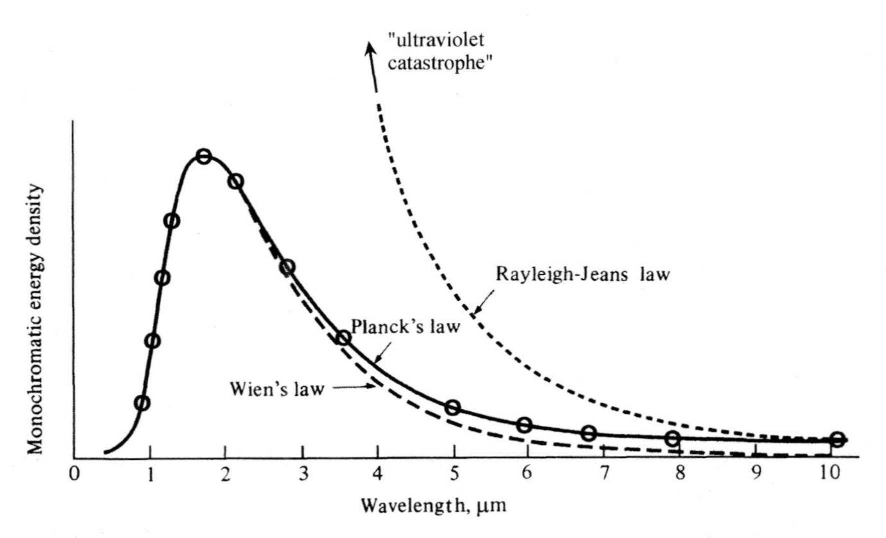

FIGURE 3-10. Planck's radiation law showing an excellent fit to measured data. Wein's law fits the data for short wavelengths; the Rayleigh–Jeans law holds for large wavelengths but exhibits ultraviolet catastrophe at short wavelengths.

Since the temperature must appear in the combination λT, or T/ν, or ν/T, ε_0 must be proportional to ν, or

$$\varepsilon_0 = h\nu$$

where h is a new universal constant, called *Planck's constant. Planck's distribution law* for thermal radiation is therefore

$$\mu_\nu = \frac{8\pi h\nu^3}{c^3}\frac{1}{e^{h\nu/kT} - 1}$$

or, in terms of wavelength,

$$\mu_\lambda = \frac{8\pi h}{\lambda^5}\frac{1}{e^{hc/\lambda kT} - 1}$$

This predictive relationship is plotted as the smooth curve in Figure 3-10 along with the curves based on theories developed by Wein and Rayleigh–Jeans. Whereas their distribution functions break down at either end of the measured spectrum, Planck's distribution law fits the experimental points exactly and is obviously the more general and correct law for emission of radiant energy. It can be shown that it encompasses Rayleigh–Jeans formula by expanding the exponential function in the denominator for $h\nu/kT \ll 1$ and Wein's formula for $h\nu/kT \gg 1$. Planck's law thus incorporates the principles observed by Wein and Rayleigh–Jeans, but its greatest significance is the key discovery that radiation is emitted in discrete units. Without it, theory and experiment did not agree, but more important, the discovery states a basic law of nature. Its further development and use led to one of the greatest periods of thought in human history and science: the use of quantum mechanics to explain the structure of matter.

Planck determined the first value for h from the experimental data and was also able to calculate the best value to that date for k. These, too, are fundamental constants of nature, especially Planck's constant, which appears in all descriptions of electromagnetic radiation and treatments of the atom by quantum mechanics. His results were close to the modern values, which are

$$h = 6.6260755 \times 10^{-34} \text{ J} \cdot \text{s}$$

$$= 4.1356692 \times 10^{-15} \text{ eV} \cdot \text{s}$$

$$k = 1.380658 \times 10^{-23} \text{ J/K}$$

The revolutionary nature of Planck's theory is contained in the postulate that the emission and absorption of radiation must be discontinuous processes. Emission can take place only when an oscillator makes a discontinuous transition from a state in which it has one particular energy to another state in which it has another energy different from the first by an amount that is an integral multiple of $h\nu$. Therefore, any physical system capable of emitting or absorbing electromagnetic radiation is limited to a discrete set of possible energy values or levels; energies intermediate between these allowed values simply do not occur. This revolutionary theory applies to all energy systems, including the energy states of particles in atoms, and because it does so, it greatly influences the structure that an atom can have.

CHECKPOINTS

- The quantum theory is a bold hypothesis that was made to bring theory and experiment into agreement.
- Discrete energy levels, which are a direct consequence of Planck's quantum hypothesis, are fundamental to understanding both atomic and nuclear phenomena.
- The energy of any electromagnetic wave, an essential parameter in radiation protection, is provided by Planck's law as $E = h\nu$.

Quantum Theory and the Photoelectric Effect

Plank himself did not immediately recognize the breadth and importance of his discovery, believing he had merely found an ad hoc formula for thermal radiation. Wrong! How could one adopt the idea that electromagnetic radiation was a bunch of quanta when the wave theory so clearly explained interference, diffraction, and polarization of light waves? Einstein's explanation of the photoelectric effect and Compton's explanation of x-ray scattering both depended on the quantum hypothesis and finally convinced physicists, including Planck, of its validity.

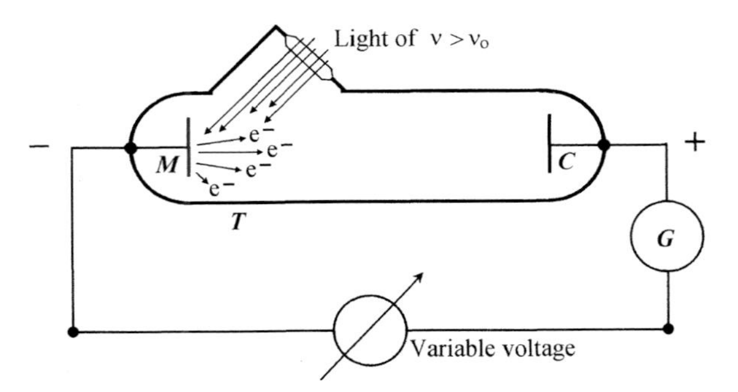

FIGURE 3-11. Experimental arrangement for measuring photoelectric phenomena.

Ironically, the photoelectric effect was discovered by Heinrich Hertz in 1887 while confirming the existence of long-wavelength electromagnetic waves as predicted by Maxwell's theories. He found that the spark discharge in his experiment was enhanced if the surface were irradiated with ultraviolet light. This *photoelectric effect* can be studied by enclosing a metallic plate and a charge-collecting plate in a glass vacuum tube as shown in Figure 3-11. When light above a certain frequency shines on the metal plate in Figure 3-11, electrons are ejected from the metallic surface to produce a current that is measured by the galvanometer *G*. A variable voltage can be applied to just stop the electrons from reaching the plate, and this voltage corresponds to the energy of the emitted photoelectrons. The following results have been obtained in such experiments:

- If the frequency of the light source is kept constant, the photoelectric current increases with the intensity of the incident radiation.
- Photoelectrons are emitted immediately (in less than 10^{-9} s) after the surface is illuminated by light.
- For a given surface, the emission of photoelectrons takes place *only* if the frequency of the incident radiation is equal to or greater than a certain minimum frequency ν_0, which has a specific value for each metal.
- The maximum kinetic energy KE_{max} of photoelectrons depends only on the frequency of the incident radiation ν, not on its intensity.
- There is a linear relation between KE_{max} and ν as shown in Figure 3-12.

According to classical electromagnetic theory there should be no minimum (threshold) frequency for emission of photoelectrons because light of any intensity would eventually build up enough energy to pull the electrons from the surface. Finding electrons with specific energy related to frequency immediately after illumination was unexplained, although clearly present. Einstein explained these phenomena by adopting and extending Planck's quantum hypothesis (for this, and other contributions obviously, he received a Nobel prize in 1921). He simply assumed that the electromagnetic radiation incident on the metallic surface consisted of Planck-like discrete bundles of energy, called *quanta*, which could interact as a "particle." If an electron absorbs a photon of energy $h\nu$, it could be ejected from

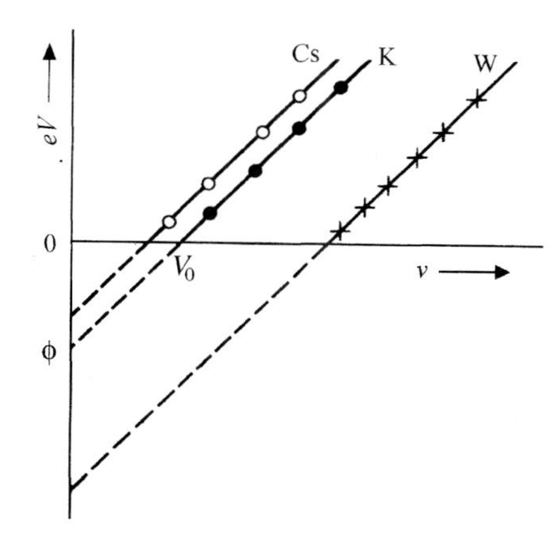

FIGURE 3-12. Kinetic energies of photoelectrons versus frequency of the incident light. $KE = h\nu - \phi$, where ϕ is the y-intercept and the threshold frequency ν_0 is the intercept of the x-axis.

the metallic surface with a kinetic energy equal to $h\nu$ minus an amount of energy ϕ, which represented the degree to which the electrons were bound in the metallic surface. Einstein called this quantity the *work function*, and it is different for each metal. The kinetic energy of the ejected electrons is

$$KE = h\nu - \phi$$

which is an equation for a straight line with a slope of h and a y-intercept of ϕ. The work function ϕ is determined by extrapolating the straight line to intercept the y-axis, and the threshold frequency ν_0 is determined by the intercept with the x-axis. Millikan used experimental data similar to that shown in Figure 3-12 to determine a value of Planck's constant that agreed with that from the blackbody radiation spectrum.

Example 3-1: Light of wavelength $\lambda = 5893$ Å incident on a potassium surface yields electrons that can be stopped by 0.36 V. Calculate (a) the maximum energy of the photoelectron, (b) the work function, and (c) the threshold frequency.

SOLUTION: (a) The maximum energy KE_{max} of the ejected photoelectrons is

$$KE_{max} = eV_0 = 0.36 \text{ eV}$$

(b) Since $KE_{max} = h\nu - \phi$,

$$eV_0 = h\nu - \phi$$

or

$$\phi = \frac{hc}{\lambda} - eV_0 = \frac{(4.13567 \times 10^{-15} \text{ eV} \cdot \text{s})(3 \times 10^8 \text{ m/s})}{5893 \times 10^{-10} \text{ m}} - 0.36 \text{ eV} = 1.7454 \text{ eV}$$

(c) The threshold frequency is one that will just overcome the work function energy of 1.7454 eV:

$$\nu_0 = \frac{\phi}{h} = \frac{1.7454 \text{ eV}}{4.13567 \times 10^{-15} \text{ eV} \cdot \text{s}} = 4.22 \times 10^{14} \text{ s}^{-1}$$

$$= 4.22 \times 10^{14} \text{ Hz}$$

Electromagnetic Spectrum

The photon hypothesis represents one of the remarkable aspects of science, the broad applicability of a basic principle once it is discovered. The photon or quantum hypothesis was developed to explain blackbody radiation, but it extends to all other electromagnetic radiation. Hertz showed the existence of low-frequency radio waves, which greatly extended the low end of the spectrum of electromagnetic energy, and Roentgen's x rays extended it on the higher end. The simple expression for the energy of electromagnetic radiation, $E = h\nu$, applies across the full electromagnetic spectrum, which, as now known and shown in Figure 3-13, covers more than 22 orders of magnitude.

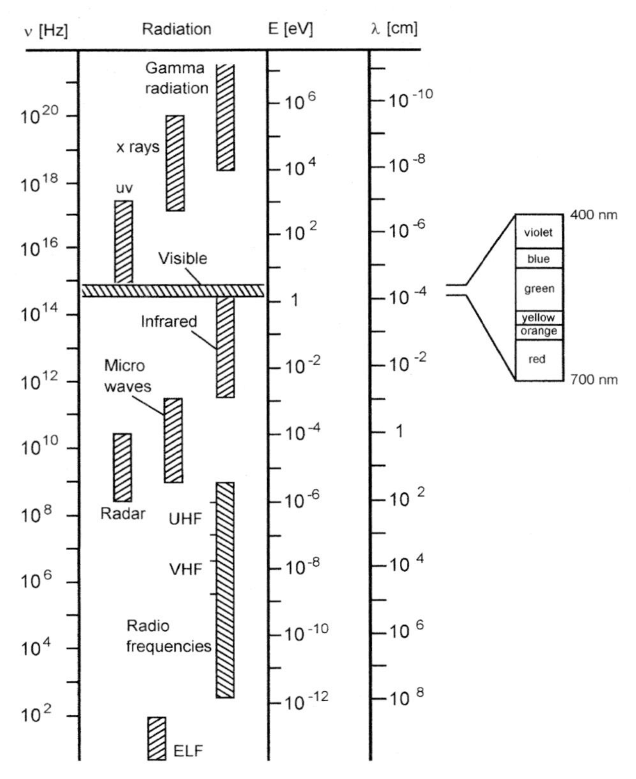

FIGURE 3-13. Electromagnetic spectrum with the wavelengths of the visible region amplified. (Planck's radiation law $E = h\nu$ applies to wave phenomena over more than 22 orders of magnitude.)

Photon Hypothesis

The successful explanation of the photoelectric effect established that the photon could be treated as a packet (or quantum) of energy, $E = h\nu$, that behaves like a particle and travels at the speed of light c. According to the special theory of relativity, the photon has a moving (not rest) mass, $E = mc^2$; thus several important relationships can be derived for photons:

$$m_0 = 0, \qquad E = h\nu, \qquad m = h\nu/c^2, \qquad p = h\nu/c$$

These characteristics of photons as particles were successful in establishing the photoelectric effect; they are also necessary for explaining the Compton effect.

Compton Effect

It took the work of A. H. Compton, who in 1922 explained the scattering of x rays in terms of particle properties of the photon, to firmly establish the photon (or quantum) hypothesis. He received a Nobel prize for describing the phenomenon known as the *Compton effect*, which was fundamental in the evolution of physics as well as interpreting interactions of x rays and gamma rays with matter. Compton found that x rays produce two frequencies when they interact with materials containing free or loosely bound electrons: the original frequency, ν, and another frequency, ν', which is less than ν. X rays that do not change frequency (or wavelength) undergo coherent scattering. Those that do experience *incoherent scattering*, or *Compton scattering*, as it is commonly called.

Compton used a collimated beam of monochromatic x rays from molybdenum to irradiate a carbon target, which has electrons so loosely bound that they are assumed to be free (the binding energy of the outer electrons in carbon is only 11 eV, which is negligible compared to x rays). The x rays were scattered at different angles with a modified wavelength, λ', as shown schematically in Figure 3-14. He then treated the incoming and scattered photons as particles and applied the laws of conservation of momentum and energy to calculate the change in wavelength of each.

The incident photon with energy $h\nu$ and momentum $h\nu/c$ interacts with an electron that has a rest-mass energy of m_0c^2. The photon scatters at an angle θ, but now has energy $h\nu'(< h\nu)$ and momentum $h\nu'/c$, which is less than $h\nu/c$. By resolving the momentum and energy components, Compton calculated the change in wavelength, $\Delta\lambda = \lambda' - \lambda$, associated with the scattered photon as

$$\Delta\lambda = \lambda' - \lambda = \frac{h}{m_0c}(1 - \cos\theta) = \frac{6.625 \times 10^{-34}\ \text{J}\cdot\text{s}}{(9.10953 \times 10^{-31}\ \text{kg})(3 \times 10^8\ \text{m/s})}(1 - \cos\theta)$$

$$= 0.002424(1 - \cos\theta)\ \text{Å}$$

The change in wavelength $\lambda' - \lambda$ depends on the rest mass of the free electron m_0 and the scattering angle θ, but not on the energy or the wavelength of the incident photon. The quantity h/m_0c has the dimensions of length and is called the *Compton wavelength*, or *Compton shift*. Figure 3-15 shows the scattered x rays of increased wavelength (decreased energy) for two different scattering angles. The points are

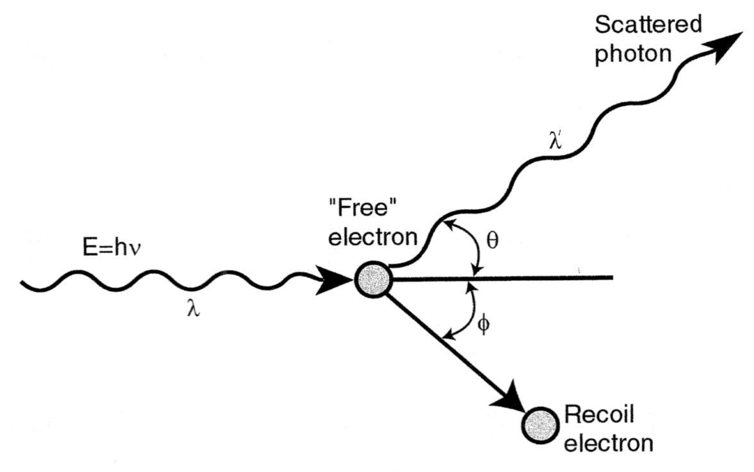

FIGURE 3-14. Compton effect for x-ray scattering represented as a collision between an incident photon and a loosely bound or free electron, producing a scattered photon of reduced energy (i.e., longer wavelength) and a recoiling electron.

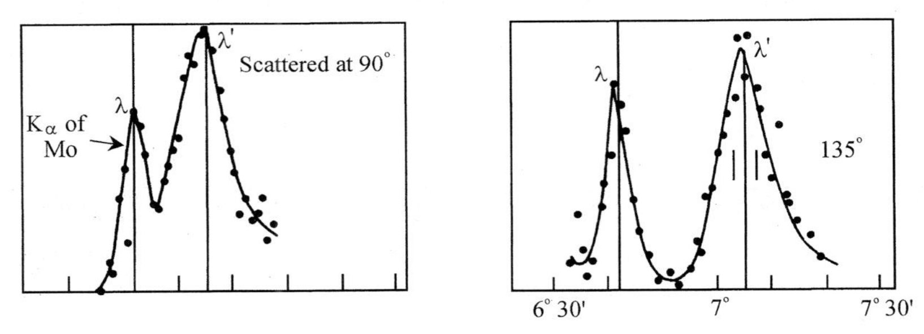

FIGURE 3-15. Intensity of Compton scattered x rays at 90° and 135° for K_α x rays from molybdenum. The modified wavelength λ' is due to a photon collision with free electrons; the unmodified wavelength λ is due to interaction of the primary K_α x ray with bound electrons in which the deflection is insignificant. (From Compton, 1923.)

the experimentally measured values, while the continuous curves are the values that Compton predicted.

Photon scattering may also take place with tightly bound inner electrons. When it does, the entire atom recoils. If we replace m_0 with the atom mass M (which is 1836 times m_0 for hydrogen, and much greater for other atoms) the Compton wavelength h/Mc is 0.0000133 Å or smaller, which is negligible even at the maximum angle $\theta = 180°$; thus the change in wavelength is minimal when photons interact with bound electrons.

The kinetic energy of the recoiling electron is useful in gamma-ray spectroscopy. This energy is

$$\mathrm{KE}_e = \frac{h\nu}{m_0c^2}\frac{h\nu(1-\cos\theta)}{1 + h\nu/m_0c^2(1-\cos\theta)}$$

which has a maximum value when the photon scattering angle θ is 180°, or

$$\mathrm{KE}_{e(\max)} = \frac{2(h\nu)^2}{m_0c^2(1 + 2h\nu/m_0c^2)}$$

The Compton electron scatters at a definite angle ϕ which is sometimes useful to know. This angle is

$$\cos\phi = \frac{E^2 - E'^2 + K^2[1 + (2E_0/K)]}{2EK\sqrt{1 + 2E_0/K}}$$

where E and E' are the energies of the incident and scattered photons, $E_0 = 511$ keV/c^2 is the electron rest-mass energy, and K is the kinetic energy of the scattered electron.

Example 3-2: For x rays of 100 keV that undergo Compton scattering, calculate (a) the energy of the x rays scattered at an angle of 30°, (b) the energy of the recoiling electron, (c) the angle of the recoiling electron, and (d) the maximum energy that the recoiling electron could have.

SOLUTION: (a) Compton's scattering equation can be modified by substituting $\lambda = hc/E$ and $\lambda' = hc/E'$, where E and E' are the energies of the incident and scattered x rays:

$$\Delta\lambda = \lambda' - \lambda = \frac{h}{m_0c}(1 - \cos\theta)$$

$$\frac{1}{E'} - \frac{1}{E} = \frac{1}{m_0c^2}(1 - \cos\theta)$$

For $E = 100$ keV,

$$E' = \frac{1}{\dfrac{1-\cos\theta}{m_0c^2} + \dfrac{1}{E}} = \frac{1}{\dfrac{1-\cos 30^\circ}{511\ \text{keV}} + \dfrac{1}{100\ \text{keV}}} = \frac{1}{\dfrac{1}{3814\ \text{keV}} + \dfrac{1}{100\ \text{keV}}} = 97.4\ \text{keV}$$

(b) Since energy is conserved, the kinetic energy of the recoiling electron is

$$E - E' = (100 - 97.4)\ \text{keV} = 2.6\ \text{keV}$$

(c) The angle of the scattered electron for $E = 100$ keV, $E' = 97.4$ keV, $K = 2.6$ keV, and $E_0 = 511$ keV is

$$\phi = \cos^{-1}\frac{(100\ \text{keV})^2 - (97.4\ \text{keV})^2 + (2.6\ \text{keV})^2[1 + (2 \times 511\ \text{keV}/2.6\ \text{keV})]}{2(100\ \text{keV})(2.6\ \text{keV})\sqrt{1 + 2 \times 511\ \text{keV}/2.6\ \text{keV}}}$$

$$= 72.1^\circ$$

(d) The maximum energy for the recoil electron is

$$\text{KE}_{\text{max}} = \frac{2(100\ \text{keV})^2}{511\ \text{keV}(1 + 200/511)}$$

$$= 28.13\ \text{keV}$$

CHECKPOINTS: QUANTIZED ENERGY

- A body that emits radiant energy (a furnace as heat, a light bulb as light, a radioactive element as gamma rays) undergoes changes that produce an electromagnetic wave that transmits energy from where it is produced (the energetic body) to some other place—all radiation is, therefore, a form of transmitted energy.
- Radiant energy emission occurs only as a pulse, not as a continuum of energy. An electron or particle in the atom can exist only at energy level 1 or 2 (see Figure 3-16) but not in between, so that when an atom emits energy it does so in an amount that is the discrete difference between the two permitted states. For an electron, say, to become more tightly bound (i.e., achieve a lower potential energy state), the atom has to spit out a pulse of energy, the photon; for it to go to a higher (less tightly bound) level, it must gain the same amount of energy (i.e., $E_2 - E_1$).
- The energy of a photon is directly proportional to its frequency. This constant of proportionality is one of the fundamental constants of nature and is, appropriately, called *Planck's constant.* All photons travel at the speed of light, which is another fundamental constant of nature; thus the energy of a photon is expressed in terms of these two fundamental constants

$$E = h\nu = \frac{hc}{\lambda}$$

That is, photons with long wavelengths have less energy than photons with very short wavelengths.

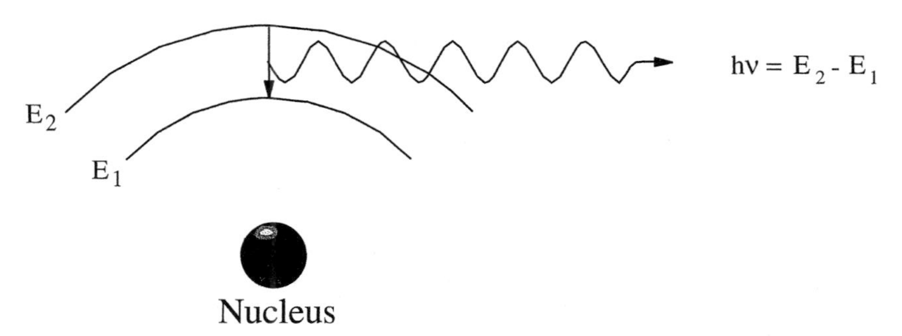

FIGURE 3-16. Discrete photon energy emission for permitted energy levels of an atomic electron; similar emissions occur for particles in the nucleus.

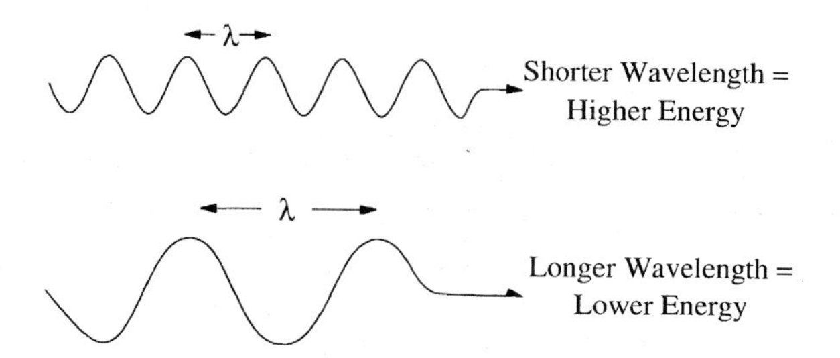

- Radiant energy (light, heat, x rays) **is** pure energy propagated along a wave in packets of energy or quanta that behave like a particle:

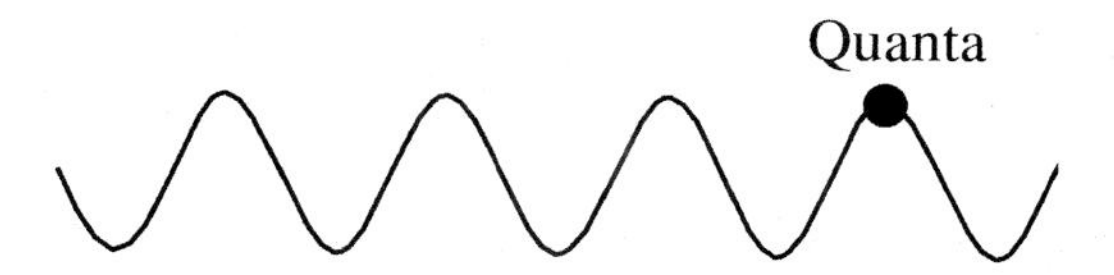

- Photons clearly have no rest mass, but their particle properties can be used to explain the photoelectric effect (Einstein, 1905), in which light of a certain frequency (wavelength) causes the emission of electrons from the surface of various metals.

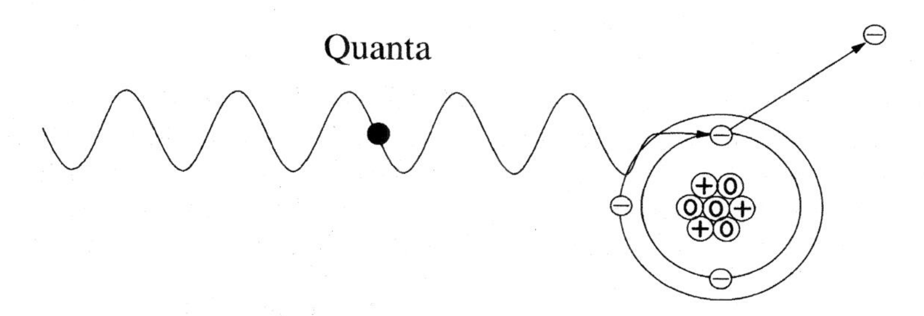

The photon, behaving like a particle, hits a bound electron and "knocks" it out of the atom (photocells depend on this principle.)

- Compton scattering of x rays can be explained as interactions between "particlelike" photons and "free" electrons in which energy and momentum are conserved, and the calculated wavelength changes agree with experimental observations.
- Atoms that emit electromagnetic radiation must have quantized, or discrete, energy states for all particle arrangements: those involving electrons around the nucleus, and also the arrangement of neutrons and protons within the nucleus.

DISCOVERY OF THE ATOM'S STRUCTURE

The discoveries associated with x rays, radioactivity, quantized energy, and special relativity led to vital insights about the atom, its ultimate structure, and the dynamics of its components. The work of Thomson and Faraday established that electrons

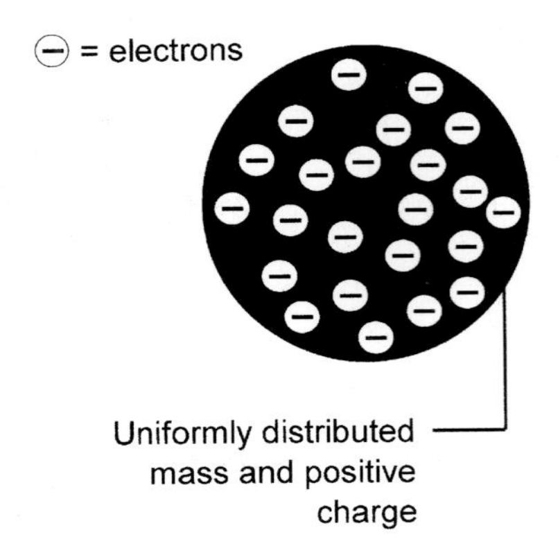

FIGURE 3-17. J. J. Thomson's plum-pudding model of the atom.

exist as a corpuscular unit that can be broken off atoms and that these are all the same regardless of the element. Obviously, the atom is made up of components, one of which is the electron and another that carries a positive charge since atoms are electrically neutral; this unit would probably be the remainder of the hydrogen atom, the lightest and simplest element, when the electron is stripped away or ionized. The expulsion of alpha particles (positively charged helium nuclei) during radioactive transformation also strongly suggests that atoms have components. An alpha particle must be a substructure of the atom, and its positive charge confirms that part of the atom carries a positive charge.

Since atoms are electrically neutral, any structure for the atom must describe a configuration whereby the negative charges of electrons and positive charges cancel each other. Thomson proposed a model in which the charge and mass are distributed uniformly over a sphere of radius about 10^{-8} cm, with electrons embedded in it (like plums in pudding) to make a neutral atom. Thomson's "plum-pudding model" can be represented as shown in Figure 3.17. Unfortunately, Thomson's configuration of the atom did not produce a stable equilibrium, nor frequencies observed in optical spectra, nor the fact that the positive charges account for only about half the atom's weight.

Rutherford's Alpha-Scattering Experiments

Geiger (inventor of the Geiger counter) and Marsden, working with Rutherford, used alpha particles from radioactive substances to literally shoot holes in Thomson's distributed atom because it could not account for the scattering observed. They irradiated thin gold foils (about 10^{-4} cm thick) with alpha particles using the apparatus shown schematically in Figure 3-18. The entire apparatus assembly was placed in a vacuum chamber to avoid absorption of the alpha particles by air. Most of the particles passed through the foil undeviated, as if there were empty spaces in the foil. But some of them collided with the atoms of the foil and were scattered at different angles, which these early researchers measured by counting the flashes of light produced by each alpha particle when it struck a screen coated with ZnS. These scintillations were observed through a magnifying glass or a microscope after they had sat in the dark for awhile to dark-adapt their eyes. Measurements revealed

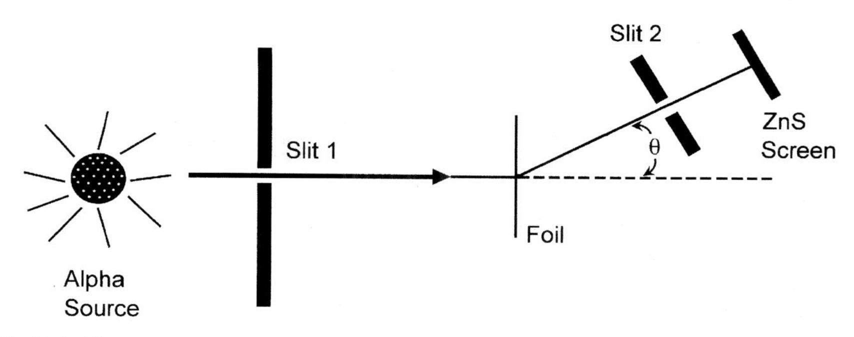

FIGURE 3-18. Experimental arrangement for studying scattering of alpha particles by thin metallic foils.

that most of the alpha particles, as expected, passed through the foil undeviated. But one out of every 8000 or so alpha particles was scattered back toward the source. The distribution of charge in the Thomson model could not account for such results.

The scattering of alpha particles observed by Geiger–Marsden can be represented as

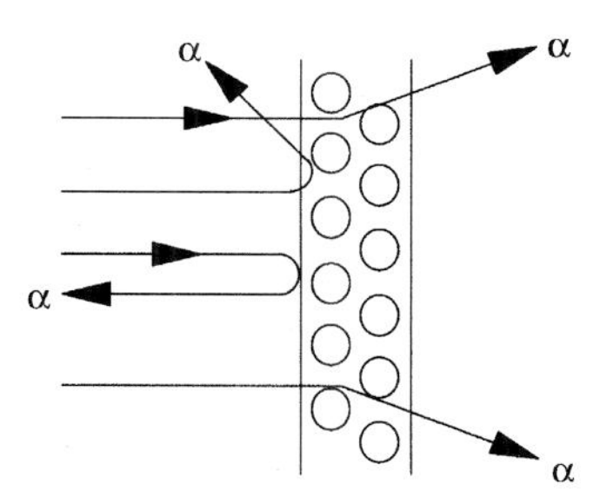

The large-angle deflection observed must be consistent with the force of repulsion between the positive charge of the alpha particle and the positive charge of the atom, or

$$F = k\frac{q_1 q_2}{r^2} = k\frac{(2e)(Ze)}{r^2}$$

Rutherford suggested that the only way that the observed large-angle scattering could be explained was if the value for r in the Coulomb force equation was very small, on the order of 10^{-12} cm. This led him to suggest that all of the positive charge—and hence almost all the mass of the atom—is confined to a very small volume at the center of the atom, which he called the *nucleus*. Since the radius of the atom is about 10^{-8} cm, electrons could be assumed to be distributed in orbits around the nucleus such that the atom is mostly empty space.

As brilliant as Rutherford's proposed model was, it had a fundamental flaw that he could not explain. According to classical electromagnetic theory, an accelerated charge must radiate electromagnetic waves. An electron circling a hydrogen nucleus experiences a continuous acceleration v^2/r, and according to classical theory, should radiate energy continuously; thus r would steadily decrease and the electron would,

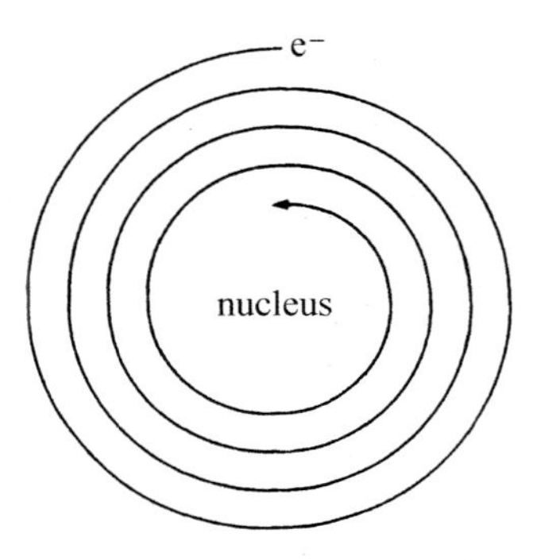

FIGURE 3-19. Rutherford's conundrum, in which classical theory calls for the electron to spiral into the nucleus.

as illustrated in Figure 13-19, quickly (in about 10^{-8} s or so) spiral into the nucleus. Such a dynamic corresponds to classical theory, but is wrong. The hydrogen atom is very stable, and it does not radiate electromagnetic waves unless stimulated to do so. Obviously, a new theory was needed, and Neils Bohr, a Danish physicist, would provide it.

Bohr Model of the Atom

In 1913, Niels Bohr applied a mixture of classical physics and the energy-quantization ideas introduced by Planck to solve Rutherford's conundrum of the electron spiraling into the nucleus. Bohr simply declared that an electron in its orbit did not radiate energy; it radiates energy only when it undergoes one of Planck's discontinous quantum changes. Bohr used three postulates to construct a model of the atom:

Postulate I: An electron in an atom can move about the nucleus in discrete stationary states without radiating.

Postulate II: The allowed stationary states of the orbiting electrons are those for which the orbital angular momentum is an integral multiple of $h/2\pi$; that is,

$$L = n\frac{h}{2\pi}$$

where $n = 1,2,3,4,\ldots$ represents the principal quantum number for discrete, quantized energy states.

Postulate III: An electron can jump from a higher-energy state E_2 to a lower-energy orbit E_2 (both of which are negative potential energy states for bound electrons) by emitting electromagnetic radiation of energy $h\nu$, where

$$h\nu = E_2 - E_1$$

Postulate I is completely arbitrary and represents Bohr's genius and his boldness —physics would just have to catch up. Postulate III comes somewhat from Planck's quantum hypothesis, but postulate II was so unfounded that it would only be explained by de Broglie's hypothesis some 13 years later. How then did Bohr come up with such a successful concept?

At first glimpse it may appear that Bohr had no firm basis for his model of the hydrogen atom, but he was able to take advantage of some significant building blocks:

1. Planck had established the discreteness of electromagnetic radiation, that is, that it appears only as integral multiples, $n = 1, 2, 3, \ldots$ of a basic unit of energy.
2. The scattering of alpha particles clearly supported Rutherford's nuclear model, in which the positive charge and mass were contained in a nucleus of about 10^{-12} cm radius at the center of the atom, with the negative charges (electrons) distributed around it out to a radius of about 10^{-8} cm.
3. Balmer had observed a regularity in the wavelengths emitted by hydrogen and had derived an empirical relationship containing an integral term, n, much like the one in Planck's quantum theory:

$$\lambda = 3645.6 \frac{n^2}{n^2 - 4} \text{ Å}$$

where $n = 3, 4, 5$.

Balmer had published his results in 1885, and in 1896, Rydberg expressed them in terms of the wave number, $1/\lambda$, as

$$\frac{1}{\lambda} = R_H \left(\frac{1}{2^2} - \frac{1}{n^2} \right)$$

where R_H is Rydberg's constant, which has a well-known value of $1.097373 \times 10^7 \text{ m}^{-1}$, and where $n = 3, 4, 5, \ldots$. Apparently, Bohr found some significance in the integral values 2 and n in devising his theory of the atom. Since Bohr's theory has provided the greatest value in explaining the emissions of electromagnetic waves from atoms, in particular hydrogen, it is useful to discuss emission and absorption spectra from elements, particularly gases.

Emission and Absorption Spectra

An emission spectrum of a gas can be obtained by exciting the gas. This is commonly done by producing an electric discharge through such a gas confined in an enclosed tube. The excitation energy causes the emission of electromagnetic radiations, which for hydrogen are in the visible region, so they were easy for Balmer and others to observe. These emitted radiations can be separated into individual wavelengths to produce an emission spectrum as shown in Figure 3-20.

As theorized by Planck, an emission spectrum from hydrogen or some other pure gas contains only specific wavelengths, not a continuous band of energies. This pattern of discrete wavelengths is called a *line spectrum* because of the way it

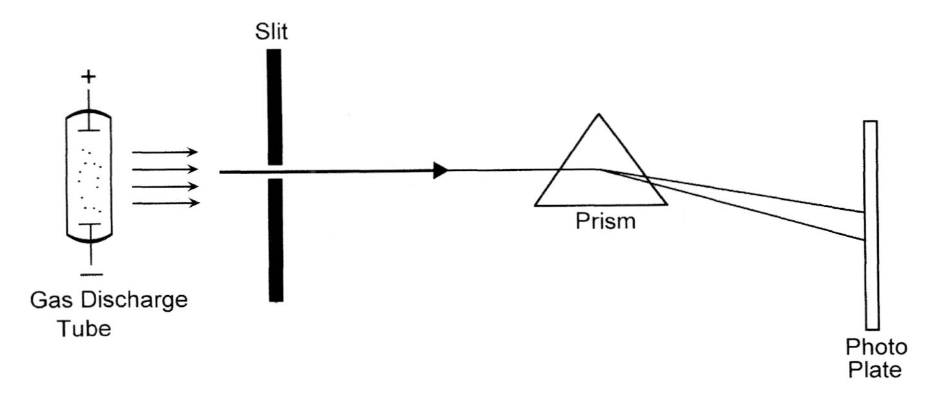

FIGURE 3-20. Wavelength studies of the emissions of light from an excited gas.

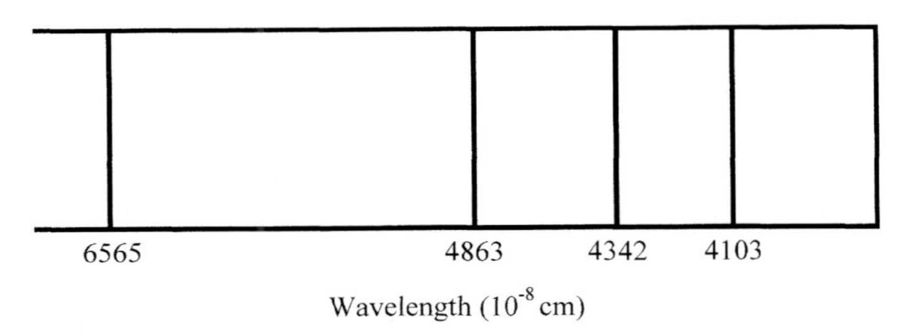

FIGURE 3-21. Line spectrum of light emitted by hydrogen in a gas discharge tube as observed by Balmer.

appears on a photographic plate, a common means of recording the emitted light photons. The line spectrum observed by Balmer for hydrogen is shown in Figure 3-21.

Absorption spectra are the inverse of emission spectra. Since a particular gas emits only specific wavelengths, it will also preferentially absorb those wavelengths when they occur in a broad spectrum of radiation. An absorption spectrum for hydrogen is obtained by placing the gas between a standard light source that emits a broad spectrum of wavelengths and observing which ones are absorbed as they pass through the gas. This can be done with the arrangement shown in Figure 3-22.

In this arrangement some of the wavelengths penetrating the gas are observed to yield a spectrum of transmitted (unabsorbed) light. The same telltale lines will appear as shown in Figure 3-23, but this time as a shadow of untransmitted (i.e., absorbed) radiation. For hydrogen, the absorption spectrum matched the emission spectrum, thus hydrogen could be said to emit only specific wavelengths of radiation or to absorb only those same wavelengths. Furthermore, these discrete patterns of emission and absorption presupposed some intrinsic nature of the hydrogen atom, which took on extra importance with Planck's quantum theory of radiation.

Bohr's model for the atom set out to explain the regularities observed for hydrogen spectra with a consistent theory. To do so he used Rutherford's nuclear atom and assumed that orbiting electrons moved in circular orbits under the influence

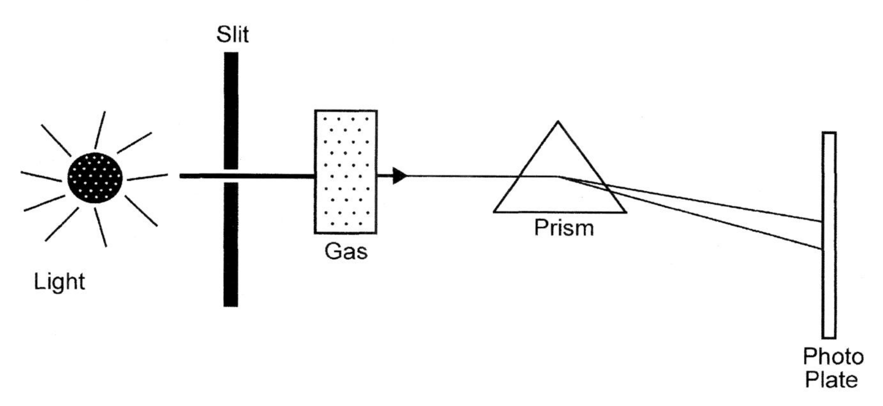

FIGURE 3-22. Experimental arrangement for observing the absorption of electromagnetic radiation in a gas.

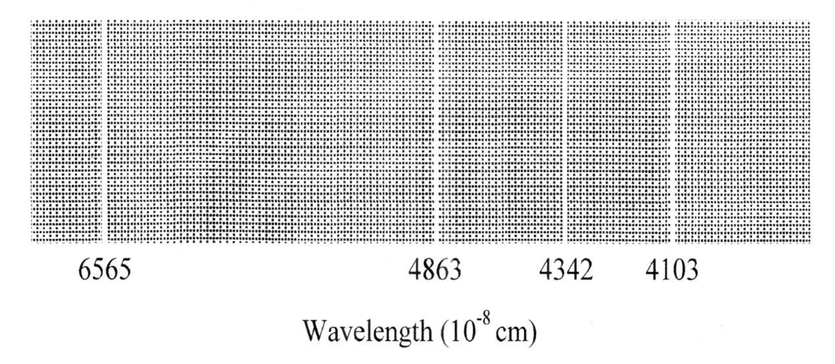

FIGURE 3-23. Absorption lines (i.e., nonpenetrating wavelengths) of hydrogen, which have the same wavelength but are the exact inverse of the emission spectrum in Figure 3-21.

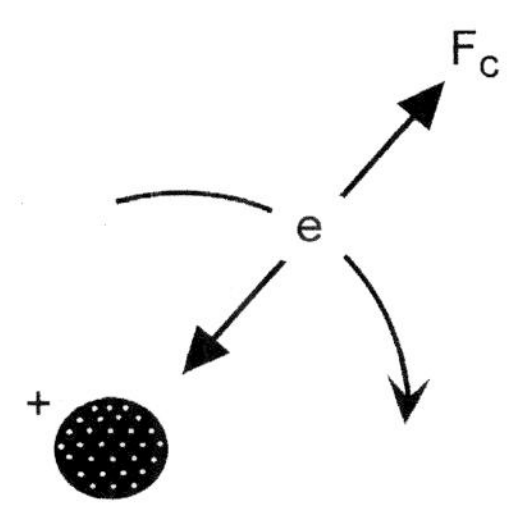

FIGURE 3-24. An electron in a Bohr orbit experiences a coulombic force of attraction toward the nucleus that is counterbalanced by the centrifugal force due to its orbital motion.

of two force fields: the Coulomb attraction (a centripetal force here) provided by the positively charged nucleus and the centrifugal force of each electron in orbital motion, as shown in Figure 3-24. Bohr assumed (postulate I) that the electron is in constant circular motion at a radius r_n with velocity v_n and that the centrifugal force and the attractive Coulomb force are equal and opposite each other:

$$\frac{mv_n^2}{r_n} = k\frac{q_1 q_2}{r_n^2}$$

The radius r_n can be calculated from postulate II,

$$mv_n r_n = n\frac{h}{2\pi}$$

Since q_1 and q_2 are unity for hydrogen:

$$r_n = \frac{(nh)^2}{(2\pi)^2 kmq^2} = n^2 r_1$$

where $n = 1,2,3,4,\ldots$ and r_1 is the radius of the first orbit of the electron in the hydrogen atom, called the *Bohr orbit*. If the values of h, k, m, and q^2 are inserted, $r_1 = 0.529 \times 10^{-8}$ cm, which agrees with the experimentally measured value, and since the quantum hypothesis limits values of n to integral values, the electron can only be in those orbits that are given by

$$r_n = r_1, 4r_1, 9r_1, 16r_1, \ldots$$

These relationships can also be used to calculate the velocity of the electron in each orbit, but the most useful relation is for the energy levels of the electron. The total energy E_n of the electron in the nth orbit is the sum of the kinetic energy and the potential energy. That is,

$$E_n = \frac{mv_n^2}{2} + \left(-\frac{ke^2}{r_n}\right)$$

Substituting the equations for v_n and r_n and simplifying, the energy for the electron in each orbit of quantum number n, E_n, is

$$E_n = -\frac{1}{n^2}\frac{(2\pi)^2 k^2 q^4 m}{2h^2} = -\frac{1}{n^2} \times 13.58 \text{ eV}$$

This is a direct calculation of the total energy of the electron in the first orbit when $n = 1$, or E_1. It is in perfect agreement with the measured value of the energy required to ionize hydrogen, which is, of course, the binding energy of the electron in the hydrogen atom. Thus, for other values of $n = 2,3,4,\ldots$, the allowed energy levels of the hydrogen atom are given by

$$E_n = -\frac{E_1}{4}, -\frac{E_1}{9}, -\frac{E_1}{16}, \ldots$$

where $E_1 = -13.58$ eV. These predicted energy levels are shown in Figure 3-25; they can be used to calculate the possible emissions (or absorption) of electromagnetic

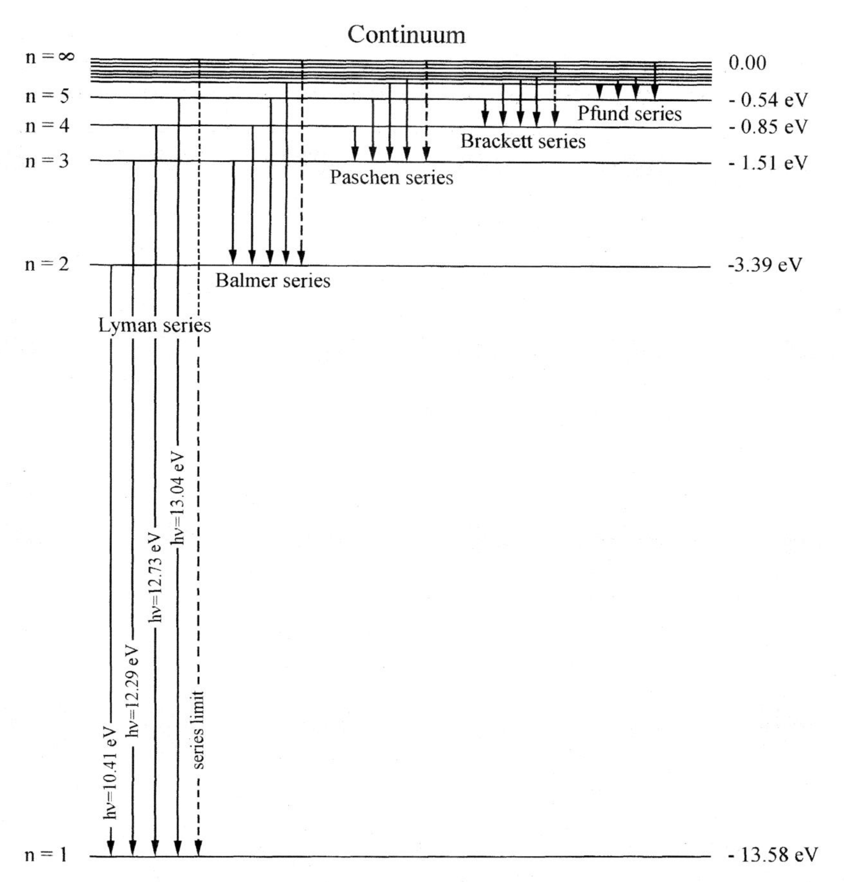

FIGURE 3-25. Relative energy states of electrons in the hydrogen atom and the corresponding radiation series named after those who discovered them.

radiation and their wavelengths for hydrogen. When Bohr did so for $n = 3$, he obtained Balmer's series of wavelengths, and since the theory holds for $n = 3$, Bohr postulated that it should also hold for other values of n. The wavelengths calculated for $n = 1, 2, 4, \ldots$ were soon found, providing dramatic proof of the theory. Bohr had put together several diverse concepts to produce a simple model of atomic structure, at least for hydrogen.

WAVE MECHANICS: A NECESSARY THEORY

Although Bohr's second postulate $[mvr = n(h/2\pi)]$ happened to pick out the stable orbits, it remained largely arbitrary even though it worked (why should mvr be directly proportional to n, for instance?). Some 13 years later, Louis de Broglie provided the key for explaining postulate II, which also provided the essential property embodied in quantum mechanics, or wave mechanics: the wave motion of particles.

De Broglie Waves

In 1926, Louis de Broglie postulated that if Einstein's and Compton's assignment of particle properties to waves was correct, why shouldn't the converse be true (i.e., that particles have wave properties)? This simple but far-reaching concept was proved by the diffraction (a wave phenomenon) of electrons (clearly particles) from a crystal containing nickel. It means that a particle such as an electron (or a car for that matter) has a wavelength λ associated with its motion of

$$\lambda = \frac{h}{p} = \frac{h}{mv}$$

and that as a wave it has momentum p with the value

$$p = \frac{h}{\lambda}$$

Simple though it appears, de Broglie's hypothesis has consequences as significant and far-reaching as Einstein's famous equation $E = mc^2$. In Einstein's equation, energy and mass are in direct proportion to each other, with c^2 the constant of proportionality (a large constant by normal standards). In de Broglie's equations, the wavelength and momentum of a particle are related to each other through Planck's constant (a small constant by normal standards). As Einstein's equation drew together two previously distinct concepts, energy and mass, so de Broglie's equation drew together apparently unrelated ideas, a wave property λ and momentum p, which is a particle property. De Broglie's wave–particle behavior of electrons provided two important contributions to modern physics: (1) it provides a direct physical basis for Bohr's second postulate, $L = nh/2\pi$, and (2) it opened the door to description of the dynamics of particles by wave mechanics, perhaps the most revolutionary development in physical thought since Einstein's special theory of relativity.

Confirmation of De Broglie's Hypothesis

Davisson and Germer inadvertently provided experimental proof of de Broglie's hypothesis in studies of the intensity of scattered electrons from a nickel target, much as Rutherford had done earlier with alpha particles and Compton with x rays. They were looking for secondary electrons, not wave phenomena. Their experimental arrangement and a plot of their observed scattering which was consistent with classical theory is shown schematically in Figure 3-26. The entire apparatus was contained inside an evacuated chamber since they were working with low-energy electrons. The electron beam caused the target to get very hot, and then the vacuum failed, which quickly caused the hot block of nickel to oxidize. To clean it, they baked it in a high-temperature oven for several hours, which oriented the microcrystals in the nickel sample such that it became one large crystal. After inserting it back into the experiment, the results changed significantly, with a peak (as shown in Figure 3-26*b*) at an angle of 65° for electrons accelerated at 54 V. Their prepared minds and de Broglie's hypothesis caused them to rethink their new observations; otherwise, they might have simply thrown the target away and started over. They had observed diffraction of electrons!

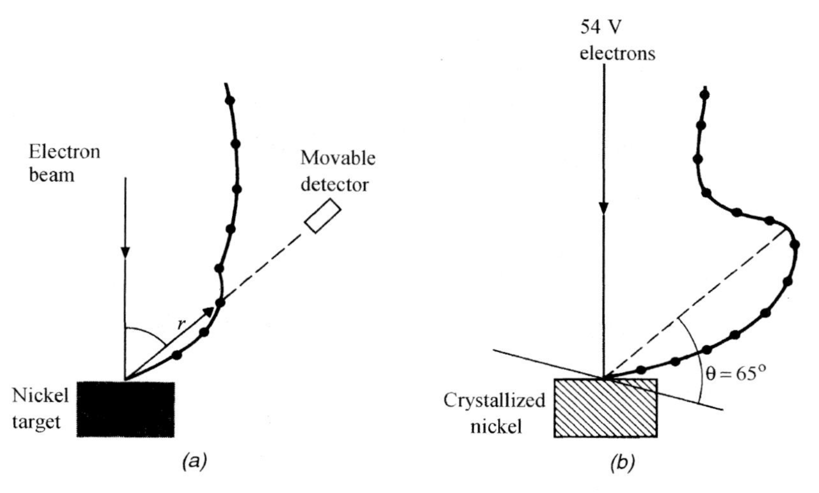

FIGURE 3-26. Distribution of electrons scattered from a nickel target before (*a*) and after (*b*) the nickel had been heated clearly demonstrates the presence of a diffraction peak.

It had long been established that crystals diffract electromagnetic waves such as x rays, a phenomenon described by *Bragg's law*, which predicts a diffraction peak at integral multiples of λ as

$$n\lambda = 2d\sin\theta$$

where λ is the wavelength incident on the crystal, d the spacing between crystalline layers, and θ the angle between the scattered beam and the plane of the crystal. When applied to the data in Figure 3-26 for the crystalline spacing in nickel (0.91 Å) and a diffraction at $\theta = 65°$,

$$\lambda = 2d\sin\theta = 2(0.91\ \text{Å})\sin 65° = 1.65\ \text{Å}$$

An electron that has been accelerated through a potential difference of V has a kinetic energy of

$$\frac{mv^2}{2} = eV$$

from which the velocity can be obtained as

$$v = \sqrt{\frac{2eV}{m}}$$

According to the de Broglie hypothesis, a 54 V electron would have a wavelength

$$\lambda = \frac{h}{p} = \frac{h}{mv} = \frac{h}{\sqrt{2\ \text{meV}}} = 1.67\ \text{Å}$$

which agrees with Bragg's law and establishes that Davisson and Germer had observed the wave motion of electrons. Similar diffraction patterns for electrons have

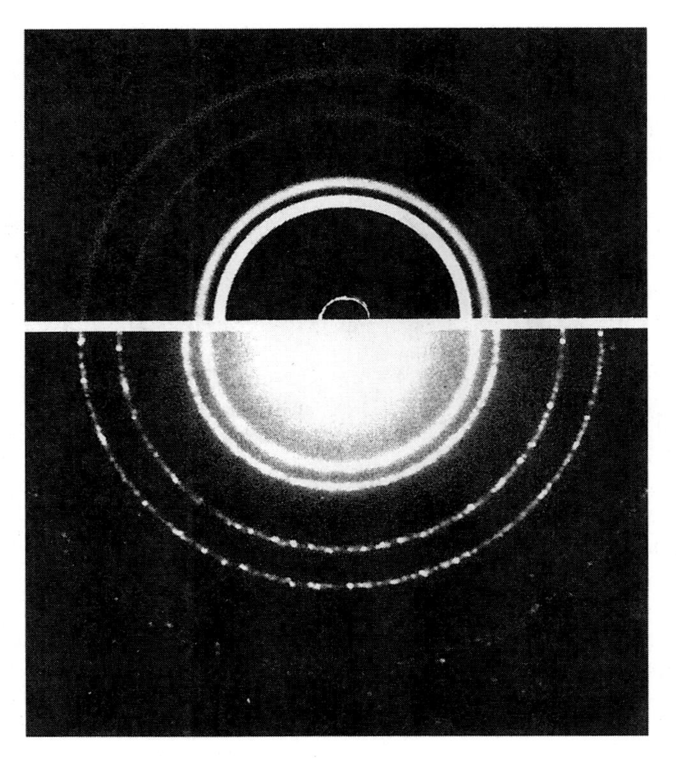

FIGURE 3-27. Diffraction pattern produced by 600 eV electrons (top) and x rays of 0.71 Å (bottom).

been demonstrated, and they match those of x rays. A typical set of matching patterns is shown in Figure 3-27.

Example 3-3: Calculate the de Broglie wavelength associated with (a) a 50 g golf ball launched from Tiger Woods' driver at 20 m/s, (b) a proton with a velocity of 2200 m/s, and (c) a 10 eV electron.

SOLUTION: (a) According to de Broglie's hypothesis,

$$\lambda = \frac{h}{p} = \frac{h}{mv} = \frac{6.625 \times 10^{-34}\ \text{J} \cdot \text{s}}{0.050\ \text{kg}(20\ \text{m/s})} = 6.625 \times 10^{-34}\ \text{m}$$

which is so small that it can't be counted on to keep the ball in the fairway.

(b) A proton has a mass of 1.67×10^{-27} kg and at a speed of 2200 m/s, the nonrelativistic expression for momentum applies:

$$\lambda = \frac{h}{mv} = \frac{6.625 \times 10^{-34}\ \text{J} \cdot \text{s}}{1.67 \times 10^{-27}\ \text{kg}(2200\ \text{m/s})} = 1.8 \times 10^{-10}\ \text{m} = 1.8\ \text{Å}$$

which, because of the very small mass, is on the order of an x-ray wavelength.

(c) A 10 eV electron can be treated nonrelativisticly since its total energy is much less than its rest mass (511 keV); thus the de Broglie wavelength is

$$\lambda = \frac{h}{p} = \frac{h}{\sqrt{2\ \mathrm{meV}}}$$

$$= \frac{6.625 \times 10^{-34}\ \mathrm{J \cdot s}}{\sqrt{2(9.11 \times 10^{-31}\ \mathrm{kg})(10\ \mathrm{eV})(1.602 \times 10^{-19}\ \mathrm{J/eV})}} = 3.88 \times 10^{-10}\ \mathrm{m} = 3.88\ \text{Å}$$

Such a wavelength is easily measured with modern equipment.

De Broglie Waves and the Bohr Model

Bohr's second postulate can be derived from the fact that an electron travels at high speed some distance, r, from a nucleus with an associated de Broglie wavelength as shown in Figure 3-28. If the electron is precisely in phase with itself, it will remain in phase as a standing wave, as shown in Figure 3-28a. The only values of r that allow it to remain in phase are when the circumference contains an integral number of wavelengths, or

$$2\pi r \equiv n\lambda$$

Since electrons around a nucleus have a de Broglie wavelength of h/mv,

$$2\pi r = n\frac{h}{mv}$$

or

$$mvr = n\frac{h}{2\pi}$$

Since the angular momentum L is equal to mvr,

$$L = n\frac{h}{2\pi}$$

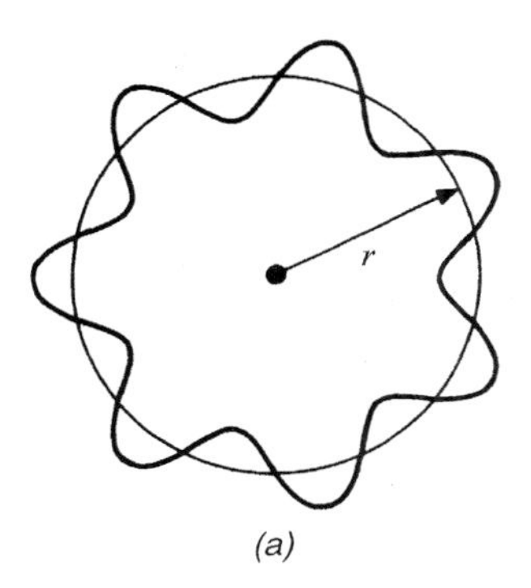

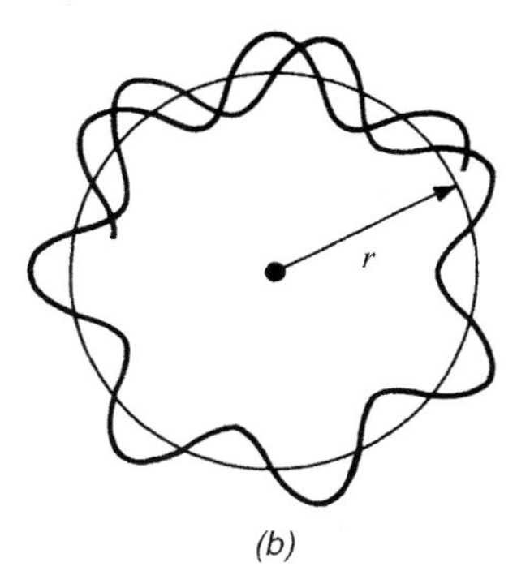

FIGURE 3-28. Orbital electron positioned (a) such that its de Broglie wavelength λ will continue to exist; at a radius such that $n\lambda \neq 2\pi r$, it exhibits destructive interference (b) and will cease to exist.

which is the second postulate that Bohr had stated on faith and insight in 1913 (some 13 years earlier) with no idea about the wave nature of electrons. Now the rule is "explained" as the result of a wave closing on itself and interfering constructively only for certain discrete radii corresponding to discrete energy states, an essential concept of atomic structure. If the electrons become more tightly bound, electromagnetic radiation equal to the difference between the two states must be emitted; to change to a less tightly bound state (a higher orbit), energy must be added.

Wave Mechanics

Confirmation of de Broglie's hypothesis allows a wave equation to be used to describe the way an electron, or any other particle, moves under a given force. Since the mechanics of the particle is described in terms of a wave, this type of mechanics is called *wave mechanics*, or since it also addresses quantum effects, *quantum mechanics*. Unfortunately, complex mathematics is required for complete understanding of wave mechanics, but the concepts necessary to radiation physics are relatively straightforward and can be discussed without performing these intricate calculations.

It has been well demonstrated that the variation in amplitude of a wave throughout space is given by

$$\frac{\partial^2 \psi}{\partial x^2} + \frac{\partial^2 \psi}{\partial y^2} + \frac{\partial^2 \psi}{\partial z^2} + \frac{4\pi^2}{\lambda^2}\psi = 0$$

where ψ is the amplitude of the wave and λ is its wavelength. This second-order differential equation can be used to calculate the dynamics of a particle with a de Broglie wavelength. First, the kinetic energy of a particle is related to its momentum by

$$\text{KE} = \frac{mv^2}{2} = \frac{m^2v^2}{2m} = \frac{p^2}{2m}$$

or

$$p = \sqrt{2m(\text{KE})}$$

and since the kinetic energy (KE) of a particle is the total energy E minus its potential energy P, the particle has a wavelength

$$\lambda = \frac{h}{p} = \frac{h}{\sqrt{2m(\text{KE})}} = \frac{h}{\sqrt{2m(E-P)}}$$

Insertion of this wavelength into the wave equation yields the time-independent Schrödinger wave equation:

$$\frac{\partial^2 \psi}{\partial x^2} + \frac{\partial^2 \psi}{\partial y^2} + \frac{\partial^2 \psi}{\partial z^2} + 4\pi^2\frac{2m(E-P)}{h^2}\psi = 0$$

By solving this equation for the energy states of electrons, Schrödinger was able in 1926 to explain completely Bohr's rules of quantization. Not only did the equation provide Bohr's results for the energy states of electrons in hydrogen, but

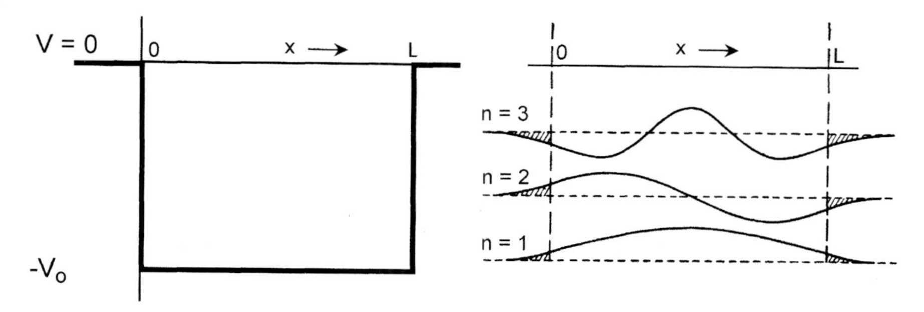

FIGURE 3-29. Wavefunction of a particle inside a square potential well of energy $-V_0$, showing probabilities of it existing outside the potential well.

it was found to be (and still is) applicable to all other elements and applicable to the nucleus as well. The solution of the Schrödinger wave equation for an electron provides its amplitude at a point in space, which is just the probability of finding the electron at that point. Such a solution is called the *wave function* or the *probability amplitude*.

One of the more interesting applications of the Schrödinger equation is for a bound particle (such as in an atom) where forces hold it in an energy "hole" where the potential energy inside, $-V_0$, is lower than the potential energy outside. Solution of the Schrödinger equation for this situation yields states in which the bound particle is sinusoidal inside the potential well, as shown in Figure 3-29, but significantly, with exponential tails outside. Thus there is some probability of finding the particles in a region where, according to classical theory, they do not have enough energy to be! Such behavior with a strongly supported theoretical basis is fundamental to explaining several important phenomena in radiation physics, especially energy states of orbital electrons and nuclear changes such as in radioactive transformation.

Bohr had arbitrarily introduced concepts to derive energy states for hydrogen; however, these same features plus others develop naturally by solving the Schrödinger equation for an electron with a de Broglie wavelength. The potential energy of the electron in the hydrogen atom is

$$P = -\frac{e^2}{(4\pi\varepsilon_0)r} = -\frac{e^2}{kr}$$

where r is the distance from the proton to the electron, which is $\sqrt{x^2 + y^2 + z^2}$. Substitution of P into the Schrödinger equation yields

$$\frac{\partial^2\psi}{\partial x^2} + \frac{\partial^2\psi}{\partial y^2} + \frac{\partial^2\psi}{\partial z^2} + \frac{8\pi^2 m}{h^2}\left(E + \frac{e^2}{4\pi\varepsilon_0\sqrt{x^2 + y^2 + z^2}}\right)\psi = 0$$

The solution of this equation, $\psi(x, y, z)$, is a function that has a definite value at each point in space around the nucleus. Where ψ is large, the electron is likely to be found; where it is small, the electron is unlikely to be found, as shown in Figure 3-30 for an electron in the hydrogen atom. The electron can be found anywhere between

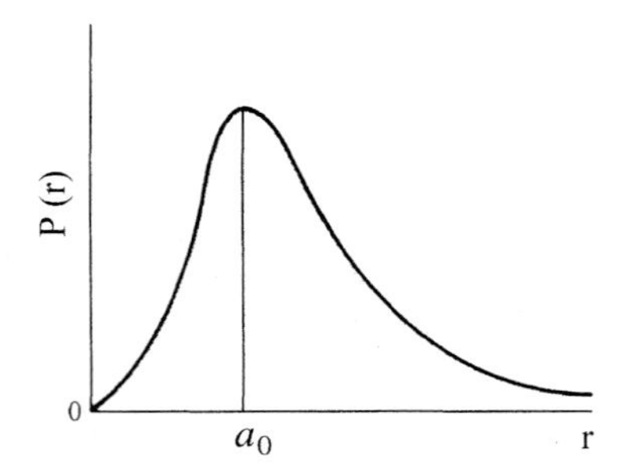

FIGURE 3-30. Plot of the probability of finding an electron in hydrogen versus r, the distance from the nucleus. Note that r_{max} corresponds to the first Bohr radius shown in Figure 3-31.

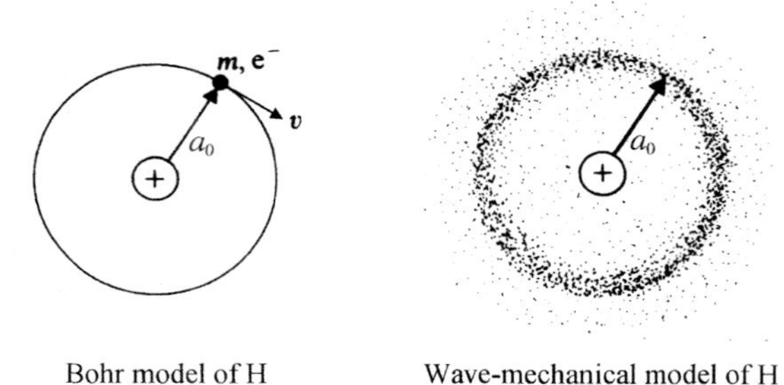

FIGURE 3-31. The Bohr model of the hydrogen atom places the electron at a_0; the wave mechanical model shows the relative probability of locating an electron, which is most likely to be found at $r = a_0$.

$r = 0$ and $r = \infty$, but it is most likely to be found at $r = a_0$ where its binding energy corresponds to the ground-state energy of -13.58 eV. This is in sharp contrast to Bohr's theory, as shown for two-dimensional space in Figure 3-31. Obviously, an electron moves around a nucleus in three-dimensional space, which produces a nebularlike cloud of electrons in atoms.

Schrödinger obtained energy eigenvalues for hydrogen as

$$E_n = -\frac{1}{n^2}\left(\frac{\mu k^2 e^4}{2\hbar^2}\right) = -\frac{E_1}{n^2}$$

where n is the principal quantum number. This result agrees with Bohr's calculations and confirms the exceptional insight of Bohr's original model of the atom for hydrogen; however, the Schrödinger equation and wave mechanics provide a more consistent and broadly applicable theory.

Exclusion Principle

As the broad principles of wave mechanics were developed, it was shown that other parameters of the atom system were also quantized, requiring additional quantum numbers to describe atomic phenomena completely. Four quantum nembers are used to account for the dynamics of electrons in atoms.

n = principal quantum number; this is the most important factor in specifying the total energy of the system: $n = 1, 2, 3, \ldots$

l = orbital quantum number, which quantizes the angular momentum of the system: $l = 0, 1, 2, 3, \ldots (n - 1)$

m = magnetic orbital quantum number, which quantizes the spatial orientation of the orbital plane in the presence of an external field; thus, only certain angles can exist between the magnetic field and the electron's orbit; thus $m = 0, \pm 1, \pm 2, \pm 3, \ldots, \pm l$

s = spin quantum number, which recognizes the inherent angular momentum of an electron spinning on its axis; it has only one magnitude, $\pm\frac{1}{2}$, but two directions, sometimes known as *parallel* and *antiparallel*

Each electron, in each energy state, is characterized by this set of four numbers. The *Pauli exclusion principle* states that no two electrons can have exactly the same set of the four basic quantum numbers. For example, if $n = 1$, l and m are zero and only two electron states are allowed (those with spins of $+\frac{1}{2}$ and $-\frac{1}{2}$). For the ground state of hydrogen, two states are possible, and two forms of atomic hydrogen are known. In elements beyond hydrogen, which require two or more orbital electrons for the atoms to be electrically neutral, the electrons will be grouped into energy levels or shells. The two electrons in helium form a closed shell (the K shell). The exclusion principle requires that the electron spins in helium be oppositely directed so they will contribute nothing to the total angular momentum. For $n = 2$, there are eight possible combinations and neon ($Z = 10$) has a closed K shell and a closed L shell, for a total of 10 electrons, just balancing Z.

The maximum number of electrons in any shell is $2n^2$. The plot of ionization potentials against atomic number in Figure 3-32 shows the tight electron binding in the closed shells of the noble gases ($Z = 2, 10, 28, \ldots$), the loose binding of the first electron added in each succeeding shell, and the gradual rise in binding energy as each shell fills.

Uncertainty Principle

Another important consequence of wave mechanics is the *Heisenberg uncertainty principle*, introduced by the German theoretical physicist Werner Heisenberg in 1927, which states that any position measurement generates a momentum uncertainty (and vice versa) such that the product of the uncertainties in the simultaneous measurements of position and momentum has a lower limit on the order of h, Planck's constant. Heisenberg derived a relationship for momentum and position from Schrödinger's equation to be

$$(\Delta x)(\Delta p) \sim \frac{h}{4\pi}$$

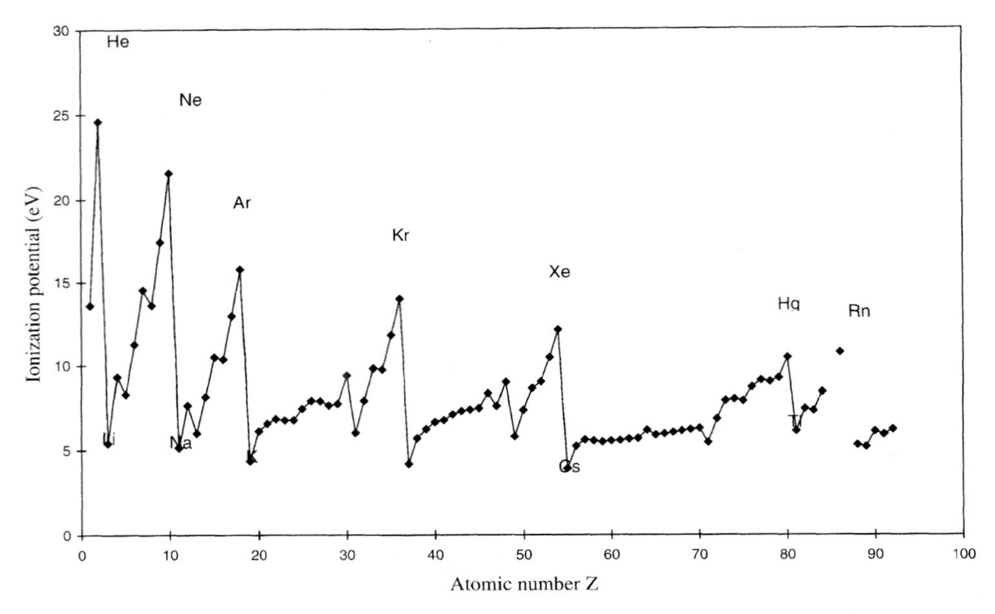

FIGURE 3-32. The periodic nature of the elements is evident from this plot of ionization potential versus Z for various elements.

An uncertainty in momentum occurs simply by "looking at" a particle, because to do so at least one photon must bounce off the particle. The photon will not locate the particle more closely than its own wavelength λ, because of the effects of diffraction. An accurate measure of position thus requires light of very short wavelength, but such photons have high momentum ($p = h/\lambda$) and in bouncing off the particle will impart some momentum to the particle, as shown by the Compton effect. Uncertainty between certain pairs of variables has to be accepted (i.e., there is a natural limit to how well position and momentum can both be known together).

The uncertainty principle provided the proof that electrons could not be inside the nucleus, which had been conjectured to account for the fact that atoms weighed approximately twice the atomic number, yet were electrically neutral. The uncertainty principle established that for an electron to be within the dimension of the nucleus, it would need a momentum that corresponded to an energy well above 20 MeV. Since the highest-energy beta particles were on the order of 1 MeV or so, free electrons could not exist in the nucleus.

Uncertainty also exists between the energy a particle is determined to have and the time over which it is measured. This uncertainty is also on the order of h and is

$$\Delta E\, \Delta t \sim \frac{h}{4\pi}$$

This uncertainty relationship means that to determine the energy of a particle exactly, it must be observed over an infinite amount of time, a clearly impossible requirement. In atomic systems, where particles move through small distances at high speed, the available time is quite small, requiring a corresponding uncertainty in determining the energy of the particle.

ATOM SYSTEMS

Although physicists have begun to delve deeper into the structure of matter and to develop and support a grand unification theory based on smaller and more energetic constituents (quarks, leptons, bosons, hadrons, etc.), the purposes of radiation physics can be represented with simplified models based on electrons, protons, neutrons, and their respective energy states.

The Neutron

Rutherford had postulated a neutral particle many years earlier to deal with the issues of mass and charge, but no evidence of its existence had been found. Rutherford's penchant for shooting alpha particles at substances not only discovered the transmutation of elements through nuclear interactions, but his discoveries led other researchers to do similar experiments. The Joliet-Curies, in particular, had irradiated all the light elements with alpha particles and discovered that they could produce radioactive substances artificially; however, their results for beryllium were puzzling. Alpha particles on beryllium yielded a highly penetrating radiation which they postulated to be high-energy gamma radiation since the emitted radiation was not deflected by a magnetic field. Their proposed reaction was

$$^{9}_{4}\mathrm{Be} + {}^{4}_{2}\mathrm{He} \rightarrow ? + \gamma$$

In an attempt to determine its energy, they used various materials to intercept it, one of which was paraffin. The interactions with paraffin yielded energetic protons that would have required gamma radiation on the order of 50 MeV. The highest-energy gamma rays observed thus far were at most a few MeV.

Chadwick repeated the Joliet-Curie experiments but with a suspicion that perhaps the product was Rutherford's neutral particle. By treating the product as a neutral particle of about the same mass as the proton, he was able to explain all the energies and momenta observed and to state the correct reaction as

$$^{9}_{4}\mathrm{Be} + {}^{4}_{2}\mathrm{He} \rightarrow [{}^{13}_{6}\mathrm{C}^{*}] \rightarrow {}^{12}_{6}\mathrm{C} + {}^{1}_{0}\mathrm{n}$$

The discovery of the *neutron*, which was made in 1932, tied together so many aspects of nuclear physics that Chadwick was quickly awarded the Nobel prize just four years later, in 1936. The Joilet-Curies had been one of the first to produce and observe neutrons; they just didn't know what they were.

The discovery also solved the puzzle of the mass of individual atoms and the presence of several isotopes of elements. Electrons could not exist in the nucleus, but with discovery of the neutron's existence, they weren't necessary. A nucleus made up of Z protons provided part of the mass (a little less than half) and all the charge required of the nucleus; the neutrons accounted for the balance of the mass, and since they had no charge, they had no effect on electrical neutrality.

All of a sudden, physicists had a new source of projectiles with which to probe the nucleus of atoms. Since they had no charge, neutrons were excellent projectiles for bombarding all elements, not just light ones, to see what happens; and marvelous things did. Fermi, in particular, set about to bombard all the known elements with neutrons, with particular emphasis on producing elements beyond uranium, *transuranic elements*. He also most certainly fissioned the uranium nucleus, one of the greatest discoveries of all time, but was so focused on producing transuranic elements that he missed it.

Nuclear Shell Model

There is no single complete model of the nucleus. Instead, a shell model is used to describe the energy levels in the nucleus, and a liquid-drop model best describes fission and other phenomena. The nucleus exhibits periodicities similar to those of atomic structure, as shown in Figure 3-33. These periodicities indicate a nuclear shell structure similar to the patterns of electron shells. Atoms that have 2, 8, 20, 28, 50, and 82 neutrons or protons and 126 neutrons are particularly stable. These values of N and Z are called *magic numbers*, and elements with them have many more stable isotopes than do their immediate neighbors. For example, Sn ($Z = 50$) has 10 stable isotopes, whereas In ($Z = 49$) and Sb ($Z = 51$) each have only 2. Similarly, for $N = 20$, there are 5 stable *isotones* (different elements with the same N number), while for $N = 19$, there is none, and for $N = 21$, there is only one. The same pattern holds for other magic numbers.

Neither the exact form of the nuclear force in the shell model nor the structure of potential energy states is currently known. It is, however, well known that the nuclear force is strongly attractive and extends only over a very short range from the center of the nucleus, going to zero within about 10^{-13} cm of the center. Rutherford's alpha-scattering experiments established the smooth variation of the coulombic force (inversely proportional to r^2) down to 10^{-12} cm or so with no perturbation by the nuclear force; thus the nuclear force is distributed only over the nucleus itself. Whatever the form of the potential, the shell model of the nucleus is based on two primary assumptions;

1. Each nucleon moves freely in a force field described by the nuclear potential, which appears to be related to the radial distance from the center of the system.
2. The Pauli exclusion principle applies (i.e., the energy levels or shells are filled according to nuclear quantum numbers).

The shell model with discrete energy states corresponds nicely with the emission of gamma rays from excited nuclei and can be thought of as similar to the emission of a photon when an orbital electron changes to an allowed quantum state of lower potential energy.

SUMMARY

Roentgen's discovery of x rays in 1895 began a series of discoveries that explained the structure of matter. Radioactivity, which was discovered shortly thereafter, es-

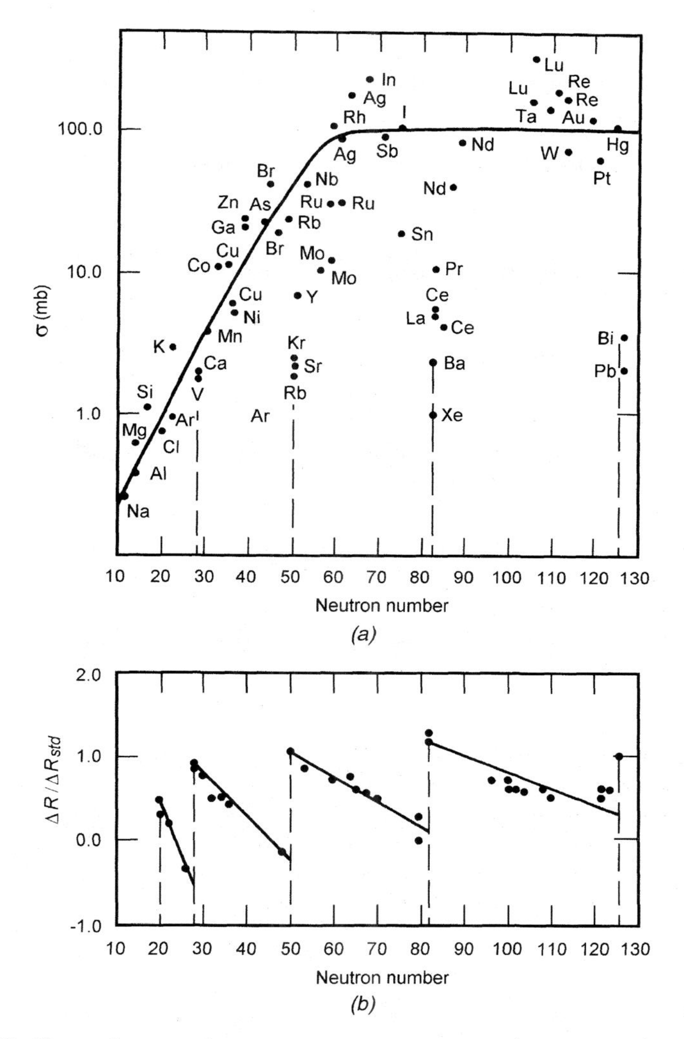

FIGURE 3-33. Sharp decrease in neutron-capture cross sections (*a*) and changes in the nuclear charge radius (*b*) for atoms with 20, 28, 50, 82, and 126 neutrons suggest a nuclear shell model. (From Shera et al., 1976.)

tablished that atoms are made up of constituent components, and it also made it possible to aim alpha particles at foils and gases. By doing so, Rutherford discovered that an atom must have a small positively charged nucleus because the observed scattering of alpha particles requires a very small scattering center at the center of the atom. Bohr used this discovery and other findings, including Planck's revolutionary hypothesis, to describe the motion of electrons and the processes of emission and absorption of radiation. Despite its simple clarity, Bohr's model of the atom was not quite precise, especially when applied to more complex atoms. Our current concept of the atom might better be called a nebular model, as shown in Figure 3-34, with electrons spread as waves of probability over the entire volume of

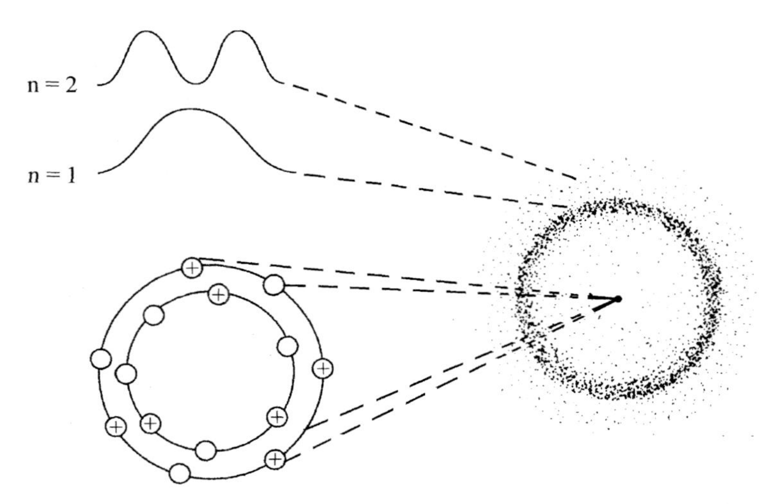

FIGURE 3-34. Very simplified model of a nebular atom consisting of an array of protons and neutrons with shell-like states within a nucleus surrounded by a cloud of electrons with three-dimensional wave patterns and also with shell-like energy states.

the atom, a direct consequence of de Broglie's discovery of the wave characteristics of electrons and other particles.

The electrons are distributed around the nucleus in energy states determined by four quantum numbers according to the exclusion principle at radii that are an equal number of de Broglie wavelengths, or $n\lambda$, where n is the principal quantum number corresponding to the respective energy shells, K, L, M,... for $n = 1, 2, 3, \ldots$. The dynamics of the electrons are described by the Schrödinger wave equation; changes between states are quantized, and discrete energies are apparent. The outer radius of the nebular cloud of electrons is about 10^{-10} m, which is some four to five orders of magnitude greater than the nuclear radius, thus the atom is mostly empty space. The nucleus has a radius of about 10^{-15} m and contains Z protons, which contribute both mass and positive charge, and neutrons (slightly heavier than protons), which make up the balance of the mass and help to distribute the nuclear force. Electrons do not, and, according to the uncertainty principle, cannot exist in the nucleus, although they can be manufactured and ejected during radioactive transformation.

Three force fields affect the behavior of particles in the nucleus (the fourth, gravitational, is also present but is negligible): the electrostatic force of repulsion between the protons; the strongly attractive nuclear force, which acts uniformly on both neutrons and protons across the nucleus but nowhere else (there is no center point toward which nucleons are attracted); and the weak force (relatively speaking), which influences radioactive transformation. No single model exists for the behavior, structure, and energy states of the nucleons under the influence of the three forces. Rather, a nuclear shell model is used to describe dynamics and changes in energy states of nuclear constituents, and a liquid-drop model is used for modeling the opposing nuclear and electrostatic forces in nuclear fission.

The nucleons in an atom have a deBroglie wavelength and their dynamics can be described by the Schrödinger wave equation, although with complex mathematics. Einstein expressed his difficulty with the probabilistic solutions of wave mechanics with the phrase "God doesn't play dice," believing, instead, that a more fundamental relationship must exist based on deterministic principles even though wave mechanics provided a mathematical tool, at least for the time being, to explain a wide range of physical phenomena. Despite Bohr's response that "Einstein shouldn't try to tell God what he could do," he spent the latter part of his life seeking a unified field theory, but was not successful. Only time will tell if even greater discoveries await the prepared minds of physicists as they develop experimental data and construct theories to explain the subtleties of nature and its atoms.

REFERENCES AND ADDITIONAL RESOURCES

Compton, A. H., *Physical Review* **22**, 411 (1923).

Shera, E. B., et al., *Physical Review C* **14**, 731 (1976).

Arya, A. P., *Elementary Modern Physics*, Addison-Wesley: Reading, MA, 1974.

PROBLEMS

3-1. What is the energy, in electron volts, of a quantum of wavelength $\lambda = 5500$ Å?

3-2. When light of wavelength 3132 Å falls on a cesium surface, a photoelectron is emitted for which the stopping potential is 1.98 V. Calculate (**a**) the maximum energy of the photoelectron, (**b**) the work function, and (**c**) the threshold frequency.

3-3. Light of wavelength 4350 Å is incident on a sodium surface for which the threshold wavelength of the photoelectrons is 5420 Å. Calculate (**a**) the work function in electron volts, (**b**) the stopping potential, and (**c**) the maximum velocity of the photoelectrons emitted.

3-4. The work function of potassium is 2.20 eV. (**a**) What should be the wavelength of the incident electromagnetic radiation so that the photoelectrons emitted from potassium will have a maximum kinetic energy of 4 eV? (**b**) Calculate the threshold frequency.

3-5. Calculate the value of Planck's constant from the following data, assuming that the electronic charge e has a value of 1.6022×10^{-19} C: A surface when irradiated with light of wavelength 5896 Å emits electrons for which the stopping potential is 0.12 V; when the same surface is irradiated with light of wavelength 2830 Å, it emits electrons for which the stopping potential is 2.20 V.

3-6. Calculate the work function and threshold frequency for the surface in Problem 3-5 (**a**) for the data provided and (**b**) with the actual value for Planck's constant of $6.6260755 \times 10^{-34}$ J · s.

3-7. Electrons in an x-ray tube are accelerated through a potential difference of 3000 V. What is the minimum wavelength of the x rays produced in a target?

3-8. X rays with energies of 200 keV are incident on a target and undergo Compton scattering. Calculate (**a**) the energy of the x rays scattered at an angle of 60° to the incident direction, (**b**) the energy of the recoiling electrons, and (**c**) the angle of the recoiling electrons.

3-9. Calculate the de Broglie wavelength associated with (**a**) an electron with a kinetic energy of 1 eV, (**b**) an electron with a kinetic energy of 511 keV, (**c**) a thermal neutron (2200 m/s), and (**d**) a 1500 kg automobile at a speed of 100 km/h.

3-10. Calculate the de Broglie wavelength associated with (**a**) a proton with 15 MeV of kinetic energy and (**b**) a neutron of the same energy.

3-11. Through what potential difference should an electron be accelerated so that the de Broglie wavelength associated with it is 0.1 Å?

3-12. Calculate Avogadro's number from the following data: the distance between the adjacent planes of a KCl crystal is 3.14 Å; the density and molecular weight of the KCl crystal are 1.98×10^3 kg/m^3 and 74.55, respectively.

4

INTERACTIONS

When the intervals, passages, connections, weights, impulses, collisions, movement, order and position of the atoms interchange, so also must the things formed from them change.

—Lucretius, ca. 100 B.C.

Much of what is known about atoms and radiation from them was learned by aiming subatomic particles at various target materials. The discovery of cathode rays provided a source of projectiles that led to the discovery of x rays and subsequently, radioactivity. Isolation of alpha-emitting radioactive substances following the discovery of radioactivity provided Rutherford and his collegues with another group of projectiles to shine on gold foils and other materials. The scattering of these particles gave the first clues and insight that the positive charges of atoms are coalesed into a very small nucleus.

The processes associated with subatomic projectiles can be discussed under the broad heading *interactions*, including a special category of nuclear interactions between projectiles and the nuclei of atoms. Interactions and the parameters that relate to them provide useful information on the physics and energetics of atoms and on the creation of new products and emanations that lead to key perspectives in defining and solving radiation protection issues related to them. Five aspects of interactions are important:

1. Production of x rays by accelerated electrons
2. The concept of cross section and how it is used to describe interactions
3. Interactions of alpha particles, protons, deuterons, neutrons, and light nuclei with various target nuclei
4. Transmutation of heavy elements
5. Fission and fusion reactions

PRODUCTION OF X RAYS

Roentgen was able to describe the behavior of x rays after his monumental discovery by conducting several experiments; however, it was not possible to explain how

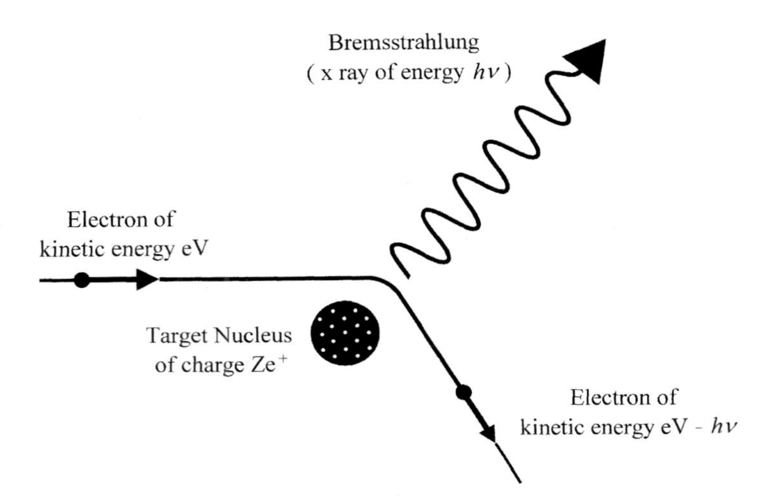

FIGURE 4-1. Production of x rays in which accelerated electrons emit bremsstrahlung.

x rays are produced until the concepts of atoms, particles, and quanta were understood. X-ray production is now known to be a special interaction between an accelerated electron and a target atom, as shown in Figure 4-1.

The negatively charged electron has a kinetic energy of eV. When it comes under the force field of the positively charged nucleus of the target atom, which is strongest for high-Z materials such as tungsten, it is deflected and accelerated as it is bent near the nucleus, causing the emission of electromagnetic radiation. This is consistent with classical electromagnetic theory because the electron is not bound. Because radiation is emitted and energy is lost in the process, the electron must slow down, so that when it escapes the force field of the nucleus, it has less energy. Overall, the electron experiences a net deceleration, and the energy after being decelerated is $eV - h\nu$, where $h\nu$ appears as electromagnetic radiation, which Roentgen named x rays to characterize their unknown status. This process of radiation being produced by an overall net deceleration of the electrons is called *bremsstrahlung*, a German word meaning *braking radiation*.

About 98% of the kinetic energy of the accelerated electrons is lost as heat because most of the interactions ionize target atoms. X-ray production occurs only when the electrons come close to a target nucleus and undergo significant deflection; thus x-ray production is a probablistic process depending on the path that any individual electron may take past a target nucleus. Since the electron can take any path, including one in which all of its energy is lost, the bremsstrahlung photons are emitted at all energies up to the accelerating energy eV, and in all directions, including absorption in the target.

Figure 4-2 shows two typical x-ray spectra plotted as relative intensity versus energy of the emitted radiation from targets of tungsten (W) and molybdenum (Mo), both of which are operated at the same voltage. The x-ray spectrum from each target has a continuous distribution of radiation of all energies up to the maximum energy of the incoming electron. The value of E_{max} does not depend on the target material, but is directly proportional to the maximum voltage, V. Other aspects of the production and uses of x rays are described in Chapter 15.

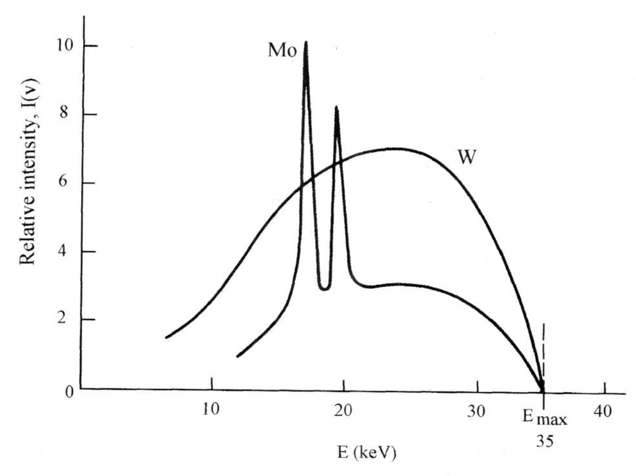

FIGURE 4-2. X-ray spectra of intensity $I(v)$ versus energy E for tungsten (W) and molybdenum (Mo) targets operated at the same voltage, 35 keV.

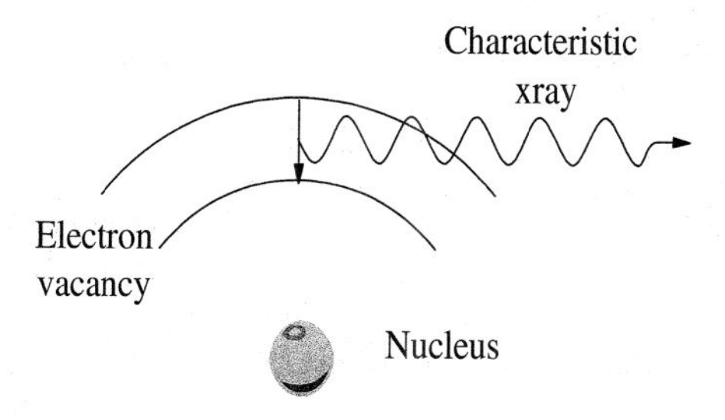

FIGURE 4-3. Emission of a characteristic x ray due to a higher-energy electron giving up energy to fill a particular shell vacancy.

CHARACTERISTIC X RAYS

Figure 4-2 shows discrete lines superimposed on the continuous x-ray spectrum for a molybdeum target but not for tungsten. The lines occur because the electrons striking the target have sufficient energy to overcome the binding energy of inner-shell electrons in the molybdenum target. Since the inner-shell electrons in tungsten are tightly bound at 69.5 keV, the electrons accelerated through 35 kV do not have sufficient energy to dislodge them, but 35 keV electrons can dislodge those in molybdenum, which have a binding energy of 20 keV. When these shell vacancies are filled by higher-energy electrons, electromagnetic radiation is emitted. The energy emitted is just the difference between the binding energy of the shell being filled and that of the shell from whence it came. The electrons in each element have unique energy states due to the energy array of the electrons in the element as dictated by quantum theory; thus these emissions of electromagnetic radiation are "characteristic" of the element, hence the term *characteristic x rays* (see Figure 4-3). They identify each element uniquely.

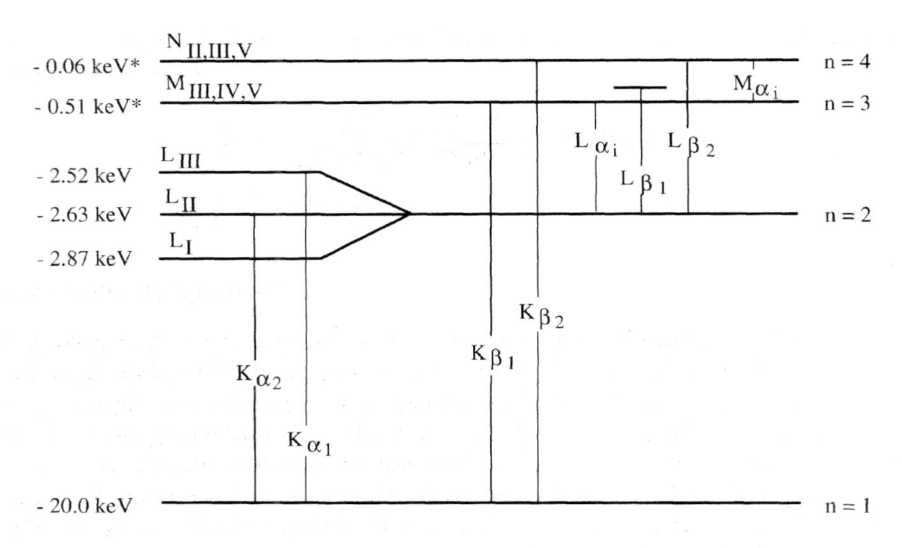

FIGURE 4-4. Emission of characteristic x rays from a Mo target due to electron vacancies in various shells followed by a higher-energy electron giving up energy to fill a particular shell vacancy.

The vacancy created by a dislodged orbital electron can be filled by an outer-shell (or free) electron changing its energy state, or as Bohr described it, jumping to a lower potential energy state. If the electron vacancy exists in the K shell, the characteristic x rays that are emitted in the process of filling this vacancy are known as K-shell x rays, or simply K x rays. Although the filling electrons can come from either the L, M, N,... shells, characteristic x rays are known by the shell that is filled. Further, the K x ray that originates from the L shell is known as the K_α x ray; if the transition is from the L_{III} subshell, it is a $K_{\alpha 1}$ x ray, and if from the L_{II} subshell, a $K_{\alpha 2}$ x ray. A transition from the L_I subshell is forbidden by the laws of quantum mechanics. In a similar fashion those from M, N, and O shells and subshells are known as K_β, K_γ, K_δ, and so on, with appropriate numerical designations for the originating subshells as shown in detail in Figure D-1 of Appendix D. Figure 4-4 illustrates these transitions in molybdenum.

L x rays are produced when the bombarding electrons knock loose an electron from the L shell; however, this occurs with lower probability than for the K shell. Similarly, electrons from higher levels drop down to fill these L-shell vacancies, producing L x rays. The lowest-energy x ray of the L series is known as L_α, and the other L x rays are labeled in order of increasing energy, which corresponds to being filled by electrons from higher-energy orbits, as shown in Figure 4-4. Characteristic x rays for M, N,... shells are designated by this same pattern. The lower probabilities of interactions in these higher-energy shells is due to the electrons being spread out over a larger volume, thus presenting a smaller target to incoming electrons.

A wide array of characteristic x rays can be observed for a given element, due to the subshells within the major shells. For example, the K_α x ray could originate from any one of the subshells of the $n = 2$ level, as shown in the left side of Figure 4-4. The energies of these different transitions will be slightly different, and emissions associated with each are possible. The energy differences are very small for most shells, but worth noting for the K and L shells, especially for the higher-Z elements. Listings of emitted radiations associated with radioactive transformations denote these as $K_{\alpha 1}$, $K_{\alpha 2}$, $L_{\alpha 1}$, $L_{\alpha 2}$, and so on (see Chapter 5).

It is relatively straightforward to determine the energies of characteristic x rays from an element if the binding energies of each of the electron shells and subshells is known, because the emitted photons are directly related to the differences in energy states for each transition that can occur. A simplified schematic of electron shell energies in molybdenum is shown in Figure 4-4, including the energies of the L subshells.

Example 4-1: From the data in Figure 4-4, compute the energy of $K_{\alpha 1}$, $K_{\alpha 2}$, K_{β}, K_{γ}, and L_{α} characteristic x rays.

SOLUTION: The emitted x rays occur because electrons change energy states to fill shell vacancies. Each characteristic x-ray energy is the difference in potential energy the electron has before and after the transition. For K x rays:

$$K_{\alpha 1} = -2.52 - (-20.0) = 17.48 \text{ keV}$$

$$K_{\alpha 2} = -2.63 - (-20.0) = 17.37 \text{ keV}$$

$$K_{\beta} = -0.50 - (-20.0) = 19.50 \text{ keV}$$

$$K_{\gamma} = 0 - (-20.0) = 20.00 \text{ keV}$$

For L-shell vacancies, it is presumed that the lowest-energy subshell is filled from the lowest-energy M-shell, or

$$L_{\alpha} = -0.5 - (-2.87) = 2.37 \text{ keV}$$

Of course, other permutations are possible which would produce a large array of discrete characteristic x rays, all of which can be resolved with modern x-ray spectrometers.

X Rays and Atomic Structure

Study of characteristic x rays led to fundamental information on atomic structure. In particular, the work of Moseley, who analyzed K_{α} x rays in considerable detail, proved that the periodic table should be ordered by increasing Z instead of by mass and that certain elements were out of order and others were not yet discovered. The relative positions of the K_{α} and K_{β} lines in Moseley's original photographic images arrayed according to atomic number Z are shown in Figure 4-5. The lines are for elements from calcium to copper, and it is clear that the wavelengths decrease in a regular way as the atomic number increases. The gap between the calcium and titanium lines represents the positions of the lines of scandium, which occurs between those two elements in the periodic table.

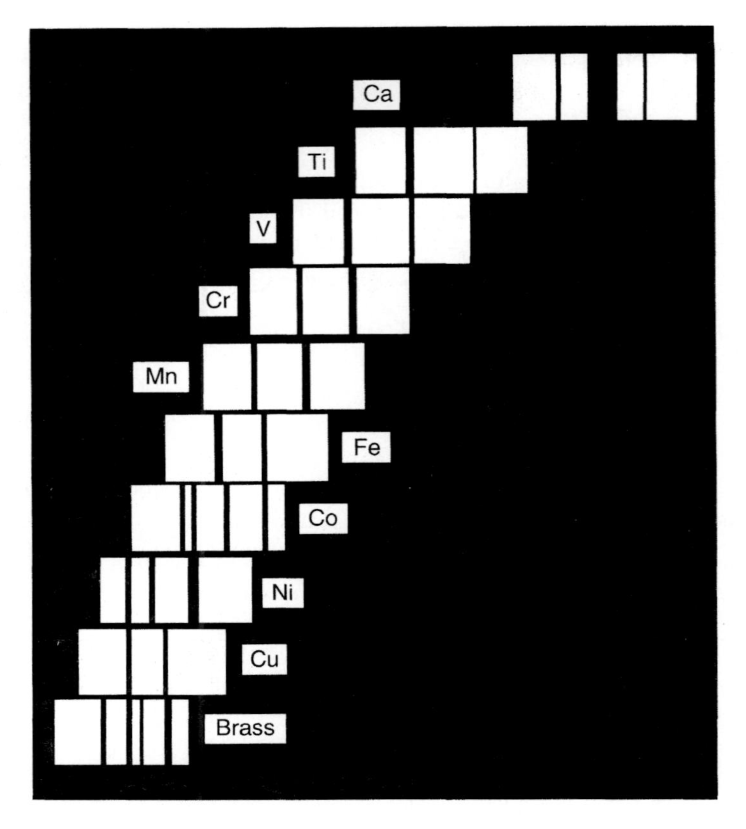

FIGURE 4-5. Moseley's photographs of K_α and K_β x-ray lines for various light elements.

Before Moseley's work the periodic table was ordered according to increasing mass, which placed certain elements in the wrong period (e.g., cobalt and nickel or iodine and tellurium). Moseley found that when the elements were ordered according to Z, the chemical properties corresponded to the proper group. He also found gaps corresponding to yet undiscovered elements; for example, the radioactive element technetium ($Z = 43$) does not exist in nature and was not known at the time, but his work clearly showed such a gap at $Z = 43$.

Auger Electrons

A characteristic x-ray photon is produced any time an electron vacancy exists in an atom and it is filled by a higher-energy electron. But not every characteristic x ray that is produced is emitted. Some of these do not exit the atom because of a phenomenon called the *Auger effect*. When this occurs, the x ray interacts with another orbital electron, causing the emission of an electron from a shell farther out. The Auger effect is thus similar to an inner photoelectric effect. One electron may fall into a lower shell and another one is emitted simultaneously. The Auger effect is represented schematically in Figure 4-6.

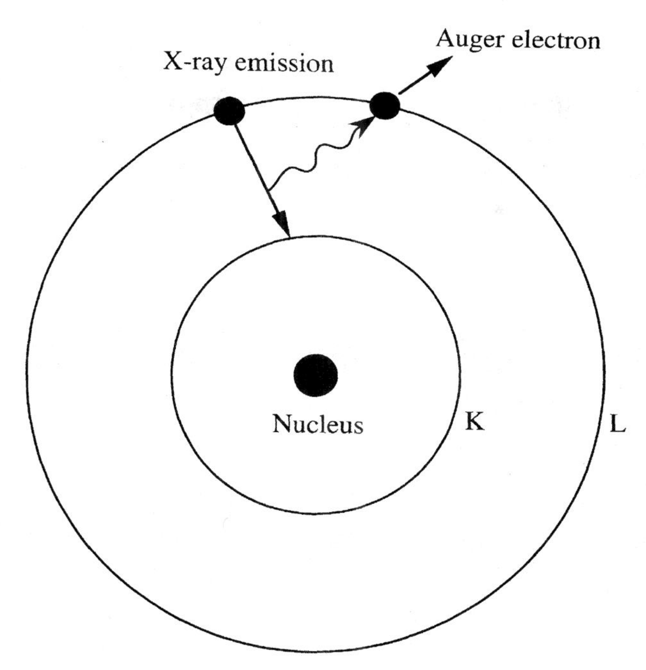

FIGURE 4-6. Auger electron emission (right) competes with emission by characteristic x rays (left).

First, a K-shell vacancy occurs, which can be filed by any higher-energy electron, and the energy released as a characteristic x ray can remove a second L electron from the L shell. The result is that the L shell loses two electrons. These are then replaced by electrons from the M shell or farther out, sometimes producing a cascade of Auger electrons. The probability of such nonradiative processes, which compete with x-ray emission, has been found to decrease with increasing nuclear charge. In light atoms with low Z the ejection of Auger electrons far outweighs characteristic x-ray emissions (Figure 4-7). The fraction of total emissions that occur by Auger electrons is denoted by $1 - \eta$. The kinetic energy of the Auger electron is determined simply by the energy $h\nu$ of the characteristic x ray (if it were to be emitted) minus the binding energy of the ejected electron in its respective shell, or

$$\text{KE}_{\text{auger}} = h\nu - BE_e$$

Example 4-2: Silver is bombarded with 59.1 keV K_α radiation from tungsten. If the binding energies of the K- and L-shell electrons in Ag are 25.4 and 3.34 keV, respectively, what are (a) the energies of electrons ejected from the K and L shells by direct ionization, and (b) the energies of Auger electrons ejected from the L shell by the K_α and K_β x rays?

SOLUTION: (a) The energy of photoelectrons from the L shell of silver is

$$59.1 - 3.34 = 55.76 \text{ keV}$$

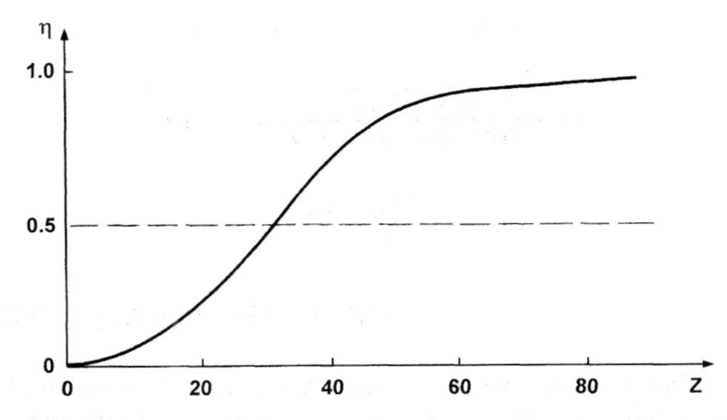

FIGURE 4-7. Ratio η of characteristic x-ray emission to Auger electron production as a function of Z.

and from the K shell is

$$59.1 - 25.4 = 33.7 \text{ keV}$$

(b) The energy of auger electrons ejected from the L shell by K_β x rays is

$$24.9 - 3.34 = 21.56 \text{ keV}$$

and by K_α x rays is

$$22.06 - 3.34 = 18.72 \text{ keV}$$

NUCLEAR INTERACTIONS

Nuclear interactions are those that involve absorption of a bombarding particle by the nucleus of a target material (in x-ray production the electron is not absorbed by the nucleus). Absorption of the projectiles produces first a compound nucleus, which then breaks up to yield the final products. Most interactions of interest use alpha particles (α), protons (p), deuterons (d), neutrons (n), or light nuclei such as tritium (^{3}H) or helium (^{3}He) as projectiles. In certain circumstances, reactions can be induced by photons.

Nuclear interactions can be represented as follows:

$$\text{X} + \text{x} \rightarrow [\text{compound nucleus}] \rightarrow \text{Y} + \text{y}$$

where X is the target nucleus, x the bombarding particle or projectile, Y the product nucleus, and y the emitted product (either a particle, a nucleus, or a photon). Such reactions are also shown in condensed form as

target (projectile, emission) product

The total charge (total Z) and the total number of nucleons (total A) are the same before and after the reaction, and momentum and energy must be conserved.

Rutherford was one of the first to bombard materials with alpha particles to observe what happened, and his results on the scattering of alpha particles led him to put the nucleus at the center of the atom. He obtained an equally startling result from bombarding nitrogen with alpha particles when he showed that nitrogen could be transmuted into oxygen and a stream of emitted protons. This famous nuclear interaction, which demonstrated the transmutation of elements, the first of its kind, is

$$ {}^{14}_{7}\text{N} + {}^{4}_{2}\text{He} \rightarrow [{}^{18}_{9}\text{F}] \rightarrow {}^{17}_{8}\text{O} + {}^{1}_{1}\text{H} $$

or in abbreviated form, $^{14}N(\alpha,p)^{17}O$.

Rutherford was a clever experimentalist. He placed a movable source of radium C′, which emits 7.69 MeV alpha particles, inside a box that he could fill with various gases. A zinc sulfide screen was placed just outside the box to detect scintillations that occurred. When the box was filled with oxygen or carbon dioxide at atmospheric pressure, no scintillations were seen on the screen with the source 7 cm or more away, which is the range of the alpha particles in these gases. However, when the box was filled with nitrogen, scintillations were observed on the screen when the source of the alpha particles was as much as 40 cm away. Since the alpha particles could not penetrate 40 cm of air, Rutherford concluded that the scintillations were caused by particles ejected from the nitrogen nuclei by the impact of the alpha particles. Measurement of the magnetic deflection of the emitted particles confirmed that they were, in fact, protons. The probability of such disintegration is very small; 1 million alpha particles are required to produce one interaction in which nitrogen is transmuted to oxygen and a proton.

Dynamics of Nuclear Interactions

The (α,p) reaction with nitrogen as well as other particle–target reactions is governed by a general set of dynamic principles. The charge, mass, and energy of the bombarding particle (or projectile) determines its interaction probability. Although it is generally assumed that the target atoms are at rest, their mass and charge also affect the likelihood of interaction. The emitted particle from the compound nucleus also has mass, charge, and energy. All of these need to be accounted for in describing how the interaction proceeds to produce a changed atom and an outgoing particle and/or energy emission. The effects of these parameters on the probability of any given interaction are collectively described by the cross section of the interaction (see below).

Three categories of interactions can occur, as illustrated in Figure 4-8: (1) scattering, in which the projectile bounces off the target nucleus with a transfer of some of its energy; (2) a high-energy interaction in which the projectile either collects (picks up) or loses (strips) nucleons; and (3) absorption into the target nucleus, forming a new atom that then undergoes change. Each type of interaction will produce recoil of the nucleus and deceleration (or stopping) of the particle, which alters its momentum and energy.

Scattering reactions produce a decrease in the energy of the projectile by elastic or inelastic scattering. If the residual nucleus is left in its lowest or ground state,

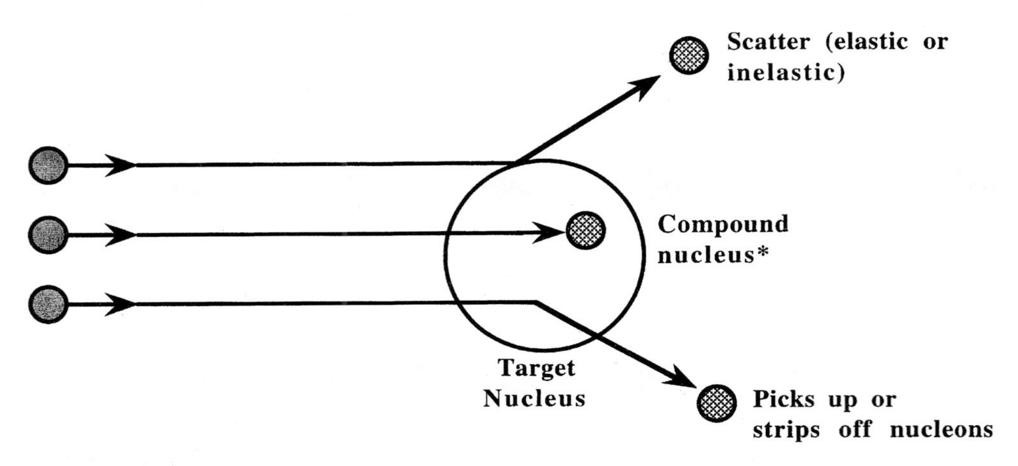

FIGURE 4-8. Nuclear interactions in which a bombarding particle is scattered by the target nucleus, absorbed to form a compound nucleus with excess energy, or picks up or strips nucleons.

the scattering is *elastic*; when the residual nucleus is left in an excited state, the scattering is called *inelastic*.

Pickup and stripping reactions usually occur when the projectile has high energy, and in such reactions the nucleon enters or leaves a definite "shell" of the target nucleus without disturbing the other nucleons in the target. Rutherford's (α, p) reaction with nitrogen may also be thought of as a stripping reaction.

Absorption reactions occur when the incident projectile is fully absorbed into a target atom to form a new compound nucleus which lives for a very short time in an excited state and then breaks up. It exists only for 10^{-16} s or so; thus, it cannot be observed directly, but this is much longer than the 10^{-21} s required to traverse the nucleus. It is therefore assumed that the compound nucleus does not "remember" how it was formed, and consequently, it can break up any number of ways, depending only on the excitation energy available. The formation and breakup of ^{64}Zn is one such example:

$$\begin{array}{lcl} ^{63}\mathrm{Cu} + {}^{1}_{1}\mathrm{H} \rightarrow & & \rightarrow {}^{63}\mathrm{Zn} + \mathrm{n} \\ & ^{64}\mathrm{Zn}^{*} & \rightarrow {}^{62}\mathrm{Cu} + \mathrm{n} + \mathrm{p} \\ ^{60}\mathrm{Ni} + {}^{4}_{2}\mathrm{He} \rightarrow & & \rightarrow {}^{62}\mathrm{Zn} + 2\mathrm{n} \end{array}$$

The relative probabilities of forming any of the three interaction products from the compound nucleus $^{64}\mathrm{Zn}^{*}$ are essentially identical, even though the initial particles are different.

Cross Section

The likelihood of an interaction between a bombarding particle and a nucleus is described by the concept of cross section denoted by σ_i, where i refers to the product

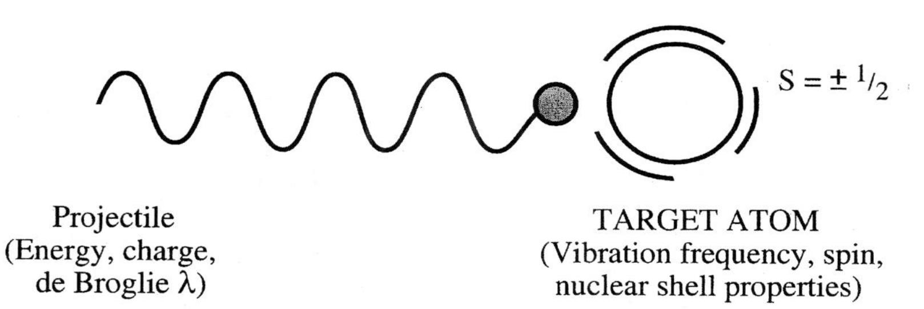

FIGURE 4-9. Parameters related to interactions of particles with target nuclei.

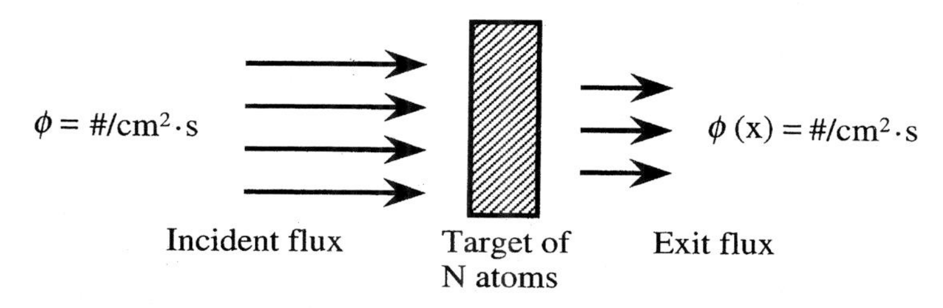

FIGURE 4-10. An incident flux ϕ_0 of projectiles on a target containing N atoms is diminished to $\phi(x)$ after traversing a thickness x.

of the interaction. There is no guarantee that a particular bombarding projectile will interact with a target nucleus to bring about a given reaction; thus σ provides only a measure of the probability that it will occur. The cross section is dependent on the target material and features of the incident "particle," which may in fact be a photon. Schematically, the cross-sectional area presented to the incoming projectile can be represented as shown in Figure 4-9.

Several properties of both the projectile and the target nucleus influence whether a given scattering or absoption interaction will occur. These include the energy, charge, mass, and de Broglie wavelength of the projectile, and the vibrational frequency, the spin, and the energy states of nucleons in the target atom. These cannot be predicted directly by nuclear theory; thus, cross sections for any given arrangement of projectile and target are usually determined by measuring the number of projectiles per unit area (i.e., the flux) before and after they impinge on a target containing N atoms (see Figure 4-10).

The reduction in the flux after passing through the target is a direct measure of the number of interactions that occur in the target; it is a function of the flux ϕ and the interaction probability σN of the target atoms, or

$$-\frac{d\phi}{dx} = \sigma N \phi$$

which can be integrated to yield

$$\phi(x) = \phi_0 e^{-\sigma N x}$$

The units of σ will be in terms of apparent area (m^2 or cm^2) per target atom. Although the cross section has the dimensions of area, it is not the physical area presented by the nucleus to the incoming particle but is better described as the sum total of those nuclear properties that determine whether or not a reaction is favorable. When such measurements and calculations were first made for a series of elements, the reseachers were surprised that the calculated values of σ were as large as observed. Apparently, someone exclaimed that it was "as big as a barn," and the name stuck. One *barn*, or b, is

$$\text{b} = 10^{-24}\ \text{cm}^2 \qquad \text{or} \qquad 10^{-28}\ \text{m}^2$$

which is on the order of the square of a nuclear radius. Cross sections vary with the energy of the bombarding particle, the target nucleus, and the type of interaction.

Example 4-3: A flux of 10^6 neutrons/s in a circular beam of 1 cm radius is incident on a foil of aluminum (density = 2.7 g/cm^3) that was measured in several places with a micrometer to average 0.5 cm in thickness. After passing through the foil the beam contained 9.93×10^5 neutrons/s. Calculate the cross section σ_γ.

SOLUTION: The number of atoms/cm^3 in the foil is

$$\frac{(2.7\ \text{g/cm}^3)(6.022 \times 10^{23}\ \text{atoms/mol})}{26.982\ \text{g/mol}} = 6.026 \times 10^{22}\ \text{atoms/cm}^3$$

$$\ln \frac{\phi(x)}{\phi_0} = -\sigma_\gamma N x = -\sigma_\gamma \times 6.026 \times 10^{22}\ \text{atoms/cm}^3 \times 0.5\ \text{cm}$$

$$\ln \frac{9.93 \times 10^5}{10^6} = -3.013 \times 10^{22}\ \sigma_\gamma$$

and $\sigma_\gamma = 0.233 \times 10^{-24}$ cm^2 or 0.233 b.

Q-Value

The Q-value represents the energy balance, or mass change, that occurs between the particles and targets in a nuclear interaction and the products of the reaction, or

$$Q = (M_X + m_x)c^2 - (M_Y + m_y)c^2$$

where m_x, M_X, m_y, and M_Y represent the masses of the incident particle, target nucleus, product particle, and product nucleus, respectively, and the E's represent their kinetic energies. Thus the Q-value is easily calculated by subtracting the masses of the products from the masses of the reactants. If the Q-value is positive, the kinetic energy of the products is greater than that of the reactants and the reaction is *exoergic*, that is, energy is gained at the expense of the mass of the reactants. If the Q-value is negative, the reactions are *endoergic* and energy must be supplied for the reaction to occur.

An amount of energy actually needed to bring about an endoergic reaction is somewhat greater than the Q-value. When the incident particle collides with the target nucleus, conservation of momentum requires that the fraction $m_x/(m_x + M_X)$ of the kinetic energy of the incident particle must be retained by the products as kinetic energy; thus, only the fraction $M_X/(m_x + M_X)$ of the energy of the incident particle is available for the reaction. The threshold energy E_{th}, which is somewhat larger than the Q-value, is the kinetic energy that the incident particle must have for the reaction to be energetically possible; it is given by

$$E_{th} = -Q\frac{M_X + m_x}{M_X}$$

Example 4-4: Calculate the Q-value for the $N^{14}(\alpha,p)O^{17}$ reaction and the threshold energy.

SOLUTION: The atomic masses from Appendix B are:

Mass of reactants

- 14.003074 u for ^{14}N
- 4.002603 u for ^{4}He

Mass of products

- 16.999131 u for ^{17}O
- 1.007826 u for ^{1}H

The mass difference is −0.001280 u; thus $Q = -1.19$ MeV, and the reaction is endoergic (i.e., energy must be supplied). The threshold energy E_{th} that the bombarding alpha particle must have for the reaction to occur is

$$E_{th} = -(-1.19 \text{ MeV})\left(\frac{14.003074 + 4.002603}{14.003074}\right) = 1.53 \text{ MeV}$$

which can be supplied by almost any natural source of alpha particles. Although in this example we used exact mass values, calculations of E_{th} using integer mass numbers yield sufficiently accurate values.

Q-value calculations are done with the masses of the neutral atoms rather than the masses of the bombarding and ejected particles. The number of electrons required for a nuclear reaction equation balances when the masses of the neutral atoms are used because the number of electrons is the same on both sides of the equation. It is necessary, however, to examine the balance equation for a given reaction to assure that this is in fact the case. Two reactions involving radioactive transformation illustrate these factors:

$$^{14}_{6}C \rightarrow {}^{14}_{7}N + {}^{0}_{-1}e + Q$$

$$^{22}_{11}Na \rightarrow {}^{22}_{10}Ne + {}^{0}_{+1}e + \text{orbital e} + Q$$

The first reaction yields a positive Q-value (as all radioactive transformations must) of 0.156 MeV, which is obtained simply by subtracting the ^{14}N mass from that of ^{14}C. It may appear that the beta-particle (electron) mass was left out, but it is in fact included because the mass of neutral ^{14}N includes seven electrons, one more than the parent ^{14}C. By using the masses of the neutral isotopes, we have effectively "waited" until the product ^{14}N picks up an additional orbital electron to balance the net increase of one proton in the nucleus due to the beta-particle transformation of ^{14}C (i.e., the electron mass is included as though the beta particle stopped and attached itself to the ionized ^{14}N product). The binding energy of the electron in ^{14}N is in fact not accounted for, but this is very small compared to the other energies involved.

This convenient circumstance does not exist, however, in the second reaction in which a positron is emitted. The proton number decreases by one unit, which requires that an electron drift off from the product atom. Using the neutral masses of the parent and product nuclei to calculate the Q-value leaves two electron masses unaccounted for (the orbital that drifts off and the emitted positron); thus it is necessary to add these to obtain a correct Q-value.

ALPHA-PARTICLE INTERACTIONS

The transmutation of a nitrogen atom by an alpha particle is one of the most important ever observed:

$$^{14}_{7}\text{N} + ^{4}_{2}\text{He} \rightarrow [^{18}_{9}\text{F}] \rightarrow ^{17}_{8}\text{O} + ^{1}_{2}\text{H}$$

When Rutherford first observed it, he was uncertain whether the alpha particle and the nitrogen nucleus had combined to form a new compound nucleus or whether a proton had just been knocked out of the nitrogen nucleus. If the alpha particle disappears, a compound nucleus is produced and transmutation occurs; if the proton is knocked out, the alpha particle would still exist and nuclear interactions could not be characterized as transmutations, but just knocking things together. Blackett performed a classic experiment, the results of which are shown schematically in Figure 4-11, which showed that in fact the proton is emitted after a compound nucleus is formed. Blackett's photograph (one of only eight from 20,000 cloud chamber photos and over 400,000 alpha tracks) shows that the only tracks observed after the

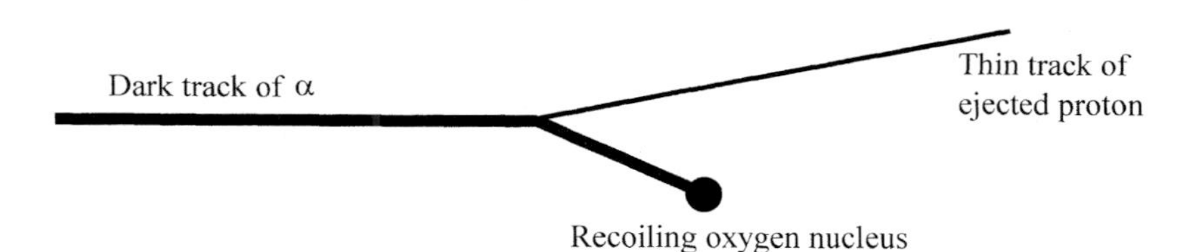

FIGURE 4-11. Schematic of P. M. S. Blackett's classic photograph showing the thick alpha track before it "vanishes" (is absorbed), the thin track of the proton, and the heavy track of the recoiling product nucleus.

collision were those of the incident alpha particle, a proton, and a recoil nucleus. The alpha particle had vanished. These three tracks (not four) proved that the alpha particle entered the nucleus of the nitrogen atom to form an unstable compound nucleus which immediately expelled a proton. Atoms had been transmuted, and Blackett received a Nobel prize in 1948 for providing the experimental proof of this fundamental precept.

Rutherford and his colleagues observed alpha-particle transmutation for all the light elements from boron to potassium, except carbon and oxygen. Some of these are

$$^{10}_{5}\mathrm{B} + {}^{4}_{2}\mathrm{He} \rightarrow [{}^{14}_{7}\mathrm{N}] \rightarrow {}^{13}_{6}\mathrm{C} + {}^{1}_{1}\mathrm{H}$$

$$^{23}_{11}\mathrm{Na} + {}^{4}_{2}\mathrm{He} \rightarrow [{}^{27}_{13}\mathrm{Al}] \rightarrow {}^{26}_{12}\mathrm{Mg} + {}^{1}_{1}\mathrm{H}$$

$$^{27}_{13}\mathrm{Al} + {}^{4}_{2}\mathrm{He} \rightarrow [{}^{31}_{15}\mathrm{P}] \rightarrow {}^{30}_{14}\mathrm{Si} + {}^{1}_{1}\mathrm{H}$$

$$^{32}_{16}\mathrm{S} + {}^{4}_{2}\mathrm{He} \rightarrow [{}^{36}_{18}\mathrm{Ar}] \rightarrow {}^{35}_{17}\mathrm{Cl} + {}^{1}_{1}\mathrm{H}$$

$$^{39}_{19}\mathrm{K} + {}^{4}_{2}\mathrm{He} \rightarrow [{}^{43}_{21}\mathrm{Se}] \rightarrow {}^{42}_{20}\mathrm{Ca} + {}^{1}_{1}\mathrm{H}$$

The ejected protons from some of these interactions were found to have more energy than the bombarding alpha particles. This extra energy was acquired by rearrangement of the bombarded nucleus, an important result in understanding atoms and how they acquire and emit energy with accompanying mass changes.

Alpha–Neutron Reactions

The discovery of the neutron is the most vivid example of (α,n) reactions and is of extraordinary importance because it provided a comprehensive structure of the atom (see Chapter 3). When beryllium was bombarded by alpha particles, no protons were observed, but several reseachers observed a very penetrating radiation. It was first assumed that the emitted radiation was a high-energy gamma ray which would have required formation of $^{13}\mathrm{C}$ with a Q-value of about 50 MeV, which was highly questionable. Rutherford and Chadwick had long suspected the existence of a neutral particle, and Chadwick turned his attention to these strange emissions from bombarding beryllium with alpha particles. The Joliet–Curies had provided an important piece of the puzzle by showing that the emissions produced a stream of protons when directed at paraffin, a hydrogenous material. Chadwick assumed that a neutral particle with mass near that of the proton was emitted and that elastic collisions occurred with the hydrogen nuclei in the paraffin. From the momenta observed, he showed that this was the correct hypothesis and that the emitted radiation was in fact a neutral particle (the neutron) with a mass very nearly equal to that of the proton. The nuclear reaction that was observed is

$$^{9}_{4}\mathrm{Be} + {}^{4}_{2}\mathrm{He} \rightarrow [{}^{13}_{6}\mathrm{C}] \rightarrow {}^{12}_{6}\mathrm{C} + {}^{1}_{0}\mathrm{n}$$

where ${}^{1}_{0}\mathrm{n}$ is the symbol for the neutron. Neutrons were highly penetrating because they had no charge and thus were not repelled by positively charged nuclei.

Once it was known that neutrons existed and were produced by alpha-particle interactions, other reactions were soon discovered. Some important ones are

$$^{7}_{3}\text{Li} + ^{4}_{2}\text{He} \rightarrow [^{11}_{5}\text{B}] \rightarrow ^{10}_{5}\text{B} + ^{1}_{0}\text{n}$$

$$^{14}_{7}\text{N} + ^{4}_{2}\text{He} \rightarrow [^{18}_{9}\text{F}] \rightarrow ^{17}_{9}\text{F} + ^{1}_{0}\text{n}$$

$$^{23}_{11}\text{Na} + ^{4}_{2}\text{He} \rightarrow [^{27}_{13}\text{Al}] \rightarrow ^{26}_{13}\text{Al} + ^{1}_{0}\text{n}$$

$$^{27}_{13}\text{Al} + ^{4}_{2}\text{He} \rightarrow [^{31}_{15}\text{P}] \rightarrow ^{30}_{15}\text{P} + ^{1}_{0}\text{n}$$

The reactions of alpha particles with ^{14}N, ^{23}Na, and ^{27}Al may, in accordance with the theory of a compound nucleus, also emit a proton. The (α,p) reaction with aluminum was found by the Joliet–Curies to be radioactive, which was the first observation of artificial radioactivity and gained them a Nobel prize in 1935.

TRANSMUTATION BY PROTONS

Although alpha particles can be used to produce a number of useful interactions, they are limited in application to light elements because the coulombic repulsion between their +2 charge and the high Z value of heavy elements is just too great to overcome. The range of possible interactions can be extended somewhat by using long-range alpha particles of 10 to 12 MeV from short-lived alpha emitters, but these are very difficult to produce and are very short lived. Protons and deuterons with a single + charge are more useful for nuclear interactions, but no natural sources are available.

Nuclear reactions with protons and deuterons became available with the Cockcroft and Walton linear accelerator, shown schematically in Figure 4-12, and the cyclotron, which was developed somewhat later (see Figure 9-7). A source of hydrogen ions (protons, deuterons, tritons) is introduced at the beginning of a series of drift tubes, and these can be accelerated by applying an alternating voltage across the gaps between each tube. The accelerated particles then enter a drift tube, where they travel just far enough to be in phase to receive another "kick" as they enter the next gap. The length of the drift tubes increases along the particle path because the

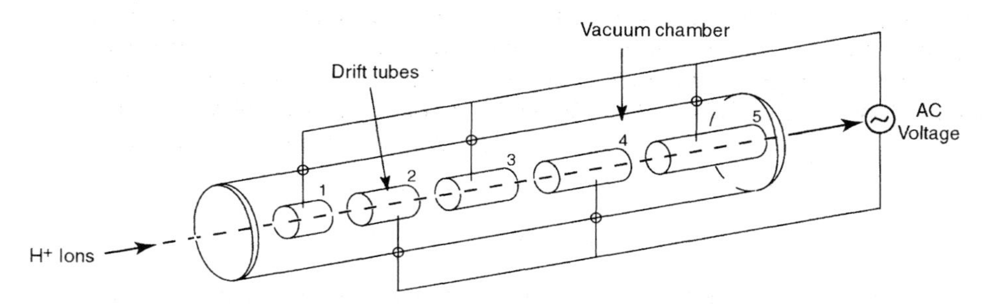

FIGURE 4-12. Schematic of a linear accelerator for protons and deuterons.

accelerated particles are traveling faster and faster as they proceed down the line. The Cockcroft and Walton machine could accelerate protons to energies of 0.1 to 0.7 MeV, which though miniscule by today's standards, led to major contributions to nuclear physics.

Cockcroft and Walton bombarded lithium with protons and showed that two alpha particles were produced, as evidenced by their tracks in a cloud chamber. The two alpha particles were emitted with equal energies in opposite directions according to the reaction

$$^{7}_{3}\text{Li} + ^{1}_{1}\text{H} \rightarrow [^{8}_{4}\text{Be}] \rightarrow ^{4}_{2}\text{He} + ^{4}_{2}\text{He}$$

This reaction has a certain historical interest because it was the first quantitative proof of the validity of the Einstein mass–energy relationship. The Q-value calculated from mass changes is 17.32 MeV, and the measured value of Q (from the energies of the incident protons and the emergent alpha particles) was found to be 17.33 MeV. This experiment thus demonstrated a genuine release of energy from the lithium atom at the expense of its mass. Many other nuclear transformations have since been studied in detail, and all validate the fundamental relationship between mass and energy.

Other proton reactions that produce alpha particles are

$$^{6}_{3}\text{Li} + ^{1}_{1}\text{H} \rightarrow [^{7}_{4}\text{Be}] \rightarrow ^{3}_{2}\text{He} + ^{4}_{2}\text{He}$$

$$^{9}_{4}\text{Be} + ^{1}_{1}\text{H} \rightarrow [^{10}_{5}\text{B}] \rightarrow ^{6}_{3}\text{Li} + ^{4}_{2}\text{He}$$

$$^{11}_{5}\text{B} + ^{1}_{1}\text{H} \rightarrow [^{12}_{6}\text{C}] \rightarrow ^{8}_{4}\text{Be} + ^{4}_{2}\text{He}$$

$$^{19}_{9}\text{F} + ^{1}_{1}\text{H} \rightarrow [^{20}_{10}\text{Ne}] \rightarrow ^{16}_{8}\text{O} + ^{4}_{2}\text{He}$$

$$^{27}_{13}\text{Al} + ^{1}_{1}\text{H} \rightarrow [^{28}_{14}\text{Si}] \rightarrow ^{24}_{12}\text{Mg} + ^{4}_{2}\text{He}$$

Proton bombardment of ^{11}B yields $^{8}_{4}\text{Be}$, which is highly unstable and immediately breaks up into two alpha particles, yielding three alpha particles overall. Proton–alpha reactions can be used with high-energy protons in a cyclotron to produce radioactive substances for use in nuclear medicine applications. Two reactions that produce positron emitters (see Chapter 5) are

$$^{14}_{7}\text{N} + ^{1}_{1}\text{H} \rightarrow [^{15}_{8}\text{O}] \rightarrow ^{11}_{6}\text{C} + ^{4}_{2}\text{He}$$

$$^{16}_{8}\text{O} + ^{1}_{1}\text{H} \rightarrow [^{17}_{9}\text{F}] \rightarrow ^{13}_{7}\text{N} + ^{4}_{2}\text{He}$$

Proton–Neutron Reactions

Protons can be used to produce neutrons in certain target elements through interactions that are mostly endoergic. The effect of these transmutations is to increase the charge on the target nucleus by one unit, moving it above the line of stability on the chart of the nuclides with no change in the mass number. Examples of such (p, n) reactions are

$$^{11}_{5}B + ^{1}_{1}H \rightarrow [^{12}_{6}C] \rightarrow ^{11}_{6}C + ^{1}_{0}n$$

$$^{18}_{8}O + ^{1}_{1}H \rightarrow [^{19}_{9}F] \rightarrow ^{18}_{9}F + ^{1}_{0}n$$

$$^{58}_{28}Ni + ^{1}_{1}H \rightarrow [^{59}_{29}Cu] \rightarrow ^{58}_{29}Cu + ^{1}_{0}n$$

$$^{65}_{29}Cu + ^{1}_{1}H \rightarrow [^{66}_{30}Zn] \rightarrow ^{65}_{30}Zn + ^{1}_{0}n$$

Proton–Gamma Reactions

Some proton interactions produce an excited state of the target nucleus, which is relieved by emission of a gamma photon. Examples of (p, γ) reactions are

$$^{7}_{3}Li + ^{1}_{1}H \rightarrow [^{8}_{4}Be] \rightarrow ^{8}_{4}Be + \gamma$$

$$^{12}_{6}C + ^{1}_{1}H \rightarrow [^{13}_{7}N] \rightarrow ^{13}_{7}N + \gamma$$

$$^{19}_{9}F + ^{1}_{1}H \rightarrow [^{20}_{10}Ne] \rightarrow ^{20}_{10}Ne + \gamma$$

$$^{27}_{13}Al + ^{1}_{1}H \rightarrow [^{28}_{14}Si] \rightarrow ^{28}_{14}Si + \gamma$$

Proton–gamma interactions with lithium are particularly important because they yield photons with an energy of 17.3 MeV, which is far more energetic than those from naturally radioactive substances. These high-energy photons can thus be used to produce nuclear interactions by a process called *photodisintegration*.

Proton–Deuteron Reactions

Proton interactions may also produce deuterons. Examples of (p, d) reactions are

$$^{9}_{4}Be + ^{1}_{1}H \rightarrow [^{10}_{5}B] \rightarrow ^{8}_{4}Be + ^{2}_{1}H$$

$$^{7}_{3}Li + ^{1}_{1}H \rightarrow [^{8}_{4}Be] \rightarrow ^{6}_{3}Li + ^{2}_{1}H$$

TRANSMUTATION BY DEUTERONS

Deuterons, which like protons also have the advantage of a single + charge, can also be used as bombarding particles by accelerating them up to energies of several MeV, typically in a cyclotron (see Chapter 9). One of the first deuteron-induced reactions studied was again that on natural lithium, which contains 7.5% ^{6}Li:

$$^{6}_{3}Li + ^{2}_{1}H \rightarrow [^{8}_{4}Be] \rightarrow ^{4}_{2}He + ^{4}_{2}He$$

Other examples of (d, α) reactions are

$$^{16}_{8}O + ^{2}_{1}H \rightarrow [^{18}_{9}F] \rightarrow ^{14}_{7}N + ^{4}_{2}He$$

$$^{27}_{13}Al + ^{2}_{1}H \rightarrow [^{29}_{14}Si] \rightarrow ^{25}_{12}Mg + ^{4}_{2}He$$

$$^{20}_{10}Ne + ^{2}_{1}H \rightarrow [^{22}_{11}Na] \rightarrow ^{18}_{9}F + ^{4}_{2}He$$

The latter reaction produces ^{18}F, which is radioactive with a 110-min half-life and is used widely in nuclear medicine. These reactions are usually exoergic (i.e., they have positive Q-values).

Deuteron–Proton and Deuteron–Neutron Reactions

Depending on the energies, it is possible for deuterons to produce protons, just as protons produce deuterons. Examples of (d,p) reactions are

$$^{12}_{6}C + ^{2}_{1}H \rightarrow [^{14}_{7}N] \rightarrow ^{13}_{6}C + ^{1}_{1}H$$

$$^{23}_{11}Na + ^{2}_{1}H \rightarrow [^{25}_{12}Mg] \rightarrow ^{24}_{11}Na + ^{1}_{1}H$$

$$^{31}_{15}P + ^{2}_{1}H \rightarrow [^{33}_{16}S] \rightarrow ^{32}_{15}P + ^{1}_{1}H$$

These transformations increase the mass of the nucleus by one unit, leaving the charge unchanged. The reactions are usually exoergic.

Deuteron bombardment can also yield neutrons as in the following (d,n) reactions:

$$^{7}_{3}Li + ^{2}_{1}H \rightarrow [^{9}_{4}Be] \rightarrow ^{8}_{4}Be + ^{1}_{0}n$$

$$^{9}_{4}Be + ^{2}_{1}H \rightarrow [^{11}_{5}B] \rightarrow ^{10}_{5}B + ^{1}_{0}n$$

$$^{12}_{6}C + ^{2}_{1}H \rightarrow [^{14}_{7}N] \rightarrow ^{13}_{7}N + ^{1}_{0}n$$

$$^{14}_{7}N + ^{2}_{1}H \rightarrow [^{16}_{8}O] \rightarrow ^{15}_{8}O + ^{1}_{0}n$$

An interesting reaction occurs when frozen D_2O (*heavy water*) is bombarded with deuterons, in effect fusing deuterium by accelerator processes. Both the (d,p) and (d,n) reactions have been observed because the excited compound nucleus (^{4}He) can break up in two ways:

$$^{2}_{1}H + ^{2}_{1}H \rightarrow [^{4}_{2}He] \rightarrow ^{3}_{1}H + ^{1}_{1}H$$

$$^{2}_{1}H + ^{2}_{1}H \rightarrow [^{4}_{2}He] \rightarrow ^{3}_{2}He + ^{1}_{0}n$$

The first reaction produces tritium with the release of a proton and 4.03 MeV of energy, and the second produces helium and a neutron and 3.27 MeV of energy. Tritium is unstable and has a half-life of about 12.3 years; helium-3 is stable and is found in nature.

NEUTRON INTERACTIONS

Neutrons have proved to be especially effective in producing nuclear transformations because they have no electric charge and are more likely to penetrate nuclei than are protons, deuterons, or alpha particles. Nuclear reactors provide copious quantities of neutrons; however, other sources are available from radioactive substances mixed with beryllium and/or deuterium or from (d,n) reactions produced in accelerators.

Neutron–Alpha Reactions

Examples of (n, α) reactions are

$$^{6}_{3}\text{Li} + ^{1}_{0}\text{n} \rightarrow [^{7}_{3}\text{Li}] \rightarrow ^{3}_{1}\text{H} + ^{4}_{2}\text{He}$$

$$^{10}_{5}\text{B} + ^{1}_{0}\text{n} \rightarrow [^{11}_{5}\text{B}] \rightarrow ^{7}_{3}\text{Li} + ^{4}_{2}\text{He}$$

$$^{27}_{13}\text{Al} + ^{1}_{0}\text{n} \rightarrow [^{28}_{13}\text{Al}] \rightarrow ^{24}_{11}\text{Na} + ^{4}_{2}\text{He}$$

The first two reactions have a relatively high yield, and both are often used to detect neutrons. In one method, an ionization chamber is lined with boron, usually BF_3, which provides efficient capture of neutrons. The liberated alpha particle produces significant ionization, which can be collected and amplified to produce an electronic signal allowing detection. The (n, α) reaction with ^{6}Li yields tritium, and ^{6}Li is a common source target for its production. Tritium is a key component of nuclear weapons and fusion energy, thus the reaction with lithium is quite important.

Neutron–Proton Reactions

In these reactions a proton in the nucleus is replaced by a neutron. The mass number is not changed, but the charge is decreased by one unit and the atom is moved below the line of stability on the *chart of the nuclides*. Some examples of (n, p) reactions are

$$^{14}_{7}\text{N} + ^{1}_{0}\text{n} \rightarrow [^{15}_{7}\text{N}] \rightarrow ^{14}_{6}\text{C} + ^{1}_{1}\text{H}$$

$$^{27}_{13}\text{Al} + ^{1}_{0}\text{n} \rightarrow [^{28}_{13}\text{Al}] \rightarrow ^{27}_{12}\text{Mg} + ^{1}_{1}\text{H}$$

$$^{64}_{30}\text{Zn} + ^{1}_{0}\text{n} \rightarrow [^{65}_{30}\text{Zn}] \rightarrow ^{64}_{29}\text{Cu} + ^{1}_{1}\text{H}$$

$$^{39}_{19}\text{K} + ^{1}_{0}\text{n} \rightarrow [^{40}_{19}\text{K}] \rightarrow ^{39}_{18}\text{Ar} + ^{1}_{1}\text{H}$$

The first of these reactions can be induced by slow neutrons; however, the latter three require more energetic neutrons. Activation of ^{39}K to produce ^{39}Ar can be used in the potassium–argon dating method (see Chapter 6).

Neutron–Neutron Reactions

Sometimes when a high-energy neutron is captured, two or more neutrons are emitted. The (n, 2n) reaction leaves the charge of the nucleus unchanged; however, an isotope of the target nucleus is produced with a mass number one unit smaller. Examples of this reaction are

$$^{27}_{13}\text{Al} + ^{1}_{0}\text{n} \rightarrow [^{28}_{13}\text{Al}] \rightarrow ^{26}_{13}\text{Al} + ^{1}_{0}\text{n} + ^{1}_{0}\text{n}$$

$$^{238}_{92}\text{U} + ^{1}_{0}\text{n} \rightarrow [^{239}_{92}\text{U}] \rightarrow ^{237}_{92}\text{U} + ^{1}_{0}\text{n} + ^{1}_{0}\text{n}$$

In the latter reaction, ^{237}U is radioactive and undergoes transformation to ^{237}Np, which is long lived and of significance in radioactive wastes. It is also a target material for neutron activation by an (n, γ) reaction to ^{238}Np, which soon transforms to ^{238}Pu, a useful radioisotope for remote power systems. The mass change in the (n,2n) reaction is always negative; thus the Q-value is negative and fast neutrons are needed to bring about this reaction. Very high energy neutrons can produce reactions in which three or four or more neutrons are emitted.

Radiative Capture Reactions

The most common neutron interaction is *radiative capture*, in which the mass number of an element is increased by one unit due to the addition of a neutron. The excitation energy induced by the extra neutron is relieved by gamma radiation, and Q-values are always positive. These reactions shift the target atom to the right of the line of stability on the chart of the nuclides, and the element remains the same since the Z doesn't change.

Numerous (n,γ) reactions are possible with slow neutrons, beginning with hydrogen to produce deuterium:

$$^{1}_{1}\mathrm{H} + {}^{1}_{0}\mathrm{n} \rightarrow [^{2}_{1}\mathrm{H}] \rightarrow {}^{2}_{1}\mathrm{H} + \gamma$$

which in turn can be bombarded with slow neutrons to produce tritium:

$$^{2}_{1}\mathrm{H} + {}^{1}_{0}\mathrm{n} \rightarrow [^{3}_{1}\mathrm{H}] \rightarrow {}^{3}_{1}\mathrm{H} + \gamma$$

Even though the (n,γ) reaction with deuterium has a very small cross section, all light water–cooled nuclear reactors will have tritium in the coolant due to these two reactions. Similarly, if air contacts the flux of neutrons in a reactor, ^{40}Ar will be converted by an (n,γ) reaction to

$$^{40}_{18}\mathrm{Ar} + {}^{1}_{0}\mathrm{n} \rightarrow [^{41}_{18}\mathrm{Ar}] \rightarrow {}^{41}_{18}\mathrm{Ar} + \gamma$$

Other typical (n,γ) reactions are

$$^{59}_{27}\mathrm{Co} + {}^{1}_{0}\mathrm{n} \rightarrow [^{60}_{27}\mathrm{Co}] \rightarrow {}^{60}_{27}\mathrm{Co} + \gamma$$

$$^{27}_{13}\mathrm{Al} + {}^{1}_{0}\mathrm{n} \rightarrow [^{28}_{13}\mathrm{Al}] \rightarrow {}^{28}_{13}\mathrm{Al} + \gamma$$

$$^{115}_{49}\mathrm{In} + {}^{1}_{0}\mathrm{n} \rightarrow [^{116}_{49}\mathrm{In}] \rightarrow {}^{116}_{49}\mathrm{In} + \gamma$$

$$^{202}_{80}\mathrm{Hg} + {}^{1}_{0}\mathrm{n} \rightarrow [^{203}_{80}\mathrm{Hg}] \rightarrow {}^{203}_{80}\mathrm{Hg} + \gamma$$

These reactions are quite probable with slow neutrons and the products are commonly radioactive. Targets made of such materials can be important sources of artificial radionuclides: for example, ^{60}Co, ^{99}Mo, ^{116}In, ^{203}Hg, and so on.

Photodisintegration Reactions

High-energy photons, through the process of photodisintegration, can also transform nuclei. Since the photon has no mass, it can supply only its kinetic energy, which must be at least as great as the binding energy of a neutron or proton in order to eject it from a target nucleus. Photodisintegration (γ,n) reactions are, therefore, endoergic, and with the exception of deuterium and beryllium, the threshold energies are on the order of 8 to 10 MeV or more.

Photodisintegration is practical for deuterium, in which the binding energy between the neutron and proton is only 2.225 MeV, and ^{9}Be, in which one neutron is loosely bound. The reactions are

$$^{2}_{1}\mathrm{H} + \gamma \rightarrow [^{2}_{1}\mathrm{H}^{*}] \rightarrow {}^{1}_{1}\mathrm{H} + {}^{1}_{0}\mathrm{n}$$

$$^{9}_{4}\mathrm{Be} + \gamma \rightarrow [^{9}_{4}\mathrm{Be}^{*}] \rightarrow {}^{8}_{4}\mathrm{Be} + {}^{1}_{0}\mathrm{n}$$

Photodisintegration of deuterium is the inverse of the (n,γ) reaction with hydrogen.

Beryllium has a Q-value of -1.67 MeV and is relatively easy to use with a source of high-energy photons to produce neutrons (Fermi and his colleagues extracted radon from a sizable radium source and used the gamma rays from the radon transformation products to generate neutrons from beryllium.) The (p,α) reaction with Li yields 17.3 MeV gamma rays, which are in turn used to produce photoneutrons, as in the (γ,n) reaction with ^{31}P:

$$^{31}_{15}\mathrm{P} + \gamma \rightarrow [^{31}_{15}\mathrm{P}] \rightarrow {}^{30}_{15}\mathrm{P} + {}^{1}_{0}\mathrm{n}$$

The (γ,p) reaction requires still higher energies, which can be obtained from a betatron.

MEDICAL ISOTOPE REACTIONS

Several (n,γ) reactions produce radionuclides used in nuclear medicine, for example:

$$^{50}_{24}\mathrm{Cr} + {}^{1}_{0}\mathrm{n} \rightarrow [^{51}_{24}\mathrm{Cr}] \rightarrow {}^{51}_{24}\mathrm{Cr} + \gamma$$

$$^{98}_{42}\mathrm{Mo} + {}^{1}_{0}\mathrm{n} \rightarrow [^{99}_{42}\mathrm{Mo}] \rightarrow {}^{99}_{42}\mathrm{Mo} + \gamma$$

$$^{132}_{54}\mathrm{Xe} + {}^{1}_{0}\mathrm{n} \rightarrow [^{133}_{54}\mathrm{Xe}] \rightarrow {}^{133}_{54}\mathrm{Xe} + \gamma$$

^{51}Cr is used for labeling red blood cells and spleen scanning; ^{99}Mo is the source of ^{99m}Tc, the most commonly used radionuclide in nuclear medicine; and ^{133}Xe is used for lung ventilation studies.

Medical isotopes can also be produced by accelerating protons, deuterons, ^{3}He, and ^{4}He nuclei toward targets. A certain threshold energy is required (generally, 5 to 30 MeV) for these reactions to overcome the repulsive coulombic forces of target nuclei. Common reactions are

$$^{16}_{8}\mathrm{O} + {}^{3}_{2}\mathrm{He} \rightarrow [^{19}_{10}\mathrm{Ne}] \rightarrow {}^{18}_{9}\mathrm{F} + {}^{1}_{1}\mathrm{H}$$

$$^{68}_{30}\mathrm{Zn} + {}^{1}_{1}\mathrm{H} \rightarrow [^{69}_{31}\mathrm{Ga}] \rightarrow {}^{67}_{31}\mathrm{Ga} + 2\mathrm{n}$$

^{18}F is used for labeling radiopharmaceuticals for position emission tomography (PET) imaging, and ^{67}Ga is widely used for soft tumor and occult abscess detection.

A radionuclide of interest may be formed indirectly from radioactive transformation of the product of a nuclear reaction. Two examples of these indirect methods are ^{123}I and ^{201}Tl:

$$ {}^{122}_{52}\text{Te} + {}^{4}_{2}\text{He} \rightarrow [{}^{126}_{54}\text{Xe}] \rightarrow {}^{123}_{54}\text{Xe} + 3\text{n} $$

$$ {}^{123}_{54}\text{Xe} \rightarrow {}^{123}_{53}\text{I} + {}^{0}_{1}\text{e}^{+} $$

$$ {}^{203}_{81}\text{Tl} + {}^{1}_{1}\text{H} \rightarrow [{}^{204}_{82}\text{Pb}] \rightarrow {}^{201}_{82}\text{Pb} + 3\text{n} $$

$$ {}^{201}_{82}\text{Pb} + {}^{0}_{-1}\text{e}^{+} \rightarrow {}^{201}_{81}\text{Tl} $$

Iodine 123 ($T_{1/2}$ = 13 h) shows promise for replacing ^{131}I in many diagnostic studies, which would produce lower patient doses, and ^{201}Tl ($T_{1/2}$ = 3.04 d) transforms by electron capture and gamma emission at energies that are useful in heart imaging studies. In most charged-particle reactions, the product radionuclide is a new element; therefore, it can be chemically separated yielding a carrier-free source. Also, since many reactions increase the proton number, several β^+ or electron capture radionuclides can be produced.

TRANSURANIUM ELEMENTS

In 1934, Fermi bombarded uranium with neutrons to make and identify new elements with nuclear charge greater than 92, or transuranic elements. Fermi undoubtedly was one of the first to fission uranium but did not recognize it because in his focus on transuranic production, he had covered the uranium target with foil, thus preventing the detection of the resulting fission fragments.

Neutron capture in ^{238}U yields

$$ {}^{238}_{92}\text{U} + {}^{1}_{0}\text{n} \rightarrow [{}^{239}_{92}\text{U}] \rightarrow {}^{239}_{92}\text{U} + \gamma $$

The ^{239}U product undergoes radioactive transformation by beta particle emission to produce ^{239}Np, which does not exist in nature. Radioactive transformation of ^{239}Np by beta-particle emission forms an isotope of another new element with $Z = 94$, or ^{239}Pu. Plutonium-239 ($T_{1/2}$ = 24,500 y) is an important nuclide because it can be fissioned in nuclear reactors and nuclear weapons.

Soon after the discovery of ^{239}Np, two other isotopes of neptunium were made by the following reactions:

$$ {}^{238}_{92}\text{U} + {}^{2}_{1}\text{H} \rightarrow [{}^{240}_{93}\text{Np}] \rightarrow {}^{238}_{93}\text{Np} + {}^{1}_{0}\text{n} + {}^{1}_{0}\text{n} $$

$$ {}^{238}_{92}\text{U} + {}^{1}_{0}\text{n} \rightarrow {}^{237}_{92}\text{U} + {}^{1}_{0}\text{n} + {}^{1}_{0}\text{n} $$

The ^{238}Np product in the first reaction is transformed by beta-particle emission (half-life of 2.0 d) to ^{238}Pu ($T_{1/2}$ = 87.7 y), which is an alpha emitter that is used in thermoelectric generators, primarily for space missions. The ^{237}U produced in the second (n, 2n) reaction undergoes radioactive transformation with a half-life of 6.8 d to ^{237}Np, which is the head of the neptunium series (see Chapter 6). It has a half-life of 2.2×10^6 y and is the longest-lived of the known transuranium nuclides.

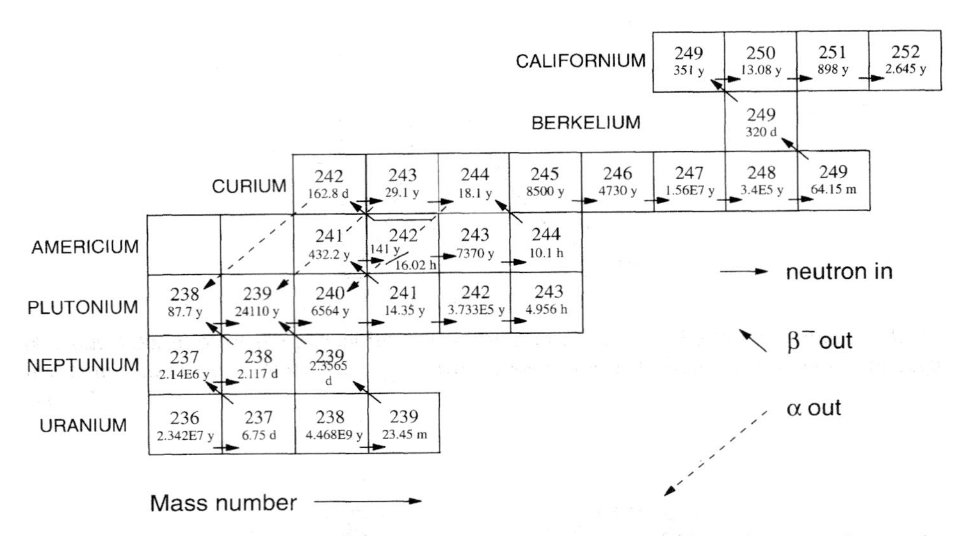

FIGURE 4-13. Transuranium element production by a combination of neutron interactions to build mass followed by radioactive transformation to new products, which may also be activated by neutrons to produce even heavier elements.

The production of transuranium elements generally starts with neutron bombardment of uranium, plutonium, americium, curium, and so on, to first build mass, then waiting for radioactive transformation to yield the desired target for further neutron bombardment to gradually reach the desired combination of Z and A. Sometimes it is necessary to use high-energy neutrons to obtain the desired product by (n, 2n) or (n, 3n) or greater reactions. Figure 4-13. shows the interrelationships of some of these paths of formation for some of the important transuranic elements.

FISSION AND FUSION REACTIONS

Fission occurs when the nucleus of a fissionable material such as ^{235}U or ^{239}Pu absorbs a neutron and splits into two fragments that are both smaller and more tightly bound than the fissioning nucleus. The products of this reaction, in addition to the two fission fragments, are two or three neutrons plus a considerable quantity of energy. Fission reactions are therefore exoergic, and because of this, can be a source of energy. Since excess neutrons are emitted, they are also sources of neutrons. A typical fission reaction with each of these components is

$$^{235}_{92}\text{U} + ^{1}_{0}\text{n} \rightarrow [^{236}\text{U}^*] \rightarrow {}^{144}\text{Nd} + {}^{90}\text{Zr} + 2\text{n} + Q$$

The Q-value of this reaction is +205.4 MeV. Numerous other reactions are possible since the uranium nucleus can split any number of ways after absorbing a neutron.

The discovery of fission occurred due to efforts by several researchers, including Enrico Fermi, to produce transuranic elements by irradiating uranium with neutrons and other particles. Products were indeed found, and many were radioactive, but the products were varied with many different half-lives. These physicists did chemical separations to better understand the interaction processes and found unexpected bariumlike radioactive products in the precipitates; however, the presence of these products was attributed to their rudimentary chemistry techniques. When Hahn, an expert chemist, and Strassmann showed that the products of neutron irradiation of uranium indeed contained radioactive barium, Hahn's former assistant, Lise Meitner, and her nephew, Otto Frisch, decided to apply Bohr's recent liquid drop model of the nucleus to the results; and there it was! Uranium had split into lighter fragments, one of which was an isotope of barium which subsequently transformed to lanthanum. They termed it *nuclear fission* because it resembled division of cells (called *fission*) in biology. As soon as this insight was announced, physicists all over the world used ion chambers to observe the high rates of ionization produced by the fission fragments. If Fermi had not wrapped his uranium target in foil, this foremost physicist would also have been the discoveror of fission, a phenomenon to which he contributed so much.

The discovery of nuclear fission not only changed physics, but has also changed the world. Fissioning of the ^{235}U nucleus yields neutrons, numerous radioactive fission products, and enormous amounts of energy. These concepts, their use in recovering the energy of nuclear fission, and the dynamics of producing radioactive fission products are discussed in detail in Chapter 10.

Fusion

Fusion reactions occur when light elements are brought together to form a heavier element in which the nucleons are more tightly bound. Such reactions, when they can be made to occur, are exoergic, yielding considerable amounts of energy because the products move up the curve of binding energy per nucleon for light elements shown in Figure 2-2. Fusion reactions are the source of the sun's tremendous energy generation and the enormous energy release in thermonuclear weapons (or hydrogen bombs). Isotopes of hydrogen are fused in the following typical reactions:

$$^{1}_{1}H + ^{1}_{1}H \rightarrow ^{2}_{1}H + e^{+}$$

$$^{1}_{1}H + ^{2}_{1}H \rightarrow ^{3}_{2}He$$

$$^{3}_{2}He + ^{3}_{2}He \rightarrow ^{4}_{2}He + 2^{1}_{1}H$$

These reactions, which are shown here as reactions between whole elements, actually occur between bare nuclei in the sun. Electrons are present to maintain electrical neutrality, but because of the sun's intense temperature, they are stripped away and do not participate in the reactions. Once a deuteron is created, it can fuse with another proton to make a nucleus of ^{3}He, and two ^{3}He nuclei fuse to form an alpha particle and two protons, which can participate in other reactions. The net result of these reactions is the fusing of four protons to produce an alpha particle, two positrons that annihilate, and 26.7 MeV of energy. This cycle of reactions is known as the *proton–proton cycle*. The sun burns 5.3×10^{16} kg/d of its hydrogen into alpha particles, thereby transforming about 3.7×10^{14} kg of its mass into energy every day.

Fusion reactions are called *thermonuclear reactions* because high temperatures are required to overcome the coloumbic repulsion between the nuclei being fused (i.e., "thermo" for the heat required and "nuclear" for the interactions that occur). For a pair of deuterons, or a deuteron and a triton, to come close enough together to fuse, their centers must be separated by not more than about 10^{-14} m. At this separation, the potential energy is

$$\mathrm{PE} = \frac{1}{4\pi e_0}\frac{e^2}{r} = \frac{(8.987 \times 10^9\ \mathrm{N/C^2})(1.6022 \times 10^{-19}\ \mathrm{C})^2}{10^{-14}\ \mathrm{m}}$$

$$= 2.3 \times 10^{-14}\ \mathrm{J} = 144\ \mathrm{keV}$$

shared between the two; therefore, each hydrogen nucleus (deuteron or triton) would need at least 72 keV of kinetic energy to come close enough to fuse. The Boltzmann distribution of the kinetic energy of molecules of a gas is

$$\overline{E} = \tfrac{3}{2}kT$$

where the Boltzmann constant $k = 1.38 \times 10^{-23}$; therefore, for a hydrogen nucleus to have an $\overline{E}$ of 72 keV (or 1.15×10^{-14} J),

$$T = 5.6 \times 10^8\ \mathrm{K}$$

The actual temperature required is not quite this extreme because some nuclei have considerably more kinetic energy than the average, and not all the nuclei need to surmount the potential energy barrier to interact. In practice, thermonuclear reactions proceed at temperatures of about $\frac{1}{10}$ this value, or 6×10^7 K. A fission bomb can provide this temperature.

The fusing of protons to form deuterons in the proton–proton cycle takes place on a very long time scale, which effectively controls the rate of energy generation in the sun. For hotter stars, which the sun will one day inevitably become, the sequence of reactions will be dominated by the carbon cycle, which is comprised of the following reactions:

$$^{12}\mathrm{C} + {}^{1}\mathrm{H} \rightarrow {}^{13}\mathrm{N} + \gamma$$

$$^{13}\mathrm{N} \rightarrow {}^{13}\mathrm{C} + \mathrm{e}^+$$

$$^{13}\mathrm{C} + {}^{1}\mathrm{H} \rightarrow {}^{14}\mathrm{N} + \gamma$$

$$^{14}\mathrm{N} + {}^{1}\mathrm{H} \rightarrow {}^{15}\mathrm{O} + \gamma$$

$$^{15}\mathrm{O} \rightarrow {}^{15}\mathrm{N} + \mathrm{e}^+$$

$$^{15}\mathrm{N} + {}^{1}\mathrm{H} \rightarrow {}^{12}\mathrm{C} + {}^{4}\mathrm{He}$$

These reactions require more thermal energy to overcome the repulsion between carbon and hydrogen nuclei, or about 2×10^7 K. Since the sun's interior temperature is about 1.5×10^7 K, a long period of time must elapse during which sufficient helium is produced by burning hydrogen. When all the hydrogen has been converted to helium, the sun will contract and heat up, which will create conditions whereby helium nuclei can fuse into ^{12}C. The presence of carbon, which is neither produced nor consumed, permits this sequence of reactions to take place at a much greater rate than the reactions in the proton–proton cycle. The net process is still the fusion of four protons to produce a helium nucleus with the release of 26.7 MeV of energy.

The primordial matter produced when the universe was formed consisted of about 75% hydrogen and 25% helium; all of the other chemical elements were formed by nuclear reactions in the interiors of stars. At about 10^8 K, three ^4He nuclei are converted into ^{12}C by a two-step process:

$$^4He + ^4He \rightarrow ^8Be$$

$$^8Be + ^4He \rightarrow ^{12}C$$

The first reaction is endothermic with a Q-value of 92 keV. The ^8Be nucleus is unstable and quickly (in about 10^{-16} s) transforms back into two alpha particles. Even so, at 10^8 K there will be a small concentration of ^8Be, which because of a particularly large cross section allows fusion with ^4He to form ^{14}C and 7.3 MeV of energy before the breakup of ^8Be can occur. Once ^{12}C is formed, it can successively fuse with other alpha particles to make even heavier elements; for example,

$$^{12}C + ^4He \rightarrow ^{16}O$$

$$^{16}O + ^4He \rightarrow ^{20}Ne$$

$$^{20}Ne + ^4He \rightarrow ^{24}Mg$$

Each of these reactions is exothermic. At still higher temperatures (10^9 K), fusion of carbon and oxygen atoms can occur to produce still heavier products:

$$^{12}C + ^{12}C \rightarrow ^{20}Ne + ^4He$$

$$^{16}O + ^{16}O \rightarrow ^{28}Si + ^4He$$

Such processes will continue with the buildup of elements according to Figure 2-2 until ^{56}Fe is reached; beyond this point no further energy can be gained by fusion.

Nitrogen is almost as abundant in nature as carbon and oxygen, which are the two most abundant elements, presumably because they are products of fusion processes involving hydrogen and helium. The likely reactions that yield nitrogen are

$$^{12}C + ^1H \rightarrow ^{13}N + \gamma$$

$$^{13}N \rightarrow ^{13}C + e^+ + \nu$$

$$^{13}C + ^1H \rightarrow ^{14}N + \gamma$$

and

$$^{16}O + {}^{1}H \rightarrow {}^{17}F + \gamma$$
$$^{17}F \rightarrow {}^{17}O + e^{+}$$
$$^{17}O + {}^{1}H \rightarrow {}^{14}N + {}^{4}He$$

The stable isotopes ^{13}C and ^{17}O are found in natural carbon and oxygen with abundances of 1.1 and 0.04%, respectively, which confirms that reactions of this sort can indeed take place.

Formation of elements beyond ^{56}Fe requires the presence of neutrons which become abundant after fusion yields atoms with an excess of neutrons such as ^{13}C, ^{17}O, and so on. Neutron captures in ^{56}Fe begin a sequence of mass building and radioactive transformation to build heavier and heavier elements,

$$^{56}Fe + n \rightarrow {}^{57}Fe + n \rightarrow {}^{58}Fe + n \rightarrow {}^{59}Fe$$

and since ^{59}Fe is radioactive, it transforms to ^{59}Co, which can absorb neutrons to become ^{60}Ni, and so on.

Thermonuclear Weapons

Fusion reactions are the source of energy in thermonuclear weapons or hydrogen bombs. Lithium hydride, made up of ^{6}Li deuteride ($^{6}Li^{2}H$), is a solid at normal temperature and makes an excellent fuel. The deuterons are held close together in this formulation, which enhances the DD reaction, and the ^{6}Li "manufactures" tritium to stoke the DT reaction. The related reactions are

$$^{2}_{1}H + {}^{2}_{1}H \rightarrow {}^{3}_{1}H + p \qquad \text{(energy release = 4.0 MeV)}$$
$$^{2}_{1}H + {}^{2}_{1}H \rightarrow {}^{3}_{2}He + n \qquad \text{(energy release = 3.3 MeV)}$$
$$^{2}_{1}H + {}^{3}_{1}H \rightarrow {}^{4}_{2}He + n \qquad \text{(energy release = 17.6 MeV)}$$
$$^{6}_{3}Li + n \rightarrow {}^{3}_{1}H + {}^{4}_{2}He \qquad \text{(energy release = 4.9 MeV)}$$

These four interlocked reactions use both tritons and neutrons which are not initially present in the fuel. The first two reactions fuse deuterium to produce energy, and tritium from the first reaction participates in DT fusion (shown as the third reaction). The neutron from the second reaction interacts with ^{6}Li according to the fourth reaction to supply additional tritium for DT fusion. The fourth reaction is not in itself a thermonuclear reaction since the neutron has no potential energy barrier to overcome to reach the ^{6}Li nucleus, and is, in fact, a fission reaction, although fission is usually associated with the heavy elements. The neutrons from DD and DT fusion are used to induce more tritium, neutrons, and fission-energy releases in fissionable materials. These reactions have been used by weapons designers in ways not known to ordinary people to produce very efficient thermonuclear weapons.

The 14 MeV neutrons produced by the DT reaction can induce fission in ^{238}U, which is often fabricated into weapons to increase the yield substantially. Fissioning

^{238}U yields several lower-energy neutrons that can stimulate the production of ^3H in ^6Li, adding more fuel and neutrons. This sequence of fission–fusion–fission–fusion can be used to multiply the yield of a fission weapon a thousandfold with cheap and readily available ^6Li, ^2H, and natural uranium. The products of fission cause radioactive fallout when detonated in the open atmosphere; fusion produces "clean" but very powerful weapons.

SUMMARY

Much of what is known about atoms and radiation from them was learned by aiming subatomic particles at various target materials. The discovery of cathode rays provided a source of projectiles that led to the discovery of x rays, which is now recognized as a special interaction between an accelerated electron and an atom in the target material of an x-ray tube that causes the emission of bremsstrahlung radiation, with energies up to that of the accelerated electrons. Electrons that impinge on the target may also dislodge orbital electrons, and when the electron vacancy is filled by an outer shield electron, discrete-energy x rays, known as *characteristic x rays*, are also emitted. Characteristic x rays that are emitted in the process of filling a K-shell vacancy are known as K-shell x rays, or simply K x rays, and similarly for other shell vacancies, as L, M,... x rays. Moseley's detailed analysis of K_α x rays established that the periodic table should be ordered by increasing Z instead of by mass, determined that certain elements were out of order, and identified gaps which were filled by elements that were discovered later.

Nuclear interactions are those that involve absorption of a bombarding particle by the nucleus of a target material (in x-ray production the electron is not absorbed by the nucleus). Absorption of the projectiles produces first a compound nucleus, which then breaks up to yield the final products; these changes are characterized by the Q-value, the energy balance between the interacting particle and the target atom (the reactants) and the products of the reaction. If the Q-value is positive, the reaction is exoergic (i.e., energy is gained at the expense of the mass of the reactants). If the Q-value is negative, the reactions are endoergic and energy must be supplied for the reaction to occur.

The likelihood of an interaction between a bombarding particle and a nucleus is described by the concept of cross section, which is only a measure of the probability that it will occur. The cross section for a specific interaction is dependent on the energy, charge, mass, and de Broglie wavelength of the projectile, and the vibrational frequency, the spin, and the energy states of nucleons in the target atom. These cannot be predicted directly by nuclear theory; thus cross sections for any given arrangement of projectile and target are usually determined experimentally.

Neutrons have proved to be especially effective in producing nuclear transformations because they have no electric charge and are not repelled by the positively charged nuclei of target atoms. Consequently, they are more likely to penetrate nuclei than are protons, deuterons, or alpha particles, and many different reactions have been produced. The most common neutron interaction is an (n, γ) or radiative capture reaction in which the mass number of the target element is increased by one unit and the excitation energy induced by the extra neutron is relieved by gamma radiation.

Protons, deuterons, tritons (^3H), and helium nuclei can also be accelerated and used to produce various products by interactions with target atoms. A certain threshold energy is required (generally, 5 to 30 MeV) for these reactions in order to overcome the repulsive coulombic forces of target nuclei. Nuclear reactions can also be used to induce fission and fusion of nuclei. Fission occurs when the nucleus of a fissionable material such as ^{235}U or ^{239}Pu absorbs a neutron and splits into two fragments that are both smaller and more tightly bound than the fissioning nucleus. The products of this reaction, in addition to the two fission fragments, are two or three neutrons plus a considerable quantity of energy. Fusion reactions occur when light elements are brought together to form a heavier element in which the nucleons are much more tightly bound. Such reactions, when they can be made to occur, are exoergic, and yield considerable amounts of energy.

PROBLEMS

4-1. Complete the following reactions by writing them in equation form to show the balance of atomic and mass numbers and the compound nucleus: **(a)** $^7Li(p,\alpha)$; **(b)** $^9Be(d,p)$; **(c)** $^9Be(p,\alpha)$; **(d)** $^{11}B(d,\alpha)$; **(e)** $^{12}C(d,n)$; **(f)** $^{14}N(\alpha,p)$; **(g)** $^{15}N(p,\alpha)$; **(h)** $^{16}O(d,n)$.

4-2. Calculate, from the masses listed in Appendix B, the Q-value for each of the reactions of Problem 4-1, and determine which reactions are exoergic and which are endoergic.

4-3. What are the Q values for the reactions **(a)** $^9Be(p,n)^9B$, **(b)** $^{13}C(p,n)^{13}N$, and **(c)** $^{18}O(p,n)^{18}F$ which have threshold energies of 2.059, 3.236, and 2.590 MeV, respectively?

4-4. Calculate, from the atomic masses in Appendix B, the threshold energies for the reactions **(a)** $^{11}B(p,n)^{11}C$, **(b)** $^{18}O(p,n)^{18}F$, and **(c)** $^{23}Na(p,n)^{23}Mg$.

4-5. If ^{12}C is bombarded with neutrons, protons, deuterons, and alpha particles, each having kinetic energy of 10.00 MeV, what would be the excitation energy of each compound nucleus if it is assumed that the recoil energy of the compound nucleus can be neglected?

4-6. With the mass data in Appendix B, calculate the binding energy of the "last neutron" in each of the following sets of isotopes: **(a)** ^{11}C, ^{12}C, ^{13}C and **(b)** ^{13}N, ^{14}N, ^{15}N.

4-7. With the mass data in Appendix B, calculate the binding energy of the "last proton" in **(a)** ^{11}C, **(b)** ^{12}C, and **(c)** ^{13}C.

4-8. What causes the binding energy for those reactions in Problems 4-6 and 4-7 to be particularly small, i.e., smaller than 5 MeV?

4-9. An (n,γ) reaction in ^{113}Cd using thermal neutrons ($E = 0.025$ eV), produces ^{114}Cd. What energy does the emitted photon have?

4-10. Find the energy of the gamma ray emitted in the (α,γ) reaction with ^7Li assuming the alpha particle energy is negligible.

4-11. Find the Q-value (and therefore the energy released) from the masses in Appendix B for the fission reaction

$$^{235}_{92}\text{U} + ^{1}_{0}\text{n} \rightarrow [^{236}_{92}\text{U}^*] \rightarrow ^{93}_{40}\text{Zr} + ^{141}_{58}\text{Ce} + ^{1}_{0}\text{n} + ^{1}_{0}\text{n} + 6\,^{0}_{-1}\beta^- .$$

4-12. Find the kinetic energy of the emitted neutron in the (d, triton) fusion reaction (*Hint*: the recoil energies must be accounted for).

4-13. Under the proper conditions two alpha particles can combine to produce a proton and a nucleus of ^{7}Li. (**a**) Write the equation for this reaction. (**b**) What minimum kinetic energy must one of the alpha particles have to make the reaction proceed if the other alpha particle is at rest?

4-14. Thermal neutrons (0.025 eV) produce an (n, α) reaction with ^{6}Li in natural lithium. Calculate the kinetic energy of the ejected alpha particle.

4-15. Two reactions that occur in the upper atomsphere due to the effects of cosmic "rays" are $^{14}_{7}$N(n, p)? and $^{14}_{7}$N(n, ?)$^{12}_{6}$C. (**a**) Complete the nuclear reaction equation for each and calculate the Q-value of each. (**b**) Is the nuclide formed in each case stable or radioactive?

4-16. Radioactive transformation is a special kind of exoergic reaction. Carbon-14 emits an electron (beta particle) to transform to ^{14}N. From the Q-value, determine the maximum energy available to the ejected electron (beta particle).

5

RADIOACTIVE TRANSFORMATION

Don't call it transmutation...they'll have our heads off as alchemists.
—Ernest Rutherford, 1902

Radioactive *transformation* of unstable nuclei is a major focus of the profession of radiation protection, and much of what has been learned about atomic and nuclear physics is based on this remarkable property by which certain nuclei transform themselves spontaneously from one value of Z and N to another. Discovery of the emissions of alpha and beta particles established that atoms are not indivisible but are made up of more fundamental particles. Use of the emitted particles as projectiles to transform nuclei led eventually to some of the greatest discoveries in physics.

The process of radioactive transformation was recognized by Rutherford as transmutation of one element to another. It is also quite common to use the term *radioactive decay*, but transformation is a more accurate description of what actually happens; decay suggests a process of disappearance, when what actually happens is that an atom with excess energy transforms itself to another atom that is either stable or one with more favorable conditions to proceed on to stability.

PROCESSES OF RADIOACTIVE TRANSFORMATION

Atoms undergo radioactive transformation because constituents in the nucleus are not arrayed in the lowest potential energy states possible; therefore, rearrangement of the nucleus occurs in such a way that this excess energy is emitted and the nucleus is transformed to an atom of a new element. The transformation of a nucleus may involve the emission of alpha particles, negatrons, positrons, electromagnetic radiation in the form of x or gamma rays, and to a lesser extent neutrons, protons, and fission fragments. Such transformations are spontaneous, and the Q-values are positive; if the array of nuclear constituents is in the lowest potential energy states possible, the transformation yields a stable atom; if not, another transformation must occur.

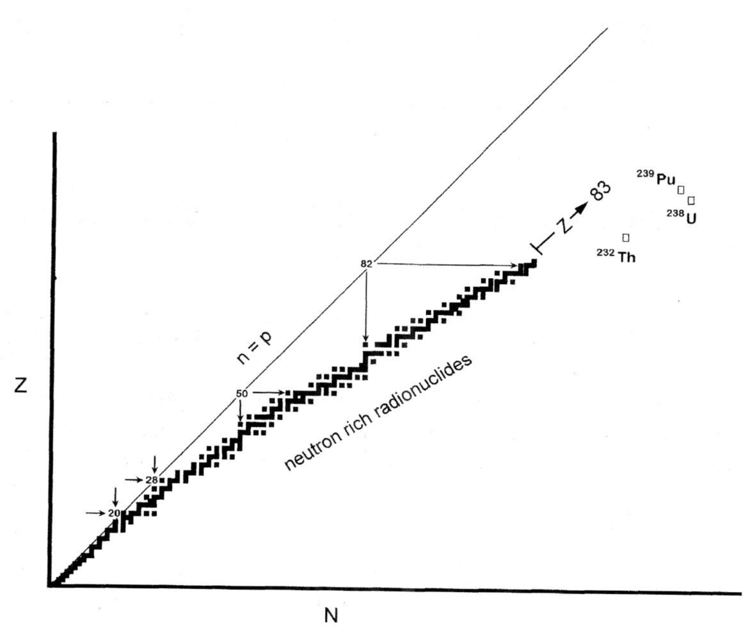

FIGURE 5-1. Line of stable nuclides on the chart of the nuclides relative to neutron-rich (below the line), proton-rich (above the line), and long-lived heavy ($Z > 83$) nuclei.

Perhaps the best way to relate the dynamics of radioactive transformations is to consider unstable, or radioactive nuclei relative to those that are stable in nature. Figure 5-1 is a plot of the number of protons (Z) and the number of neutrons (N, or $A - Z$) in all stable nuclei. This plot uses the same axes as the chart of the nuclides. Stable nuclei are shown as black squares extending from $Z = 1$ for hydrogen up to $Z = 83$ for ^{209}Bi. Also shown in Figure 5-1 are very long-lived ^{238}U and ^{232}Th, which are quite prevalent in nature even though they are radioactive. These nuclides and their transformation products exist well above the end of the line of stable nuclei. Also shown is ^{239}Pu, a very important artificial radionuclide produced in nuclear reactors; it is above uranium (or transuranic). These long-lived radionuclides produce long chains of radioactive products before they become stable.

Although the plot of the stable nuclei is called the *line of stability*, it is not a line but more a zigzag array with some gaps. For light nuclei, the neutron and proton numbers are roughly equal. However, for heavy nuclei, the coulombic repulsion between the protons is substantial, and extra neutrons are needed to supply additional binding energy to hold the nucleus together. Thus all heavy stable nuclei have more neutrons than protons. Interestingly, there are no stable nuclei with $A = 5$ or $A = 8$. A nucleus with $A = 5$, such as ^{5}He or ^{5}Li, will quickly (in 10^{-21} s or so) disintegrate into an alpha particle and a neutron or proton. A nucleus with $A = 8$, such as ^{8}Be, will quickly break apart into two alpha particles or two helium nuclei. The helium nucleus, $^4_2\text{He}^{2+}$, is particularly stable, with a binding energy per nucleon of 7.07 MeV; the energetics of existing separately as two helium nuclei is more favorable than being joined as ^{8}Be.

Regardless of their origin, radioactive nuclides surround the zigzag line of stable nuclei. These can be grouped into three major categories that will determine how they must undergo transformation to become stable:

1. Neutron-rich nuclei, which lie below the zigzag line of stable elements
2. Proton-rich nuclei, which are above the line
3. Heavy nuclei with $Z > 83$

The basic energy changes represented by these groupings comprise several transformation modes of importance to health physics for they govern how a particular nuclide undergoes transformation and the form of the emitted energy. The characteristics of these emissions in turn establishes how much energy is available for potential absorption in a medium to produce biological change(s) and how they may be detected. We examine first the energetics of these transformations, followed by the mathematical laws describing the rate of radioactive transformation.

Transformation of Neutron-Rich Radioactive Nuclei

Neutron-rich nuclei fall below the line of stable nuclei as shown in Figure 5-2 for ^{14}C, an activation product produced by an (n,p) reaction with stable nitrogen thus reducing the proton number from 7 to 6 and increasing the neutron number from 7 to 8. For neutron-rich nuclei to become stable, they need to reduce the number of neutrons in order to become one of the stable nuclides which are diagonally up and to the left on the chart of the nuclides. In simplest terms, this requires a reduction of one negative charge (or addition of a positive one) in the nucleus; since these nuclei have excess energy (mass), the transformation can occur if the nucleus emits a negatively charged electron (or beta particle). This results in an increase in the charge on the nucleus and a slight diminution of mass (equal to the electron mass and the mass equivalence of emitted energy) and has the effect of conversion of a neutron to a proton. The negatively charged electron is emitted with high energy. The mass change is small; the only requirement is to reduce the ratio of neutrons to protons.

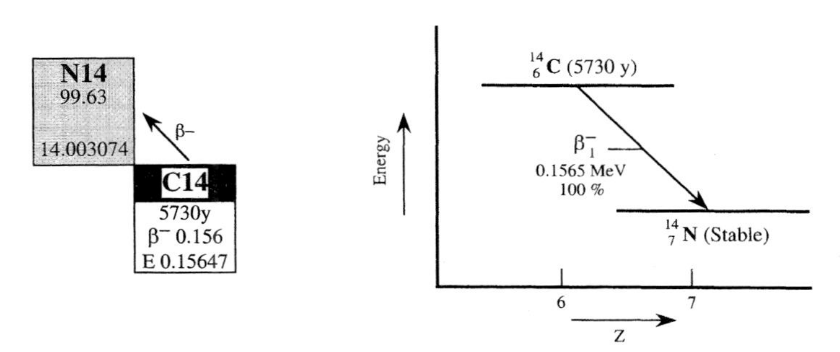

FIGURE 5-2. Position of ^{14}C relative to its transformation product ^{14}N on the chart of the nuclides, and its decay scheme.

The transformation of ^{14}C (see Figure 5-2) by beta-particle emission can be shown by a reaction equation as follows:

$$^{14}_{6}C \rightarrow ^{14}_{7}N + ^{0}_{-1}e + \overline{\nu} + Q$$

The Q-value for this transformation is 0.156 MeV and is positive, which it must be for it to be spontaneous. This energy is distributed between the recoiling product nucleus (negligible) and the ejected electron, which has most of it. When ^{14}C undergoes transformation to ^{14}N, the atomic number Z increases by 1, the neutron number N decreases by 1, and the mass number A remains the same. The transition should be thought of as a total nuclear change producing a decrease in the neutron number (or an increase in the proton number); this change is shown in Figure 5-2 as a shift upward and to the left on the chart of the nuclides to stable ^{14}N. Radioactive transformations are typically displayed in decay schemes, which are a diagram of energy (vertical axis) versus Z (horizontal axis). Because the net result is an increase in atomic number and a decrease in total energy, transformation by β^- emission is shown by an arrow down and to the right on a plot of energy versus Z, also shown in Figure 5-2.

Somewhat more complex transformations of neutron-rich radionuclides are shown in Figure 5-3 for ^{60}Co and ^{137}Cs along with their positions on the chart of the nuclides. These neutron-rich nuclei fall below the line of stability; some are fission products (e.g., ^{137}Cs), but they may also be the product of nuclear interactions (e.g., ^{60}Co, an activation product) which result in an increase in the N number. Both of these radionuclides undergo tranformation by the emission of two beta particles with different probabilities, and gamma emission follows one of the routes to relieve the excitation energy that remains because the shell configuration of the neutrons in the nucleus has not achieved the lowest potential energy states possible due to the beta emission.

Decay schemes for simplicity do not typically show the vertical and horizontal axes, but it is understood that the transformation product is depicted below the radioactive nuclide (i.e., with less energy) and that the direction of the arrow indicates the change in atomic number Z. For similar reasons, gamma emission (Figure 5-3) is indicated by a vertical line (no change in Z), often with a wave motion.

Unlike alpha particles, which are monoenergetic from a given source, beta particles are emitted with a range of energies ranging from just above 0 MeV to the maximum energy (denoted as $E_{\beta,\max}$) available from the mass change for the particular radioactive nuclide. These energies form a continuous spectrum, as illustrated in Figure 5-4 for ^{14}C.

When it was first observed that beta particles are emitted with a spectrum of energies, it was most perplexing to physicists, since conservation of energy requires that all beta transformations from a given source undergo the same energy change since the mass change is constant. For example, in the transformation of ^{14}C to ^{14}N, the energy of the nucleus decreases by 0.156 MeV because of the decrease in mass that occurs during the transformation. Some of the emitted beta particles have the maximum energy $E_{\beta,\max}$ but most do not, being emitted at some smaller value with no observable reason since the "missing" energy is not detectable. The missing energy posed an enigma to physicists: Is the well-established law of conservation of energy wrong, is Einstein's mass–energy relation wrong, or is there a

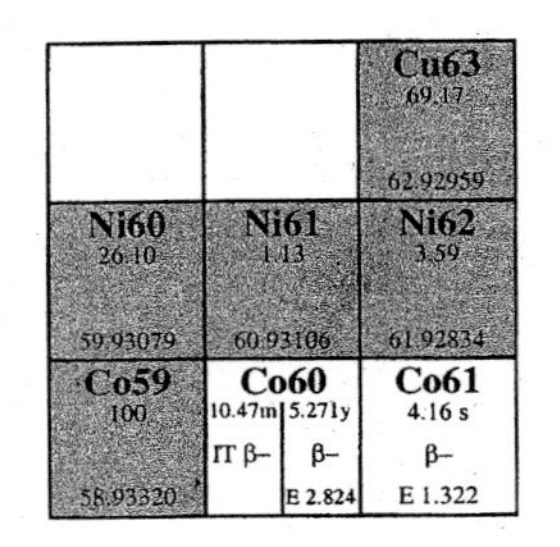

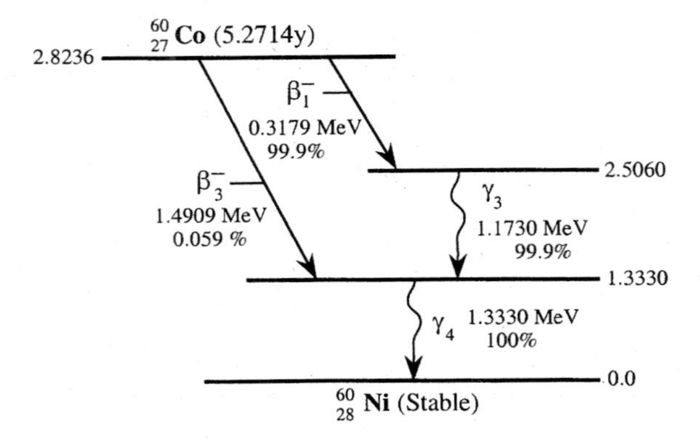

	La138 0.090 1.05E11 y β+ EC, E 1.04 137.90711	La139 99.910 β 138.90635
Ba137 2.552 m 11.23 136.90582	Ba138 71.70 137.90524	Ba139 1.396 h β- E 2.314
Cs136 19 s 13.16 d IT β- E 1.322	Cs137 30.17 y β⁻ 0.514 γ 661.7D E 1.176	Cs138 2.9 m 32.2 m IT β- β- E 1.322

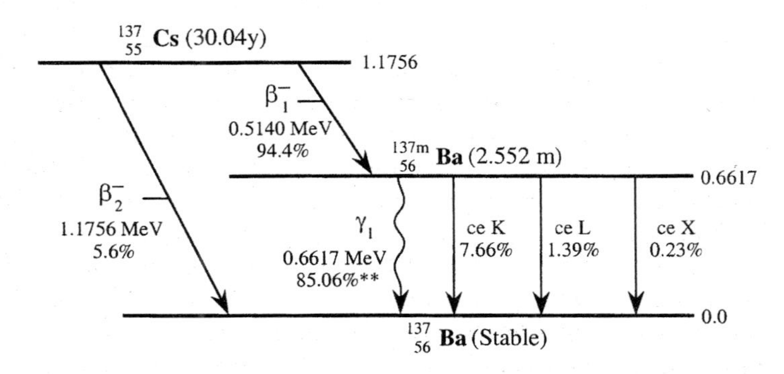

** Percent transformation of the ^{137}Cs parent

FIGURE 5-3. Artificial radionuclides of neutron-rich ^{60}Co and ^{137}Cs relative to stable nuclides and their transition by β^- emission. (Note that the excerpt from the chart of the nuclides lists a gamma ray energy for ^{137}Cs of 661.7 D, indicating that it is delayed, which occurs through ^{137m}Ba.)

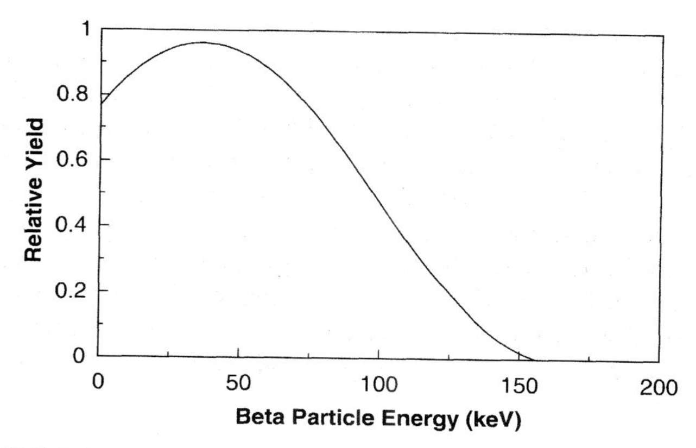

FIGURE 5-4. Continuous spectrum of beta-particle emission from ^{14}C.

way to explain the missing energy? The suggestion was made by Pauli and further developed by Fermi in 1934 that the emission of a beta particle is accompanied by the simultaneous emission of another particle, a *neutrino* (Fermi's "little one") with no charge and essentially no mass, that would carry off the rest of the energy. The neutrino (actually, an *antineutrino*, $\overline{\nu}$, from ^{14}C transformation to conserve spin) does possess momentum and energy; however, it is so evasive that its existence wasn't confirmed until 1956, when Reines and Cowan obtained direct evidence for it using an extremely high beta field from a nuclear reactor and an ingenious experiment, which is described later in this chapter.

Radioactive transformation is a process that involves the entire atom, which is demonstrated by the transformation of ^{187}Re to ^{187}Os by negative beta-particle emission (i.e., the entire atom must "participate"). The Q-value for $^{187}Re \rightarrow {}^{187}Os + \beta^-$ is 2.64 keV; however, the binding energy of atomic electrons in Os is 15.3 keV greater than in Re. This means that a Re nucleus weighs less than an Os nucleus plus an electron by 12.7 keV (15.3 − 2.6 keV) and the ^{187}Re nucleus alone cannot supply the mass–energy to yield ^{187}Os. The entire ^{187}Re atom must supply the energy for the transformation from ^{187}Re to ^{187}Os by an adjustment of all of the constituent particles and energy states; this circumstance influences the very long half-life of 4.3×10^{10} y of this primordial radionuclide.

Double-Beta Transformation

A very rare transformation of neutron-rich radionuclides is $\beta\beta$ transformation, in which two negatively charged beta particles are emitted in cascade. Double-beta transformation can occur in cases where only one beta transformation would be energetically impossible. For example, ^{128}Te to ^{128}I would require a Q-value of −1.26 MeV, which cannot occur spontaneously, but $\beta\beta$ transformation from ^{128}Te to ^{128}Xe has a Q-value of 0.87 MeV and is therefore energetically possible.

Since two beta particles must be emitted for $\beta\beta$ transformation to occur, it would be expected to be highly improbable and the half-life would be very long. A classic example of $\beta\beta$ transformation is ^{130}Te to ^{130}Xe associated with the primordial radionuclide ^{130}Te, which has a half-life of 2.5×10^{21} y (see the chart of the nuclides). The long half-life for $\beta\beta$ transformation of ^{130}Te was determined by mass spectroscopy. The method was based on observing an excess abundance of Xe (relative to its abundance in atmospheric Xe) in a tellurium-bearing rock. If the rock is T years old and short compared to the $\beta\beta$ half-life of ^{130}Te, the number of Xe atoms from $\beta\beta$ transformation is

$$N(\mathrm{Xe}) = N(\mathrm{Te})(1 - e^{-\lambda t}) \simeq N(\mathrm{Te})\lambda(\mathrm{Te})$$

Thus the half-life of ^{130}Te is

$$T_{1/2}(\mathrm{Te}) = \ln 2 \frac{N(\mathrm{Te})}{N(\mathrm{Xe})}$$

The number of Te and Xe atoms measured by the mass spectrometer can be used directly to determine the $\beta\beta$ half-life for $^{130}\mathrm{Te} \rightarrow {}^{130}\mathrm{Xe}$, which is 2.5×10^{21} y. Similarly, the $\beta\beta$ half-life for $^{82}\mathrm{Se} \rightarrow {}^{82}\mathrm{Kr}$ has been measured at 1.4×10^{20} y. The long half-lives of $\beta\beta$ transformation preclude direct detection since 1 mol of a sample would produce just a few transformations per year, and this rate would be virtually impossible to distinguish from natural radioactivity or cosmic rays.

Transformation of Proton-Rich Nuclei

Proton-rich nuclei have charge and mass such that they are above the line of stable nuclei (i.e., they are unstable due to an excess of protons). These nuclei are typically produced by interactions of protons or deuterons with stable target materials, thereby increasing the number of protons in target nuclei. Such interactions commonly occur with cyclotrons, linear accelerators, or other particle accelerators, and there are numerous (p,n) reactions that occur in nuclear reactors that produce them. Proton-rich nuclei achieve stability by a total nuclear change in which a positively charged electron mass, or positron, is emitted or an orbital electron is captured by the unstable nucleus. Both processes transform the atom to one with a lower Z value.

Some of the common proton-rich radionuclides—^{11}C, ^{13}N, ^{18}F, and ^{22}Na—are shown in Figure 5-5 as they appear on the chart of the nuclides relative to stable nuclei. These nuclei will transform themselves to the stable elements ^{11}B, ^{13}C, ^{18}O, and ^{22}Ne, respectively, which are immediately down and to the right (i.e., on the diagonal) on the chart. The mass number does not change in these transitions, only the ratio Z/N.

Positron Emission

Radioactive transformation by positron emission is often described as a proton being converted to a neutron and a positively charged electron, which is not quite correct because the proton cannot supply the mass necessary to produce a neutron (which is heavier than the proton) plus the electron mass and the associated kinetic energy carried by the positron. Rather, the process occurs as a total change in the nucleus where the protons and neutrons rearrange themselves to become more tightly bound, thus supplying the energy needed to yield a neutron and create a positron, and the kinetic energy given to it when it is emitted.

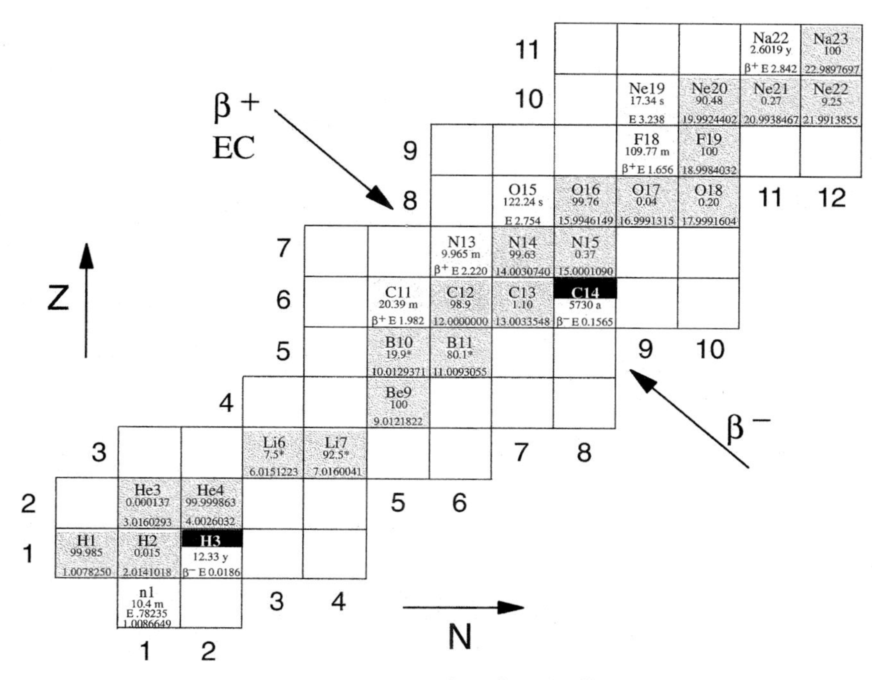

FIGURE 5-5. Proton-rich radionuclides ^{11}C, ^{13}N, ^{18}F, and ^{22}Na relative to stable nuclei and neutron rich ^{3}H and ^{14}C and the radioactive free neutron. (Adapted from GE, 1996.)

Two examples of radioactive transformation of proton-rich nuclides by positron emission are ^{18}F and ^{22}Na. These are shown in Figure 5-6 above the line of stable nuclei; they undergo transformation by the following reactions:

$$^{18}_{9}\mathrm{F} \xrightarrow{\beta^+} {}^{18}_{8}\mathrm{O} + {}^{0}_{1}\mathrm{e} + \nu + Q + \text{orbital electron}$$

$$^{22}_{11}\mathrm{Na} \xrightarrow{\beta^+} {}^{22}_{10}\mathrm{Ne} + {}^{0}_{1}\mathrm{e} + \nu + Q + \text{orbital electron}$$

It is particularly important to note that two electron masses are necessary on the product side to balance these reaction equations. One electron mass is the emitted positron; the other occurs because of the reduced charge on the ^{18}O or ^{22}Ne product nucleus. Therefore, for a nucleus to undergo transformation by positron emission, it must have enough excess mass to be able to supply these two electron masses over and above the mass of the product nucleus. In simpler terms, if the radioactive proton-rich nucleus is not two electron masses (i.e., 1.022 MeV) heavier than the product nucleus, transformation by positron emission cannot occur. When positron emission is possible, the positron is emitted with a maximum energy, $E_{\beta^+,\max}$, that is equal to the Q-value of the transition minus 1.022 MeV that is consumed in producing two electron masses. Since the available energy is shared between the positron and a neutrino, a spectrum of energies is produced.

The spectrum of positron energies is skewed to the right because the Coulomb force field of the positively charged nucleus gives the positively charged electron an

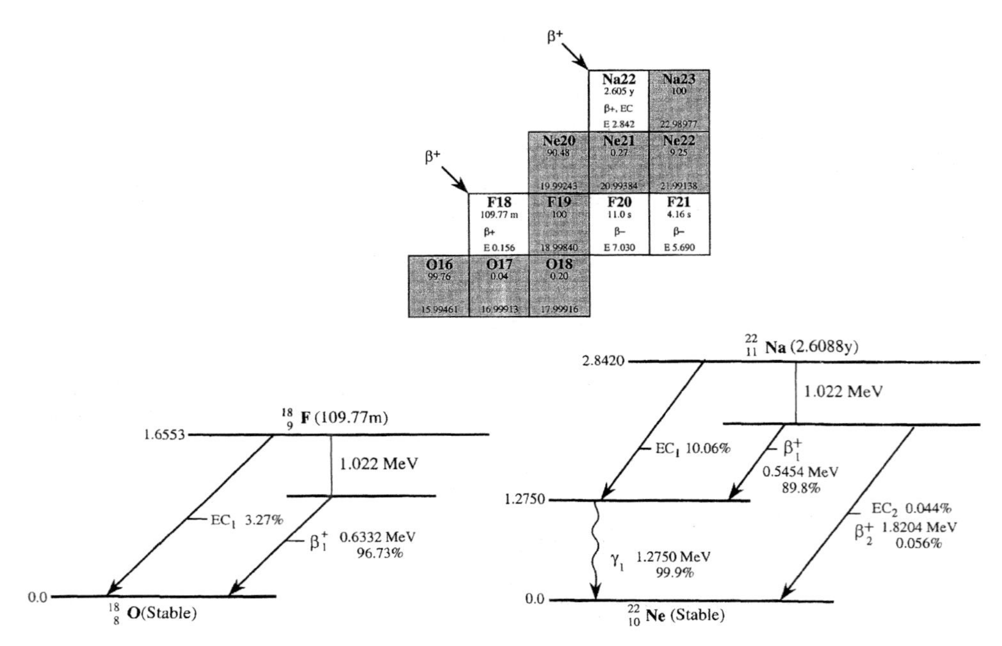

FIGURE 5-6. Proton-rich ^{18}F and ^{22}Na relative to stable nuclei and their transformation by β^+ emission, electron capture, and gamma-ray emission.

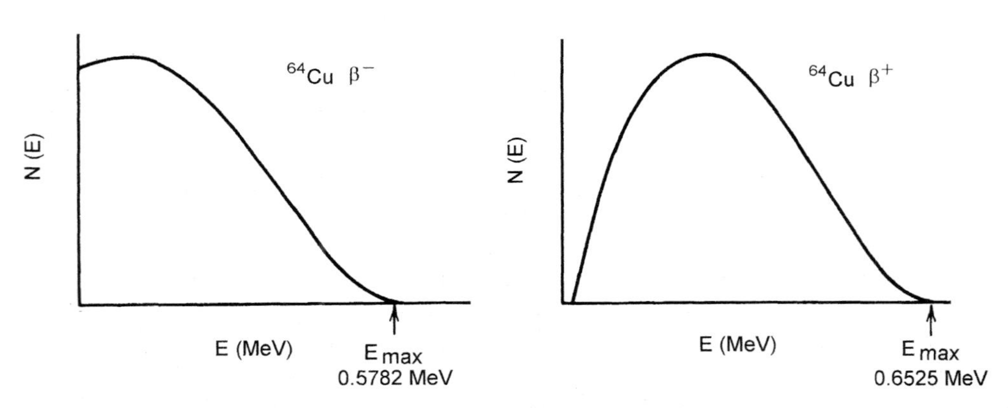

FIGURE 5-7. Comparison of β^- and β^+ energy spectra from ^{64}Cu transformation, showing the effect of the positively charged nucleus on spectral shape.

added "kick" as it leaves the nucleus. Figure 5-7 shows the different-shaped spectra for β^- and β^+ emission from radioactive transformation of ^{64}Cu. The average β^+ and β^- energy is about $\frac{1}{3}E_{max}$; however, the true average energy requires a weighted sum of the energies across the spectrum, and these are tabulated in the listings for beta-emitting and positron-emitting nuclides in Appendix D.

Annihilation radiation always accompanies positron emission because the emitted positively charged electrons are antimatter that will interact with and be annihilated by negative electrons, which are so freely distributed throughout matter. Annihi-

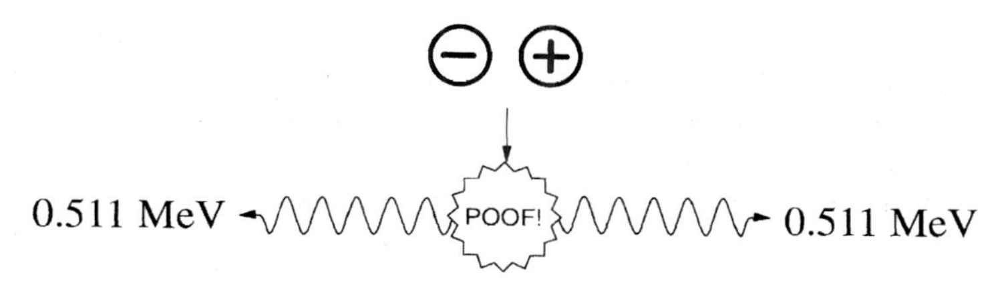

FIGURE 5-8. Annihilation of a positron and a negatron, which converts the electron masses to electromagnetic radiation.

lation of the positron with a negatron results in complete conversion of the two electron masses into pure energy, which almost always occurs as the simultaneous emission of two 0.511 MeV photons by the reaction

$$e^+ + e^- \rightarrow h\nu_1 + h\nu_2$$

This interaction is shown in Figure 5-8. The positron and negatron come together for a fleeting moment to form positronium, which consists of the two electron masses with unique spin, charge neutralization, and energetics before the complete conversion of the electrons to pure energy. The 0.511 MeV photons produced are emitted back to back or 180° from each other; they are commonly called gamma rays, but they do not originate from the nucleus, and it is more accurate to refer to them as annihilation photons, or just 0.511 MeV photons.

Average Energy of Negatron and Positron Emitters

A general relationship for the average kinetic energy of the β^- particles emitted in radioactive transformation is

$$E_{\beta^-,\mathrm{avg}} = \tfrac{1}{3}E_{\beta^-,\mathrm{max}}\left(1 - \frac{\sqrt{Z}}{50}\right)\left(1 + \frac{\sqrt{E_{\beta^-,\mathrm{max}}}}{4}\right)$$

where $E_{\beta^-,\mathrm{avg}}$ and $E_{\beta^-,\mathrm{max}}$ are in MeV. A reasonable rule of thumb for $E_{\beta^-,\mathrm{avg}}$ is $\frac{1}{3}E_{\beta^-,\mathrm{max}}$. The general relationship for the average kinetic energy for the β^+ particles emitted by positron emitters is

$$E_{\beta^+,\mathrm{avg}} = \tfrac{1}{3}E_{\beta^+,\mathrm{max}}\left(1 + \frac{\sqrt{E_{\beta^+,\mathrm{max}}}}{4}\right)$$

where $E_{\beta^+,\mathrm{avg}}$ and $E_{\beta^+,\mathrm{max}}$ are in MeV. Although the β^+ energy spectrum has a different shape from the β^- spectrum, a reasonable rule of thumb for $E_{\beta^+,\mathrm{avg}}$ is also $\frac{1}{3}E_{\beta^+,\mathrm{max}}$. The best values of $E_{\beta^-,\mathrm{max}}$ and $E_{\beta^+,\mathrm{avg}}$ are weighted averages based on the beta spectrum for each radionuclide that emits beta particles or positrons. These weighted values of β^- and β^+ energies are listed in the tables of radiation emissions provided in Appendix D.

Example 5-1: Determine the average beta energy for negatron emission of ^{64}Cu, which by a Q-value calculation has a maximum beta energy of 0.5787 MeV. How does this compare with the $\frac{1}{3}E_{\beta^-,\text{max}}$ rule-of-thumb calculation?

SOLUTION:

$$E_{\beta^-,\text{avg}} = \tfrac{1}{3} \times 0.5787 \left(1 - \frac{\sqrt{29}}{50}\right)\left(1 + \frac{\sqrt{0.5787}}{4}\right)$$

$$= 0.2049 \text{ MeV}$$

By rule of thumb,

$$E_{\beta^-,\text{max}} = \frac{0.5787}{3} = 0.1929 \text{ MeV}$$

which is roughly 94.2% of the more careful calculation; the accepted value based on spectrum weighting (Appendix D) is 0.1902 MeV.

Copper-64 also emits a positron with an $E_{\beta^+,\text{max}}$ value of 0.6531 MeV. Its average energy, $E_{\beta^+,\text{avg}}$, is 0.2335 MeV by the detailed calculation and 0.2177 MeV by the rule-of-thumb determination; both are considerably lower (83.9 and 78.3%, respectively) than the value of 0.2782 MeV listed in Appendix D. These examples are adequate for most radiation protection considerations, but the more exact values listed in Appendix D and the NNDC database are perhaps more appropiate for beta dose determinations.

Electron Capture

The decay scheme for ^{22}Na shows that only 90% of the transformations of ^{22}Na occur by positron emission. The remaining 10% is by electron capture, which is a competing mechanism by which proton-rich nuclei reduce the number of protons. Electron capture competes with positron emission only when positron emission is energetically possible, as in ^{22}Na; when sufficient energy is not available for positron emission, the number of protons in proton-rich nuclei can be reduced only by capture of an orbital electron. Since both positron emission and electron capture reduce the number of protons to achieve stability, they compete with each other, but only if both can occur.

Electron capture transformation occurs because the wave motion of orbital electrons can bring them close enough to the unstable nucleus that capture of the electron results in the reduction of the number of protons in the nucleus (as illustrated in Figure 5-9). Typical reactions for electron capture are represented by the transformation of ^{59}Ni and ^{55}Fe, neither of which have sufficient excess mass to supply two electron masses (i.e., $Q < 1.022$ MeV); thus electron capture is the only means by which such nuclei can undergo transformation to more stable products by the following reactions:

$$^{59}_{28}\text{Ni} + {}^{0}_{-1}\text{e} \rightarrow {}^{59}_{27}\text{Co} + \nu + Q$$

$$^{55}_{26}\text{Fe} + {}^{0}_{-1}\text{e} \rightarrow {}^{55}_{25}\text{Mn} + \nu + Q$$

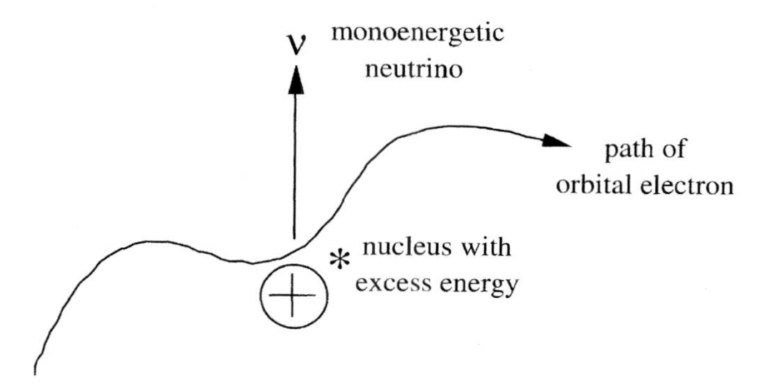

FIGURE 5-9. Capture of an orbital electron in electron capture transformation of proton-rich radionuclides to reduce the nuclear charge. The de Broglie wave pattern of the orbiting electrons causes them to come close to or pass through the nucleus, allowing capture to occur.

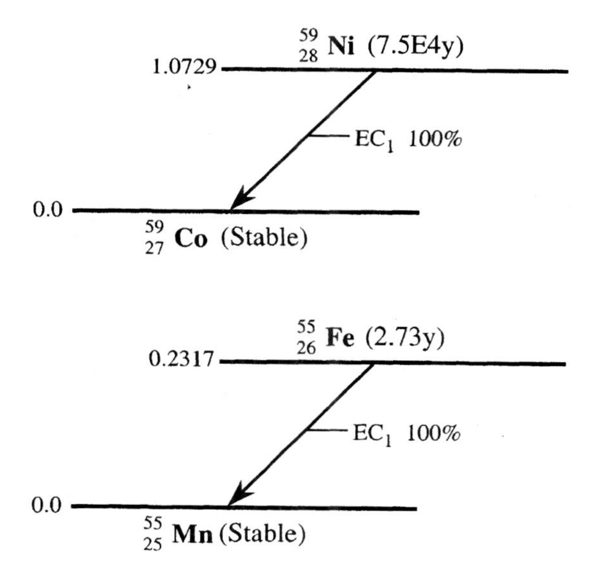

FIGURE 5-10. Decay scheme of ^{55}Fe and ^{59}Ni. Each is followed by emission of characteristic x rays (as shown in Figure 5-11 for $^{55}\text{Fe} \rightarrow {}^{55}\text{Mn}$).

The accompanying decay schemes for these transitions are shown in Figure 5-10.

Since K-shell electrons are closest to the nucleus to begin with, most electron capture events (on the order of 90%) involve K-capture; however, it is probable for L-shell electrons to come close enough for capture, and some 10% of the captured electrons are L-shell electrons. M-shell electrons can also be captured, but with much lower (about 1%) probability. In electron capture transformations, Z decreases by 1, N increases by 1, and A, the mass number, remains the same. However, the capture of an electron from the K, L, or M shell leaves a vacancy that is filled immediately by electrons from higher energy levels. This process is accompanied

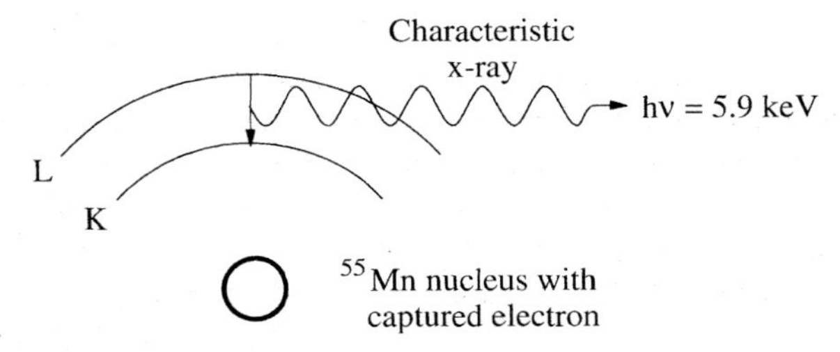

FIGURE 5-11. Emission of a 5.9 keV characteristic x ray by filling the K-shell electron vacancy from the L shell in an atom of ^{55}Mn, the product of K-shell electron capture transformation of a nucleus of ^{55}Fe.

by the emission of x-rays characteristic of the product nuclide. For example, as shown in Figure 5-11, radioactive transformation of ^{55}Fe by electron capture results in the emission of x rays that are due to energy changes between the electron shells of ^{55}Mn, the transformed atom; hence the emitted x rays are characteristic of Mn. The energy of the characteristic x rays emitted by ^{55}Mn is about 5.9 keV. Therefore, the only way that electron capture transformations produce energy emission or deposition (radiation dose) is by these very weak x rays or Auger electrons associated with them (see Chapter 4). They also represent the only practical means for detecting these radioactive nuclides.

Electron capture transformation is accompanied by emission of a monoenergetic neutrino in contrast to the spectrum of neutrino energies in β^- and β^+ transformations. Electron capture is also referred to as a form of beta transformation because an electron is involved and the mass number of the transforming nucleus does not change, only the ratio of protons to neutrons. The energy of the monoenergetic neutrino is just the Q-value for the transformation minus any excitation energy that may be emitted in the form of gamma rays to relieve excitation energy retained by the transformed nucleus.

Example 5-2: What is the energy of the neutrino emitted in electron capture transformation of ^{22}Na to ^{22}Ne?

SOLUTION: From Figure 5-6, electron capture of ^{22}Na is followed by emission of a 1.28 MeV gamma photon. The Q-value of the reaction is 2.84 MeV; thus

$$E_{\text{neutrino}} = 2.84 - 1.28 = 1.56 \text{ MeV}$$

Radioactive Transformation of Heavy Nuclei by Particle Emission

A number of unstable nuclei exist above $Z = 83$ (bismuth). These nuclei include the long-lived naturally occurring elements of ^{238}U, ^{235}U, and ^{232}Th and their transformation products, and they also include artificially produced transuranic radionuclides such as ^{239}Pu and ^{241}Am. ^{238}U and ^{232}Th are so long-lived that they still exist after formation of the solar system some 4 to 14 billion years ago. For them to achieve stability, they must convert themselves to one of the stable isotopes of lead (^{206}Pb for ^{238}U or ^{208}Pb for ^{232}Th), each of which is considerably lighter than

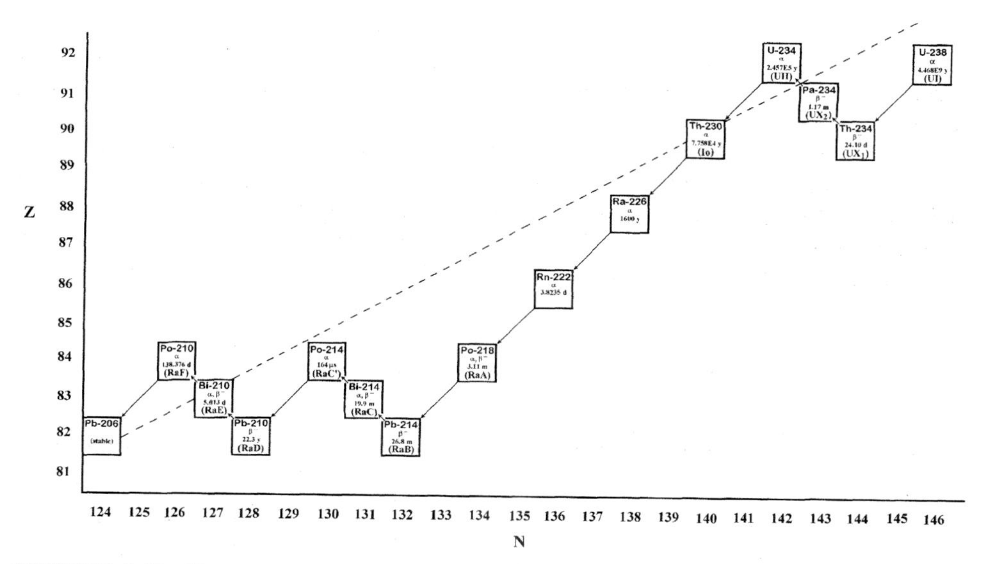

FIGURE 5-12. Change in proton number (Z) versus neutron number (N) for the uranium series relative to an extrapolated line of stability.

the parent; ^{238}U atoms must lose 32 mass units, including 10 protons ($Z = 92$ to $Z = 82$) and 22 neutrons, and ^{232}Th must lose 24 mass units, or 8 protons ($Z = 90$ to $Z = 82$) and 16 neutrons. The most efficient way for these nuclides to reduce both mass and charge is through the emission of several $^4He^{2+}$ nuclei (or alpha particles), each of which will reduce the proton and the neutron number by 2 and the mass number by 4. Reduction of A, Z, and N for these heavy nuclides yields a long chain of transformation products (called *series decay*) before a stable end product is reached.

Figure 5-12 shows the series transformation of ^{238}U through a long chain of radioactive products before the atoms are transformed to stable ^{206}Pb. A hypothetical dashed *line of stability* is shown as an extrapolation from the stable nuclei below to indicate how the path of transformation crosses back and forth before a stable configuration of N and Z is found at ^{206}Pb. The intermediate alpha emissions yield unstable nuclei, and some of these emit beta particles if the neutron/proton ratio is "too high." A zigzag path of transformation prevents the transformation to nuclei that are too far from a projection (no actual line of stability exists above bismuth) of the line of stability.

The alpha particle is a helium nucleus (i.e., a helium atom stripped of its electrons), and it consists of two protons and two neutrons. It is, on the nuclear scale, a relatively large particle. For a nucleus to release such a large particle, the nucleus itself must be relatively large. Alpha particles resulting from a specific nuclear transformation are monoenergetic, and they have large energies, generally above about 4 MeV. A typical example is ^{226}Ra:

$$^{226}_{88}\text{Ra} \xrightarrow{\alpha} {}^{222}_{86}\text{Rn} + {}^{4}_{2}\text{He} + 4.87\ \text{MeV}$$

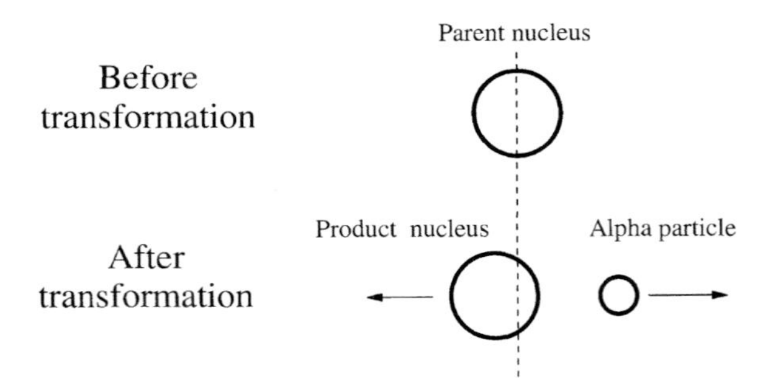

FIGURE 5-13. Parent nucleus at rest before and after transformation; the emitted alpha particle, and the product nucleus recoil in opposite directions to conserve linear momentum.

where both Z and N of the transforming atom decrease by 2 while the mass number A decreases by 4. Although the total energy change for the reaction is 4.87 MeV, the energy of the emitted alpha particle is only 4.78 MeV. The difference of 0.09 MeV is the energy of recoil imparted to the newly formed radon nucleus. As shown in Figure 5-13, the parent nucleus is essentially at rest before the alpha particle is emitted, and the mass of the emitted alpha particle is large enough that the product nucleus undergoes recoil in order to conserve momentum.

Calculation of the Q-value for alpha-particle emission yields the total disintegration energy, which is shared between the emitted alpha particle and the recoiling product nucleus. The kinetic energy of the ejected alpha particle will be equal to the Q-value of the reaction minus the recoil energy imparted to the product nucleus. For example, the Q-value of ^{226}Ra transformation by alpha-particle emission is

$$\begin{aligned} Q &= \text{mass of } ^{226}\text{Ra} - (\text{mass of } ^{222}\text{Rn} + \text{mass of } ^{4}\text{He}) \\ &= 226.025406 \text{ u} - (222.017574 + 4.002603) \text{ u} \\ &= 0.005227 \text{ u} \times 931.502 \text{ MeV/u} \\ &= 4.87 \text{ MeV} \end{aligned}$$

The Q-value energy is shared between the product nucleus and the alpha particle, both of which can be considered nonrelativistically in terms of their velocity and mass, or

$$Q = \tfrac{1}{2}Mv_{\text{p}}^2 + \tfrac{1}{2}mv_{\alpha}^2$$

since energy is conserved. Momentum is also conserved such that $Mv_{\text{p}} = mv_{\alpha}$, thus the kinetic energy, E_{α}, imparted to the alpha particle is

$$E_{\alpha} = \frac{Q}{1 + m_{\alpha}/M_{\text{p}}}$$

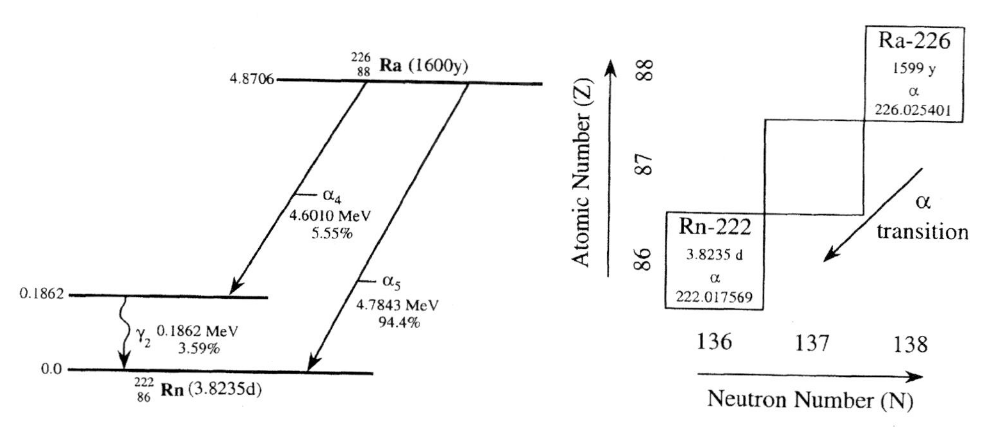

FIGURE 5-14. Decay scheme of ^{226}Ra plotted as a change in E and Z and its transition to ^{222}Rn on the chart of the nuclides.

where m_α and M_p are the masses (approximated by the mass numbers) of the alpha particle and the recoiling product nucleus, respectively.

Example 5-3: Calculate the kinetic energy of the alpha particle emitted when ^{226}Ra is transformed directly to the ground state of ^{222}Rn. What is the recoil energy of the ^{222}Rn product nucleus?

SOLUTION: The Q-value of the transformation reaction is 4.87 MeV; thus

$$E_\alpha = \frac{4.87 \text{ MeV}}{1 + 4/222} = 4.78 \text{ MeV}$$

and the recoil energy is

$$E_R = 4.87 \text{ MeV} - 4.78 \text{ MeV} = 0.09 \text{ MeV}.$$

Alpha-particle transformation is shown schematically in the decay scheme (or the E–Z diagram) as a decrease in energy due to the ejected mass and a decrease in the atomic number by two units. Figure 5-14 shows the transformation of ^{226}Ra to ^{222}Rn, and the change in position of the transforming nucleus and the product nucleus on the chart of the nuclides.

The decay scheme shows two alpha particles of different energies from the radioactive transition of ^{226}Ra. The first (4.78 MeV) occurs in 94.4% of the transformations and goes directly to the ground state of ^{222}Rn; the other (4.6 MeV) occurs 5.6% of the time and leaves the product nucleus in an excited state. The excited nucleus reaches the ground state by emitting a gamma ray of 0.186 MeV with a yield of 3.59% (internal conversion accounts for the other 2.01% by this route). The existence of these two different alpha particle energies and the gamma ray demonstrates the existence of discrete nuclear energy states. The alpha particles for a given transition are monoenergetic (i.e., they all have the same energy).

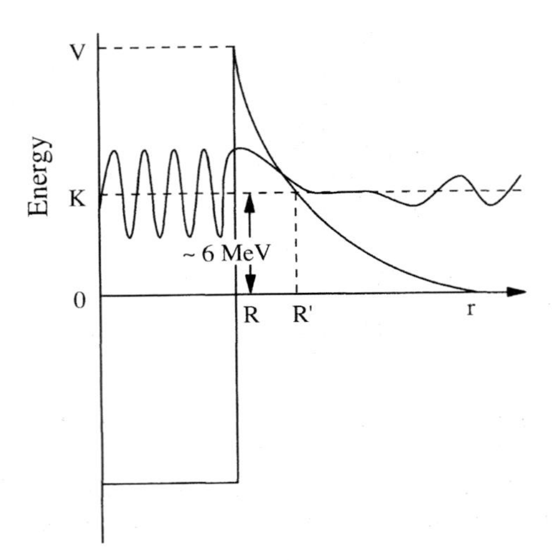

FIGURE 5-15. Barrier penetration of a heavy nucleus by an alpha particle.

Theory of Alpha-Particle Transformation

Alpha-particle transformation cannot be explained by classical physics because the alpha particle doesn't have enough energy to penetrate the large potential energy barrier of heavy nuclei such as radium. This potential energy barrier can be as much as 30 to 40 MeV, whereas most emitted alpha particles have energies well below 10 MeV. When considered as a wave-mechanical phenomenon, however, the de Broglie wavelength of an alpha particle in the nucleus has a probability of existing beyond the barrier, and in fact does so. An alpha particle confined to the nucleus can be thought of as two protons and two neutrons coming together temporarily to make up the particle. An alpha particle of 4 to 5 MeV will have a velocity such that it will bounce back and forth within the potential well of the nucleus some 10^{22} times per second. This large number of attacks on the potential barrier of the nucleus combined with the de Broglie wavelength (Figure 5-15) eventually leads to its "tunneling through" the barrier.

The probability of emission (tunnelling) increases with alpha-particle energy as shown in Figure 5-16, which is a plot of the half-lives of various alpha emitters versus the energies of the emitted alpha particles. Clearly, the higher the alpha-particle energy, the shorter the half-life, indicating that the probability of transformation is markedly higher. This can be explained by the "thickness" of the potential energy barrier in Figure 5-15. The higher-energy particles interact with the thinner part of the barrier and have a greater likelihood of tunneling through, whereas the lower-energy particles intersect the thicker part of the potential barrier and are less likely to penetrate. From Figure 5-16 it is startling that a twofold change in alpha energy results in a change in half-life of 10^{22}. This effect, which is quite uniform over typical alpha particle energies, is termed the *Geiger–Nuttall rule* after the scientists who first deduced the relationship. It is often plotted as the log of the disintegration constant, λ, or of half-life, each of which shows an approximate straight-line relationship.

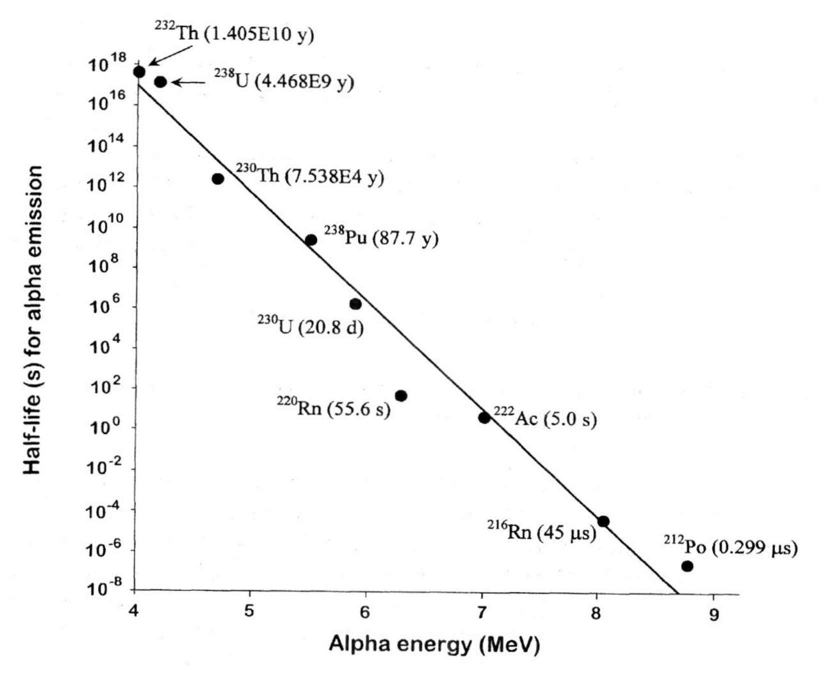

FIGURE 5-16. Geiger–Nuttall relationship between probability of alpha-particle emission (i.e., half-life), and kinetic energy of the emitted alpha particle—a twofold change in alpha particle energy changes the half-life by a factor of 10^{22}.

Transuranic Radionuclides

Transuranic (TRU) elements, which exist above $Z = 92$, also emit a large number of alpha and beta particles to reduce mass and charge as they undergo transformation to a stable isotope of bismuth or lead. They represent a special group of heavy elements, all of which are radioactive and are produced primarily by bombarding ^{238}U with neutrons. Transformation of TRUs typically joins one of the naturally radioactive series in the process of achieving a path to stability. Selected TRUs are shown in Figure 5-17. All of these are well above the end of the curve of stable nuclei and are above uranium in the chart, hence the name *transuranic nuclides*. The arrows in Figure 5-17 provide a general path of the formation of some of the major TRU nuclides by neutron activation and beta transformation from ^{238}U to produce ^{239}Pu and further neutron activation of ^{239}Pu to yield increasingly heavier TRUs, many of which undergo radioactive transformation by alpha-particle emission. Most TRU nuclides undergo radioactive transformation by alpha emission. For example, ^{240}Pu (half-life of 6560 years) emits two alpha particles with energies of 5.16 and 5.12 MeV, as shown in Figure 5-18.

Gamma Emission

Radioactive transformation is a chaotic process and often leaves the transforming nucleus in an excited state in which the protons and neutrons in the shells of the nucleus are not in the most tightly bound state possible. This excitation energy will

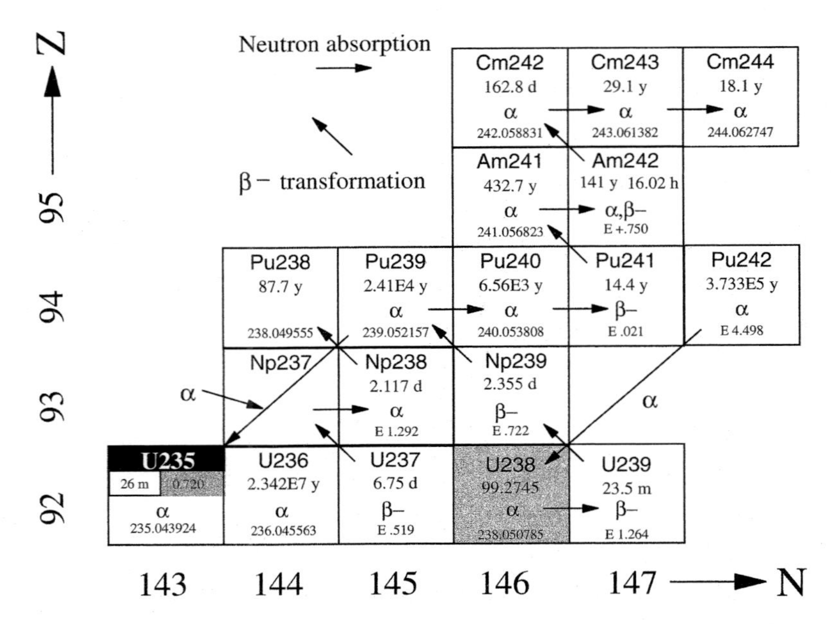

FIGURE 5-17. Transuranic radionuclides (TRUs) relative to naturally occurring uranium with arrows showing routes of production via neutron absorption and/or beta transformation. (Adapted from GE, 1996.)

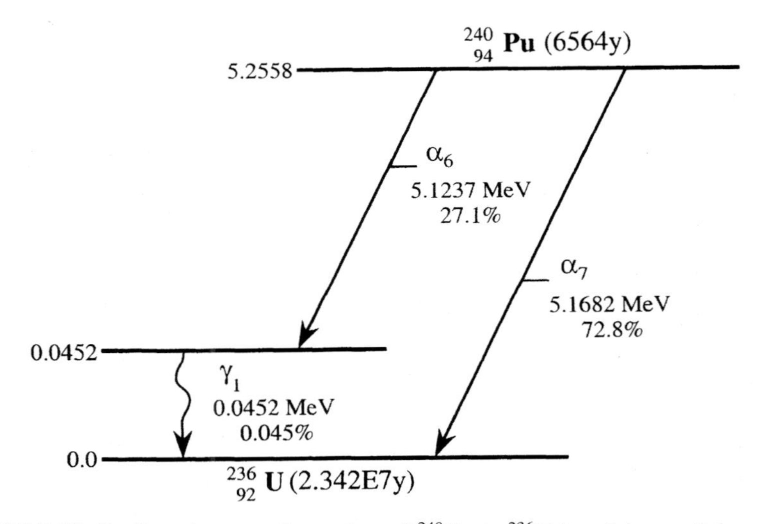

FIGURE 5-18. Radioactive transformation of ^{240}Pu to ^{236}U by alpha-particle emission.

be emitted as electromagnetic radiation as the protons and neutrons in the nucleus rearrange themselves to the desired lowest-energy state. The shell model of the nucleus suggests discrete energy states for neutrons and protons; it is this difference between energy states that is emitted as a gamma photon when rearrangement takes place. Thus the emitted gamma ray is characteristic of that particular nucleus.

As important as it is in radiation protection, gamma emission occurs only after radioactive transformation has occurred by alpha-particle emission, negatron emission, positron emission, or electron capture, all of which change the unstable atom to another atom. It is incorrect to refer to this process as gamma decay because gamma rays are produced only to relieve excitation energy (i.e., the atom is still the same atom after the gamma ray is emitted). The decay schemes in Figure 5-3 for ^{60}Co and ^{137}Cs illustrate how gamma emission occurs to relieve excited states after a radioactive transformation occurs.

Internal Transition: Metastable or Isomeric States

Most excited states in atoms are relieved in less than 10^{-9} s or so by gamma emission; however, some excited states may exist long enough to be readily measurable, and in these cases the entire process of radioactive transformation may be thought of as two separate events. Such a delayed release of excitation energy represents a metastable (or isomeric) state; when radioactive transformation is interrupted by a metastable state, it is known as *internal transition.*

Figure 5-19 illustrates the radioactive transformation of ^{99}Mo, with about 15% of the beta transitions omitted for clarity; the transitions shown yield a metastable state of ^{99m}Tc at an energy level of 0.1405 MeV above the ground state of ^{99}Tc. The metastable state, ^{99m}Tc, is delayed considerably with a 6-h half-life, and it is relieved by the emission of gamma radiation (or internal transition) to the ground state of ^{99}Tc. The only change that occurs in ^{99m}Tc (i.e., the excited ^{99}Tc nucleus) is the relief of the excitation energy, and although the gamma-emission rate of ^{99m}Tc is often stated in curies or becquerels, this is not strictly correct because these units define the quantity of a radionuclide undergoing radioactive transformation to another element, which ^{99m}Tc does not do.

The production of ^{99m}Tc by ^{99}Mo has found widespread use in nuclear medicine because a ^{99}Mo source with its 66-h half-life can be fabricated into a generator of ^{99m}Tc, which provides a relatively pure source of 140 keV gamma rays without the confounding effects of beta particles. The ^{99m}Tc can be extracted periodically by eluting the ion-exchange column with saline solution. This selectively strips the ^{99m}Tc; however, some ^{99}Mo may also be eluted. The gamma-emission rate of the eluted ^{99m}Tc is directly dependent on the activity of the ^{99}Mo parent, the fraction (82.84%) of beta-particle transformations that produce the metastable state of Tc, and the period of ingrowth that occurs between elutions. These relative activities represent series transformation.

Internal Conversion

Many nuclei are left in an excited state following radioactive transformation. This excitation energy is usually relieved by the emission of a gamma ray, but the wave motion of the orbital electrons can bring them close enough to the nucleus such that the excitation energy can be transferred directly to one of the orbital electrons, ejecting it from the atom; i.e., they are internally converted to an ejected electron to relieve the excitation energy remaining in the nucleus. The kinetic energy of the conversion electron is discrete; it is the difference between that available for emis-

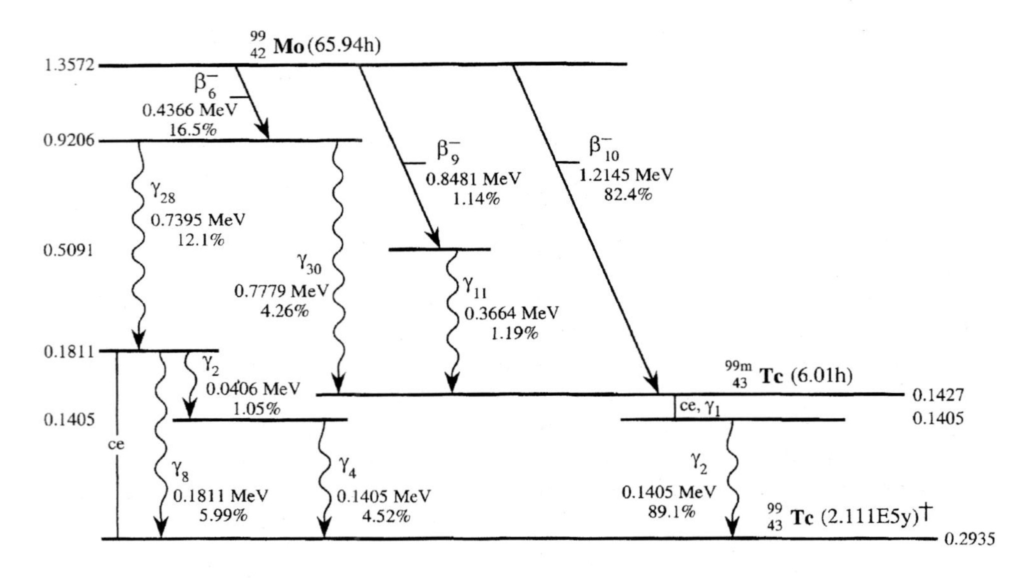

† NOTE: 87.907% of transitions of ^{99}Mo produce the 0.1427 and in turn the 0.1405 level of ^{99m}Tc. Metastable ^{99m}Tc is relieved in 89.9% of its transitions by a 0.1405 MeV gamma ray; therefore, the 0.1405 MeV γ is produced in 78.32% of the transformations of ^{99}Mo. However, if the ^{99}Mo parent and the ^{99m}Tc product are together and in equilibrium, the 0.1405 MeV gamma is produced in 82.84 % of transformations of ^{99}Mo (this accounts for the 4.52% contribution through β^-_6 and γ_4).

FIGURE 5-19. Decay scheme of ^{99}Mo through ^{99m}Tc to ^{99}Tc. The gamma-ray yields are for isomeric transition of ^{99m}Tc and thus total 100%; however, the emission rate of 0.1405 MeV gamma rays is only 82.84 per 100 transformations of ^{99}Mo. The ^{99}Tc end product is effectively stable.

sion as a gamma photon minus the binding energy of the electron in the particular shell from which it is converted (i.e., from a bound electron to an ejected electron with kinetic energy). This process competes with gamma emission to relieve the excitation energy left in the nucleus after a transformation occurs by particle emission or electron capture. Since internal conversion occurs when an orbital electron and an excited nucleus meet, it is more prevalent for metastable nuclides in which the nucleus retains the excitation energy for a longer period, but it can occur with any excited nucleus that could emit gamma radiation, just with lower (often negligible) probability. The process of internal conversion is shown schematically in Figure 5-20.

Energetically, the process of internal conversion can be thought of as similar to an internal photoelectric effect, except that the photon does not appear, only the energy. The gamma photon is not emitted because the excitation energy in the nucleus is transferred directly to the orbital electron. The kinetic energy of the ejected electron is the energy available for gamma emission minus the binding energy of the electron in its particular orbit; thus it is similar to the work function energy which must be overcome when photons undergo photoelectric interactions with bound orbital electrons in target atoms. Conversion of K-shell electrons is most probable because the K electrons spend more time close to the nucleus than do any of the outer orbitals, but internal conversions can occur for L, M, and on

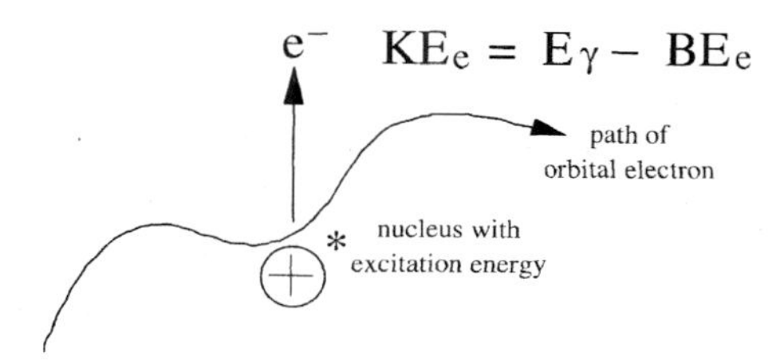

FIGURE 5-20. Relief of excitation energy by internal conversion. The excitation energy of the nucleus is transferred directly to an orbital electron, which due to its wave motion, can exist close to or in the nucleus, where it can "pick up" the energy and be ejected. The kinetic energy of the internally converted electron is equal to that available for gamma emission minus the orbital binding energy of the ejected electron.

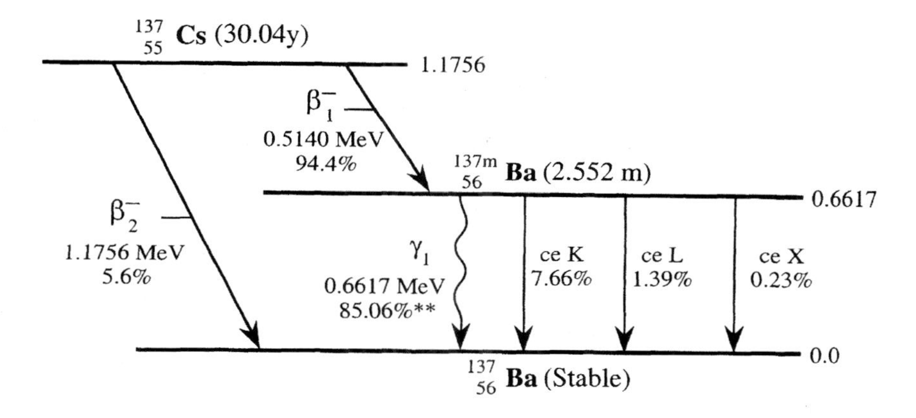

FIGURE 5-21. Gamma emission and internal conversion of ^{137m}Ba in the radioactive transformation of ^{137}Cs through ^{137m}Ba to ^{137}Ba.

rare occasions, the more distant orbital electrons. In contrast to beta emission, the electron is not created in the transformation process, and no neutrinos are involved. For this reason internal conversion rates can be altered slightly by changing the chemical environment of the atom, thus changing somewhat the atomic orbits. It is not a two-step process; that is, a photon is not emitted to knock loose an orbiting electron as in the photoelectric effect. The probability of such an event would be negligible.

A common example of internal conversion occurs in the transformation of ^{137}Cs through ^{137m}Ba, as shown in Figure 5-21. When ^{137}Cs atoms undergo transformation, 94.4% go to an excited state of ^{137}Ba, indicated as ^{137m}Ba, with a metastable half-life of 2.55 min. The excited nuclei of ^{137m}Ba then emit γ-rays in 90.1% of emissions, 8.11% by internal conversion of K-shell electrons ($\alpha_K = 0.0811$) and 1.72% by electron conversion from other shells (i.e., for each 100 transformations of ^{137}Cs, 85.1 gamma photons, 7.66 internally converted electrons from the

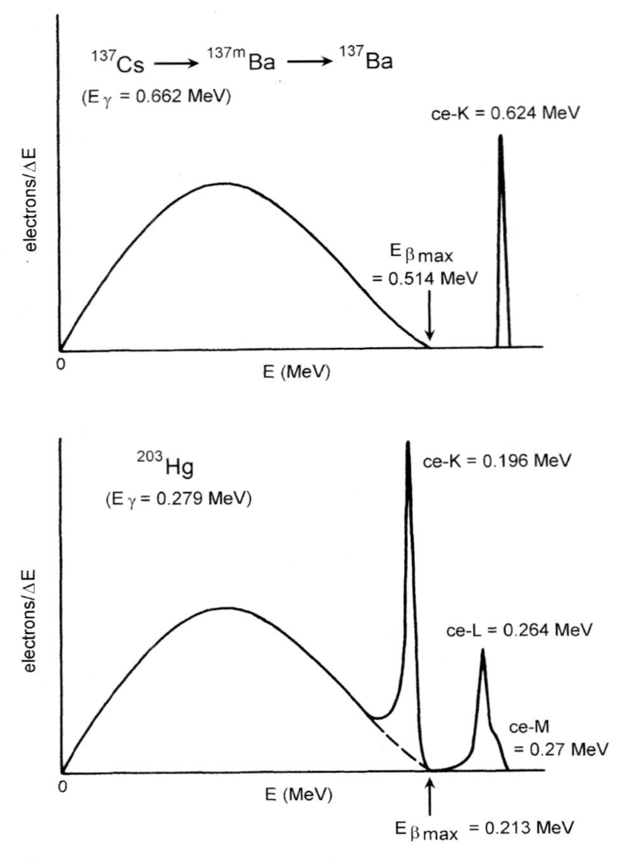

FIGURE 5-22. Internal conversion electron emission superimposed upon continuous beta spectra of ^{137}Cs and ^{203}Hg, illustrating the effect of the available excitation energy in the transformed nucleus on the position of the discrete energy peaks.

K-shell, and 1.62 electrons from other shells will occur through relief of the excitation energy retained in ^{137m}Ba).

Conversion electrons are emitted monoenergetically because all of the energy states involved in internal conversion have definite values. The excitation energy left in a transformed nucleus is discrete, as is the electron binding energy associated with each shell; therefore, electrons emitted due to internal conversion are expected to have discrete energies. Beta spectra of beta/gamma-emitting radionuclides will have the usual energy distribution but will also show the internally converted electrons as lines superimposed on the beta continuum. Figure 5-22 shows beta spectra of ^{137}Cs and ^{203}Hg with internal conversion electrons appearing as discrete lines associated with the electron shells from which they are ejected. The internal conversion electrons from ^{137}Cs appear as a sharp peak at 0.624 MeV and above the 0.514 MeV endpoint energy of ^{137}Cs because the internally converted electrons have a kinetic energy of 0.662 MeV (the gamma-excitation energy) minus the orbital binding energy of 0.038 or 0.624 MeV, which happens to be larger than $E_{\beta^-,\max}$ of

^{137}Cs beta particles. This sharp peak of internally converted electrons makes ^{137}Cs a useful source for energy calibration of a beta spectrometer.

The energy peaks of internal conversion electrons on beta spectra vary considerably. As shown in Figure 5-22, the peaks of internally converted electrons of ^{203}Hg are superimposed on the spectrum of beta particles and serve to extend it somewhat. Also notable is the combined peak for converted L- and M-shell electrons. These electron spectra indicate that a beta source that also emits gamma rays has a number of individual and discrete components. Furthermore, characteristic x rays follow internal conversion to fill electron shell vacancies; thus gamma spectra from a radioactive source in which internal conversion is prominent may also contain peaks due to these x rays, usually in the low-energy range.

Conversion electrons are labeled according to the electronic shell from which they come: K, L, M, and so on. These are designated in the listings of radiations (see, e.g., Figure 5-21 for ^{137}Cs) and Appendix D as ce-K, ce-L, and so on. Some listings also include the substructure corresponding to the individual electrons in the shell (e.g., L_I, L_{II}, L_{III}).

Since internal conversion *competes* with gamma emission, an internal conversion coefficient, α, has been defined as the fraction of deexcitation that occurs by internal conversion. This fraction is listed in some decay tables and can be used to compute the number of gamma emissions that actually occur versus those that could occur, or similarly, the number of internally converted electrons. The coefficient is denoted α_K, α_L, α_M, and so on, where the subscripts designate the electron shell from which the conversion electron originates. More recent compilations (see Appendix D) don't list the conversion coefficient but instead list the fraction of such electrons per transformation and the fractional emission of the competing gamma ray(s).

Internal conversion coefficients have a number of interesting features. First, they increase as Z^3, so the conversion process is more important for heavy nuclei than for light nuclei. For example, α_K for the 1.27 MeV transition in $^{22}_{10}$Ne has a value of 6.8×10^{-6}, and the 1.22 MeV transition in $^{187}_{74}$W has an α_K value of 2.5×10^{-3}; thus the ratio of converted electrons is very nearly equal to $(10/74)^3$, as expected. The conversion coefficient also decreases rapidly with increasing transition energy, and for higher atomic shells ($n > 1$) it decreases as $1/n^3$ such that for a given transition, the ratio of $\alpha_K/\alpha_L \approx 8$.

Example 5-4: What are the energies of internally converted electrons when ^{203}Hg undergoes transformation to ^{203}Tl, in which the residual excitation energies are emitted as a single gamma ray of energy 279.190 keV or by internal conversion from the K, L, or M_I shells?

SOLUTION: Binding energies of electrons in Tl are (from Appendix C)

$$E_b(\text{K}) = 85.529 \text{ keV}$$

$$E_b(\text{L}_\text{I}) = 15.347 \text{ keV}$$

$$E_b(\text{L}_\text{II}) = 14.698 \text{ keV}$$

$$E_b(\text{L}_\text{III}) = 12.657 \text{ keV}$$

$$E_b(\text{M}_\text{I}) = 3.704 \text{ keV}$$

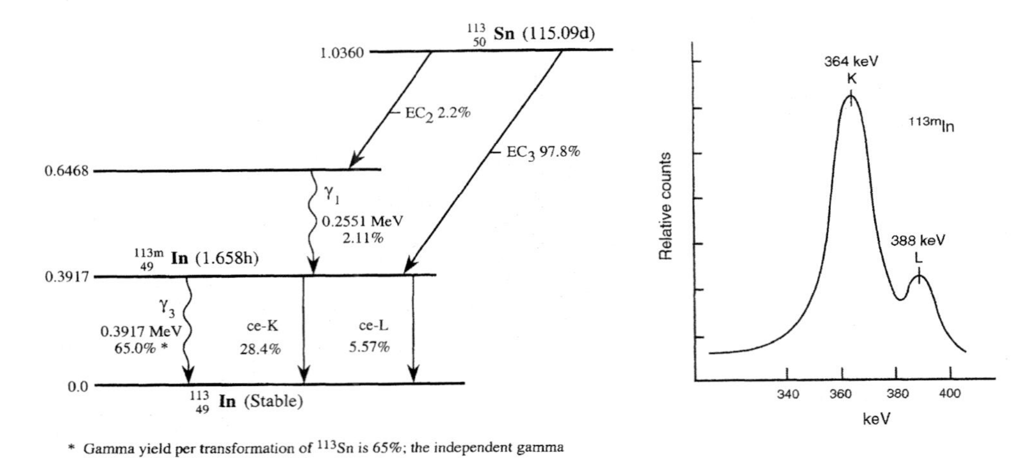

FIGURE 5-23. Transformation of ^{113}Sn to ^{113}In by electron capture through metastable ^{113m}In, and the spectrum of conversion electrons ejected from the K and L shells of ^{113m}In.

Consequently, the conversion electrons emitted from the excited nuclei of ^{203}Tl after β^- transformation of ^{203}Hg will have the following energies:

$$\text{ce-K} = 279.190 - 85.529 \text{ keV} = 193.661 \text{ keV}$$

$$\text{ce-L}_{\text{I}} = 279.190 - 15.347 \text{ keV} = 263.843 \text{ keV}$$

$$\text{ce-L}_{\text{II}} = 279.190 - 14.698 \text{ keV} = 264.492 \text{ keV}$$

$$\text{ce-L}_{\text{III}} = 279.190 - 12.657 \text{ keV} = 266.533 \text{ keV}$$

$$\text{ce-M}_{\text{I}} = 279.190 - 3.704 \text{ keV} = 275.486 \text{ keV}$$

and so on, for higher-energy shells. The designations for these conversion electrons (ce-K, etc.) are the same as those listed in Appendix D.

The transformation of ^{113}Sn through ^{113m}In ($T_{1/2}$ = 1.66 h) to ^{113}In is another example of internal conversion. The metastable atoms of ^{113m}In have a half-life of 1.66 h and thus can exist in an excited state for a relatively longer time, during which orbital electrons passing by can receive the excitation energy. In this case, 98% of the electron capture transformations of ^{113}Sn go to the 392 keV excitation state of ^{113m}In. The energy of the internally converted electrons will be 392 keV minus 27.93 keV, the binding energy of the K-shell electron, or 4.25 keV for the L-shell electron. Figure 5-23 shows the discrete energies of the K- and L-shell converted electrons at 364 and 388 keV and their average yields. Indium characteristic x rays are also emitted as the vacant electron shells are filled.

Example 5-5: What will be the kinetic energy of internally converted electrons from the K and L shells of ^{113}In due to radioactive transformation of ^{113}Sn through ^{113m}In to ^{113}In?

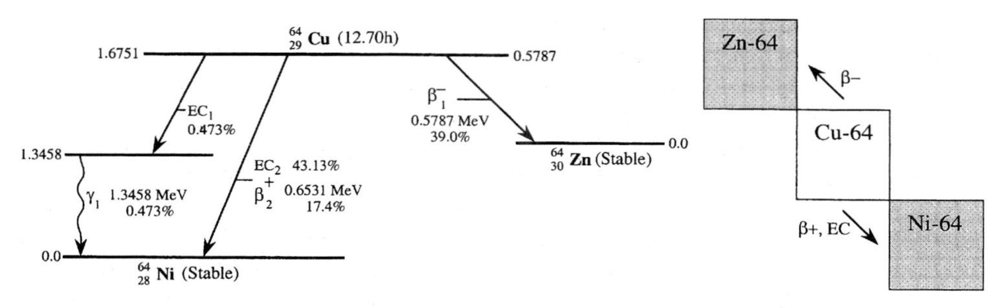

FIGURE 5-24. Decay scheme of ^{64}Cu for β^- transformation to ^{64}Zn and β^+ and electron capture transformation to ^{64}Ni.

SOLUTION: The binding energy of K- and L-shell electrons in ^{113}In are 27.93 and 4.25 keV, respectively. Since ^{113m}In has excitation energy of 392 keV above the ground state of ^{113}In, it can be relieved either by gamma emission or internal conversion. The energy of K-shell converted electrons is therefore

$$KE_{eK} = (392 - 27.93)\text{ keV} = 364\text{ keV}$$

and for the L-shell conversion,

$$KE_{eL} = (392 - 4.25)\text{ keV} = 388\text{ keV}$$

Multiple Modes of Radioactive Transformation

Some nuclei [e.g., ^{64}Cu (Figure 5-24)], are so configured that they can undergo transformation by either β^+, β^-, or electron capture, without a change in the mass number. As shown in Figure 5-24, the position of ^{64}Cu on the chart of the nuclides is such that it could , if energetics allow, transform to ^{64}Zn or ^{64}Ni. The test of whether any given radionuclide can undergo a given type of transformation is whether sufficient energy is available for the transformation, which is determined (1) by whether a positive Q-value exists, and (2) if it meets the conditions that must be satisfied for the respective transformation to occur (Example 5-6).

Example 5-6: Determine whether ^{64}Cu can undergo radioactive transformation by β^-, β^+, and electron capture processes.

SOLUTION: The reactions are

$${}^{64}_{29}\text{Cu} \rightarrow {}^{64}_{30}\text{Zn} + \beta^- + \overline{\nu} + Q$$

$${}^{64}_{29}\text{Cu} \rightarrow {}^{64}_{28}\text{Ni} + \beta^+ + \nu + Q$$

$${}^{64}_{29}\text{Cu} + {}^{0}_{-1}\text{e} \rightarrow {}^{64}_{28}\text{Ni} + \nu + Q$$

and the atomic masses of the three isotopes are

$${}^{64}_{29}\text{Cu}(63.929768\text{ u}), \qquad {}^{64}_{28}\text{Ni}(63.9279698\text{ u}), \qquad {}^{64}_{30}\text{Zn}(63.929147\text{ u})$$

For transformation by negative beta-particle emission, the Q-value need only be positive, which is determined from the nuclide masses:

$$Q(\beta^-) = 63.929768 \text{ u} - 63.929147 \text{ u}$$
$$= (0.000621 \text{ u})(931.5 \text{ MeV/u}) = 0.58 \text{ MeV}$$

Therefore, β^- transformation can occur. For transformation by positron emission or β^+, the Q-value must be positive *and* larger than the energy equivalent of two electron masses (1.022 MeV), or

$$Q(\beta^+) = 63.929768 \text{ u} - 63.927970 \text{ u} = 0.001798 \text{ u}$$
$$= (0.001798 \text{ u})(931.5 \text{ MeV/u}) = 1.675 \text{ MeV}$$

which is greater than 1.022 MeV by 0.655 MeV, and thus transformation by positron emission can occur.

For electron capture transformation to occur the Q-value for $^{64}\text{Cu} \rightarrow {}^{64}\text{Ni}$ need only be positive, which it is; therefore, it too can occur (whether it actually does occur is a random probabilistic process since it compete with β^+ transformation). It is possible, therefore, for $^{64}_{29}\text{Cu}$ to undergo radioactive transformation by all three processes: 39% by β^- to stable ^{64}Zn, 17% by β^+ to stable ^{64}Ni, and about 44% by electron capture to stable ^{64}Ni.

Transformation by Delayed Neutron Emission

A few nuclides, usually fission products, emit neutrons, and these emissions can be very important to reactor physics. This process does not change the element, but it does produce a transformation of the atom to a different isotope of the same element; for example, the radioactive transformation of ^{87}Br, a fission product, produces $^{87\text{m}}\text{Kr}$ by β^- emission in 70% of its transformations, which in turn emits gamma radiation or a 0.3 MeV neutron to stable ^{86}Kr, as shown in Figure 5-25. The neutron emission of $^{87\text{m}}\text{Kr}$ is delayed entirely by the 56-s half-life of the ^{87}Br parent; therefore, these delayed neutrons can be considered as having a 56-s half-life in nuclear reactor physics and criticality control procedures.

Transformation by neutron emission is a process that competes with transformation by beta-particle emission, which is a considerably slower process of radioactive transformation.

Transformation by Spontaneous Fission

Another means of neutron emission is spontaneous fission in which a heavy nucleus has so much excess energy that it splits on its own (i.e., it undergoes transformation without the addition of energy or a bombarding particle). The process yields the usual two fission fragments and two to four neutrons, depending on the distribution probability of each fission. Radioactive transformation by emission of an alpha or beta particle may compete with spontaneous fission; for example, ^{252}Cf emits an

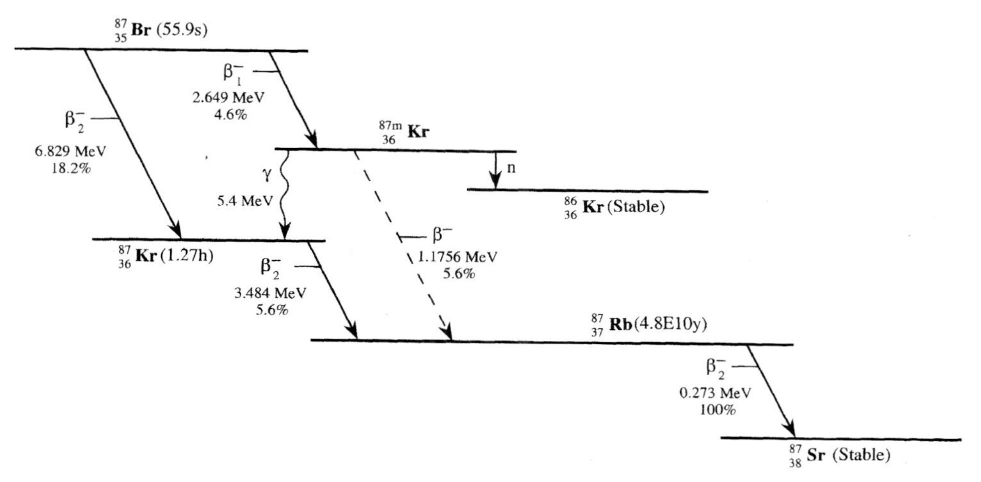

FIGURE 5-25. Decay scheme of ^{87}Br and its products, including neutron emission from ^{87m}Kr.

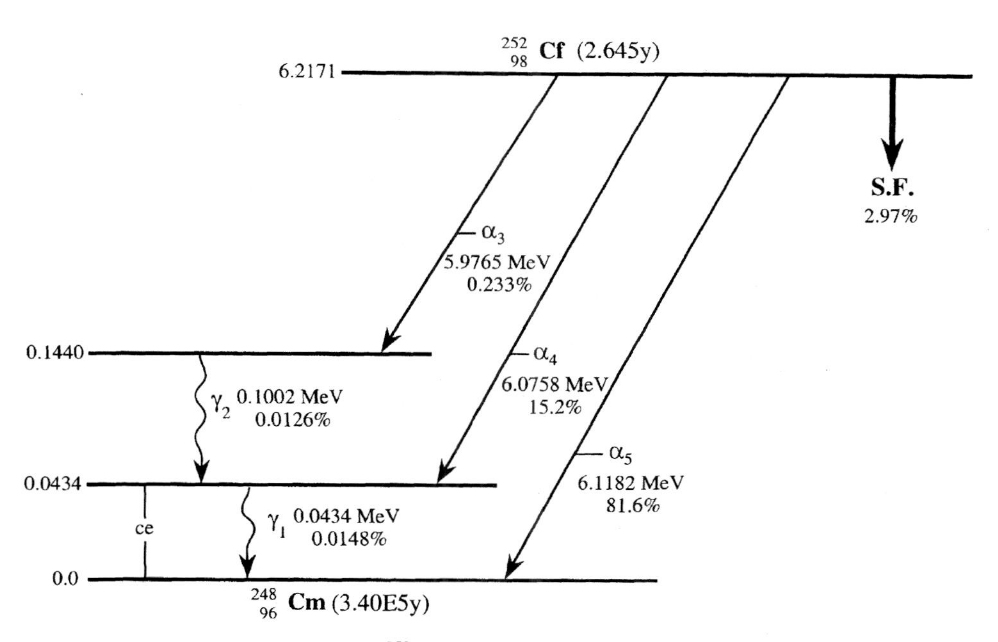

FIGURE 5-26. Decay scheme of ^{252}Cf, showing spontaneous fission in competition with alpha transition.

alpha particle in about 97% of its transformations and neutrons in 2.97% as shown in Figure 5-26. The neutron yield is 2.3×10^{12} neutrons/s per gram of ^{252}Cf, or 4.3×10^{9} neutrons/s per Ci of ^{252}Cf. Although the spontaneous fission half-life is 87 years, the overall half-life of ^{252}Cf is 2.645 y because alpha transformation to ^{248}Cm is more probable than spontaneous fission.

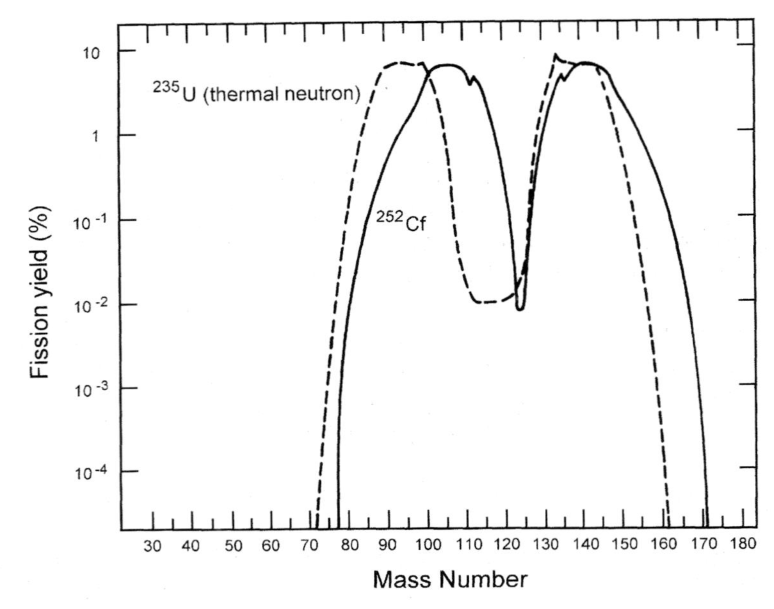

FIGURE 5-27. Mass number distribution of ^{252}Cf spontaneous fission fragments compared to thermal neutron fission of ^{235}U.

The distribution of fission fragments from ^{252}Cf fission is shown in Figure 5-27 compared to that for ^{235}U. The fission product yield from spontaneous fission of ^{252}Cf is quite different from thermal neutron fission of ^{235}U, being somewhat skewed for lighter fragments and containing a few products of slightly heavier mass. Other examples of nuclides that undergo spontaneous fission are ^{256}Fm ($T_{1/2}$ = 1.30 h) and ^{254}Cf ($T_{1/2}$ = 60.5 d). Isotopes of plutonium, uranium, thorium, and protactinium also experience spontaneous fission, as do many other heavy nuclei. Most of these are not useful as neutron sources, but ^{252}Cf is commonly produced to provide a source of neutrons with a spectrum of energies representative of those of fission.

Proton Emission

A few radionuclides far above the line of stability are so proton-rich that they may emit protons as they undergo radioactive transformation. These are rare and of little, if any, interest in radiation protection, but can occur physically because of so much excess mass. An example of proton emission is $^{73}_{36}$Kr, which undergoes transformation by positron emission, 0.7% of which yields an excitation state of $^{73m}_{36}$Br that is relieved by proton emission to $^{72}_{34}$Se. The decay scheme is shown in Figure 5-28 with two end products: $^{72}_{34}$Se due to protons emitted from $^{73m}_{35}$Br, or $^{73}_{35}$Br when positrons are emitted. For proton emission the mass number decreases by one unit; for positron emission it remains the same. Another example is ^{59}Zn, and there are others, although all are only of academic interest in radiation protection.

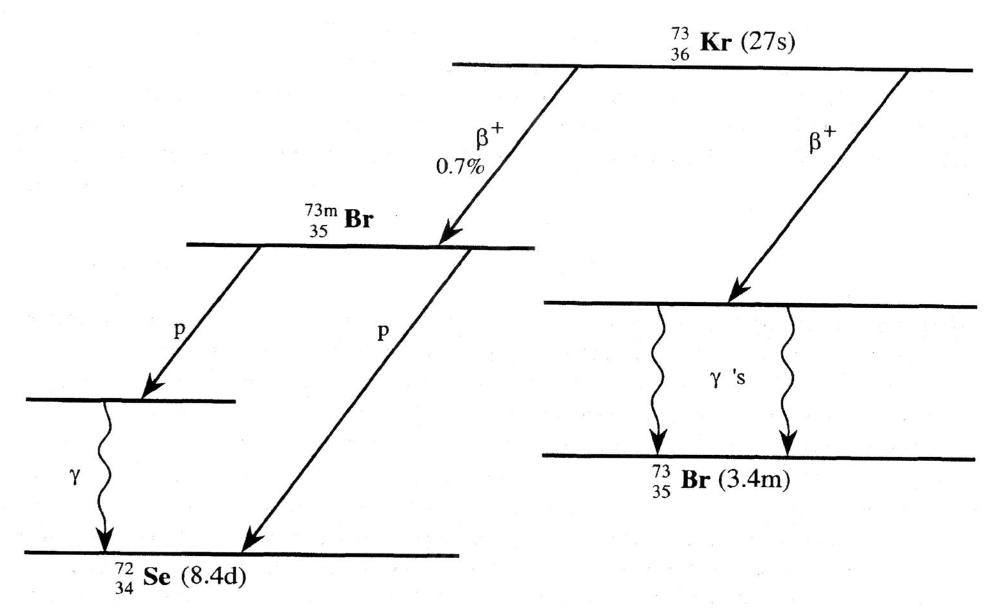

FIGURE 5-28. Proton emission in the radioactive transformation of ^{73}Kr in which 0.7% of β^+ transitions are to the metastable state of ^{73m}Br, which is relieved by emission of protons to ^{72}Se (greatly simplified).

DECAY SCHEMES

The probability that a radionuclide will undergo transformation by a particular route is constant, and it is possible to predict, on the average, the percentage of transformations that will occur by any given mode. Several compilations of the various modes of radioactive transformation have been made with varying degrees of completeness. The chart of the nuclides provides very basic information on whether an isotope is radioactive, and if so, the primary modes of radioactive transformation, the total transition energy, and the predominant energies of the emitted "radiations." Examples of these for ^{11}C, ^{13}N, and ^{14}C are shown in Figure 5-29. Each is denoted as radioactive by having no shading; the half-life is provided just below the isotopic symbol; the principal (but not necessarily all) mode(s) of transformations and the energies are shown just below the half-life; and the disintegration energy in MeV is listed at the bottom of each block. The emitted energies for particles are listed in MeV; however, gamma energies are listed in keV presumably because gamma spectroscopy uses this convention.

The most complete information on radioactive transformation is displayed in diagrams such as those for ^{51}Cr and ^{131}I shown in Figures 5-30 and 5-31. Such figures, which are commonly called *decay schemes*, provide a visual display of the modes of transformation and the principal radiations involved, but much more detailed information on the frequency and energy of emitted particles and related electromagnetic energy is necessary, and it is useful to provide these data in a detailed listing. Figures 5-30 and 5-31 contain such listings for ^{51}Cr and ^{131}I; these

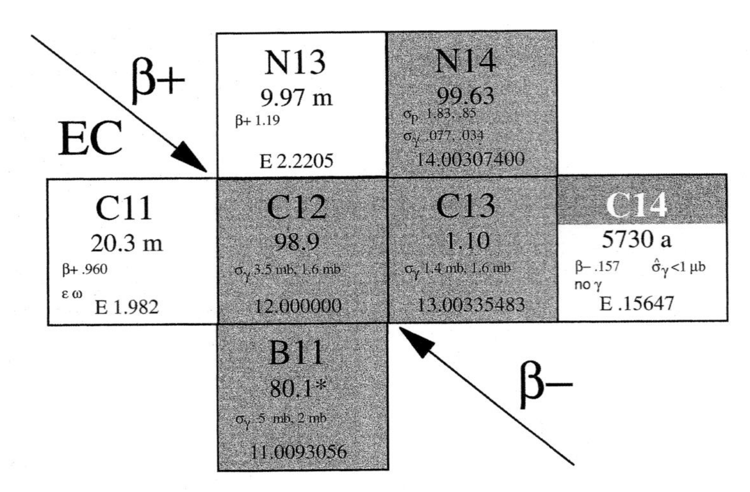

FIGURE 5-29. Excerpt from the chart of the nuclides showing basic transformation modes with principal energies, half-lives, and disintegration energies of ^{11}C, ^{13}N, and ^{14}C. Also shown are stable isotopes and their percent abundances, activation cross sections, and rest masses (u). (Adapted from GE, 1996.)

lists and diagrams are typical of those provided in Appendix D for many of the common radionuclides of interest to radiation protection.

Two other compilations on radioactive transformation data are those published by the Medical Internal Radiation Dosimetry (MIRD) Committee of the Society of Nuclear Medicine (SNM, 1989) and Report 38 of the International Commission on Radiological Protection (ICRP-38) (ICRP, 1978). Both of these data sets were compiled for internal dosimetry calculations and contain all the radiation information necessary for such calculations. They are available from the National Nuclear Data Center (NNDC), which is operated for the U.S. Department of Energy by Brookhaven National Laboratory and is an invaluable national resource for such data. The ICRP-38 report lists just about all the radionuclides; however, even though it is very comprehensive, it is quite old (published in 1978). The MIRD compilation (published in 1989) is more recent but lists only those radionuclides of interest to nuclear medicine.

The decay schemes and listed data in Appendix D were compiled from current NNDC data and use the same designators used by MIRD, ICRP-38, and NNDC, although somewhat simplified for radiation protection needs. The abbreviations used in the decay scheme tables are described in Table D-1. Also shown in Figures 5-30, and 5-31 and in Appendix D are the mode of production of the radioisotope (upper left of the table of radiations) and the date the data were last reviewed by NNDC (at the upper right). Data for other radionuclides can be obtained from the MIRD and ICRP-38 reports, or if up-to-date information is desired, from the National Nuclear Data Center.

Iodine-131 is an example of a nuclide that has an especially complex decay scheme, as shown in Figure 5-31. Four distinct beta-particle transitions occur (plus others of lesser yield) and each has its own particular yield and transformation

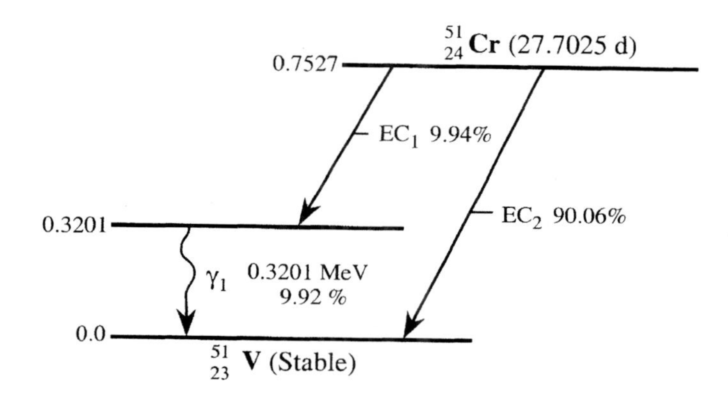

^{50}Cr(n,γ) 17-JUN-97

Radiation	Y_i (%)	E_i (MeV)
γ_1	9.92	0.3201
ce-K, γ_1	0.0167	0.3146
ce-L, γ_1	0.0016	0.3195[a]
ce-M, γ_1	2.58E-04	0.3200[a]
$K\alpha_1$ X-ray	13.1	0.0050
$K\alpha_2$ X-ray	6.6	0.0049
$K\beta$ X-ray	2.62	0.0054*
L X-ray	0.334	0.0005*
Auger-K	66.9	0.0044*

*Average Energy
[a]Maximum Energy For Subshell

FIGURE 5-30. Decay scheme of ^{51}Cr and a listing of principal radiations emitted.

energy, referred to as the *disintegration energy*. The most important of these transitions is beta-particle emission in 89.9% of transformations, with a maximum beta energy of 0.61 MeV, followed by immediate gamma emission through two routes, including the prominent gamma-ray energy of 0.364 MeV. This 0.364 MeV gamma ray, which occurs for 81.7% of the transformations of ^{131}I, is often used to quantitate sources of ^{131}I. Several other gamma-ray energies also appear from other beta transitions.

RATE OF RADIOACTIVE TRANSFORMATION

The rate of transformation of radioactive atoms is not affected by natural processes such as burning, freezing, solidifying, dilution, or making chemical compounds

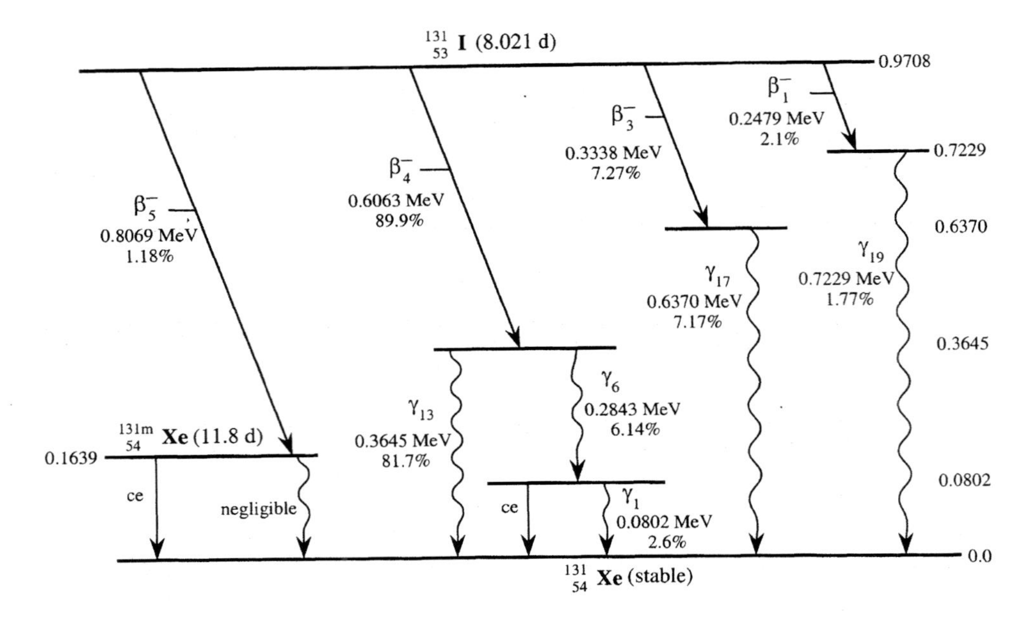

Radiation	Y_i (%)	E_i (MeV)
β^-_1	2.10	0.0694*
β^-_3	7.27	0.0966*
β^-_4	89.9	0.1916*
β^-_5	1.18	0.8069*
γ_1	2.62	0.0802
ce-K, γ_1	3.54	0.0456
ce-L, γ_1	0.464	0.0747[a]
γ_6	6.14	0.2843
ce-K, γ_6	0.252	0.2497
γ_{13}	81.7	0.3645
ce-K, γ_{13}	1.55	0.3299
ce-L, γ_{13}	0.246	0.3590[a]
ce-M+, γ_{13}	0.051	0.3633[a]
γ_{17}	7.17	0.6370
γ_{19}	1.77	0.7229
$K\alpha_1$ X-ray	2.56	0.0298
$K\alpha_2$ X-ray	1.38	0.0295

*Average Energy
[a]Maximum Energy For Subshell
^{131m}Xe product, yield 1.18%; ^{131}Xe yield 98.8%

FIGURE 5-31. Decay scheme of ^{131}I and a listing of major radiations emitted and their yield.

with them. The presence or absence of oxygen or other substances has no effect; thus radioactive substances can be used as energy sources in outer space or in other hostile environments. One small exception is a chemical change that increases the electron density near the nucleus, which in turn can have a small influence on the probability of electron capture or internal conversion, both of which are related to the likelihood of interaction with orbital electrons. The effect has been observed to increase the half-life of electron capture transformation of $^{7}BeF_2$ by 0.08% as compared with ^{7}Be metal. The half-life of photon emission from $^{99m}Tc_2S_7$ has been observed to be 0.27% greater than in the pertechnetate form because of chemical alteration of the electron density and the corresponding effect on the probability of internal conversion. Other metastable nuclei also exhibit this small effect in half-life; however, the effect is of minimal consequence in most radiation protection circumstances.

Radioactive transformation requires definition of three important concepts: (1) activity, or transformation rate, (2) a unit to describe the rate of transformation, and (3) a mathematical law that allows calculation of the amount and rate of radioactive transformation over time.

Activity of a Radioactive Source

Up to this point, radioactive transformation has been discussed in terms of those features of atoms that cause energy emission to occur, the particles and radiations that are emitted, and the processes that cause them to do so. The types of particles and radiations emitted and their energy are of utmost importance for radiation protection because they determine how much is available to produce various radiation effects. Equally important to these concepts is the rate of radioactive transformation or the activity of a radioactive source. Regardless of which mode of transformation occurs (β^-, β^+, γ, electron capture, etc.) radioactive atoms of a given species have a constant probability of transformation, designated by the disintegration constant λ; therefore, the activity of a source containing N atoms is defined as

$$\text{activity} \equiv \lambda N$$

The number of radioactive atoms, N, is determined from the mass of the radioactive isotope by Avogadro's number (see Chapter 1); and the disintegration constant, λ, has been determined experimentally for each radionuclide as the fraction of the number of atoms in a radionuclide will undergo transformation in a given amount of time, usually per second, minute, hour, and so on.

Units of Radioactive Transformation

Activity, or *intensity of radioactive emission*, is the number of atoms that undergo transformation to new atoms per unit time. For many years the standard unit of radioactivity has been the *curie* (Ci), first defined as the emission rate of 1 g of radium (assumed to be 3.7×10^{10} transformations per second) at the request of Madame Curie, the discoverer of radium, to honor her husband, Pierre. As measurements of the activity of a gram of radium became more precise, this definition led to a standard that, unfortunately, varied. This variability, which is unacceptable for a standard unit, was eliminated by defining the curie as that *quantity* of *any* radioactive material that produces 3.7×10^{10} transformations per second (t/s):

$$\text{transformation rate per curie} \equiv 3.7 \times 10^{10} \text{ t/s}$$

Unfortunately, the curie is a very large unit for most samples and sources commonly encountered in radiation protection. Consequently, activity is often stated in terms of decimal fractions of a curie, especially for laboratory or environmental samples. The *millicurie* (mCi), which is one-thousandth of a curie, and the *microcurie* (μCi) (equal to one-millionth of a curie) correspond to amounts of radioactive materials that produce 3.7×10^7 (37 million) and 3.7×10^4 (37,000) t/s, respectively. Environmental levels of radioactivity are often reported in *nanocuries* (1 nCi = 37 t/s) or *picocuries* (1 pCi = 0.037 t/s, or 2.22 t/min). Despite its familiarity to radiation physicists, the curie is a somewhat inconvenient unit because of the necessity to state most measured decay rates in μCi, nCi, or even pCi. The *rutherford* (Rd), defined as 10^6 disintegrations per second (d/s), was used for a time to provide a smaller decimalized unit, but it is rarely used as a unit although it is subtly included in SI units for radioactivity since the designation MBq, or 10^6 t/s, is used frequently.

International units have been defined by the Systeme International d'Unités (SI), which attempts to use basic quantities (kg, m, s, etc.) and decile and decimal multiples of each. It has defined the *becquerel* (Bq) in honor of Henri Becquerel, the discoverer of radioactivity, as that quantity of any radioactive material that produces 1 transformation per second:

$$\text{transformation rate per Bq} \equiv 1 \text{ t/s}$$

The SI unit for radioactivity suffers from being small compared to most radioactive sources measured by radiation physicists, which in turn requires the use of a different set of prefixes: kBq, MBq, GBq, and so on, for kilobecquerel (10^3 t/s), megabecquerel (10^6 t/s), gigabecquerel (10^9 t/s), and so on, respectively. Neither the curie nor the becquerel is easy to use despite the tradition of paying respect to great and able scientists or the SI goal of pure units. Nonetheless, both are used with appropriate prefixes and it is necessary for radiation protection to adapt. It is perhaps useful to reflect that both units are just surrogates for listing the number of emissions that occur per unit of time, and when used with the energy for each emission, an energy deposition rate or the total energy deposition can be determined for use in measurement of radiation and/or determination of its effects. In this context, the number of transformations per second (t/s) or disintegrations per minute (d/m) are useful starting points; therefore, the most practical expressions and uses of activity may be to convert Ci and Bq to t/s or d/m whenever they are encountered. The becquerel, which has units of t/s, may offer some advantages in this respect, but the curie, the conventional unit for activity, is still widely used. These considerations are illustrated in Example 5-7.

Example 5-7: A radioactive source with an activity of 0.2 μCi is used to determine the counting efficiency of a radiation detector. (a) If the response of the detector is 66,600 counts per minute (c/m), what is the detector efficiency? (b) What is the efficiency if the source activity is 20 kBq?

SOLUTION: (a) The 0.2 μCi source produces an emission rate of

$$A = 0.2\ \mu\text{Ci} \times 2.22 \times 10^6\ \text{d/m} \cdot \mu\text{Ci}$$
$$= 4.44 \times 10^5\ \text{d/m}$$

$$\text{counting efficiency} = \frac{\text{counts/min}}{\text{activity}}$$
$$= \frac{66{,}600\ \text{c/m}}{4.44 \times 10^5\ \text{d/m}} = 0.15 \quad \text{or} \quad 15\%$$

(b) The 20 kBq source produces an emission rate of

$$A = 20\ \text{kBq} \times 10^3\ \text{Bq/kBq} \times 1\ \text{t/s} \cdot \text{Bq} = 2 \times 10^4\ \text{t/s}$$

$$\text{counting efficiency} = \frac{\text{counts/min}}{\text{activity}}$$
$$= \frac{66{,}600\ \text{c/m}}{2 \times 10^4\ \text{t/s} \times 60\ \text{s/m}}$$
$$= 5.55 \times 10^{-2} \quad \text{or} \quad 5.55\%$$

Mathematics of Radioactive Transformation

The activity of a radioactive source is proportional to the number of radioactive atoms present; therefore, it can be written mathematically as a differential change in N in a differential unit of time as

$$\text{activity} \equiv -\frac{dN}{dt} = \lambda N$$

where the constant of proportionality λ is the disintegration constant, and $-dN/dt$ is the rate of decrease of the number N of radioactive atoms at any time t. Rearranging the equation gives an expression that can be integrated directly between the limits N_0 at $t = 0$ and $N(t)$, the number of atoms for any other time t:

$$\int_{N_0}^{N(t)} \frac{dN}{N} = -\lambda \int_0^t dt$$

Integration and evaluation of the limits yields

$$\ln N(t) - \ln N_0 = -\lambda t$$

or

$$\ln N(t) = -\lambda t + \ln N_0$$

which is an equation of a straight line with slope of $-\lambda$ and a y-intercept of $\ln N_0$. By applying the law of logarithms, this can also be written as

$$\ln \frac{N(t)}{N_0} = -\lambda t$$

and since the logarithm of a number is the exponent to which the base (in this case e) is raised to obtain the number, the expression above is literally

$$\frac{N(t)}{N_0} = e^{-\lambda t}$$

or

$$N(t) = N_0 e^{-\lambda t}$$

If both sides are multiplied by λ, and recalling that activity = λN, then

$$\lambda N(t) = \lambda N_0 e^{-\lambda t}$$

or

$$A(t) = A_0 e^{-\lambda t}$$

In other words, the activity $A(t)$, at some time t, of a source of radioactive atoms, all of the same species with a disintegration constant λ, is equal to the initial activity A_0, multiplied by the exponential $e^{-\lambda t}$, where e is the base of the natural logarithm. If the activity $A(t)$ is plotted against time, the exponential curve in Figure 5-32*a* is obtained. A plot of the natural logarithm of the activity with time yields the straight line shown in Figure 5-32*b*. Neither curve goes to zero except when t is infinite.

Half-Life

The half-life (or half-period) of a radioative substance is used to describe the exponential behavior of radioactive transformation since it is more meaningful than the disintegration constant with reciprocal units of time. The *half-life* is the amount of time it takes for half of the atoms in a radioactive source to undergo transformation; this special value of time, $T_{1/2}$, is calculated as the value of t that corresponds to $A(t)/A_0 = \frac{1}{2}$ [or $N(t)/N_0$ if one prefers] as follows:

$$\frac{A(t)}{A_0} = \frac{1}{2} = e^{-\lambda T_{1/2}}$$

which can be solved by taking the natural logarithm of both sides, or

$$\ln 1 - \ln 2 = -\lambda T_{1/2}$$

such that

$$T_{1/2} = \frac{\ln 2}{\lambda}$$

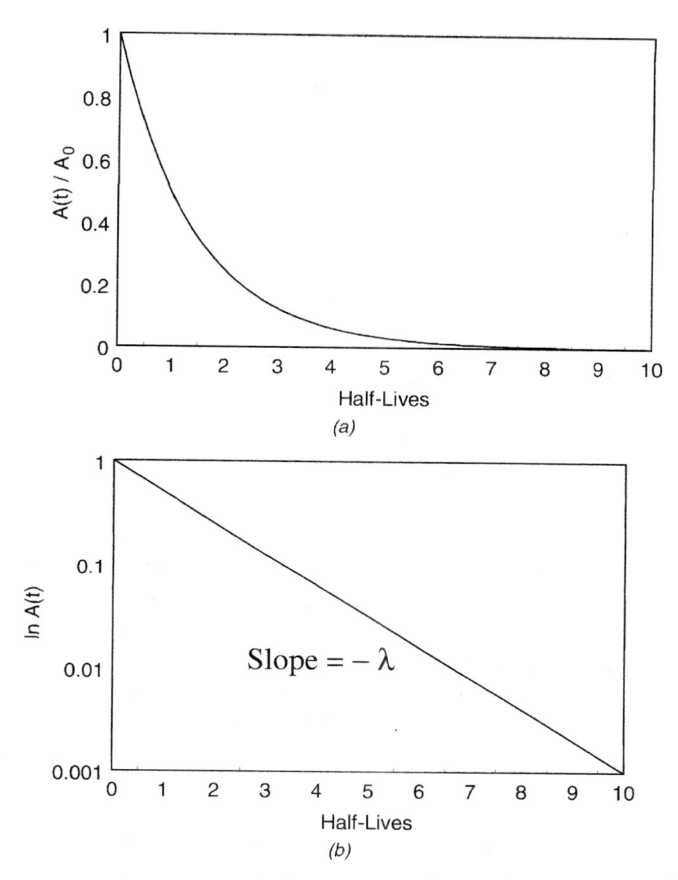

FIGURE 5-32. Radioactive transformation versus the number of elapsed half-lives plotted as (*a*) a linear function and (*b*) the natural log of activity.

The disintegration constant λ, which is required for calculating activity, follows directly from this relationship as

$$\lambda = \frac{\ln 2}{T_{1/2}}$$

The disintegration constant has units of reciprocal time (s^{-1}, min^{-1}, h^{-1}, y^{-1}) and is a value much less than 1.0. Consequently, it is also not a familiar quantity; thus radioactive isotopes are characterized in terms of their half-life instead of λ, which makes communication about radioactive transformation much easier. Half-life, which is given in amounts of time, is easier to relate to, but one needs to be mindful that the process is still exponential. It doesn't matter how much activity you start with or when you start the clock, only that you know how much you have (i.e., A_0), when you start the clock; after one half-life, you will have only half the activity. Then you must start again, and so on (i.e., the decrease is exponential).

Since λ is derived from the known half-life of a radioactive substance, the expression $\lambda = (\ln 2)/T_{1/2}$ can be used in the general relationship for activity to provide an equation for straightforward calculations of activity with time, or

$$A(t) = A_0 e^{[-(\ln 2)/T_{1/2}]t}$$

The exponential nature of radioactive transformation must be kept in mind when dealing with radioactive sources. As shown in Figure 5-32*b*, the activity of a source will diminish to 1% ($A/A_0 = 0.01$) after about 6.7 half-lives and to 0.1% ($A/A_0 = 0.001$) after 10 half-lives. Although these are good rules of thumb for most sources, some activity will, because of exponential removal, remain after these periods. The common practice of assuming that a source has decayed away after 10 half-lives should be used with caution because in fact it has only diminished to $0.001A_0$, which may or may not be of significance.

Mean Life

The mean life of each atom in a radioactive source can be useful for determining the total number of emissions of radiation from the source. The mean life is the average time it takes each atom to transmute, recognizing that some will transmute right away, some will last an infinite time, and others will have lifetimes in between. The mean life can be deduced from the relationship between N_0, dN, dt, and λ.

As shown in Figure 5-33, the number of atoms in a source will decrease exponentially with time, which will affect the average time each persists. The area under the curve will have units of atom · seconds if t is in seconds, although other units

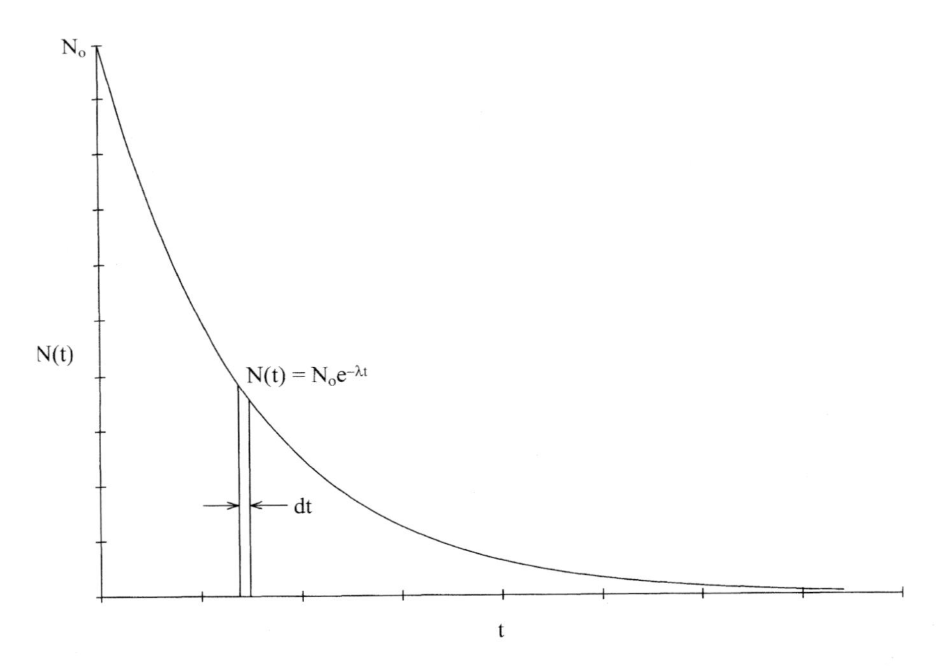

FIGURE 5-33. The mean life τ of a radioactive substance is obtained by determining a value τN_0 that equals the total area under the curve or the sum of the differential areas $N(t)dt$ from $t = 0$ to $t = \infty$.

of t could also be used. The area under the curve is obtained by integrating each differential unit of area, which has a width of dt and a height of $N(t) = N_0 e^{-\lambda t}$, from $t = 0$ to $t = \infty$, or

$$\text{area} = \int_0^t N(t)\,dt = \int_0^\infty N_0 e^{-\lambda t}\,dt$$
$$= \frac{N_0}{\lambda}$$

which has units of atom · seconds. If each of the initial atoms N_0 in the source has a mean life of τ, the quantity $N_0\tau$ will represent the same total area, so that

$$N_0\tau = \frac{N_0}{\lambda}$$

The mean life of the atoms in a particular radioisotope is, therefore,

$$\tau = \frac{1}{\lambda}$$

where λ is actually the fraction of the total number of atoms of the given radioelement disintegrating in unit time. The expression for mean life can also be simplified in terms of the half-life of the substance, or

$$\tau = \frac{T_{1/2}}{\ln 2} = 1.4427 T_{1/2}$$

which states that the mean life of the atoms in a radionuclide is about 44.3% longer than the half-life. Any given atom may disintegrate immediately or it may last forever; it is impossible to tell in advance when a given atom will do so, but each atom will, on average, exist for a time τ. On the other hand, for a large number of radioactive atoms of the same type, a definite fraction λ will decay in unit time. Radiation physicists are, therefore, somewhat like insurance companies that cannot foretell the fate of any individual, but by the laws of statistics they can accurately predict the life expectancy of a given set and respond accordingly. Mean life has a practical use in dosimetry because it provides an average time that can be assigned to all of the atoms in a source that may deposit energy in a medium. Although the mean life of atoms is analogous to the average life expectancy of persons, a given radioactive atom might never transform and atoms don't age.

Effective Half-Life

Other processes may serve to reduce radioactivity in a source separate and distinct from that due to atom transformations. For example, if a radioactive material is located in a physiological system in which it is removed by biological processes, it may be necessary to account for both in assessing the number of atoms that transform over some period of time. To do so, we use an effective removal constant that is obtained by adding the biological rate constant, λ_b, and other removal constants,

λ_i (with similar units, of course), to the radioactivity constant. The effective removal constant for a biological system is

$$\lambda_{\text{eff}} = \lambda_r + \lambda_b + \text{other removal constants}$$

The amount of radioactivity in an organ of the body is often described by an effective removal constant, λ_{eff}, when biological removal is a factor. An effective half-life can be derived from the effective removal constant as

$$\frac{\ln 2}{T_{\text{eff}}} = \frac{\ln 2}{T_r} + \frac{\ln 2}{T_b}$$

where T_{eff}, T_r, and T_b are the effective, radioactive, and biological half-lives, respectively. Solving for T_{eff} yields

$$T_{\text{eff}} = \frac{T_r T_b}{T_r + T_b}$$

Example 5-8: Determine the effective removal constant and the effective half-life of iodine-131 ($T_{1/2} = 8.02$ d) in the human thyroid if it is removed with a biological half-life of 120 days.

SOLUTION:

$$\lambda_{\text{eff}} = \lambda_r + \lambda_b = \frac{\ln 2}{8.02\ \text{d}} + \frac{\ln 2}{120\ \text{d}} = 0.0922\ \text{d}^{-1}$$

$$T_{\text{eff}} = \frac{8.02\ \text{d} \times 120\ \text{d}}{8.02 + 120} = 7.52\ \text{d}$$

The effective removal constant could also be obtained from the effective half-life if it were known, and vice versa.

RADIOACTIVITY CALCULATIONS

Many useful calculations can be made for radioactive substances using the relationships just derived, for example:

- What activity does a given mass of a radioactive substance have, or vice versa?
- From a set of activity measurements made at different times, what is the half-life and disintegration constant of a substance?
- How radioactive will a substance be at some time t in the future?
- If the current activity is known, what is the activity at a previous time, such as when the source was shipped or stored?
- How long will it take for a source to diminish to a given level, perhaps one that will no longer be of concern?

These calculations can be made with some basic starting information and the equation for activity $A(t)$ and its decrease with time:

$$A(t) = A_0 e^{[-(\ln 2)/T_{1/2}]t}$$

which relates A and t in terms of A_0 and the half-life $T_{1/2}$. Calculations with this relationship are mathematically precise, but familiarity with exponential relationships is necessary; these are now relatively straightforward using scientific calculators that have a natural logarithm function [i.e., a key labeled ln(x), ln x, or ln].

Since the half-life is a fundamental part of this exponential relationship, the change in the activity of a radioactive source can be calculated just by dividing the initial activity by 2 a sufficient number of times to account for the half-lives elapsed. Mathematically, this is given as

$$A(t) = \frac{A_0}{2^n}$$

where n is the number of half-lives that have elapsed during time t. If n is an integer, the activity after an elapsed time is calculated simply by dividing A_0 by 2 n times. If n is not an integer, it is necessary to solve the relationship by other means, one of which is a ratio, $A(t)/A_0$, versus the number of half-lives that have elapsed. Figure 5-34 is such a plot that is generally applicable, subject of course to the uncertainty in reading such curves. The use of such a curve is demonstrated in Examples 5-9 and 5-10.

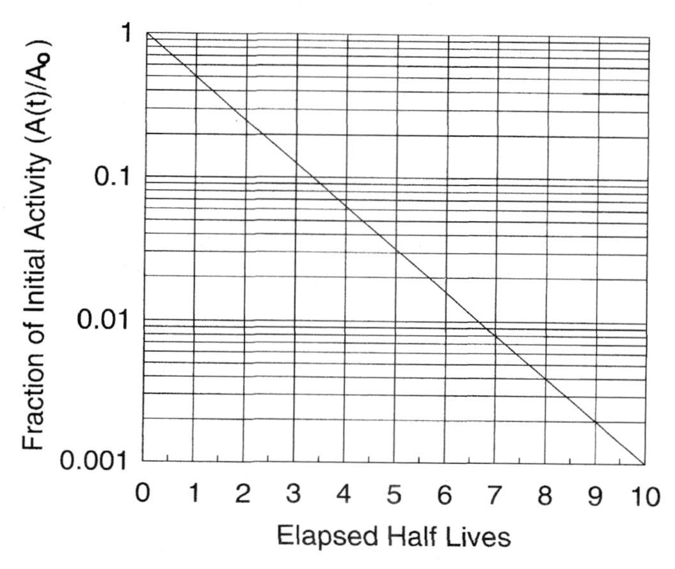

FIGURE 5-34. Semilog plot of the fraction of activity, $A(t)/A_0$, remaining versus the number of elapsed half-lives.

Example 5-9: If a reactor part contains 1000 Ci of cobalt-60 ($T_{1/2} = 5.27$ y), how much will exist in 40 years?

SOLUTION: Figure 5-34 can be used. The number of elapsed half-lives in 40 years is 7.59. The fractional activity from Figure 5-34 is approximately 0.0052, and the activity at 40 years is

$$A(40 \text{ y}) = 1000 \text{ Ci} \times 0.0052 = 5.2 \text{ Ci}$$

Example 5-10: Assume that a source of ^{137}Cs ($T_{1/2} = 30.07$ y) can be managed safely if it contains less than 0.01 Ci. How long would it take for a source containing 5 Ci of ^{137}Cs to reach this value?

SOLUTION: First we determine what fraction 0.01 Ci is of the original 5 Ci [i.e., $A(t)/A_0 = 0.002$]. From Figure 5-34, the number of elapsed half-lives that corresponds to 0.002 is 8.97. The time required for 5.0 Ci to diminish to 0.01 Ci is $8.97 \times 30.07 \text{ y} = 269.7$ y.

Example 5-11: A waste shipment contains 800 Ci of iodine 125 ($T_{1/2} = 60$ d). How much will be left in (a) one year and (b) 11 months?

SOLUTION: (a) Since a year is 12 months, the elapsed time will be 6 half-lives. Dividing 800 by 2 six times yields

$$A(1 \text{ y}) = \frac{800 \text{ Ci}}{2^6} = 12.5 \text{ Ci}$$

(b) At 11 months, or 5.5 half-lives, the activity can be obtained directly by solving the general expression for activity with time:

$$A(11 \text{ months}) = 800e^{[-(\ln 2)/2]11} = 17.68 \text{ Ci}$$

The activity at 11 months can also be solved in terms of half-lives. Since 5.5 half-lives would have elapsed, the activity is between 25 Ci (5 half-lives) and 12.5 Ci (6 half-lives) (note that it is not halfway between because radioactive transformation is exponential). The solution for 5.5 half-lives can be represented as

$$A(11 \text{ months}) = \frac{800 \text{ Ci}}{2^{5.5}}$$

which can be solved by one of at least four methods:

1. By solving $2^{5.5}$ by y^x and completing the arithmetic, the activity A is

$$A = \frac{800}{45.2548} = 17.68 \text{ Ci}$$

2. By taking the natural logarithm of both sides and finding the corresponding natural antilogarithm:

$$\ln A = \ln 800 - 5.5 \ln 2 = 2.872$$

$$A = e^{2.8723} = 17.68 \text{ Ci}$$

3. By taking $\log_{10}$ of both sides and finding the corresponding antilogarithm$_{10}$:

$$\log A = \log 800 - 5.5 \log 2 = 1.247425$$

$$A = 10^{1.247425} = 17.68 \text{ Ci}$$

4. By using the graph in Figure 5-34 to determine the ratio $A(t)/A_0$ at 5.5 half-lives as

$$\frac{A(t)}{A_0} = 0.022$$

from which $A(t)$ is calculated to be 17.68 Ci. (*Note*: The accuracy of this method is limited by reading the graph.)

Half-Life Determination

The half-life of an unknown radioactive substance can be determined by taking a series of activity measurements over several time intervals and plotting the data. If the data are plotted as ln $A(t)$ versus t and a straight line is obtained, we can be reasonably certain that the source contains only one radioisotope. The slope of the straight line provides the disintegration constant, and once the disintegration constant is known, the half-life can be determined directly.

Example 5-12: What is the half-life of a sample of ^{55}Cr that has activity at 5-min intervals of 19.2, 7.13, 2.65, 0.99, and 0.27 t/s?

SOLUTION: The general relationship between activity and time can be expressed as

$$\ln A(t) = -\lambda t + \ln A$$

which is a straight line with slope of $-\lambda$ where $\ln A_0$ is the y-intercept. Figure 5-35 is plotted as a least-squares fit of the data shown in Table 5-1. (Note that it is necessary to take the natural logarithm of each activity rate.) From the curve in Figure 5-35, the slope $-\lambda$ is $-0.197\,\text{min}^{-1}$ and $\lambda = 0.197\,\text{min}^{-1}$; therefore,

$$T_{1/2} = \frac{\ln 2}{0.197 \text{ min}^{-1}} = 3.52 \text{ min}$$

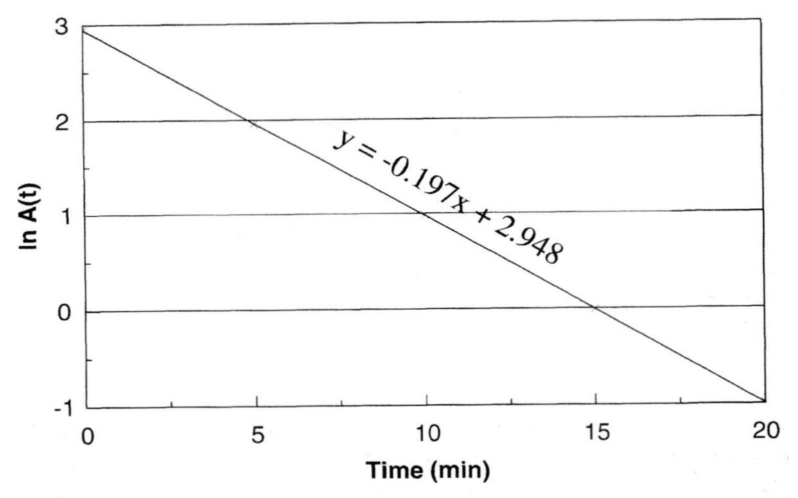

FIGURE 5-35. Linear least-squares fit of $\ln A(t)$ versus time for a sample of ^{55}Cr.

TABLE 5-1. Data for Example 5-12

t(min)	$\ln A(t)$
0	2.95
5	1.96
10	0.98
15	−0.01
20	−1.31

The technique used in Example 5-12 is also applicable to a source that contains more than one radionuclide with different half-lives. This can readily be determined by plotting the $\ln A(t)$ of a series of activity measurements with time as shown in Figure 5-36 for two radionuclides. If only one radionuclide were present, the semilog plot would be a straight line; however, when two or more are present, the line will be curved as in Figure 5-36. The straight-line portion at the far end represents the longer-lived component, and thus can be extrapolated back to time zero and subtracted from the total curve to yield a second straight line. The slopes of these two lines establish the disintegration constants, λ_1 and λ_2, from which the half-life of each can be calculated for possible identification of the two radionuclides in the source.

ACTIVITY–MASS RELATIONSHIPS

The activity of a radioisotope is directly related to the number of atoms, or

$$\text{activity} \equiv \lambda N$$

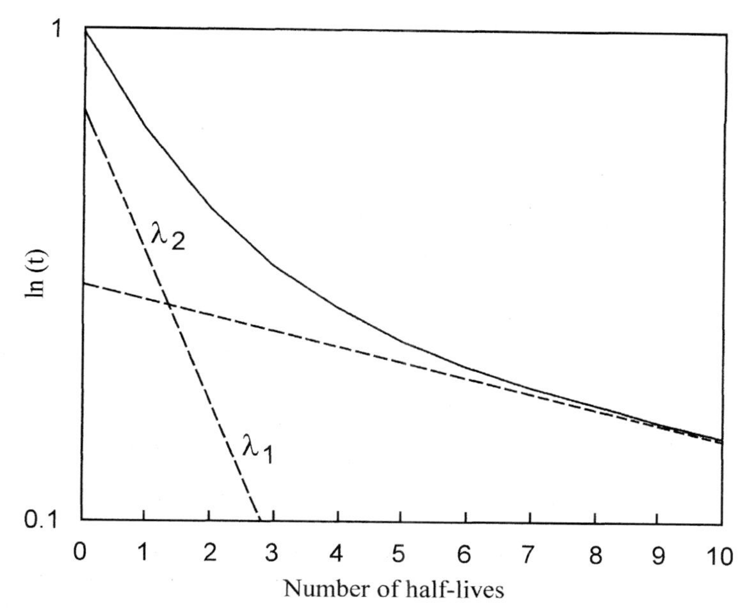

FIGURE 5-36. The $\ln A(t)$ plotted as a function of time for two radionuclides with two different half-lives.

and since N is determined from the mass m of the sample by Avogadro's number, mass can be used directly as a surrogate for activity, or

$$m(t) = m_0 e^{-\lambda t}$$

Example 5-13: If a sample of ^{24}Na ($T_{1/2} = 15$ h) originally contains 48 g, how many hours will be required to reduce the ^{24}Na concentration in this sample to 9 g?

SOLUTION: Since activity = λN, and N is determined from the mass of the radioelement, the activity of ^{24}Na is directly proportional to its mass. Therefore,

$$9 \text{ g} = (48 \text{ g})e^{-\lambda t}$$

$$\ln \frac{9 \text{ g}}{48 \text{ g}} = -\lambda t = \frac{\ln 2}{15 \text{ h}} t$$

$$t = 36.2 \text{ h}$$

Specific Activity

The basic equation for the radioactivity of a substance is

$$A(t) = \lambda N = \frac{\ln 2}{T_{1/2}} N$$

Since the number N of radioactive atoms in a radioactive element and its half-life $T_{1/2}$ are specific for every isotope of a radioactive element, the activity per unit mass of a pure radionuclide is a constant that is defined as the specific activity of the element. The *specific activity* of a radioelement is obtained by determining the number of atoms in a unit mass (e.g., 1 gram) of the element, multiplying by the disintegration constant λ to obtain its transformation rate, and then expressing the transformation rate in activity units (Ci, Bq, etc.) per unit mass, as shown in Example 5-14.

Example 5-14: What is the specific activity of 1 g of ^{226}Ra if the radioactive half-life is 1600 y?

SOLUTION: The number of atoms per gram of ^{226}Ra is

$$N = \frac{1\ \text{g} \times 6.02214 \times 10^{23}\ \text{atoms/mol}}{226\ \text{g/mol}} = 2.665 \times 10^{21}\ \text{atoms}$$

The disintegration constant λ is

$$\lambda = \frac{\ln 2}{T_{1/2}} = \frac{\ln 2}{1600\ \text{y}(3.1557 \times 10^{7}\ \text{s/y})} = 1.373 \times 10^{-11}\ \text{s}^{-1}$$

and the activity A is

$$A = \lambda N = 1.373 \times 10^{-11}\ \text{s}^{-1} \times 2.665 \times 10^{21}\ \text{atoms/g} = 3.66 \times 10^{10}\ \text{t/s} \cdot \text{g}$$

This value is slightly less than the current official definition of the curie (3.700×10^{10} t/s) primarily because the half-life measurement has become more precise.

The specific activity (SpA) of pure radioelements is often given in units of Ci/g. A general expression of specific activity in Ci/g for radioelements of atomic mass A (g/mol) where half-life is given in seconds is

$$\text{SpA(Ci/g)} = \frac{1.12824 \times 10^{13}}{T_{1/2}(\text{s})A(\text{g/mol})}$$

Example 5-15: Calculate the specific activity of ^{131}I, which has a half-life of 8.0207 d.

SOLUTION:

$$\text{SpA(Ci/g)} = \frac{1.12824 \times 10^{13}\ \text{s} \cdot \text{Ci/mol}}{T_{1/2}(\text{s})A(\text{g/mol})} = \frac{1.12824 \times 10^{13}\ \text{s} \cdot \text{Ci/mol}}{8.0207\ \text{d}(86{,}400\ \text{s/d})(131\ \text{g/mol})}$$
$$= 1.243 \times 10^{5}\ \text{Ci/g}$$

Table 5-2 lists values of specific activity for several radionuclides that were calculated with half-lives from the National Nuclear Data Center. Integer mass numbers

TABLE 5-2. Specific Activity of Selected Pure Radionuclides

Radionuclide	$T_{1/2}$	SpA (Ci/g)	Radionuclide	$T_{1/2}$	SpA (Ci/g)
Hydrogen-3	12.3 y	9.64×10^3	Ruthenium-106	373.59 d	3.30×10^3
Carbon-14	5730 y	4.46	Cadmium-109	462.0 d	2.59×10^3
Nitrogen-16	7.13 s	9.89×10^{10}	Iodine-123	13.27 h	1.92×10^6
Sodium-22	2.6088 y	6.23×10^3	Iodine-125	59.402 d	1.76×10^4
Sodium-24	14.959 h	8.73×10^6	Iodine-131	8.0207 d	1.24×10^5
Phosphorus-32	14.26 d	2.86×10^5	Barium-133	10.52 y	262.00
Sulfur-35	87.51 d	4.27×10^4	Cesium-134	2.0648 y	1.29×10^3
Chlorine-36	3.01×10^5 y	3.30×10^{-2}	Cesium-137	30.07 y	86.90
Argon-41	109.43 m	4.19×10^7	Barium-140	12.752 d	7.32×10^4
Potassium-42	12.36 h	6.04×10^6	Lanthanum-140	40.22 h	5.57×10^5
Calcium-45	162.61 d	1.79×10^4	Cerium-141	32.5 d	2.85×10^4
Chromium-51	27.704 d	9.25×10^4	Cerium-144	284.9 d	3.18×10^3
Manganese-54	312.12 d	7.76×10^3	Praseodymium-144	17.28 m	7.56×10^7
Iron-55	2.73 y	2.39×10^3	Promethium-147	2.6234 y	928.00
Manganese-56	2.5785 h	2.17×10^7	Europium-152	13.542 y	174.00
Cobalt-57	271.79 d	8.44×10^3	Tantalum-182	114.43 d	6.27×10^3
Iron-59	44.503 d	4.98×10^4	Iridium-192	73.831 d	9.21×10^3
Nickel-59	7.6×10^4 y	7.99×10^{-2}	Gold-198	2.69517 d	2.45×10^5
Cobalt-60	5.2714 y	1.13×10^3	Mercury-203	46.612 d	1.38×10^4
Nickel-63	100.1 y	56.80	Thallium-204	3.78 y	4.64×10^2
Copper-64	12.7 h	3.86×10^6	Thallium-208	3.053 m	2.96×10^8
Zinc-65	244.26 d	8.23×10^3	Polonium-210	138.4 d	4.49×10^3
Gallium-72	14.1 h	3.09×10^6	Polonium-214	164 μs	3.13×10^{14}
Arsenic-76	1.0778 d	1.60×10^6	Radium-226	1600 y	0.99
Bromine-82	35.3 h	1.08×10^6	Thorium-232	1.41×10^{10} y	1.10×10^{-7}
Rubidium-86	18.631 d	8.16×10^4	Uranium-233	1.592×10^5 y	9.64×10^{-3}
Strontium-89	50.53 d	2.91×10^4	Uranium-235	7.0E $\times 10^8$ y	2.16×10^{-6}
Strontium-90	28.74 y	138.00	Uranium-238	4.468×10^9 y	3.36×10^{-7}
Yttrium-90	64.1 h	5.44×10^5	Plutonium-239	24,110 y	6.21×10^{-2}
Molybdenum-99	65.94 h	4.80×10^5	Plutonium-241	14.35 y	103.00
Technetium-99m	6.01 h	5.27×10^6	Americium-241	432.2 y	3.43

were used rather than actual masses, except for ^{3}H, where the exact mass was used. It is very important to note that specific activities for the radionuclides listed in Table 5-2 are for pure samples of the radioisotope, although it is recognized that no radioactive substance is completely pure since transformations produce new atoms, which generally remain in the source.

It is common practice to measure a source containing a radioactive substance after weighing it, to report the measured value in units of activity/gram, and to refer to this value as its specific activity; however, this use of the term is not consistent with the definition of specific activity, which refers to a pure sample. It is more appropriate to refer to measured values of activity per unit mass in a nonpure sample as having a concentration of so many Bq or Ci per gram of a sample or source; otherwise, the impression may be given that the sample is pure when, in fact, it is a mixture.

RADIOACTIVE SERIES TRANSFORMATION

The radioactive transformation of many radionuclides often yields a product that is also radioactive. The radioactive product in turn undergoes transformation to produce yet another radioactive product, and so on, until stability is achieved (see, e.g., Figure 5-12 for $^{238}U \rightarrow {}^{206}Pb$). Occasionally, the product radionuclide may be the more important for radiation protection considerations. For example, ^{222}Rn, the product of the radioactive transformation of ^{226}Ra, can migrate while its parent (^{226}Ra) remains fixed in soil. The transformation of ^{90}Sr produces ^{90}Y, which is also radioactive, and since it has a much higher beta energy than the ^{90}Sr parent, it is a larger contributor to radiation dose than is ^{90}Sr itself. Because of its higher energy, ^{90}Y is also used to measure ^{90}Sr indirectly.

Series Decay Calculations

The number of atoms of each member of a radioactive series at any time t can be obtained by solving a system of differential equations that relates each product, $N_1, N_2, N_3, \ldots, N_i$ with corresponding disintegration constants $\lambda_1, \lambda_2, \lambda_3, \ldots, \lambda_i$. Each series begins with a parent nuclide, N_1, which has a rate of transformation

$$\frac{dN_1}{dt} = -\lambda_1 N_1$$

The second nuclide in a radionuclide series will be produced at a rate of $\lambda_1 N_1$ due to the transformation of N_1, but as soon as atoms of N_2 exist, they too can undergo transformation if they are radioactive; thus the rate of change of atoms of N_2 is the rate of production minus the rate of removal of N_2 atoms, or

$$\frac{dN_2}{dt} = \lambda_1 N_1 - \lambda_2 N_2$$

Similarly, for atoms of N_3, which are produced by transformation of N_2 atoms and subject to removal as a function of the disintegration constant λ_3,

$$\frac{dN_3}{dt} = \lambda_2 N_2 - \lambda_3 N$$

and so on, up to the ith member of the series,

$$\frac{dN_i}{dt} = \lambda_{i-1} N_{i-1} - \lambda_i N_i$$

If the end product is stable, the atoms of the stable end product appear at the rate of the last radioactive precursor, and of course are not removed since they are stable.

It is useful to solve this system of equations for the first four members of a radioactive series in order to demonstrate their direct applicability to radon and its progeny in a later section. The number of atoms of N_1 is

$$N_1(t) = N_1^0 e^{-\lambda_1 t}$$

where N_1^0 is the number of atoms of the parent at $t = 0$. This expression for N_1 can be inserted into the equation for dN_2/dt to give

$$\frac{dN_2(t)}{dt} = \lambda_1 N_1^0 e^{-\lambda_1 t} - \lambda_2 N_2$$

Collecting terms, we have

$$\frac{dN_2}{dt} + \lambda_2 N_2 = \lambda_1 N_1^0 e^{-\lambda_1 t}$$

This type of equation can be converted into one that can be integrated directly by multiplying through by an appropriate integrating factor, which for this form is always an exponential with an exponent that is equal to the constant in the second term multiplied by the variable in the denominator of the derivative, or in this case $e^{\lambda_2 t}$; thus

$$e^{\lambda_2 t}\frac{dN_2}{dt} + e^{\lambda_2 t}\lambda_2 N_2 = \lambda_1 N_1^0 e^{(\lambda_2 - \lambda_1)t}$$

Multiplying through by $e^{\lambda_2 t}$ converts the left side of the equation to the time derivative of $N_2 e^{\lambda_2 t}$, which can be demonstrated by differentiating the expression. It also yields an exponential expression multiplied by a constant on the right side, or

$$\frac{d}{dt}(N_2 e^{\lambda_2 t}) = \lambda_1 N_1^0 e^{(\lambda_2 - \lambda_1)t}$$

which can be integrated directly to give

$$N_2 e^{\lambda_2 t} = \frac{\lambda_1}{\lambda_2 - \lambda_1} N_1^0 e^{(\lambda_2 - \lambda_1)t} + C$$

where C, the constant of integration, is determined by stating the condition that when $t = 0$, $N_2 = 0$; thus

$$C = -\frac{\lambda_1}{\lambda_2 - \lambda_1} N_1^0$$

Therefore, the solution for N_2 as a function of time is

$$N_2 = \frac{\lambda_1}{\lambda_2 - \lambda_1} N_1^0 (e^{-\lambda_1 t} - e^{-\lambda_2 t})$$

The number of atoms of the third kind is found by inserting this expression for N_2 into the equation for the rate of change of N_3, which as before is the rate of production of N_3 by transformation of atoms of N_2 (or $\lambda_2 N_2$) minus the rate of removal of N_3 by radioactive transformation (or $\lambda_3 N_3$),

$$\frac{dN_3}{dt} = \lambda_2 N_2 - \lambda_3 N_3$$

After the integration is performed and the constant of integration is evaluated, the equation for the number of atoms of N_3 with time is

$$N_3(t) = \lambda_1 \lambda_2 N_1^0 \left[\frac{e^{-\lambda_1 t}}{(\lambda_2 - \lambda_1)(\lambda_3 - \lambda_1)} + \frac{e^{-\lambda_2 t}}{(\lambda_1 - \lambda_2)(\lambda_3 - \lambda_2)} + \frac{e^{-\lambda_3 t}}{(\lambda_1 - \lambda_3)(\lambda_2 - \lambda_3)} \right]$$

In a similar fashion, for the number of atoms of the fourth kind, the expression for $N_3(t)$ is inserted into the equation for dN_4/dt, which is integrated to obtain the number of atoms of N_4 with time, or

$$N_4(t) = \lambda_1 \lambda_2 \lambda_3 N_1^0 \left[\frac{e^{-\lambda_1 t}}{(\lambda_2 - \lambda_1)(\lambda_3 - \lambda_1)(\lambda_4 - \lambda_1)} + \frac{e^{-\lambda_2 t}}{(\lambda_1 - \lambda_2)(\lambda_3 - \lambda_2)(\lambda_4 - \lambda_2)} + \frac{e^{-\lambda_3 t}}{(\lambda_1 - \lambda_3)(\lambda_2 - \lambda_3)(\lambda_4 - \lambda_3)} + \frac{e^{-\lambda_4 t}}{(\lambda_1 - \lambda_4)(\lambda_2 - \lambda_4)(\lambda_3 - \lambda_4)} \right]$$

These equations yield the number of atoms of each of the first four members of a radioactive series that begins with a pure radioactive parent (i.e., there are no transformation products at $t = 0$).

A radioactive series typically ends at a stable nuclide or one with a very large half-life such that it is not unreasonable to terminate the production of radioactive atoms. In this case, the disintegration constant, λ_i, for the end product will be zero, or at least very small. For example, if the third element in a series is stable (i.e., $\lambda_3 = 0$), the number of atoms of N_3 will be

$$N_3 = N_1^0 \left(1 - \frac{\lambda_2}{\lambda_2 - \lambda_1} e^{-\lambda_1 t} - \frac{\lambda_1}{\lambda_1 - \lambda_2} e^{-\lambda_2 t} \right)$$

Recursive Kinetics: Bateman Equations

The solutions for the numbers of atoms in the first four members of a radioactive series yields a recursion of similar terms, which has been generalized into a series of expressions known as the *Bateman equations*. If it is assumed, as Bateman did, that at $t = 0$ only the parent substance is present, then the number of atoms of any member of the chain at a subsequent time t is given by

$$N_n(t) = C_1 e^{-\lambda_1 t} + C_2 e^{-\lambda_2 t} + C_3 e^{-\lambda_3 t} + \cdots + C_n e^{-\lambda_n t}$$

where

$$C_1 = \frac{\lambda_1 \lambda_2 \cdots \lambda_{n-1}}{(\lambda_2 - \lambda_1)(\lambda_3 - \lambda_1)\cdots(\lambda_n - \lambda_1)} N_1^0$$

$$C_2 = \frac{\lambda_1 \lambda_2 \cdots \lambda_{n-1}}{(\lambda_1 - \lambda_2)(\lambda_3 - \lambda_2)\cdots(\lambda_n - \lambda_2)} N_1^0$$

$$\vdots$$

$$C_n = \frac{\lambda_1 \lambda_2 \cdots \lambda_{n-1}}{(\lambda_1 - \lambda_n)(\lambda_2 - \lambda_n)\cdots(\lambda_{n-1} - \lambda_n)} N_1^0$$

The constants also require the condition that if the value $(\lambda_n - \lambda_n)$ appears, it be set equal to 1.0; otherwise, the denominator would be zero and the constant (and the number of atoms) would be infinite.

These relationships can be further simplified using product $\prod$ notation in the following general expression, which holds for any member of the series:

$$N_i(t) = N_1^0 \prod_{i=1}^{n-1} \lambda_i \sum_{i=1}^{n} \frac{e^{-\lambda_i t}}{\prod_{n=1}^{n} (\lambda_n - \lambda_i)}$$

again with the provision that $\lambda_n - \lambda_n = 1.0$. It is left to the reader to become familiar with this expression by using it to write down the equations for the number of atoms for the first four members of a radioactive series, as developed above. The product expression applies to any member of a radioactive series that begins with a pure parent. It can readily be programmed for computer solutions, but most radioactive series calculations as a practical matter rarely extend past the fourth member of the series. Radon progeny calculations (see Chapter 6) are one of the most important and can be calculated as a subseries of four transformation products.

RADIOACTIVE EQUILIBRIUM

The relative activities of a radioactive parent and its radioactive product (commonly referred to as the *daughter*) can be determined from the equation for the number of atoms for the second member of a series by multiplying both sides by λ_2, with the following result:

$$A_2(t) = \frac{\lambda_2}{\lambda_2 - \lambda_1} A_1^0 (e^{-\lambda_1 t} - e^{-\lambda_2 t})$$

Calculations with this equation can be made somewhat simpler by use of the identity

$$\frac{\lambda_2}{\lambda_2 - \lambda_1} = \frac{T_1}{T_1 - T_2}$$

Thus

$$A_2(t) = \frac{T_1}{T_1 - T_2} A_1^0 (e^{-\lambda_1 t} - e^{-\lambda_2 t})$$

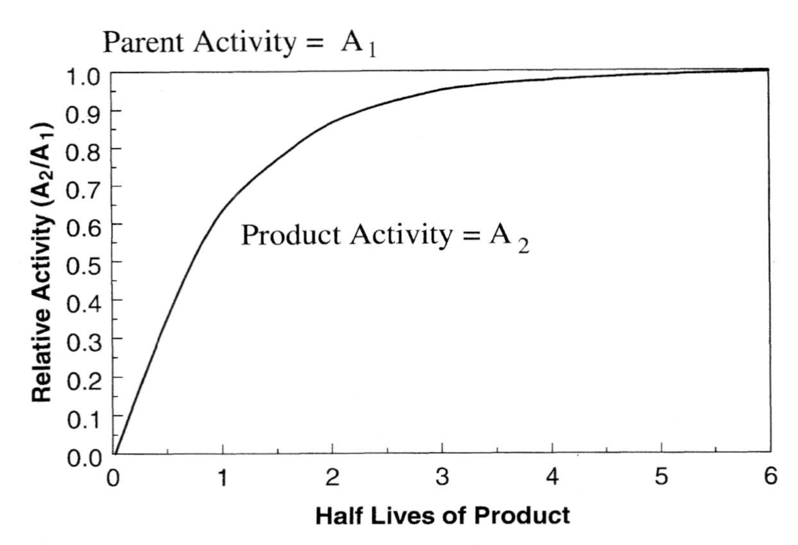

FIGURE 5-37. Example of secular equilibrium in which the parent activity remains essentially unchanged while the short-lived product (activity = 0 at $t = 0$) activity builds up to 99% of the parent activity in about 6.7 half-lives.

Time of Maximum Activity

The time of maximum activity of $A_0(t)$ can be determined exactly for all cases by differentiating the equation for $A_2(t)$, setting it equal to zero and solving for t_m as follows:

$$\frac{d(A_2)}{dt} = -\lambda_1 e^{-\lambda_1 t_m} + \lambda_2 e^{-\lambda_2 t_m} = 0$$

or

$$t_m = \frac{\ln(\lambda_2/\lambda_1)}{\lambda_2 - \lambda_1}$$

where λ_1 and λ_2 must be expressed in the same units. The value of t_m can also be expressed in terms of half-lives as

$$t_m = 1.4427 \times \frac{T_1 T_2}{T_1 - T_2} \times \ln \frac{T_1}{T_2}$$

The time of maximum activity occurs at the same time that the activities of the parent and daughter are equal (see Figures 5-37 and 5-38) if, and only if, the parent has only one radioactive product. As a general rule of thumb t_m occurs in 7 to 10 half-lives for a long-lived parent and a relatively short-lived product; however, the best practice is to calculate it. Once the time of maximum of activity is known, it can, in turn, be used in the equation for activity of the second species to determine the maximum amount of activity of the product.

Example 5-16: If 10 kBq of ^{132}Te ($T_{1/2}$ = 3.2 d = 76.8 h) is used as a generator of ^{132}I ($T_{1/2}$ = 2.28 h), what will be the maximum activity of ^{132}I that will grow into the source, and when will it occur?

SOLUTION: First determine the time of maximum activity t_m:

$$t_m = 1.4427 \times \frac{76.8 \times 2.28}{76.8 - 2.28} \times \ln \frac{76.8 \text{ h}}{2.28 \text{ h}} = 11.92 \text{ h}$$

Since the maximum activity of the product occurs at t_m = 11.92 h,

$$A_2(11.92 \text{ h}) = \frac{T_1}{T_1 - T_2} A_1^0 (e^{[-(\ln 2)/T_1]T_m} - e^{[-(\ln 2)/T_2]T_m})$$

$$A_2(11.92 \text{ h}) = \frac{76.8}{76.8 - 2.28} \times 10 \text{ kBq} (e^{[-(\ln 2)/76.8]11.92} - e^{[-(\ln 2)/2.28]11.92})$$
$$= 8.98 \text{ kBq}$$

Therefore, at t_m = 11.92 h, the activity of ^{132}I in the source will have a transformation rate of 8980 t/s.

Secular Equilibrium

If the period of observation (or calculation) is such that the activity of the parent nuclide remains essentially unchanged, the activity of the radioactive product (the radioactive daughter) in the equation for the second member of a radioactive series can be simplified as follows:

$$A(t) = A_1^0(1 - e^{-\lambda_2 t})$$

It should be noted that this is accurate only if the parent is much longer lived than the product. When this condition is satisfied, the product activity will grow to a level that is essentially identical to that of the parent, as shown in Figure 5-37, and the activities of the parent and the product show no appreciable change during many half-lives of the product. This condition is called *secular equilibrium*. The buildup of ^{90}Y from the transformation of ^{90}Sr is a good example of secular equilibrium, as is the ingrowth of ^{222}Rn from the transformation of ^{226}Ra. In both instances the activity of the product will increase until it is the same as that of the parent (i.e., it is in equilibrium with the parent); the activity of the product reaches a value of 99% of the parent activity after about 6.7 half-lives of the product nuclide and is effectively in equilibrium with it at this and subsequent times.

The simplified equation for secular equilibrium is very convenient and easy to use, but it is necessary to assure that the conditions for its appropriate use are met. If any uncertainty exists, the general solution for the second member of a series will always give the correct result for a radioactive parent undergoing transformation to a radioactive product.

Example 5-17: ^{90}Sr ($T_{1/2} = 28.78$ y) is separated from a radioactive waste sample and measured to contain 2 μCi of ^{90}Sr. If measured again 10 days later, how much ^{90}Y ($T_{1/2} = 2.67$ d) will be in the sample?

SOLUTION: This is a clear case of secular equilibrium; therefore,

$$A_2(t) = A_1^0(1 - e^{-\lambda_2 t})$$

or

$$A_2(t) = 2\ \mu\text{Ci}(1 - e^{[-(\ln 2)/2.67\ \text{d}]10\ \text{d}}) = 1.85\ \mu\text{Ci}\ {}^{90}\text{Y}$$

Transient Equilibrium

A condition called *transient equilibrium* results between a radioactive parent and a radioactive product if the parent is longer-lived than the product ($\lambda_1 < \lambda_2$), but the half-life of the parent is such that its activity diminishes appreciably during the period of consideration. If no product activity exists at $t = 0$, the activity of the product is given by

$$A_2(t) = \frac{T_1}{T_1 - T_2} A_1^0(e^{-\lambda_1 t} - e^{-\lambda_2 t})$$

For values of t above 6.7 half-lives or so, $e^{-\lambda_2 t}$ becomes negligible compared with $e^{-\lambda_1 t}$ and the activity of the product is essentially

$$A_2(t) = \frac{T_1}{T_1 - T_2} A_1^0 e^{-\lambda_1 t}$$

Thus the product eventually diminishes with the same half-life as the parent. When this condition exists, the two nuclides are said to be in transient equilibrium, and the product activity is greater than that of the parent by the factor $T_1/(T_1 - T_2)$. Since $A_1^0 e^{-\lambda_1 t}$ is just the activity of the parent at time t, the product activity can be obtained easily by determining the parent activity $A_1(t)$ at time t and mulitplying it by $T_1/(T_1 - T_2)$ *if* it is known that transient equilibrium has been established. This is done by determining that the ingrowth time has exceeded the time when the product activity has reached its maximum value (see the earlier discussion on time of maximum activity).

An example of transient equilibrium is shown graphically in Figure 5-38 for ^{132}Te ($T_{1/2} = 76.8$ h) undergoing transformation to ^{132}I ($T_{1/2} = 2.28$ h). Transient equilibrium is achieved just after, but not before, the activity of the product nuclide ^{132}I crosses over the activity curve for ^{132}Te. Once transient equilibrium is established, the product activity can be obtained simply by multiplying the existing parent activity by the ratio $T_1/(T_1 - T_2)$.

When the parent has a shorter half-life than the product ($\lambda_1 > \lambda_2$), no state of equilibrium is attained. If the parent and product are separated initially, then as the parent undergoes transformation, the number of product atoms will increase, pass through a maximum, and the product activity will eventually be a function of its own unique half-life as calculated by the equation for $A_2(t)$.

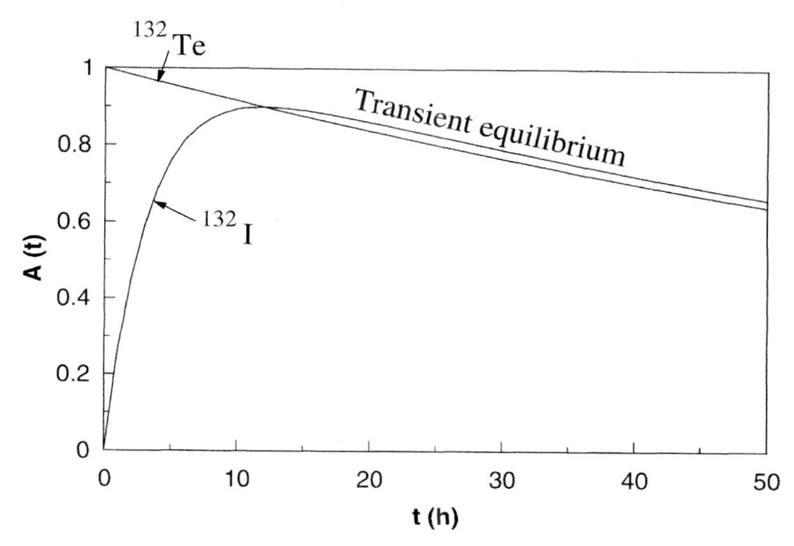

FIGURE 5-38. Activity of ^{132}I produced from initially pure ^{132}Te undergoing radioactive transformation to ^{132}I, showing the establishment of transient equilibrium.

Radionuclide Generators

Various radionuclide generators are fabricated to produce relatively short-lived radionuclides from a long-lived parent; for example:

$$\underset{(66\ \text{h})}{^{99}\text{Mo}} \longrightarrow \underset{(6\ \text{h})}{^{99\text{m}}\text{Tc}} \longrightarrow \underset{(> 10^5\ \text{y})}{^{99}\text{Tc}}$$

$$\underset{(115\ \text{d})}{^{113}\text{Sn}} \longrightarrow \underset{(1.67\ \text{h})}{^{113\text{m}}\text{In}} \longrightarrow \underset{(\text{stable})}{^{113}\text{In}}$$

$$\underset{(275\ \text{d})}{^{68}\text{Ge}} \longrightarrow \underset{(1.1\ \text{h})}{^{68}\text{Ga}} \longrightarrow \underset{(\text{stable})}{^{68}\text{Zn}}$$

The transformation rate of the parent radionuclide determines the activity of the product radionuclide, which is produced continuously such that a state of equilibrium often occurs. Once equilibrium has been achieved between the parent and daughter radioactivities, it can be disturbed only by chemical separation (referred to as *milking*) of the two radionuclides which will be followed by regrowth of the product radioactivity as new atoms are produced by transformation of the parent. For example, the ^{99m}Tc activity in a ^{99}Mo–^{99m}Tc generator will achieve a state of transient equilibrium with ^{99}Mo, and the activity of ^{99m}Tc in the generator at any given time is governed by the 66-h half-life of ^{99}Mo rather than the 6-h half-life of ^{99m}Tc. In transient equilibrium the activities of the two radionuclides maintain a constant ratio with time even though the half-lives of the two radionuclides are quite different. Once ^{99m}Tc is removed (eluted) from the generator its activity is, of course, governed by its 6-h half-life. Transient equilibrium is reestablished in about 7 half-lives of the product, although at a new level, due to the decrease in activity of the parent. In practical terms, a fresh supply of the product radionuclide is available in about 2 to 4 half-lives; however, the generator can be remilked

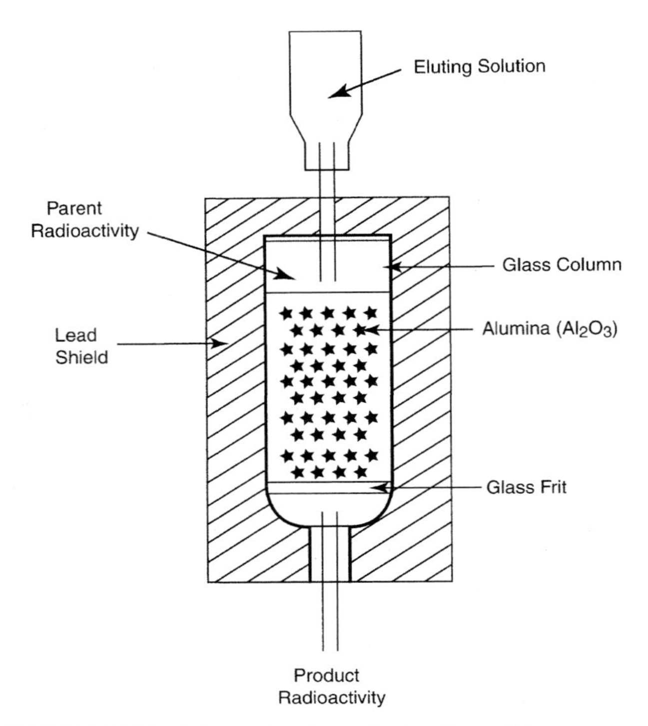

FIGURE 5-39. Schematic of a typical radionuclide generator.

sooner, and the extracted activity determined by the parent source strength and the ingrowth period. The growth and removal of the product activity in a generator can be predicted exactly using the radioactive series transformation equations.

In a typical ^{99}Mo–^{99m}Tc generator, ^{99}Mo in the form of sodium molybdedate is absorbed onto alumina, and when eluted with 0.9% sodium chloride solution (or saline), technetium is extracted as sodium pertechnetate (Na $^{99m}TcO_4$). A typical generator consists of a glass column filled with a suitable exchange material such as alumina (Al_2O_3), held in place with a porous glass disk and enclosed in a lead shield (Figure 5-39). The parent radionuclide is firmly absorbed on the top of the alumina and the product is also retained in the matrix until it is separated (eluted or milked) from the parent radionuclide by a solution that preferentially leaches the product radionuclide without removing the parent. The ^{99}Mo used in the generator can be produced by neutron activation of ^{98}Mo or by nuclear fission of ^{235}U, which yields a carrier-free product with very high activity; the activity for neutron activation of ^{98}Mo is generally much lower.

Transient Equilibrium: Branching

Transformation branching affects the relative activity of a radionuclide in a series. For example, the decay scheme for ^{99}Mo–^{99m}Tc indicates (as shown in Figure 5-19) that the ^{99m}Tc product is produced in only 82.84% of the transformations, and it is necessary to account for this branching fraction in calculations of the activity

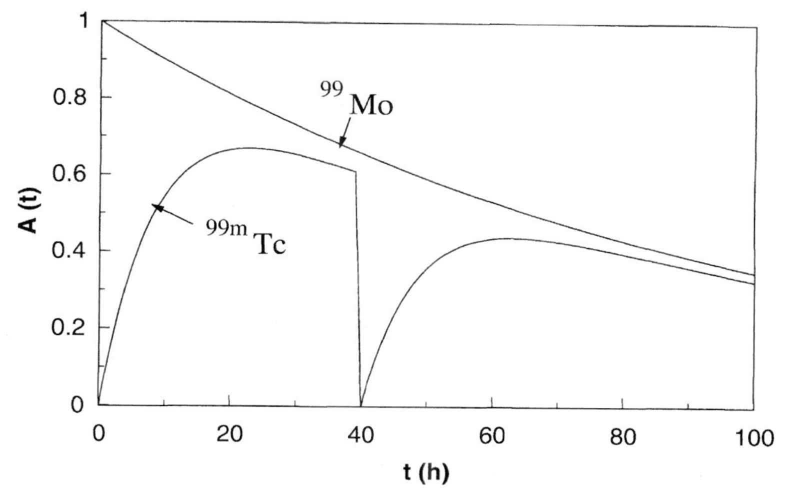

FIGURE 5-40. Transient equilibrium for branched radioactive transformation of ^{99}Mo to ^{99m}Tc.

of ^{99m}Tc produced from ^{99}Mo; that is,

$$A(^{99m}\text{Tc}) = (0.8284)\frac{T_1}{T_1 - T_2}(e^{-\lambda_1 t} - e^{-\lambda_2 t})$$

where T_1 and T_2 are in the same units. This relationship is plotted in Figure 5-40 for the ingrowth of ^{99m}Tc (in gamma transitions) for several extractions of ^{99m}Tc produced in the generator. The gamma-transition rate of ^{99m}Tc that grows into the generator never reaches the parent activity (as it does for ^{132}I in Figure 5-38) because only 82.84% of ^{99}Mo transformations produce ^{99m}Tc, and thus the recovery curve does not rise above the parent even though a state of transient equilibrium is established.

TOTAL NUMBER OF TRANSFORMATIONS

Many circumstances in radiation dosimetry or measurement require knowledge of the total number of transformations that occur over a time interval. If the time interval is short and the activity remains essentially constant over the interval, the number of transformations is just activity × time, and fortunately many situations can be so represented (a count taken over a few minutes, a detector reading, a dose-rate measurement, etc.). In other situations, the activity will change significantly over the time of interest, and determination of the number of transformations must account for the change. One such circumstance is clearance of a radioactive substance that has been deposited in a biological tissue (e.g., a human organ), and removal is not only by radioactive transformation but biological processes as well (i.e., with an effective half-life). If an initial activity, A_0, is deposited in a biological system, it will vary with time as shown in Figure 5-41.

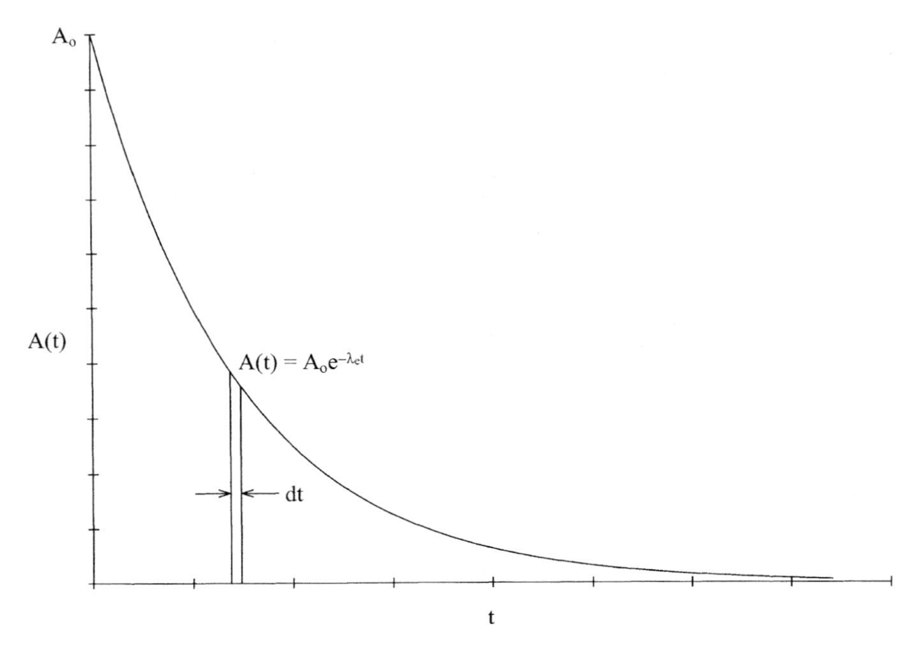

FIGURE 5-41. Variation of activity in a system due to multiple removal processes, where λ_e accounts for both biological λ_b, and radiological, λ_r removal processes.

The number of transformations that occur over a given time interval from $t = 0$ to $t = t$ is obtained by determining the area under the curve. The area will have units of activity · seconds (or other time units if preferred) and can be obtained by integration of each of the differential units of area $A(t)\,dt$, recognizing that both are variables, or

$$\text{area (activity} \cdot \text{seconds)} = \int_0^t A(t)\,dt = \int_0^t A_0 e^{-\lambda_e t}\,dt$$

$$= \frac{A_0}{\lambda_e}(1 - e^{-\lambda_e t}) = 1.4427 T_e A_0 (1 - e^{-\lambda_e t})$$

which yields the total number of transformations that occur in a time t, because activity has units of transformations per unit time and T_e is in units of time. If $t \to \infty$, which it effectively does in 7 to 10 effective half-lives, the total number of transformations is

$$\text{activity} \cdot \text{seconds} = \frac{A_0}{\lambda_e} = 1.4427 T_e A_0$$

Example 5-18: For a radiopharmaceutical containing ^{32}P ($T_{1/2} = 14.3$ d) with an activity of 10^4 d/m that is deposited in the liver of a person from which it clears with a biological half-life of 18 days, (a) how many transformations will occur in 7 d, and (b) how many total transformations will occur?

SOLUTION: First, determine the effective half-life and the effective removal constant:

$$T_e = \frac{14.3 \times 18}{14.3 + 18} = 7.97 \text{ d} = 1.15 \times 10^4 \text{ min}$$

(a) The number of transformations occurring in 7 d is

$$\begin{aligned}\text{transformations} &= 1.4427 \times 1.15 \times 10^4 \text{ m} \times 10^4 \text{ t/m}[1 - e^{(\ln 2/7.97 \text{ d})(7 \text{ d})}] \\ &= 7.55 \times 10^7 \text{ transformations}\end{aligned}$$

(b) The total number of transformations as $t \to \infty$ is

$$\begin{aligned}\text{transformations} &= 1.4427 \times 1.15 \times 10^4 \text{ m} \times 10^4 \text{ t/m} \\ &= 1.66 \times 10^8 \text{ transformations}\end{aligned}$$

The results of Example 5-18 can be used to determine the amount of energy deposited in the tissue by assigning an energy to each transformation and determining the fraction of each energy unit that is absorbed. For ^{32}P in the liver, the average energy per transformation is 0.695 MeV (see Appendix D) and the absorbed fraction is 1.0; therefore, the total energy deposited is

$$\begin{aligned}E_{\text{dep}} &= 1.66 \times 10^8 \text{ t} \times 0.695 \text{ MeV/t} \\ &= 1.154 \times 10^8 \text{ MeV} = 1.85 \times 10^2 \text{ ergs}\end{aligned}$$

which can be converted to radiation absorbed dose by dividing by the tissue mass (see Chapter 7).

DISCOVERY OF THE NEUTRINO

Discovery of the neutrino was accomplished by Reines and Cowan some 20 years after it was first postulated. They used the intense radiation field from a high-powered nuclear reactor at the Savannah River site to determine whether the inverse of β^- transformation occurred in a tank containing 1400 L of proton-rich triethyl benzene, as shown schematically in Figure 5-42. The theoretical reaction being tested was for the inverse of proton transformation, in which the large flux of postulated antineutrinos would produce positrons and neutrons in a proton-rich solution, or

$$\bar{\nu} + {}^1_1\text{p} + 1800 \text{ keV} \to {}^0_1\beta^+ + {}^1_0\text{n}$$

which is the inverse of β^- transformation. Antineutrino absorption by a proton should produce a positron and a neutron, and the reaction would require a Q-value of 780 keV plus the energy of two electon masses, for a total of 1800 keV.

Coincidences between positrons and neutrons in the liquid scintillator were used as the primary means of detecting the reactions. If antineutrino reactions occurred with protons, both a positron and a neutron would be produced. The postulated

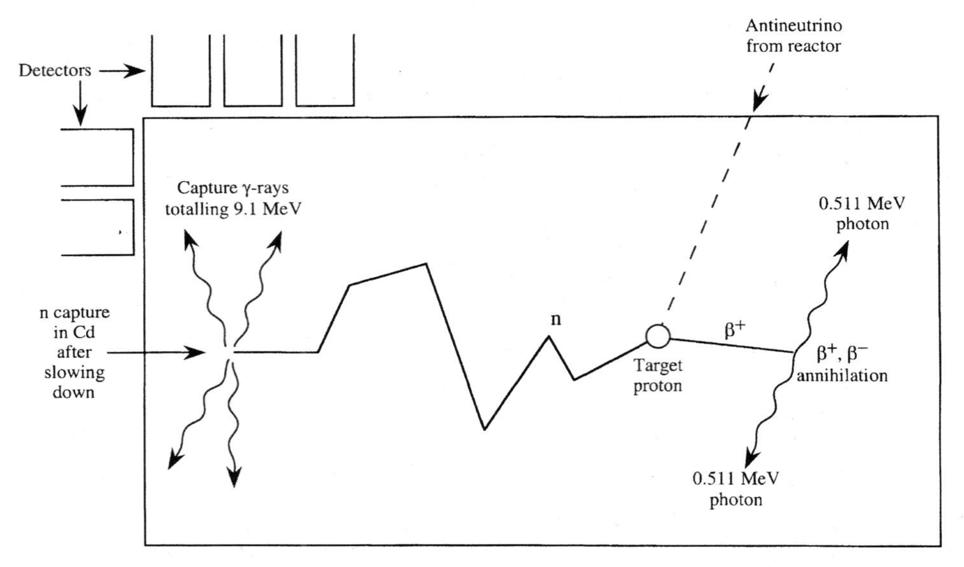

FIGURE 5-42. Tank of liquid scintillating triethyl benzene used to detect the neutrino by inducing antineutrino capture by protons yielding a prompt signal (interactions of the positron produced in the interaction and its annihilation) and a delayed signal when the neutron produced by the reaction is captured in cadmium. The two signals were detected some 30 μs apart by scintillation detectors surrounding the tank.

positron would quickly ionize the medium and then annihilate with an available electron, and the neutron would slow down and be captured yielding a capture gamma ray; therefore, $CdCl_2$ was dissolved in the scintillator to enhance detection. Detectors surrounding the tank of liquid scintillator would produce two signals: a *prompt* signal from the ionization produced by absorption of the emitted positron and the two 0.511 MeV annihilation photons in the scintillator, and a *delayed* signal produced by 9.1 MeV of gamma radiation emitted when the neutron was captured in cadmium, which due to the slowing-down time of the neutron, should occur in about 30 μs. Reines and Cowan (1953) measured 36 ± 4 events associated with the reactor being on and calculated the cross section for neutrino capture to be $(11 \pm 2.6) \times 10^{-44}$ cm^2, thus confirming its existence.

ACKNOWLEDGMENTS

This chapter was compiled with the able and patient assistance of Arthur Ray Morton III and Chul Lee, both graduates of the University of Michigan Radiological Health Program, and Patricia Ellis, Research Associate in the University of Michigan Radiological Health Program.

REFERENCES AND ADDITIONAL RESOURCES

Krane, K. S., *Introductory Nuclear Physics*, Wiley, New York, 1988.

Evans, R. D., *The Atomic Nucleus*, McGraw-Hill, New York, 1955.

GE, Chart of the nuclides, *Nuclides and Isotopes*, 15th ed., General Electric Company, San Jose, CA, 1996.

ICRP, *Radionuclide Transformations: Energy and Intensity of Emissions*, Report 38, Pergamon Press, Elmsford, NY, 1978.

National Nuclear Data Center, Brookhaven National Laboratory, Upton, NY. Data resources are accessible through the Internet at *www.nndc.bnl.gov*.

SNM, *MIRD: Radionuclide Data and Decay Schemes*, Society of Nuclear Medicine, New York, 1989.

Reines, F., and C. L. Cowan, Detection of the free neutrino, *Physical Review* **92**, 8301 (1953).

Robson, J. M., The radioactive decay of the neutron, *Physical Review* **83**, 349–358 (1951).

PROBLEMS

5-1. It is noted that a sample of ^{134}La with a half-life of 6.5 min has an activity of 4 Ci. What was its activity 1 h ago?

5-2. Determine by Q-value analysis whether ^{65}Zn can undergo radioactive transformation by β^+, electron capture, or β^- transition.

5-3. A sample of material decays from 10 Ci to 1 Ci in 6 h. What is the half-life?

5-4. An isotope of fission is ^{151}Sm, with a half-life of 90 y. How long will it take 0.05 g to decay to 0.01 g?

5-5. (**a**) How many days will it take to reduce 4 Ci of ^{210}Po ($T_{1/2}$ = 138 d) to 1 Ci? (**b**) How many grams of ^{210}Po does it take to produce 4 Ci of activity?

5-6. (**a**) What is the initial activity of 40 g of ^{227}Th? (**b**) How many grams will be in the sample 2 y later?

5-7. In August 1911, Marie Curie prepared an international standard of activity containing 21.99 mg of $RaCl_2$. Calculate the original activity of the solution and its activity as of August 1996 using constants from the chart of the nuclides.

5-8. (**a**) Calculate the specific activity of ^{210}Pb. (**b**) What is the activity per gram of a pure sample that has now aged 3 y?

5-9. A reference person weighs 70 kg and contains, among other elements, 18% carbon and 0.2% potassium. How many microcuries of ^{14}C and of ^{40}K will be present in such a person?

5-10. What must be the activity of radiosodium (^{24}Na) compound when it is shipped from Oak Ridge National Laboratory so that upon arrival at a hospital 24 h later its activity will be 100 mCi?

5-11. 750 mg of ^{226}Ra is used as a source of radon which is pumped off and sealed into tiny seeds or needles. (**a**) How many millicuries of radon will be available at each pumping if this is done at weekly intervals? (**b**) How much time will be required to accumulate 700 mCi of the gas?

5-12. An assay of an equilibrium ore mixture shows an atom ratio for $^{235}U/^{231}Pa$ of 3.04×10^6. Calculate the half-life of ^{235}U from the assay data and the known half-life of ^{231}Pa (3.28×10^4 y).

5-13. Regulations permit use of ^3H in luminous aircraft safety devices up to 4 Ci. (**a**) How many grams of ^3H could be used in such a device? (**b**) What change in luminosity would occur after 8 y of use?

5-14. (**a**) What is the energy of neutrino(s) from electron capture transformation of ^{55}Fe? (**b**) What energy distribution do these neutrinos have?

5-15. Artificially produced ^{47}Ca ($T_{1/2} = 4.54$ d) is a useful tracer in studies of calcium metabolism, but its assay is complicated by the presence of its radioactive product ^{47}Sc ($T_{1/2} = 3.35$ d). Assume an assay based on beta-particle counting with equal counting efficiencies for the two nuclides, and calculate the data for and plot a correction curve that can be used to determine ^{47}Ca from a total beta count.

5-16. (**a**) What mass of ^3H will be needed to replace 0.08 Ci of ^{210}Po in a luminous instrument dial if each is equally effective in producing scintillations? (**b**) How much ^3H would be needed for equal luminosity of ^{210}Po 1 y after fabrication?

5-17. A source of radioactivity was measured to have 40, 14, 5.5, 2, and 0.6 d/m at 8-min intervals. Use a plot of the data to determine the half-life of the radionuclide in the source.

5-18. A mixed source was measured to produce the following activity levels at 3-min intervals: 60, 21, 8, 1, and 0.3 d/m. What are the half-lives of the radionuclides in the source?

5-19. Ten millicuries of ^{90}Sr is separated from a fission product mixture and allowed to set for ^{90}Y to grow in. What is the activity of ^{90}Y at (**a**) 20 h and (**b**) 600 h?

5-20. (**a**) What is the maximum activity of ^{90}Y that will ever be present in the sample in Problem 5-19? (**b**) At approximately what time after separation does it occur?

5-21. If ^{140}Ba ($T_{1/2} = 12.75$ d) is freshly separated from a fission product mixture and found to contain 200 mCi: (**a**) What is the activity of ^{140}La ($T_{1/2} = 1.678$ d) that will grow into the sample in 12 h? (**b**) What is the maximum activity of ^{140}La that will ever be present in the sample, and when will it occur?

5-22. Derive from first principles the equation for the third member of a radioactive series that begins with a radioactive parent containing N_1^0 atoms.

5-23. The steel compression ring for the piston of an automobile engine has a mass of 30 g. The ring is irradiated with neutrons until it has an activity of 10 μCi of ^{59}Fe ($T_{1/2} = 44.529$ d). The ring is installed in an engine and removed 30 d later. If the crankcase oil has an average activity due to ^{59}Fe of 0.126 d/m per cubic centimeter, how much iron was worn off the piston ring? (Assume that the total volume of the crankcase oil is 20 L.)

5-24. The mass of the human thyroid is about 20 g and it clears iodine deposited in it with a biological half-life of 80 d. If 2 μCi of ^{131}I ($T_{1/2}$ = 8.02 d) is present in a person's thyroid: (**a**) How many transformations will occur in a period of 6 d? (**b**) How many will occur over a period of 1 y, or effectively infinite time? (**c**) If the principal beta of $E_{\beta^-,\mathrm{max}} = 0.605$ MeV and $E_{\beta^-,\mathrm{avg}} = 0.192$ MeV occurs in 89.9% of transformations, how much energy will be deposited per gram of thyroid tissue?

6

NATURALLY OCCURRING RADIATION AND RADIOACTIVITY

The majority of a person's environmental radiation dose is due to the natural radiation background to which humans have shown remarkable adaptability; it provides, therefore, a reasonable baseline for judging the significance of radiation exposures from other human activities.

—Floyd L. Galpin, 1999

A natural radiation background exists everywhere and every natural substance contains some amount of radioactive material. Natural radiation and radioactivity are of importance to radiation physics for at least four reasons:

1. They presaged and stimulated perhaps the greatest period of scientific discovery and thought in human history and the development of basic models of the atom, nuclear structure, and radioactive transformation.
2. The natural sources represent a continuous exposure of beings on the earth and thus are a benchmark for consideration of levels of radiation protection not only for enhanced sources of natural radiation and radioactivity, but other sources as well.
3. Human activities affect exposure to these sources, and understanding of such processes, including discovery and uses of uranium, thorium, and radon and their transformation products, is fundamental to assessing and controlling their radiological impacts.
4. Every measurement of a radiation level or the radioactivity in a sample or source must take into account the background associated with naturally occurring radioactive material and radiation sources.

The natural radiation environment consists of cosmic rays and naturally radioactive materials. Some of the materials are cosmogenic, others are primordial, and others exist naturally because of the radioactive transformations of substances produced by these processes. The radiological significance of naturally occurring radioactive materials and radiation sources is closely linked to the physical behavior of the materials in the source and how they change with time. The purpose of this chapter is to describe several, but not all, typical circumstances.

DISCOVERY AND INTERPRETATION OF RADIOACTIVITY

In reporting his discovery of x rays Roentgen had made the observation "that the place of most brilliant fluorescence on the wall of the (glass) discharge tube is the chief seat whence the x rays originate." This association between fluorescence and the penetrating rays dominated the thinking of Henri Becquerel, who set out to find the source of the rays. He placed a fluorescent crystal of potassium uranyl sulfate, inherited from his father's laboratory, on a photographic plate and exposed it to sunlight because it produced the brightest fluorescence; an image of the crystal appeared on the photographic plate, clearly penetrating the wrapping. At almost the same time, Silvanus Thompson in London exposed several fluorescent compounds on thin sheets of aluminum atop sealed photographic plates and noted that only the uranium compound produced a silhouette of the crystals on the developed photographic plate. On February 24, 1986, Becquerel reported his finding to the Academy; Thompson's much broader observations were made on the 26th, and he wrote right away to the president of the Royal Society, who on February 29 encouraged him to publish his findings, only to write again two days later that Becquerel had preceded him. Close, but not close enough.

Becquerel had asserted that the invisible emissions from uranium were a form of fluorescence that emitted radiation produced only after exposure to an external source of energy, the sun. He was premature, and wrong, which he recognized several days later. He had placed a couple of wrapped photographic plates with uranium crystals in place into a drawer because the sun was not out when he prepared them, nor did it shine again for several days. He decided to develop them anyway, probably thinking that the plates would not show an image and serve as a control, but it is reported that even he did not know why he did so. Instead, the silhouettes of the uranium appeared as intense as ever, which Becquerel declared to be "very important...quite outside the range of phenomena one might expect" and that the photographic images were due to "invisible radiations emitted by phosphorescence." Although the radiation from uranium had nothing to do with either fluorescence or phosphorescence, March 1, 1896 is established as the date that radioactivity was discovered; even though Thompson's results were the more definitive, Becquerel, who published first, is credited with the discovery. The discovery attracted little attention at the time; x rays were much more in vogue, and the rays from uranium were too weak to produce photographic images of the skeleton. It was the persistence of the Curies and the seminal work of Rutherford and Soddy that established the science and importance of substances with, to use Becquerel's phrase, "radiation activité."

The work of the Curies is quite remarkable when contrasted to the times, in particular that of Marie, who in her doctoral research observed that the intensity of radiation from unprocessed uranium ore was greater than from purified uranium salts and realized that something else in the ore must be producing it. To find it, she processed a ton of pitchblende ore from which most of the uranium had been removed and discovered in two different precipitates the element polonium, and a second element with even more activity that she named radium.

The Curies claimed that radioactivity is "an atomic property of substances," that "each atom of a radioactive substance functions as a constant source of energy," and that "this activity does not vary with time" (i.e., a radioactive atom absorbs energy continuously from an external source and releases the energy as radiation at a steady rate) (Frame, 1996). Although the phenomenon was believed to be atomic in nature, models of the atom were not available to develop the necessary understanding of the process. It took the work of Rutherford and Soddy to sort the matter out. Rutherford had studied the ionization of air by thorium, discovered to be radioactive by Schmidt in 1898, but his results were spurious, which he traced to a short-lived substance that deposited on his apparatus some distance away from the thorium source. He concluded that thorium must be producing a substance he called *emanation* (now known as ^{220}Rn, or thoron) that migrated away from the source and then produced a short-lived radioactive substance that deposited on the surface of nearby objects, thus making them radioactive. Rutherford enlisted the help of Frederick Soddy, who quickly demonstrated that the emanation was unchanged by heating and it condensed at low temperatures. It was clearly a gas, and since it could pass unaffected through a series of solutions and metals, it was also a noble gas—a unique element quite distinct from thorium. If the emanation were a noble gas, it could not have been present in the original thorium sample, which had been of the highest available purity. There was but one explanation: Thorium was spontaneously producing a noble gas, a different element, by transmutation of atoms.

Crookes sent Rutherford some extremely pure thorium nitrate, and a copy of a paper by Becquerel that described how the activity of a precipitate of uranium (called UX) decreased while the activity of the purified uranium recovered. Suspecting that something similar might occur for thorium, he and Soddy performed chemical separations on the thorium nitrate and discovered a substance (^{224}Ra) that in keeping with Becquerel's designation, they called thorium X (ThX). Most of the radioactivity originally in the thorium ended up in the precipitate containing ThX, and as expected, the thorium gradually recovered its activity. More important, they noted that the activity of the isolated ThX decreased exponentially with a half period of 4 days, while the activity of the purified thorium recovered at precisely the same rate! Figure 6-1 contains the classic and historical plots of their data on the depletion of ThX activity and its recovery in thorium as well as the more careful measurements they made of uranium and the UX precipitate observed by Becquerel. These results, together with the evidence that their thorium samples were also producing a noble gas, clearly established that radioactivity was a phenomenon of transmutation; however, Rutherford was reluctant, stating "don't call it transmutation or they'll have our heads off as alchemists." Their paper, "The Cause and Nature of Radioactivity," was quickly published in 1902. In their words: "The disintegration of the atom and the expulsion of a...charged particle leaves behind a new system lighter than before and possessing physical and chemical properties quite different from those of the original parent element. The disintegration process, once started, proceeds from state to state with measurable (rates) in each case. Since...radioactivity...is an atomic phenomenon...in which new types of matter are produced, these changes must be occurring within the atom. The results that have so far been obtained, which indicate that the (rate) of the (radioactive) reaction is unaffected by the conditions, make it clear that the changes in question are different in character from any that have been before dealt with in chemistry.... Radioactivity may therefore be considered as a manifestation of subatomic change." The atom could no longer be considered permanent.

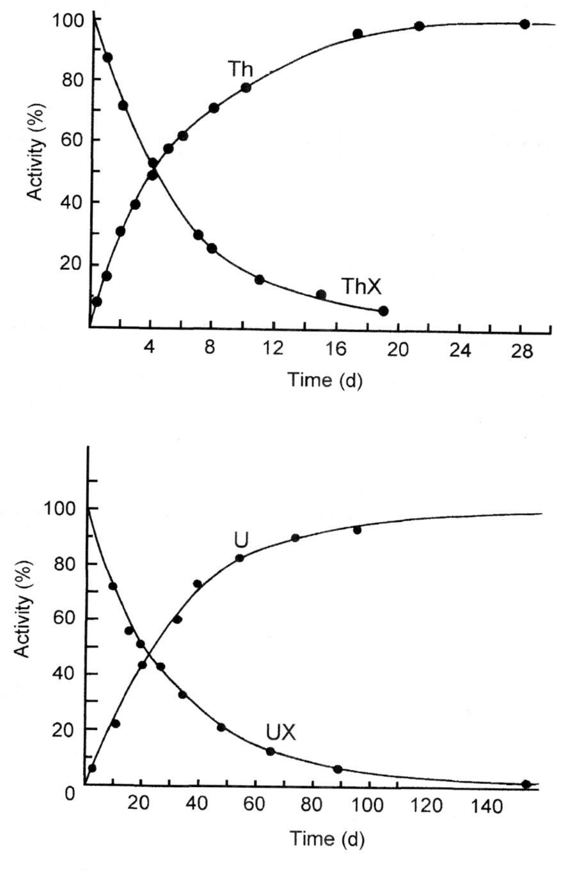

FIGURE 6-1. Recovery of thorium and uranium activity after separation and loss of ThX (^{224}Ra; $T_{1/2} = 3.66$ d) and UX (^{234}Th; $T_{1/2} = 24.1$ d) as measured by Rutherford and Soddy.

The shapes of the curves in Figure 6-1 show that activity diminishes exponentially and that the number of atoms which transition in a unit interval of time is related to the total number of atoms present by a constant which Rutherford and Soddy called the *radioactive constant* of the element under consideration (now commonly referred to as the disintegration or decay constant λ). The radioactive constant λ is a definite and specific property of a given radioelement. Its value depends only on the nature of the species, is independent of the physical or chemical conditions of the atoms, and the exponential equations represent very accurately the rates of transformations for radioelements for values of λ ranging over 24 orders of magnitude (i.e., the equations are applicable to very short-lived species as well as those with very long half-lives).

BACKGROUND RADIATION

Creatures on the earth are continually exposed to external radiation from cosmic rays and a number of naturally occurring radionuclides that produce external gamma

TABLE 6-1. Average Effective Dose Equivalent Rates from Various Sources of Natural Background Radiation in the United States

Source	Dose Equivalent Rate (mrem/y)
Cosmic radiation	27
Cosmogenic nuclides	1
External terrestrial	28
Nuclides in body	39

Source: Adapted from NCRP, 1987.

radiation or can be incorporated into the body to produce internal radiation dose. These sources, of course, vary widely depending on location and the surrounding environment. Average values of exposure data from the various sources of natural radiation are listed in Table 6-1, and the particular features of each source are discussed in the sections that follow.

The doses in Table 6-1 do not include radon and its products, although it is common to do so. Published estimates of total actual background exposure that are on the order of 300 mrem/y usually include an average for the uncertain radon product lung dose adjusted to a whole body effective dose so it can be added to the other common sources which expose the whole body. Although it is proper to recognize radon exposure, it should be done warily.

COSMIC RADIATION

The earth is bombarded continuously by radiation originating from the sun and from sources within and beyond the galaxy. This cosmic radiation slams into the earth's upper atmosphere, which provides an effective shield for beings below. Cosmic rays consist of high-energy atomic nuclei, some 87% of which are protons. About 11% are alpha particles, approximately 1% are heavier atoms which decrease in importance with increasing atomic number, and the remaining 1% are electrons. These "rays" have very high energies, some as high as 10^{14} MeV, but most are in the range 10 MeV to 100 GeV.

Cosmic rays are extraterrestrial radiations. They consist of *galactic particles*, which originate outside the solar system, and *solar particles*, emitted by the sun. As they strike the atmosphere they produce cascades of nuclear interactions that yield many secondary particles which are very important in production of cosmogenic radionuclides. The distribution of cosmic ray components and their energy spectra are shown in Figure 6-2. Galactic cosmic radiation produces various spallation reactions in the upper atmosphere which yield secondary neutrons and protons. Many pions are also produced and their subsequent disintegration results in electrons, photons, neutrons, and muons. Muon disintegrations, in turn, lead to secondary electrons, as do coulombic scattering interactions of charged particles in the atmosphere. As shown in Figure 6-3, the number and mix of these particles varies with altitude, as does the resulting tissue absorbed dose rate.

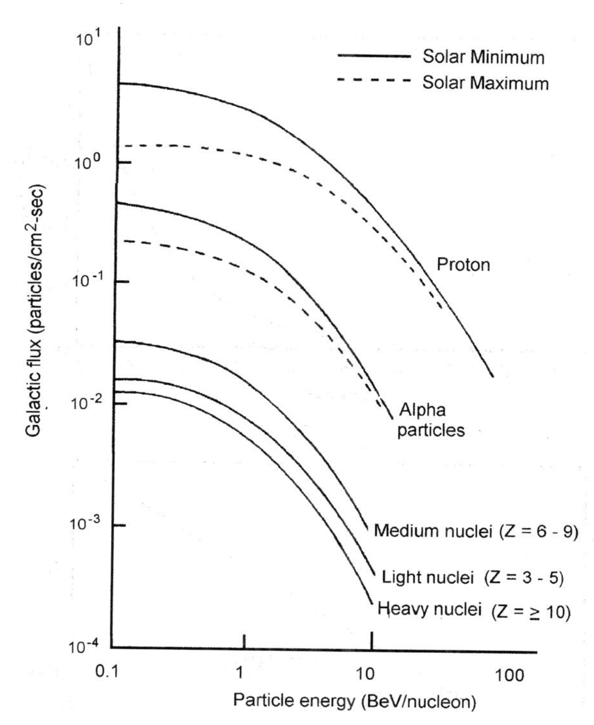

FIGURE 6-2. Components of galactic cosmic radiation and their energies. (Adapted from NCRP, 1987, with permission.)

Radiation dose rates at the earth's surface due to cosmic radiation are caused largely by muons and electrons, and both vary with elevation and with latitude. The atmospheric shield is a column of air that weighs 1033 g/cm^2 at sea level; the effectiveness of this shield is related to the atmospheric "thickness" or "depth" in units of g/cm^2 of overlying air. Consequently, the total dose rate from cosmic rays will increase with altitude as the thickness of the atmosphere decreases, as shown in Figure 6-3. At geomagnetic latitude 55°N, for example, the absorbed dose rate in tissue approximately doubles with each 2.75 km (9000 ft) increase in altitude, up to about 10 km (33,000 ft). The neutron component of the dose equivalent rate increases more rapidly with altitude than does the directly ionizing component and dominates at altitudes above about 8 km (UN, 1988).

Cosmic rays of solar origin are mainly hydrogen and helium nuclei of relatively low energy (about 1 keV), but solar flares can generate particles of several GeV. Their energies contribute little to radiation doses at the surface of the earth; however, they perturb the earth's magnetic field, which in turn deflects galactic cosmic rays that might otherwise reach the earth's atmosphere and the surface. Maximum solar flare activity leads to decreases in dose rates, and vice versa. Consequently, at any given location, cosmic ray doses may vary in time by a factor of about 3. At sea level they can vary with geomagnetic latitude by as much as a factor of 8, being greatest at the poles and least at the equator. The global average cosmic ray dose equivalent rate at sea level is about 24 mrem (240 μSv) per year for the directly

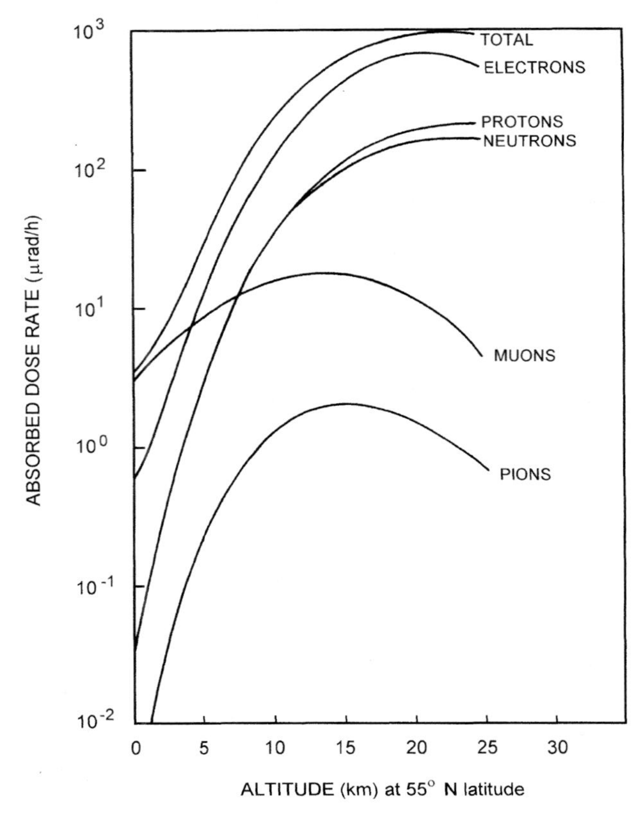

FIGURE 6-3. Absorbed dose rates from various components of cosmic radiation at solar minimum at 5-cm depth in a 30-cm slab of tissue. (To convert to μGy/h, divide by 100.)

ionizing component and 2 mrem (20 μSv) per year for the neutron component. The annual tissue dose rate from cosmic rays increases with altitude as shown in Figure 6-4, which contains data averaged over geomagnetic latitudes between 43° and 55° and over two periods of solar activity.

COSMOGENIC RADIONUCLIDES

Several radionuclides of cosmogenic origin are produced when high-energy protons (87% of cosmic radiation) interact with constituents of the atmosphere. Showers of secondary particles, principally neutrons, from such interactions yield a number of such radionuclides, in particular ^{3}H, ^{7}Be, ^{14}C, and ^{22}Na (see Table 6-2), which are produced at relatively uniform rates. The high-energy (ca. MeV) secondary neutrons that produce these interactions originate in the atmosphere from cosmic ray interactions, rather than coming from outer space. Their concentration rises with altitude to a maximum around 40,000 ft and then decreases, and since free neutrons are radioactive with a life of only about 12 min, they do not last long enough to travel from outer space.

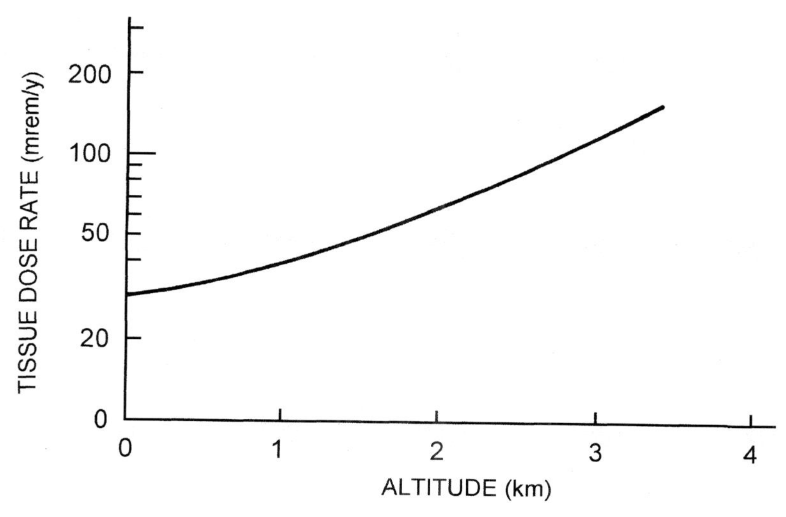

FIGURE 6-4. Long-term average outdoor total dose equivalent rate (charged particles plus neutrons) due to cosmic radiation at 5-cm depth in a 30-cm-thick slab of tissue. (To convert to mSv/y, divide by 100.)

TABLE 6-2. Global Distribution of Cosmogenic Radionuclides

	^{3}H	^{7}Be	^{14}C	^{22}Na
Global inventory (pBq)	1300	37	8500	0.4
Distribution (%)				
Stratosphere	6.8	60	0.3	25
Troposphere	0.4	11	1.6	1.7
Land surface/biosphere	27	8	4	21
Mixed ocean layer	35	20	2.2	44
Deep ocean	30	0.2	92	8
Ocean sediments	—	—	0.4	—

Cosmic ray interactions can also occur with constituents of the sea or the earth, but atmospheric interactions dominate. The total neutron flux density at sea level is only about 30 neutrons/cm$^2 \cdot$h of which only 8 are thermal neutrons; therefore, production of radioactive materials by neutron capture in the earth's crust or the sea is minor. An important exception is the production of long-lived ^{36}Cl (NCRP, 1987) by neutron activation of ^{35}Cl in the earth's crust.

Oxygen in the atmosphere is essentially transparent to neutrons, but (n,p) interactions with ^{14}N to form ^{14}C are very probable. Neutron interactions with ^{14}N yield ^{14}C 99% of the time and tritium through (n,t) reactions 1% of the time. Carbon-14 from ^{14}N(n,p)^{14}C reactions exist in the atmosphere as CO_2, but the main reservoir is the ocean. It has a half-life of 5730 years and undergoes radioactive transformation back to ^{14}N by beta-particle emission with a maximum energy of 157 keV (average energy of 49.5 keV). The natural atomic ratio of ^{14}C to stable carbon is 1.2×10^{-12} (corresponding to 0.226 Bq of ^{14}C per gram of carbon) (NCRP, 1987). For carbon weight fractions of 0.23, 0.089, 0.41, and 0.25 in the soft tissues, gonads, red marrow, and skeleton, annual average absorbed doses in those tissues are respectively 1.3, 0.5, 2.3, and 1.4 millirem.

Tritium is produced mainly from $^{14}N(n,t)^{12}C$ and $^{16}O(n,t)^{14}N$ reactions in the atmosphere and exists in nature almost exclusively as HTO, but in foods may be partially incorporated into organic compounds. Tritium has a half-life of 12.3 years and, upon transformation, releases beta particles with maximum energy of 18.6 keV (average energy 5.7 keV). The average concentration of cosmogenic tritium in environmental waters is 100 to 600 Bq/m^3 (3 to 16 pCi/L) based on a seven-compartment model (NCRP, 1987). If it is assumed that ^{3}H exists in the body, which contains 10% hydrogen, at the same concentration as surface water (i.e., 400 Bq/m^3 or 12 pCi/L) the average annual absorbed dose in the body is 1.2 μrem (0.012 μSv).

Except as augmented by human-made sources, ^{14}C and ^{3}H have existed for eons in the biosphere in equilibrium (i.e., the rate of production by neutron interactions is equal to its subsequent removal by radioactive transformation). The cosmogenic content of ^{14}C in the environment has been diluted the past century or so by combustion of fossil fuels and the emission of CO_2 not containing ^{14}C. On the other hand, atmospheric nuclear-weapons tests and other human activities have added to the natural inventories of ^{3}H and ^{14}C, and these nuclides no longer exist in natural equilibria in the environment; however, they are now returning to natural levels, since atmospheric nuclear testing stopped for the most part in 1962.

Beryllium-7, with a half-life of 53.4 d, and ^{10}Be ($T_{1/2} = 1.6 \times 10^6$ y), to a lesser extent, are also produced by cosmic ray interactions with nitrogen and oxygen in the atmosphere. ^{7}Be undergoes radioactive transformation by electron capture to ^{7}Li, with 10.4% of the captures resulting in the emission of a 478 keV gamma ray, which makes its quantitation relatively straightforward. Also, being a naturally occurring radionuclide, it is often observed in gamma-ray spectra of environmental samples. Environmental concentrations in temperate regions are about 3000 Bq/m^3 in surface air and 700 Bq/m^3 in rainwater (UN, 1982). Atmospheric concentrations of ^{7}Be are noticeably influenced by seasonal changes in the troposphere as shown in Figure 6-5, and by latitude as shown in Figure 6-6. Average annual absorbed doses in the adult are 1.2 mrem to the walls of the lower large intestine, 0.12 mrem to the red marrow, and 0.57 mrem to the gonads.

Sodium-22 is produced by spallation interactions between atmospheric argon and high-energy secondary neutrons from cosmic rays. It has a half-life of 2.60 y and is transformed by positron emission (90%) and electron capture (10%) to ^{22}Ne. The positron has a maximum energy of 546 keV (average energy 216 keV), and essentially all transformations are accompanied by emission of a 1.28 MeV gamma ray from the excited ^{22}Ne product. A large fraction of ^{22}Na remains in the stratosphere where it is produced, but nearly half the natural inventory is in the mixed layer of the ocean (Table 6-2). Annual average absorbed doses to the adult are about 1 mrad to the soft tissues, 2.2 mrad to red marrow, and 2.7 mrad to bone surfaces.

A whole host of other radionuclides (^{26}Al, ^{35}Cl, ^{38}Cl, ^{39}Cl, ^{31}Si, ^{32}Si, ^{32}P, ^{35}S, ^{38}Mg, ^{24}Na, etc.) are produced by spallation reactions with atmospheric ^{40}Ar, including ^{39}Ar ($T_{1/2} = 269$ y) and ^{37}Ar ($T_{1/2} = 35$ d). Since the energies of cosmic ray particles are so high, they can cause an argon nucleus to break up in many different ways to yield these diverse products. Although these spallation products

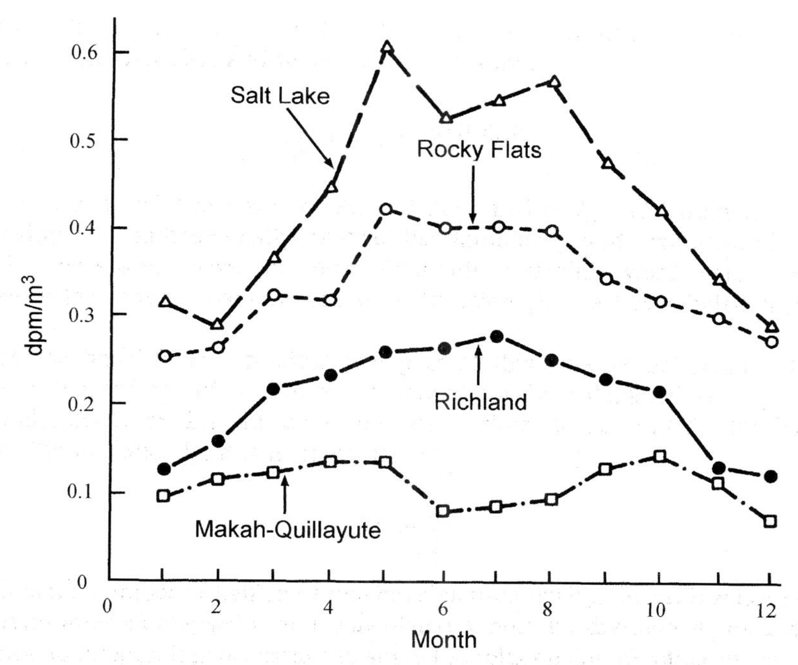

FIGURE 6-5. Variation of atmospheric ^{7}Be concentrations at Makah-Quillayute, Washington (48°N), Richland, Washington (46°N), Salt Lake City, Utah (41°N), and Rocky Flats, Colorado (40°N).

are detectable in environmental media to varying degrees, most are of minimal consequence in radiation protection.

Finally, neutrons of thermal energies interact with ^{80}Kr to produce ^{81}Kr ($T_{1/2}$ = 2.1×10^5 y) and with ^{84}Kr to produce ^{85}Kr ($T_{1/2}$ = 10.7 y), both in relatively small amounts. The quantities of cosmogenic ^{81}Kr and ^{85}K produced by these interactions are of little interest in radiation protection.

NATURALLY RADIOACTIVE SERIES

Many of the naturally occurring radioactive elements are members of one of four long chains, or radioactive series, stretching through the last part of the chart of the nuclides. These series are named the uranium (^{238}U), actinium (^{235}U), thorium (^{232}Th), and neptunium (^{237}Np) series, according to the radionuclides that serve as progenitor (or parent) of all the series products. With the exception of neptunium, each of the parent radionuclides is primordial in origin because they are so long lived that they still exist some 4.5 billion years after the solar system was formed.

The members of each of the three naturally radioactive series and the neptunium series are listed in Tables 6-3 to 6-6 along with the principal emissions, the half-life, and the maximum energy and frequency of occurrence of emitted particles and electromagnetic radiations. Each table contains a diagram of the progression of the series, including information on branching ratios that occur. Only the modern symbol is shown, despite the historical significance of the earlier designations;

TABLE 6-3. ^{238}U Series and Radiations of Yield Y_i Greater Than 1%[a]

		α		β⁻		γ	
Nuclide	$T_{1/2}$	E (MeV)	Y_i (%)	E (MeV)[b]	Y_i (%)	E (keV)	Y_i (%)
^{238}U	4.468×10^9 y	4.15	21				
α ↓		4.2	79				
^{234}Th	24.10 d			0.08	2.9	63.3	4.8
				0.1	7.6	92.4	2.8
β⁻ ↓				0.1	19.2	92.8	2.8
				0.2	70.3		
^{234m}Pa[c]	1.17 m			2.27	98.2	766	0.3
IT ↓						1001	0.84
^{234}Pa	6.75 h			22β⁻s		1313	18
				E_{avg} = 0.224			
β⁻ ↓				E_{max} = 1.26			
^{234}U	2.457×10^5 y	4.72	28.4				
α ↓		4.78	71.4				
^{230}Th	7.538×10^4 y	4.62	23.4				
α ↓		4.69	76.3				
^{226}Ra	1600 y	4.6	5.55			186.2	3.6
α ↓		4.79	94.5				
^{222}Rn	3.8235 d	5.49	99.9			510	0.08
α ↓							
^{218}Po[c]	3.11 m	6	100.0				
α ↓							

^{214}Pb	26.8 m			0.19	2.35	53.2	1.11
				0.68	46.0	242	7.5
β^-				0.74	40.5	295.2	18.5
				1.03	9.3	351.9	35.8
$^{214}Bi^c$	19.9 m			0.79	1.45	609.3	44.8
				0.83	2.74	768.4	4.8
				1.16	4.14	934.1	3.03
				1.26	2.9	1120.3	14.8
				1.26	1.66	1238.1	5.86
				1.28	1.38	1377.7	3.92
β^-				1.38	1.59	1408.9	2.8
				1.43	8.26	1729.6	2.88
				1.51	16.9	1764.5	15.4
				1.55	17.5	2204.2	4.86
				1.73	3.05	2447.9	1.5
				1.9	7.18	Nine other γs	
				3.27	19.9		
^{214}Po	164.3 μs	7.69	99.99				
α							
^{210}Pb	22.3 y			0.02	84	46.5	4.25
β^-				0.06	16		
$^{210}Bi^c$	5.013 d			1.16	100	1764.5	15.4
β^-							
^{210}Po	138.376 d	5.3	100.0				
α							
^{206}Pb	Stable						

Source: NNDC, 1995.

[a] X rays, conversion electrons, and Auger electrons are not listed.

[b] Maximum beta energy.

[c] Branching occurs in 0.13% of ^{234m}Pa to ^{234}Pa by IT; 0.02% of ^{218}Po to ^{218}At by β^-; 0.02% of ^{214}Bi to ^{210}Tl by α; 0.00013% of ^{210}Bi to ^{206}Tl by α.

TABLE 6-4. ^{235}U Series and Radiations of Yield Y_i Greater Than 1%[a]

Nuclide	$T_{1/2}$	α		β^-		γ	
		E (MeV)	Y_i (%)	E (MeV)[b]	Y_i (%)	E (keV)	Y_i (%)
^{235}U	703.8 × 10⁶ y	4.2–4.32	11			143.8	11
		4.37	17			163.3	5.1
↓ α		4.4	55			185.7	57.2
		4.5–4.6	10.9			205.3	5
^{231}Th	25.52 h			0.21	12.8	25.6	14.5
				0.29	12	84.2	6.6
↓ β^-				0.31	35		
^{231}Pa	3.276 × 10⁴ y	4.95	22.8			27.4	10.3
		5.01	25.4			283.7	1.7
↓ α		5.03	20			300.1	2.5
		5.06	11			330.1	1.4
^{227}Ac	21.773 y	4.94	0.55	0.02	10		
β^-		4.95	0.66	0.03	35		
98.62% → ^{227}Th; α 1.38% → ^{223}Fr				0.04	54		
^{227}Th	18.72 d	5.76	20.4			50	8
		5.98	23.5			236	12.3
		6.04	24.2			300	2.3
						304	1.2
α → ^{223}Ra						330	2.7
^{223}Fr	21.8 m			0.91	10.1	50	36
				1.07	16	79.7	9.1
↓ β^-				1.09	67	234.9	3
^{223}Ra	11.435 d	5.43	2.27			144	3.2

α		5.54	9.2			154	5.6
		5.61	25.7			269	13.7
		5.72	52.6			324	3.9
		5.75	9.2			338	2.8
^{219}Rn	3.96 s	6.43	7.5			271	10.8
		6.55	12.9			401.8	6.4
α		6.82	79.4				
$^{215}Po^{c}$	1.781 ms	7.39	100				
α							
^{211}Pb	36.1 m			0.97	1.5	405	3.8
				1.38	91.3	427	1.8
β^-						832	3.5
^{211}Bi	2.14 m	6.28	16.2	0.58	0.28	351	12.9
α 99.7%; β^- 0.276%		6.62	83.5				
^{211}Po	0.516 s	7.45	98.9				
^{207}Tl (^{211}Po α)	4.77 m			1.43	99.7	898	0.26
β^-							
^{207}Pb	Stable						

Source: NNDC, 1995.

[a] X rays, conversion electrons, and Auger electrons are not listed.

[b] Maximum beta energy.

[c] ^{215}Po has minor branching (0.00023%) by β^- emission to ^{215}At followed by α emission.

TABLE 6-5. ^{232}Th Series and Radiations of Yield Y_i Greater Than 1%[a]

		α		β^-		γ	
Nuclide	$T_{1/2}$	E (MeV)	Y_i (%)	E (MeV)[b]	Y_i (%)	E (keV)	Y_i (%)
^{232}Th	14.05×10^9 y	3.95	22.1				
α ↓		4.01	77.8				
^{228}Ra	5.75 y			0.02	40		
β^- ↓				0.04	60		
^{228}Ac[c]	6.15 h			0.45	2.6	209.3	3.88
				0.5	4.18	270.2	3.43
				0.61	8.1	328	2.95
β^- ↓				0.97	3.54	338.3	11.3
				1.02	5.6	463	4.44
				1.12	3	794.9	4.34
				1.17	31	835.7	1.68
				1.75	11.6	911.2	26.6
				2.08	10	964.8	5.11
						1588	3.3
^{228}Th	1.9131 y	5.34	28.2			84.4	1.27
α ↓		5.42	71.1				
^{224}Ra	3.62 d	5.45	5.06			241	3.97
α ↓		5.69	94.9				
^{220}Rn	55.6 s	6.29	99.9				
α ↓							
^{216}Po	0.145 s	6.78	100				
α ↓							
^{212}Pb	10.64 h			0.16	5.17	238.6	43.3
				0.34	82.5	300.1	3.28
β^- ↓				0.57	12.3		
^{212}Bi	60.55 m	6.05	25.13	0.63	1.87	727.3	6.58
α 35.9% ↙ ↘ β^- 64.1%		6.09	9.75	0.74	1.43	785.4	1.1
				1.52	4.36	1621	1.49
				2.25	55.5		
^{212}Po	0.299 μs	8.79	100				
^{208}Tl	3.053 m			1.03	3.1	277.4	6.3
				1.29	24.5	510.8	22.6
				1.52	21.8	583.2	84.5
β^- (^{208}Tl) ↓, α (^{212}Po) ↓				1.8	48.7	763.1	1.8
						860.6	12.4
						2615	99.2
^{208}Pb	Stable						

Source: NNDC, 1995.

[a] X rays, conversion electrons, and Auger electrons are not listed.

[b] Maximum beta energy.

[c] Only γ with yield greater than 2% listed.

TABLE 6-6. ^{237}Np Series and ^{241}Pu and ^{241}Am Precursors

		α		β^-		γ	
Nuclide	$T_{1/2}$	E (MeV)	Y_i (%)	E (MeV)[b]	Y_i (%)	E (keV)	Y_i (%)
^{241}Pu	14.35 y	3.95	0.0003	0.021	~100	149	0.0002
α ↓		4.9	0.002				
^{241}Am	432.2 y	5.44	10			26.3	2.4
		5.49	84.5			59.5	35.9
α ↓						26.3	2.4
^{237}Np[c]	2.14×10^6 y	4.64	6.18			29.4	15
		4.766	8			86.5	12.4
α ↓		4.771	47				
^{233}Pa	26.976 d			0.16	27.7	300	6.62
				0.17	16.4	312	38.6
β^- ↓				0.23	40	341	4.47
				0.26	17		
^{233}U	1.592×10^5 y	4.78	13.2			42.4	0.09
α ↓		4.82	84.4			97	0.3
^{229}Th	7340 y	4.82	9.3			31.5	1.19
		4.84	5			86.4	2.57
		4.85	56.2			137	1.18
α ↓		4.9	10.2			156	1.19
		4.97	5.97			194	4.41
		4.98	3.17			211	2.8
		5.05	6.6				
^{225}Ra	14.9 d			0.33	69.5	40	30
β^- ↓				0.37	30.5		
^{225}Ac	10.0 d	5.73	8			99.6	0.62
		5.79	18.1			157	0.36
α ↓		5.83	50.7			188	0.54
^{221}Fr	4.9 m	6.13	15.1			219	11.6
α ↓		6.34	83.4			411	0.14
^{217}At	0.0323 s	7.07	~100			259	0.06
α ↓							
^{213}Bi	45.59 m	5.87	1.94	0.98	31	293	0.43
β^- ↓		5.55	0.15	1.42	65.9	440	26.1
^{213}Po	4.2 μs	8.38	~100				
α ↓							
^{209}Pb	3.253 h			0.644	100		
β^- ↓							
^{209}Bi	Stable						

Source: NNDC, 1995.
[a]X rays, conversion electrons, and Auger electrons are not listed.
[b]Maximum beta energy.
[c]Beginning of ^{237}Np series.

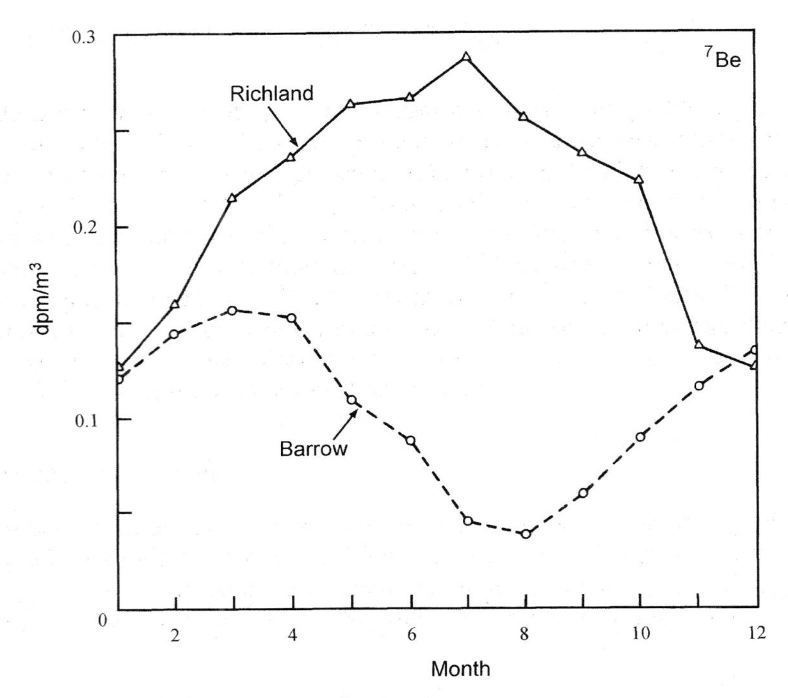

FIGURE 6-6. Variation of atmospheric ^{7}Be concentration by month and latitude. Richland, Washington is at 46°N and Barrow, Alaska at 70°N latitude.

however, the names of many of the nuclides are derived from their historical names [e.g., radium, radon (radium emanation), thorium (named after the god Thor), etc.].

Primordial sources of ^{237}Np no longer exist because its half-life is only 2.1 million years. Table 6-6, which contains the nuclides of the ^{237}Np series, is shown as starting with the radioactive transformation of ^{241}Pu to ^{241}Am, the two parent radionuclides produced in nuclear reactors that are now common sources of ^{237}Np. Were it not for large-scale uses of nuclear fission, ^{237}Np and its series products would be insignificant. There are notable similarities in radioactive transformations in the radioactive series and also some striking differences. The uranium, thorium, and actinium series each have an intermediate gaseous isotope of radon, and each ends in a stable isotope of lead. The neptunium series has no gaseous product and its stable end product is ^{209}Bi instead of an isotope of lead. Other common factors exist, especially in the sequences of transformation, as follows:

$$^{238}\mathrm{U}\text{---}\alpha\text{-}\beta\text{-}\beta\text{-}\alpha\text{-}\alpha\text{-}\alpha\text{-}\alpha\text{-}\alpha\text{-}\beta\text{-}\beta\text{-}\alpha\text{-}\beta\text{-}\beta\text{-}\alpha\text{--}\,^{206}\mathrm{Pb} \qquad \Delta Z = 10 \qquad \Delta A = 32$$

$$^{232}\mathrm{Th}\text{--}\alpha\text{-}\beta\text{-}\beta\alpha\text{-}\alpha\text{-}\alpha\text{-}\alpha\text{-}\beta\text{-}\beta\text{-}\alpha\text{----------}\,^{208}\mathrm{Pb} \qquad \Delta Z = 8 \qquad \Delta A = 24$$

$$^{235}\mathrm{U}\text{---}\alpha\text{-}\beta\text{-}\alpha\text{-}\beta\text{-}\alpha\text{-}\alpha\text{-}\alpha\text{-}\alpha\text{-}\beta\text{-}\alpha\text{-}\beta\text{--------}\,^{207}\mathrm{Pb} \qquad \Delta Z = 10 \qquad \Delta A = 28$$

$$^{237}\mathrm{Np}\text{--}\alpha\text{-}\beta\text{-}\alpha\text{-}\alpha\text{-}\beta\text{-}\alpha\text{-}\alpha\text{-}\alpha\text{-}\beta\text{-}\alpha\text{-}\beta\text{--------}\,^{209}\mathrm{Bi} \qquad \Delta Z = 10 \qquad \Delta A = 28$$

The long-lived parent of each series undergoes transformation by alpha-particle emission, which is followed by one or more beta transitions. Near the middle of each series there is a sequence of three to five alpha emissions, which are followed

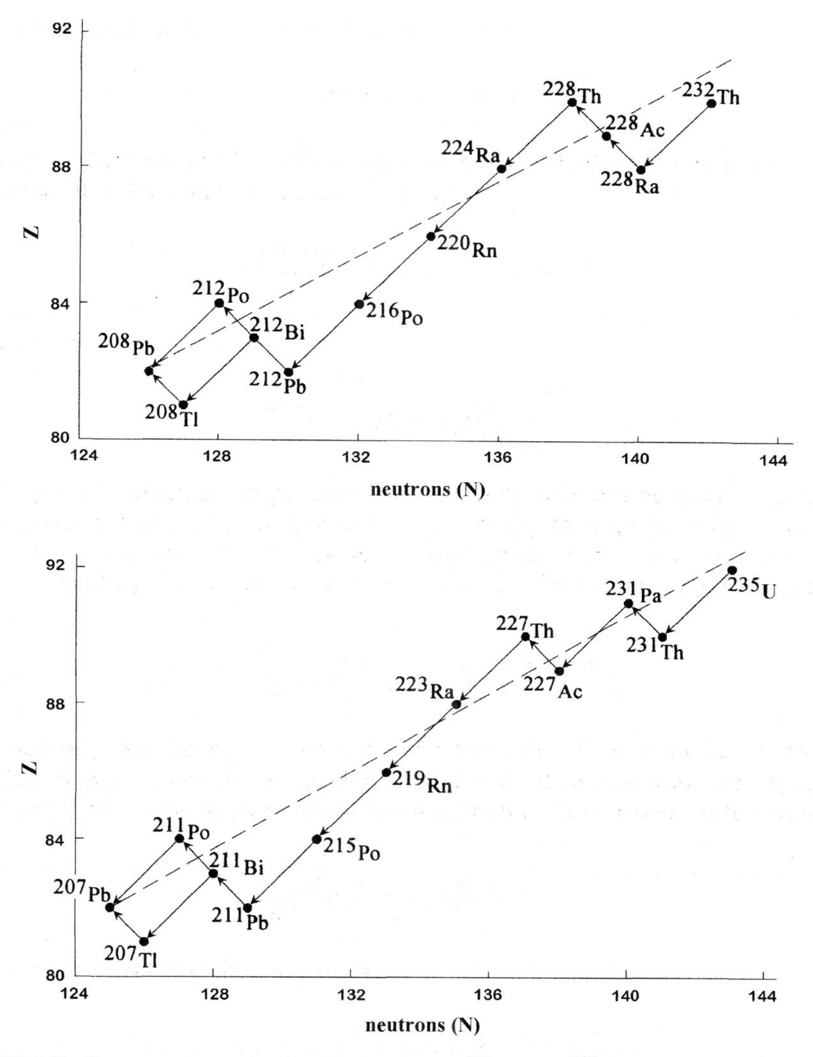

FIGURE 6-7. Comparative series transformation of ^{232}Th and ^{235}U relative to an extrapolated line of stability.

by one or two beta transformations; and there is at least one more alpha emission before stability is reached. Figure 6-7 illustrates these similarities graphically for the ^{232}Th and ^{235}U series using the same sequences as those in the chart of the nuclides. These patterns produce a zigzag path back and forth across a stability line (there are no stable nuclides above ^{209}Bi; thus such a line is only theoretical) extrapolated from the zigzag line of stable nuclides below ^{208}Pb or ^{209}Bi.

The zigzag patterns in the ^{238}U series shown in Figure 6-8 suggest some explanation for the types of transformation that occur. In the ^{238}U series the first alpha-particle emission produces a product that is quite a bit less stable and apparently farther away from the line. Beta emissions from ^{234}Th and ^{234}Pa bring the product

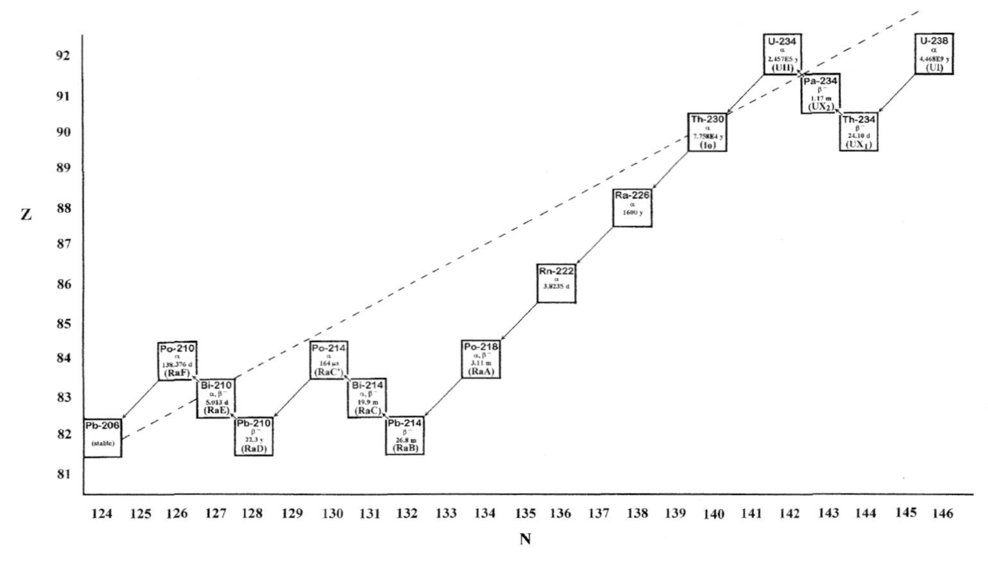

FIGURE 6-8. Uranium series with radioactive transformation products relative to an extrapolated line of stability (minor branching of ^{218}Po to ^{218}At and ^{214}Bi to ^{210}Tl not included).

nuclei back across the stability line, apparently in search of a stable endpoint. The resulting product is once again well away from the line of stability, and a sequence of five alpha-particle emissions occurs from ^{234}U to ^{214}Pb, again producing product nuclei that move steadily away from the line. This departure is reflected in the half-life sequence in the ^{238}U series: 2.5×10^5 y, 8×10^4 y, 1.6×10^3 y, 3.82 d, and 3.11 min. Beta-particle emissions then occur to bring the nuclei back towards the stability line, followed by alpha and beta emissions to ^{206}Pb. Somewhat similar sequences are observed in the other series (see Figure 6-7).

In practical terms, the uranium series contains several subseries in addition to ^{238}U itself. The more important ones are ^{230}Th, ^{226}Ra, ^{222}Rn, and ^{210}Pb and their subsequent products. ^{230}Th is strongly depleted from its precursors in seawater and enhanced in bottom sediments, thus it can also be used to date such sediments. It has also been found in raffinates from uranium processing, and as such, represents a separated source that produces ^{226}Ra and ^{222}Rn and their products. ^{226}Ra is often found separated from its precursors and is rarely found in sufficient concentrations to precipitate in the presence of anions, for which it has a strong affinity, particularly sulfate, although ^{226}Ra "hot spots" remain where uranium deposits have contacted sulfuric acid from natural oxidation of ferrous sulfide. Once released into natural waters, ^{226}Ra and other radium isotopes are mobile until scavenged or coprecipitated (e.g., the cementlike calcium carbonate "sinters" around some hot springs). Radium can also be separated from its precursors by precipitation and recrystallization as either the chloride or the bromide. When freshly prepared and free of its decay products, ^{226}Ra produces minimal gamma radiation (a 0.19 MeV gamma ray is emitted in 3.6% of ^{226}Ra transformations). However, as ^{222}Rn and its succeeding products achieve equilibrium, the gamma intensity increases substantially.

Radon-222 is also a subseries when separated from ^{226}Ra, as it often is because it is a noble gas and the only gaseous member of the ^{238}U series. It is produced from ^{226}Ra and grows into any radium source in a few days. Since it is a noble gas, it can migrate away from ^{226}Ra and its other long-lived parents or be readily separated from them, and when it becomes separated it is a subseries. Radon-222 can be removed from radium sources, and because of its high intrinsic specific activity can be concentrated into tubes and needles for implantation in malignant tissues for treatment of disease. For similar reasons, ^{222}Rn was obtained from ^{226}Ra generators by early researchers to be mixed with beryllium to produce neutrons. Radon-222 has a half-life of 3.82 d, and its principal transformation products are ^{218}Po, ^{214}Pb, ^{214}Bi, and ^{214}Po, which have half-lives ranging from 26.8 min to 164 μs. The gamma rays of ^{214}Bi have the highest yield and are the most energetic of the uranium series, which makes this subseries important with respect to external radiation.

Lead-210, a long-lived radon transformation product, heads the final subseries of ^{238}U. Lead-210 itself and its ^{210}Bi and ^{210}Po products can be observed in significant concentrations in the atmosphere. The stable end product of this subseries, and the end of the ^{238}U series, is ^{206}Pb.

The thorium series may also be considered in three subseries: ^{232}Th itself, ^{228}Ra, and ^{220}Rn. Thorium-232 is the least mobile of the series radionuclides. It exists naturally in the tetravalent state as a very stable oxide or in relatively inert silicate minerals, and it is strongly adsorbed on silicates. The subseries headed by ^{228}Ra yields ^{228}Ac, ^{228}Th, and ^{224}Ra, which are generally in radioactive equilibrium. The third subseries is headed by ^{220}Rn (thorium emanation, or thoron) which has a 56 s half-life and which quickly forms transformation products down to stable ^{208}Pb, the longest lived of which is 10.6-h ^{212}Pb. Although generally less important than ^{222}Rn in the uranium series, ^{220}Rn and its products can occasionally present radiation exposure situations if high concentrations of ^{228}Ra exist. An important product of ^{220}Rn (and ^{228}Ra itself because of the short half-life of ^{220}Rn) is ^{208}Tl, which emits 2.62 MeV gamma rays, which can produce significant gamma exposures from residues of thorium recovery.

SINGLY OCCURRING PRIMORDIAL RADIONUCLIDES

Naturally occurring primordial radioactive nuclides exist that are not members of any of the four series. Most of the major ones are listed in Table 6-7. Each of these nonseries nuclides has an extremely long half-life and a correspondingly low specific activity, which makes detection and identification of the emitted radiations very difficult. Some nuclides, now thought to be stable, may be added in the future because of detectable radioactivity. Bismuth-209, now considered to be stable and the endpoint of the neptunium series, has been on and off this list; it is currently believed to be stable ($T_{1/2} > 10^{18}$ y). So has ^{192}Pt, which is also now considered stable.

There is no obvious pattern to the distribution of long-lived nonseries radioactive nuclei. Some of the half-lives are surprisingly long in view of the energies available in the transitions. The double-beta transformation of ^{130}Te to ^{130}Xe is unique and has been used in connection with the properties of the neutrino. All of the half-lives

TABLE 6-7. Naturally Occurring Primordial Radionuclides

Nuclide	% Abundance	$T_{1/2}$ y	Emissions	Q (MeV)
^{40}K	0.0117	1.27×10^9	β^-,EC,γ	1.505, 1.311
^{50}V	0.25	1.4×10^{17}	β^-,EC,γ	2.208
^{87}Rb	27.25	4.9×10^{10}	β^-	0.283
^{113}Cd	12.22	9.3×10^{15}	β^-	0.316
^{115}In	95.7	4.4×10^{14}	β^-	0.495
^{123}Te	0.91	$> 1.3 \times 10^{13}$	EC	0.052
^{130}Te	33.87	1.25×10^{20}	$\beta^-\beta^-$	0.42
^{138}La	0.09	1.05×10^{11}	EC,β^-,γ	1.737, 1.044
^{142}Ce	11.13	5.0×10^{16}	α	?
^{144}Nd	23.8	2.1×10^{15}	α	1.905
^{147}Sm	15	1.06×10^{11}	α	2.31
^{152}Gd	0.2	1.08×10^{14}	α	2.205
^{174}Hf	0.162	2.0×10^{15}	α	2.496
^{176}Lu	2.59	3.8×10^{10}	β^-,γ	1.192
^{180}Ta	0.012	$> 1.2 \times 10^{15}$	EC,β^+,γ	0.853, 0.708
^{186}Os	1.58	2.0×10^{15}	α	2.822
^{187}Re	62.6	4.4×10^{10}	β^-	0.00264
^{190}Pt	0.01	6.5×10^{11}	α	3.249

Source: NNDC, 1995.

listed in Table 6-7 are compatible with the formation of the elements at about the time of the Big Bang, some 4.5 billion years ago.

The primordial radionuclides of ^{40}K and ^{87}Rb are of particular interest to radiation protection because of their presence in environmental media and their contribution to human exposure. Potassium-40 contributes about 40% of the exposure that humans receive from natural radiation. It is present in natural potassium with an isotopic abundance of 0.0117% and has a half-life of 1.28×10^9 y. It undergoes radioactive transformation both by electron capture (10.67%) and beta-particle emission (89.23%). The beta particle has a maximum energy of 1.312 MeV (average energy 0.51 MeV). Electron capture is followed by emission of a 1461 keV gamma ray in 10.7% of the transformations of ^{40}K, conversion electrons (0.3% of ^{40}K transformations), and the usual low-energy Auger electrons and characteristic x rays. The average elemental concentration of potassium in reference man is 2%, which produces annual doses of 14 mrem to bone surfaces, 17 mrem on average to soft tissue, and 27 mrem to red marrow (UN, 1982). It also contributes in a major way to external exposure due to an average soil concentration of 12 pCi/g. This concentration results in an annual whole-body dose equivalent due to external gamma radiation of 12 mrem (UN, 1982).

Rubidium-87 has a half-life of 4.8×10^{10} y and emits beta particles of maximum energy of 292 keV (average energy about 79 keV). Its natural isotopic abundance is 27.84% and the mass concentrations in reference man range from 6 ppm in the thyroid to 20 ppm in the testes (ICRP, 1987). It contributes an annual dose to body tissues of 1.4 mrem to bone-surface cells (UN, 1982) and an annual effective dose equivalent of about 0.6 mrem.

RADIOACTIVE ORES AND BY-PRODUCTS

Many ores are processed for their mineral content, which enhances either the concentration of the radioactive elements in process residues or increases their environmental mobility; these processes result in materials that are no longer purely "natural." The presence of uranium and/or thorium, ^{40}K, and other naturally radioactive materials in natural ores or feedstocks is often the result of the same geochemical conditions that concentrated the main mineral-bearing ores (e.g., uranium in bauxite and phosphate ore) or that led to separation of decay products such as radium into groundwaters for later concentration onto water treatment filters or pipes and equipment used in oil and gas recovery. These technologically enhanced sources of naturally occurring radioactive material represent radiation sources that can pose generally low-level radiation exposures of the public, or in some localized areas exposures significantly above the natural background.

Various sites contain residues of naturally occurring radionuclides, principally the series products of ^{238}U and ^{232}Th. For example, processing of uranium-rich ores for the uranium required for the Manhattan Project in World War II created residues containing thorium, radium, and other transformation products of uranium. These materials are characterized in radiation protection as NORM (naturally occurring radioactive material) or technologically enhanced NORM (TENORM). A similar categorization is NARM (naturally occurring and accelerator-produced radioactive material), which captures those materials not regulated under the Atomic Energy Act of 1954, as amended.

Enhanced levels of NORM may be difficult to discern because natural concentrations of ^{226}Ra, ^{40}K, and ^{232}Th and its products vary widely, especially in minerals and extracts, as shown in Table 6-8.

Resource Recovery

Resource recovery of metals and oil and gas has produced huge inventories of NORM radionuclides that are discarded because they are not of value or interest to the recovery industry. For example, vanadium mining in Colorado produced residues that contained uranium; these residues (or tailings) were later reprocessed for uranium, producing the usual uranium tailings with elevated concentrations of ^{230}Th and ^{226}Ra and their transformation products. Similar circumstances exist to varying degrees for aluminum, iron, zironium, and other metals.

Removing a mineral from ore can concentrate the residual radioactivity, especially if the ore is rich (i.e., it contains a high fraction of the metal of interest). This process of concentration is called *beneficiation*. For example, bauxite is rich in aluminum, and its removal yields a residue in which uranium and radium originally present in the ore can be concentrated to 10 to 20 pCi of ^{226}Ra per gram in the residues, or "red mud," so called because its iron content (also concentrated) has a red hue. Similarly, processing phosphate rock in furnaces to extract elemental phosphorus produces a vitrified waste called *slag*, which contains 10 to 60 pCi/g of ^{226}Ra. Slag has been commonly used as an aggregate in making roads, streets, pavements, residential structures, and buildings. Such uses of high-bulk materials from ore processes raise risk/benefit trade-offs of the values of such uses and whether exposure conditions exist that warrant attention. High volumes and relatively low, but

TABLE 6-8. Radiological Constituents of Selected Minerals and Extracts

Mineral	Nuclides	Concentration (pCi/g)
Aluminum ore	U	6.76
Bauxitic lime, soil	Th	0.8–3.5
	Ra	3–10
Tailings	Ra	20–30
Iron	U + products	Varies
	Th + products	Varies
Molybdenum	U + products	Varies
Tailings	U + products	Varies
Monazite	Th + products	4–8%
	U + products	150–500
Natural gas		
United States	^{222}Rn	0.1–500 pCi/L
Canada	^{222}Rn	0.01–1500 pCi/L
Scale	^{210}Pb	3–1500
Oil		
Brines	^{226}Ra	3–3000
Sludges	^{226}Ra	30–2000
Scales	^{226}Ra	300–100,000
Phosphate ore	U + products	3000–100,000
	Th + products	400–4000
	^{226}Ra	15–100
Potash	Th + products	Varies
	^{40}K	Varies
Rare earths	U + products	Varies
	Th + products	Varies
Vanadium	U + products	Varies
Zirconium	U	~ 100
	Th	~ 15
	Ra	100–200

Source: NCRP, 1993.

enhanced concentrations make such decisions challenging. For example, radium-contaminated soil from radium recovery in Montclair, New Jersey and Denver, Colorado caused higher-than-normal direct gamma radiation exposure levels. The use of elemental phosphorus slag to construct streets, roads, and parking lots in Soda Springs and Pocatello, Idaho has doubled the natural background radiation levels in some areas. In Mississippi and Louisiana, the use of pipes contaminated with radium scale in playgrounds and welding classes has resulted in some radiation exposures to students.

Uranium Ores

Prior to Becquerel's discovery, uranium had been mined for use as a coloring agent in the glass industry. Most of the material used in this way came from the Joachistimal mines in Czechoslovakia, which was also the source of pitchblende, a brown-black ore that contains as much as 60 to 70% uranium as the oxide U_3O_8, commonly called *yellowcake* because of its golden color when purified. The U.S. form is primarily yellow carnotite (hydrated potassium uranium vanadate), which was first processed for vanadium, then later to recover uranium when it became valuable.

The principal radioactive products associated with uranium residues are the tailings from milling the ores to extract uranium oxide. These tailings contain ^{230}Th and ^{226}Ra in concentrations directly related to the richness of uranium in the original ores. Radon gas is continually produced by ^{226}Ra in the tailings, which requires that they managed to preclude radon problems as well as ^{226}Ra exposures. Some milling processes contain a large fraction of ^{230}Th; for example, raffinate streams, which contained much of the ^{230}Th, were sluiced to separate areas for storage and/or further processing. These ^{230}Th residues represent minor radiation problems in themselves since ^{230}Th is a pure alpha emitter, but its radioactive transformation produces ^{226}Ra, which will slowly grow in to create a future source of external gamma radiation and radon gas. Modern management of uranium tailings is done under regulations that require stabilization for up to 1000 y in such a way that emissions of radon and its products are controlled. Older abandoned mill sites are being remediated to achieve the same requirements.

Water Treatment Sludge

Most water treatment wastes are believed to contain ^{226}Ra in concentrations comparable to those in typical soils. However, some water supply systems, primarily those relying on groundwater sources, may generate sludge with higher ^{226}Ra levels, especially if their process equipment is effective in removing naturally occurring radionuclides from the water. A typical drinking water source contains about 8 pCi/L of ^{226}Ra, which yields an average ^{226}Ra concentration of about 16 pCi/g in sludge after processing. About 700 water utilities in the United States generate and dispose of 300,000 metric tons (MT) of sludge, spent resin, and charcoal beds in landfills and lagoons or by application to agricultural fields.

Phosphate Industry Wastes

Phosphate deposits occurred by geochemical processes in ancient seas that were also effective in depositing uranium; thus some phosphate deposits, principally in Florida (see Figure 6-9), contain significant amounts of uranium and its radioactive progeny. Uranium in phosphate ores found in the United States ranges in concentration from 20 to 300 ppm (or about 7 to 100 pCi/g), while thorium occurs at essentially ambient background concentrations, between 1 and 5 ppm (or about 0.1 to 0.6 pCi/g).

Phosphate rock is processed to produce phosphoric acid and elemental phosphorus. Some 5 million MT of phosphate rock is used each year to produce fertilizer which contains 8 pCi/g of ^{226}Ra; thus 20 y of repeated fertilizer applications would increase the soil concentration by about 0.002 pCi/g above a natural soil concentration of 0.1 to 3 pCi/g. On the other hand, about 86% of the uranium and 70% of the thorium originally present in the ore are found in the phosphoric acid because the process used to extract phosphorus also entrains these elements.

Phosphogypsum is the principal waste by-product generated from the production of phosphoric acid. It is stored in waste piles called *stacks* and contains

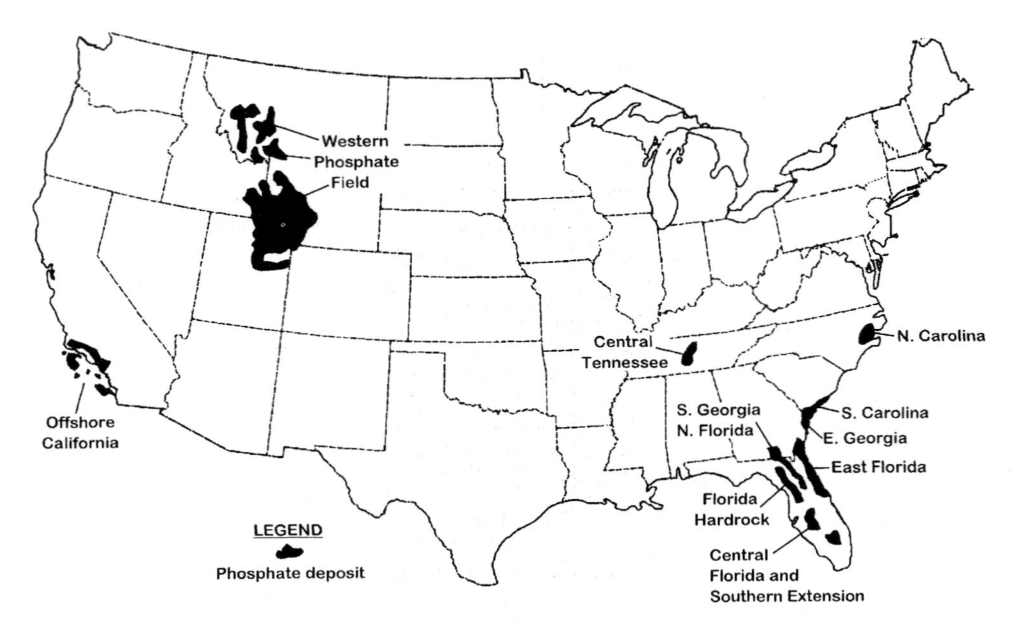

FIGURE 6-9. Major uraniferous phosphate deposits in the United States.

about 80% of the ^{226}Ra originally present in the phosphate rock. A huge inventory of phosphogypsum totaling some 8 billion MT now exists, which is being increased at at rate of about 40 million MT per year. Phosphogypsum stacks are huge, ranging from 2 to 300 ha (750 acres) in area and from 3 to 60 m high. They are typically covered with water in ponds, beaches, and ditches. Sixty-three large stacks and three relatively smaller ones exist, mostly in Florida. Phosphogypsum could be used in wallboard or applied as a soil conditioner, but such use is very small because it contains radium, which most consumers would expect to be removed.

Elemental Phosphorus

About 10% of mined phosphate ore goes into production of elemental phosphorus, which is used in the production of high-grade phosphoric acid, phosphate-based detergents, and organic chemicals used in cleaners, foods, baking powder, dentifrice, animal feed, and so on. There are only eight elemental phosphorus plants in the United States, and these are located in Florida, Idaho, Montana, and Tennessee; only three are active, two in Idaho and one in Montana. Production of elemental phosphorus has declined from a peak of 4 million MT in 1979 to 310,000 MT in 1988, a trend that is expected to continue due to rising energy costs and decreasing demand.

Slag is the principal waste by-product of elemental phosphorus, and ferrophosphorus is also a by-product. Because of furnace temperatures, the slag is a glassy like material that seals the radionuclides in a vitrified matrix which reduces their leachability. Some 90% of phosphate rock ore processed into elemental phosphorus ends up in slag, and large inventories (some 200 to 400 million MT) of slag are stored at operating and closed elemental phosphorus plants. Some of this inventory has, until recently, been used in road building, paving of airport runways and streets, and backfill which may cause exposure to the public due to external γ-radiation (radon emanation is minimal because of the glassy matrix of slag).

Neptunium Series Radionuclides

Primordial sources of ^{237}Np ($t_{1/2} = 2.14 \times 10^6$ y) have long since decayed away; however, once produced it can be a challenge to provide long-term management as a waste product. Two important precursors of ^{237}Np are ^{241}Pu and ^{241}Am, both of which are fairly abundant because ^{241}Pu is produced by neutron activation of ^{239}Pu through ^{240}Pu in nuclear reactors. ^{241}Pu is transformed by β-particle emission to ^{241}Am, which emits alpha particles to become ^{237}Np. ^{241}Am is used in smoke detectors, and if disposed in landfills can accumulate a source of ^{237}Np and its series products.

Manhattan Project Wastes

During World War II a major effort, called the Manhattan Project, required considerable amounts of uranium to develop the atomic bomb. The African Metals Co. provided the U.S. Government over 1200 tons of pitchblende from the Belgian Congo in the early 1940s which contained as much as 65% uranium. This material was processed at a number of nongovernment sites in the 1940s, and to some extent into the 1950s and 1960s. As a result, many private properties, including former laboratories, processing facilities, and waste disposal facilities, contain residues of naturally radioactive materials from storage, sampling, assaying, processing, and metal machining. Over 2 million cubic yards of soil and soil-like material that contain various concentrations of ^{238}U, ^{234}U, ^{230}Th, ^{226}Ra, ^{232}Th, ^{228}Ra, and their associated transformation products remain at these sites. In some cases, contaminated material from these properties spread to surrounding properties, including many residential properties, through erosion and runoff or improper management. Most of the radioactivity concentrations are less than 50 pCi/g; however, some materials contain thousands of pCi/g of ^{226}Ra or ^{230}Th.

The major radionuclides in Manhattan Project residues are ^{232}Th and ^{226}Ra since the parent ores were processed to remove uranium which was scarce and very valuable for development of nuclear weapons. It is also fortunate that African Metals Co. required the U.S. government to remove most of the ^{226}Ra, which was considered valuable at the time; it is equally unfortunate that the ^{226}Ra materials (labeled K-65) were poorly handled after African Metals Co. decided it no longer wanted them. Thorium-230 transforms into ^{226}Ra, which has a 1600-y half-life; thus if the ^{226}Ra has been removed (as it was in the Manhattan Project), the residues pose little exposure because ^{226}Ra and its gamma-emitting transformation products are absent and ^{230}Th by itself does not produce significant gamma radiation. However, after several hundred years, ^{226}Ra and its radioactive products will grow back in.

The Belgian Congo pitchblende also contained significant amounts of ^{235}U and its transformation products even though the abundance of ^{235}U in natural uranium is only 0.72%. The richness of the ores accounts for two other interesting circumstances associated with the process residues. Whereas ^{222}Rn is effectively a

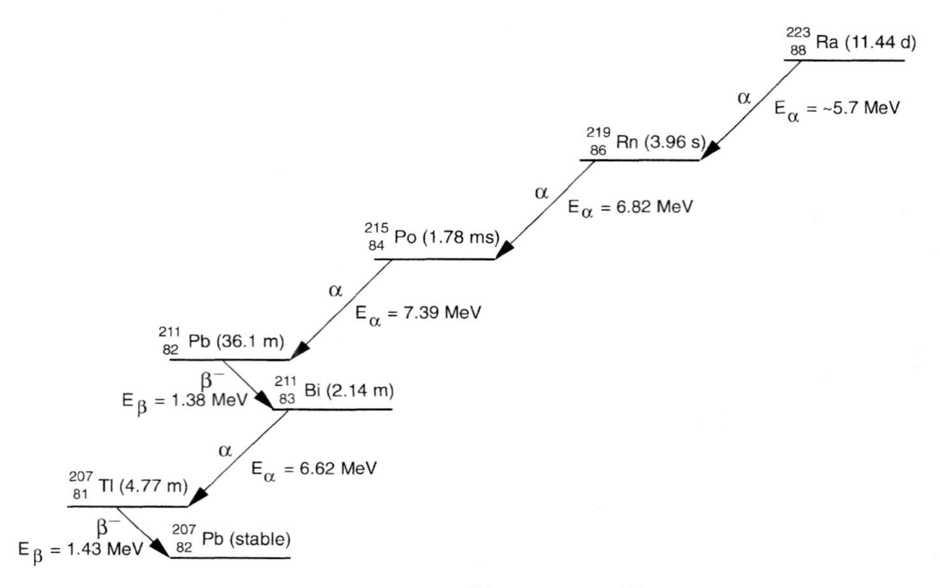

FIGURE 6-10. Radioactive transformation of ^{223}Ra to stable ^{207}Pb, a subseries of the actinide (^{235}U) series.

subseries in the ^{238}U series, ^{219}Rn, its counterpart in the actinium (^{235}U) series, is rarely observed in nature because the concentration of ^{235}U in most ores is so low and the half-life of ^{219}Rn is so short (3.96 s). However, ^{219}Rn is noticeably present in the process residues left over from the Belgian Congo ores; it is however not a health threat due to its short half-life.

In a similar way, ^{227}Ac, a transformation product in the actinium series, is also present in significant concentrations, due to the large amounts of ^{235}U in the rich ores. This nuclide can be separated and used in interesting ways. Since it is fairly long lived ($t_{1/2}$ = 21.773 y), ^{227}Ac was used for a period as the alpha emitter of choice to be mixed with beryllium to produce neutron initiators for nuclear weapons. It has recently found application as a generator of ^{223}Ra ($t_{1/2}$ = 11.435 d), which can be tagged onto a monoclonal antibody or injected as a colloid into a tumor site. Purified ^{227}Ac is captured on a resin column and the ^{233}Ra product that grows in after a few weeks is in turn eluted off for processing. As such, ^{223}Ra and its subsequent alpha-emitting transformation products, which rapidly reach radioactive equilibrium, can deliver a whopping radiation dose to cancer cells. Figure 6-10 shows schematically that each transformation of ^{223}Ra can yield about 28 MeV of deposited energy if it can be assumed that ^{219}Rn does not diffuse away from the cancer site.

Although ^{227}Ac exists as a subseries of the ^{235}U series, it is perhaps more practical to produce it by an (n, γ) reaction with ^{226}Ra in reactors rather than by separating it from K-65 or other actinium ore residues. Such production requires high neutron fluxes, which are available in large reactors, and neutron activation to ^{227}Ac may be a useful way to convert considerable amounts of ^{226}Ra to more useful products; ^{226}Ra itself is now problematic since earlier uses of the material have been replaced by artificially produced radionuclides.

Thorium Ores

Thorium ores have been processed since about 1900 to recover rare earth elements and thorium, a soft silvery metal used in specialized applications such as welding electrodes, gas lantern mantles, ultraviolet photoelectric cells, and in certain glasses and glazes. Thorium-232 is also a "fertile" material that can be transmuted by neutron irradiation into ^{233}U, a fissionable material.

Thorium-232 is a naturally occurring radionuclide which heads the thorium series of radioactive transformation products (see Table 6-4). Concentrations of 1 pCi/g are typical of natural soils and minerals, but some areas of the world contain significant thorium; for example, Monazite sands in Brazil and India contain 4 to 8% thorium. Thorium extraction does not appear to have been as thorough as that for uranium, or perhaps not as much care was exercised because sources are more available. And, of course, processing for rare earths would not preferentially remove the thorium, which would become a waste by-product. Regardless, residues from processing monazite and other thorium-rich ores may still contain significant amounts of ^{232}Th, and huge inventories of these residual materials exist. Thorium-232 emits alpha particles to become ^{228}Ra ($t_{1/2}$ = 5.75 y), which will exist in secular equilibrium with ^{232}Th some 40 y after ^{232}Th is purified. Radium-228 has a number of short-lived, gamma-emitting decay products; thus, even if radium is removed in processing, ^{232}Th will produce an external gamma flux after a few decades. Concentrations of 5 pCi/g of ^{232}Th in equilibrium with its transformation products and spread over a large area with no soil cover will produce an estimated exposure of 15 mrem/y to an average person standing on the area.

Residues that contain significant concentrations of ^{232}Th represent long-term radiological source terms because each of the transformation products will achieve equilibrium in about 40 y and will continue indefinitely because they are driven by the ^{232}Th parent with a half-life of 14 billion years. The series transformation products emit significant gamma radiation, primarily from ^{208}Tl at 2.6 MeV (see Table 6-4). If, however, ^{232}Th were removed, the remaining products would constitute a subseries headed by ^{228}Ra ($T_{1/2}$ = 5.75 y) and through series decay would be effectively eliminated in about 40 years.

Whereas radon exposures can be significant for uranium wastes if elevated ^{226}Ra concentrations exist in soil and a structure is present, ^{220}Rn (thoron) in the ^{232}Th series is of minor significance because it has such a short (56-s) half-life. In general, concentrations of thoron are quite low and generally nondetectable.

RADIOACTIVITY DATING

Carbon Dating

Carbon dating of objects that were previously alive depends on an assumption that cosmic ray production of ^{14}C and its natural distribution has remained constant throughout the time interval being considered. In general, this assumption is valid, with the exception of the past 150 years or so in which the ^{14}C pool has been perturbed by burning fossil fuels and the atmospheric testing of nuclear weapons.

Cosmic ray interactions in the atmosphere produce copious quantities of neutrons with MeV energies. Oxygen in the atmosphere is essentially transparent to these neutrons, but they interact readily with ^{14}N to form ^{14}C through (n,p) interactions about 99% of the time and tritium in the remaining 1% by (n,t) reactions. It is well established that the neutrons that produce these interactions are secondary products of cosmic ray interactions in the atmosphere for two reasons: (1) since free neutrons are radioactive with a life of only about 12 min, they would not last long enough to have come from outer space; and (2) their concentration rises up to an altitude of about 40,000 ft and then decreases.

Atmospheric ^{14}C exists primarily as $^{14}CO_2$ which diffuses throughout the atmosphere and comes into equilibrium with the carbonate pool of the earth; thus it will be acquired by any living organism that metabolizes carbon. While alive, the organisms will contain carbon in a constant $^{14}C/^{12}C$ ratio, which, before the advent of atmospheric testing of nuclear devices, was about 15.3 d/m per gram of carbon. When an organism (e.g., a plant or animal) dies, ^{14}C incorporation ceases and the ^{14}C activity per gram of carbon will then decrease with a half-life of 5730 y; thus measurement of the activity per gram can be used to fix the time of death. Organic materials can be dated by ^{14}C measurements to about 50,000 years because of the half-life of ^{14}C and the concentration of carbon in materials of interest.

The accuracy of radiocarbon dating is influenced by human activity, principally the fossil-fuel effect (or Suess effect) and ^{14}C production by nuclear weapon tests. Since about 1850, the combustion of coal and oil has released large quantities of ^{14}C-free CO_2 into the environment since coal and oil are millions of years old and the ^{14}C in these hydrocarbons has long since decayed away. Injecting this "old" carbon into the atmosphere dilutes the radiocarbon activity of recently grown samples by about 2% compared to samples prior to 1850. Early dating laboratories used recently grown samples as a standard of ^{14}C activity before the importance of the Suess effect was recognized, and dates based on such recent samples are too young by 100 to 200 years. New standards prepared by the U.S. National Institute of Standards and Technology (NIST) and Heidelberg University should remove these inaccuracies. Offsetting somewhat the fossil-fuel effect since 1954 has been the production of ^{14}C by hydrogen bomb explosions. The radioactivity of plants grown since that time is almost 1% higher than in the pre-bomb era.

Since ^{14}C is heavier than ^{12}C, it is concentrated somewhat higher in ocean carbonate in the exchange reaction between the atmosphere and the ocean. The ^{14}C concentration in terrestrial plant life resulting from the photosynthesis of atmospheric carbon dioxide is lower, and modern wood should be about 4% lower in ^{14}C than air. Marine plants, however, are less affected than wood because they are fed by bicarbonate already enriched in ^{14}C relative to atmospheric $^{14}CO_2$. Also modern carbonate is enriched in ^{14}C, and there is a calculated 5% ^{14}C excess in the shell of marine organisms compared with wood. This 5% enrichment means that the date for old shells would be 400 years older relative to modern wood.

Dating by Primordial Radionuclides

For ages ranging in millions of years, several methods have been developed that are based upon the radioactive transformation of elements occurring naturally in the earth's crust at its formation. Of these, the most useful are:

^{238}U ($T_{1/2}$ = 4.47 billion years) to ^{206}Pb
^{232}Th ($T_{1/2}$ = 14.0 billion years) to ^{208}Pb
^{230}Th ($T_{1/2}$ = 75,380 years) to ^{226}Ra
^{87}Rb ($T_{1/2}$ = 48 billion years) to ^{87}Sr
^{40}K ($T_{1/2}$ = 1.28 billion years) to ^{40}Ar

The age of a rock sample can be determined by measuring the relative amounts of the primordial "parent" radionuclide and its related transformation product, usually by means of a mass spectrometer. For example, the greater the amount of ^{206}Pb relative to ^{238}U in a sample, the greater is the age of that sample. This was one of the earliest applications of radioactive dating because the complete transformation of ^{238}U is equivalent to

$$^{238}\mathrm{U} \rightarrow {}^{206}\mathrm{Pb} + 8\alpha + 6\beta + \gamma\text{'s}$$

The longest intermediate half-life is 75,380 y ^{230}Th; however, after a few million years, the holdup at this point will be negligible compared to the amount of ^{206}Pb that has been formed. If one can be sure that no ^{206}Pb was formed by other processes, an assay for the two nuclides in question will serve to determine the time at which the ^{238}U was incorporated into the mineral. Times determined by this method agree well with those obtained from purely geological considerations. An assay for ^{4}He may also be used when there is some doubt as to the radiogenic purity of the ^{206}Pb; however, helium can diffuse out of porous structures and caution is in order when age determinations are based on ^{4}He gas analyses.

Potassium–Argon Dating

Radioactive dating can also be done with ^{40}Ar, a radioactive transformation product of ^{40}K, a primordial radionuclide, which undergoes radioactive transformation by two competing modes:

$$^{40}\mathrm{K} \rightarrow {}^{40}\mathrm{Ca} + \beta^- + \overline{\nu}$$

$$^{40}\mathrm{K} + \mathrm{e}^- \rightarrow {}^{40}\mathrm{Ar} + \gamma$$

Eighty-nine percent of the transformations of ^{40}K are by beta emission, and 11% produce ^{40}Ar by electron capture. Both of the product elements are found in nature with well-established abundances. If radiogenic ^{40}Ar is present, measurement with a mass spectrometer will show an abnormal $^{39}Ar/^{40}Ar$ ratio, and the excess ^{40}Ar can be attributed to transformation of ^{40}K.

The potassium–argon method has been used successfully to date samples up to about 2 million years old. Volcanic ash beds interlayered with fossils are ideal sources because they can be defined by stratigraphy. The potassium-bearing mineral laid down by molten lava would contain no ^{40}Ar since, as a gas, it would have boiled away. After the lava layer cooled, however, all ^{40}Ar atoms produced by ^{40}K transformation would be entrapped in the matrix. The number of ^{40}Ar atoms locked therein is a direct measure of the time required to produce them.

Use of the potassium–argon method relies on the ratio of ^{40}K atoms to ^{40}Ar atoms rather than the more abundant yield of ^{40}Ca atoms because ^{40}Ar atoms can be distinguished from other isotopes of argon. It is nearly impossible to distinguish ^{40}Ca from naturally occurring ^{40}Ca atoms. The method is a two-step process. First, it is necessary to remove surface-layer ^{40}Ar atoms that may have been entrained from the air since only the ^{40}Ar atoms within the sample are wanted. These are removed by heating the sample in a vacuum furnace at about 750°F for 12 to 48 h. After the surface-layer Ar is removed, the sample is heated to about 2200°F to melt the rock and release the entrapped ^{40}Ar atoms from inside the rock matrix. These are absorbed on porous charcoal cooled to −40°F and their number is determined by mass spectrometry or neutron activation analysis. The mass spectrometer uses a magnetic field to separate the lighter and heavier argon atoms into separate circular paths and aims them at an electronic target where they are registered. A separate sample of the rock is analyzed for the gamma-emission rate to quantify ^{40}K, and the relative amounts of each are used to date the sample.

The potassium–argon technique requires separate analyses of potassium in a sample and the entrapped argon. A modification of the technique called argon–argon dating is to place the sample in a flux of fast neutrons to produce ^{39}Ar by (n,p) interactions with ^{39}K in natural potassium at 93.3% abundance. The rock sample is then heated, the argon collected, and a mass spectrogram ratio of ^{39}Ar/^{40}Ar obtained. After suitable correction for irradiation time, depletion of ^{40}Ar to ^{41}Ar by neutron activation, and fractional transformation of ^{40}K by electron capture to ^{40}Ar, the age of the sample is related directly to the argon ratio. The method can be used to test various layers of rock by differential heating and has the advantage that experimental errors cancel out in determining the ^{39}Ar/^{40}Ar ratio, which can be done precisely with mass spectroscopy.

Ionium (^{230}Th) Method

The ionium, or ^{230}Th ($T_{1/2}$ = 75,380 y), method can be used to date some deep ocean sediments. Ionium is formed continually in seawater and in rocks by the radioactive transformation of uranium (see Table 6-8). The chemistry of the ocean causes ^{230}Th to be deposited preferentially in sediments where its radioactivity decreases according to its own half-life rather than the half-life of uranium which remains in the seawater. The transformation product of ^{230}Th is ^{226}Ra ($T_{1/2}$ = 1600 y); thus, measuring the radium concentration, which can be done with gamma spectroscopy, yields the amount of ^{230}Th. The ionium method can be used alone only if the sediment has been deposited in an undisturbed condition and is of uniform mineralogical composition over its thickness; however, this restriction can be overcome by measuring the activity of ^{231}Pa, a transformation product of ^{235}U, which is also deposited preferentially along with ionium from seawater. Since its half-life is only 38,800 years, the ratio of the activities of ^{231}Pa and ^{230}Th is a measure of the age of the sediment, independent of the rate of deposition.

Lead-210 Dating of Sediments

Lead-210, with a half-life of 22.6 y, can be used to date recent sediments that have remained relatively stable, for example in lakes and estuaries. ^{210}Pb is the end product of the relatively short-lived ^{222}Rn decay series. Radon-222 ($T_{1/2}$ = 3.82 d) is fairly uniformly dispersed within the atmosphere from decay of uranium and radium in soils and minerals, and produces a relatively steady ubiquitous fallout of long-lived ^{210}Pb, which is transferred from the atmosphere to the lithosphere in about a month. If the buildup of sediment layers can be assumed to be uniform and steady, a not unreasonable assumption, the activity of ^{210}Pb will decrease with depth in sediments, allowing the age of a given sediment layer to be estimated. This dating method is based on the assumption that there is a uniform flux of ^{210}Pb to the sediments and that the sediments are not disturbed after deposition occurs. Further, core samples need to be collected carefully to prevent commingling of sediment layers, and if significant amounts of ^{226}Ra exist in the sediment, the samples need to be corrected for the in situ formation of ^{210}Pb from its transformation.

RADON AND ITS PROGENY

Radon is a radioactive transformation product of ^{238}U, ^{235}U, and ^{232}Th, and it exists in various concentrations in all soils and minerals. Radon-220, or thoron from the ^{232}Th series, and ^{219}Rn from ^{235}U have very short half-lives (55.6 s and 3.96 s, respectively), and they are of minor significance compared to ^{222}Rn in the ^{238}U series. Uranium has a half-life of 4.5 billion years, and its intermediate transformation products, ^{230}Th and ^{226}Ra, the immediate parent of ^{222}Rn, have half-lives of 75,380 y and 1600 y, respectively. In natural soils, ^{230}Th and ^{226}Ra are in radioactive equilibrium with uranium; thus a perpetual source of radon exists naturally even though its half-life is only 3.82 d. Even if ^{230}Th and ^{226}Ra are separated from uranium, each (especially, ^{230}Th), will still represent long-term sources of radon.

Whereas uranium and its intermediate products are solids and remain in the soils and rocks where they originate, ^{222}Rn is a radioactive noble gas that migrates through soil to zones of low pressure such as homes. Its 3.82-day half-life is long enough for it to diffuse into and build up in homes unless they are constructed in ways that preclude entry of radon gas, or provisions are made to remove the radon. Once radon accumulates in a home, it will undergo radioactive transformation; however, the resulting transformation products are no longer gases but are solid particles which, due to an electrostatic charge, become attached to dust particles that are inhaled by occupants, or the particles can be inhaled directly. Because they are electrically charged, the particles readily deposit in the lung, and since they have half-lives on the order of minutes or seconds, their transformation energy is almost certain to be deposited in lung tissue. Radon transformation products or progeny (commonly called daughters) emit alpha particles with energies ranging from 6 to 7.69 MeV, and because alpha particles (as helium nuclei) are massive and highly charged, this energy is delivered in a huge jolt to the cellular structure of the surface of the bronchi and the lung, which damages and kills these cells. The body can tolerate and replace killed cells, but damaged lung tissue cells can replicate, and cellular defects may eventually lead to lung cancer.

Unlike many pollutants which are controlled by standards, lung cancer has been observed directly in humans exposed to radon. Over 300 years ago, before radioactivity was known to science, it was common for feldspar miners in Germany and Czechoslovakia to contract and die from a mysterious disease—it is now known that

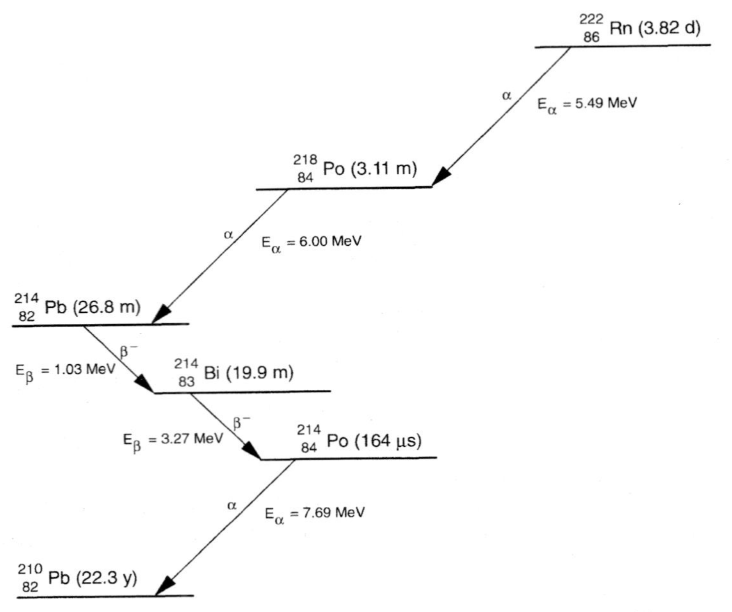

FIGURE 6-11. Radioactive series transformation of ^{222}Rn to long-lived ^{210}Pb (minor branching of ^{218}Po to ^{218}At and ^{214}Bi to ^{210}Th not included).

this mysterious disease was lung cancer caused by the short-lived alpha-emitting transformation products of radon continuously emanating from the uranium in the rocks. Similariy, in the United States and Canada there is a direct epidemiological association between uranium miner exposure to radon and lung cancer, evidence so compelling that vigorous controls have been used to reduce the concentrations of radon and its transformation products in uranium mines.

Radon Subseries

The radioactive transformation of radon and its radioactive products is a practical example of series decay. As shown in Figure 6-11, ^{222}Rn undergoes transformation by alpha-particle emission to produce ^{218}Po (RaA), which in turn emits 6.0 MeV alpha particles with a half-life of 3.11 min to ^{214}Pb (RaB). Beta-particle emissions from RaB (^{214}Pb, $T_{1/2}$ = 26.8 min) and ^{214}Bi (RaC, $T_{1/2}$ = 19.9 min) produce ^{214}Po (RaC′), which quickly ($T_{1/2}$ = 164 μs) produces the end product ^{210}Pb (RaD) by emission of 7.69 MeV alpha particles. The alpha transformation of RaC′ is effectively an isomeric alpha transformation of RaC because it occurs so quickly after RaC′ is formed. The end product ^{210}Pb (RaD) is effectively stable with a half-life of 22.3 y and is treated as such in most calculations of radon and its progeny. Properties of this important segment of the uranium series are shown in simplified form in Figure 6-11 and Table 6-9.

The Bateman equations for series transformation can be used to determine the activity and the number of atoms of each of the products of radon through ^{214}Po (RaC′). Since RaC′ is so short lived, its activity will be exactly the same as that of RaC, so determination of the amounts of the primary energy-emitting products

TABLE 6-9. Radon and Its Radioactive Products

Name	Isotope	Radiation	$T_{1/2}$	Atoms in 100 pCi
Radon	^{222}Rn	α	3.82 d	1,770,000
Radium A	^{218}Po	α	3.11 min	996
Radium B	^{214}Pb	β, γ	26.8 min	8,583
Radium C	^{214}Bi	β, γ	19.9 min	6,374
Radium C′	^{214}Po	α	164 μs	0.0008
Radium D	^{214}Pb	β	22.3 y	3.7×10^9

as a function of time and radon concentration only involves equations for the first four members of a radioactive series, starting with ^{222}Rn. The activity of each of the transformation products for time t in minutes is

- For Rn:

$$A_{Rn} = A_1^0$$

- For RaA (A_2):

$$A_2(t) = A_1^0(1 - e^{-t/4.49})$$

- For RaB (A_3):

$$A_3(t) = A_1^0(1 + 0.1313e^{-t/4.49} - 1.1313e^{-t/38.66})$$

- For RaC (A_4):

$$A_4(t) = A_1^0(1 - 0.0243e^{-t/4.49} - 4.394e^{-t/38.66} + 3.4183e^{-t/28.71})$$

- For RaC′ (A_5):

$$A_5 = A_4(t)$$

Figure 6-12 shows a plot of the *activity* of each radon transformation product obtained from these equations. The activity of each product approaches equilibrium with the ^{222}Rn activity when t is about 3 h or more.

A total quantity of 100 pCi of ^{222}Rn in 1 L of standard air contains 1.77×10^6 atoms (N_1^0), which will remain essentially unchanged over a period of 3 h or so, the period of interest for ^{222}Rn calculations. When the numerical values of the half-lives of Rn, RaA, RaB, and RaC are substituted into the Bateman equations, the number of atoms of each of the products at any time t is obtained by dividing the activity by λ for each product; that is, the number of atoms of Rn ($N_{Rn} = N_1^0$) is

$$N_{Rn} = N_1^0$$

The number of atoms of RaA, $N_2(t)$, is

$$N_2(t) = N_1^0(5.657 \times 10^{-4}e^{-t/7936} - 5.657 \times 10^{-4}e^{-t/4.49}) \tag{6-1}$$

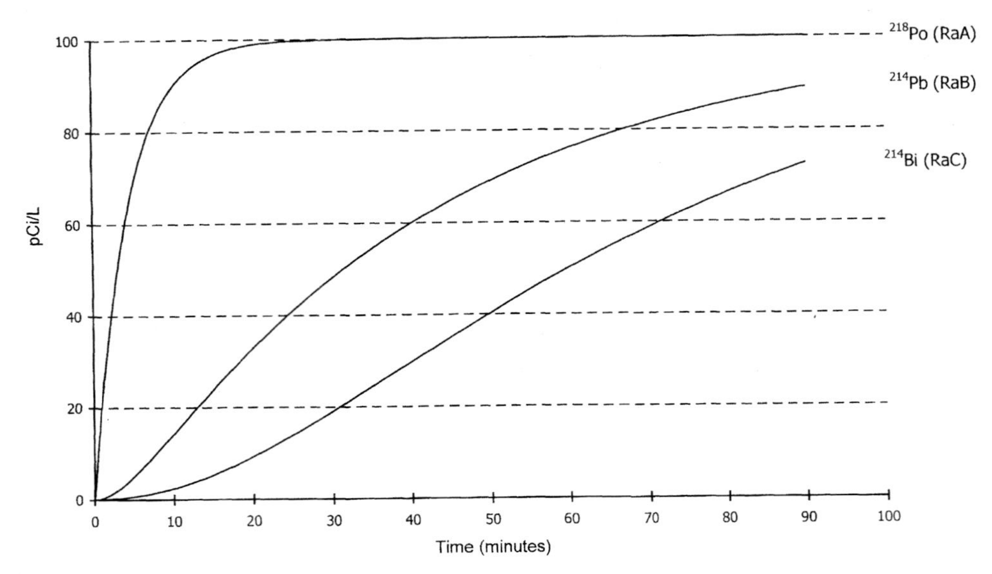

FIGURE 6-12. Ingrowth of activity of ^{222}Rn transformation products from 100 pCi of ^{222}Rn.

of RaB, $N_4(t)$,

$$N_3(t) = N_1^0(4.875 \times 10^{-3}e^{-t/7936} + 6.4 \times 10^{-4}e^{-t/4.49} - 5.547 \times 10^{-3}e^{-t/38.66}) \tag{6-2}$$

and of RaC, $N_4(t)$,

$$N_4(t) = N_1^0(3.6176 \times 10^{-3}e^{-t/7936} - 8.81 \times 10^{-5}e^{-t/4.49} - 1.59 \times 10^{-2}e^{-t/38.66} + 1.237 \times 10^{-2}e^{-t/28.71}) \tag{6-3}$$

Since RaC′ can be thought of as an alpha-isomeric state of RaC, $N_4(t)$, the number of atoms that produce RaC′ emissions, $N_5(t)$, will be the same as $N_4(t)$:

$$N_5(t) = N_4(t) \tag{6-4}$$

Working Level for Radon Progeny

The *working level* (WL) has been defined to deal with the special conditions of exposure to radon and its decay products. It was first defined for exposure of uranium miners as 100 pCi of ^{222}Rn in 1 L of standard air in equilibrium with its transformation products RaA, RaB, RaC, and RaC′; however, this condition rarely exists in working environments or in homes. Since radiation exposure of the human lung is caused by deposition of alpha-particle energy from the radioactive transformation of the particulate atoms of RaA, RaB, RaC, and RaC′ inhaled into and deposited

TABLE 6-10. Determination of Ultimate Alpha Energy Due to a Concentration of 100 pCi of Radon in Equilibrium with Its Four Principal Decay Products, or One Working Level

Nuclide	MeV/t	$T_{1/2}$	Atoms in 100 pCi	α Energy per Atom (MeV)	Total Energy (MeV/100 pCi)	Energy Fraction
^{222}Rn	5.5	3.82 d	1.77×10^6	Excluded	None	None
^{218}Po (RaA)	6	3.11 min	996	13.69	1.36×10^4	0.11
^{214}Pb (RaB)	0^a	26.8 min	8583	7.69	6.60×10^4	0.51
^{214}Bi (RaC)	0^a	19.9 min	6374	7.69	4.90×10^4	0.38
^{214}Po (RaC′)	7.69	164 μs	0.0008	7.69	~ 0	~ 0
				Total	1.3×10^5	1

[a] β and γ emissions excluded in working level because of minimal energy deposition in the lung.

in the lung, and not ^{222}Rn itself, the working level is defined as any combination of the short-lived decay products of radon (RaA, RaB, RaC, and RaC′) in 1 L of air that will result in the ultimate emission by them of 1.3×10^5 MeV of alpha-ray energy (to lung tissue).

Radon is not deposited in the lung because it is an inert gas, and since it does not interact with lung tissue, it is quickly exhaled. Beta-particle emissions and gamma rays emitted by RaB and RaC make a negligible contribution to the amount of energy deposited in the lung when compared to the alpha particles emitted by RaA and RaC′.

Calculation of the working level requires first a determination of the number of atoms of each of the radon transformation products and the ultimate alpha energy that each of the product atoms will produce as it undergoes radioactive series transformation to the end product ^{210}Pb. The number of atoms of each is directly related to the "age" of the radon in the air being considered. For 100 pCi of ^{222}Rn in equilibrium with its transformation products, the ultimate alpha energy emitted by the number of atoms at equilibrium will be 1.3×10^5 MeV, as shown in Table 6-10.

The data in Table 6-10 (for 100 pCi of ^{222}Rn in equilibrium with its progeny) show that each atom of RaA can deliver not only its own characteristic 6.00 MeV energy but also the 7.69 MeV alpha emission from RaC′ since each RaA atom will transition through RaB and RaC and eventually into RaC′ as well (see Figure 6-11). Hence the "ultimate" or "potential" alpha energy associated with each atom of RaA is $6.00 + 7.69 = 13.69$ MeV. Similarly, every atom of RaB and RaC, even though they are beta emitters, will undergo transformation to RaC′, which will emit 7.69 MeV of alpha energy when it rapidly undergoes radioactive transformation.

Table 6-10 also shows that RaA, in equilibrium with ^{222}Rn, contributes only about 11% of the ultimate or potential alpha energy, and that RaB contributes about 51%, because in equilibrium with 100 pCi of ^{222}Rn it constitutes the largest number of atoms (because it has the longest half-life) that will eventually produce 7.69 MeV alpha emissions through RaC′ transformations. Analogously, the atoms in 100 pCi of RaC, the other beta and gamma emitter, supply 38% of the ultimate alpha energy through RaC′. RaC′ produces ^{210}Pb ($t_{1/2} = 22.3$ y), which is unlikely to remain in the lung; thus the alpha energy of its subsequent transformation product ^{210}Po (RaF) is excluded.

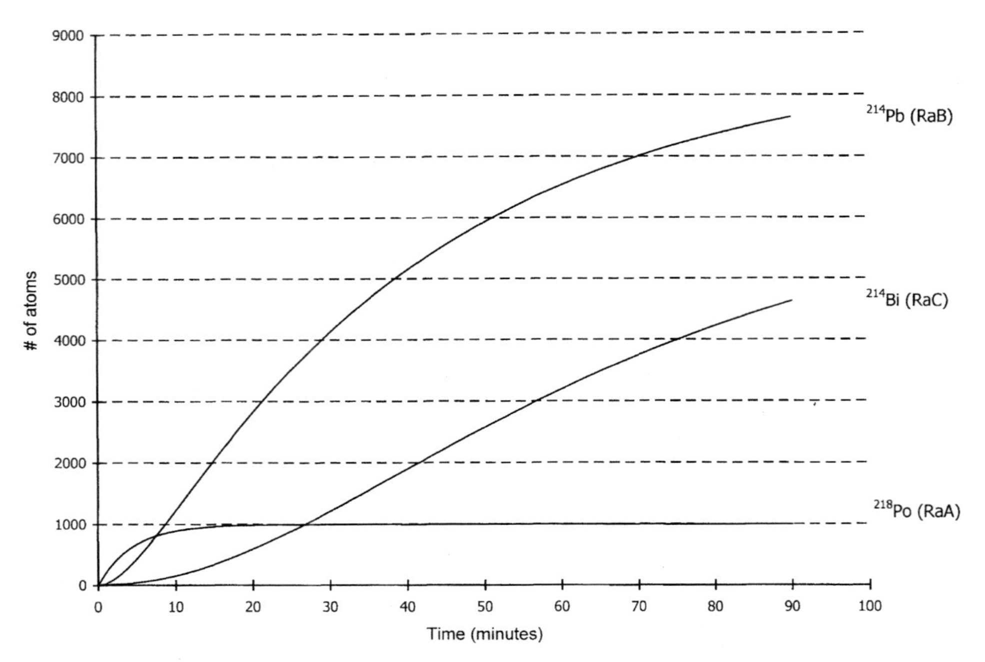

FIGURE 6-13. Graphical representation of the number of atoms of RaA, RaB, and RaC (also equal to RaC′) in a source containing 100 pCi of ^{222}Rn, the activity of which does not change appreciably over the period of observation.

Radioactive equilibrium between ^{222}Rn and its short-lived transformation products takes well over an hour, and in a practical sense such equilibrium rarely exists. Consequently, determination of the working level cannot rely on determination of the ^{222}Rn concentration alone. Since the particulate progeny of Rn determine the working level, it is always necessary to determine the number of atoms of these products, which in turn is determined directly by the amount of ^{222}Rn present and the time period over which ingrowth occurs, or in effect the age of the radon–progeny mixture. The number of atoms of each product can be calculated from equations (6-1) to (6-4), or they can be determined from the curves in Figure 6-13, which is a plot of the number of atoms of RaA, RaB, and RaC at any given time due to radioactive transformation of a pure source of 100 pCi of ^{222}Rn.

As shown in Figure 6-14, it takes 40 to 60 min to achieve 0.5 to 0.7 of the equilibrium value of 1.0 WL. During the first few minutes the WL is due almost entirely to the rapid but limited ingrowth of RaA. Between 4 and 20 min the increase in the WL is approximately linear and is due primarily to the ingrowth of RaB. The contribution from RaC is delayed considerably, due to the 26.8- and 19.9-min half-lives of RaB and RaC; thus, the RaB contribution becomes significant only after the radon source is more than 15 min old and the RaC contribution only after about 40 min. Clearly, the age of the air has a significant effect on the WL, thus it is desirable to breathe young air. This can be achieved by passing the air through an efficient filter shortly before it is to be inhaled or by using ventilation to mix in fresh air free of radon and its products.

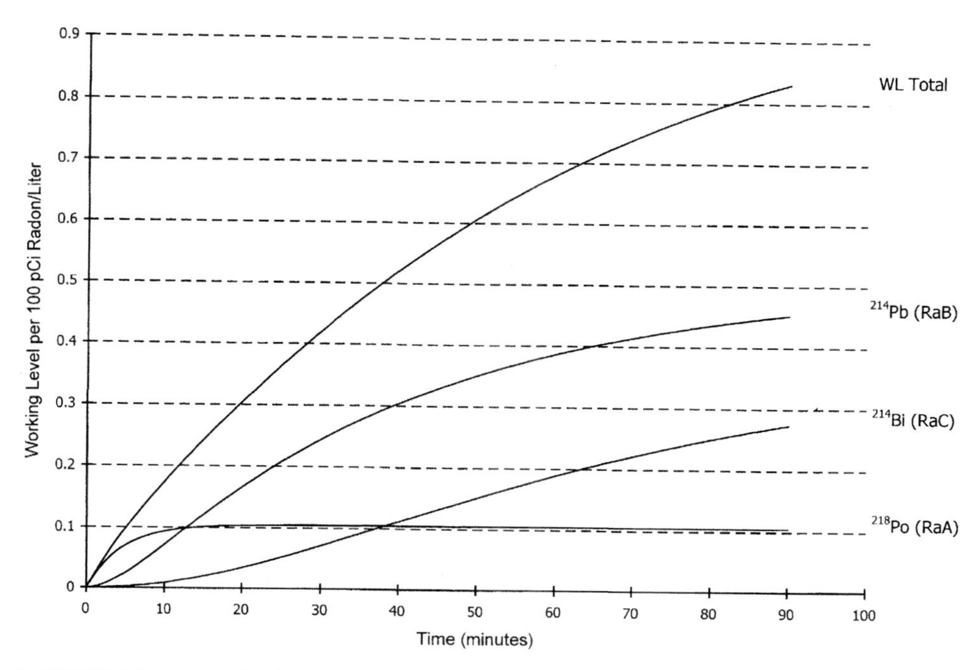

FIGURE 6-14. Contribution of each of the transformation products of ^{222}Rn to the working level.

TABLE 6-11. Results for Example 6-1

Nuclide	Atoms in 100 pCi	α Energy per Atom (MeV)		α Energy (MeV/100 pCi)
^{222}Rn	1.77×10^6	Excluded		None
^{218}Po (RaA)	876	13.69		1.2×10^4
^{214}Pb (RaB)	1219	7.69		9.4×10^3
^{214}Bi (RaC)	114	7.69		8.8×10^2
^{214}Po (RaC′)	~ 0	7.69		0
			Total	2.23×10^4

Example 6-1: Determine the WL for air containing 100 pCi/L of radon 10 min after filtration.

SOLUTION: Following the example for defining the WL, the number of atoms present after 10 min of ingrowth are calculated by equations (6-1) through (6-4) or alternatively, read from the curves in Figure 6-13. Tabulated values are listed in Table 6-11. The working level is

$$\text{WL} = \frac{2.23 \times 10^4 \text{ MeV}}{1.3 \times 10^5 \alpha \text{ MeV/WL}} = 0.17 \text{ WL}$$

Although the working level unit has shortcomings, it is the most practical single parameter for describing the effect of radon and its transformation products. Since the WL is based on radon, it is, like the roentgen (for x rays), a unit of exposure, not radiation dose. Controls are based on limiting human exposure by keeping airborne levels of radon and its products below a specified number of units of WL.

The working-level month (WLM) is used to describe cumulative radon exposure. The average number of exposure hours in a month is 173 ($40 \times 52/12 = 173$); thus a WLM would be exposure to 1 WL for 1 month. The WLM is calculated by multiplying the exposure in WL by the number of hours of exposure and dividing by 173 h/month. For example, exposure to 1 WL for 173 h = 1 WLM; similarly, exposure to 10 WL for 173 h would be 1730 WL-h, or 10 WLM.

Example 6-2: A person is exposed to air containing radon and transformation products which is determined ultimately to emit 10^5 MeV of alpha energy per liter of air. If a person is exposed to this atmosphere for 1000 h, what is the cumulative exposure in WLM?

SOLUTION:

$$\frac{10^5 \text{ MeV}}{1.3 \times 10^5 \text{ MeV/WL}} = 0.77 \text{ WL}$$

$$\frac{0.77 \text{ WL} \times 1000 \text{ h}}{173 \text{ h/month}} = 4.45 \text{ WLM}$$

Measurement of Radon

The two principal measurement methods for radon are (1) collecting the particles of the transformation products and determining the WL directly, or (2) absorbing the radon, a noble gas, on an adsorbent and measuring it to determine its concentration in air. Since the radon concentration and the WL are related, each can be inferred from the other if certain parameters (e.g., fraction of equilibrium) are known or assumed.

Working environments such as mines or uranium mills are usually characterized by a WL measurement, which is made by collecting a short-term (5 to 10 min) particulate sample on a filter paper and measuring the alpha-disintegration rate. Home environments are usually sampled over a period of several days (2 to 7) by adsorbing radon onto activated charcoal, which is then measured by counting the gamma-emission rate from RaC (^{214}Bi), which is the same as the radon transformation rate when secular equilibrium exists. The first (workplace) requires a particulate air sampler and a field method for measurement; the second (residences) uses passive adsorption and a laboratory technique.

Particulate sampling for radon involves a field method that can be summarized as follows:

1. Collect a 5- to 10-min air sample at a flow rate of 10 ft^3/min or greater, and record the liters (L) of air sampled.

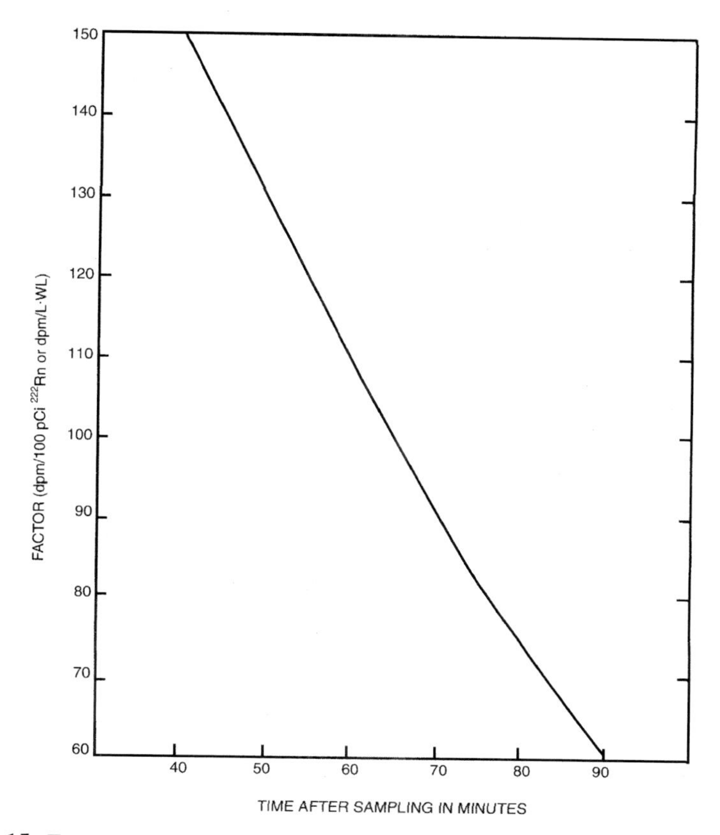

FIGURE 6-15. Factors of (d/m)/(L · WL) (or d/m per 100 pCi of ^{222}Rn) for times after collection of alpha-emitting particulate progeny of radon.

2. Measure the alpha activity (in d/m) on the filter 40 to 90 min later and adjust it by dividing by the appropriate factor from Figure 6-15 to obtain d/m/(L · WL) for the sample.
3. Calculate the WL as

$$\text{WL} = \frac{\alpha \text{ d/m (measured)}}{\text{volume (L)} \times \text{C.F. (d/m per L per WL)}}$$

This method is often referred to as the *Kusnetz method* after the person who devised and published the correction factors in 1956. The Kuznetz method is illustrated in Example 6-3.

Example 6-3: A particulate air sample was collected for 5 min at 2 ft^3/min in a tent erected over an excavation site at an old uranium processing site. The filter was measured 50 min later with a portable ZnS alpha scintillation detector and found to contain alpha activity of 10,000 d/m. The air filter is assumed to have 100% collection efficiency and no self-absorption of the emitted alpha particles. (a) What was the WL for the tent atmosphere? (b) If a worker worked inside the tent 40 hours per week for 25 weeks, what would be the accumulated exposure?

SOLUTION: (a) The WL in the tent is based on the activity per liter of air

$$\text{volume (L)} = 5 \text{ m} \times 2 \text{ ft}^3/\text{m} \times 28.32 \text{ L/ft}^3$$
$$= 283.2 \text{ L}$$

From Figure 6-15, the Kusnetz correction factor at 50 min after sample collection is 129 (d/m)/(L · WL) and

$$\text{WL} = \frac{10{,}000 \text{ d/m}}{283.2 \text{ L} \times 129 \text{ (d/m)/(L} \cdot \text{WL)}}$$
$$= 0.274 \text{ WL}$$

which is just below the recommended exposure guide for uranium miners.

(b) The accumulated exposure in WLM for the worker in the tent is

$$\text{WLM} = \frac{0.274 \text{ WL} \times 25 \text{ wk} \times 40 \text{ h/wk}}{173 \text{ WL} \cdot \text{h/WLM}}$$
$$= 1.58 \text{ WLM}$$

The radon absorption method uses activated charcoal to absorb radon directly from the air. Since this procedure is passive (i.e., no air sampler is used to force air through the charcoal) it requires a standard sampler, usually a small canister that contains about 75 g of charcoal that is calibrated in a known radon atmosphere. The absorption of radon on the charcoal of the reference canister occurs at a constant rate and is a function of the exposure time; thus the activity of the absorbed radon is directly proportional to the concentration of radon (pCi/L) in the air surrounding the canister and the exposure period, as given in the following relationship for a 75-g canister (Gray, 1989):

$$\text{absorption [(d/m)} \cdot \text{L/pCi} \cdot \text{min]} = 35.8511 \times 10^{-3} - (5.1353 \times 10^{-5} \times t)$$

where t is exposure time in hours.

The absorbed radon is determined by counting ^{214}Bi (RaC) gamma rays to determine the ^{214}Bi disintegration rate, and since ^{222}Rn and ^{214}Bi are in equilibrium after 90 min or so, their activities are the same. This activity is then decay-corrected back to the midpoint of the sampling period and reported as pCi of radon per liter of air.

Example 6-4: A 75-g charcoal canister was exposed in the author's home for 189 h (11,340 min) with doors and windows closed. A 10-min count taken 40 h later yielded a net gamma count rate of 119.3 counts per minute (c/m) of ^{214}Bi (RaC) gamma rays after subtraction of background, and the detector efficiency for ^{214}Bi was 26.2%. What was the radon concentration (pCi/L)?

SOLUTION: Since the radon transformation products were in equilibrium, the measured activity of ^{214}Bi and ^{222}Rn is

$$\frac{119.3 \text{ c/m}}{0.262 \text{ (c/m)/(d/m)}} = 455.34 \text{ d/m}$$

Since ^{222}Rn is the parent of ^{214}Bi, both of which would be in equilibrium on the charcoal, the activity rate of ^{222}Rn is also 455.34 d/m. The activity of ^{222}Rn at the midpoint of the sample collection period is a good value of the average ^{222}Rn concentration in the home; therefore, this value should be decay-corrected back to the midpoint, which occurred 5.85 d before the measurement. The ^{222}Rn activity at the midpoint of the sampling period is

$$455.34 \text{ d/m} = A_0 e^{-(\ln 2/3.82 \text{ d})(5.85 \text{ d})}$$

$$A_0 = 1254 \text{ d/m}$$

The absorbed radon on the canister is

$$\text{absorption} = 35.85 \times 10^{-3} - (5.1353 \times 10^{-5} \times 189 \text{ h})$$
$$= 0.0261 \text{ (d/m)} \cdot \text{L/pCi} \cdot \text{min}$$

The average radon concentration over the sampling period is

$$\text{Rn (pCi/L)} = \frac{1254 \text{ d/m}}{11{,}340 \text{ min} \times 0.0261 \text{ (d/m)} \cdot \text{L/pCi} \cdot \text{min}} = 4.23 \text{ pCi/L}$$

The working level (WL) can be calculated from the measured radon concentration if the degree of radioactive equilibrium between radon and its transformation products is known. For a residence, this equilibrium fraction is about 50% due to the number of air changes that occur; for energy-efficient homes it can be as high as 75 to 80%, and for some older homes it could be as low as 20 to 30%. The working level can be defined in terms of the equilibrium fraction as

$$\text{WL} = \frac{\text{Rn (pCi/L)}}{100 \text{ pCi/L} \cdot \text{WL}} \times F$$

where F is the fractional equilibrium between radon and its alpha-emitting transformation products.

Example 6-5: The recommended action level guideline for a residence is 4 pCi of radon per liter of air. (a) If the fractional equilibrium between ^{222}Rn and its particulate, alpha-emitting transformation products is assumed to be 50%, what is the corresponding WL? (b) If 70% occupancy of the residence is assumed, what is the cumulative WLM associated with this exposure level in 1 y?

SOLUTION: (a)

$$\text{WL} = \frac{4\ \text{pCi/L}}{100\ \text{pCi/L} \cdot \text{WL}} \times 0.5 = 0.02\ \text{WL}$$

(b) At 70% occupancy, the annual exposure period is

$$E(\text{h}) = 365\ \text{d/y} \times 24\ \text{h/d} \times 0.7 = 6132\ \text{h}$$

$$\text{total exposure} = \frac{6132\ \text{h} \times 0.02\ \text{WL}}{173\ \text{WL} \cdot \text{h/WLM}} = 0.71\ \text{WLM}$$

SUMMARY

Natural radiation and radioactivity represent a continuous exposure of beings on the earth and produce a natural background that must be considered in every measurement of a radiation level or the radioactivity in a sample or source. The natural radiation environment is influenced by a number of natural processes, especially extraterrestrial cosmic ray activity, the distribution of soils and minerals, and various human activities.

The natural radiation environment consists of cosmic rays and naturally radioactive materials that are either cosmogenic, primordial, or the products of these substances that occur in one of naturally radioactive series, stretching through the last part of the chart of the nuclides. These series are named the uranium (^{238}U), actinium (^{235}U), thorium (^{232}Th), and neptunium (^{237}Np) series according to the radionuclides that serve as progenitor (or parent) of the series products. With the exception of neptunium, each of the parent radionuclides is primordial in origin because they are so long lived that they still exist some 4.5 billion years after the solar system was formed.

Many ores are processed for their mineral content, which enhances either the concentration of the radioactive elements in process residues or increases their environmental mobility; these processes result in materials that are no longer purely "natural." These materials are called NORM (naturally occurring radioactive material), or more recently, TENORM (technologically enhanced NORM).

Radon, an important transformation product of ^{238}U, ^{235}U, and ^{232}Th, exists in various concentrations in all soils and minerals. Radon-222, which is a product of the ^{238}U series, is the most important radioisotope of radon because of its 3.82-d half-life and the 4.5 billion year half-life of its uranium parent. The intermediate transformation products, ^{230}Th and ^{226}Ra (the immediate parent of ^{222}Rn), have half-lives of 75,380 y and 1600 y; thus they are in radioactive equilibrium with uranium and represent a perpetual source of radon. Whereas uranium and its intermediate products are solids and remain in the soils and rocks where they originate, ^{222}Rn is a radioactive noble gas that migrates through soil to zones of low pressure such as homes. Its 3.82-d half-life is long enough for it to diffuse into and build up in homes unless they are constructed in ways that preclude its entry, or unless provisions are made to remove it.

Naturally occurring radiation and radioactive materials account for a major portion of radiation dose received by members of the public, on the order of 100 mrem (1 mSv) per year to the average person in the United States. Cosmic rays contribute about 28 mrem/y, terrestrial radiation about 27 mrem/y, and radionuclides in the body, principally ^{40}K, about 35 mrem/y. Exposure to radon and its products produces doses, primarily to lung tissue, which vary widely with location and residence and can be substantial, representing a radiation risk well above that from other natural sources.

ACKNOWLEDGMENTS

This chapter was compiled with substantial help from Chul Lee and Rebecca L. Pintar of the University of Michigan Radiological Health Program.

REFERENCES AND ADDITIONAL RESOURCES

NCRP, *Radiation Exposure of the Population in the United States and Canada from Natural Background Radiation*, Report 94, National Council on Radiation Protection and Measurements, Bethesda, MD, 1987.

NCRP, *Radiation Protection in the Mineral Extraction Industry*, Report 118, National Council on Radiation Protection and Measurements, Bethesda, MD, 1993.

Evans, R. D., Engineers guide to behavior of radon and daughters, *Health Physics* **17**, no. 2 (August 1969).

Frame, Paul W., Radioactivity: conception to birth, The Health Physics Society 1995 Radiology Centennial Hartman Oration, *Health Physics*, April 1996.

Kusnetz, H. L., Radon daughters in mine atmospheres, *Ind. Hygiene Qtr.* **17**, 85–88 (March 1956).

Gray, D., and S. T. Windham, Standard Operating Procedures for Radon-222 Measurement Using Charcoal Canisters, U.S. Env. Prof. Agency, Nat'l. Air and Radiation Env. Laboratory (NAREL), Montgomery, AL, March 1987.

PROBLEMS

6-1. When Becquerel separated uranium X (^{234}Th; $T_{1/2}$ = 24.1 d) from his uranium phosphor, its radioactivity essentially vanished. When he measured it again 4 months later, the uranium had regained activity. What fraction of the original radioactivity due to UX had returned?

6-2. Rutherford and Soddy precipitated thorium nitrate and observed a phenomenon similar to that for uranium and UX. They found that essentially all the activity was removed in the precipitate that they called thorium X (^{224}Ra; $T_{1/2}$ = 3.66 d) but that the thorium regained its ThX activity much quicker. How long did it take for ThX to reach 99.9% of its original activity in thorium?

6-3. A sample of RaE (^{210}Bi) is freshly isolated and found to contain 1 μCi. What is the maximum activity of RaF (^{210}Po) that will exist in this naturally occurring mixture, and when will it occur? (*Hint*: Ignore the short-lived ^{206}Tl.)

6-4. The transition of ^{238}U to ^{206}Pb can be used to determine the age of minerals by determining the weight ratio of Pb to U by mass spectroscopy. (**a**) Show that if ^{238}U transformation is neglected, the age of the mineral is the ratio (Pb/U $\times$ 7.45 $\times$ 10^9 y). (**b**) What is the age of a rock that contains 0.1 g of ^{206}Pb per gram of U? and (**c**) What would be the age if corrected for the radioactive transformation of ^{238}U?

6-5. If 10 μCi of ^{223}Ra is eluted from an ^{227}Ac generator and tagged to monoclonal antibodies that in turn are incorporated into a 5 g cancerous mass, how much alpha-particle energy would ultimately be delivered to the cancer? (Assume that ^{219}Rn does not diffuse out of the cancer and that all product atoms transform in the cancerous mass.)

6-6. Some of the carbon atoms in trees and lumber are ^{14}C. Why are there none among the carbon atoms in petroleum products?

6-7. If the activity of ^{14}C in wood grown prior to 1850 (before significant fossil-fuel burning) was 15.3 d/m per gram of carbon, what would be the age of cinders found in an ancient fire pit that had an activity of 10 d/m per gram of carbon when discovered?

6-8. An old rock of volcanic origin is found and measured to contain 0.1 g of potassium. The rock is then heated and the ^{40}Ar from transformation of ^{40}K is collected and measured by mass spectroscopy to be 1.66 $\times$ 10^{-10} g. What is the age of the rock?

6-9. The bones of "Lucy" were found in 1974 in sedimentary rock containing trapped potassium with a typical sample yielding a total of 0.1 g. A fraction (0.0000118) of the trapped potassium was ^{40}K, which decays into ^{40}Ar. This sample contained 2.39 $\times$ 10^{-9} g of ^{40}Ar, of which 7.25 $\times$ 10^{-10} g was contamination from the air. What is Lucy's approximate age?

6-10. Water from a deep well is measured and found to contain tritium that is only 40% of the tritium in fresh rainwater. If it can be assumed that the well water is isolated from surface flow, what is the approximate age of the water in the well (i.e., how long did it take to diffuse from the surface to the aquifer)?

6-11. The average human being is made up of 18% by weight of natural carbon. (**a**) If a 70 kg person contains 0.1 μCi, what fraction of carbon in the body is ^{14}C? (**b**) How does the activity per gram of body mass compare to that of living wood which contains 15.3 d/m per gram of carbon?

6-12. Regulations require that the concentration of ^{226}Ra in uranium tailings left at mill sites not exceed 15 pCi/g for a period of 1000 y. For soil that contains ^{230}Th but no ^{226}Ra, what is the maximum concentration of ^{230}Th that can be left in the soil today such that the regulations are not exceeded in 1000 y?

6-13. A settling basin near an abandoned uranium processing site contains sediment with concentrations of 100 pCi/g of ^{230}Th and essentially no ^{226}Ra. (**a**) If the external gamma-exposure rate is 2.5 mrem/y for 1 pCi/g of ^{226}Ra in equilibrium with its transformation products for the geometry of the sediment, what would be the exposure rate for 100% occupancy of the site 500 y hence? (**b**) If roughly 5 pCi/g of ^{226}Ra corresponds to 0.02 WL of ^{222}Rn and its products in a structure built upon soils containing ^{226}Ra, what would be the radon level in homes built on the site in 200 y? In 500 y?

6-14. (**a**) If ^{230}Th and ^{231}Pa are preferentially precipitated from seawater into silt with a concentration of 10 and 2 pCi/g, respectively, what is the age of a sediment sample at the 1-m depth if it contains 7 pCi/g of ^{226}Ra which is not brought down in the sediment? (**b**) What would be the corresponding concentration of ^{223}Ra?

6-15. Determine the working level for 100 pCi/L of thoron (^{220}Ra) continuously in equilibrium with its short-lived decay products.

6-16. The average age of ^{222}Rn in a mine shaft is determined to be 20 min. What would be the WL in the mine shaft for a measured concentration of 100 pCi/L of ^{222}Rn?

6-17. Calculate the working level for air that contains 100 pCi/L of radon that is only 3 min old.

7

INTERACTIONS OF RADIATION WITH MATTER

Much of what we know of the Universe comes from information transmitted by photons.
—John Hubbell, Ann Arbor, Michigan, 1995

When radiation is emitted, regardless of what type it is, it produces various interactions that deposit energy in the medium that surrounds it. This deposition of energy is characterized as radiation dose, and if it occurs in living human tissue, the endpoint effects will be biological changes, most of which are undesirable. Understanding these interactions leads naturally to the determination of radiation exposure and dose and the units used to define them.

The mechanisms by which various radiations interact in an absorbing medium are fundamental to describing the amount of deposited energy, how the characteristics of shields or other absorbers modify and affect radiation exposure and dose, and the design of detectors to measure the various types of radiation based on their respective interaction principles. The goals of this chapter are first to describe interaction processes that attenuate and absorb charged particles and photons and then to apply these concepts to the calculation of deposited energy (i.e., radiation exposure and dose). The design of radiation shields is provided in Chapter 8, and the detection of radiation is discussed in Chapter 12. A similar treatment is provided in Chapter 14 for neutrons.

RADIATION DOSE AND UNITS

The term *radiation dose*, or simply *dose*, is defined carefully in terms of two key concepts: (1) the energy deposited per gram in an absorbing medium, principally tissue, which is the absorbed dose, and (2) the damaging effect of the radiation type, which is characterized by the term *effective dose equivalent*. A related term is *radiation exposure*, which applies to air only and is a measure of the amount of ionization produced by x- and γ-radiation in air. Each of these is defined in the conventional system of units, which in various forms and refinements has been used for several decades, and in the newer SI system, which is gradually replacing the conventional units. This presentation uses both sets, but for the most part, emphasizes conventional units, which are firmly embedded in governmental standards and regulations, at least in the United States.

Radiation Absorbed Dose

The *absorbed dose* is defined as the amount of energy deposited per unit mass. The conventional unit for absorbed dose, the *rad* (radiation absorbed dose), is equal to the absorption of 100 ergs of energy in 1 g of absorbing medium, typically tissue:

$$1 \text{ rad} = 100 \text{ ergs/g of medium}$$

The SI unit of absorbed dose, the *gray* (Gy), is defined as the absorption of 1 J of energy per kilogram of medium, or

$$\begin{aligned} 1 \text{ Gy} &= 1 \text{ J/kg} \\ &= 100 \text{ rad} \end{aligned}$$

A milligray (mGy) is 100 mrad, which is about the amount of radiation a person receives in a year from natural background, excluding radon. This is a convenient relationship for translating between the two systems of units since the annual radiation dose due to natural background (equal to 1 mGy or 100 mrad) is a convenient reference point for radiation dose received by a person. The rate at which an absorbed dose is received is quite often of interest. Common dose rates are rad/s, mrad/h, and so on; in SI units, dose rates may be expressed as Gy/s, mGy/h, and so on, and because the Gy is such a large unit compared to many common circumstances, the unit μGy/h is often used.

Radiation Dose Equivalent

The definition of dose equivalent is necessary because different radiations produce different amounts of biological damage even though the deposited energy may be the same. If the biological effects of radiation were directly proportional to the energy deposited by radiation in an organism, the radiation absorbed dose would be a suitable measure of biological injury, but this is not the case. Biological effects depend not only on the total energy deposited, but also on the way in which it is distributed along the path of the radiation. Radiation damage increases with the linear energy transfer (LET) of the radiation; thus for the same absorbed dose, the biological damage from high-LET radiation (e.g., α particles, neutrons, etc.) is much greater than from low-LET radiation (β particles, γ rays, x rays, etc.).

The *dose equivalent*, denoted by H, is defined as the product of the absorbed dose and a factor, Q, the quality factor, that characterizes the damage associated with each type of radiation, or

$$H(\text{dose equivalent}) = D(\text{absorbed dose}) \times Q(\text{quality factor})$$

In the conventional system of units, the unit of dose equivalent is the rem, which is calculated from the absorbed dose as

$$\text{rem} = \text{rad} \times Q$$

The value of Q varies with the type of radiation; $Q = 1.0$ for x rays, gamma rays, and electrons, $Q = 20$ for alpha particles and fission fragments, and $Q = 2$ to 10 for neutrons of different energies.

The SI unit of dose equivalent is the sievert (Sv), or

$$\text{sieverts} = \text{Gy} \times Q$$

The sievert, as is the gray, is a very large unit, corresponding to 100 rem in conventional units, and it is often necessary to use mSv, or in some cases μSv, to describe the radiation dose equivalent (or rate) received by workers and the public.

Radiation Exposure

The term *exposure* is used to describe the quantity of ionization produced when x or gamma rays interact in air because it can conveniently be measured directly by collecting the electric charge, whereas that which occurs in a person cannot be. The roentgen is the unit of radiation exposure; it is defined only for air and applies only to x and gamma rays up to energies of about 3 MeV. A milliroentgen (mR), is 0.001 R. Exposure rates are often expressed as roentgens per unit time, for example, R/s, mR/h, and so on. Since the roentgen is determined in air, it is not a radiation dose, but with appropriate adjustment it can be converted to a dose.

The roentgen was originally defined as that amount of x or gamma radiation such that the associated corpuscular emission produces in 0.001293 g of air (1 cm^3 of air at atmosphere pressure and 0°C) 1 electrostatic unit (1 esu = 3.336×10^{-10} C) of charge of either sign. The ionization produced by the associated corpuscular emissions is due to photoelectric, Compton, and if applicable, pair production interactions. In terms of charge produced per unit weight of air, the R corresponds to

$$\begin{aligned} 1\ \text{R} &= \frac{1\ \text{esu} \cdot 3.336 \times 10^{-10}\ \text{C/esu}}{0.001293\ \text{g}} \times 10^3\ \text{g/kg} \\ &= 2.58 \times 10^{-4}\ \text{C/kg of air} \end{aligned}$$

The modern definition of the roentgen (R) is based on this value (i.e., 1 R is that amount of x or gamma radiation that produces 2.58×10^{-4} C of charge in 1 kg of air).

The roentgen is not included in the SI system of units; the SI unit for exposure is the *X unit*, defined as the production of 1 C of charge in 1 kg of air, or

$$X = 1\ \text{C/kg air}$$

The X unit corresponds to deposition of 33.97 J in 1 kg of air and is equal to 3876 R. The X unit is a huge unit; consequently, most exposure measurements are made and reported in R, which seems appropriate for the discoverer of x rays. Exposure and exposure rate apply *only* to x and gamma rays and only in *air*, and neither R nor the X unit is appropriate for describing energy deposition from particles or for energy deposition in the body.

Energy deposition per roentgen is an important relationship because the deposition of energy in air is readily calculated. The amount of charge produced in air varies with energy but is approximately linear for photons between about 70 keV and 3 MeV such that the amount of energy required to produce an ion pair is 33.97 eV on average. The roentgen thus corresponds to an energy deposition in air, the absorbing medium, of 87.64 ergs/g of air (see Example 7-1).

Example 7-1: From the original definition of the roentgen, determine the energy deposition in air (a) in ergs per gram of air for an exposure of 1 R and (b) in joules per kilogram of air for 1 R.

SOLUTION: (a) Since 1 R = 1 esu in air at STP (0.001293 g),

$$\text{exposure (ergs/g)} = \frac{1 \text{ esu}}{0.001293 \text{ g}} \times 33.97 \text{ eV/ion pair}$$

$$\times \frac{1.6022 \times 10^{-12} \text{ erg/eV}}{1.6022 \times 10^{-19} \text{ C/ion pair}} \times 3.336 \times 10^{-10} \text{ C/esu}$$

$$= 87.64 \text{ ergs/g} \cdot \text{R}$$

(b) In SI units, 1 R = 2.58×10^{-4} C/kg:

$$\text{exposure (J/kg)} = 2.58 \times 10^{-4} \text{ C/kg} \times 33.97 \text{ J/C}$$

$$= 8.764 \times 10^{-3} \text{ J/kg} \cdot \text{R}$$

RADIATION DOSE CALCULATIONS

Radiation dose can be calculated by a three-step procedure as follows:

1. Establish the number of radiation pulses (particles or photons) per unit area entering a volume of medium of known density.
2. Establish the mass of medium in which the energy is dissipated. For particles, this is just the depth of penetration; for photons, it is necessary to use a unit depth (e.g., 1 cm) due to the probabilistic pattern of interactions.
3. From the pattern(s) of interaction probabilities, determine the amount of energy deposited.

All emitted radiations must be considered in this process, and adjustments should be made for any attenuating medium between the source and the point of interest.

Since different radiations penetrate to different depths in tissue, radiation dose has been specified in regulations for three primary locations:

1. The shallow dose, which is just below the dead layer of skin, which has a density thickness of 7 mg/cm^2 or an average thickness of 70 μm

2. The eye dose just below the lens of the eye, with a density thickness of 300 mg/cm^2
3. The deep dose, which is located at a depth of 1 cm in tissue or at a density thickness of 1000 mg/cm^2, primarily to account for highly penetrating radiation such as x or gamma rays or neutrons

A precise determination of absorbed energy at these depths requires an adjustment for any diminution in the flux due to energy losses in the overlying tissue layer.

Since energy deposition is dependent on the radiation type, its energy, and the absorbing medium, it will, for the sake of clarity, be discussed in the context of the interaction mechanisms of each type of radiation. Shielding of radiation, which is related to the same considerations, is discussed in the same manner in Chapter 8.

Inverse-Square Law

The number of radiation pulses that enter and/or traverse a medium is governed by source strength and the geometry between the source and the medium of interest, a relationship that can often be conveniently expressed as a fluence (number/cm^2) or a fluence rate or flux (number/cm$^2 \cdot$ s). A radioactive point source of activity S (t/s) emits radiation uniformly in all directions, and at a distance r will pass through an area equal to that of a sphere of radius r; therefore, the flux ϕ is

$$\phi \text{ (no./cm}^2 \cdot \text{s)} = \frac{S(\text{t/s})f_i}{4\pi r^2}$$

where f_i is the fractional yield per transformation of each emitted radiation. This relationship is the *inverse-square law*, which states that the flux of radiation emitted from a point source is inversely proportional to r^2, as shown in Example 7-2.

Example 7-2: What is the photon flux produced by a 1-mCi point source of ^{137}Cs at a distance of (a) 100 cm and (b) 400 cm?

SOLUTION: (a) ^{137}Cs ($T_{1/2}$ = 30.07 y) emits 0.662 MeV gamma rays through ^{137m}Ba ($T_{1/2}$ = 2.52 m) in 85% of its transformations. The gamma flux is

$$\begin{aligned}\phi(\gamma/\text{cm}^2 \cdot \text{s}) &= \frac{1 \text{ mCi} \times 3.7 \times 10^7 \text{ t/s} \cdot \text{mCi} \times 0.85 \ \gamma/\text{t}}{4\pi(100 \text{ cm})^2} \\ &= 2.5 \times 10^2 \ \gamma/\text{cm}^2 \cdot \text{s}\end{aligned}$$

(b) At 400 cm

$$\begin{aligned}\phi(\gamma/\text{cm}^2 \cdot \text{s}) &= \frac{2.5 \times 10^2 \ \gamma/\text{cm}^2 \cdot \text{s}(100 \text{ cm}^2)}{(400 \text{ cm})^2} \\ &= 15.625 \ \gamma/\text{cm}^2 \cdot \text{s}\end{aligned}$$

Once the flux of radiation is known, it can be converted to an energy flux by multiplying the flux by the average energy of each radiation emitted by the source.

The energy flux due to particle emissions (betas, alphas, deuterons, neutrons, etc.) is calculated in a similar way using the average energy of each.

INTERACTION PROCESSES

Radiation, either in the form of emitted particles or electromagnetic radiation, has properties of energy, mass, momentum, and charge which combine to determine how it interacts with matter. Various absorption and/or scattering interactions produce ionization or excitation of the medium, or the radiation may be converted into yet another type (e.g., photons from positron–electron annihilation). In general, charged particles lose considerable energy by ionization, whereas photons and neutrons give up energy by scattering and absorption reactions. Table 7-1 summarizes various properties of the major types of radiation.

Several terms are used to describe the changes in energy of a particle and the absorbing medium. The stopping power S is defined as the loss of energy *from* a particle over a path length, dx:

$$S = -\frac{dE}{dx}$$

Linear energy transfer (LET), which is similar to stopping power but has a distinct property, is defined as the energy imparted *to the medium*, or

$$\text{LET} = -\frac{dE}{dx}$$

Although the expression for LET appears to be the same as the stopping power, the LET is the energy imparted to the medium at or near the site of the collision. When bremsstrahlung is produced, as in beta absorption, LET will be different from S because the bremsstrahlung photons carry some of the energy lost by the particles away from the collision site.

Specific ionization (SI) is yet another energy-loss term. It is the number of ion pairs formed per unit path length and is represented by

$$\text{SI} = \frac{dN}{dx}$$

TABLE 7-1. Interaction Properties of Radiation

Radiation	Charge	Energy	Range in Air	Range in H_2O
α particles	+2	3–10 MeV	2–10 cm	20–125 μm
β^+, β^- particles	± 1	0–3 MeV	0–10 m	< 1 cm
Neutrons	0	0–10 MeV	0–100 m	0–1 m
X rays	0	0.1–100 KeV	m–10 m	mm–cm
Gamma rays	0	0.01–10 MeV	cm–100 m	mm–10s of cm

where N is the number of ion pairs (or atoms ionized) produced per unit path length along the path of the radiation. The average amount of energy expended to create an ion pair in air is currently accepted to be 33.97 eV, which is often rounded to 34 eV/ion pair for most calculations; in other gases it ranges from 32 to 35 eV.

Example 7-3: Calculate the total number of ion pairs produced in air by a 4.78 MeV alpha particle emitted by ^{226}Ra.

SOLUTION:

$$\text{Ion pairs} = \frac{E_\alpha}{34} = \frac{4{,}780{,}000 \text{ eV}}{34 \text{ eV/ion pair}} = 140{,}600 \text{ ion pairs}$$

Example 7-3 illustrates the large number of energy-transferring interactions when alpha particles traverse a medium. In tissue, all of these occur in a few tens of micrometers.

Relative stopping power S_{rel} is used to compare the energy loss in various substances to a reference medium such as air. When the relative stopping power is known, the range R_m in a medium can be determined by

$$S_{\text{rel}} = \frac{(dE/dx)_m}{(dE/dx)_{\text{air}}} = \frac{R_a}{R_m}$$

where R_a and R_m represent the ranges in air and in the medium, respectively. An approximate relative stopping power, S_{rel}, can be calculated from the densities; for example, the relative stopping power for aluminum relative to air is

$$S_{\text{Al}} = \frac{\rho_{\text{Al}}}{\rho_{\text{air}}} = \frac{2.669}{1.293 \times 10^{-3}} = 2064$$

Therefore, the stopping power of alpha particles in aluminum is almost 2100 times greater than that of air, or alternatively, the range in aluminum is about 2100 times less.

Kerma (kinetic energy released in material) is also used to describe energy loss in a medium. It is a unit of exposure, expressed in rad, that represents the kinetic energy transferred to charged particles per unit mass of irradiated medium when indirectly ionizing (uncharged) radiations such as photons or neutrons traverse the medium. Kerma is thus the starting point for determining the energy deposition by a given type of radiation in an absorbing medium and varies according to radiation type and absorption medium. For this reason it will be discussed as each type of radiation is presented.

INTERACTIONS OF ALPHA PARTICLES AND HEAVY NUCLEI

Alpha particles, heavy recoil nuclei, and fission fragments are highly charged and interact by ionization in traversing a medium. Because of their high charge, their paths of interaction are characterized by very dense patterns of ionized atoms, the

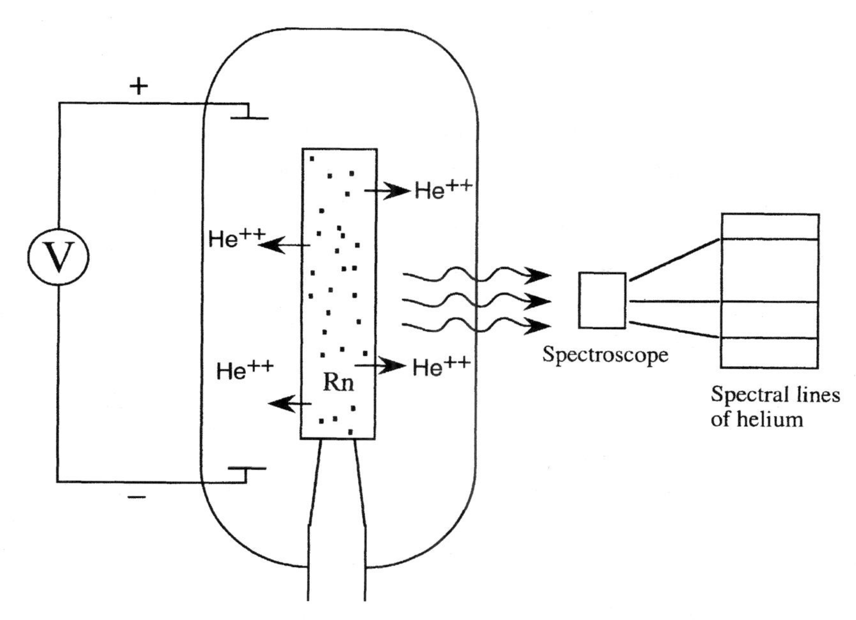

FIGURE 7-1. Apparatus used to identify alpha particles as helium nuclei.

energy transfer per millimeter of path length is large, and their depth of penetration in most absorbing media is on the order of micrometers; even in air the maximum range is only a few centimeters. Kerma and absorbed dose from alpha particles and heavy nuclei are essentially equal since the energy transferred is deposited very near the sites of interactions to produce absorbed dose.

An alpha particle does not have orbital electrons when it is ejected from the nucleus of a radioactive atom: it is a helium nucleus with a charge of +2. Sir Ernest Rutherford conducted an ingenious experiment which demonstrated that alpha particles are in fact helium nuclei. He placed radon in a tube with walls so thin that the alpha particles emitted by the radon could penetrate into a second tube surrounding the first (see Figure 7-1). All the air was removed from the second tube (that vacuum pump again) before the experiment began, and after several days, Rutherford induced an electrical discharge across the secondary chamber and observed the classic emission spectrum of helium, thus proving that alpha particles were indeed helium nuclei stripped of their electrons.

Alpha particles are emitted with several MeV of kinetic energy, typically 4 to 10 MeV, and because of their relatively large mass, the product nucleus recoils with a significant amount of energy. Consequently, both the alpha particle with its +2 charge and the recoiling nucleus, which is also charged, produce considerable energy deposition in a very short distance. Such energy losses occur primarily by ionization and gradually reduce the velocity of the particle, which allows it to spend more time in the vicinity of the target atoms with a greater probability of producing ionization. When it slows down and stops, it picks up orbital electrons to become a neutral helium atom. This variation in specific ionization is shown in Figure 7-2 for highly ionizing alpha particles of ^{210}Po (E_α = 5.304 MeV) and radium C′ or ^{214}Po (E_α = 7.687 MeV). These particles produce on average about 50,000 ion pairs/cm

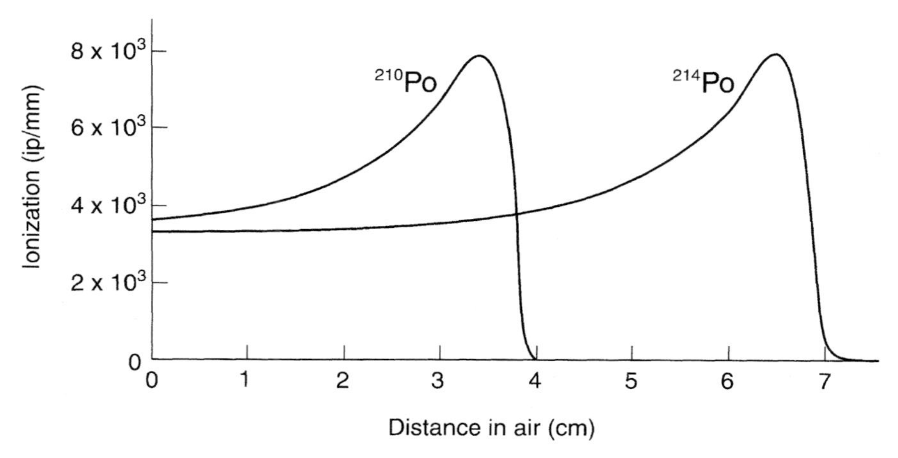

FIGURE 7-2. Bragg curves of specific ionization (ion pairs/cm) of alpha particles from ^{210}Po and radium C′ (^{214}Po) in air at 760 mm and 15°C.

and range up to about 80,000 ion pairs/cm near the end of their path. Such curves with increased ionization per unit path length near the end of the particle track are called *Bragg curves* and are characterized by their shape as shown in Figure 7-2. Beta particles produce a similar curve, but with a much reduced rate of ionization per millimeter, typically about 50 ion pairs/cm on average.

Recoil Nuclei and Fission Fragments

Recoiling nuclei from nuclear fission also produce very short range energy deposition. The two fission fragments share about 170 MeV of energy, which is usually apportioned as about 70 MeV of kinetic energy to a light fragment ($A \approx 94$) and about 100 MeV to a heavier one ($A \approx 140$). The fissioning nucleus supplies so much energy that the newly formed fragments repel each other with such force that they literally break loose from their respective electron clouds. Consequently, each fragment carries a charge of about +20 as they move through matter, and because of this excessive charge, their range in most media is just a few micrometers. In uranium fuel rods, where they are commonly produced, their range is about 7 to 14 μm, and considerable heat is produced as the tremendous energy of the fission fragments is absorbed in the fuel matrix. Similar patterns occur for plutonium fission fragments and recoil nuclei from high-energy nuclear interactions.

Range of Alpha Particles

Alpha particles are emitted monoenergetically, and each particle will typically have the same range in air (or some other medium) except for some straggling about a mean range, $\overline{R}$, as shown in Figure 7-3. The straggling is statistical and forms a normal distribution about the mean range $\overline{R}$. The range of an alpha particle is usually measured in and expressed as centimeters of air, which is roughly the same as its energy in MeV, as shown in Figure 7-4. An empirical fit of the data in Figure

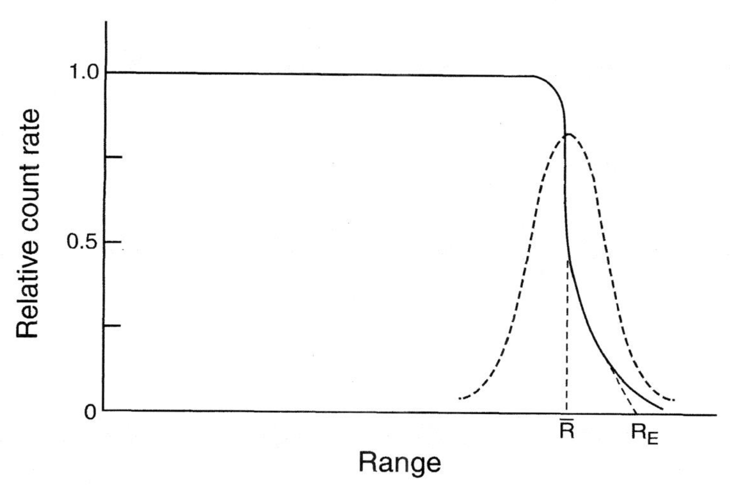

FIGURE 7-3. Range in air of a collimated monoenergetic source of alpha particles showing straggling that is normally distributed about the mean range at $\overline{R}$. An extrapolated range, R_E, can be obtained by extending the straight-line portion of the curve to the x-axis.

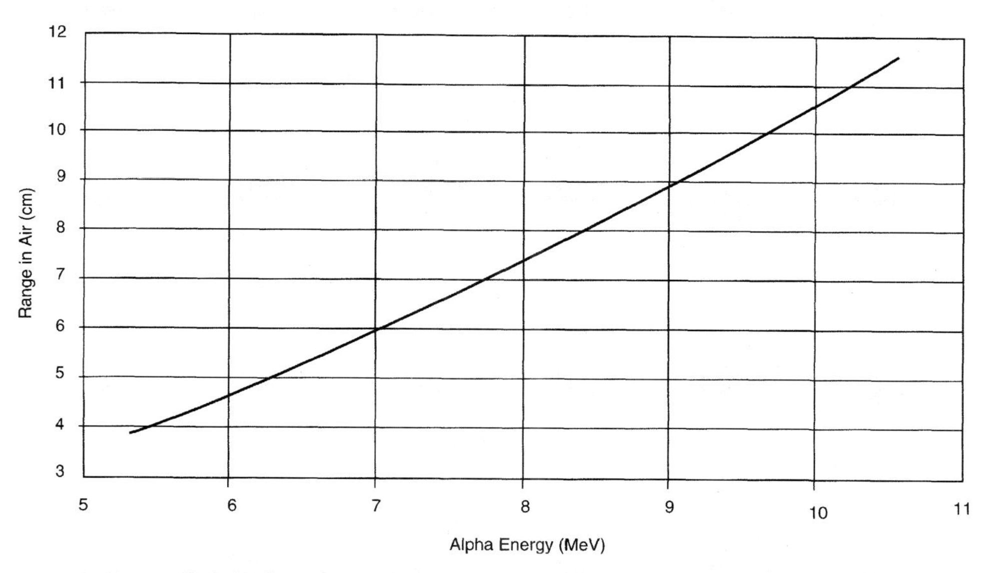

FIGURE 7-4. Range versus energy of alpha particles in air.

7-4 shows a relationship between range and energy as

$$R = 0.325E^{3/2}$$

or

$$E = 2.12R^{2/3}$$

where R is in centimeters of standard air and E is in MeV.

The range of an alpha particle in air can be used to determine its range in any other medium by the *Bragg–Kleeman rule*:

$$R_m = \frac{\rho_a}{\rho_m} R_a \sqrt{\frac{M}{M_a}}$$

The most useful relationship for the range of alpha particles in a medium is in terms of their range in air since these values have been well established. By inserting the molecular weight and density of air at 20°C and 1 atm,

$$R_m = 3.2 \times 10^{-4} \times \frac{\sqrt{M}}{\rho_m} R_a$$

where R_m is the range in the medium in centimeters, R_a is the range of the alpha particle in centimeters of air, ρ_m is the density of the medium, and M is the atomic weight of the medium.

Example 7-4: What thickness of aluminum ($\rho = 2.7$) is required to attenuate a source of 5 MeV alpha particles completely?

SOLUTION: The range of a 5 MeV alpha particle in air is first obtained from Figure 7-4 or the empirical relationship $R = 0.325E^{3/2}$ to be 3.6 cm. The range in Al is

$$R_{\mathrm{Al}} = 3.2 \times 10^{-4} \frac{\sqrt{27}}{2.7} \times 3.6 \text{ cm}$$
$$= 22.2 \times 10^{-4} \text{ cm} = 22.2\ \mu\text{m}$$

For tissue, the mass stopping power is almost the same as in air since the composition of tissue is similar. The range of an alpha particle in tissue can be determined from the range in air, R_a, by adjusting it by the ratio of the densities of each:

$$R_t = \frac{\rho_a}{\rho_t} R_a$$

Thus the 5.0 MeV alpha particle in Example 7-4 would have a range in tissue of 4.65×10^{-3} cm, or about 46.5 μm. The range of alpha particles in tissue is of particular interest in radiation protection, as illustrated in Example 7-5.

Example 7-5: If ^{218}Po is uniformly distributed on the lining of the bronchi (bronchial epithelium) of the human lung with an activity of 100 pCi/cm^2, what is the energy deposition rate per unit mass in the bronchial epithelium?

SOLUTION: ^{218}Po emits 6.0 MeV alpha particles in 100% of its transformations, of which 50% would, due to geometry considerations, penetrate the bronchial epithelium, creating a flux of

$$\phi_\alpha = 100 \text{ pCi/cm}^2 \times 2.22 \text{ (d/m)/pCi} \times 0.5 = 111\ \alpha/\text{cm}^2 \cdot \text{min}$$

These α particles would dissipate their energy over the mean range in tissue ($\rho =$ 1.0), which is determined from Figure 7-4 or from the equation for the range in air:

$$R_a = 0.325E^{3/2} = 4.78 \text{ cm}$$

from which the range in tissue can be calculated; that is,

$$R_t = \frac{1.293 \times 10^{-3}}{1} \times 4.78 \text{ cm} = 6.2 \times 10^{-3} \text{ cm}$$

The energy deposition rate in rad/h over this depth of bronchial epithelium is

$$E_{\text{dep}} = \frac{111\alpha/\text{cm}^2 \cdot \text{min} \times 6.0 \text{ MeV}/\alpha \times 1.6022 \times 10^{-6} \text{ erg/MeV} \times 60 \text{ min/h}}{6.2 \times 10^{-3} \text{ cm} \times 1 \text{ g/cm}^3 \times 100 \text{ ergs/g} \cdot \text{rad}}$$
$$= 0.104 \text{ rad/h}$$

Or, if the condition were to persist for 1 y due to continuous exposure to radon, the dose to the bronchial epithelium due to continuous emission of ^{218}Po alpha particles would be 911 rad (or 9.11 Gy, the unit of radiation absorbed dose in SI units).

BETA PARTICLE INTERACTIONS AND DOSE

A beta particle is a high-speed electron that is so labeled because it originates from the nucleus of a radioactive atom. It has a rest mass m_0 of only 9.1×10^{-28} g and a charge Q of 1.6022×10^{-19} C. Its energy is dependent on its velocity, and conversely, and due to their small mass beta particles with energies in the MeV range have velocities that approach the speed of light. Beta particles lose energy to a medium in four ways: direct ionization, delta rays from electrons ejected by ionization, production of bremsstrahlung, and Čerenkov radiation. Although each mechanism can occur, the most important ones are direct ionization and bremsstrahlung production.

Energy Loss by Ionization

The kinetic energy of beta particles and their negative charge is such that coulombic forces dislodge orbital electrons in the absorbing medium to create ion pairs. Similar interactions occur for positrons. The particles thus lose energy by ionizing the absorbing medium, and the energy carried by these excited electrons is absorbed essentially at the interaction site. If the beta particles eject K-, L-, or M-shell electrons, characteristic x rays will also be emitted as these vacancies are filled. These, too, are likely to be absorbed nearby.

Since beta particles and orbital electrons are about the same size, beta particles are deflected through a rather tortuous path, as shown in Figure 7-5*a*. The path length is much longer, and both linear energy transfer and stopping power are relatively lower for electrons. A 3 MeV β^- particle has a range in air of over 1000 cm and produces only about 50 ion pairs per centimeter of path length.

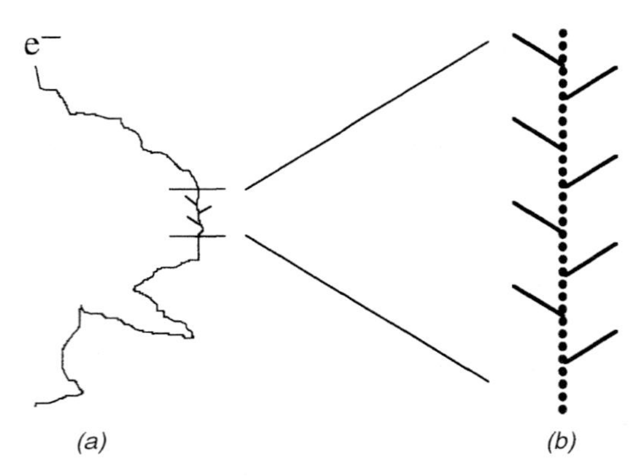

FIGURE 7-5. (*a*) Range and ionization path of a beta particle in an absorbing medium; (*b*) amplified segment of ionization track showing delta ray tracks produced by ejected electrons.

Delta rays are often formed along the ionization tracks of beta particles because some of the ionized electrons are ejected from target atoms with so much energy (on the order of 1 keV or so) that they can ionize other target atoms. These secondary ionizations form a short trail of ionization extending outward from the main path of the beta particle as shown in Figure 7-5*b*.

Energy Loss by Bremsstrahlung

Since beta particles are high-speed electrons, they produce bremsstrahlung, or "braking radiation," in passing through matter, especially if the absorbing medium is a high-*Z* material. As electrons pass near a nucleus they experience acceleration due to the deflecting force and give up energy in producing bremsstrahlung, which in turn reduces the speed of the beta particle by an amount that corresponds to the energy lost to the bremsstrahlung photon. Whether a given beta particle converts part (or all) of its energy to radiation emission depends on the path it takes toward a target nucleus and the amount of deflection that occurs. The deflecting force is directly proportional to the nuclear charge (or *Z*) of the target, and since the beta particles can approach the nucleus of a target atom from many different angles, a source of beta particles will produce a spectrum of bremsstrahlung forming a continuous band of energies that extends up to the maximum energy of the beta particle. The process is the same one that occurs in an x-ray tube and is illustrated in Figure 7-6.

The yield or fraction of bremsstrahlung produced is proportional to the atomic number of the target (or absorbing) material and the energy of the electrons striking the target, which of course rapidly decreases as the particles traverse the target material. Describing the process analytically is difficult; consequently, best-fit models have been constructed of empirical data on the fraction of the beta energy from a source that is converted to photons in order to account for them in dosimetry or radiation shielding. Unfortunately, the measured data are for monoenergetic electrons, which complicate their applicability for beta particles that are emitted as a spectrum of energies from a source. Berger and Seltzer have tab-

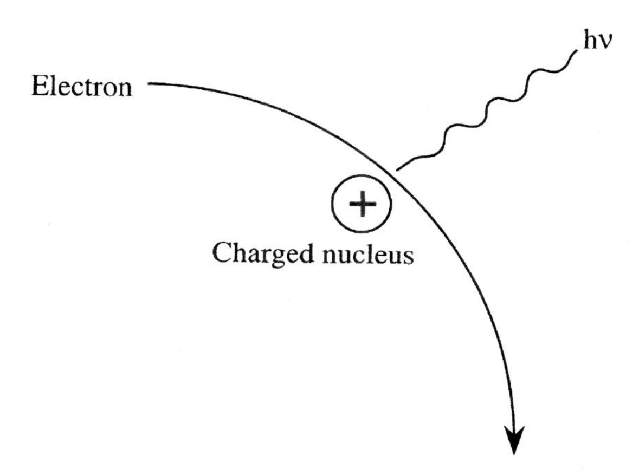

FIGURE 7-6. Schematic of bremsstrahlung production by a high-speed electron.

ulated fractional yields for monoenergetic electrons (see Chapter 8), and these can be used with some discretion to estimate photon production for beta emitters. As much as 10% or so of the energy of 2-MeV particles can be converted to bremsstrahlung in heavy elements such as lead ($Z = 82$), an effect that must be recognized in choosing shield material (see Chapter 8). When beta particles are absorbed in tissue (which has a low atomic number), less than 1% of the interactions produce bremsstrahlung, and many of those that do are likely to escape the tissue medium because their probability of interaction is also low in this low-Z medium.

Čerenkov Radiation

High-speed beta particles (and other high-energy particles as well) can cause the emission of visible radiation with a blue tint, *Čerenkov radiation*, named after the pioneer scientist who studied the effect. Čerenkov radiation occurs because charged particles moving through a medium experience deflections (and accelerations) which yield photons of electromagnetic radiation that must traverse the same medium. The velocity of these photons is equal to c/n, where n is the index of refraction, which is specific for each medium ($n = 1.4$ for water). Since a medium such as water will refract electromagnetic radiation traveling in it, it is common for high-speed beta particles to have a velocity in the medium $\geq c/n$; when this happens, photons of electromagnetic energy are produced that cannot outrun the fast-moving beta particle. These photons will lag behind the beta particle, which will produce other photons up ahead, and as the trailing photons overlap the new ones, they constructively interfere with each other. The effect can be thought of as a charged-particle sonic boom, analogous to the compression wave created by an aircraft which, due to its physical constraint in traveling through air, simply cannot keep up with the aircraft that creates the shockwave (a similar analogy is the bow wave produced by a speedboat).

The constructive interference produced by high-speed beta particles (see Figure 7-7) yields photons in the ultraviolet region of the visible light spectrum and

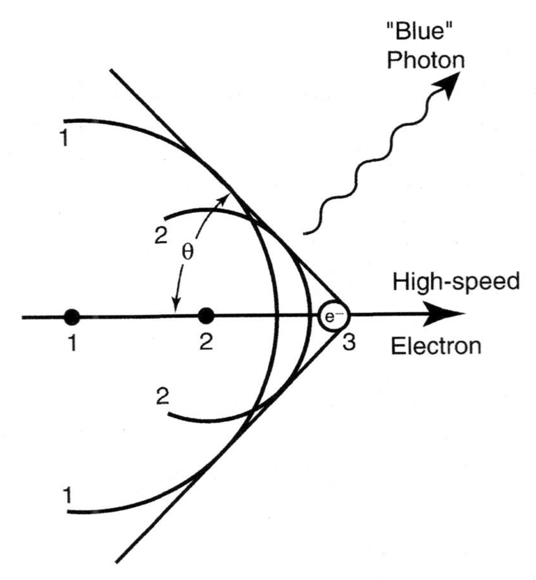

FIGURE 7-7. Wavefronts produced by a particle with velocity $\geq c/n$ constructively interfere to produce blue photons of light, Čerenkov radiation.

accounts for the blue glow that occurs from highly radioactive sources (e.g., multicurie cobalt sources and highly radioactive reactor fuel elements). Čerenkov radiation around nuclear reactor cores and other high-intensity radiation sources is fascinating to observe, but it accounts for little energy loss compared to that from ionization and bremsstrahlung production.

Attenuation of Beta Particles

Beta particles are attenuated in an absorption medium by the various interaction processes, primarily through ionization and radiative energy losses. The degree of absorption (or attenuation) is determined by observing the change in source strength due to insertion of different thickness of absorber (usually, aluminum in units of mg/cm^2) in the beam.

As shown in Figure 7-8, the counting rate, when plotted on a logarithmic scale, decreases as a straight line, or very nearly so, over a large fraction of the absorber thickness, eventually tailing off into another straight-line region represented by the background, which is always present. The point where the beta absorption curve meets the background is the range $R_{\beta,\max}$ traversed by the most energetic particles emitted, and curves of this type can be used to determine $R_{\beta,\max}$ and $E_{\beta,\max}$ for a beta source as described in Chapter 8.

The measured range in mg/cm^2 in aluminum or similar absorber is then used to determine the energy by an empirical range–energy curve, as shown in Figure 7-9. Aluminum is the absorber medium most often used for determining ranges of beta particles, but most absorber materials will have essentially the same electron density since, with the exception of hydrogen, the Z/A ratio varies only slowly

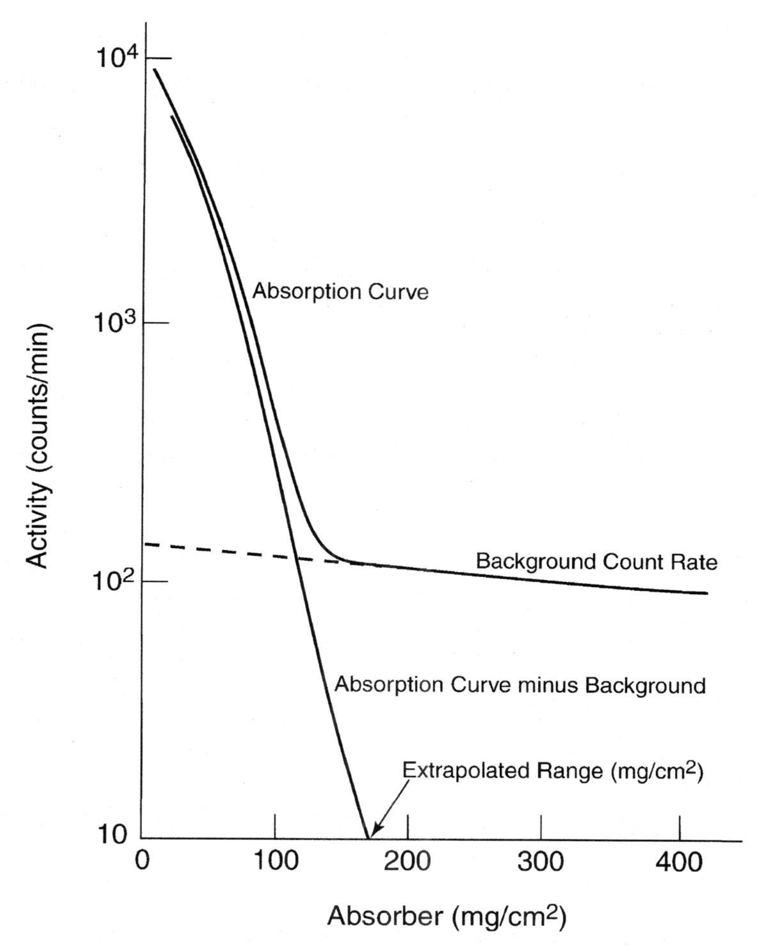

FIGURE 7-8. Decrease in measured activity of a beta particle source versus mg/cm^2 of absorber thickness that trails off into the background of the detector system. Subtraction of the background portion from the total curve yields a curve that can be extrapolated to estimate the maximum range (in mg/cm^2) of the beta source.

with Z. Consequently, the ranges in Figure 7-9 can be considered generic for most light absorbers, with a notable exception for air (see Figure 8-1), as long as the equivalent mg/cm^2 of absorber is known.

Empirical relations have been developed from experimental data to relate range to beta-particle energy. The range R in mg/cm^2 versus energy in MeV has been empirically derived for $0.01 \le E \le 2.5$ MeV as

$$R = 412E^{1.265-0.0954\ln E}$$

or alternatively for $R < 1200$ mg/cm^2,

$$\ln E = 6.63 - 3.2376(10.2146 - \ln R)^{1/2}$$

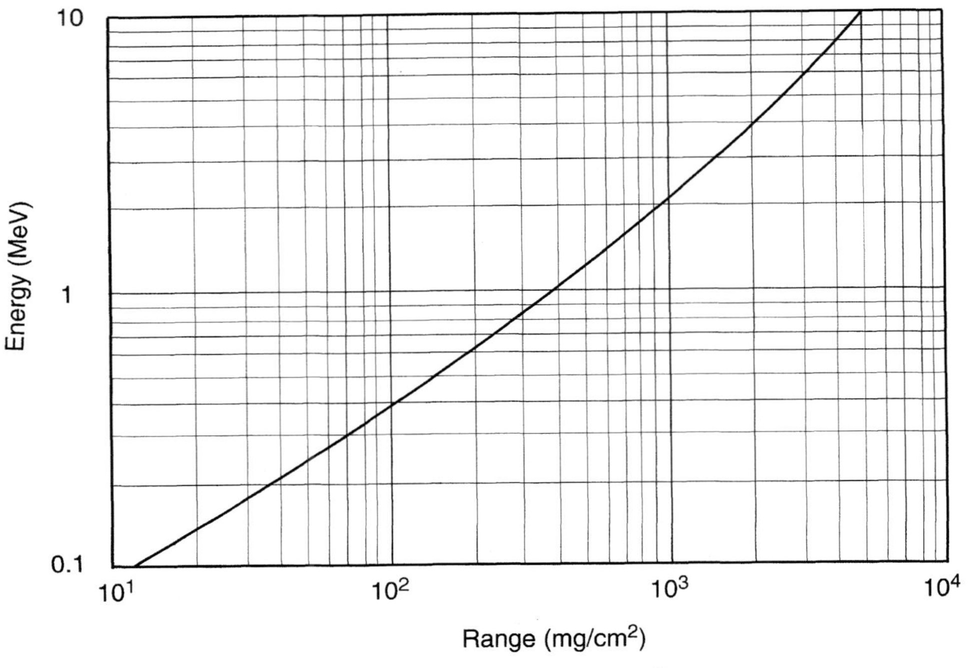

FIGURE 7-9. Equivalent range of electrons in mg/cm^2 of low-Z absorbers.

Another empirical relationship between E in MeV and R in mg/cm^2 is Feather's rule for beta particles of $E \geq 0.6$ MeV:

$$R = 542E - 133$$

Although these relationships can be used to calculate ranges and energies for low-energy electrons, they can generally be obtained more accurately from the empirical range–energy curve in Figure 7-9, where range is determined in units of mg/cm^2.

The exponential decrease in the number of beta particles counted versus absorber thickness (Figure 7-8) is represented, to a good approximation, for the straight-line part of the curve as

$$N(x) = N_0 e^{-\mu_\beta(\rho x)}$$

where N_0 is the number of beta particles counted with zero thickness of absorber, $N(x)$ the number observed for an absorber of thickness x, μ_β the beta absorption coefficient in cm^2/g, and ρx the density thickness (g/cm^2) of the absorber. Since the units of μ_β for various beta sources are determined (and expressed) in cm^2/g, it is necessary to specify the absorber as a density thickness with units of g/cm^2, which is obtained by multiplying the absorber thickness x by its density ρ.

The absorption of beta particles is not truly exponential, or probabilistic, as the expression for absorption would suggest. The exponential form of the curve is an artifact produced by a combination of the continually varying spectrum of beta energies and the scattering of the particles by the absorber. It can, however, be used with reasonable accuracy for density thicknesses less than the maximum range of the beta particles; there is no practical value in evaluating thicknesses greater than $R_{\beta,\text{max}}$.

Approximate relationships for the beta absorption coefficient have been determined for air and tissue specifically because of their utility in radiation protection. Appropriate values of μ_β in cm^2/g can be determined for air and tissue as follows:

$$\mu_{\beta,\text{air}} = 16(E_{\beta,\text{max}} - 0.036)^{-1.4}$$

$$\mu_{\beta,\text{tissue}} = 18.6(E_{\beta,\text{max}} - 0.036)^{-1.37}$$

For any other medium,

$$\mu_{\beta,i} = 17(E_{\beta,\text{max}})^{-1.14}$$

where values of μ_β are in units of cm^2/g (based on measurements with aluminum absorbers) and $E_{\beta,\text{max}}$ is in units of MeV. It should be noted that μ_β is defined uniquely with units of cm^2/g instead of the more traditional units of cm^{-1}, or as μ/ρ, because the most practical uses of the beta absorption coefficients are in calculations of beta dose or beta attenuation.

Radiation Dose from Beta Particles

Radiation dose calculations for beta particles are based on the number traversing a medium (usually tissue) per unit area; their range, which is energy dependent; and the energy deposition fraction per unit mass μ_β. When it is possible to establish a flux of beta particles impinging on a mass of medium, dose calculations are straightforward if the beta absorption coefficient and the average beta-particle energy are known. The average beta-particle energy can be obtained by equations in Chapter 5 or, more accurately, by the listed values for each radionuclide contained in Appendix D. The rule of thumb, $\overline{E} = \frac{1}{3}E_{\beta,\text{max}}$, can also be used as a first approximation, but for accurate calculation the weighted average values of $\overline{E}$ for each beta emitter should be used.

Once the beta particle energy flux ($\text{MeV/cm}^2 \cdot \text{s}$) is known, the energy absorption rate in a medium (or dose) can be calculated by using the beta absorption coefficient for the medium and the particular beta-particle energy. The beta-radiation dose rate $\dot{D}_\beta$ for beta particles of energy $\overline{E}$ (MeV) is

$$\dot{D}_\beta(\text{rad/h}) = \frac{\phi_\beta \times \overline{E} \times 1.6022 \times 10^{-6}\ \text{erg/MeV} \times \mu_\beta(\text{cm}^2/\text{g}) \times 3600\ \text{s/h}}{100\ \text{ergs/g} \cdot \text{rad}}$$

or

$$\dot{D}_\beta(\text{rad/h}) = 5.768 \times 10^{-5}\phi_\beta\overline{E}\mu_\beta$$

Example 7-6: What is the beta dose rate at the surface of a person 1 m from a 1 Ci source of ^{32}P ($E_{\beta,\max} = 1.710$ MeV and $\overline{E}_\beta \simeq 0.695$ MeV) assuming no attenuation of the beta particles as they traverse the medium?

SOLUTION: The beta flux at 100 cm is

$$\phi_{\beta,E} = \frac{3.7 \times 10^{10}\ \text{t/s} \cdot \text{Ci}}{4\pi(100\ \text{cm})^2}$$
$$= 2.944 \times 10^5\ \beta\text{s/cm}^2 \cdot \text{s}$$

The beta-absorption coefficient in tissue for ^{32}P beta particles is

$$\mu_{\beta,\text{tissue}} = 18.6(1.71 - 0.036)^{-1.37} = 9.183\ \text{cm}^2/\text{g}$$

and the beta dose rate due to beta particles with $\overline{E} = 0.695$ MeV is

$$\dot{D}_\beta = 5.768 \times 10^{-5} \times 2.944 \times 10^5\ \beta/\text{cm}^2 \cdot \text{s}$$
$$\times\, 0.695\ \text{MeV}/\beta \times 9.183\ \text{cm}^2/\text{g}$$
$$= 108.4\ \text{rads/h}$$

Most practical problems of beta radiation dose require adjustments of the energy flux. In Example 7-6, for instance, the dose rate from a point source would be more accurate if absorption of beta particles in the 1-m thickness of air were also considered since some absorption occurs in the intervening thickness of air before reaching the person. This is done by adjusting the energy flux for air attenuation using $\mu_{\beta,\text{air}}$ for the 1.71 MeV beta particles of ^{32}P, which is

$$\mu_{\beta,\text{air}} = 16(1.71 - 0.036)^{-1.4} = 7.78\ \text{cm}^2/\text{g}$$

The beta dose rate for ^{32}P beta particles after traversing a thickness x of air is then

$$\dot{D}_\beta(x) = \dot{D}^0_\beta e^{-\mu_{\beta,\text{air}}(\rho x)}$$
$$= 108.4(0.3657) = 39.64\ \text{rad/h}$$

A further adjustment to Example 7-6 is appropriate to account for absorption of beta particles in the dead layer of skin with a density thickness of 0.007 g/cm^2 (i.e., to calculate the dose where it matters, to living tissue). The dose rate just below the 7-mg/cm^2 skin layer is known as the *shallow dose*. To make this adjustment, the beta dose rate is again calculated considering air attenuation and the density thickness of tissue (0.007 g/cm^2), or

$$\dot{D} = 39.64e^{-\mu_{\beta,t}(\rho x)}$$
$$= 39.64e^{(-9.183\ \text{cm}^2/\text{g})(0.007\ \text{g/cm}^2)} = 37.2\ \text{rad/h}$$

TABLE 7-2. Stopping Powers and Attenuation of Beta-Particle Dose, Relative to Air, for Low-*Z* Materials

Medium	Z^a	Stopping Power	Attenuation Relative to Air
Polyethylene	4.75	1.205	1.12
Polystyrene	5.29	1.118	1.04
Carbon	6.00	1.007	0.96
Mylar	6.24	1.066	1.03
Water	6.60	1.150	1.12
Muscle	6.65	1.136	1.11
Air	7.36	1.00	1.00
LiF	7.50	0.965	0.97
Teflon (CF_2)	8.25	1.032	1.05
Bone	8.74	1.063	1.09

Source: Adapted from *The Health Physics and Radiological Health Handbook*, with permission.
[a] Values of *Z* for compounds are weighted based on chemical composition.

To summarize, if ϕ_β, the flux of beta particles, is known, the general expression for the beta dose rate at the shallow depth in tissue after traversing a thickness x of air is

$$\dot{D}_\beta(\text{rad/h}) = 5.768 \times 10^{-5} \phi_\beta \mu_\beta \overline{E}[e^{-\mu_{\beta,a}(\rho x)}][e^{-\mu_{\beta,t}(0.007\ \text{g/cm}^2)}]$$

where ϕ_β is in units of $\beta/\text{cm}^2 \cdot \text{s}$, $\overline{E}$ is MeV/t, $\mu_{\beta,a}$ and $\mu_{\beta,t}$ are in units of cm^2/g, ρx is the density thickness of air, and 0.007 g/cm^2 is the density thickness of the dead skin layer.

Energy deposition of beta particles in a medium can be represented relative to air, as shown in Table 7-2. These factors can be useful for determining energy deposition (dose) or attenuation for a given medium from a beta dose rate measured in air. For example, when LiF is used as a dosimeter, its response can be estimated by multiplying the beta dose rate in air by 0.97 (from Table 7-2).

Beta Dose from Contaminated Surfaces

If a floor, wall, or other solid surface is uniformly contaminated with a beta emitter, the dose rate due to the emitted beta particles can be determined by establishing the beta flux and then proceeding as demonstrated in Example 7-7. For contaminated solid surfaces, the flux will be determined by the area contamination level (e.g., μCi/cm^2), a geometry factor (usually $\frac{1}{2}$), and a backscatter factor to account for beta particles that penetrate the surface but are scattered back out to increase the beta flux reaching a receptor near the surface. Table 7-3 contains backscatter factors versus beta energy for various materials. It is noticeable that the backscatter factor is highest for low-energy beta particles on dense (or high-*Z*) materials.

Example 7-7: What is the beta dose rate to skin (a) at the shallow depth just above the surface (i.e., at contact) of a large copper shield uniformly contaminated with 10 μCi/cm^2 of ^{32}P, and (b) at a distance of 30 cm?

TABLE 7-3. Backscatter Factors (BSF) for Beta Particles on Thick Surfaces

E_β (MeV)	Carbon	Aluminum	Concrete[a]	Iron	Copper	Silver	Gold
0.1	1.040	1.124	1.19	1.25	1.280	1.38	1.5
0.3	1.035	1.120	1.17	1.24	1.265	1.37	1.5
0.5	1.025	1.110	1.15	1.22	1.260	1.36	1.5
1.0	1.020	1.080	1.12	1.18	1.220	1.34	1.48
2.0	1.018	1.060	1.10	1.15	1.160	1.25	1.40
3.0	1.015	1.040	1.07	1.12	1.125	1.20	1.32
5.0	1.010	1.025	1.05	1.08	1.080	1.15	1.25

Source: Adapted from Tabata, Ito, and Okabc, 1971.
[a]BSF values for concrete based on arithmetic average of BSF values for aluminum and iron.

SOLUTION: (a) The beta flux is presumed to be equal to the area emission rate, and since half of the beta particles would be directed into the surface, the geometry factor is 0.5. The backscatter factor for ^{32}P beta particles ranges between 1.22 and about 1.26 based on the average energy, which by interpolation yields a backscatter factor of 1.24. The surface-level beta dose rate at the air–tissue interface due to beta particles with $\overline{E} = 0.695$ MeV/t is

$$\begin{aligned}\dot{D}_\beta(\text{rad/h}) &= 5.768 \times 10^{-5}\phi_\beta\overline{E} \times 0.5 \times \mu_{\beta,t}\\ &= (5.768 \times 10^{-5})(10\ \mu\text{Ci/cm}^2 \times 3.7 \times 10^4\ \text{t/s} \times 1.24\\ &\quad \times 0.5 \times 0.695\ \text{MeV/t})(9.183\ \text{cm}^2/\text{g})\\ &= 84.45\ \text{rads/h}\end{aligned}$$

and the beta dose at the shallow depth in tissue is

$$\dot{D}_{\beta,\text{sh}} = 84.45e^{(-9.183\ \text{cm}^2/\text{s})(0.007\ \text{g/cm}^2)} = 79.2\ \text{rad/h}$$

(b) At 30 cm, the beta particles will be attenuated by 30 cm of air, $\mu_{\beta,a} = 7.78$ cm^2/g, $\rho_a = 0.001205$, and the beta dose rate at the skin is

$$\begin{aligned}\dot{D}_{\beta,\text{skin}} &= 84.45\ \text{rad/h} \cdot e^{-\mu_{\beta,a}\rho_{\text{air}}(30\ \text{cm})}\\ &= 63.75\ \text{rad/h}\end{aligned}$$

and at the shallow depth, a dose of

$$\begin{aligned}\dot{D}_{\beta,\text{sh}} &= 63.75e^{(-9.183\ \text{cm}^2/\text{g})(0.007\ \text{g/cm}^2)}\\ &= 59.78\ \text{rad/h}\end{aligned}$$

Beta Contamination on Skin or Clothing

A shallow dose from beta particles can occur from direct contamination of the skin or from contaminated protective clothing in contact with the skin. If an area energy flux can be established, calculation of the shallow dose is straightforward by assuming a geometry factor of 0.5 (one-half the activity goes outward) and adjusting the energy flux to account for beta-energy absorption in the dead layer of skin. There will be some backscatter of beta particles by the tissue layer below the skin, but since tissue is a low-Z material, this is only a few percent and can be ignored for most radiation protection situations.

Example 7-8: Estimate the shallow dose rate from 5 mL of solution containing 10 μCi/mL of ^{32}P ($E_{\beta,\max} = 1.71$ MeV) spilled onto the sleeve of a worker's lab coat and distributed uniformly over an area of about 50 cm^2.

SOLUTION: Uniform absorption of the solution onto the lab coat would produce an area contamination of 1 μCi/cm^2. Assuming no attenuation of beta particles in the fabric, one half of the beta particles emitted would impinge on the skin, and the energy flux just reaching the basal skin layer would be

$$\begin{aligned}\phi_{\beta,E} &= 1\ \mu\text{Ci/cm}^2 \times 3.7 \times 10^4\ \text{t/s} \cdot \mu\text{Ci} \times 0.695\ \text{MeV/t} \times 0.5e^{-\mu_{\beta,t}0.007} \\ &= 1.21 \times 10^4\ \text{MeV/cm}^2 \cdot \text{s}\end{aligned}$$

and the beta dose rate at the shallow depth of tissue is

$$\dot{D}_{\beta,\text{sh}} = (5.768 \times 10^{-5})(1.21 \times 10^4\ \text{MeV/cm}^2 \cdot \text{s})(9.183\ \ \text{cm}^2/\text{g}) = 6.39\ \text{rad/h}$$

A similar calculation can be made for direct contamination of the skin, although in most practical situations it is more difficult to determine the area contamination. The skin contamination may be estimated by careful measurements with a thin-window detector or by washing the wet area with a swab or liquid and measuring the activity removed.

Beta Dose from Hot Particles

Small, but very radioactive, particles that emit beta particles and gamma rays are occasionally found on the skin or clothing of persons in radiation areas. Because of their small size and intense radioactivity, small areas of tissue can receive significant energy deposition before the particles are detected and removed. Various computer codes, such as VARSKIN, have been developed to deal with the unusual and often diverse conditions presented by hot particles; however, it is possible to calculate the resultant beta doses by applying several general principles. The activity of the particle establishes the number of beta particles being emitted from a point, and since the particles are emitted uniformly in all directions from a point source, the emitted energy will be deposited into a hemispherical volume of tissue of radius equal to the range of the beta particles in tissue, as shown in Figure 7-10.

To determine the beta dose from a hot particle, it can be assumed that the average energy of the beta particles is distributed fairly uniformly throughout the tissue mass in the hemispherical volume of radius R_β, where R_β is based on the maximum

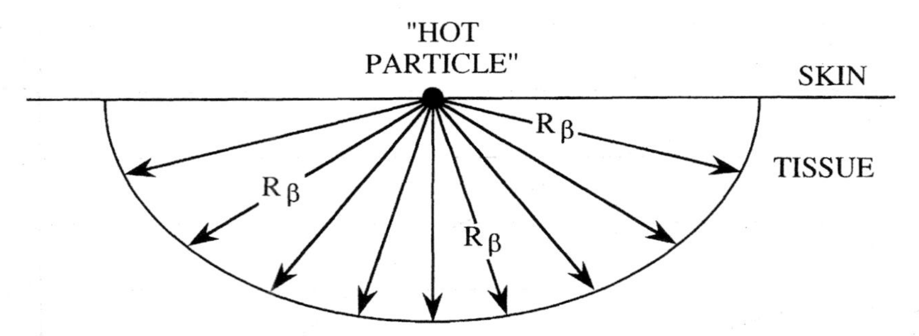

FIGURE 7-10. Schematic of energy deposition of beta particles emitted by a point-source "hot particle" on the surface of the skin. The energy of the beta particles is deposited uniformly in a three-dimensional hemispherical mass of tissue of radius equal to the maximum range R_β of the beta particles.

energy of the emitted beta particles. The average energy is used to determine energy deposited rather than the maximum energy even though R_β is based on $E_{\beta,\max}$. The interplay of the continuous energy spectrum of a beta emitter, the maximum depth of penetration of each particle, and the Bragg absorption pattern for each beta particle combine to provide fairly uniform distribution of beta energy throughout the hemispherical mass. This is a reasonable determination since the continuous energy spectrum of beta particles and the random attenuation combine to produce an approximate exponential absorption pattern (see Figure 7-8). VARSKIN uses a value of $0.9R_\beta$ to account for these variables, thus conservatively assuming a smaller mass (and a somewhat higher absorbed dose) in which the energy deposition occurs. The beta dose rate to tissue, without correcting for absorption in the dead layer of skin, from a hot particle of activity $q(\mu\text{Ci})$, one-half of which impinges on the skin, is

$$\dot{D}_\beta = \frac{0.5q(\mu\text{Ci}) \times 3.7 \times 10^4 \text{ t/s}\cdot\mu\text{Ci} \times \overline{E} \text{ MeV/t} \times 1.6022 \times 10^{-6} \text{ erg/MeV} \times 3600 \text{ s/h}}{\frac{4}{3}\pi R_\beta^3 \times \frac{1}{2} \times 1 \text{ g/cm}^3 \times 100 \text{ ergs/g}\cdot\text{rad}}$$

$$= \frac{0.51\overline{E}_\beta \times q(\mu\text{Ci})}{R_\beta^3} \text{ rad/h}$$

where q is the activity of the hot particle in μCi, R_β the range in centimeters of the beta particles of maximum energy, and $\overline{E}$ is the average energy in MeV of each beta particle emitted.

Example 7-9: Estimate the radiation dose, ignoring the gamma emissions, due to a hot particle containing 1 μCi of ^{60}Co ($E_{\beta,\max} = 0.318$ MeV; $\overline{E} = 0.106$ MeV) on a person's skin.

SOLUTION: The beta dose due to 1 μCi of ^{60}Co is assumed to be produced by deposition of the average beta energy of 0.106 MeV/t (Appendix D) uniformly in a hemispherical mass of radius R_β. The range R_β of 0.318 MeV beta particles is determined from Figure 7-9 to be 84 mg/cm^2, which corresponds to a distance in tissue of unit density of 0.084 cm. The radiation dose due to beta particles emitted by the hot particle is

$$D_\beta(\text{rad/h}) = \frac{0.51 \times 1\ \mu\text{Ci} \times 0.106\ \text{MeV/t}}{(0.084)^3}$$

$$= 91.1\ \text{rad/h}$$

PHOTON INTERACTIONS

Describing a photon is difficult, even though its roles in radiation protection are fairly well understood. Its more obvious properties are no rest mass, it always travels at the speed of light, and it can interact as a particle even though it is also a wave. The makeup or composition of a photon is unknown, but like the electron, it is a true "point"—it has no physical size and cannot be taken apart to yield subcomponents.

The energy of a photon is $h\nu$, its momentum is h/λ, and its energy can be described in terms of its momentum as $E = pc$. Einstein showed that a photon feels the pull of gravity as if it were a particle even though it has no rest mass. Photons are fundamental to physics because they transmit the electromagnetic force; two electric charges are believed to interact by "exchanging" photons (photons are emitted by one charge and absorbed by the other), even though these photons exist only in the mathematical framework of theoretical physics.

Photons originate in interesting ways; they simply appear when it is necessary to carry off excess energy (such as bremsstrahlung production, in radioactive transformation, or in nuclear interactions). The dichotomy of being both a wave and a particle is intriguing, but both properties are required to explain photon interactions in media. The origin and appearance of interference and diffraction phenomena are clearly wave properties, and its absorption to deposit energy and impart momentum requires a particle description as so effectively deduced by Einstein (photoelectric effect) and Compton (in explaining photon scattering). It is difficult to accept that both could exist simultaneously, yet both properties are required to describe photons completely. Neither the wave nor the particle theory is wholly correct all the time, but the two, as defined by Bohr, are complementary to each other.

The principal modes by which photons interact with matter to be attenuated and to deposit energy are the photoelectric effect, the Compton effect, and pair production. Photons also undergo Rayleigh scattering, Bragg scattering, photodisintegration, and nuclear resonance scattering; however, these result in negligible attenuation or energy deposition and can generally be ignored for purposes of radiation protection.

Photoelectric Interactions

A low-energy photon can, by a process known as the *photoelectric effect*, collide with a bound orbital electron and eject it from the atom. The electron is ejected with an energy equal to that of the incoming photon, $h\nu$, minus the binding energy of the electron in its particular orbit, an energy that must be overcome to free the electron from the atom. The interaction must occur with a bound electron since the entire atom is necessary to conserve momentum, and it often occurs with one of the

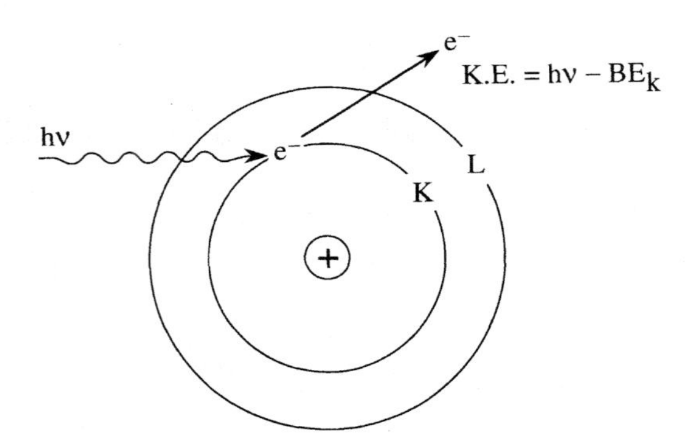

FIGURE 7-11. Schematic of the photoelectric effect.

inner-shell electrons. Since a vacancy is created in the electron shell, a characteristic x ray, typically from filling the K shell, will also be emitted. The kinetic energy of the ejected electron is almost always absorbed in the medium where photoelectric absorption occurs. Characteristic x rays that are produced are also very likely to be absorbed in the medium, typically by another photoelectric interaction or by the ejection and absorption of Auger electrons produced from them (see Figure 7-11).

The *photoelectric absorption coefficient* τ is a function of the atomic number Z of the absorbing material (generally related to the density, ρ, of the absorbing medium) and the energy of the radiation as follows:

$$\tau \simeq \text{constant}\frac{Z^5}{E^3}$$

It is evident that photoelectric absorption is most pronounced in high-Z materials and for low-energy photons (less than 0.5 MeV). In a high-Z material such as lead, L x rays and M x rays can also be prominent emissions from target atoms, and these will either be absorbed in the absorbing medium or will contribute to the photon fluence. The photoelectric effect can be summarized as follows:

- It occurs only with bound electrons because the entire atom is necessary to conserve momentum.
- The interaction coefficient is greatest when the photon energy just equals the amount to overcome the binding energy of the orbital electron, causing it to be ejected from its shell.
- The photoelectric absorption coefficient is directly proportional to Z^5 and inversly proportional to $(h\nu)^3$ on average.
- In tissue the absorbed energy $E_{\text{ab}} \approx h\nu$, and the transfer, absorption, and attenuation coefficients are nearly equal.

Compton Interactions

Compton scattering interactions are especially important for gamma rays of medium energy (0.5 to 1.0 MeV), and for low-Z materials such as tissue can be the dominant

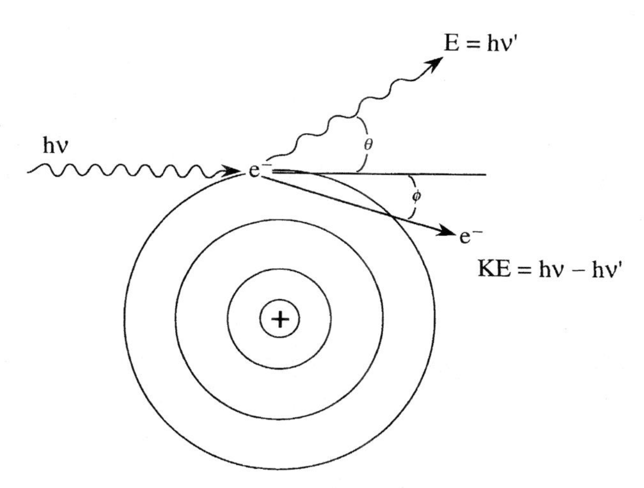

FIGURE 7-12. Compton scattering with a "free" electron.

interaction mechanism down to about 0.1 MeV. Compton scattering involves a collision between a photon and a "free" or very loosely bound electron in which a part of the energy of the photon is imparted to the electron, as shown in Figure 7-12. Both energy and momentum are conserved in the collision.

The Compton-scattered photon emerges from the collision in a new direction and with reduced energy and increased wavelength. The change in wavelength, $\lambda' - \lambda$, commonly referred to as the *Compton shift*, is

$$\lambda' - \lambda = \frac{h}{m_0 c}(1 - \cos\theta) = 0.024264(1 - \cos\theta)\ \text{Å}$$

It is notable that the change in wavelength (and decrease in energy) of the photon is determined only by the scattering angle. The term h/m_0c, often called the *Compton wavelength*, has the value 2.4264×10^{-10} cm.

Energy transfer to the recoiling electron is the most important consequence of Compton interactions since it will be absorbed locally to produce radiation dose. This is a variable quantity and can range from zero up to a maximum value for electrons ejected in the forward direction. The fraction of the photon energy $h\nu$ that is transferred to the Compton electron is shown in Figure 7-13; the value E_{tr} is dependent on the scattering angle, where $\overline{E}_{tr}$ is the average value due to random processes and $E_{tr,max}$ is the maximum value for a scattering angle of 180°.

The *Compton interaction coefficient* σ consists of two components:

$$\sigma = \sigma_a + \sigma_s$$

where σ is the total Compton interaction coefficient, σ_a the Compton absorption coefficient for photon energy lost by collisions with electrons, and σ_s the loss of energy due to the scattering of photons out of the beam. The Compton interaction coefficient σ is determined by electron density, which is directly related to Z and

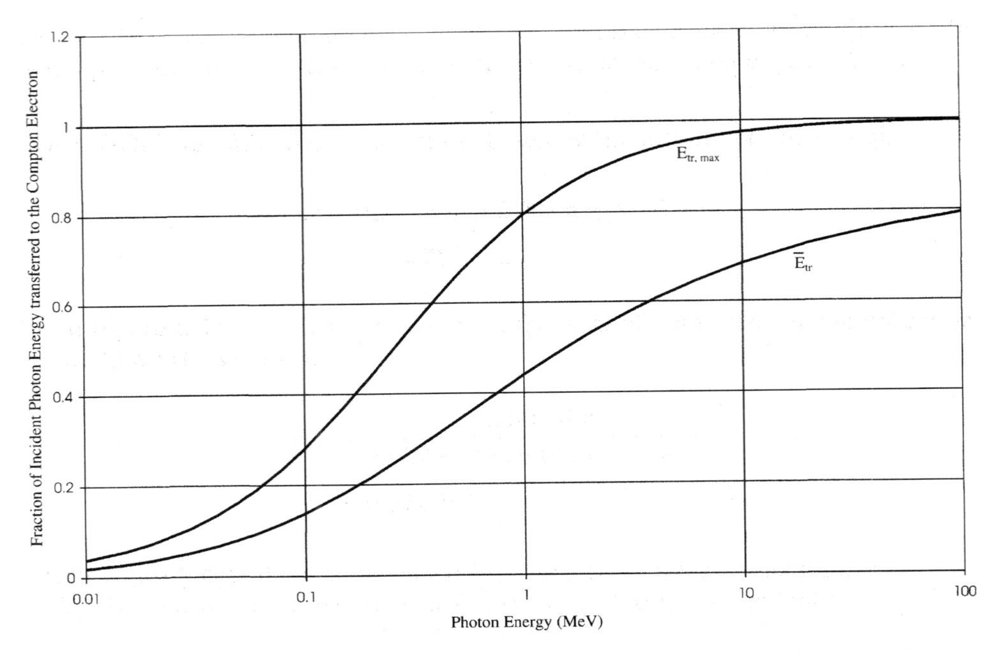

FIGURE 7-13. Fraction of incident photon energy ($h\nu$) transferred to the Compton electron.

inversely proportional to E as follows:

$$\sigma \simeq \text{constant}\frac{Z}{E}$$

Compton scattering interactions can be summarized as follows:

- A Compton interaction occurs between a photon and a free electron, producing a recoiling electron and a scattered photon of reduced energy.
- Kinetic energy transferred to the electron is directly proportional to the scattering angle of the scattered photon and, on average, increases with photon energy.
- The Compton interaction coëfficient decreases with increasing energy and is almost independent of atomic number.
- For photons with energies above 100 keV, Compton interactions in soft tissue (low-Z material) are much more important than either the photoelectric or pair production interactions.

Pair Production

When a high-energy (> 1.022 MeV) photon interacts with the strong electromagnetic field surrounding a nucleus as shown in Figure 7-14, its energy can be converted into a pair of electron masses, one of which is negatively charged (the electron) and the other, the *positron*, positively charged. Pair production can also occur

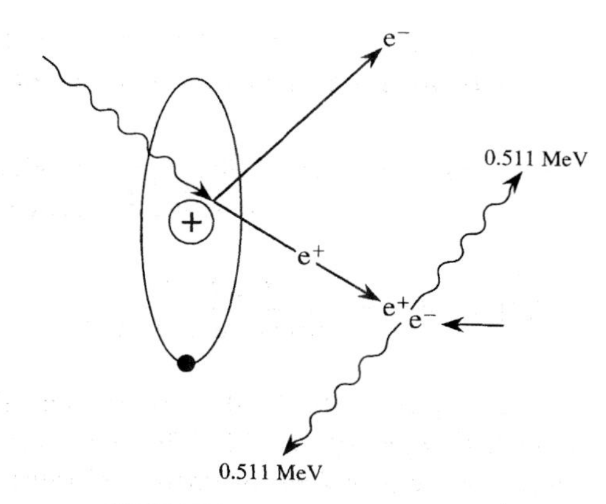

FIGURE 7-14. Pair production.

in the field of an electron yielding a triplet consisting of a positron, a negatron, and the recoiling electron.

Pair production is a classic example of Einstein's special theory of relativity in which the pure energy of the photon is converted into two electron masses, and since energy is conserved, the positron and electron share the energy left over ($h\nu - 1.022$) after the electron masses have been formed. This remaining energy appears as kinetic energy of the e^+ and e^- pair but is not shared equally. The positively charged nucleus repels the positively charged positron, which provides an extra "kick" while the negatron is attracted and thus slowed down with a decrease in its kinetic energy. Because of these circumstances, the positron should receive a maximum of about $0.0075Z$ more kinetic energy than the average negatron. The slight difference in energy shared by the positron and the electron in pair production interactions is of little consequence to radiation dosimetry or detection since the available energy, $h\nu - 1.022$ MeV, will be absorbed in the medium with the same average result, regardless of how it is shared.

Example 7-10: What energy sharing will exist for positrons and electrons when 2.622 MeV gamma rays from ThC′ produce a positron–electron pair?

SOLUTION: The kinetic energy to be shared is $2.622 - 1.022$ MeV = 1.6 MeV. If the energy were shared equally, the positron and negatron would each possess 0.80 MeV, but the positron energy should be 0.8 MeV + $0.0075Z$ = 1.06 MeV. The negatron energy is $1.6 - 1.06 = 0.54$ MeV.

Pair production interactions are also accompanied by the emission of two annihilation photons of 0.511 MeV each, which are also shown in Figure 7-14. The positron will exist as a separate particle as long as it has momentum and kinetic energy. However, when it has been fully absorbed, being antimatter in a matter world, it will interact with a negatively charged electron, forming for a brief moment a "neutral particle" of "positronium," which then vanishes, yielding two 0.511 MeV photons (i.e., mass becomes energy). The absorption of high-energy photons thus yields a complex pattern of energy emission and absorption in which the pure energy of the photon produces an electron and a positron which deposit $h\nu - 1.022$ MeV of kinetic energy along a path of ionization, followed in turn by positron annihilation with a free electron to convert mass back into energy. The absorption of a high-energy photon by pair production thus yields two new photons of 0.511 MeV, which may or may not interact in the medium, and an intermediate pair of electron masses, which almost certainly do.

The *pair production interaction coefficient* κ is proportional to the square of the atomic number Z for photons with energy greater than 2×0.511 MeV (the energy required to form an electron–positron pair) and has the following relationship:

$$\kappa \simeq \text{constant } Z^2(E - 1.022)$$

where Z is the atomic number and E is the photon energy in MeV. Pair production interactions can be summarized as follows:

- They occur for photons with $h\nu \geq 1.022$ MeV primarily in the field of the nucleus to produce two electron masses with kinetic energy, but can also occur with an orbital electron yielding a triplet of electron masses.
- The kinetic energy shared between the positron and the electron is $h\nu - 1.022$ MeV.
- The positron annihilates with a free electron after dissipating its kinetic energy to produce two 0.511 MeV annihilation photons.
- The absorption coefficient increases rapidly with energy above the 1.022 MeV threshold and varies approximately as Z^2 of the absorbing medium.

Photodisintegration

Photodisintegration interactions can also deplete a beam of photons if the photon energy is sufficiently large. From a practical standpoint, photodisintegration can generally be neglected in calculating the energy removed from a photon beam; however, such reactions are quite sharp at 1.66 MeV for ^{9}Be(γ,n) interactions and 2.225 MeV for ^{2}H(γ,n) interactions. Except for these two light elements, photodisintegration reactions require photon energies of 8 MeV or more. For photon energies above 20 MeV, the cross sections for photodisintegration are sufficiently large and must be accounted for in shield designs. Large neutron fields are created at these energies, and these neutrons will be mixed with a large number of unabsorbed photons, both of which must be considered in radiation dosimetry and shield designs.

PHOTON ATTENUATION AND ABSORPTION

Photon interactions with matter are very different from those of charged particles. When x or gamma rays traverse matter, some are absorbed, some pass through without interaction, and some are scattered as lower-energy photons in directions that are quite different from those in the primary beam. The attenuation of a photon beam by an absorber is characterized as occurring in good geometry or poor geometry, as shown in Figure 7-15. Good geometry exists when every photon that

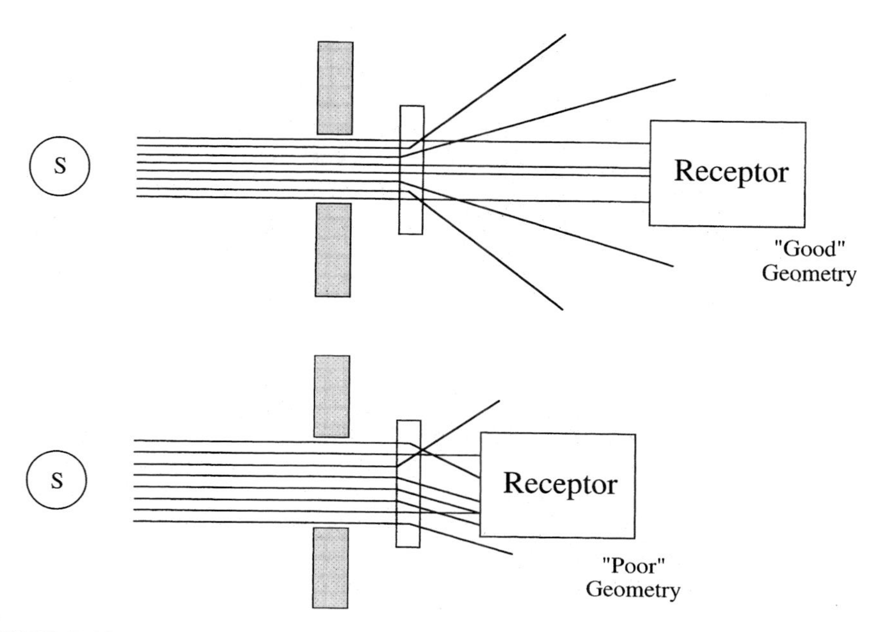

FIGURE 7-15. Attenuation of a photon beam in good geometry, in which all interactions reduce the number of photons reaching a small receptor, and in poor geometry, in which the relationship between the beam and the receptor is such that the scattered photons reach the receptor. Attenuation coefficients for photons of a particular energy are determined under good geometry (or narrow beam) conditions.

interacts is either absorbed or scattered out of the primary beam such that it will not impact a small receptor some distance away. When *good geometry* exists, only those photons that have passed through the absorber without any kind of interaction will reach the receptor, and each of these photons will have all its original energy. This situation exists when the primary photons are confined to a narrow beam and the receptor (e.g., a detector) is small and sufficiently far away that scattered photons have a sufficiently large angle with the original narrowly focused beam that they truly leave the beam and do not reach the receptor. Because of this condition, good geometry is also characterized as *narrow-beam geometry*. Readings taken with and without the absorber in place will yield the fraction of photons removed from the narrow beam, by whatever process.

In *poor geometry* or *broad-beam geometry*, a significant fraction of the scattered photons will also reach the receptor of interest, in addition to those transmitted without interaction. Poor geometry exists in most practical conditions. In addition to the configuration shown in Figure 7-15, other typical poor geometry configurations are a source enclosed by an absorber (e.g., a point source in a lead pig), a shielded detector, or any other condition where a broad beam of photons strikes an absorber. Such conditions often exist when tissue is exposed or a shield is used to attenuate a photon source, and each scattered photon will be degraded in energy according to the angle through which it is scattered. The photons reaching a receptor will then have a complex energy spectrum, including scattered photons of many energies and

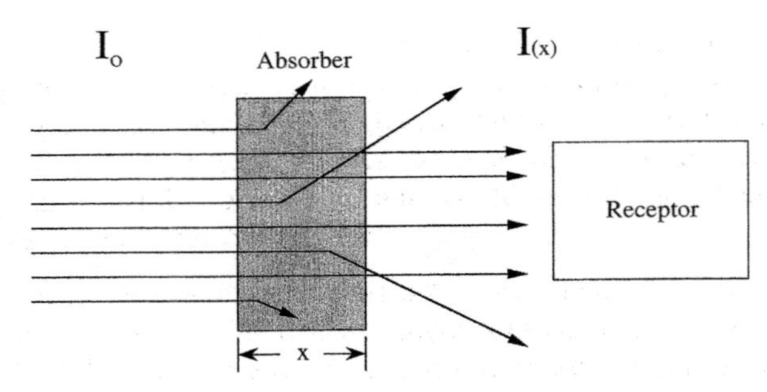

FIGURE 7-16. Alteration of a beam of photons by attenuation processes.

unscattered photons originally present in the primary beam. The pattern of energy deposition in the receptor (e.g., a person or a detector) will be equally complex and governed by the energy distribution and the exact geometrical arrangement used. The amount of energy deposited in an absorber or a receptor under such conditions is very difficult to determine analytically.

The change in intensity of a photon in good geometry (as illustrated in Figure 7-16) is expressed mathematically as a decreasing function with thickness of absorber, or

$$-\frac{dI}{dx} = \mu I$$

where the constant of proportionality μ is the total attenuation coefficient of the medium for the photons of interest. If all the photons possess the same energy (i.e., the beam is monoenergetic) and if the photons are attenuated under conditions of good geometry (i.e., the beam is narrow and contains no scattered photons), the number or intensity $I(x)$ of photons penetrating an absorber of thickness x (i.e., without interaction in the medium) is found by rearranging and integrating

$$\int_{I_0}^{I(x)} \frac{dI}{I} = \int_0^x -\mu\, dx$$

to yield

$$\ln I(x) - \ln I_0 = -\mu x$$

or

$$\ln I(x) = -\mu x + \ln I_0$$

which is an equation of a straight line with a slope of $-\mu$ and a y-intercept (i.e., with no absorber) of $\ln I_0$, as shown in Figure 7-17a. This can be simplified by the law of logarithms to

$$\ln \frac{I(x)}{I_0} = -\mu x$$

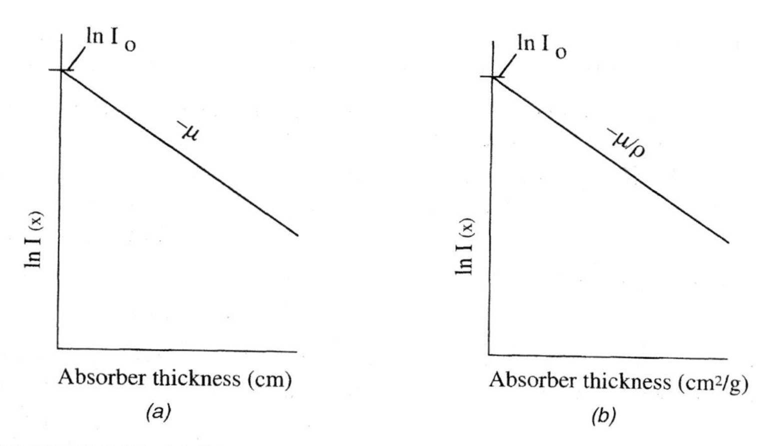

FIGURE 7-17. (*a*) Linear attenuation and (*b*) mass absorption of photons.

and since the natural logarithm of a number is the exponent to which the base e is raised to obtain the number, this expression translates to

$$\frac{I(x)}{I_0} = e^{-\mu x}$$

or

$$I(x) = I_0 e^{-\mu x}$$

where I_0 is the intensity of the incident beam, $I(x)$ is the intensity after traversing a distance x through the absorbing medium, and μ, the *linear attenuation coefficient*, is the probability of interaction per unit distance in an absorbing medium. It is synonymous with the radioactive disintegration constant λ, which expresses the probability of transformation of radioactive atoms per unit time. The exponential relationship for photon absorption suggests that theoretically, complete absorption of a beam of photon radiation never really occurs, but in a practical sense, exponential attenuation and/or absorption can be used to reduce most beam intensities to imperceptable levels.

Attenuation (μ) and Energy Absorption (μ_{en}) Coefficients

The relationship between beam intensity and the attenuation coefficient is valid for photoelectric, Compton, and for photons of sufficient energy, pair production interactions. It also holds for each mode of attenuation separately, or in combination with each other; that is, the total probability that a narrow beam of photons will interact with an absorber of thickness x and be depleted from the beam is obtained by multiplying the individual probabilities of each, or

$$I(x) = I_0(e^{-\tau x} \cdot e^{-\sigma x} \cdot e^{-\kappa x}) = I_0 e^{-(\tau+\sigma+\kappa)x} = I_0 e^{-\mu x}$$

where, μ, the total attenuation coefficient, is the sum of the coefficients of each of the principal modes of photon interaction in a medium,

$$\mu = \tau + \sigma + \kappa$$

where τ, σ, and κ are photoelectric, Compton, and pair production coefficients, respectively.

It is somewhat surpising that μ is a constant since it is the sum of the coefficients of individual interaction processes, each of which is a function of gamma-ray energy, atomic number, and the mass and density of the absorbing medium; however, measured photon intensities versus absorber thickness consistently produce straight lines when plotted on semilog paper. If other processes (e.g., photodisintegration) were to have a significant effect on the absorption of a beam of photons, these would need to be accounted for by an increase in μ or accomplished by a separate exponential term. Photodisintegration and Rayleigh (or coherent) scattering usually increase μ by only a small amount and are ignored.

Radiation exposure or absorbed dose from photons is determined by the amount of energy deposited by the various photon interactions as they traverse a medium such as tissue. Since some interactions produce radiant energy that carries energy out of the medium, the attenuation coefficient μ cannot be used to determine energy deposition in a medium. Consequently, a linear energy absorption coefficient μ_{en} has been defined that accounts for this loss. It has the value

$$\mu_{\text{en}} = \mu - (\sigma_s + \text{other low-probability interactions})$$

Processes that produce radiant energy that can escape the medium are Compton-scattered photons that don't interact, bremmstrahlung from high-energy recoil electrons, annihilation radiation if $h\nu > 1.022$ MeV, and characteristic x rays that don't interact. These depleting mechanisms and their effect on μ are shown in Figure 7-18.

The mass energy absorption coefficient, μ_{en}/ρ, with units of cm^2/g, is the most useful form for determining radiation exposure or dose when a flux of x- or gamma

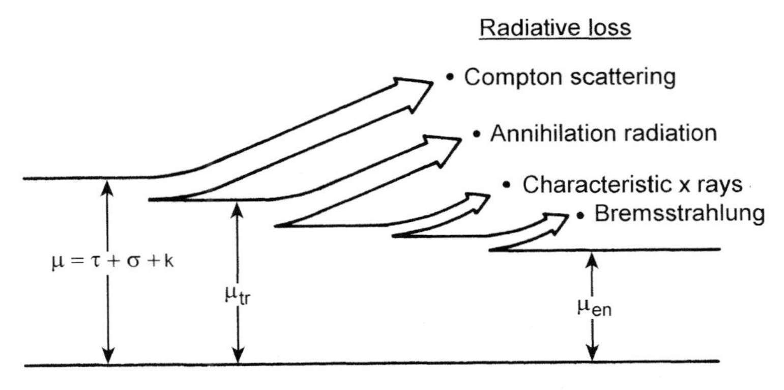

FIGURE 7-18. Relationship of the photon attenuation coefficient μ, the energy transfer coefficient $\mu_{\text{tr}} = \mu - \sigma_s$, and the energy absorption coefficient μ_{en}, and the radiative loss processes that propagate energy out of the absorbing medium.

rays is known or can be determined. Values of μ_{en}/ρ are provided together with values of μ and μ/ρ in Table 7-4 for air, tissue, muscle, and bone for determination of radiation exposure in air and radiation dose in tissue. These values are believed to be appropriate for a wide range of photon absorption problems, but the terminology for describing the various attenuation and absorption coefficients is not uniform, and care must be exercised to be sure that the proper coefficient is selected for a particular radiation protection problem. Values for various other media which are useful for radiation shielding and energy deposition in various materials (e.g., dosimeters) are included in Table 8-2.

Effect of *E* and *Z* on Photon Attenuation and Absorption

Most radiation protection work is done with the total attenuation coefficient μ (for radiation shielding) or the total mass energy absorption coefficient μ_{en}/ρ (for radiation dose calculations) instead of coefficients for the individual interactions; however, the features of each interaction process determine which is most important at various energies in different materials. Attenuation (or absorption) of photons varies considerably with photon energy and the Z of the absorbing medium, as shown in Figure 7-19*a* and *b* for water (similar to tissue) and lead, because of the influence of E and Z on the individual interaction coefficients. At low energies (< 15 keV), the photoelectric effect accounts for virtually all of the interactions in water. As the photon energy increases, τ drops rapidly and Compton scattering interactions (σ) become dominant in water at about 100 keV and at about 500 keV in lead. The Compton scattering effect remains dominant up to several hundred keV, where it decreases with energy and continues to do so until pair production becomes the dominant process.

In lead, a high-Z material, the photoelectric effect is the dominant interaction at low energies. Photoelectric absorption decreases rapidly with increasing photon energy but rises abruptly when the photon energy is sufficient to eject a photoelectron from the K shell of the atom (Figure 7-19*b*). For photon energies above a few hundred keV, Compton-scattering interactions dominate and continue to do so until photon energies are well above the 1.022 MeV pair production threshold. Figure 7-20 also shows the effect of increasing Z on the photon attenuation coefficient, which is relatively larger for lead (a high-Z material) than for iron and aluminum (a relatively low Z material).

The relationship between values of μ and μ_{en} for lead is shown in Figure 7-21. They are similar at the intermediate energies because of the Z/E effect on Compton interactions, but μ_{en} is lower for energies above the K-absorption edge and remains so above the pair production threshold. Both μ and μ_{en} for lead are quite large over the entire photon energy range, illustrating its effectiveness as a photon shield. The μ and μ_{en} values for air and tissue are almost the same as those for H_2O, as would be expected because they consist of similar elements.

Absorption Edges

When the energy of the incoming photon is the exact amount required to remove a K electron entirely from the atom, strong resonance absorption occurs. The attenuation (or absorption) cross section increases sharply (see Figures 7-20 and 7-21)

TABLE 7-4. Photon Attenuation (μ), Mass Attenuation (μ/ρ), and Mass Energy-Absorption (μ_{en}/ρ) Coefficients for Air, Tissue, Muscle, and Bone

	Dry Air (Sea Level) (ρ = 0.001205 g/cm^3)			Tissue (ICRU) (ρ = 1.00 g/cm^3)			Muscle (ICRP) (ρ = 1.05 g/cm^3)			Cortical Bone (ICRP) (ρ = 1.92 g/cm^3)		
Energy (keV)	μ (cm^{-1})	μ/ρ (cm^2/g)	μ_{en}/ρ (cm^2/g)	μ (cm^{-1})	μ/ρ (cm^2/g)	μ_{en}/ρ (cm^2/g)	μ (cm^{-1})	μ/ρ (cm^2/g)	μ_{en}/ρ (cm^2/g)	μ (cm^{-1})	μ/ρ (cm^2/g)	μ_{en}/ρ (cm^2/g)
10	0.0062	5.120	4.742	4.937	4.937	4.564	5.6238	5.356	4.964	54.7392	28.51	26.80
15	0.0019	1.614	1.334	1.558	1.558	1.266	1.7777	1.693	1.396	17.3414	9.032	8.388
20	0.0009	0.7779	0.5389	0.7616	0.7616	0.5070	0.8615	0.8205	0.564	7.6819	4.001	3.601
30	4.26×10^4	0.3538	0.1537	0.3604	0.3604	0.1438	0.3972	0.3783	0.1610	2.5555	1.331	1.070
40	2.99×10^4	0.2485	0.0683	0.2609	0.2609	0.0647	0.2819	0.2685	0.0719	1.2778	0.6655	0.4507
50	2.51×10^4	0.2080	0.0410	0.2223	0.2223	0.0399	0.2375	0.2262	0.0435	0.8145	0.4242	0.2336
60	2.26×10^4	0.1875	0.0304	0.2025	0.2025	0.0305	0.2150	0.2048	0.0326	0.6044	0.3148	0.1400
70[a]	2.10×10^4	0.1744	0.0255	0.1899	0.1899	0.0263	0.2006	0.1910	0.0276	0.4866	0.2534	0.0905
80	2.00×10^4	0.1662	0.0241	0.1813	0.1813	0.0253	0.1914	0.1823	0.0262	0.4280	0.2229	0.0690
100	1.86×10^4	0.1541	0.0233	0.1688	0.1688	0.0250	0.1778	0.1693	0.0254	0.3562	0.1855	0.0459
150	1.63×10^4	0.1356	0.0250	0.1490	0.1490	0.0273	0.1567	0.1492	0.0275	0.2842	0.1480	0.0318
200	1.49×10^4	0.1233	0.0267	0.1356	0.1356	0.0294	0.1426	0.1358	0.0294	0.2513	0.1309	0.0300
300	1.29×10^4	0.1067	0.0287	0.1175	0.1175	0.0316	0.1235	0.1176	0.0316	0.2137	0.1113	0.0303
400	1.15×10^4	0.0955	0.0295	0.1051	0.1051	0.0325	0.1105	0.1052	0.0325	0.1902	0.0991	0.0307
500	1.05×10^4	0.0871	0.0297	0.0959	0.0959	0.0327	0.1008	0.0960	0.0327	0.1732	0.0902	0.0307
600	9.71×10^5	0.0806	0.0295	0.0887	0.0887	0.0325	0.0932	0.0887	0.0325	0.1600	0.0833	0.0305
662[a]	9.34×10^5	0.0775	0.0293	0.0853	0.0853	0.0323	0.0896	0.0853	0.0323	0.1538	0.0801	0.0303
800	8.52×10^5	0.0707	0.0288	0.0779	0.0779	0.0318	0.0818	0.0779	0.0318	0.1403	0.0731	0.0297
1,000	7.66×10^5	0.0636	0.0279	0.0700	0.0700	0.0307	0.0736	0.0701	0.0307	0.1261	0.0657	0.0288
1,173[a]	7.05×10^5	0.0585	0.0271	0.0644	0.0644	0.0299	0.0677	0.0644	0.0299	0.1160	0.0604	0.0279
1,250	6.86×10^5	0.0569	0.0267	0.0626	0.0626	0.0294	0.0658	0.0627	0.0294	0.1127	0.0587	0.0275
1,333[a]	6.62×10^5	0.0550	0.0263	0.0605	0.0605	0.0290	0.0636	0.0605	0.0290	0.1090	0.0568	0.0271
1,500	6.24×10^5	0.0518	0.0255	0.0570	0.0570	0.0281	0.0599	0.0570	0.0281	0.1026	0.0535	0.0262
2,000	5.36×10^5	0.0445	0.0235	0.0489	0.0489	0.0258	0.0514	0.0490	0.0258	0.0885	0.0461	0.0242
3,000	4.32×10^5	0.0358	0.0206	0.0393	0.0393	0.0226	0.0413	0.0393	0.0226	0.0719	0.0375	0.0215
4,000	3.71×10^5	0.0308	0.0187	0.0337	0.0337	0.0204	0.0354	0.0337	0.0205	0.0625	0.0326	0.0198
5,000	3.31×10^5	0.0275	0.0174	0.0300	0.0300	0.0189	0.0315	0.0300	0.0190	0.0566	0.0295	0.0186
6,000	3.04×10^5	0.0252	0.0165	0.0274	0.0274	0.0179	0.0288	0.0274	0.0179	0.0525	0.0273	0.0179
6,129[a]	3.01×10^5	0.0250	0.0164	0.0271	0.0271	0.0178	0.0285	0.0271	0.0178	0.0520	0.0271	0.0178
7,000[a]	2.83×10^5	0.0235	0.0159	0.0255	0.0255	0.0171	0.0267	0.0255	0.0170	0.0495	0.0258	0.0174
7,115[a]	2.82×10^5	0.0234	0.0158	0.0253	0.0253	0.0170	0.0265	0.0253	0.0170	0.0492	0.0256	0.0173
10,000	2.46×10^5	0.0205	0.0145	0.0219	0.0219	0.0155	0.0230	0.0219	0.0155	0.0444	0.0231	0.0164

Source: Adapted from Hubbell and Seltzer, 1995.

[a]Coefficients for these energies were interpolated using polynomial regression.

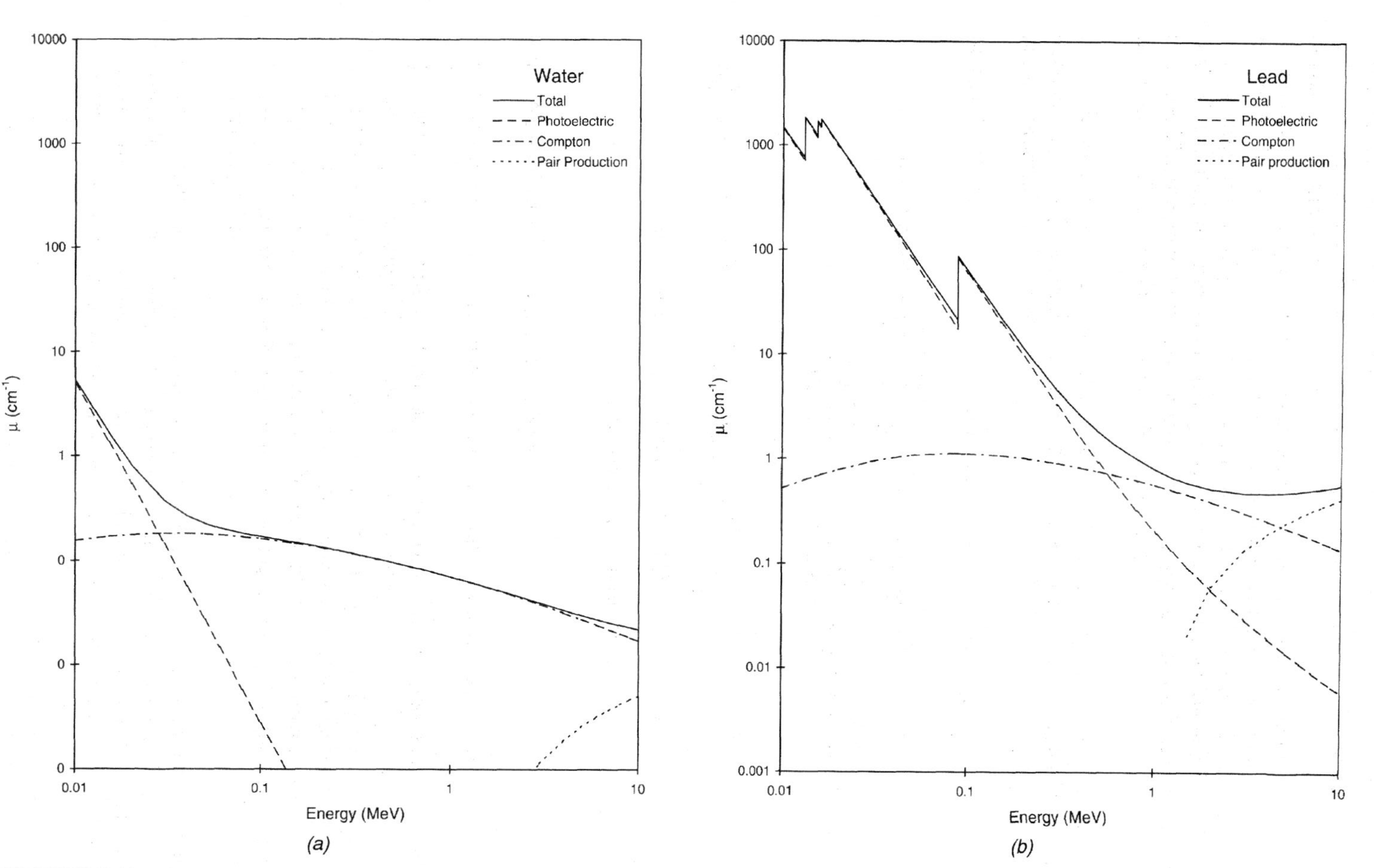

FIGURE 7-19. Photon attenuation coefficients for low-Z water (a) and high-Z lead (b), and the relative effects of the photoelectric, Compton, and pair production interactions versus energy.

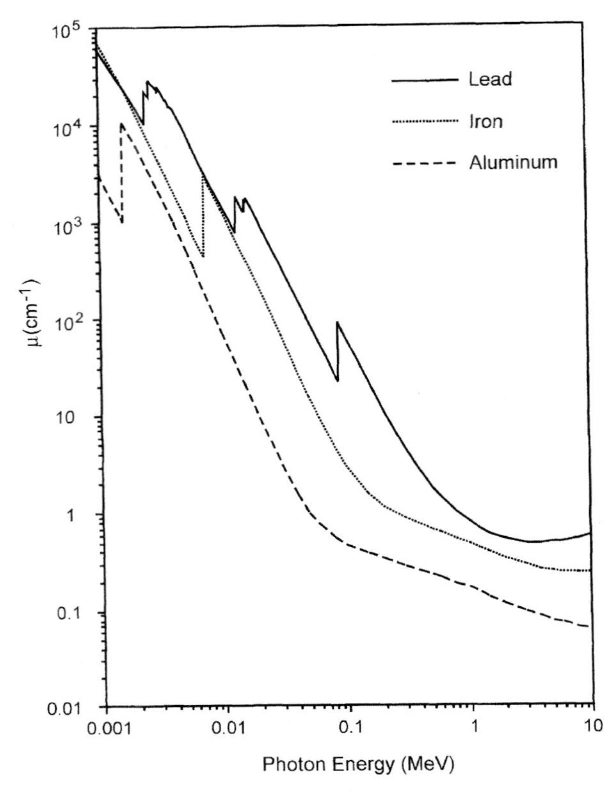

FIGURE 7-20. Variation of μ versus Z, which also illustrates the effect of Z on resonance absorption to create absorption "edges" in Pb ($Z = 82$), Fe ($Z = 26$), and Al ($Z = 13$).

at this energy, producing what is called an *absorption edge.* As long as the photon energy is below this amount, there can be no K resonance absorption, and the absorption cross sections due to photoelectric and Compton interactions will decrease with increasing photon energy. Even at the exact $K_{\alpha 1}$ energy, resonance absorption is still impossible, since there is only enough energy to move a K-shell electron up to the L shell, which by the Pauli principle is full; therefore, resonance absorption occurs only when the photon energy equals the electron binding energy. As photon energies increase above the absorption edge, the attenuation coefficient again decreases. Similar interactions produce L-absorption edges, which are also shown in detail in Figure 7-21 for lead (these edges also occur in other materials but are difficult to observe). M-absorption edges (shown in gross detail in Figure 7-21) also exist for lead, but observing their detail requires special techniques; they occur at such low energies for most materials that detection is difficult.

The energies at which absorption edges occur have practical value since they are determined uniquely by the atomic number of the absorber. This allows absorbers made of elements with adjacent atomic numbers to be used to "shape" photon beams by permitting the transmission of only those photons with a narrow band

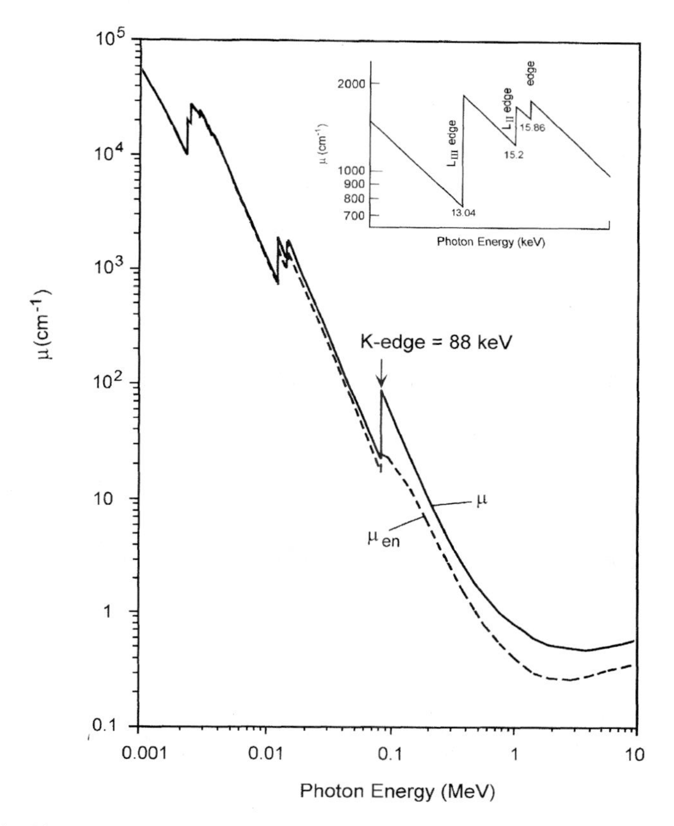

FIGURE 7-21. Photon attenuation (μ) and energy (μ_{en}) absorption coefficients versus energies in lead; the K-absorption edge is prominent at 88 keV and the substructures of the L shell (shown in detail) and the M shell are also visible.

of energies, a circumstance that is very useful in x-ray beam quality (see Chapter 14).

Characteristic x rays are emitted following the absorption of photons at and above the absorption edge of a material as the photoelectron vacancies are filled. The energies of these emissions are below the absorption edge of the material in which they are produced and hence are poorly absorbed by it. They can, however, be effectively removed by an absorber of slightly lower atomic number, and the characteristic x rays produced can, in turn, be absorbed in materials with even lower atomic numbers. This principle is used to "harden" the beams generated by x-ray tubes by removing the low-energy portion of the x-ray emission spectrum. For example, a copper filter of 0.2 to 0.5 mm in an x-ray tube operating above 100 V will effectively remove a large fraction of the lower energies, but in doing so will produce a considerable amount of 8 keV K_α x rays from copper. These can in turn be absorbed by an aluminum ($Z = 13$) filter, but photoelectric absorption in aluminum will produce K_α photons of 1.5 keV, which can in turn be absorbed in a bakelite filter ($Z = 6$). To be effective, filter combinations must be inserted in the proper order.

CHECKPOINTS

Several things are noteworthy about the exponential relationship for photon absorption:

- Since the radiation exposure rate or absorbed dose is proportional to flux, the attenuation coefficient and thickness of an absorber determine how much reduction occurs in the incident flux and in turn the exposure or dose.
- The value of μ depends on the absorbing medium and the energy of the photons; thus, extensive tabulations of μ are needed (as provided in Tables 7-4 and 8-2) for calculations of flux changes.
- The values of μ are given for good geometry [i.e., a narrow beam of photons where scattered photons are removed from the beam and do not reach a receptor of interest (either a detector or a person)]; most real-world situations do not meet these ideal conditions and are characterized as poor geometry.
- Values of μ_{en} are based only on the energy absorbed in the medium; therefore, energy losses due to Compton-scattered photons, bremsstrahlung, and other radiative processes following interaction have been subtracted because they are very likely to leave the medium.
- The most useful forms of the photon coefficients are μ_{en}/ρ (cm^2/g) for determining energy deposited in the absorbing medium and μ (cm^{-1}) for radiation shielding calculations.

ENERGY TRANSFER AND ABSORPTION BY PHOTONS

Energy transfer and absorption from a beam of photons takes place in two stages, as shown in Figure 7-22: first, a photon interaction, which is probabilistic, produces an electron or electrons with kinetic energy; and second, each ejected electron transfers energy to the medium through excitation and ionization all along its path. The first process is characterized by kerma, which is defined as *k*inetic *e*nergy *r*eleased in *ma*terial (or matter); the second is the absorbed dose. Kerma corresponds to energy transferred E_{tr}, and absorbed dose corresponds to the energy absorbed E_{ab}. Both quantities are shown for air, various tissues, and Pb for different photon energies in Table 7-5; their physical difference is illustrated in Example 7-11.

Kerma is directly related to the photon fluence and the likelihood of interaction, $(\mu/\rho)_{med}$, which causes energy to be released at the site of each interaction; it is calculated from the transfer of energy for a fluence ϕ of photons as

$$K_{med} = \phi \left(\frac{\mu}{\rho}\right)_{med} \overline{E}_{tr}(h\nu) \times 1.6022 \times 10^{-6} \text{ erg/MeV}$$

where ϕ is the fluence (photons/cm^2), $(\mu/\rho)_{med}$ the mass attenuation coefficient for the medium, and $\overline{E}_{tr}(h\nu)$ the average amount of the photon energy transferred to electrons in the medium.

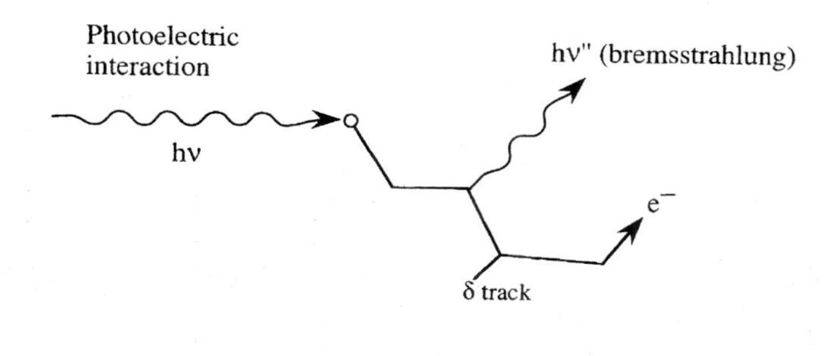

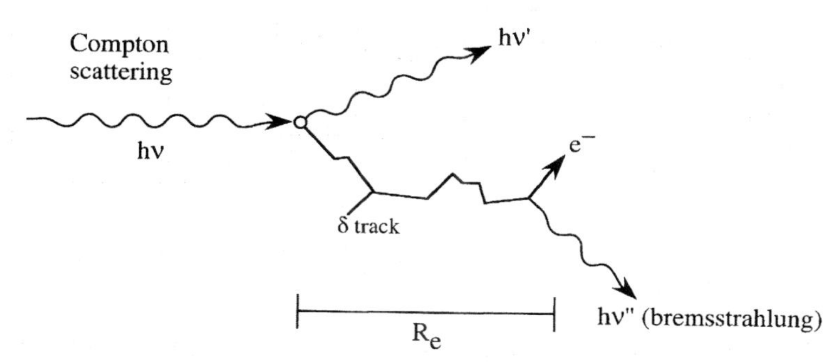

FIGURE 7-22. Energy transfer and absorption is a two-stage process: first, a photon ($h\nu$) interacts with an atom of the medium, transferring all (photoelectric effect) or some of its energy (Compton scattering) to an electron, or in the case of pair production to the created electron and positron. The kinetic energy of these electrons is then deposited in the medium mostly by ionization, although some may be lost as bremsstrahlung and other radiative processes. The transfer of energy, or kerma, occurs at the point of interaction; the absorbed dose occurs, however, along the path of the recoiling electrons extending over their entire range R_e in the medium.

Example 7-11: For a fluence of $10^{10}\ \gamma/\text{cm}^2$ of 10 MeV photons on muscle tissue, calculate (a) the kerma and (b) the absorbed dose.

SOLUTION: (a) From Table 7-4, $\mu/\rho = 0.0219\ \text{cm}^2/\text{g}$ for muscle tissue, and the average energy $\overline{E}_{tr}$ transferred by 10 MeV photons (Table 7-5) is 7.32 MeV.

$$\text{kerma} = \frac{10^{10}\gamma}{\text{cm}^2} \times 0.0219\frac{\text{cm}^2}{\text{g}} \times 7.32\ \text{MeV} \times 1.6022 \times 10^{-6}\ \text{erg/MeV}$$
$$= 2.58 \times 10^3\ \text{ergs/g of muscle}$$

(b) Absorbed dose is calculated in the same manner as kerma except that $\overline{E}_{ab}(h\nu)$ (from Table 7-5) is used instead of $E_{tr}(h\nu)$. Since E_{ab} is 7.07 MeV, the absorbed dose is

$$\text{dose} = 2.49 \times 10^3\ \text{ergs/g}$$

which is somewhat less than the value of kerma because only 7.07 MeV of the 7.32 MeV transferred is absorbed along the electron track. The difference of

TABLE 7-5. Average Energy Transferred E_{tr} and Absorbed E_{ab} by Photons Through Interactions in Various Media[a]

	Air		Water		Muscle		Bone		Fat		Lead	
$h\nu$ (keV)	E_{tr} (keV)	E_{ab} (keV)	E_{tr} (keV)	E_{ab} (keV)	E_{tr} (keV)	E_{ab} (keV)	E_{tr} (keV)	E_{ab} (keV)	E_{tr} (keV)	E_{ab} (keV)	E_{tr} (keV)	E_{ab} (keV)
10	9.23		9.25		9.24		9.37		9.70		9.55	
20	13.5		13.2		13.4		17.5		18.0		16.0	
50	9.31		8.82		9.13		22.9		24.6		42.2	42.0
80	11.4		11.2		11.3		20.4		21.7		—	—
88	—		—		—		—		—		69.2[b]	68.5[b]
100	15.0		14.8		14.9		21.4		22.3		36.2	35.7
200	43.4		43.3		43.4		45.2		45.2		122	119
300	80.8		80.8		80.8		81.6		81.4		191	185
400	124		124		124		124		124		247	239
500	171		171		171		171		171		298	286
662	252		252		252		253		253		376	360
800	327		327		327		327		327		444	423
1,000	440		440		440		440		440		550	520
1,250	588	586	588	586	588	586	588	586	588	584	693	649
1,500	741	739	741	739	741	739	741	738	741	735	—	—
2,000	1,060	1,050	1,060	1,060	1,060	1,060	1,060	1,060	1,060	1,050	1,130	1,040
3,000	1,740	1,720	1,740	1,730	1,740	1,730	1,740	1,720	1,750	1,720	1,860	1,660
4,000	2,460	2,430	2,460	2,430	2,460	2,430	2,470	2,430	2,480	2,430	2,700	2,350
5,000	3,220	3,170	3,210	3,160	3,210	3,160	3,230	3,160	3,270	3,170	3,600	3,060
10,000	7,370	7,100	7,330	7,070	7,320	7,070	7,430	7,100	7,610	7,140	8,450	6,420
20,000	16,600	15,500	16,500	15,300	16,500	15,300	16,800	15,300	17,200	15,200	18,500	12,000

Source: Adapted from Johns and Cunningham, 1983.

[a]Where the two are essentially the same only one entry is made.

[b]Energy transferred and absorbed at the K edge in Pb; for photons with energy just below the K edge, the values of E_{tr} and E_{ab} are 24.8 and 24.7 keV, respectively.

0.25 MeV is lost by radiative processes, principally bremsstrahlung. The 7.07 MeV actually absorbed is, at 34 eV/ion pair, sufficient to ionize about 208,000 atoms (i.e., although only one atom was ionized in the initial transfer of energy by a photon interaction, some 208,000 atoms were disrupted along the track of this energetic electron as it traversed the medium).

As illustrated in Figure 7-22, for photoelectric and Compton interactions, energy is transferred to an electron but not all of it is deposited in the medium since bremsstrahlung may be produced and radiated away (a small fraction of energy may also be lost in the form of characteristic x rays that are not absorbed after photoelectric interactions) over the entire range R_e of the ejected electrons in the medium. Any pair production interactions that occur yield a positron and an electron which deposit their kinetic energy ($h\nu - 1.022$ MeV) by similar processes, and if both annihilation photons are absorbed in the medium, most of the initial energy $h\nu$ is deposited. Because the length of the electron tracks may be appreciable, kerma and absorbed dose do not take place at the same location. The units of absorbed dose are joules per kilogram (or ergs/g in conventional units); kerma has the same units, but there is no special designation (such as the gray for absorbed dose) for kerma. It is just kinetic energy released or transferred per unit mass of medium.

An *energy transfer coefficient* μ_{tr} that represents the amount of energy transferred to a medium by photons of energy $h\nu$ can be calculated from listed values of μ_{en} as

$$\mu_{tr} = \frac{\mu_{en}}{1 - g}$$

where values of g, the fraction of energy lost by radiative processes, is numerically equal to $(E_{tr} - E_{ab})/E_{tr}$. Similarly, a *mass energy transfer coefficient* (μ_{tr}/ρ) can be calculated by dividing μ_{tr} by the density of the absorbing medium. Since values of E_{tr} and E_{ab} are identical below 1 MeV (see Table 7-5) and are essentially the same up to about 1.5 MeV, μ_{tr} and μ_{en} for a given medium are either identical or very nearly so for most photons emitted from radionuclides commonly encountered in radiation protection. They differ, however, for high-energy photons produced around accelerators or in nuclear interactions, and it is essential to determine the appropriate value of E_{tr} (and μ_{tr} if it is used) for calculations of kerma or absorbed dose.

The value $(1 - g)$ slowly decreases with increasing photon energy as shown in Table 7-6. Because of their importance as gamma calibration sources for dosimetry,

TABLE 7-6. Photon Energy and the Value $(1 - g)$

E (MeV)	$1 - g$
0.662	0.9984
1.25	0.9968
1.5	0.996
2.0	0.995
3.0	0.991
5.0	0.984
10.0	0.964

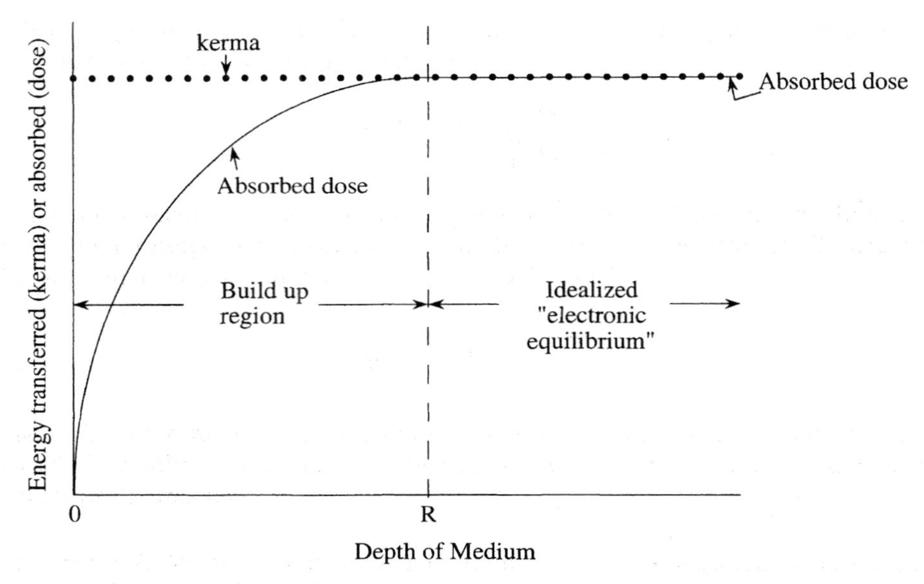

FIGURE 7-23. Idealized conditions of electronic equilibrium.

exact values of $(1 - g)$ have been determined for ^{137}Cs photons ($h\nu = 0.662$ MeV) and ^{60}Co photons ($h\nu = 1.332$ and 1.172 MeV) to be 0.9984 and 0.9968.

Electronic Equilibrium

Absorbed dose cannot be calculated unless a state of electronic equilibrium exists which, as shown in Figure 7-23, occurs only some distance into the medium as determined by the range of the ejected electrons. The curve in Figure 7-23 represents, for purposes of illustration, an idealized condition in which there is no attenuation of the photons as they pass through the medium. The deposition of energy by ionization along the path of the ejected electron starts at zero and reaches its maximum over a depth R called the *buildup region*; beyond R as many electrons stop in any volume as are set in motion in it. A condition known as *electronic equilibrium* is established where the kerma is constant with depth, and if no bremsstrahlung losses occur, the absorbed dose is equal to the kerma after electronic equilibrium is established.

The idealized conditions of Figure 7-23 do not exist in reality because, as shown in Figure 7-24, photons are in fact attenuated as they traverse the medium. Consequently, true electronic equilibrium is not attained, and kerma decreases with depth because the primary radiation is attenuated exponentially. The absorbed dose increases at first, reaches a maximum at R, and then decreases exponentially along with kerma. If bremsstrahlung losses are minimal, the absorbed dose curve will be above the kerma curve because the absorbed energy will be due to kerma processes that occurred at the shallower depths. Even this more realistic condition is oversimplified because a spectrum of electrons of all energies is always set in motion, and these travel in many different directions. Even more complexity exists when two

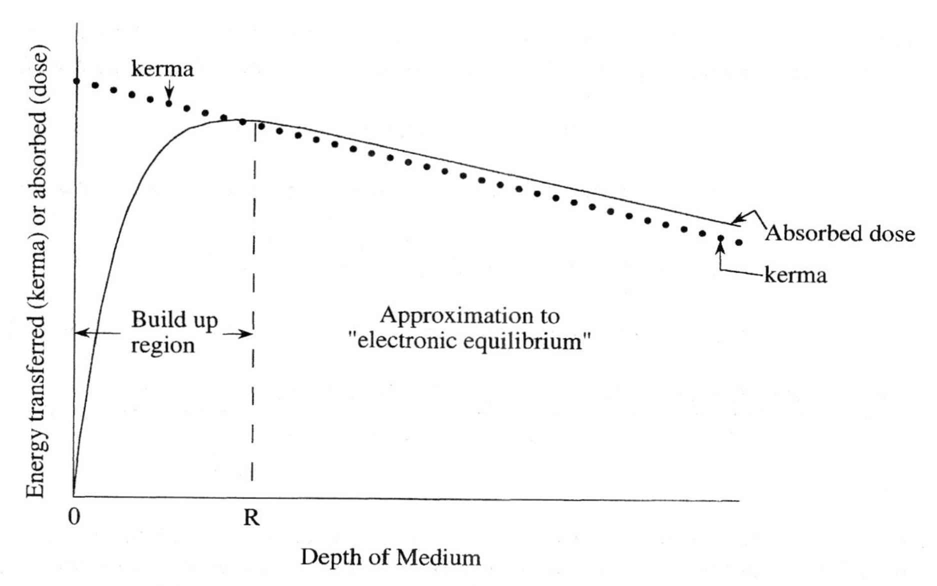

FIGURE 7-24. Realistic equilibrium of kerma and absorbed dose where the primary photons are attenuated as they traverse the medium; thus both kerma and absorbed dose decrease with depth after equilibrium occurs.

different media are involved, such as for bone in soft tissue. Electrons set in motion in one material deposit their energy in the other, and kerma and absorbed dose will not be in equilibrium since path lengths of absorbed energy are determined by the range of the electrons in each.

Even though true electronic equilibrium does not occur, it is a common and reasonable practice to treat the region beyond R as being in equilibrium and to calculate absorbed dose using $\overline{E}_{ab}$ as follows:

$$D_{med} = \phi \left(\frac{\mu}{\rho}\right)_{med} \overline{E}_{ab}(h\nu) \times 1.6022 \times 10^{-6} = K(1 - g)$$

where $\overline{E}_{ab}(h\nu)$ is the part of the average kinetic energy transferred to electrons that contributes to ionization (it excludes energy lost by radiative processes, principally bremsstrahlung) and K is the kerma. Although the absorbed dose beyond equilibrium is slightly larger than kerma, the calculated value of kerma is a reasonably accurate determination of absorbed dose for photons less than 5 MeV or so.

Bragg–Gray Theory

Bragg–Gray theory is dependent on the same factors that relate kerma and energy absorption. A true measurement of absorbed dose can only be done with calorimetry, but Bragg–Gray theory allows absorbed dose to be determined from the ionization produced by a beam of photons in a small gas cavity, usually air, embedded in a tissue-equivalent medium of sufficient thickness to establish electronic equilibrium. Under these conditions ionization will be produced in the cavity by the electrons ejected by photon interactions in the tissue medium, and under electronic equilibrium, gas-produced electrons that are lost to the wall medium are compensated for by electrons produced in the medium and absorbed in the gas. The charge liberated in the gas is collected, and since the average energy required to cause one ionization in the gas is constant over a wide range of gas pressures and electron energies, the collected charge is directly proportional to the energy deposited in the gas. The dose absorbed in the enclosed gas cavity is therefore related to the ionization produced in the gas as

$$D_{gas} = \frac{Q}{m_{gas}} W$$

where Q is expressed in coulombs, m_{gas} in grams, and W has a value of 33.97 eV per ion pair, or 33.97 J/C.

Example 7-12: Determine the energy deposited in air if 3.336×10^{-10} C of charge is collected in a 1 cm^3 cavity of air at 0°C and 1 atm (STP).

SOLUTION: Since the density of air at STP is 1.293×10^{-3} g, the air dose is

$$\begin{aligned} D_{air} &= \frac{3.336 \times 10^{-10}\ \text{C/cm}^3}{1.293 \times 10^{-3}\ \text{g/cm}^3} \times 33.97\ \text{J/C} \times 10^7\ \text{ergs/J} = 87.64\ \text{ergs/g} \\ &= 0.8764\ \text{rad (or 8.764 mGy)} \end{aligned}$$

The charge liberated for the conditions of Example 7-12 and the energy deposited is that produced by an exposure of 1 roentgen.

The Bragg–Gray relationship gives the energy imparted to the gas in the cavity, but the quantity of most interest is the energy deposited in the surrounding medium. Since the air cavity is assumed to be so small that it does not alter the electron spectrum, the gas in the cavity will "see" the same electron fluence as exists in the medium. The dose in the medium is related to the dose in air by the relative stopping power, or

$$D_{med} = S_{rel} D_{air}$$

For tissue, Bakelite, Lucite, carbon, and most tissue-equivalent materials, S_{rel} is curently accepted to be 1.12, although 1.14 has been used extensively.

EXPOSURE AND DOSE CALCULATIONS

Calculations of radiation exposure or dose are straightforward if the source of photons is characterized in terms of a fluence, fluence rate, or flux (photons/$cm^2 \cdot s$), from which the energy fluence (MeV/cm^2) or energy fluence rate (MeV/$cm^2 \cdot s$) can be determined. The energy fluence can, in turn, be used to determine exposure in air or energy deposition in an absorbing medium.

The exposure rate is directly proportional to photon flux. This relationship is plotted in Figure 7-25 as the flux at different energies that produces an exposure

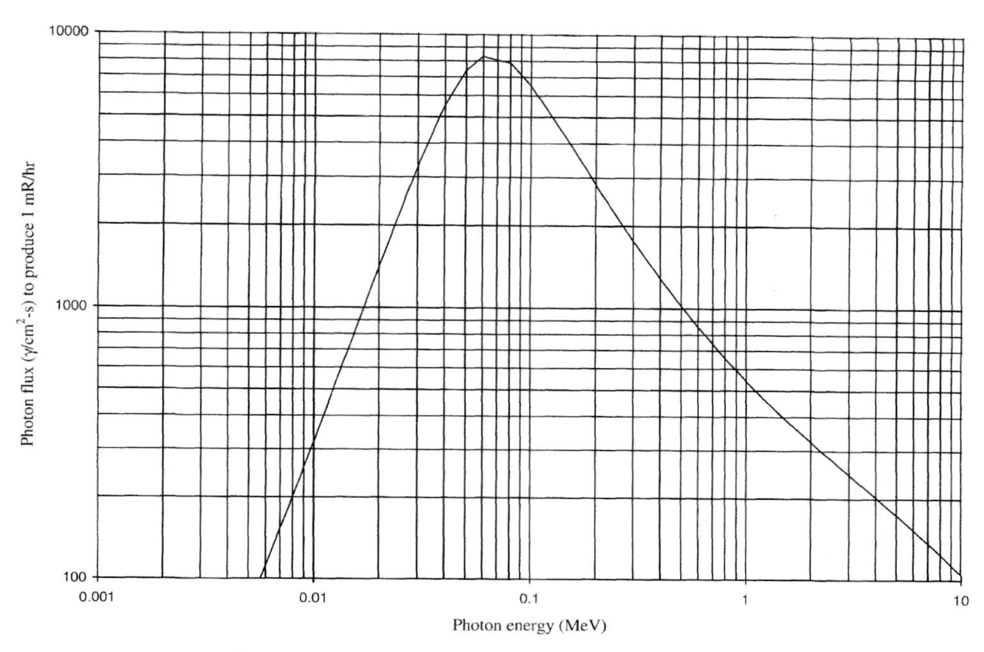

FIGURE 7-25. Photon flux ($\gamma/\text{cm}^2 \cdot \text{s}$) to produce an exposure rate of 1 mR/h.

rate of 1 mR/h. It is noteworthy that the curve in Figure 7-25 is quite linear from about 70 keV up to about 2 MeV. Over this energy range μ_{en}/ρ for air varies little and for air at STP is about 0.0275 cm^2/g. This range of approximate linearity allows two straightforward calculational tools: an inverse-square exposure rate formula for point sources based on source strength, and a gamma constant for point sources of each of the gamma-emitting radionuclides.

Point Sources

Since a point source emits photons uniformly in all directions, the number passing through a unit area is inversely proportional to the square of the distance r (the inverse-square law) and the point-source exposure rate in R/h per curie at a distance r of 1 m is

$$\text{exposure} = \frac{\begin{array}{r}\text{Ci} \times 3.7 \times 10^{10}\ \text{t/s} \times f \times E\ (\text{MeV}) \times 1.6022 \\ \times 10^{-6}\ \text{erg/MeV} \times 0.0275\ \text{cm}^2/\text{g} \times 3600\ \text{s/h}\end{array}}{4\pi r^2\ (100\ \text{cm/m})^2 (87.64\ \text{ergs/g} \cdot \text{R})}$$

$$= \frac{0.533 \cdot \text{Ci} \cdot f \cdot E}{r^2}\ \text{R/h}$$

where Ci is in curies, r is the distance in meters from the point source, f the number of photons emitted per transformation, and E the photon energy between 0.07 and 3 MeV. The point-source exposure rate is very useful since many sources can be represented as points. If distance is measured in feet, this can be converted to the familiar and easy-to-remember approximation

$$\text{R/h at 1 ft} = 6CE$$

where C is in curies and E is MeV of gamma energy per transformation. When this expression is used, it is necessary to include the photon yield f in determining E (e.g., $f = 0.85$ for 0.662-MeV gamma rays from ^{137}Cs such that $E = 0.85 \times 0.662 = 0.563$ MeV/t and for the two gamma rays from ^{60}Co, $E = 1.33$ and $1.17 = 2.50$ MeV/t). Each of these expressions can be used to obtain the exposure rate at any other distance from a point source by dividing the result by the square of the actual distance (m^2 or ft^2).

Gamma-Ray Constant

The exposure rate from a point source of each of several gamma-emitting radionuclides (as illustrated in Example 7-13) can readily be calculated because the photon emission rate and energy are unique for each; i.e., each can be represented as a *gamma-ray constant* (Γ) for a unit of activity (usually Ci or Mbq) and at a unit distance (usually 1 m) as tabulated in Table 7-7. A separate calculation was performed for each gamma ray emitted by the nuclide and the results were summed. For example, ^{60}Co has two gamma rays, one at 1.332 MeV in 100% of transformations and one at 1.173 MeV in 99.9% of transformations; Γ was determined for each gamma ray and the two were added together (see Problem 7-10).

Example 7-13: Determine Γ, the exposure gamma-ray constant (R/hr · mCi at 1 cm) for a point source of ^{137}Cs.

SOLUTION: The yield of 0.662 MeV gamma rays from ^{137}Cs is 0.85/t, and the photon flux produced by the 1 mCi ^{137}Cs cource at 1 cm is

$$\begin{aligned}\phi &= \frac{1 \text{ mCi} \times 3.7 \times 10^7 \text{ t/s} \cdot \text{mCi} \times 0.85\ \gamma/\text{t}}{4\pi(1 \text{ cm})^2}\\ &= 2.503 \times 10^6 \text{ photons/cm}^2 \cdot \text{s}\end{aligned}$$

The mass energy absorption coefficient for air is 0.0293 cm^2/g, and

$$\begin{aligned}E_{\text{ab}} &= \phi \times E(\text{MeV}/\gamma)\mu_{\text{en}}/\rho\\ &= (2.503 \times 10^6\ \gamma/\text{cm}^2 \cdot \text{s})(0.662 \text{ MeV}/\gamma)\\ &\quad \times (1.602 \times 10^{-6} \text{ erg/MeV})(0.0293 \text{ cm}^2/\text{g})(3600 \text{ s/h})\\ &= 2.8 \times 10^2 \text{ ergs/g} \cdot \text{h}\end{aligned}$$

and since 1 R = 87.64 ergs/g of air,

$$\text{exposure} = 3.193 \text{ R/h for 1 mCi of } ^{137}\text{Cs at 1 cm or 0.319 R/h per Ci at 1 m}$$

The gamma constant Γ is generally given in units of R/h at 1 m per curie but has also been tabulated as R/h at 1 cm per mCi.

TABLE 7-7. Gamma-Ray Constant, Γ (R/h · Ci at 1 m) for 1 Ci of Selected Radionuclides

Nuclide	Γ	Nuclide	Γ	Nuclide	Γ
Antimony-122	0.304	Europium-155	0.067	Platinum-197	0.021
Antimony-124	1.067	Gallium-67	0.111	Potassium-40	0.082
Antimomy-125	0.380	Gallium-72	1.456	Praseodymium-144[a]	0.017
Arsenic-72	1.165	Gold-198	0.292	Radium-226	0.012
Arsenic-76	0.545	Hafnium-181	0.393	Rhodium-106[a]	0.138
Barium-133	0.455	Iodine-123	0.277	Rubidium-86	0.054
Barium-140	0.165	Iodine-125	0.275	Scandium-46	1.167
Beryllium-7	0.034	Iodine-130	1.403	Scandium-47	0.080
Bromine-82	1.619	Iodine-131	0.283	Selenium-75	0.860
Cadmium-115m	0.013	Iodine-132	1.427	Silver-110m	1.652
Calcium-47	0.585	Iridium-192	0.592	Sodium-22	1.339
Carbon-11	0.717	Iron-59	0.662	Sodium-24	1.938
Cerium-141	0.073	Krypton-85	0.002	Strontium-85	0.759
Cerium-144	0.023	Lanthanum-140[a]	1.253	Tantalum-182	0.772
Cesium-134	0.999	Magnesium-28	0.879	Tellerium-132	0.279
Cesium-137	0.319	Manganese-54	0.511	Thallium-208[a]	1.704
Chlorine-38	0.719	Manganese-56	0.924	Tin-113	0.179
Chromium-51	0.023	Mercury-197	0.069	Tungsten-187	0.329
Cobalt-56	1.926	Mercury-203	0.253	Vanadium-48	1.701
Cobalt-57	0.151	Molybdenum-99	0.113	Xenon-133	0.010
Cobalt-58	0.614	Neodymium-147	0.139	Xenon-133m	0.112
Cobalt-60	1.370	Nickel-65	0.297	Xenon-135	0.189
Copper-64	0.132	Niobium-95	0.480	Yttrium-88	1.783
Europium-152	0.744	Nitrogen-13	0.708	Zinc-65	0.330
Europium-154	0.756	Nitrogen-16	1.474	Zirconium-95	0.465

[a]Short-lived transformation product in equilibrium with longer-lived parent.

Exposure and Absorbed Dose

The point-source formulas and the gamma-ray constant are based on exposure, usually in R/h per curie at 1 m. The absorbed dose in rad to a person is usually of primary interest, however, and this can be obtained from exposure by the relative stopping power, S_{rel}, of photons in air and tissue, which has the value of 1.12. To use S_{rel} to obtain absorbed dose, it is necessary to convert exposure rate to an energy deposition rate in air since S_{rel} relates energy loss per unit path length:

$$\text{dose(rad)} = S_{\text{rel}} \frac{\text{exposure(R)} \times 87.64 \text{ ergs/g} \cdot \text{R}}{100 \text{ ergs/g} \cdot \text{rad}}$$

or

$$\text{tissue dose(rad)} = 1.12 D_a$$

where D_a is the air dose in rad.

The absorbed dose in tissue can also be calculated directly from the energy deposition rate in the medium of interest (usually, tissue) by using the μ_{en}/ρ value for the tissue medium of interest and the appropriate photon energy. In this direct method, one proceeds as before to determine an energy flux impinging on a person, which is multiplied by the appropriate μ_{en}/ρ value to establish the energy deposition rate and the absorbed dose, as illustrated in Example 7-14.

Example 7-14: Determine the dose to muscle tissue at 1 m from a 1 mCi point source of ^{137}Cs (a) by the indirect method (i.e., relative stopping power) and (b) by the direct method.

SOLUTION: (a) The gamma-ray constant for ^{137}Cs is 0.319 R/h · Ci at 1 m:

$$D_a = \frac{0.319 \text{ R/h} \cdot \text{Ci} \times 0.001 \text{ Ci} \times 87.64 \text{ ergs/g} \cdot \text{R}}{100 \text{ ergs/g} \cdot \text{rad}}$$

$$= 2.8 \times 10^{-4} \text{ rad/h} = 0.28 \text{ mrad/h}$$

The dose to muscle tissue is obtained from the relative stopping power of tissue to air of 1.12, or

$$D = 1.12 \times D_a = 1.12 \times 0.28 \text{ mrad/h} = 0.31 \text{ mrad/h}$$

(b) ^{137}Cs emits 0.662 MeV gamma rays in 85% of its transformations and produces a flux at 1 m of 2.5×10^2 g/cm^2 · s. The energy flux is

$$\phi_{en} = 2.5 \times 10^2 \ \gamma/\text{cm}^2 \cdot \text{s} \times 0.662 \text{ MeV} = 1.66 \times 10^2 \text{ MeV/cm}^2 \cdot \text{s}$$

The mass energy absorption coefficient μ_{en}/ρ for muscle is 0.0323 cm^2/g; therefore, the energy deposition rate in muscle tissue is determined directly as

$$D = \frac{1.66 \times 10^2 \text{ MeV/cm}^2 \cdot \text{s} \times 0.0323 \text{ cm}^2\text{/g} \times 1.6022 \times 10^{-6} \text{ erg/MeV} \times 3600 \text{ s/h}}{100 \text{ ergs/g} \cdot \text{rad}}$$

$$= 3.1 \times 10^{-4} \text{ rad/h} = 0.31 \text{ mrad/h}$$

Exposure, Kerma, and Absorbed Dose

The absorbed dose at various tissue depths is dependent on the photon flux reaching the depth, the interaction fraction that occurs by the various attenuation/absorption mechanisms, and the paths and distribution of charged particles released in the tissue layer, including those backscattered from interactions in deeper layers. A very practical approach for radiation dosimetry has been developed in which all these factors are incorporated into a C_k factor that relates air kerma K_a and the absorbed dose D_i at various depths as

$$D_i = C_{ki} K_a$$

where C_{ki} is a coefficient for a particular depth in tissue and has units of Sv/Gy (or rem/rad). Values of C_{ki} are provided in Table 7-8 for tissue depths of 7, 300,

TABLE 7-8. Factors of C_k to Convert Air Kerma K_a to Tissue Dose Equivalent in a 30-cm-Thick Semi-infinite Tissue Slab at 7 mg/cm^2 (Shallow Dose), 300 mg/cm^2 (Eye Dose), and 1000 mg/cm^2 (Deep Dose), Also Defined as H(10) by ICRU (1992)

		C_k		
Photon Source	$\overline{E}_{avg}$ (keV)	Shallow Dose	Eye Dose	Deep Dose
X rays[a] (mm Al, Cu, Sn)				
M30 (0.5, 0, 0)	20	1.02		0.42
M60 (1.5, 0, 0)	35	1.21		1.00
M100 (5.0, 0, 0)	53	1.49		1.52
M150 (5.0, 0.25, 0)	73	1.64		1.78
H150 (4.0, 4.0, 1.5)	118	1.60		1.71
Monoenergetic photons (keV)				
10	10	0.949	0.251	0.010
20	20	1.040	0.910	0.616
25	25	1.124	1.071	0.886
30	30	1.228	1.221	1.115
50	50	1.657	1.737	1.803
70	70	1.821	1.913	2.026
90	90	1.798	1.879	1.990
120	120	1.702	1.766	1.852
150	150	1.616	1.670	1.731
300	300	1.405	1.431	1.459
500	500	1.299	1.312	1.325
^{137}Cs	662	1.210	1.210	1.210
1000	1000	1.205	1.209	1.208
^{60}Co	1250	1.180	1.180	1.170
2000	2000	1.140	1.149	1.140

[a]X-ray sources are standard beam designations established by the National Institute of Standards and Technology (NIST) for x-ray spectra obtained with noted filtration; M denotes moderate filtration, H denotes heavy filtration, and the number is the x-ray tube voltage in kilovolts.

and 1000 mg/cm^2. The absorbed doses at each of these tissue depths is determined by first calculating or measuring the air kerma at the interface of a tissue of sufficient thickness for electronic equilibrium to exist and then multiplying it by the appropriate C_{ki} value.

Air kerma can be calculated from the incident photon flux as

$$K_a = 1.6022 \times 10^{-10} \phi \left(\frac{\mu_{\text{en}}}{\rho}\right)_{\text{air}} \frac{E_{h\nu}}{1-g}$$

where K_a is in units of gray (100 rad), ϕ is the photon flux ($h\nu/\text{cm}^2 \cdot \text{s}$), μ_{en}/ρ or μ_{tr}/ρ is in units of cm^2/g, and $E_{h\nu}$ is in MeV. The quantity g is unitless and is negligible for photons below 1.5 MeV or so. A mass energy transfer coefficient μ_{tr}/ρ can also be used to calculate K_a; it is more convenient to use the mass energy absorption coefficient μ_{en}/ρ and to adjust it by the value of $1/(1-g)$ since values of μ_{en}/ρ are more readily available.

The *air kerma/exposure relationship* represents a practical means of determining absorbed dose in tissue because the exposure or exposure rate in free air can be readily measured. The air kerma rate is calculated from the exposure rate measurement as

$$K_a = \text{exposure(R)}\frac{W}{e}\frac{1}{1-g} = \frac{\text{exposure(R)(33.97 J/C)}}{1-g} \times 2.58 \times 10^{-4}\ \text{C/kg}\cdot\text{R}$$

$$= \frac{87.64 \times 10^{-3}\ \text{exposure(R)}}{1-g}$$

where K_a is in units of gray, exposure(R) is the measured exposure in roentgens, and g is unitless. The value of K_a is then multiplied by the appropriate C_{ki} value (from Table 7-8) to obtain the absorbed dose. These relationships are particularly useful in calibrating dosimeters or other devices in terms of absorbed dose.

SUMMARY

Radiation interacts in an absorbing medium to deposit energy, which is defined as the absorbed dose, and when weighted according to the damaging effect of the radiation type, as the effective dose equivalent. A related term is radiation exposure, which applies to air only and is a measure of the amount of ionization produced by x and gamma radiation in air.

Radiation dose can be calculated if three things are known: (1) the mass of the medium being irradiated, (2) the number of radiations per unit area (the flux) that impinge on the mass, and (3) the amount or rate of energy deposition in the mass specified. For particles, the mass in which the energy is dissipated is just the depth of penetration; for photons, it is necessary to use a unit depth (e.g., 1 cm) due to the probabilistic pattern of interactions represented by the mass energy absorption coefficient μ_{en}/ρ. All emitted radiations must be considered in this process, and adjustments should be made for any attenuating medium between the source and the point of interest.

ACKNOWLEDGMENTS

This chapter was inspired by many forms of interactions, in particular those with R. C. Riley, A. P. Jacobson, D. M. Hamby, T. W. Philbin, R. F. Fleming, and John H. Hubbell. Its compilation is due in large measure to the patient, careful, and untiring efforts of Chul Lee, a graduate of the University of Michigan Radiological Health Program.

REFERENCES AND ADDITIONAL RESOURCES

Schleien, B., *The Health Physics and Radiological Health Handbook*, Revised Ed., Scinta, Inc., Silver Spring, MD, 1992.

Hubbell, J. H., and S. M. Seltzer, *Tables of X-ray Attenuation Coefficients and Mass Absorption Coefficients* 1 keV *to* 20 MeV *for Elements* Z = 1 *to* 92 *and* 48 *Additional Substances of Dosimetric Interest*, NISTIR 5632, National Institute of Standards and Technology, Gaithersburg, MD, 1995.

Johns, H. E., and J. R. Cunningham, *Physics of Radiology*, 4th ed., Charles C Thomas, Springfield, IL, 1983.

Lamarsh, J. R., *Introduction to Nuclear Engineering*, 2nd ed., Addison-Wesley, Reading, MA, 1983, Chapters 5 and 9.

Lapp, R. E., and H. L. Andrews, *Nuclear Radiation Physics*, 4th ed., Prentice Hall, Upper Saddle River, NJ, 1972.

Tabata, T., R. Ito, and S. Okabe, Backscatter Factors for Beta-Particles, *Nucl. Instrum. Meth.* 94, 509, 1971.

International Commission on Radiological Units, Measurement of Dose Equivalent from External Photon and Electron Radiations, ICRU Report 47, Bethesda, MD, 1992.

PROBLEMS

7-1. What energy must (**a**) an alpha particle and (**b**) a beta particle have to just penetrate the dead layer of a person's skin?

7-2. The binding energies of electrons in different shells of an element may be determined by measuring the transmission of a monoenergetic beam of photons through a thin foil of the element as the energy of the beam is varied. Explain how this method works.

7-3. A γ ray of 2.75 MeV from ^{24}Na undergoes pair production in a lead shield. How much kinetic energy is shared by the two electron masses, and how is it distributed among them?

7-4. A ^{90}Sr–^{90}Y point source contains an activity due to ^{90}Sr of 10^6 t/s (i.e., 1 Mbq). Determine (**a**) the beta-particle flux at 40 cm, (**b**) the energy flux due to ^{90}Y at 40 cm, (**c**) the absorbed dose rate in tissue at 40 cm if air attenuation is not a factor, and (**d**) the effect of absorbed dose rate due to air attenuation.

7-5. A "hot particle" containing 1 μCi of ^{192}Ir was found on a person's lab coat which was next to the skin. What was the absorbed dose rate in rad/h?

7-6. The following data were recorded for identical samples mounted on different holders: a "weightless" Mylar mount, 2038 c/m; a silver disk, 3258 c/m. Compute the backscatter factor for silver.

7-7. (**a**) What is the kinetic energy of the Compton electron for photons scattered at 45° during a Compton interaction if the energy of the incident photon is 150 keV? (**b**) What effect does an increase in the photon scattering angle have on the scattered photon?

7-8. A narrow beam of 40 keV photons which initially contains 500,000 photons is passed through three different filters. One filter is 5 mm of Al, one is 3 mm of Cu, and one is 1 mm of Pb. How many photons interact in each of the three filters?

7-9. For ^{54}Mn, determine (**a**) the gamma-ray constant (R/h at 1 m) and (**b**) the absorbed dose rate in tissue at 50 cm.

7-10. Determine Γ for 1 mCi of ^{60}Co which emits 1.17 MeV (μ_{en} in air of 3.43×10^{-5} cm^{-1}) and 1.33 MeV γ rays assuming that each is emitted 100% of the time.

7-11. A solution of 10 μCi of ^{32}P was spilled on a wooden tabletop but was not discovered until it dried. (**a**) If the area of contamination was a circle 30 cm in diameter and the solution dried uniformly over it, what would be the beta-particle flux, the energy flux, and the dose rate just over the tabletop? (**b**) How would these change if the tabletop were stainless steel?

7-12. An exposure rate was measured some distance from a point source as 100 mR/h. (**a**) What would be the absorbed dose rate (rad/h) to a person who stood at the exact spot? (**b**) What would be the dose equivalent rate (mrem/h)?

7-13. A source of 1 MeV gamma rays produces a uniform flux of 10^6 $\gamma/\text{cm}^2 \cdot \text{s}$ at a distance of 150 cm. (**a**) What is the absorbed dose rate to a person at that distance? (**b**) What would it be at 200 cm?

8

RADIATION SHIELDING

Eureka! I have found it.
—Archimedes, ca. 250 B.C.; W. A. Mills, 1970

Various materials, placed between radiation sources and a receptor, can affect the amount of radiation reaching the receptor. Such effects are due to attenuation and absorption of the emitted radiation in the source itself, in material used for encapsulation of the source, or in a shielding barrier. Regardless of how it occurs, shielding is an important aspect of radiation protection since it can be a form of radiation control; therefore, the features of shields and their design, use, and effectiveness warrant specific consideration.

The presentation in this chapter builds on the interaction principles described in Chapter 7. It is noteworthy that even though various aspects are straightforward, radiation shielding is a very complex discipline for many radiation sources and the many geometric configurations in which they may occur. Therefore, this presentation focuses on straightforward configurations of point sources, line sources, and area and volume sources, which fortunately can be used conservatively to address most of the situations encountered in practical radiation protection. Many shielding problems can be treated in terms of one of these configurations, and in general the exposure will be slightly overestimated such that shield designs would be conservative.

SHIELDING OF ALPHA-PARTICLE SOURCES

Alpha particles are easy to shield since their relatively large mass and charge limit their range in most media to a few tens of micrometers and to just a few centimeters in air. Only the most energetic ones are capable of penetrating the 7 mg/cm^2 dead layer of human skin. However, inside the body, they do considerable damage because of these same properties.

Most shielding problems for alpha emitters involve fixing the material in place so it cannot become a source of contamination taken into the body by touch or as airborne particles. A layer of paint of other fixative can be used to reduce even highly contaminated areas to nondetectable levels. Caution and continuous vigilance through inspections and surveys are required to assure that the fixative remains intact, especially on floors or high-use areas.

In a similar vein, the very short range of alpha particles makes their detection very difficult because they must be able to penetrate detector coverings or windows to reach the sensitive volume of the detector and be registered. It is easy to miss highly contaminated areas by failing to get close enough (most alpha particles are absorbed in a few centimeters of air), use of the wrong detector, or use of one with a window that is too thick.

SHIELDING OF BETA-PARTICLE SOURCES

Beta particles have a limited range in most media because they are charged particles and will dissipate all their energy as they traverse a medium by ionization and bremsstrahlung production. The path of ionization can be rather long and tortuous because of the smallness of the electrons, but limited nonetheless. The mean range of most beta particles is no more than a few meters in air (see Figure 8-1); thus air can significantly attenuate beta particles and shielding properties of other media are even better since the range of most beta particles is only a few millimeters in dense materials. Because of this behavior, all of the beta particles emitted by a source can be stopped by determining the maximum range of the highest-energy beta particles and choosing a thickness of medium that matches or exceeds the range. Most beta shields are based on this very practical approach, usually with a little more thickness just to be sure.

Attenuation of Beta Particles

The exponential decrease in the number of beta particles, as shown in Figure 7-8, can be used to determine beta-particle attenuation as long as the absorber thickness (mg/cm^2) is less than the maximum range of the beta particles in the particular medium i; therefore the intensity of a beta source is

$$I(x) = I_0 e^{-\mu_{\beta,i}(\rho x)}$$

where $\mu_{\beta,i}$(g/cm^2) is a function of maximum beta-particle energy, I_0 the intensity (or number) of beta particles, $I(x)$ the intensity (or number) observed for an absorber of thickness x, and ρx is the density thickness (g/cm^2) of the absorber. Since the units of μ_β for various beta sources are determined (and expressed) in cm^2/g, it is necessary to specify the absorber as a density thickness with units of g/cm^2, which is obtained by multiplying the absorber thickness x by its density ρ.

As developed in Chapter 7, beta absorption coefficients for air, tissue (or water), and solid materials as a function of $E_{\beta,\max}$, respectively, are

$$\mu_{\beta,\text{air}} = 16(E_{\beta,\max} - 0.036)^{-1.4}$$

$$\mu_{\beta,\text{tissue}} = 18.6(E_{\beta,\max} - 0.036)^{-1.37}$$

$$\mu_{\beta,i} = 17(E_{\beta,\max})^{-1.14}$$

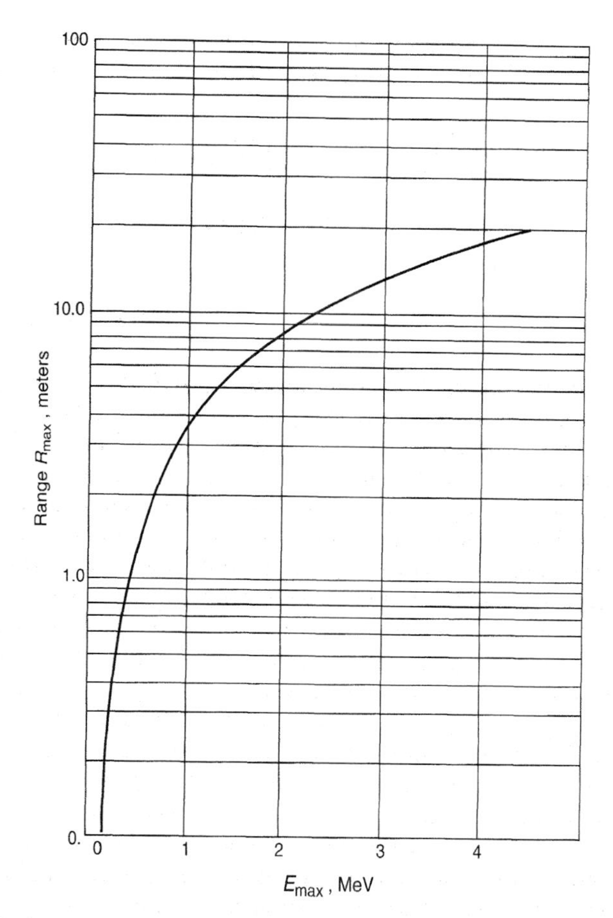

FIGURE 8-1. Maximum range $R_{\beta,\max}$ versus maximum energy $E_{\beta,\max}$ of beta particles in air at 20°C and 1 atm.

where values of μ_β are in units of $\mathrm{cm^2/g}$ and $E_{\beta,\max}$ is in units of MeV. It should be noted that μ_β is uniquely defined with units of $\mathrm{cm^2/g}$ instead of the more traditional units of $\mathrm{cm^{-1}}$ or as μ/ρ because the most practical uses of the beta absorption coefficients are in calculations of beta attenuation in materials with density thickness ρx.

Example 8-1: Beta particles with $E_{\beta,\max}$ of 2.0 MeV produce a flux of 1000 beta particles/$\mathrm{cm^2 \cdot s}$ incident on an aluminum ($\rho = 2.7$ g/cm^3) absorber 0.1 mm thick. What will be the flux of beta particles reaching a receptor just beyond the foil?

SOLUTION: The beta-absorption coefficient μ_β is

$$\mu_\beta = 17 \times (2)^{-1.14} = 17(0.45376)$$

$$= 7.714\ \mathrm{cm^2/g}$$

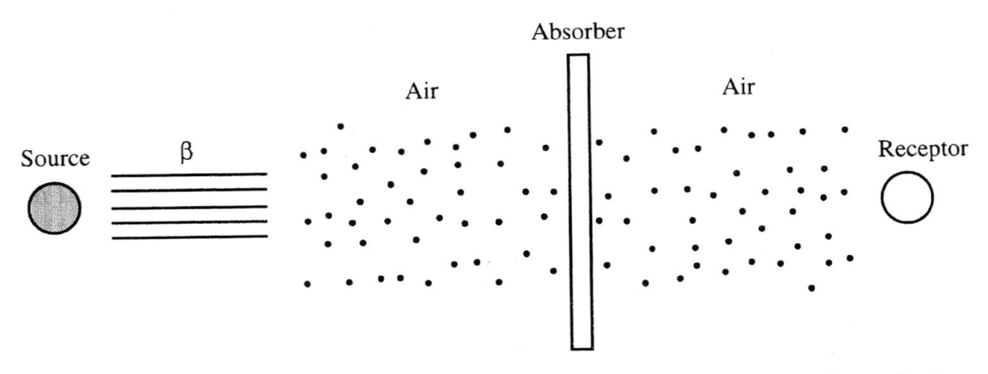

FIGURE 8-2. Schematic of various absorbing materials between a source that emits beta particles and a layer or volume of interest.

The density thickness of the aluminum absorber is

$$2.7\ \text{g/cm}^3 \times 0.01\ \text{cm} = 0.0274\ \text{g/cm}^2$$

Therefore, the flux of beta particles penetrating the foil and striking the receptor is

$$\begin{aligned} I(x) = I_0 e^{-\mu_\beta(\rho x)} &= 1000 e^{(-7.714\ \text{cm}^2/\text{g})\times(0.0274\ \text{g/cm}^2)} \\ &= 809.5\ \beta\text{s/cm}^2 \cdot \text{s} \end{aligned}$$

It is important to consider all the attenuating materials that may be present between the beta source and the point of interest. For example, as shown in Figure 8-2, a given configuration may contain in addition to the shield material enclosing the beta emitter, a thickness of air, and yet another layer between the air gap and the sensitive volume of interest (e.g., a dead layer of skin or a detector window). Of course, geometry factors for a given configuration and a backscatter factor, if appropriate, also influence the dose rate from a source. With the exception of backscatter, these tend to add additional conservatism in the effectiveness of beta shields, but not always if bremsstrahlung is considered.

Bremsstrahlung Effects for Beta Shielding

Since beta particles are high-speed electrons, they produce bremsstrahlung, or braking radiation, in passing through matter, especially if the absorbing medium is a high-Z material. Consequently, the design of shields for beta-particle sources or accelerated monoenergetic electrons need to consider the fraction of the beta-particle energy emitted by a source that is converted to photons in order to account for them. The most accurate data on the fraction of the beta energy that is converted to bremsstrahlung have been provided by Berger and Seltzer (1983) for monoenergetic electrons; these are listed in Table 8-1 and plotted in Figure 8-3 and illustrate the effects of changes in electron energy and Z of absorber on the energy fraction converted to photons. Since beta particles emitted from a radionuclide exhibit a

TABLE 8-1. Percent Radiation Yield for Electrons of Initial Energy E on Different Absorbers

	Absorber (Z)					
E (MeV)	Water	Air	Al (13)	Cu (29)	Sn (50)	Pb (82)
0.100	0.058	0.066	0.135	0.355	0.658	1.162
0.200	0.098	0.111	0.223	0.595	1.147	2.118
0.300	0.133	0.150	0.298	0.795	1.548	2.917
0.400	0.166	0.187	0.368	0.974	1.900	3.614
0.500	0.198	0.223	0.435	1.143	2.224	4.241
0.600	0.229	0.258	0.501	1.307	2.530	4.820
0.700	0.261	0.293	0.566	1.467	2.825	5.363
0.800	0.293	0.328	0.632	1.625	3.111	5.877
0.900	0.325	0.364	0.698	1.782	3.391	6.369
1.000	0.358	0.400	0.764	1.938	3.666	6.842
1.250	0.442	0.491	0.931	2.328	4.340	7.960
1.500	0.528	0.584	1.101	2.720	4.998	9.009
1.750	0.617	0.678	1.274	3.113	5.646	10.010
2.000	0.709	0.775	1.449	3.509	6.284	10.960
2.500	0.897	0.972	1.808	4.302	7.534	12.770
3.000	1.092	1.173	2.173	5.095	8.750	14.470

spectral effect (where the average beta energy is about $\frac{1}{3}E_{\beta,\max}$), it is more difficult to determine the bremsstrahlung yield for beta sources.

Empirical relations have been developed for the fraction of the total electron energy that is converted to photon production; however, these, too, are based on the experimental data for monoenergetic electrons. One such relation, expressed as an energy yield fraction Y_i for monoenergetic electrons is

$$Y_i = \frac{6 \times 10^{-4} EZ}{1 + 6 \times 10^{-4} EZ}$$

where $E_{\beta,\max}$ is the maximum energy of the electrons in MeV before striking a target and Z is the atomic number of the target material.

For a beta-particle source it is appropriate to multiply the value of Y_i for monoenergetic electrons of energy $E = E_{\beta,\max}$ on an absorber by a factor of 0.3 to obtain the total energy converted to photons. The total energy converted to photons can then be divided by an energy value assumed for each bremsstrahlung photon to obtain a beta flux for use in determining beta dose. This can be done conservatively by assuming that each bremsstrahlung photon has an energy equal to the maximum energy of the beta particles emitted by the source; however, it may be more reasonable but less conservative to assign the average beta-particle energy to each bremsstrahlung produced.

Beta-shield designs must account for the yield of bremsstrahlung photons. High-Z materials such as lead (Z = 82) should not be used to shield high activity sources such as ^{32}P($E_{\beta,\max}$ = 1.71 MeV) and ^{90}Sr-Y($E_{\beta,\max}$ = 2.28 MeV) because 8 to 12% of the emitted energy could be in the form of bremsstrahlung photons. Since the bremsstrahlung yield is higher in high-Z materials, the most practical approach

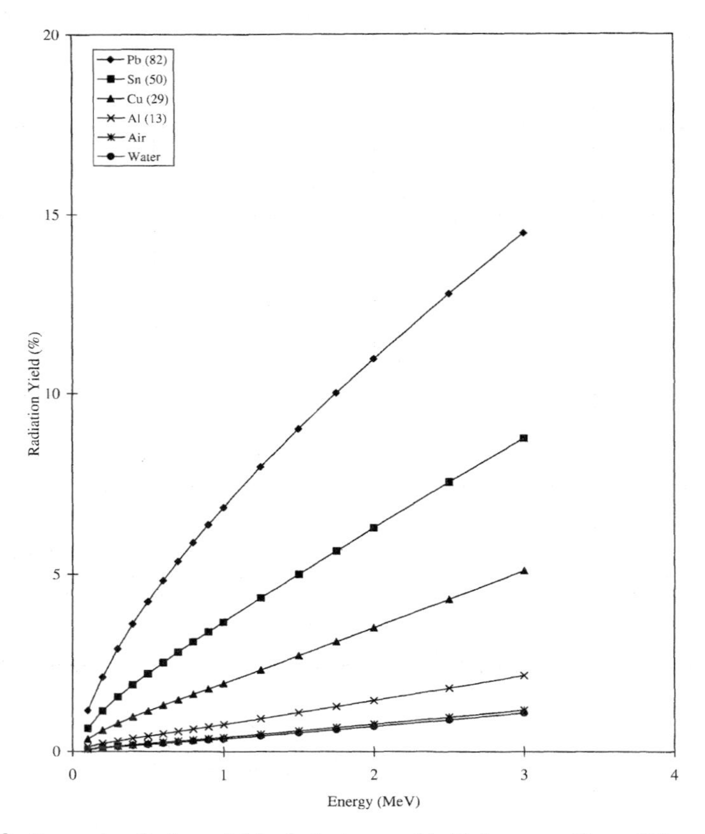

FIGURE 8-3. Percent radiation yield of electrons of initial energy E on different absorbers.

for shielding beta sources is to use a layer of plastic, aluminum, or other low-Z material to absorb all the beta particles and minimize bremsstrahlung production, then to add a layer of lead or other dense material to absorb any bremsstrahlung and characteristic x rays that are produced in the beta shield, as shown in Example 8-2.

Example 8-2: One curie of $^{32}P(E_{\beta,\text{max}} = 1.71$ MeV) is dissolved in 50 mL of water for an experiment. If it is to be kept in a polyethylene ($\rho = 0.93$) bottle, how thick should the wall be to stop all the beta particles emitted by ^{32}P, and what thickness of lead would be required to assure that the dose equivalent rate due to bremsstrahlung photons will be less than 2 mrem/h at 1 m?

SOLUTION: From Figure 7-9, the maximum range of 1.71 MeV beta particles is 810 mg/cm^2 or 0.81 g/cm^2; therefore, the thickness of polyethylene required to absorb the 1.71 MeV beta particle is

$$\frac{0.81\ \text{g/cm}^2}{0.93\ \text{g/cm}^3} = 0.87\ \text{cm}$$

Essentially all of the bremsstrahlung can be assumed to be produced in the water solution since the thin polyethylene walls are of similar density. From Table 8-1, about 0.6% of 1.71 MeV beta energy (fractional yield of about 6×10^{-3}) is converted to photons due to absorption in water, and since the average beta energy for ^{32}P (Appendix D) is 0.695 MeV, the photon energy emission rate is

$$\begin{aligned} E_{\text{rad}} &= 3.7 \times 10^{10} \text{ t/s} \times 0.695 \text{ MeV/t} \times 6 \times 10^{-3} \\ &= 1.55 \times 10^{8} \text{ MeV/s due to bremsstrahlung production} \end{aligned}$$

If this beta energy converted to bremsstrahlung is assumed to produce photons equal to the maximum beta energy of 1.71 MeV, a conservative assumption, the photon emission rate is

$$\frac{1.55 \times 10^{8}}{1.71 \text{ MeV/photon}} = 9.1 \times 10^{7} \text{ photons/s}$$

or 7.24×10^{2} photons/cm^{2} · s at 1 m.

From Table 8-2, the *mass energy* absorption coefficient, by interpolation for 1.71 MeV photons in tissue, is 0.027 cm^2/g; therefore, this flux produces an energy absorption rate in tissue of

$$\begin{aligned} E_{ab} &= 7.24 \times 10^{2} \text{ photons/cm}^{2} \cdot \text{s} \times 1.71 \text{ MeV} \times 0.027 \text{ cm}^{2}/\text{g} \\ &\quad \times 1.6022 \times 10^{-6} \text{ erg/MeV} \times 3600 \text{ s/h} \\ &= 0.193 \text{ erg/g} \cdot \text{h} \end{aligned}$$

which corresponds to an absorbed dose rate of 1.93 mrad/h, or a dose equivalent rate of 1.93 mrem/h. The attenuation coefficient μ for 1.71 MeV photons in lead is given in Table 8-2 as 0.565 cm^{-1}; thus the thickness of lead needed to attenuate the photon flux to 1 mrem/h is

$$1 \text{ mrem/h} = (1.93 \text{ mrem/h})e^{-0.565x}$$

and $x = 1.16$ cm. As will be shown later, it is usually necessary to account for scattered photons that are produced in the absorber by a buildup factor; however, since it is assumed that all the bremsstrahlung is produced with an energy of $E_{\beta,\text{max}}$, a conservative assumption, buildup can reasonably be ignored. Consideration of buildup due to Compton scattering could result in a somewhat thicker shield.

SHIELDING OF PHOTON SOURCES

As discussed in Chapter 7, photon interactions with matter are very different from those of charged particles. When x or gamma rays traverse matter, some are absorbed, some pass through without interaction, and some are scattered as lower-energy photons in directions that are quite different from those in the primary beam. The attenuation of a photon beam by an absorber is characterized, as shown in Figure 7-16, as occurring in good geometry or poor geometry. In good geometry attenuation, every photon that interacts is either absorbed or scattered out of the primary beam such that those that reach the receptor have all their original energy. In poor geometry (also called broad-beam geometry), a significant fraction of scattered photons will also reach the receptor of interest, and the energy spectrum will be quite complex, with multiple scattered photon energies in addition to unattenuated photons that retain all of their initial energy. Poor geometry exists in most practical conditions when tissue is exposed or a shield is used to attenuate a photon source and it is necessary to account for these scattered photons.

Shielding of Good Geometry Photon Sources

The attenuation of photons by various absorbing materials under ideal narrow beam conditions satisfies the relationship

$$I(x) = I_0 e^{-\mu x}$$

where I_0 is the initial photon intensity (usually expressed as a fluence or flux), $I(x)$ the photon intensity after passing through an absorber of thickness x in narrow-beam geometry, and μ (cm^{-1}) the total attenuation coefficient, which accounts for all interaction processes, including scattering reactions, that remove photons from the beam. The attenuation coefficient μ is dependent on the particular absorber medium and the photon energy; therefore, extensive listings are required as provided in Table 8-2 along with mass attenuation and mass–energy absorption coefficients. The values of μ listed in Table 8-2 were obtained with collimated narrow beams of monoenergetic photons incident on the different absorbers with a detector placed behind each absorber such that scattered photons were nondetectable.

Values of μ generally increase as the Z of the absorber increases (see Figure 7-20) because photoelectric interactions are increased in high-Z materials, especially for low energy photons, and high-Z materials yield increases in pair production interactions for high-energy photons. Because of the high-Z effect, lead is often used to line the walls of x-ray rooms, made into lead aprons for personnel protection, and incorporated into leaded glass, and $BaSO_4$ is incorporated into concrete (called *barite* or *barytes concrete*) to increase its effectiveness as a photon shield.

Photon attenuation coefficients in various materials can be calculated from measurements of the intensity of a narrow beam of photons of a given energy, as shown in Example 8-3.

Example 8-3: If a narrow beam of 2000 monoenergetic photons of 1.0 MeV is reduced to 1000 photons by a slab of copper 1.31 cm thick, determine (a) the total linear attenuation coefficient of the copper slab for these photons and (b) the total mass attenuation coefficient.

SOLUTION: (a) Since $I/I_0 = e^{-\mu x}$, μ is determined by taking the natural logarithm of each side of the equation, or

$$\ln 0.5 = -\mu(1.31 \text{ cm})$$

$$\mu = \frac{0.69315}{1.31 \text{ cm}} = 0.5287 \text{ cm}^{-1}$$

TABLE 8-2. Photon Attenuation (μ), Mass Attenuation (μ/ρ), and Mass Energy Absorption (μ_{en}/ρ) Coefficients for Selected Elements, Compounds, and Mixtures

Energy (keV)	Dry Air (Sea Level) (ρ = 0.001205 g/cm^3)			Water, Liquid (ρ = 1.00 g/cm^3)			Aluminum (ρ = 2.699 g/cm^3)			Iron (ρ = 7.874 g/cm^3)		
	μ (cm^{-1})	μ/ρ (cm^2/g)	μ_{en}/ρ (cm^2/g)	μ (cm^{-1})	μ/ρ (cm^2/g)	μ_{en}/ρ (cm^2/g)	μ (cm^{-1})	μ/ρ (cm^2/g)	μ_{en}/ρ (cm^2/g)	μ (cm^{-1})	μ/ρ (cm^2/g)	μ_{en}/ρ (cm^2/g)
10	0.0062	5.120	4.742	5.329	5.329	4.944	70.795	26.23	25.43	1343.304	170.6	136.90
15	0.0019	1.614	1.334	1.673	1.673	1.374	21.4705	7.955	7.487	449.4479	57.08	48.96
20	0.0009	0.7779	0.5389	0.8096	0.8096	0.5503	9.2873	3.441	3.094	202.2043	25.68	22.60
30	4.26×10^4	0.3538	0.1537	0.3756	0.3756	0.1557	3.0445	1.128	0.8778	64.3778	8.176	7.251
40	2.99×10^4	0.2485	0.0683	0.2683	0.2683	0.0695	1.5344	0.5685	0.3601	28.5747	3.629	3.155
50	2.51×10^4	0.2080	0.0410	0.2269	0.2269	0.0422	0.9935	0.3681	0.1840	15.4173	1.958	1.638
60	2.26×10^4	0.1875	0.0304	0.2059	0.2059	0.0319	0.7498	0.2778	0.1099	9.4882	1.2050	0.9555
70[a]	2.10×10^4	0.1744	0.0255	0.1948	0.1948	0.0289	0.6130	0.2271	0.0713	6.2318	0.7914	0.5836
80	2.00×10^4	0.1662	0.0241	0.1924	0.1924	0.0272	0.5447	0.2018	0.0551	4.6866	0.5952	0.4104
100	1.86×10^4	0.1541	0.0233	0.1707	0.1707	0.0255	0.4599	0.1704	0.0379	2.9268	0.3717	0.2177
150	1.63×10^4	0.1356	0.0250	0.1505	0.1505	0.0276	0.3719	0.1378	0.0283	1.5465	0.1964	0.0796
200	1.49×10^4	0.1233	0.0267	0.1370	0.1370	0.0297	0.3301	0.1223	0.0275	1.1496	0.1460	0.0483
300	1.29×10^4	0.1067	0.0287	0.1186	0.1186	0.0319	0.2812	0.1042	0.0282	0.8654	0.1099	0.0336
400	1.15×10^4	0.0955	0.0295	0.1061	0.1061	0.0328	0.2504	0.0928	0.0286	0.7402	0.0940	0.0304
500	1.05×10^4	0.0871	0.0297	0.0969	0.0969	0.0330	0.2279	0.0845	0.0287	0.6625	0.0841	0.0291
600	9.71×10^5	0.0806	0.0295	0.0896	0.0896	0.0328	0.2106	0.0780	0.0285	0.6066	0.0770	0.0284
662[a]	9.34×10^5	0.0775	0.0293	0.0862	0.0862	0.0326	0.2024	0.0750	0.0283	0.5821	0.0739	0.0280
800	8.52×10^5	0.0707	0.0288	0.0787	0.0787	0.0321	0.1846	0.0684	0.0278	0.5275	0.0670	0.0271
1,000	7.66×10^5	0.0636	0.0279	0.0707	0.0707	0.0310	0.1659	0.0615	0.0269	0.4720	0.0600	0.0260
1,173[a]	7.05×10^5	0.0585	0.0271	0.0650	0.0650	0.0301	0.1526	0.0565	0.0261	0.4329	0.0550	0.0251
1,250	6.86×10^5	0.0569	0.0267	0.0632	0.0632	0.0297	0.1483	0.0550	0.0257	0.4213	0.0535	0.0247
1,333[a]	6.62×10^5	0.0550	0.0263	0.0611	0.0611	0.0292	0.1435	0.0532	0.0253	0.4071	0.0517	0.0243
1,500	6.24×10^5	0.0518	0.0255	0.0575	0.0575	0.0283	0.1351	0.0501	0.0245	0.3845	0.0488	0.0236
2,000	5.36×10^5	0.0445	0.0235	0.0494	0.0494	0.0261	0.1167	0.0432	0.0227	0.3358	0.0427	0.0220
3,000	4.32×10^5	0.0358	0.0206	0.0397	0.0397	0.0228	0.0956	0.0354	0.0202	0.2851	0.0362	0.0204
4,000	3.71×10^5	0.0308	0.0187	0.0340	0.0340	0.0207	0.0838	0.0311	0.0188	0.2608	0.0331	0.0199
5,000	3.31×10^5	0.0275	0.0174	0.0303	0.0303	0.0192	0.0765	0.0284	0.0180	0.2477	0.0315	0.0198
6,000	3.04×10^5	0.0252	0.0165	0.0277	0.0277	0.0181	0.0717	0.0266	0.0174	0.2407	0.0306	0.0200
6,129[a]	3.01×10^5	0.0250	0.0164	0.0274	0.0274	0.0180	0.0713	0.0264	0.0173	0.2403	0.0305	0.0200
7,000[a]	2.83×10^5	0.0235	0.0159	0.0258	0.0258	0.01723	0.0683	0.0253	0.0170	0.2370	0.0301	0.0202
7,115[a]	2.82×10^5	0.0234	0.0158	0.0256	0.0256	0.0172	0.0680	0.0252	0.0170	0.2368	0.0301	0.0203
10,000	2.46×10^5	0.0205	0.0145	0.0222	0.0222	0.0157	0.0626	0.0232	0.0165	0.2357	0.0299	0.0211

(Continued)

TABLE 8-2. Photon Attenuation (μ), Mass Attenuation (μ/ρ), and Mass Energy Absorption (μ_{en}/ρ) Coefficients for Selected Elements, Compounds, and Mixtures *(Continued)*

	Copper ($\rho = 8.96$ g/cm^3)			Lead ($\rho = 11.35$ g/cm^3)			Polyethylene ($\rho = 0.93$ g/cm^3)			Concrete, Ordinary ($\rho = 2.3$ g/cm^3)		
Energy (keV)	μ (cm^{-1})	μ/ρ (cm^2/g)	μ_{en}/ρ (cm^2/g)	μ (cm^{-1})	μ/ρ (cm^2/g)	μ_{en}/ρ (cm^2/g)	μ (cm^{-1})	μ/ρ (cm^2/g)	μ_{en}/ρ (cm^2/g)	μ (cm^{-1})	μ/ρ (cm^2/g)	μ_{en}/ρ (cm^2/g)
10	1934.464	215.9	148.4	1482.31	130.6	124.7	1.9418	2.088	1.781	47.04	20.45	19.37
15	663.4880	74.05	57.88	1266.66	111.6	91.0	0.6930	0.7452	0.4834	14.61	6.351	5.855
20	302.7584	33.79	27.88	980.1860	86.36	68.99	0.4013	0.4315	0.1936	6.454	2.806	2.462
30	97.8432	10.92	9.349	344.1320	30.32	25.36	0.2517	0.2706	0.0593	2.208	0.9601	0.7157
40	43.5635	4.862	4.163	162.9860	14.36	12.11	0.2116	0.2275	0.0320	1.163	0.5058	0.2995
50	23.4125	2.613	2.192	91.2654	8.041	6.740	0.1938	0.2084	0.0244	0.7848	0.3412	0.1563
60	14.2733	1.5930	1.2900	56.9884	5.0210	4.1490	0.1832	0.1970	0.0224	0.6118	0.2660	0.0955
70[a]	9.2401	1.0313	0.7933	35.0670	3.0896	2.6186	0.1754	0.1886	0.0221	0.5131	0.2231	0.0638
80	6.8365	0.7630	0.5581	27.4557	2.4190	1.9160	0.1695	0.1823	0.0227	0.4632	0.2014	0.0505
100	4.1073	0.4584	0.2949	62.9812	5.5490	1.9760	0.1599	0.1719	0.0242	0.3997	0.1738	0.0365
150	1.9864	0.2217	0.1027	22.8589	2.0140	1.0560	0.1427	0.1534	0.0279	0.3303	0.1436	0.0290
200	1.3969	0.1559	0.0578	11.3330	0.9985	0.5870	0.1304	0.1402	0.0303	0.2949	0.1282	0.0287
300	1.0026	0.1119	0.0362	4.5752	0.4031	0.2455	0.1132	0.1217	0.0328	0.2523	0.1097	0.0297
400	0.8434	0.0941	0.0312	2.6366	0.2323	0.1370	0.1013	0.1089	0.0337	0.2250	0.0978	0.0302
500	0.7492	0.0836	0.0293	1.8319	0.1614	0.0913	0.0925	0.0995	0.0339	0.2050	0.0892	0.0303
600	0.6832	0.0763	0.0283	1.4165	0.1248	0.0682	0.0855	0.0920	0.0338	0.1894	0.0824	0.0302
662[a]	0.6555	0.0732	0.0279	1.2419	0.1094	0.0587	0.0823	0.0885	0.0335	0.1822	0.0792	0.0299
800	0.5918	0.0661	0.0268	1.0067	0.0887	0.0464	0.0751	0.0808	0.0330	0.1662	0.0723	0.0294
1,000	0.5287	0.0590	0.0256	0.8061	0.0710	0.0365	0.0675	0.0726	0.0319	0.1494	0.0650	0.0284
1,173[a]	0.4840	0.0540	0.0246	0.7020	0.0619	0.0315	0.0621	0.0668	0.0310	0.1374	0.0598	0.0276
1,250	0.4714	0.0526	0.0243	0.6669	0.0588	0.0299	0.0604	0.0650	0.0305	0.1336	0.0581	0.0272
1,333[a]	0.4553	0.0508	0.0239	0.6369	0.0561	0.0285	0.0584	0.0628	0.0300	0.1291	0.0561	0.0268
1,500	0.4303	0.0480	0.0232	0.5927	0.0522	0.0264	0.0550	0.0591	0.0291	0.1216	0.0529	0.0260
2,000	0.3768	0.0421	0.0216	0.5228	0.0461	0.0236	0.0471	0.0506	0.0268	0.1048	0.0456	0.0240
3,000	0.3225	0.0360	0.0202	0.4806	0.0423	0.0232	0.0376	0.0405	0.0233	0.0851	0.0370	0.0212
4,000	0.2973	0.0332	0.0199	0.4764	0.0420	0.0245	0.0320	0.0344	0.0209	0.0740	0.0322	0.0195
5,000	0.2847	0.0318	0.0200	0.4849	0.0427	0.0260	0.0283	0.0305	0.0192	0.0669	0.0291	0.0184
6,000	0.2785	0.0311	0.0203	0.4984	0.0439	0.0274	0.0257	0.0276	0.0179	0.0620	0.0270	0.0176
6,129[a]	0.2781	0.0310	0.0203	0.5002	0.0441	0.0276	0.0254	0.0273	0.0178	0.0616	0.0268	0.0175
7,000[a]	0.2757	0.0308	0.0206	0.5141	0.0453	0.0287	0.0236	0.0254	0.0169	0.0585	0.0254	0.0171
7,115[a]	0.2755	0.0307	0.0207	0.5161	0.0455	0.0289	0.0235	0.0252	0.0168	0.0582	0.0253	0.0170
10,000	0.2780	0.0310	0.0217	0.5643	0.0497	0.0318	0.0199	0.0215	0.0151	0.0524	0.0228	0.0162

Energy (keV)	Concrete, Barite (Type BA) ($\rho = 3.35$ g/cm^3)			Leaded Glass ($\rho = 6.22$ g/cm^3)			Borosilicate Glass ($\rho = 2.23$ g/cm^3)			Gypsum ($\rho = 2.69$ g/cm^3)		
	μ (cm^{-1})	μ/ρ (cm^2/g)	μ_{en}/ρ (cm^2/g)	μ (cm^{-1})	μ/ρ (cm^2/g)	μ_{en}/ρ (cm^2/g)	μ (cm^{-1})	μ/ρ (cm^2/g)	μ_{en}/ρ (cm^2/g)	μ (cm^{-1})	μ/ρ (cm^2/g)	μ_{en}/ρ (cm^2/g)
10	357.45	106.70	99.60	640.04	102.90	98.21	38.02	17.05	16.42	124.62	42.1000	39.7700
15	120.63	36.01	33.63	532.25	85.57	69.91	11.63	5.217	4.828	39.3384	13.2900	12.4700
20	55.44	16.55	15.27	408.53	65.68	52.52	5.122	2.297	1.995	17.2598	5.8310	5.3520
30	18.60	5.551	4.912	143.37	23.05	19.27	1.781	0.7987	0.5684	5.5530	1.8760	1.5870
40	39.70	11.85	4.439	67.98	10.93	9.198	0.9680	0.4341	0.2361	2.6427	0.8928	0.6645
50	22.35	6.671	3.206	38.15	6.134	5.118	0.6739	0.3022	0.1235	1.5928	0.5381	0.3403
60	13.88	4.143	2.266	23.9035	3.843	3.152	0.5390	0.2417	0.0765	1.1212	0.3788	0.2001
70[a]	9.04	2.699	1.630	15.0524	2.420	1.930	0.4623	0.2073	0.0523	0.8594	0.2903	0.1256
80	6.593	1.968	1.211	11.6252	1.869	1.458	0.4215	0.1890	0.0423	0.7323	0.2474	0.0929
100	3.759	1.122	0.7138	26.2235	4.216	1.498	0.3695	0.1657	0.0321	0.5808	0.1962	0.0573
150	1.482	0.4423	0.2659	9.641	1.550	0.8042	0.3097	0.1389	0.0273	0.4402	0.1487	0.0344
200	0.8603	0.2568	0.1369	4.8640	0.7820	0.4508	0.2779	0.1246	0.0276	0.3827	0.1293	0.0306
300	0.4891	0.1460	0.0641	2.0507	0.3297	0.1934	0.2384	0.1069	0.0289	0.3220	0.1088	0.0298
400	0.3698	0.1104	0.0447	1.2340	0.1984	0.1114	0.2127	0.0954	0.0295	0.2857	0.0965	0.0299
500	0.3119	0.0931	0.0372	0.8888	0.1429	0.0768	0.1939	0.0870	0.0296	0.2598	0.0878	0.0299
600	0.2762	0.0825	0.0334	0.7078	0.1138	0.0592	0.1792	0.0804	0.0294	0.2398	0.0810	0.0296
662[a]	0.2635	0.0787	0.0325	0.6602	0.1061	0.0551	0.1723	0.0773	0.0292	0.2293	0.0775	0.0294
800	0.2324	0.0694	0.0295	0.5238	0.0842	0.0425	0.1573	0.0705	0.0287	0.2102	0.0710	0.0288
1,000	0.2048	0.0611	0.0274	0.4301	0.0691	0.0347	0.1413	0.0634	0.0277	0.1888	0.0638	0.0279
1,173[a]	0.1851	0.0553	0.0256	0.3803	0.0611	0.0307	0.1300	0.0583	0.0269	0.1743	0.0589	0.0270
1,250	0.1810	0.0540	0.0254	0.3624	0.0583	0.0293	0.1264	0.0567	0.0265	0.1687	0.0570	0.0266
1,333[a]	0.1744	0.0521	0.0248	0.3467	0.0557	0.0281	0.1221	0.0548	0.0261	0.1633	0.0552	0.0262
1,500	0.1647	0.0492	0.0240	0.3239	0.0521	0.0264	0.1151	0.0516	0.0253	0.1537	0.0519	0.0254
2,000	0.1439	0.0430	0.0223	0.2841	0.0457	0.0237	0.0992	0.0445	0.0234	0.1328	0.0449	0.0235
3,000	0.1231	0.0368	0.0208	0.2539	0.0408	0.0228	0.0805	0.0361	0.0207	0.1089	0.0368	0.0210
4,000	0.1135	0.0339	0.0204	0.2449	0.0394	0.0234	0.0700	0.0314	0.0190	0.0957	0.0323	0.0196
5,000	0.1085	0.0324	0.0205	0.2438	0.0392	0.0244	0.0633	0.0284	0.0180	0.0874	0.0295	0.0187
6,000	0.1059	0.0316	0.0207	0.2462	0.0396	0.0254	0.0587	0.0263	0.0172	0.0820	0.0277	0.0181
6,129[a]	0.1056	0.0315	0.0207	0.2468	0.0397	0.0255	0.0581	0.0261	0.0171	0.0815	0.0275	0.0181
7,000[a]	0.1047	0.0313	0.0210	0.2504	0.0403	0.0263	0.0553	0.0248	0.0167	0.0783	0.0265	0.0178
7,115[a]	0.1046	0.0312	0.0211	0.2510	0.0403	0.0264	0.0549	0.0246	0.0166	0.0779	0.0263	0.0178
10,000	0.1051	0.0314	0.0221	0.2671	0.0430	0.0288	0.0496	0.0222	0.0158	0.0719	0.0243	0.0173

Source: Adapted from Hubbell and Seltzer, 1995.

[a] Coefficients for these energies were interpolated using polynomial regression.

(b) Copper has a density of 8.96 g/cm^3; therefore, the mass attenuation coefficient μ/ρ is

$$\frac{\mu}{\rho} = \frac{0.5287 \text{ cm}^{-1}}{8.96 \text{ g/cm}^3} = 0.0590 \text{ cm}^2/\text{g}$$

Half- and Tenth-Value Layers

As was done in Chapter 5 for radioactivity, it is also useful to express the exponential attenuation of photons in terms of a half-thickness, $x_{1/2}$, or *half-value layer* (HVL). The HVL (sometimes called the *half-value thickness*) is the thickness of absorber required to decrease the intensity of a beam of photons to one-half its initial value, or

$$\frac{I(x)}{I_0} = \frac{1}{2} = e^{-\mu x_{1/2}}$$

which can be solved for $x_{1/2}$ to yield

$$x_{1/2} = \text{HVL} = \frac{\ln 2}{\mu}$$

Since a thickness of 1.31 cm of copper reduces the beam intensity of 1.0 MeV photons in Example 8-3 by one-half, it is also the HVL of copper for 1.0 MeV photons.

Similarly, the *tenth-value layer* (TVL) is

$$\text{TVL} = \frac{\ln 10}{\mu} = \frac{2.3026}{\mu}$$

Half-value layers for various materials are listed in Table 8-3. These can be used in calculations for photon attenuation in much the same way that half-life is used for radioactive transformation (see Example 8-4).

Example 8-4: If the HVL for iron is 1.47 cm for 1 MeV photons (see Table 8-3) and the exposure rate from a source is 800 mR/h, calculate (a) μ (cm^{-1}), (b) the thickness of iron required to reduce the exposure rate to 200 mR/h, and (c) the thickness of iron required to reduce it to 150 mR/h.

SOLUTION: (a) The photon attenuation coefficient is

$$\mu = \frac{\ln 2}{\text{HVL}} = \frac{0.69135}{1.47 \text{ cm}} = 0.47 \text{ cm}^{-1}$$

(b) It is observed that $800/2^n = 200$ when $n = 2$; therefore, 2 HVLs, or 2.94 cm of iron, will reduce an 800 mR/h exposure rate to 200 mR/h.

TABLE 8-3. Half-Value Layers (in cm) Versus Photon Energy for Various Materials[a]

Energy (MeV)	Lead (11.35 g/cm^3)	Iron (7.874 g/cm^3)	Aluminum (2.699 g/cm^3)	Water (1.00 g/cm^3)	Air (0.001205 g/cm^3)	Stone Concrete (2.30 g/cm^3)
0.1	0.011	0.237	1.507	4.060	3.726×10^3	1.734
0.3	0.151	0.801	2.464	5.843	5.372×10^3	2.747
0.5	0.378	1.046	3.041	7.152	6.600×10^3	3.380
0.662	0.558	1.191	3.424	8.039	7.420×10^3	3.806
1.0	0.860	1.468	4.177	9.802	9.047×10^3	4.639
1.173	0.987	1.601	4.541	10.662	9.830×10^3	5.044
1.332	1.088	1.702	4.829	11.342	1.047×10^4	5.368
1.5	1.169	1.802	5.130	12.052	1.111×10^4	5.698
2.0	1.326	2.064	5.938	14.028	1.293×10^4	6.612
2.5	1.381	2.271	6.644	15.822	1.459×10^4	7.380
3.0	1.442	2.431	7.249	17.456	1.604×10^4	8.141
3.5	1.447	2.567	7.813	19.038	1.747×10^4	8.828
4.0	1.455	2.657	8.270	20.382	1.868×10^4	9.366
5.0	1.429	2.798	9.059	22.871	2.094×10^4	10.361
7.0	1.348	2.924	10.146	26.860	2.449×10^4	11.846
10.0	1.228	2.940	11.070	31.216	2.817×10^4	13.227

Source: Data from Hubbell and Seltzer, 1995.

[a]Calculated from mass attenuation coefficients listed in Table 7-4.

(c) Since $I = I_0 e^{-\mu x}$, the thickness of iron required to reduce an 800 mR/h exposure rate to 150 mR/h in good geometry conditions is

$$150 = 800e^{-0.47x}$$

$$\ln \frac{150}{800} = -1.674 = -0.47x$$

$$x = 3.55 \text{ cm}$$

This can also be solved by determining the number of HVLs (1.47 cm) necessary to reduce 800 mR/h to 150 mR/h, or

$$150 = \frac{800}{2^n} \qquad \text{or} \qquad 2^n = \frac{800}{150}$$

Taking the logarithm of both sides yields

$$n \ln 2 = \ln \frac{800}{150} = 1.674$$

$$n = \frac{1.674}{\ln 2} = 2.4156 \text{ HVL}$$

and since $x = \text{nHVL}$,

$$x = 2.4156 \text{ HVL} \times 1.47 \text{ cm/HVL} = 3.55 \text{ cm}$$

Shielding of Poor Geometry Photon Sources

When a significant absorbing medium such as a metal shield is placed between a photon source and a receptor, the photon flux (or fluence) will be altered significantly because of Compton-scattered photons produced in the absorber, many of which will reach the receptor. The scattered photons are also reduced in energy, and the flux of photons reaching the receptor becomes a complicated function of beam size, photon-energy distribution, absorber material, and geometry. The conditions that include these complexities, called poor geometry, represent most practical situations in radiation protection. A calculated value of $I(x)$ based on the attenuation coefficient μ which is determined in good geometry conditions will thus underestimate the number of photons reaching the receptor, which implies that absorption is greater than what actually occurs, and a shield designed on this basis will not be thick enough.

The effect of scattered photons, in addition to unscattered primary photons, is best dealt with by a buildup factor, B, which is greater than 1.0 to account for photons scattered toward the receptor from regions outside the primary beam. When buildup is included, the radiation intensity is

$$I(x) = I_0 B e^{-\mu x}$$

Experimentally determined values of B for photons of different energies absorbed in various media are listed in Table 8-4 for point sources (earlier compilations included buildup factors for broad beams, but these are out of date).

TABLE 8-4. Exposure Buildup Factors for Photons of Energy E Versus μx for Various Absorbers

	Energy (MeV)									
μx	0.1	0.5	1	2	3	4	5	6	8	10
Aluminum										
0.5	1.91	1.57	1.45	1.37	1.33	1.32	1.28	1.26	1.22	1.19
1.0	2.86	2.28	1.99	1.78	1.68	1.62	1.54	1.49	1.41	1.35
2.0	4.87	4.07	3.26	2.66	2.38	2.19	2.04	1.94	1.76	1.64
3.0	7.07	6.35	4.76	3.62	3.11	2.78	2.54	2.37	2.11	1.93
4.0	9.47	9.14	6.48	4.64	3.86	3.38	3.04	2.81	2.46	2.22
5.0	12.1	12.4	8.41	5.72	4.64	3.99	3.55	3.26	2.82	2.52
6.0	14.9	16.3	10.5	6.86	5.44	4.61	4.08	3.72	3.18	2.83
7.0	18.0	20.7	12.9	8.05	6.26	5.24	4.61	4.19	3.55	3.14
8.0	21.3	25.7	15.4	9.28	7.1	5.88	5.14	4.66	3.92	3.46
10.0	28.7	37.6	21.0	11.9	8.83	7.18	6.23	5.61	4.68	4.12
15.0	51.7	78.6	37.7	18.9	13.4	10.5	9.03	8.09	6.64	5.87
20.0	81.1	137	57.9	26.6	18.1	14.0	11.9	10.7	8.68	7.74
25.0	117	213	81.3	34.9	23.0	17.5	14.9	13.3	10.8	9.74
30.0	159	307	107	43.6	28.1	21.0	18.0	16.0	13.0	11.8
Iron										
0.5	1.26	1.48	1.41	1.35	1.32	1.3	1.27	1.25	1.22	1.19
1.0	1.4	1.99	1.85	1.71	1.64	1.57	1.51	1.47	1.39	1.33
2.0	1.61	3.12	2.85	2.49	2.28	2.12	1.97	1.87	1.71	1.59
3.0	1.78	4.44	4	3.34	2.96	2.68	2.46	2.3	2.04	1.86
4.0	1.94	5.96	5.3	4.25	3.68	3.29	2.98	2.76	2.41	2.16
5.0	2.07	7.68	6.74	5.22	4.45	3.93	3.53	3.25	2.81	2.5
6.0	2.2	9.58	8.31	6.25	5.25	4.6	4.11	3.78	3.24	2.87
7.0	2.31	11.7	10.0	7.33	6.09	5.31	4.73	4.33	3.71	3.27
8.0	2.41	14.0	11.8	8.45	6.96	6.05	5.38	4.92	4.2	3.71
10.0	2.61	19.1	15.8	10.8	8.8	7.6	6.75	6.18	5.3	4.69
15.0	3.01	35.1	27.5	17.4	13.8	11.9	10.7	9.85	8.64	7.88
20.0	3.33	55.4	41.3	24.6	19.4	16.8	15.2	14.2	12.9	12.3
25.0	3.61	79.9	57.0	32.5	25.4	22.1	20.3	19.3	18.2	18.1
30.0	3.86	108	74.5	40.9	31.7	27.9	25.9	25.1	24.5	25.7
Tin										
0.5	1.35	1.32	1.33	1.27	1.29	1.28	1.31	1.31	1.33	1.31
1.0	1.38	1.61	1.69	1.57	1.56	1.51	1.55	1.54	1.6	1.57
2.0	1.41	2.15	2.4	2.17	2.07	1.96	1.97	1.94	2.04	2.05
3.0	1.43	2.68	3.14	2.82	2.64	2.45	2.43	2.38	2.51	2.61
4.0	1.45	3.16	3.86	3.51	3.25	3.0	2.54	2.87	3.05	3.27
5.0	1.47	3.63	4.6	4.23	3.92	3.6	3.52	3.43	3.69	4.09
6.0	1.49	4.14	5.43	5.03	4.68	4.29	4.19	4.09	4.45	5.07
7.0	1.5	4.64	6.27	5.87	5.48	5.04	4.93	4.83	5.34	6.26
8.0	1.52	5.13	7.11	6.74	6.32	5.84	5.74	5.65	6.36	7.69
10.0	1.54	6.13	8.88	8.61	8.19	7.65	7.63	7.63	8.94	11.5
15.0	1.58	8.74	13.8	14.0	13.8	13.5	14.1	14.9	19.7	29.6
20.0	1.61	11.4	19.1	20.1	20.5	21.1	23.5	26.4	40.7	72.1
25.0	1.64	14.0	24.5	26.9	28.1	30.6	36.2	43.9	79.7	168
30.0	1.66	16.5	30.0	34.2	36.6	42.1	53.0	69.3	150	377

(*Continued*)

TABLE 8-4. Exposure Buildup Factors for Photons of Energy E Versus μx for Various Absorbers (*Continued*)

	Energy (MeV)									
μx	0.1	0.5	1	2	3	4	5	6	8	10
Lead										
0.5	1.51	1.14	1.2	1.21	1.23	1.21	1.25	1.26	1.3	1.28
1.0	2.04	1.24	1.38	1.4	1.4	1.36	1.41	1.42	1.51	1.51
2.0	3.39	1.39	1.68	1.76	1.73	1.67	1.71	1.73	1.9	2.01
3.0	5.6	1.52	1.95	2.14	2.1	2.02	2.05	2.08	2.36	2.63
4.0	9.59	1.62	2.19	2.52	2.5	2.4	2.44	2.49	2.91	3.42
5.0	17.0	1.71	2.43	2.91	2.93	2.82	2.88	2.96	3.59	4.45
6.0	30.6	1.8	2.66	3.32	3.4	3.28	3.38	3.51	4.41	5.73
7.0	54.9	1.88	2.89	3.74	3.89	3.79	3.93	4.13	5.39	7.37
8.0	94.7	1.95	3.1	4.17	4.41	4.35	4.56	4.84	6.58	9.44
10.0	294	2.1	3.51	5.07	5.56	5.61	6.03	6.61	9.73	15.4
15.0	5800	2.39	4.45	7.44	8.91	9.73	11.4	13.7	25.1	50.8
20.0	1.33×10^5	2.64	5.27	9.98	12.9	15.4	19.9	26.6	62.0	161
25.0	3.34×10^6	2.85	5.98	12.6	17.5	23.0	32.9	49.6	148	495
30.0	8.87×10^7	3.02	6.64	15.4	22.5	32.6	52.2	88.9	344	1470
Uranium										
0.5	1.04	1.11	1.17	1.19	1.2	1.19	1.23	1.24	1.28	1.27
1.0	1.06	1.19	1.31	1.35	1.35	1.32	1.37	1.38	1.48	1.49
2.0	1.08	1.3	1.53	1.65	1.64	1.6	1.64	1.66	1.85	1.97
3.0	1.1	1.39	1.73	1.95	1.95	1.89	1.94	1.98	2.27	2.56
4.0	1.11	1.45	1.9	2.25	2.28	2.21	2.27	2.33	2.78	3.31
5.0	1.12	1.52	2.07	2.56	2.62	2.55	2.63	2.74	3.39	4.26
6.0	1.13	1.58	2.23	2.88	2.99	2.93	3.04	3.19	4.11	5.43
7.0	1.14	1.63	2.38	3.19	3.38	3.33	3.49	3.71	4.96	6.9
8.0	1.14	1.68	2.52	3.51	3.78	3.76	3.99	4.28	5.97	8.73
10.0	1.16	1.77	2.78	4.17	4.64	4.72	5.14	5.68	8.61	13.9
15.0	1.18	1.96	3.35	5.84	7.06	7.72	9.1	11.0	20.8	43.4
20.0	1.2	2.11	3.82	7.54	9.8	11.6	15.1	20.1	48.6	131
25.0	1.22	2.23	4.23	9.27	12.8	16.5	23.7	35.4	110	385
30.0	1.23	2.33	4.59	11.0	16.0	22.5	36.0	60.4	244	1100
Water										
0.5	2.37	1.6	1.47	1.38	1.34	1.31	1.28	1.27	1.23	1.2
1.0	4.55	2.44	2.08	1.83	1.71	1.63	1.56	1.51	1.43	1.37
2.0	11.8	4.88	3.62	2.81	2.46	2.24	2.08	1.97	1.8	1.68
3.0	23.8	8.35	5.5	3.87	3.23	2.85	2.58	2.41	2.15	1.97
4.0	41.3	12.8	7.68	4.98	4	3.46	3.08	2.84	2.46	2.25
5.0	65.2	18.4	10.1	6.15	4.8	4.07	3.58	3.27	2.82	2.53
6.0	96.7	25.0	12.8	7.38	5.61	4.68	4.08	3.7	3.15	2.8
7.0	137	32.7	15.8	8.65	6.43	5.3	4.58	4.12	3.48	3.07
8.0	187	41.5	19.0	9.97	7.27	5.92	5.07	4.54	3.8	3.34
10.0	321	62.9	26.1	12.7	8.97	7.16	6.05	5.37	4.44	3.86
15.0	938	139	47.7	20.1	13.3	10.3	8.49	7.41	5.99	5.14
20.0	2170	252	74.0	28	17.8	13.4	10.9	9.42	7.49	6.38
25.0	4360	403	104	36.5	22.4	16.5	13.3	11.4	8.96	7.59
30.0	7970	594	139	45.2	27.1	19.7	15.7	13.3	10.4	8.78

TABLE 8-4. ***(Continued)***

	Energy (MeV)									
μx	0.1	0.5	1	2	3	4	5	6	8	10
Air										
0.5	2.35	1.6	1.47	1.38	1.34	1.31	1.29	1.27	1.23	1.2
1.0	4.46	2.44	2.08	1.83	1.71	1.63	1.57	1.52	1.43	1.37
2.0	11.4	4.84	3.6	2.81	2.46	2.25	2.09	1.97	1.8	1.68
3.0	22.5	8.21	5.46	3.86	3.22	2.85	2.6	2.41	2.15	1.97
4.0	38.4	12.6	7.6	4.96	4	3.46	3.11	2.85	2.5	2.26
5.0	59.9	17.9	10.0	6.13	4.79	4.07	3.61	3.28	2.84	2.54
6.0	87.8	24.2	12.7	7.35	5.6	4.69	4.12	3.71	3.17	2.82
7.0	123	31.6	15.6	8.61	6.43	5.31	4.62	4.14	3.51	3.1
8.0	166	40.1	18.8	9.92	7.26	5.94	5.12	4.57	3.84	3.37
10.0	282	60.6	25.8	12.6	8.97	7.19	6.13	5.42	4.49	3.92
15.0	800	134	47.0	20	13.4	10.3	8.63	7.51	6.08	5.25
20.0	1810	241	72.8	27.9	17.9	13.5	11.1	9.58	7.64	6.55
25.0	3570	385	103	36.2	22.5	16.7	13.6	11.6	9.17	7.84
30.0	6430	567	136	45	27.2	19.9	16.1	13.6	10.7	9.11
Concrete										
0.5	1.89	1.57	1.45	1.37	1.33	1.31	1.27	1.26	1.22	1.19
1.0	2.78	2.27	1.98	1.77	1.67	1.61	1.53	1.49	1.41	1.35
2.0	4.63	4.03	3.24	2.65	2.38	2.18	2.04	1.93	1.76	1.64
3.0	6.63	6.26	4.72	3.6	3.09	2.77	2.53	2.37	2.11	1.93
4.0	8.8	8.97	6.42	4.61	3.84	3.37	3.03	2.8	2.45	2.22
5.0	11.1	12.2	8.33	5.68	4.61	3.98	3.54	3.25	2.81	2.51
6.0	13.6	15.9	10.4	6.8	5.4	4.6	4.05	3.69	3.16	2.8
7.0	16.3	20.2	12.7	7.97	6.2	5.23	4.57	4.14	3.51	3.1
8.0	19.2	25.0	15.2	9.18	7.03	5.86	5.09	4.6	3.87	3.4
10.0	25.6	36.4	20.7	11.7	8.71	7.15	6.15	5.52	4.59	4.01
15.0	44.9	75.6	37.2	18.6	13.1	10.5	8.85	7.86	6.43	5.57
20.0	69.1	131	57.1	26.0	17.7	13.9	11.6	10.2	8.31	7.19
25.0	97.9	203	80.1	33.9	22.5	17.4	14.4	12.7	10.2	8.86
30.0	131	290	106	42.2	27.4	20.9	17.3	15.2	12.2	10.6

The *buildup factor B* is dependent on the absorbing medium, the photon energy, the attenuation coefficient for specific energy photons in the medium, and the absorber thickness x. The latter two are depicted in Table 8-4 as μx, which is dimensionless and is commonly referred to as the number of mean free paths or the number of relaxation lengths, the value that reduces the initial flux (or exposure) by $1/e$ (the *mean free path* can be thought of as the mean distance that a photon travels in an absorber before it undergoes an absorption or scattering interaction that removes it from the initial beam). It is also clear from Table 8-4 that B can be quite large, especially for low-energy photons, and that calculations of the radiation exposure associated with a beam of photons would be significantly in error if it were not included.

The buildup factors presented in Table 8-4 are for exposure in air after penetration through the absorber or shielding material. Other types of buildup factors also exist, in particular energy absorption buildup factors for energy deposition in an absorbing medium and dose buildup factors for absorbing media. Since a primary assessment in radiation protection is the exposure field before and after use of a radiation shield, exposure buildup factors (as provided in Table 8-4) are of most general use with appropriate adjustments of the air exposure to obtain absorbed dose.

Use of Buildup Factors

Estimates of photon fields under poor geometry conditions are made by first using good geometry conditions and then adjusting the results to account for the buildup of scattered photons, as follows:

1. Determine the total attenuation of the beam in good geometry by calculating the change in intensity for the energy–absorber combination as

$$I(x) = I_0 e^{-\mu x}$$

   where $I(x)$ is the unscattered intensity (flux, exposure, etc.) and μ is the linear attenuation coefficient (cm^{-1}).
2. Multiply the unscattered intensity by the buildup factor for the particular photon energy–absorber combination

$$I_b(x) = BI_0 e^{-\mu x}$$

   where B is obtained for the absorber in question, the photon energy, and the particular value of μx (or number of mean free paths). It is usually necessary to interpolate between the energies and the μx values (sometimes both) to obtain the proper value.

This procedure is illustrated in Examples 8-5 and 8-6.

Example 8-5: A beam of 1.0 MeV gamma rays is emitted from a point source and produces a flux of 10,000 $\gamma/cm^2 \cdot s$. If 2 cm of iron is placed in the beam, what is the best estimate of the flux after passing through the shield?

SOLUTION: First determine the flux based on narrow-beam conditions. The attenuation coefficient for 1 MeV photons in iron is 0.472 cm^{-1} and the value of μx for 2 cm of iron is 0.944; therefore, the attenuated unscattered flux would be

$$\begin{aligned} I(x) = I_0 e^{-\mu x} &= 10{,}000\ \gamma/cm^2 \cdot s(e^{-0.472(2\ cm)}) \\ &= 3890\ \gamma/cm^2 \cdot s \end{aligned}$$

This needs to be adjusted by the buildup factor, which is determined by interpolation from Table 8-4 as about 1.8, and the best estimate of $I_b(x)$ is

$$I_b(x) = 1.8 \times 3890 \simeq 7000\ \gamma/cm^2 \cdot s$$

This same approach is used if $I(x)$ is in units of exposure rate, energy flux, or absorbed dose rate, since these quantities are based on the photon flux, which is the quantity that is measured in determining values of μ for different absorbers.

Example 8-6: A fluence of $10^5\ \gamma/\text{cm}^2$ of 1.5 MeV photons strikes a 2-cm-thick piece of lead. What is the best estimate of the total energy that reaches a receptor beyond the lead shield?

SOLUTION: The linear attenuation coefficient μ for 1.5 MeV photons in lead is 0.59247 cm^{-1}, and for good geometry,

$$I(x) = 10^5\ \gamma/\text{cm}^2(e^{-(0.59247)(2)})$$
$$= 3.06 \times 10^4\ \gamma/\text{cm}^2$$

From Table 8-4, the buildup factor for 1.5 MeV photons in lead for $\mu x = 1.185$ is found by interpolation to be 1.45, and the buildup fluence is

$$I_b(x) = BI_0e^{-\mu x} = 1.45(3.06 \times 10^4) = 4.43 \times 10^4\ \gamma/\text{cm}^2$$

which contains primary beam photons and scattered photons of lower energy; however, despite the presence of lower-energy scattered photons, the best estimate of the energy fluence is, conservatively,

$$I_b(x)_E = 4.43 \times 10^4\ \gamma/\text{cm}^2 \times 1.5\ \text{MeV} = 6.65 \times 10^4\ \text{MeV/cm}^2$$

Effect of Buildup on Shield Thickness

The exposure rate in air for unscattered photons of energy E is

$$\text{exposure (mR/h)} = \frac{\phi E \times 1.6022 \times 10^{-6}\ \text{erg/MeV} \times (3600\ \text{s/h}) \times (\mu_{en}/\rho)_{air}}{87.64\ \text{ergs/g} \cdot \text{R} \times 10^{-3}\ \text{R/mR}}$$

or

$$\text{exposure (mR/h)} = 0.0658\phi E(\mu_{en}/\rho)_{air}$$

where ϕ is the flux of photons/cm$^2 \cdot$ s and $(\mu_{en}/\rho)_{air}$ is the energy absorption coefficient in air for photons of energy E. If a shield or other absorber is placed betwen the photon source and a receptor, it is necessary to introduce a buildup factor B to account for an increase in the exposure due to buildup of scattered photons, or

$$\text{exposure (mR/h)} = 0.0658\phi BE(\mu_{en}/\rho)_{air}e^{-\mu x}$$

such that the exposure is directly proportional to $(\mu_{en}/\rho)_{air}$, B, and $e^{-\mu x}$ of the absorber, all of which vary with photon energy. The buildup factor B and the exponential attenuation factor $e^{-\mu x}$ are also a function of μx, which can complicate calculations of the shield thickness x to reduce a photon intensity from I_0 to $I(x)$.

Determining a thickness for a given reduction usually requires successive approximations, and the process becomes one of homing in on the correct value of x by iterative calculations. For most circumstances B increases slowly with x while the exposure rate decreases exponentially, and at a faster rate; these various complexities are illustrated in Example 8-7.

Example 8-7: Determine the thickness of an iron shield needed to reduce the exposure rate from a point source that emits 10^8 1 MeV photons to 1 mR/h at 60 cm.

SOLUTION: First, determine the thickness x of a shielding material for narrow-beam geometry with $B = 1$; this underestimates the shield thickness, but it can then be increased by adding thicknesses of absorber until the appropriate values of x and B are obtained such that the exponential function yields the desired reduction in I_0.

The value of $(\mu_{en}/\rho)_{air}$ for 1 MeV photons in air from Table 8-2 is 0.0636 cm^2/g and the flux that produces 1 mR/h is

$$\phi = \frac{1 \text{ mR/h}}{0.0658 \times 1 \text{ MeV} \times 0.0636} = 239 \ \gamma/\text{cm}^2 \cdot \text{s}$$

The flux at 60 cm from a point source that emits 10^8 photons (i.e., without shielding) is

$$\phi = \frac{10^8 \ \gamma/\text{cm}^2 \cdot \text{s}}{4\pi(60)^2} = 2.21 \times 10^3 \ \gamma/\text{cm}^2 \cdot \text{s}$$

The linear attenuation coefficient for 1 MeV photons in iron is 0.472 cm^{-1}, and the value of x required without a buildup factor is determined as

$$239 = 2.21 \times 10^3 e^{-0.472 \text{ cm}^{-1} x}$$

or

$$x = 4.713 \text{ cm}$$

and

$$\mu x = 2.22$$

The buildup factor is found by interpolation in Table 8-4 to be 3.17, and the calculated exposure based on good geometry attenuation is, as expected, well above 1 mR/h. If the amount of iron is increased to 8.4 cm, μx becomes 3.965, B is about 5.38, and the flux is

$$\begin{aligned}\phi &= 2.2 \times 10^3 \times 5.38 e^{-3.965} \\ &= 225 \ \gamma/\text{cm}^2 \cdot \text{s}\end{aligned}$$

which is just below the flux (239 $\gamma/\text{cm}^2 \cdot \text{s}$) that produces 1 mR/h; thus the appropriate shield thickness is about 8.4 cm of iron.

The iterative process in Example 8-7 can be made more exact by a computer program or by plotting ϕ versus μx on semilog paper. Such plots produce a nearly-straight line from which the appropriate value of μx that corresponds to the desired flux can be determined.

Mathematical Formulations of the Buildup Factor

Buildup factors can be calculated by using one of several mathematical approximations, each of which is developed as an equation of fit for the experimental data. The primary use of the mathematical formulation is in analytical solutions for complex geometries, where it is useful to include a formulation of B as a varying quantity in the solution; if, however, the conditions of the problem are fixed, it is easier and more accurate to use tables of B. Two approximations, known as the Berger form or Taylor forms after their developers, are widely used. The Berger form is a bit complex to use even though it is one of the more accurate formulations for B. The Taylor form which is based on a sum of exponentials, however, has found wide use for problems in which values of x, which influences both attenuation and buildup, are to be determined. The Taylor form of the buildup factor as a function of photon energy E and μx is

$$B(E, \mu x) = Ae^{-\alpha_1, \mu x} + (1 - A)e^{-\alpha_2 \mu x}$$

where A, α_1, and α_2 are fitting parameters for the particular absorber and photon energy; these are listed in Table 8-5. The Taylor form of the buildup factor is generally accurate but needs to be used with caution for low-energy photons in low-Z material. It should also be remembered that the fitting parameters are based on older data, whereas the buildup factors listed in Table 8-5 are much more recent.

A *single-term Taylor formulation of the buildup factor* in which a single exponential term accounts for absorber/thickness combinations is

$$B(E, \mu x) \approx A_1 e^{-\alpha_x \mu x}$$

in which A_1 and α_x are fitting parameters (shown in Table 8-6) that depend on the initial gamma-ray energy, E, the attenuating medium, and μx. This approximation is almost exact for μx values of between 3 and 8 and within a few percent up to $\mu x = 12$. Below $\mu x = 3$, the one-term Taylor approximation overestimates the exposure by as much as a factor of 1.5.

A *linear formulation* of the buildup factor has also been developed and is expressed as

$$B(E, \mu x) = 1 + \alpha_\ell(\mu x)$$

where E is the incident photon energy, μx is the number of mean free paths in the absorber material, and α_ℓ is a fitting parameter that is sometimes taken to be unity, a practice that should be eschewed except in special instances. Values of α_ℓ for water, aluminum, concrete, iron, tin, and lead are listed in Table 8-7 for values of $\mu x = 7$ and 20; however, since the expression is linear, these may be plotted to interpolate values of α_ℓ for other values of μx. This simpler formulation is not quite as accurate as the Taylor or Berger formulations, but is less cumbersome to use and generally produces acceptable results.

TABLE 8-5. Values of Fitting Parameters for the Taylor Form of the Buildup Factor

Material	E (MeV)	A	α_1	α_2
Water	0.5	100.845	−0.12687	−0.10925
	1	19.601	−0.09037	−0.02520
	2	12.612	−0.53200	0.01932
	3	11.110	−0.03550	0.03206
	4	11.163	−0.02543	0.03025
	6	8.385	−0.01820	0.01640
	8	4.635	−0.02633	0.07097
	10	3.545	−0.02991	0.08717
Aluminum	0.5	38.911	−0.10015	−0.06312
	1	28.782	−0.06820	−0.02973
	2	16.981	−0.04588	0.00271
	3	10.583	−0.04066	0.02514
	4	7.526	−0.03973	0.03860
	6	5.713	−0.03934	0.04347
	8	4.716	−0.03837	0.04431
	10	3.999	−0.03900	0.04130
Concrete[a]	0.1	73.80	−0.39400	−0.01450
	0.2	144.0	−0.07410	−0.05980
	0.5	62.0	−0.06880	−0.04240
	1	97.0	−0.03960	−0.02710
	2	38.70	−0.02500	−0.00227
	5	10.42	−0.02440	0.02680
	10	5.10	−0.02690	0.04500
	15	4.04	−0.02670	0.03930
Iron	0.5	31.379	−0.06842	−0.03742
	1	24.957	−0.06086	−0.02463
	2	17.622	−0.04627	−0.00526
	3	13.218	−0.04431	−0.00087
	4	9.624	−0.04698	0.00175
	6	5.867	−0.06150	−0.00186
	8	3.243	−0.07500	0.02123
	10	1.747	−0.09900	0.06627
Lead	0.5	1.677	−0.03084	0.30941
	1	2.984	−0.03503	0.13486
	2	5.421	−0.03482	0.04379
	3	5.580	−0.05422	0.00611
	4	3.897	−0.08468	−0.02383
	6	0.926	−0.17860	−0.04635
	8	0.368	−0.23691	−0.05684
	10	0.311	−0.24024	0.02783
Tin	0.5	11.440	−0.01800	0.03187
	1	11.426	−0.04266	0.01606
	2	8.783	−0.05349	0.01505
	3	5.400	−0.07440	0.02080
	4	3.496	−0.09517	0.02598
	6	2.005	−0.13733	−0.01501
	8	1.101	−0.17288	−0.01787
	10	0.708	−0.19200	0.01552

Source: Adapted from Morgan and Turner, 1967.

[a]Values for concrete from Chilton, 1977, with permission, American Nuclear Society, La Grange Park, IL.

TABLE 8-6. Fitting Parameters for the Single-Term Taylor Form of the Buildup Factor for $3 \leq \mu x \leq 8$

E_0 (MeV)	Water A_1	Water α_x	Concrete A_1	Concrete α_x	Iron A_1	Iron α_x	Lead A_1	Lead α_x
0.2	5.5	−0.40	3.8	−0.28	2.2	−0.15	—	—
0.5	3.7	−0.31	3.0	−0.27	2.4	−0.22	1.4	−0.06
1.0	2.9	−0.24	2.6	−0.23	2.3	−0.21	1.5	−0.10
1.5	2.6	−0.21	2.4	−0.20	2.2	−0.19	1.5	−0.11
2.0	2.4	−0.18	2.2	−0.18	2.1	−0.18	1.5	−0.13
5.0	1.8	−0.13	1.8	−0.14	1.6	−0.15	1.1	−0.17

TABLE 8-7. Values of the Fitting Parameter, α_ℓ, for the Linear Buildup Factor Approximation $B(E, \mu x) \simeq 1 + \alpha_\ell(\mu x)$ for Different Absorbers

Energy (MeV)	Values of α_ℓ in Water	Concrete	Aluminum	Iron	Tin	Lead
			$\mu x = 7$			
0.5	4.680	3.744	2.646	1.428	0.5153	0.1549
1.0	1.995	1.906	1.609	1.237	0.7199	0.2990
2.0	1.030	1.023	0.9686	0.8556	0.6731	0.3796
3.0	0.7397	0.7303	0.7197	0.6691	0.5837	0.3810
4.0	0.5884	0.5736	0.5663	0.5403	0.5146	0.3523
6.0	0.4321	0.4329	0.4334	0.4297	0.4153	0.3034
8.0	0.3406	0.3376	0.3476	0.3391	0.3317	0.2419
10	0.2877	0.2923	0.2847	0.2681	0.2550	0.1933
			$\mu x = 20$			
0.5	13.093	5.012	5.737	2.377	0.5090	0.1043
1.0	3.479	2.992	2.539	1.864	0.8495	0.2549
2.0	1.255	1.233	1.193	1.119	0.8521	0.3947
3.0	0.7863	0.7857	0.8061	0.8446	0.8509	0.5123
4.0	0.5951	0.5942	0.6075	0.6942	0.8643	0.6378
6.0	0.4030	0.4145	0.4626	0.6134	1.079	1.125
8.0	0.3085	0.3200	0.3697	0.5245	1.171	1.417
10	0.2584	0.2737	0.3087	0.4759	1.108	1.237

Source: ORNL, 1966.

The various mathematical formulations can be quite useful for calculations of complex geometries such as discs, slabs, and areal and volume sources, as discussed in the next section.

GAMMA FLUX FOR DISTRIBUTED SOURCES

Many calculations of radiation exposure/dose from photon sources are straightforward once the flux is known. A useful formulation is the flux at r from an attenuated

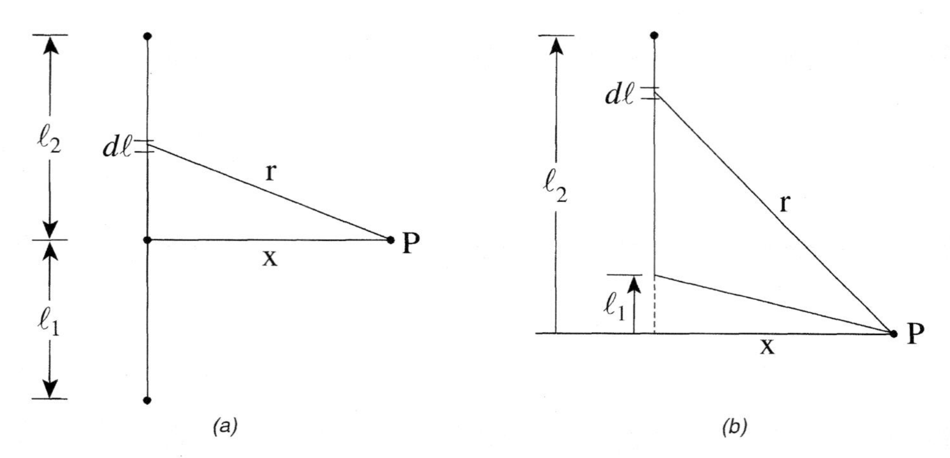

FIGURE 8-4. Schematic of a line source of radioactive material that emits S_L gamma rays per centimeter over a finite length (ℓ_1 to ℓ_2) or is infinitely long ($\ell = \infty$) to produce a flux at P with (a) coordinates along the line, or (b) some distance from one end of the line source.

point source

$$\phi(x) = \phi_0 \frac{e^{-\mu x}}{4\pi r^2}$$

The expression $e^{-\mu x}/4\pi r^2$ is referred to as the *point kernel*, which is the response at a point r from a source of unit strength. The point kernel is used extensively in developing relationships between flux and exposure for various source geometries and absorbing media.

Although many radiation sources can, with its ease and utility, be represented as a point or an approximate point source, many real-world exposure conditions cannot. Typical examples are a long pipe or tube containing radioactive material that approximates a line source, a contaminated area that is representative of a disc or infinite planar source, and various volume sources. Practical approaches can be used to determine the photon flux from point kernels spread over such geometries, and once the flux has been determined, it can be applied in the usual way to calculate radiation exposure.

Line Sources

Line sources can often be practically considered as infinitely long with respect to a point P, located a distance x away from the lineal source as shown in Figure 8-4a). The differential flux $d\phi_\ell$ at a point P located a distance r from a point-sized element $d\ell$ that emits S_L gamma rays per centimeter of length is

$$d\phi_\ell = \frac{S_L}{4\pi r^2} d\ell$$

which can be integrated by substituting $\ell^2 + x^2$ for r^2. For an infinite line source the limits of integration are $-\infty$ and $+\infty$ and the flux at P is

$$\phi_{\ell,\infty}(\gamma\text{-rays/cm}^2 \cdot \text{s}) = \frac{S_L}{4x}$$

The photon flux for a finite line source is obtained by integration between $-\ell_1$ and $+\ell_2$, or

$$\phi_\ell = \frac{S_L}{4\pi x}\left(\tan^{-1}\frac{\ell_2}{x} + \tan^{-1}\frac{\ell_1}{x}\right)$$

where the $\tan^{-1}$ solutions are expressed in radians. For the geometry shown in Figure 8-4*b*, the solution is such that the second term is $-\tan^{-1}(\ell_1/x)$ since in this case ℓ_1 is a fictitious source and thus must be subtracted.

Thus the fluence or fluence rate from line sources of photon emitters varies as $1/x$, a very useful general rule.

Example 8-8: A worker must enter a 100-ft pipe tunnel to repair a valve 30 ft from one end. The pipe is estimated to emit 1.0 MeV gamma rays at a lineal rate of 10 γ/cm · s. (a) What is the flux at 1 m from the pipe at the location of the valve? (b) How much does the flux change if the pipe is assumed to be an infinite line source?

SOLUTION: (a) The gamma flux at 30 ft from one end is

$$\phi_\ell = \frac{10\ \gamma/\text{cm}\cdot\text{s}}{4\pi(100\ \text{cm})}\left(\tan^{-1}\frac{70\ \text{ft}}{3.281\ \text{ft}} + \tan^{-1}\frac{30\ \text{ft}}{3.281\ \text{ft}}\right)$$

$$= 7.96\times10^{-3}(87.32^\circ + 83.76^\circ)$$

and since 1 rad = 57.3°,

$$\phi_\ell = 7.96\times10^{-3}\times2.986\ \text{rad}$$

$$= 2.38\times10^{-2}\ \gamma/\text{cm}^2\cdot\text{s}$$

(b) An infinite line source with the same lineal emission rate would produce a flux of

$$\phi_{\ell,\infty} = \frac{S_L}{4x} = \frac{10\ \gamma/\text{cm}\cdot\text{s}}{4\times100\ \text{cm}} = 2.5\times10^{-2}\ \gamma/\text{cm}^2\cdot\text{s}$$

which overestimates the flux by about 5%.

Shielding of a line source is often done by placing a sheet of metal close to a pipe or rod containing the radioactive material or constructing an annular ring around the lineal source. Interposing shielding material between the line source and the receptor point of interest not only attenuates the photons emitted but introduces a scattered component that must be accounted for by an appropriate buildup factor. An *approximation* to the shielded photon flux at a point P away from the lineal source can be made by assuming that all photons penetrate the shield in a perpendicular direction even though the true direction from most points along the line will be along an angular (and longer) path through the shield. This approximation overestimates the flux and the resultant exposure rate by only a few percent and is reasonable to use in lieu of the more complex integration necessary to account for an angular distance that constantly varies.

Exact calculations of shield thicknesses for line sources and other distributed sources involve the Sievert integral function, $F(\theta_i, \mu x)$ which adds more complexity than is necessary for most shielding problems in radiation protection. Such detailed calculations are justified in the design of shields for very high radiation areas such as hot cells and reactor walls, where more exact solutions may represent considerable cost savings for building materials and supporting structures. LaMarsh (1983) has provided the elements of such calculations as has Morgan and Turner (1967), and computer programs are available from the Radiation Shielding Information Center at Oak Ridge National Laboratory.

Example 8-9: What would be the photon flux in Example 8-8 if the 100-ft line source were to be enclosed in a 2-cm-thick annular shield constructed of lead?

SOLUTION: As shown in part (b) of Example 8-8, an infinite line source can be assumed. If all photons are emitted perpendicular to the pipe, which simplifies the consideration of buildup

$$\phi_s = \phi_{us} B e^{-\mu x}$$

where B is determined from Table 8-4 to be 1.56 for 1 MeV photons that penerate 1.61 mean free path lengths of lead ($\mu x = 0.8061\ \text{cm}^{-1} \times 2$ cm). The shielded flux is

$$\begin{aligned}\phi_s &= (2.38 \times 10^{-2}\ \gamma/\text{cm}^2 \cdot \text{s})(1.56)(e^{-1.61}) \\ &= 7.4 \times 10^{-4}\ \gamma/\text{cm}^2 \cdot \text{s}\end{aligned}$$

Since the primary beam contains 1 MeV photons, it can be conservatively estimated that the combination of scattered and unscattered photons yields an energy flux of 7.4×10^{-4} MeV/cm$^2 \cdot$ s. The actual energy flux will be a mixture of 1 MeV photons and many lower-energy photons.

Ring Sources

Some sources are or can be modeled as a linear source in the shape of a ring of radius R and length $\ell = 2\pi R$. Such a ring source, if thin enough, is simply a curved line source, and if the photon emission rate is S_L γ/s per centimeter of length the differential flux $d\phi_\ell(P)$ at a point P in the center of the ring at a distance r from each differential segment $d\ell$ along the line is

$$d\phi_\ell(P) = \frac{S_L}{4\pi r^2} d\ell$$

and the total flux at P is

$$\phi_\ell(P) = \frac{S_L}{4\pi r^2} \ell$$

and since ℓ is $2\pi R$,

$$\phi_\ell(P) = \frac{S_L}{2r^2}R$$

where $r^2 = x^2 + R^2$.

Example 8-10: A 5-cm-diameter pipe containing ^{60}Co with a lineal emission rate of 10^6 photons/cm · s curves halfway around the ceiling of a circular room 6 m in diameter. What is the flux 1.5 m below the pipe in the center of the room?

SOLUTION: Since the pipe is a half-circle, the flux can be computed for a ring source of radius R = 3 m, and the estimated flux for the half-circle will be one-half this value; thus

$$\phi_\ell(P = 1.5\text{ m}) = \frac{1}{2}\,\frac{10^6\ \gamma/\text{cm}\cdot\text{s} \times 3000\ \text{cm}}{2(1500^2 + 3000^2)}$$

$$= 66.7\ \gamma/\text{cm}^2\cdot\text{s}$$

Disc and Planar Sources

Spill areas on floors and/or contaminated sites can produce a flux of gamma rays and an exposure field at points above them. These area sources can be modeled as a disc source made up of a series of annular rings as shown in Figure 8-5. If the activity is spread uniformly over the area such that gamma rays are emitted isotropically as S_A $\gamma/\text{cm}^2\cdot\text{s}$, the differential flux contributed by each ring at a point P a distance x away from the center of the disc is

$$d\phi = S_A\frac{2\pi R\,dR}{4\pi r^2} = S_A\frac{R\,dR}{2r^2}$$

The total unscattered flux at P is obtained by integrating over all annuli encompassed in the disc of radius R:

$$\phi(P) = \frac{S_A}{2}\int_0^R \frac{R\,dR}{r^2}$$

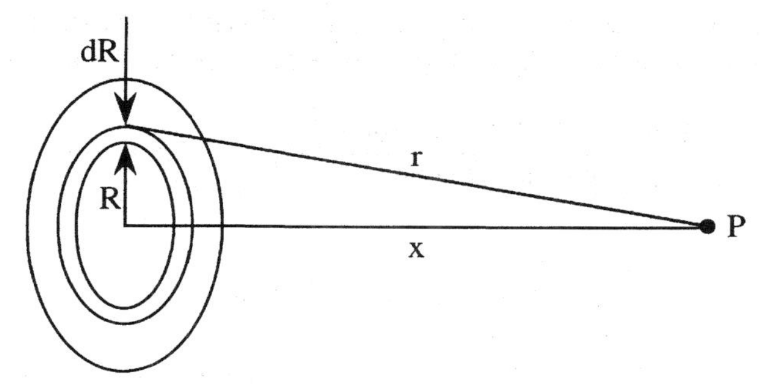

FIGURE 8-5. A disc source with a uniform emission rate of S_A photons per unit area produces a differential flux at point P a distance r from a thin ring of width dR and area $2\pi R\,dR$.

and since $r^2 = x^2 + R^2$,

$$\phi_u(P)_A = \frac{S_A}{4} \ln\left(1 + \frac{R^2}{x^2}\right)$$

This is the general solution for all values of x, which is the distance of P from the center of a disc-shaped source with radius R. When the area source is very much larger than the distance x, as it typically is, the flux is

$$\phi_u(P)_A = \frac{S_A}{2} \ln \frac{R}{x}, \qquad R \gg x$$

Therefore, for large area sources the gamma flux (and radiation exposure) decreases as $1/x$, and once the flux is determined, it is used in the usual way for determining radiation exposure or absorbed dose.

Shield Designs for Area Sources

Shielding an area source with a slab to reduce gamma exposure is often the more practical problem since the unshielded exposure rate is usually measured after contamination occurs rather than calculated (an exposure level may also be forecast for a disposal location or a work area). In such cases it may be necessary to determine the required thickness of a layer of concrete or other material that may be used to reduce an exposure rate to a desired level or perhaps a layer of soil to be placed atop a contaminated site for the same purpose.

If a shield of thickness a is placed between the disc source and the point P shown in Figure 8-5, the flux, minus any buildup effect, is obtained as before by integrating over a series of annular rings of radius R and width dR for S_A $\gamma/\text{cm}^2 \cdot \text{s}$ emitted isotropically by the disc. The integration must also account for the exponential attenuation of the gamma rays by the shield; therefore, the unscattered flux at P from a ring of width dR and radius R with the shield in place is

$$\phi_u(P) = S_A \frac{2\pi R\,dR}{4\pi r^2} e^{-\mu r_s} = S_A \frac{R\,dR}{2r^2} e^{-\mu r_s}$$

where r_s is that segment of r that is taken up by the shield. The shield path length, r_s, equals a only when P is directly over the center of the disc; for all other paths through the shield $r_s > a$. It can be shown that r_s is related to ar, and since $r^2 = x^2 + R^2$ and $r\,dr = R\,dR$, the integral becomes

$$\phi_u(P)_{A,\infty} = \frac{S_A}{2} \int_0^\infty \frac{e^{-\mu r_s} R\,dR}{r^2} = \frac{S_A}{2} \int_0^\infty \frac{e^{-\mu r_s}\,dr}{r}$$

The integral, although simplified in terms of r, cannot be evaluated analytically. Such integrals are known as exponential integrals, and E-function tables of its solutions for values of x have been prepared for integer values of n where

$$E_n(x) = x^{n-1} \int_x^\infty \frac{e^{-t}}{t}\,dt$$

A *shielded infinite area source* can be expressed in terms of the E_1 function since the integral in the relationship for the uncollided flux $\phi_u(P)_{A,\infty}$ matches the case where $n = 1$ and $x = \mu a$; therefore,

$$\phi_u(P)_{A,\infty} = \frac{S_A}{2} E_1(\mu a)$$

where $E_1(\mu a)$ is obtained for the value $x = \mu a$ from a table of the function. Plots of $E_n(x)$ can also be used but tables are more accurate, especially when it is necessary to interpolate between values of $E_n(x)$ as x changes. Values of the $E_1(x)$ function, which is encountered in many shielding problems, are contained in Table 8-8 along with the $E_2(x)$ function, which is also encountered in such problems.

Values of $E_1(\mu x)$ when μx is very small or very large are often needed, and these can be obtained fairly accurately by a series calculation. For very small values of μx (≤ 0.1), the following approximation can be used:

$$E_1(\mu x) = -0.577216 - \ln \mu x$$

For large values of μx (≥ 10),

$$E_1(\mu x) = e^{-\mu x} \left[\frac{1}{\mu x + 1} + \frac{1}{(\mu x + 1)^3} \right]$$

These approximations are accurate to a few percent, however, the best accuracy is obtained from the tabulations of $E_1(x)$ in Table 8-8.

A *finite shielded area source* can also be represented in terms of the $E_1(x)$ function. An area source of radius R will produce an unscattered (uncollided) photon flux through a shield of thickness a

$$\phi_u(P)_{A,R} = \frac{S_A}{2} [E_1(\mu a) - E_1(\mu r)]$$

where r is the distance from P to the edge of the disc. The value of r can be calculated from $r^2 = x^2 + R^2$ or as $x \sec \theta$ (or $a \sec \theta$), where θ is the angle between x and r.

A *shielded infinite area source with a buildup factor* is used when scattered photons, which can be quite significant, combine with unscattered photons to produce a buildup flux $\phi_b(P)$ at P. An analytical solution of this situation requires one of the approximations for $B(\mu x)$ such as the Taylor form, which contains two exponential terms. This formulation yields an expression for the buildup flux at P for an infinite disc or planar source in terms of E_1 functions, or

$$\phi_b(P)_{A,\infty} = \frac{S_A}{2} \{A E_1[(1 + \alpha_1)\mu a] + (1 - A) E_1[(1 + \alpha_2)\mu a]\}$$

where A, α_1, and α_2 are selected from Table 8-5 for the particular absorber and photon energy, and values of the E_1 function are obtained from Table 8-8.

Example 8-11: A large isotropic planar source emits 1 MeV gamma rays with a fluence rate of 10^7 $\gamma/\text{cm}^2 \cdot \text{s}$. If a 30-cm layer of concrete ($\rho = 2.35$) is poured over the area, what would be the adjusted fluence rate?

TABLE 8-8. Tabulations of the $E_1(x)$ and $E_2(x)$ Functions for Photon Shielding Calculations

x	$E_1(x)$	$E_2(x)$	x	$E_1(x)$	$E_2(x)$	x	$E_1(x)$	$E_2(x)$	x	$E_1(x)$	$E_2(x)$
0.00	0.000	1.000	0.29	9.309(−1)	4.783(−1)	0.58	4.732(−1)	2.855(−1)	0.87	2.742(−1)	1.804(−1)
0.01	4.038(1)[a]	9.497(−1)	0.30	9.057(−1)	4.691(−1)	0.59	4.636(−1)	2.808(−1)	0.88	2.694(−1)	1.777(−1)
0.02	3.355(1)	9.131(−1)	0.31	8.815(−1)	4.602(−1)	0.60	4.544(−1)	2.762(−1)	0.89	2.647(−1)	1.750(−1)
0.03	2.959(1)	8.817(−1)	0.32	8.583(−1)	4.515(−1)	0.61	4.454(−1)	2.717(−1)	0.90	2.602(−1)	1.724(−1)
0.04	2.681(1)	8.535(−1)	0.33	8.361(−1)	4.430(−1)	0.62	4.366(−1)	2.673(−1)	0.91	2.557(−1)	1.698(−1)
0.05	2.468(1)	8.278(−1)	0.34	8.147(−1)	4.348(−1)	0.63	4.280(−1)	2.630(−1)	0.92	2.513(−1)	1.673(−1)
0.06	2.295(1)	8.040(−1)	0.35	7.942(−1)	4.267(−1)	0.64	4.197(−1)	2.587(−1)	0.93	2.470(−1)	1.648(−1)
0.07	2.151(1)	7.818(−1)	0.36	7.745(−1)	4.189(−1)	0.65	4.115(−1)	2.546(−1)	0.94	2.429(−1)	1.623(−1)
0.08	2.027(1)	7.610(−1)	0.37	7.554(−1)	4.112(−1)	0.66	4.036(−1)	2.505(−1)	0.95	2.387(−1)	1.599(−1)
0.09	1.919(1)	7.412(−1)	0.38	7.371(−1)	4.038(−1)	0.67	3.959(−1)	2.465(−1)	0.96	2.347(−1)	1.576(−1)
0.10	1.823(1)	7.225(−1)	0.39	7.194(−1)	3.965(−1)	0.68	3.883(−1)	2.426(−1)	0.97	2.308(−1)	1.552(−1)
0.11	1.737(1)	7.048(−1)	0.40	7.024(−1)	3.894(−1)	0.69	3.810(−1)	2.387(−1)	0.98	2.269(−1)	1.530(−1)
0.12	1.660(1)	6.878(−1)	0.41	6.859(−1)	3.824(−1)	0.70	3.738(−1)	2.349(−1)	0.99	2.231(−1)	1.507(−1)
0.13	1.589(1)	6.715(−1)	0.42	6.700(−1)	3.756(−1)	0.71	3.668(−1)	2.312(−1)	1.00	2.194(−1)	1.485(−1)
0.14	1.524(1)	6.560(−1)	0.43	6.546(−1)	3.690(−1)	0.72	3.599(−1)	2.276(−1)	1.01	2.157(−1)	1.463(−1)
0.15	1.464(1)	6.410(−1)	0.44	6.397(−1)	3.626(−1)	0.73	3.532(−1)	2.240(−1)	1.02	2.122(−1)	1.442(−1)
0.16	1.409(1)	6.267(−1)	0.45	6.253(−1)	3.562(−1)	0.74	3.467(−1)	2.205(−1)	1.03	2.087(−1)	1.421(−1)
0.17	1.358(1)	6.128(−1)	0.46	6.114(−1)	3.500(−1)	0.75	3.403(−1)	2.171(−1)	1.04	2.052(−1)	1.400(−1)
0.18	1.310(1)	5.995(−1)	0.47	5.979(−1)	3.440(−1)	0.76	3.341(−1)	2.137(−1)	1.05	2.019(−1)	1.380(−1)
0.19	1.265(1)	5.866(−1)	0.48	5.848(−1)	3.381(−1)	0.77	3.280(−1)	2.104(−1)	1.06	1.986(−1)	1.360(−1)
0.20	1.223(1)	5.742(−1)	0.49	5.721(−1)	3.323(−1)	0.78	3.221(−1)	2.072(−1)	1.07	1.953(−1)	1.340(−1)
0.21	1.183(1)	5.622(−1)	0.50	5.598(−1)	3.266(−1)	0.79	3.163(−1)	2.040(−1)	1.08	1.922(−1)	1.321(−1)
0.22	1.145(1)	5.505(−1)	0.51	5.478(−1)	3.211(−1)	0.80	3.106(−1)	2.009(−1)	1.09	1.890(−1)	1.302(−1)
0.23	1.110(1)	5.393(−1)	0.52	5.362(−1)	3.157(−1)	0.81	3.050(−1)	1.978(−1)	1.10	1.860(−1)	1.283(−1)
0.24	1.076(1)	5.283(−1)	0.53	5.250(−1)	3.104(−1)	0.82	2.996(−1)	1.948(−1)	1.11	1.830(−1)	1.264(−1)
0.25	1.044(1)	5.177(−1)	0.54	5.140(−1)	3.052(−1)	0.83	2.943(−1)	1.918(−1)	1.12	1.801(−1)	1.246(−1)
0.26	1.014(1)	5.074(−1)	0.55	5.034(−1)	3.001(−1)	0.84	2.891(−1)	1.889(−1)	1.13	1.772(−1)	1.228(−1)
0.27	9.849(−1)	4.974(−1)	0.56	4.930(−1)	2.951(−1)	0.85	2.840(−1)	1.860(−1)	1.14	1.743(−1)	1.211(−1)
0.28	9.573(−1)	4.877(−1)	0.57	4.830(−1)	2.902(−1)	0.86	2.790(−1)	1.832(−1)	1.15	1.716(−1)	1.193(−1)

x	$E_1(x)$	$E_2(x)$	x	$E_1(x)$	$E_2(x)$	x	$E_1(x)$	$E_2(x)$	x	$E_1(x)$	$E_2(x)$
1.16	1.688(−1)	1.176(−1)	1.45	1.078(−1)	7.829(−2)	1.74	7.049(−2)	5.287(−2)	2.3	3.250(−2)	2.550(−2)
1.17	1.662(−1)	1.160(−1)	1.46	1.062(−1)	7.722(−2)	1.75	6.949(−2)	5.217(−2)	2.4	2.844(−2)	2.246(−2)
1.18	1.635(−1)	1.143(−1)	1.47	1.046(−1)	7.617(−2)	1.76	6.850(−2)	5.148(−2)	2.5	2.491(−2)	1.980(−2)
1.19	1.609(−1)	1.127(−1)	1.48	1.030(−1)	7.513(−2)	1.77	6.753(−2)	5.080(−2)	2.6	2.185(−2)	1.746(−2)
1.20	1.584(−1)	1.111(−1)	1.49	1.015(−1)	7.411(−2)	1.78	6.658(−2)	5.013(−2)	2.7	1.918(−2)	1.541(−2)
1.21	1.559(−1)	1.095(−1)	1.50	1.000(−1)	7.310(−2)	1.79	6.564(−2)	4.947(−2)	2.8	1.686(−2)	1.362(−2)
1.22	1.535(−1)	1.080(−1)	1.51	9.854(−2)	7.211(−2)	1.80	6.471(−2)	4.882(−2)	2.9	1.482(−2)	1.203(−2)
1.23	1.511(−1)	1.065(−1)	1.52	9.709(−2)	7.113(−2)	1.81	6.380(−2)	4.817(−2)	3.0	1.305(−2)	1.064(−2)
1.24	1.487(−1)	1.050(−1)	1.53	9.567(−2)	7.017(−2)	1.82	6.290(−2)	4.754(−2)	3.1	1.149(−2)	9.417(−3)
1.25	1.464(−1)	1.035(−1)	1.54	9.426(−2)	6.922(−2)	1.83	6.202(−2)	4.691(−2)	3.2	1.013(−2)	8.337(−3)
1.26	1.441(−1)	1.020(−1)	1.55	9.288(−2)	6.828(−2)	1.84	6.115(−2)	4.630(−2)	3.3	8.939(−3)	7.384(−3)
1.27	1.419(−1)	1.006(−1)	1.56	9.152(−2)	6.736(−2)	1.85	6.029(−2)	4.569(−2)	3.4	7.891(−3)	6.544(−3)
1.28	1.397(−1)	9.920(−2)	1.57	9.019(−2)	6.645(−2)	1.86	5.945(−2)	4.509(−2)	3.5	6.970(−3)	5.802(−3)
1.29	1.376(−1)	9.781(−2)	1.58	8.887(−2)	6.555(−2)	1.87	5.862(−2)	4.450(−2)	3.6	6.160(−3)	5.146(−3)
1.30	1.355(−1)	9.645(−2)	1.59	8.758(−2)	6.467(−2)	1.88	5.780(−2)	4.392(−2)	3.7	5.448(−3)	4.567(−3)
1.31	1.334(−1)	9.510(−2)	1.60	8.631(−2)	6.380(−2)	1.89	5.700(−2)	4.335(−2)	3.8	4.820(−3)	4.054(−3)
1.32	1.313(−1)	9.378(−2)	1.61	8.506(−2)	6.295(−2)	1.90	5.620(−2)	4.278(−2)	3.9	4.267(−3)	3.600(−3)
1.33	1.293(−1)	9.247(−2)	1.62	8.383(−2)	6.210(−2)	1.91	5.542(−2)	4.222(−2)	4.0	3.779(−3)	3.198(−3)
1.34	1.274(−1)	9.119(−2)	1.63	8.261(−2)	6.127(−2)	1.92	5.465(−2)	4.167(−2)	4.1	3.349(−3)	2.842(−3)
1.35	1.254(−1)	8.993(−2)	1.64	8.142(−2)	6.045(−2)	1.93	5.390(−2)	4.113(−2)	4.2	2.969(−3)	2.527(−3)
1.36	1.235(−1)	8.868(−2)	1.65	8.025(−2)	5.964(−2)	1.94	5.315(−2)	4.059(−2)	4.3	2.633(−3)	2.247(−3)
1.37	1.216(−1)	8.746(−2)	1.66	7.909(−2)	5.884(−2)	1.95	5.241(−2)	4.007(−2)	4.4	2.336(−3)	1.999(−3)
1.38	1.198(−1)	8.625(−2)	1.67	7.796(−2)	5.806(−2)	1.96	5.169(−2)	3.955(−2)	4.5	2.073(−3)	1.779(−3)
1.39	1.180(−1)	8.506(−2)	1.68	7.684(−2)	5.729(−2)	1.97	5.098(−2)	3.903(−2)	4.6	1.841(−3)	1.583(−3)
1.40	1.162(−1)	8.389(−2)	1.69	7.574(−2)	5.652(−2)	1.98	5.027(−2)	3.853(−2)	4.7	1.635(−3)	1.410(−3)
1.41	1.145(−1)	8.274(−2)	1.70	7.465(−2)	5.577(−2)	1.99	4.958(−2)	3.803(−2)	4.8	1.453(−3)	1.255(−3)
1.42	1.128(−1)	8.160(−2)	1.71	7.359(−2)	5.503(−2)	2.0	4.890(−2)	3.753(−2)	4.9	1.291(−3)	1.118(−3)
1.43	1.111(−1)	8.048(−2)	1.72	7.254(−2)	5.430(−2)	2.1	4.261(−2)	3.297(−2)	5.0	1.148(−3)	9.965(−4)
1.44	1.094(−1)	7.938(−2)	1.73	7.151(−2)	5.358(−2)	2.2	3.719(−2)	2.898(−2)	5.1	1.021(−3)	8.881(−4)

(Continued)

TABLE 8-8. Tabulations of the $E_1(x)$ and $E_2(x)$ Functions for Photon Shielding Calculations *(Continued)*

x	$E_1(x)$	$E_2(x)$	x	$E_1(x)$	$E_2(x)$	x	$E_1(x)$	$E_2(x)$	x	$E_1(x)$	$E_2(x)$
5.2	9.086(−4)	7.917(−4)	8.1	3.370(−5)	3.057(−5)	11.0	1.400(−6)	1.298(−6)	13.9	6.193(−8)	5.821(−8)
5.3	8.086(−4)	7.060(−4)	8.2	3.015(−5)	2.738(−5)	11.1	1.256(−6)	1.166(−6)	14.0	5.566(−8)	5.234(−8)
5.4	7.198(−4)	6.296(−4)	8.3	2.699(−5)	2.453(−5)	11.2	1.127(−6)	1.047(−6)	14.1	5.002(−8)	4.706(−8)
5.5	6.409(−4)	5.617(−4)	8.4	2.415(−5)	2.198(−5)	11.3	1.012(−6)	9.398(−7)	14.2	4.496(−8)	4.232(−8)
5.6	5.708(−4)	5.012(−4)	8.5	2.162(−5)	1.969(−5)	11.4	9.080(−7)	8.439(−7)	14.3	4.042(−8)	3.805(−8)
5.7	5.085(−4)	4.473(−4)	8.6	1.936(−5)	1.764(−5)	11.5	8.150(−7)	7.578(−7)	14.4	3.633(−8)	3.422(−8)
5.8	4.532(−4)	3.992(−4)	8.7	1.733(−5)	1.581(−5)	11.6	7.315(−7)	6.805(−7)	14.5	3.266(−8)	3.077(−8)
5.9	4.039(−4)	3.564(−4)	8.8	1.552(−5)	1.417(−5)	11.7	6.566(−7)	6.112(−7)	14.6	2.936(−8)	2.767(−8)
6.0	3.601(−4)	3.183(−4)	8.9	1.390(−5)	1.270(−5)	11.8	5.895(−7)	5.490(−7)	14.7	2.640(−8)	2.489(−8)
6.1	3.211(−4)	2.842(−4)	9.0	1.245(−5)	1.138(−5)	11.9	5.292(−7)	4.931(−7)	14.8	2.373(−8)	2.238(−8)
6.2	2.864(−4)	2.539(−4)	9.1	1.115(−5)	1.020(−5)	12.0	4.751(−7)	4.429(−7)	14.9	2.134(−8)	2.013(−8)
6.3	2.555(−4)	2.268(−4)	9.2	9.988(−6)	9.149(−6)	12.1	4.266(−7)	3.979(−7)	15.0	1.919(−8)	1.811(−8)
6.4	2.279(−4)	2.027(−4)	9.3	8.948(−6)	8.203(−6)	12.2	3.830(−7)	3.574(−7)	15.1	1.725(−8)	1.629(−8)
6.5	2.034(−4)	1.811(−4)	9.4	8.018(−6)	7.356(−6)	12.3	3.440(−7)	3.211(−7)	15.2	1.551(−8)	1.465(−8)
6.6	1.816(−4)	1.619(−4)	9.5	7.185(−6)	6.596(−6)	12.4	3.089(−7)	2.885(−7)	15.3	1.395(−8)	1.318(−8)
6.7	1.621(−4)	1.447(−4)	9.6	6.439(−6)	5.916(−6)	12.5	2.774(−7)	2.592(−7)	15.4	1.255(−8)	1.186(−8)
6.8	1.448(−4)	1.294(−4)	9.7	5.771(−6)	5.306(−6)	12.6	2.491(−7)	2.329(−7)	15.5	1.128(−8)	1.067(−8)
6.9	1.293(−4)	1.157(−4)	9.8	5.173(−6)	4.759(−6)	12.7	2.238(−7)	2.093(−7)	15.6	1.015(−8)	9.595(−9)
7.0	1.155(−4)	1.035(−4)	9.9	4.637(−6)	4.269(−6)	12.8	2.010(−7)	1.881(−7)	15.7	9.126(−9)	8.633(−9)
7.1	1.032(−4)	9.259(−5)	10.0	4.157(−6)	3.830(−6)	12.9	1.805(−7)	1.690(−7)	15.8	8.208(−9)	7.767(−9)
7.2	9.219(−5)	8.283(−5)	10.1	3.727(−6)	3.436(−6)	13.0	1.622(−7)	1.519(−7)	15.9	7.383(−9)	6.988(−9)
7.3	8.239(−5)	7.411(−5)	10.2	3.342(−6)	3.083(−6)	13.1	1.457(−7)	1.365(−7)	16.0	6.640(−9)	6.287(−9)
7.4	7.364(−5)	6.632(−5)	10.3	2.997(−6)	2.767(−6)	13.2	1.309(−7)	1.227(−7)	16.1	5.973(−9)	5.657(−9)
7.5	6.583(−5)	5.935(−5)	10.4	2.687(−6)	2.483(−6)	13.3	1.176(−7)	1.103(−7)	16.2	5.373(−9)	5.090(−9)
7.6	5.886(−5)	5.312(−5)	10.5	2.410(−6)	2.228(−6)	13.4	1.057(−7)	9.914(−8)	16.3	4.834(−9)	4.581(−9)
7.7	5.263(−5)	4.756(−5)	10.6	2.162(−6)	2.000(−6)	13.5	9.495(−8)	8.912(−8)	16.4	4.348(−9)	4.122(−9)
7.8	4.707(−5)	4.258(−5)	10.7	1.939(−6)	1.795(−6)	13.6	8.532(−8)	8.011(−8)	16.5	3.912(−9)	3.709(−9)
7.9	4.210(−5)	3.812(−5)	10.8	1.740(−6)	1.611(−6)	13.7	7.667(−8)	7.202(−8)	16.6	3.519(−9)	3.338(−9)
8.0	3.767(−5)	3.414(−5)	10.9	1.561(−6)	1.446(−6)	13.8	6.890(−8)	6.475(−8)	16.7	3.166(−9)	3.004(−9)

x	$E_1(x)$	$E_2(x)$	x	$E_1(x)$	$E_2(x)$	x	$E_1(x)$	$E_2(x)$	x	$E_1(x)$	$E_2(x)$
16.8	2.849(−9)	2.704(−9)	18.9	3.119(−10)	2.975(−10)	21.0	3.453(−11)	3.308(−11)	23.1	3.859(−12)	3.710(−12)
16.9	2.563(−9)	2.433(−9)	19.0	2.808(−10)	2.679(−10)	21.1	3.110(−11)	2.980(−11)	23.2	3.477(−12)	3.344(−12)
17.0	2.306(−9)	2.190(−9)	19.1	2.528(−10)	2.413(−10)	21.2	2.802(−11)	2.685(−11)	23.3	3.133(−12)	3.014(−12)
17.1	2.075(−9)	1.971(−9)	19.2	2.276(−10)	2.173(−10)	21.3	2.524(−11)	2.419(−11)	23.4	2.824(−12)	2.716(−12)
17.2	1.867(−9)	1.774(−9)	19.3	2.049(−10)	1.957(−10)	21.4	2.273(−11)	2.180(−11)	23.5	2.544(−12)	2.448(−12)
17.3	1.680(−9)	1.597(−9)	19.4	1.845(−10)	1.762(−10)	21.5	2.048(−11)	1.964(−11)	23.6	2.293(−12)	2.206(−12)
17.4	1.512(−9)	1.437(−9)	19.5	1.661(−10)	1.587(−10)	21.6	1.845(−11)	1.769(−11)	23.7	2.066(−12)	1.988(−12)
17.5	1.361(−9)	1.294(−9)	19.6	1.496(−10)	1.429(−10)	21.7	1.662(−11)	1.594(−11)	23.8	1.862(−12)	1.792(−12)
17.6	1.225(−9)	1.165(−9)	19.7	1.347(−10)	1.287(−10)	21.8	1.497(−11)	1.436(−11)	23.9	1.678(−12)	1.615(−12)
17.7	1.102(−9)	1.049(−9)	19.8	1.213(−10)	1.159(−10)	21.9	1.349(−11)	1.294(−11)	24.0	1.512(−12)	1.456(−12)
17.8	9.920(−10)	9.439(−10)	19.9	1.092(−10)	1.044(−10)	22.0	1.215(−11)	1.166(−11)	24.1	1.363(−12)	1.312(−12)
17.9	8.928(−10)	8.498(−10)	20.0	9.836(−11)	9.405(−11)	22.1	1.095(−11)	1.051(−11)	24.2	1.228(−12)	1.183(−12)
18.0	8.036(−10)	7.650(−10)	20.1	8.857(−11)	8.471(−11)	22.2	9.861(−12)	9.467(−12)	24.3	1.107(−12)	1.066(−12)
18.1	7.233(−10)	6.887(−10)	20.2	7.976(−11)	7.630(−11)	22.3	8.884(−12)	8.531(−12)	24.4	9.977(−13)	9.611(−13)
18.2	6.511(−10)	6.201(−10)	20.3	7.183(−11)	6.873(−11)	22.4	8.004(−12)	7.687(−12)	24.5	8.992(−13)	8.664(−13)
18.3	5.860(−10)	5.582(−10)	20.4	6.469(−11)	6.191(−11)	22.5	7.212(−12)	6.927(−12)	24.6	8.105(−13)	7.810(−13)
18.4	5.275(−10)	5.027(−10)	20.5	5.826(−11)	5.577(−11)	22.6	6.498(−12)	6.242(−12)	24.7	7.305(−13)	7.040(−13)
18.5	4.749(−10)	4.526(−10)	20.6	5.247(−11)	5.023(−11)	22.7	5.854(−12)	5.625(−12)	24.8	6.584(−13)	6.346(−13)
18.6	4.275(−10)	4.075(−10)	20.7	4.726(−11)	4.525(−11)	22.8	5.275(−12)	5.069(−12)	24.9	5.934(−13)	5.721(−13)
18.7	3.848(−10)	3.669(−10)	20.8	4.257(−11)	4.077(−11)	22.9	4.753(−12)	4.568(−12)	25.0	5.349(−13)	5.157(−13)
18.8	3.464(−10)	3.304(−10)	20.9	3.834(−11)	3.672(−11)	23.0	4.283(−12)	4.117(−12)			

[a]Each value is multiplied by 10 raised to the power shown in ( ): i.e., 4.038(1) = 40.38; 9.489(−1) = 9.489 × 10^{-1} = 0.9849.

SOLUTION: Since the planar source is large, it can be treated as a shielded infinite plane in which the buildup flux above the concrete layer using the Taylor approximation is

$$\phi_b(P)_{A,\infty} = \frac{S_A}{2}\{AE_1[(1+\alpha_1)\mu a] + (1-A)E_1[(1+\alpha_2)\mu a]\}$$

From Table 8-5, $A = 97.0$, $\alpha_1 = -0.0396$, and $\alpha_2 = -0.02710$; and from Table 8-2 $\mu/\rho = 0.0635\ \text{cm}^2/\text{g}$ (or $\mu = 0.14923\ \text{cm}^{-1}$), and $\mu a = 4.4768$. Therefore

$$\begin{aligned}
\phi_b(P)_{A,\infty} &= \frac{10^7}{2}[97.0E_1[(1-0.0396)4.4768] - 96.0E_1[(1-0.02710)4.4768]] \\
&= 5\times10^6[97.0E_1(4.30) - 5\times10^6\times 96.0E_1(4.355)] \\
&= 5\times10^6[(97.0\times2.633\times10^{-3}) - (96.0\times2.47\times10^{-3})] \\
&= 5\times10^6(0.2554 - 0.2371) = 9.15\times10^4\ \gamma/\text{cm}^2\cdot\text{s}
\end{aligned}$$

As shown in Example 8-11, it is relatively straightforward to obtain $\phi_b(P)_{A,\infty}$ when the shield thickness and the emission rate for photons of a given energy are known; it is much more complex to obtain the shield thickness required to yield a predetermined flux (or exposure rate). The latter requires iterative calculations that are facilitated by plotting solutions of the equation on semilog paper for incremental values of μx. Iterative calculations of $\phi_b(P)_{A,\infty}$ for values of μx can also be performed by computer; in either case, once the appropriate value of μx is determined, the shield thickness x can be obtained directly.

A *shielded finite area source of radius R with buildup* also uses the Taylor approximation to obtain the buildup flux, which though more complex, is

$$\begin{aligned}
\phi_b(P)_{A,R} &= \frac{S_A}{2}A\{E_1[(1+\alpha_1)\mu a] - E_1[(1+\alpha_1)\mu r]\} \\
&\quad + \frac{S_A}{2}(1-A)\{E_1[(1+\alpha_2)\mu a] - E_1[(1+\alpha_2)\mu r]\}
\end{aligned}$$

where r can be obtained from $r^2 = x^2 + R^2$ or $r = x\sec\theta$ where θ is the angle subtended between the line x perpendicular from the center of the area and the line from P to the outside edge. In essence this more detailed calculation is the same expression as for the infinite area source minus the portion that lies outside the radius R. Fortunately, most practical problems can be considered "infinitely large" and this refinement can often be neglected. Exposures calculated on the basis of ininite planar or area sources will be somewhat conservative for radiation protection purposes (i.e., higher than what actually occurs).

Gamma Exposure from Thick Slabs

Many sources are not strictly area sources because they may be a thick source (e.g., a slab) that contains a uniformly distributed gamma-emitting isotope and the effective surface activity is very much a function of the amount of self-absorption within the source. A thick slab is essentially a volume source that emits $S_V\gamma/\text{cm}^3\cdot\text{s}$;

however, this volume source can be adjusted to a surface source by accounting for self-absorption of radiation in the slab. If the linear absorption coefficient of the slab material is μ, the activity on the surface due to radioactivity in a differential layer dx at a depth of x in the slab is

$$dS_V = S_V\, dx\, e^{-\mu x}$$

which can be integrated over the total thickness t of the slab to obtain an effective surface activity of

$$S_A = \int_0^t S_V e^{-\mu x}\, dx = \frac{S_V}{\mu}(1 - e^{-\mu t})$$

This expression can then be used in the relationship above for the gamma flux at a point P above the slab, or

$$\phi(P)_{V,t} = \frac{S_V}{4\mu}(1 - e^{-\mu t}) \ln\left(1 + \frac{R^2}{x^2}\right)$$

or for very large slabs where $R \gg x$,

$$\phi(P)_{V,t} = \frac{S_V}{2\mu}(1 - e^{-\mu t}) \ln\frac{R}{x}, \qquad R \gg x$$

where $\phi(P)_{V,t}$ is the unscattered flux ($\gamma/\text{cm}^2 \cdot \text{s}$) that exits from the thick slab. It does not, however, account for scattered photons produced within the slab itself, which requires a buildup factor again based on one of the mathematical formulations of B.

An *infinite slab source* is representative of a number of typical sources encountered in radiation protection. Such sources can be represented as a homogeneous concentration in a slab of material, many of which are large enough to be considered as infinitely thick and infinite in area. Of course they are not, but if they are several inches thick and a few tens of square feet in area, their detailed solution approaches one with infinite dimensions in which reasonable accuracy is obtained and the mathematics is much easier. Contaminated soils at a site are generally representative of this condition.

Schiager (1974) published a classic paper in which the unscattered flux above large-area thick slabs of gamma-emitting radioactive material can be represented in terms of the soil-volume concentration, S_V, as

$$\phi(\gamma/\text{cm}^2 \cdot \text{s}) = \frac{S_V}{2\mu}[1 - E_2(\mu x)]$$

where ϕ is the emission rate of gamma rays per unit area ($\gamma/\text{cm}^2 \cdot \text{s}$), S_V the soil-volume emission rate ($\gamma/\text{cm}^3 \cdot \text{s}$), μ the linear attenuation coefficient, and x the slab thickness. The term in brackets has a rather difficult solution because $E_2(\mu x)$ is the second-order exponential integral (also listed in Table 8-8); however, when plotted as shown in Figure 8-6 for this special case of a very thick infinite plane, its use is rather straightforward.

An *infinite slab buildup factor* is also necessary to obtain the total exposure rate because more scattered photons exit a large slab than uncollided primary photons. Unfortunately, buildup factors are available only for point sources; however,

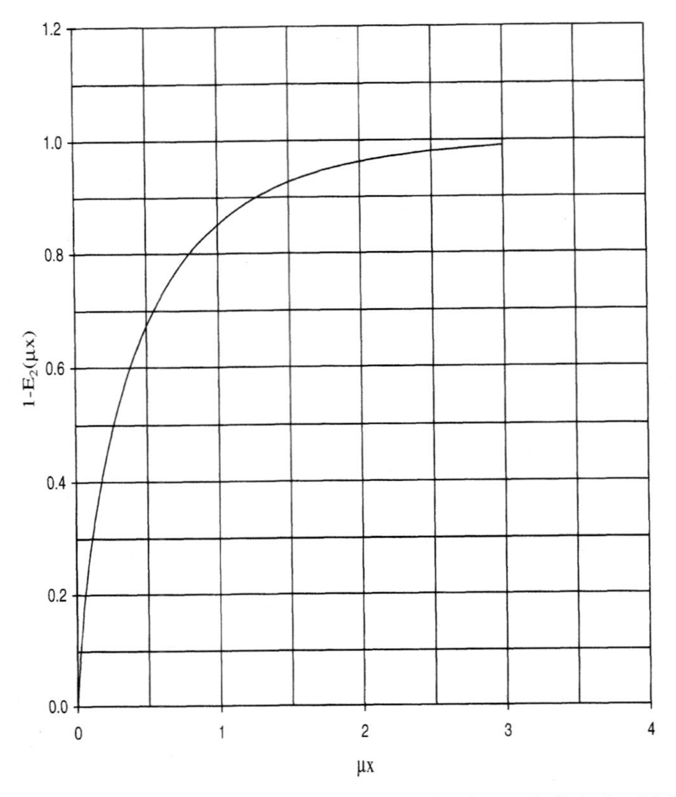

FIGURE 8-6. Plot of the function $(1 - E_2)$ versus μx for large, infinitely thick slab sources of gamma-emitting radioactive material.

a buildup factor for slab sources based on empirical data can be approximated as

$$B_{\text{slab}} = e^{\mu x/1+\mu x}$$

With these considerations, the total flux, including scattered photons, is

$$\phi_s(\gamma/\text{cm}^2 \cdot \text{s}) = \frac{S_V}{2\mu}[1 - E_2(\mu x)]e^{\mu x/1+\mu x}$$

where the relationship $[1 - E_2(\mu x)]$ is readily obtained from Figure 8-6 (or calculated using the data in Table 8-8).

The value of μ for dry or moist soils ($\rho = 1.6$) is about 0.11 cm^{-1} for 0.8 MeV photons and about 0.16 cm^{-1} for concrete ($\rho = 2.35$) and varies very slowly for thicknesses of these materials where x is greater than 30 to 60 cm. Many site conditions are typical of such depths, and for such conditions both $1 - E_2(\mu x)$ and B are essentially unity and the expression for the flux reduces to

$$\phi_s(\gamma/\text{cm}^2 \cdot \text{s}) = \frac{S_V}{2\mu}e$$

This is not the case, however, when the thickness of the radioactive layer (typically, soil or sediment) is less than 20 cm or so, and it is necessary to use the general equation with the factor $1 - E_2(\mu x)$ and the approximation for the slab buildup factor.

Radium-contaminated soils are a special case, although a fairly common one. The radioactive progeny of ^{226}Ra produce 2.184 photons per transformation of ^{226}Ra, and these photons have an average energy of 0.824 MeV (Schiager, 1974). The gamma constant for radium progeny photons is 0.84 R/h · Ci of ^{226}Ra at 1 m and 2300 photons/cm^2 yield 1 μR; therefore, 1 pCi/g of ^{226}Ra will produce an unscattered flux of

$$\phi = 0.59\ \gamma/\text{cm}^2 \cdot \text{s per pCi/g of } ^{226}\text{Ra}$$

The exposure rate due to the unscattered photons is $0.92C_{\text{Ra}}(\mu\text{R/h})$, which needs to be adjusted by a buildup factor B to account for the importance of scattered photons. Since B can be approximated by $B = e^{\mu x/1+\mu x}$, the exposure rate for a ^{226}Ra concentration C_{Ra} (pCi/g) is

$$\text{exposure } (\mu\text{R/h}) = 0.92C_{\text{Ra}}e^{\mu x/1+\mu x}$$

When the thickness of the contaminated soil layer is 30 to 60 cm or more, μx becomes very large such that B changes very little and $B \approx e$; therefore,

$$\text{exposure}_{\text{Ra}}\ (\mu\text{R/h}) = 0.92eC_{\text{Ra}} = 2.5C_{\text{Ra}}$$

where C_{Ra} is in pCi/g. If, however, the radium soil layer is less than 30 cm or so, the practical approach would be first to calculate the exposure rate for an infinitely thick layer and adjust it by the factor exposure(x)/exposure(∞) obtained from Figure 8-7. It is important to note in Figure 8-7 the rapid change of exposure(x)/exposure(∞) for thin layers of contaminated soils; therefore, exposure determinations depend very much on accurate determinations, usually by field measurements, of the thickness of the contaminated layer. These determinations have more influence on the calculated results than the analytical solutions used. If the contamination layer is covered with soil, H_2O, concrete, or other material, it is necessary to determine the exposure rate as though the shielding material were absent and then calculate the exposure rate as a shielded area source (see above) where the shield is the thickness of cover material.

Volume Sources

Volume sources such as large drums or tanks of radioactive material produce scattered photons due to self-absorption by the medium in which they are produced. Radiation exposure calculations for these various cylindrical and spherical geometries is fairly complex; however, good information can be obtained for such geometries by dividing them up into several point-source subdivisions and summing the contributions of each. Such calculations are generally conservative in that they tend to overestimate exposure, but considerable simplification of the calculations is

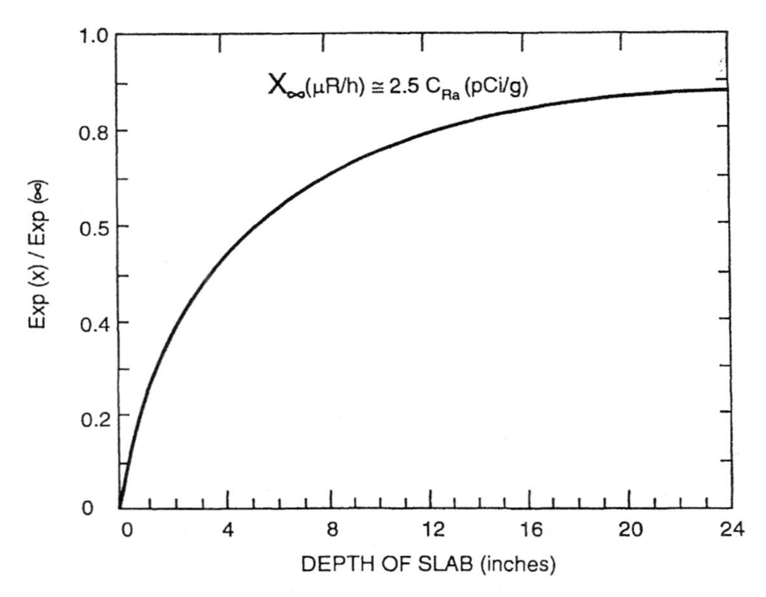

FIGURE 8-7. Ratio of exposure rate exposure(x) for a slab of thickness x of gamma-emitting material and exposure(∞) for an infinitely thick and infinitely large slab of the same concentration.

obtained, and errors in the estimates are not large. Detailed integration over many differential volume sources is of course more precise, and computerized calculations can be performed for numerous small volume elements to increase accuracy and reduce the calculational burden.

Example 8-12: Estimate the exposure rate in air at 1 m (a) from the center of a 10-cm-diameter plastic pipe if the pipe is 1.5 m long and contains 5 Ci of ^{137}Cs in H_2O and (b) with a 3-cm-thick lead shield between the pipe and the receptor.

SOLUTION: The 1.5-m pipe is divided into five segments of 30 cm, each of which contains 1 Ci, as shown in Figure 8-8. The activity of each subvolume is assumed to be concentrated at a point in the center and the exposure at P is calculated for the point sources at 1, 2, and 3, and because of symmetry the exposures for 4 and 5 are the same as 1 and 2. The individual exposures are added to obtain the total. It is necessary to determine the distance between each point source and P, the angular thickness of water and Pb to be penetrated, the attenuation of flux by each of these thicknesses, and the buildup flux for each thickness. These determinations yield the following results (see Table 8-9): (a) exposure at P without the Pb shield is 1350 mR/h due to photon absorption and the buildup effect in the H_2O solution; (b) with the lead shield in place, the exposure rate is reduced to 38.1 mR/h. Better accuracy and somewhat less conservatism (slightly lower total exposure) could be obtained by further subdividing the segments, perhaps into two lengthwise segments and then further dividing each of these into nine segments.

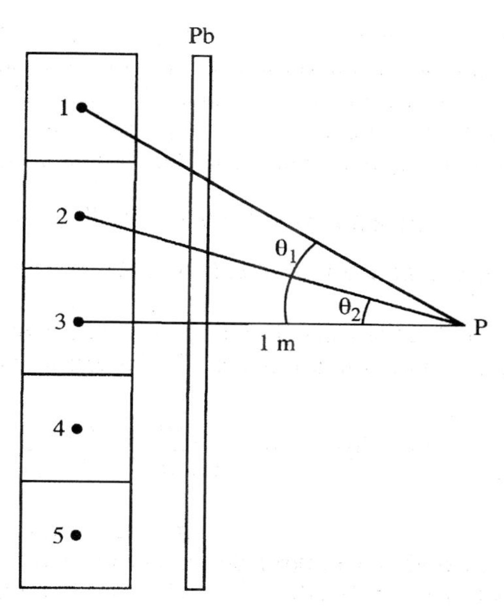

FIGURE 8-8. Volume source of 5 Ci of ^{137}Cs divided into five segments and modeled as five 1-Ci point sources, producing exposure at *P* with and without a lead shield.

Buildup Factors for Layered Absorbers

Example 8-12 is also illustrative of the influence of two different absorber mediums on photon absorption and transmission. When different photon absorbers are used in tandem, even more complexity occurs because of the different effect presented by each. The most important consideration is the production of Compton-scattered photons, which can be quite large for low- to medium-energy photons in low-*Z* absorbers, factors which in turn influence the buildup factor.

There are no precise methods for dealing with the complexities of absorber materials of different *Z*, but a few general principles can be applied. First, a low-*Z* material (such as H_2O) will result in a higher fraction of scattered photons than high-*Z* materials (e.g., Pb); therefore, the low-*Z* absorber should be placed closest to the source and the high-*Z* material placed outside it to absorb the Compton-scattered photons produced in the first absorber. If, for example, a source is shielded with a layer of lead and then enclosed in water, the buildup flux would be considerable higher than the inverse because the photons entering H_2O would produce relatively more scattered photons (due to the dominance of the Compton interaction coefficient in H_2O for low- to medium-energy photons (Figure 7-20), and these would not be absorbed in the water because photoelectric absorption is relatively low in a low-*Z* medium. If, however, the H_2O shield is placed closest to the source, the scattered photons produced in it will be highly absorbed in Pb by photoelectric interactions since it is a high-*Z* material. Although the buildup factor is higher in Pb for low- to intermediate-energy photons, this is more than offset by its greater effect on attenuation and absorption. Therefore, when there is a choice, put the low-*Z* material first.

TABLE 8-9. Parameters and Exposure Calculations for a 10-cm-Diameter Pipe Modeled as Five 1-Ci Point Sources[a]

	Distance (cm)			μx		$B(\mu x)$		Attenuated Flux at P (γ-rays/cm$^2\cdot$s)			Exposure in Air (mR/h) $\times B$	
Point Source	To P	In H_2O	In Air	For H_2O	For Pb	For H_2O	For Pb	By Air	By H_2O	By Pb	H_2O Only	H_2O + Pb
1	116.6	5.83	3.50	0.503	4.51	1.58	1.76	1.84×10^5	1.11×10^5	2.02×10^3	224	4.53
2	104.4	5.22	3.13	0.450	4.03	1.56	1.71	2.29×10^5	1.46×10^5	4.07×10^3	290	8.88
3	100	5.00	3.00	0.631	3.87	1.55	1.70	2.5×10^5	1.62×10^5	5.21×10^3	321	11.30
4	104.4	5.22	3.13	0.450	4.03	1.56	1.71	2.29×10^5	1.46×10^5	4.07×10^3	290	8.88
5	116.6	5.83	3.50	0.503	4.51	1.58	1.76	1.84×10^5	1.11×10^5	2.02×10^3	224	4.53
										Total	1349	38.12

[a]Exposure rate in air for 0.662 MeV gamma rays is 1.276×10^{-6} R/h $\cdot\,\gamma$.

Similarly, the choice of a buildup factor for layered absorbers can be generalized as follows:

1. If the atomic numbers of the two media do not differ by more than 5 to 10, use the buildup factor of the medium for which this factor is the larger, but compute the overall buildup factor for the sum of both thicknesses since (except at low energy) the buildup factors do not vary rapidly with the Z of the medium.
2. If the media are different, with the low-Z medium first, then use the buildup factor of the second medium as if the first medium were not there because the Compton-scattered radiation from the first is of lower energy and can be expected to be absorbed in the second.
3. If the media are significantly different, with the high-Z medium first, the procedure to be followed in this case depends on whether the gamma-ray energy is above or below the minimum in the μ-curve, which occurs at about 3 MeV for heavy elements. If $E < 3$ MeV, then

$$B(Z_1 + Z_2) = B_{Z1}(\mu_1 a_1) \times B_{Z2}(\mu_2 a_2)$$

This is because the energy of the photons emerging from a high-Z shield is little different from that of the source, and in the second medium the photons can be treated as if they were source γ-rays. If, on the other hand, $E > 3$ MeV, then

$$B(Z_1 + Z_2) = B_{Z1}(\mu_1 a_1) \times B_{Z2}(\mu_2 a_2)_{\min}$$

where $B_{Z2}(\mu_2 a_2)_{\min}$ is the value of B_{Z2} at 3 MeV. This assumes that the gamma rays penetrating the first layer have energies clustered about the minimum in μ, so that their penetration through the second layer is determined by this energy rather than by the actual source energy.

SHIELDING OF PROTONS AND LIGHT IONS

Beams of protons, deuterons, tritons, and helium ions can be present around accelerators. Since all of these are charged particles with considerable mass, they will ionize atoms in a shielding material and can be stopped completely in a short distance after giving up all their energy. Since attenuation occurs by ionization (and not by probabilistic exponential processes), the range is definitive. Therefore, shielding for these particles follows the same principles as for beta and alpha particles (i.e., by determining the maximum range of the most energetic particles in the beam). Being charged particles, some bremsstrahlung may be produced, but this is very minor except perhaps for highly energetic particles (say above several tens of MeV).

Shield designs for high-energy protons are based on their range in matter, as shown in Figure 8-9 for selected materials. Example 8-13 demonstrates the utility of these data.

Example 8-13: What thickness of (a) lead and (b) aluminum is necessary to attenuate completely a beam of 20-MeV protons?

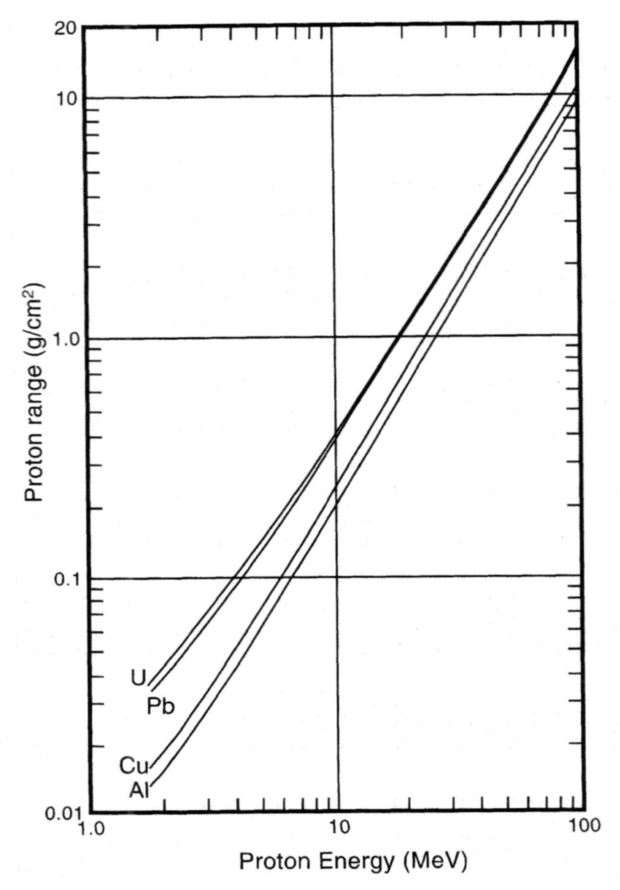

FIGURE 8-9. Range versus energy of protons in uranium, lead, copper, and aluminum. (From NCRP, 1977.)

SOLUTION: (a) From Figure 8-9, the range of 20 MeV protons in lead is 0.9 g/cm^2; therefore,

$$x = \frac{R}{\rho} = \frac{0.9\ \text{g/cm}^2}{11.34\ \text{g/cm}^2} = 0.079\ \text{cm}$$

(b) For aluminum, the range of 20 MeV protons is approximately 0.5 g/cm^2, and

$$x = \frac{R}{\rho} = \frac{0.5\ \text{g/cm}^2}{2.7\ \text{g/cm}^2} = 0.185\ \text{cm}$$

Obviously, even light materials are effective in shielding energetic protons. One must be careful, however, to consider the production of other radiations by proton absorption interactions in the material chosen.

Shield designs for light ions are based on a reference thickness of penetration for a proton of a given energy. This reference value is then adjusted to account for the different mass and charge for deuterons, tritons, and helium ions. The range of deuterons, tritons, and doubly charged helium ions of a given energy E are related to the range of protons in an absorber as follows:

$$R(^2\text{H}^+) = 2 \times (R_p \text{ at } E/2)$$

$$R(^3\text{H}^+) = 3 \times (R_p \text{ at } E/3)$$

$$R(\text{He}^{2+}) = (R_p \text{ at } E/4)$$

where the range of the respective particle of energy E is obtained by determining the proton range R_p at a lower energy and adjusting it by the appropriate factor, as shown in Example 8-14.

Example 8-14: What thickness of copper is required to attenuate 30 MeV tritons completely?

SOLUTION: First determine the range of protons in copper at $E/3 = 10$ MeV, which is 0.23 g/cm^2. The range of 30 MeV tritons in copper is

$$R(^3\text{H}^+) = 3R_p = 0.69 \text{ g/cm}^2$$

which is provided by a thickness of

$$x = \frac{0.69 \text{ g/cm}^2}{8.96 \text{ g/cm}^2} = 0.077 \text{ cm}$$

SUMMARY

Radiation shielding, which is a very complex discipline, is a form of radiation protection for many radiation sources and the many geometric configurations in which they may occur. Alpha particles and other light ions are easy to shield, usually by a sealer, since their relatively large mass and charge limit their range in most media to a few tens of micrometers. Shielding of beta particles from a source is also straightforward by choosing a thickness of medium that matches or exceeds the maximum range; however, since high-energy beta particles can produce bremsstrahlung, especially in high-Z materials, beta shields should use plastic, aluminum, or other low-Z material followed by lead or some other dense material to absorb any bremsstrahlung and characteristic x rays that are produced in the beta shield.

Photon shields, unlike those for charged particles, are governed by the exponential, or probabilistic, attenuation of electromagnetic radiation, and the flux of shielded photons is a complex mixture of scattered and unscattered photons characterized as poor geometry. A calculated value of $I(x)$ based on the attenuation coefficient μ, which is determined in good geometry conditions, will thus underestimate the number of photons reaching the receptor, which implies that absorption is greater than that which actually occurs. A buildup factor, $B > 1.0$, is used to correct good geometry calculations to more accurately reflect actual, or poor geometry conditions.

Many radiation sources can, with ease and utility, be represented as a point or an approximate point source; however, many real-world situations cannot; for example, long pipes or tubes (typical of a line source), a contaminated area (representative of a disc or infinite planar source), and various volume sources. Fortunately, various practical calculations, some of which are fairly complex, can be used to determine the photon flux, which can then be applied in the usual way to calculate radiation exposure. Such calculations are generally conservative in that they tend to overestimate exposure, but considerable simplification of the calculations is obtained and errors in the estimates are not large.

ACKNOWLEDGMENTS

Many of the data resources in this chapter came from the very helpful people at the National Institute of Standards and Technology, in particular John H. Hubbell and his colleagues M. J. Berger and S. M. Seltzer. Its compilation is due in large measure to the patient, careful, and untiring efforts of Chul Lee, a graduate of the University of Michigan Radiological Health Program.

REFERENCES AND ADDITIONAL RESOURCES

Chilton, A. B., *Nuclear Science and Engineering* **64**, 799–800 (1977).

Hubbell, J. H., and S. M. Seltzer, *Tables of X-ray Attenuation Coefficients and Mass Absorption Coefficients 1 keV to 20 MeV for Elements Z = 1 to 92 and 48 Additional Substances of Dosimetric Interest*, NISTIR 5632, National Institute of Standards and Technology, Gaithersburg, MD, 1995.

Lamarsh, J. R., *Introduction to Nuclear Engineering*, 2nd ed., Addison-Wesley, Reading, MA, 1983, Chapters 5 and 9.

Schiager, K. J., Analysis of radiation exposures on or near uranium tailings piles, *Radiation Data and Reports*, pp. 411–425, July 1974.

Berger, M. J., and S. M. Selzer, Stepping powers and ranges of electrons and positrons. NBSIR 82-2550A National Bureau of Standards, Washington, DC, 1983.

Morgan, K. Z., and J. E. Turner, eds., *Principles of Radiation Dosimetry*, Wiley, New York, 1967.

NCRP, *Radiation Protection Design Guidelines for 0.1–100 MeV Particle Accelerator Facilities*, Report 51, National Council on Radiation Protection and Measurements, Bethesda, MD, March 1977.

Oak Ridge National Laboratory, Radiation Shielding and Information Center Report 10, Oak Ridge, TN 1966.

U.S. Academic Energy Commission, Reactor Shielding for Nuclear Engineers, Editor, Schaeffer, NM, Report TID 25951, 1973.

PROBLEMS

8-1. A surface is contaminated with alpha-emitting ^{239}Pu. If it is to be sealed with fiberglass resin ($\rho \approx 1.0$), how thick must it be to absorb all the alpha particles emitted?

8-2. A point source that emits beta particles at a rate of 10^7 β/s is to be shielded. If the maximum beta energy is 3.0 MeV, what thickness of aluminum would be required to attenuate the beta particles, and what thickness of lead would need to be included to reduce the photon flux at 50 cm by a factor of 10^3?

8-3. A 30-mL solution containing 2 Ci of ^{90}Sr in equilibrium with ^{90}Y is to be put into a small glass bottle, which will then be placed in a lead container having walls 1.5 cm thick. (**a**) How thick must the walls of the glass bottle be to prevent any beta ray from reaching the lead? (**b**) Estimate the bremsstrahlung dose rate at a distance of 1.5 m from the center of the lead container.

8-4. A small vial (approximately a point source) containing 200 Ci of ^{32}P in aqueous solution is enclosed in an 8 mm-thick aluminum can. Calculate the thickness of lead shielding needed to reduce the exposure rate to 2.0 mR/h at a distance of 2 m from the can.

8-5. Show that the tenth-value layer for photons equals $2.30/\mu$, where μ is the total linear attenuation coefficient.

8-6. Assume that the exponent μx in the equation $I(x) = I_0 e^{-\mu x}$ is equal to or less than 0.1. Show that with an error of less than 1%, the number of photons transmitted is $I_0(1 - \mu x)$ and the number attenuated is $I_0 \mu x$. (*Hint*: Expand the term $e^{-\mu x}$ into a series.)

8-7. By what fraction will 2 cm of aluminum reduce a narrow beam of 1.0 MeV photons?

8-8. What thickness of copper is required to attenuate a narrow beam of 500 keV photons to one-half of the original number?

8-9. Calculate the thickness of lead shielding needed to reduce the exposure rate 2.5 m from a 16-Ci point source of ^{137}Cs to 1.0 mR/h if scattered photons are not considered (i.e., without buildup).

8-10. Recalculate the exposure rate for the shield design in Problem 8-9 when the buildup of scattered photons is considered.

8-11. How thick must a spherical lead container be to reduce the exposure rate 1 m from a small 100-mCi ^{24}Na source to 2.0 mR/h?

8-12. A beam of 500 keV photons is normally incident on a uranium sheet that is 1.5 cm thick. If the exposure rate in front of the sheet is 60 mR/h, what is it behind the sheet?

8-13. What thickness of lead shielding is needed around a 2000-Ci point source of ^{60}Co to reduce the exposure rate to 10 mR/h at a distance of 2 m?

8-14. An ion-exchange column that contains radioactive materials that emits 5.5×10^6 1 MeV gammas per centimeter of length rests on the floor of a plant. This column is 9 m long and has a diameter of 0.5 m. What is the unshielded exposure rate at a point 12 m away and 1 m above the floor (ignore air attenuation and scatter from the floor and walls) if the gamma-ray constant is 1.55 R/h per Ci at 1 m?

8-15. For Problem 8-14, estimate the approximate exposure rate behind a 2-cm-thick lead shield placed around the ion-exchange column.

8-16. A solution of ^{131}I containing an activity of 10^7 t/s is spilled on an approximate circular area 1 m in diameter. Determine the exposure rate 1 m above the center of the contaminated area.

8-17. A solution of ^{137}Cs with a concentration of 0.16 μCi/mL is to be mixed with concrete in a cylindical plastic container 30 cm in diameter and 28 cm long. What exposure rate will the ^{137}Cs concrete mixture ($\rho = 2.35$) produce at 1 m from the 30-cm-diameter surface?

8-18. Radioactive fresh fission products from an atomic explosion contaminate the surface at 1 μCi/cm^2 with an average gamma-ray energy of 0.7 MeV per transformation. A person enters a fallout shelter that has a 30-cm-thick concrete roof. What is the exposure rate inside the shelter?

8-19. A shallow circular impoundment roughly 10 m in diameter was drained, exposing a thin sediment layer containing ^{137}Cs, which dried. If the estimated activity of the thin sediment layer is 1 mCi/cm^2, what is the gamma dose rate 1 m above the center?

8-20. Residues from processing thorium ores in monazite sands and from which all the ^{232}Th has been removed are to be stored in a cylindrical tank 40 cm in diameter and 2 m high. The ^{228}Ra concentration of 20 pCi/g maintains an equilibrium concentration of ^{208}Tl, which emits photons of 0.583 MeV in 4.5% and 2.614 MeV in 94.16% of transformations. Perform a simplified volume source estimation of the exposure rate 1 m from the tank due to the 2.614-MeV gamma rays by segmenting the volume into two parallel subvolumes each of which contains five segments.

8-21. A very large basin has a 10-cm sediment layer ($\rho = 1.6$) containing ^{137}Cs that emits 10^3 0.662 MeV gamma rays per second per gram. What is the exposure rate (**a**) 1 m above the exposed sediment layer and (**b**) 1 m above the area if it is covered with 60 cm of water?

8-22. An old radium site contains a large area of debris with an average concentration of 5 pCi/g of ^{226}Ra. It is about 50 cm thick but is covered with a layer of soil. If the soil layer is removed, what would be the exposure rate above the radium debris?

9

PRODUCTION OF ACTIVATION PRODUCTS

Wow! It's as big as a barn.
—Anonymous, ca. 1932

Activation products are common sources of radioactive materials that are of concern to radiation protection; they occur from bombarding target materials with various particles, such as neutrons, protons, deuterons, and from various other reactions (as developed in Chapter 4). The quantity and types of products produced depend on the energy and type of the bombarding particle, the target material, and the interaction probabilities of the various combinations.

ACTIVATION CROSS SECTIONS

Cross section denotes the target size or apparent area that an atom of a particular element presents to a bombarding particle such that a given reaction may take place. It is not specifically an area, but is better described as a measure of the probability that a reaction will take place. Cross sections are expressed in barns, which is 10^{-24} cm^2. Since cross sections are highly dependent on particle energy and target material, the most comprehensive information is provided by a plot of the cross section versus incident particle energy as shown in Figure 9-1 for neutrons on cadmium and boron. Such plots show a general fall-off of σ with increasing neutron energy (i.e., a $1/v$ dependence of σ on neutron speed) and often contain resonance energies where the cross section is very high. The National Nuclear Data Center at Brookhaven National Laboratory is the authoritative source for similar information on cross-section data for a host of nuclear interactions, including protons, deuterons, alpha particles, and other projectiles, as well as neutrons.

NEUTRON REACTIONS

Many activation products are produced by neutron irradiation, which occurs most commonly in nuclear reactors. Other important sources of neutrons are (α, n) sources

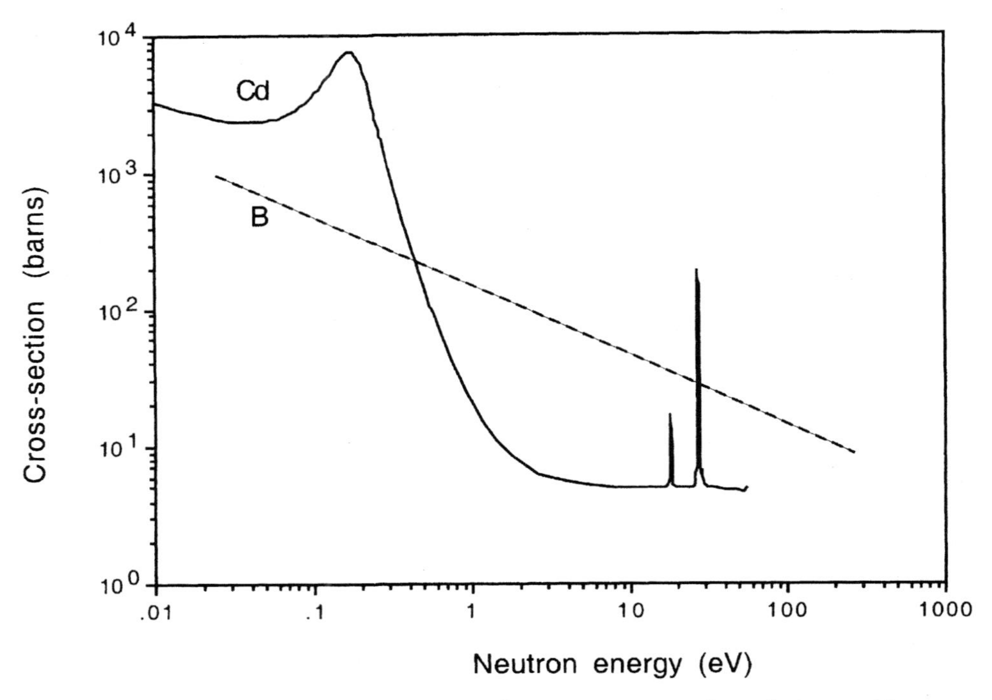

FIGURE 9-1. Neutron activation cross sections versus energy for cadmium and boron.

such as polonium or plutonium, mixed with beryllium, energetic gamma sources or x rays on deuterium or beryllium to create photodisintegration neutron sources, and spontaneous fission nuclides such as ^{252}Cf. These sources yield fast neutrons that can be used as produced, moderated to intermediate energies, or moderated to thermal energies.

Neutron activation products can be produced intentionally by placing the appropriate target material in a neutron flux, or such production can occur unintentionally as the by-product of operating a source that emits neutrons. For example, the structural materials in a nuclear reactor contain various metals, depending on the neutron cross section, which may be activated to produce radioactive products. Cobalt, which is added to steel to give it desired tensile properties, is often a troublesome by-product of reactor operation, as is the iron in steel. Cobalt-60 is produced from stable ^{59}Co by the reaction

$$^{59}\text{Co} + {}^{1}_{0}\text{n} \rightarrow [^{60}\text{Co}^{*}] \rightarrow {}^{60}\text{Co} + \gamma + Q$$

This and similar reactions in which the excited nucleus emits its excess energy as γ-radiation are known as *radiative capture* (n,γ) *reactions*. Other (n,γ) activation products produced in nuclear reactors are ^{58}Ni$(n,\gamma)^{59}$Ni, ^{53}Mn$(n,\gamma)^{54}$Mn, and ^{50}Cr$(n,\gamma)^{51}$Cr.

Another general category is *charged particle emission* (CPE), in which the compound nucleus that is formed breaks up by the emission of a charged particle such as a proton (n,p) or alpha (n,α) particle. These reactions are rare with low-energy neutrons because the expulsion of a charged particle usually requires a considerable amount of energy; however, one important (n,p) reaction that occurs with a low-energy neutron to produce ^{14}C is

$$^{14}_{7}N + ^{1}_{0}n \rightarrow [^{15}_{7}N^{*}] \rightarrow ^{14}_{6}C + ^{1}_{1}H + Q$$

Cross sections versus energy for CPE, (n,2n), (n,n′) and (n,fission) reactions are in Appendix F; or for thermal neutrons on the chart of the nuclides designated as σ_p, σ_α, etc. where the subscript denotes the emitted particle. Radiative capture (n,γ) cross sections are designated by σ_γ, and σ_f denotes that the isotope fissions after absorption of a neutron.

ACTIVATION PRODUCTS IN NUCLEAR REACTORS

Important sources of activation products in reactors are structural components containing cobalt, manganese, chromium, and other elements used to harden steel, and natural constituents and coolant additives (i.e., boron and lithium compounds) for controlling the water chemistry of the plant. Typical activation products are ^{51}Cr, ^{54}Mn, ^{56}Mn, ^{55}Fe, ^{59}Fe, ^{58}Co, ^{60}Co, ^{95}Zr, ^{97}Zr, ^{65}Zn, and ^{95}Nb produced by neutron activation of corrosion products as they circulate through the various systems. Corrosion products, as well as long-lived fission products, tend to collect in piping bends, joints, and other low points, creating significant hot spots (high radiation sources); contamination occurs if they leak out of the systems.

Water itself contains oxygen and hydrogen, both of which can be activated. Oxygen-16 (99.8% of naturally occurring oxygen) can be activated to form radioactive nitrogen (N) by the following reactions:

$$^{16}_{8}O + ^{1}_{0}n \rightarrow [^{16}_{8}O^{*}] \rightarrow ^{16}_{7}N + ^{1}_{1}p$$

$$^{16}_{8}O + ^{1}_{1}p \rightarrow [^{17}_{9}F^{*}] \rightarrow ^{13}_{7}N + ^{4}_{2}He$$

After formation, the nitrogen atoms usually combine with oxygen and hydrogen in the coolant to form N_2 ions or compounds such as NO_2^-, NO_3^-, NH_4^+, NO, and NO_2. Nitrogen-16 has a relatively large formation rate and emits extremely high energy gamma rays (6.13 and 7.12 MeV); therefore, it represents a major contributor to radiation levels near primary system piping containing either water or steam. Fortunately, the half-life of ^{16}N is only 7.1 s, so that it is no problem once the activation process terminates (i.e., after a reactor is shut down). Nitrogen-13 has a 9.97-min half-life and undergoes radioactive transformation by positron emission which in itself represents little concern; however, positron annihilation photons can contribute to gamma-exposure fields.

During normal operation of boiling water reactors (BWRs), ^{16}N is the major contributor to the radiation levels around the turbine equipment since it travels in steam to the turbine. In a pressurized water reactor (PWR) the coolant is contained in a primary loop and the nitrogen activity is confined to the reactor water. The contribution to plant radiation levels from ^{13}N is completely masked by the predominance of ^{16}N, and in PWR systems it has no practical significance at all. However, in BWR systems the half-life of ^{13}N (9.97 min) is long enough for some of it to be discharged to the environment in gaseous releases to the atmosphere, and it can also be detected in steam leaks.

Fluorine-18 is produced by proton interaction with ^{18}O, which makes up about 0.2% of natural oxygen, by the following reaction:

$$^{18}_{8}O + ^{1}_{1}p \rightarrow [^{19}_{9}F] \rightarrow + ^{18}_{9}F + ^{1}_{0}n + Q$$

Flourine-18 is radioactive and undergoes transformation by positron emission. Sometimes the ^{18}F and ^{13}N are confused since they both emit positrons, which ultimately yield 0.511 MeV annihilation photons.

Tritium (^{3}H), with maximum and average beta energies of 0.0186 and 0.006 MeV, can be produced in nuclear reactors in large quantities primarily by activation of boron or lithium, both of which have been used as additives to reactor coolant systems to control plant conditions. The reactions, which are different for thermal (n_{th}) and fast (n_f) neutrons, are

$$^{10}_{5}B + ^{1}_{0}n_{th} \rightarrow ^{7}_{3}Li + ^{4}_{2}He$$

$$^{10}_{5}B + ^{1}_{0}n_{f} \rightarrow (^{7}_{3}Li)^{*} + ^{4}_{2}He$$

where the excited lithium nucleus further breaks up to yield ^{3}H by the reaction

$$(^{7}_{3}Li)^{*} \rightarrow ^{3}_{1}H + ^{4}_{2}He$$

Boron becomes a source of tritium in reactors because it is often injected into the coolant of PWRs in concentrations up to 4000 ppm to provide a chemical shim for control of reactivity; therefore, PWRs that add boron to coolant produce more tritium than BWRs that do not require it.

Lithium hydroxide is often used to control the pH of reactor coolant. Fast neutron reactions with ^{7}Li and thermal neutron reactions with ^{6}Li form tritium by

$$^{7}_{3}Li + ^{1}_{0}n_{f} \rightarrow ^{3}_{1}H + ^{4}_{2}He + ^{1}_{0}n$$

$$^{6}_{3}Li + ^{1}_{0}n_{th} \rightarrow ^{3}_{1}H + ^{4}_{2}He$$

Natural lithium contains 92.5% ^{7}Li and 7.5% ^{6}Li; however, it is the presence of ^{6}Li in the reactor coolant that constitutes the most significant source of ^{3}H because of its high thermal neutron cross section and the preponderance of thermal neutrons. Tritium production in PWRs can be reduced considerably by use of ^{6}LiH, which is generally available. Tritium is also produced by *ternary fission* of ^{235}U:

$$^{235}_{92}U + ^{1}_{0}n_{th} \rightarrow ^{140}_{54}Xe + ^{91}_{37}Rb + ^{3}_{1}H + 2^{1}_{0}n + Q$$

In ternary fission, three fission products are produced, yielding one atom of tritium for every 12,500 fissions. However, over 99% of ternary tritium is retained within the fuel rods. Ternary fission also produces ^{14}C but at a lower yield, and it too remains in the fuel rods.

ACTIVATION PRODUCT CALCULATIONS

Activation production formation is shown schematically in Figure 9-2 for a flux of ϕ neutrons/cm$^2 \cdot$ s incident upon a thin target containing N_1^0 atoms. It is assumed that the target thickness is such that the flux of neutrons remains essentially the same as it passes through. When these conditions exist, absorption reactions will produce new atoms of N_2, at a rate

$$\text{production of } N_2 = \phi\sigma_1 N_1^0$$

where ϕ is the flux, σ_1 the absorption cross section in barns for projectiles (commonly neutrons) of a given energy incident on the target, and N_1^0 the number of target atoms. As soon as atoms of N_2 are formed, they can be removed by radioactive transformation and/or activation form to a new product; thus

$$\text{removal of } N_2 = \lambda_2 N_2 + \phi\sigma_2 N_2$$

If activation of N_2 can be ignored, which is usually (but not always) the case, the rate of removal of N_2 atoms is due only to radioactive transformation, and for thin foils containing N_1^0 atoms, the rate of change of new atoms of N_2 with time is the rate of production minus the rate of removal, or

$$\frac{dN_2}{dt} = \phi\sigma_1 N_1^0 - \lambda_2 N_2$$

which can be solved by multiplying through by the integrating factor $e^{\lambda_2 t}$ and performing the integration to obtain

$$N_2(t) = \frac{\phi\sigma_1 N_1^0}{\lambda_2}(1 - e^{-\lambda_2 t})$$

or in terms of activity,

$$A_2(t) = \lambda_2 N_2 = \phi\sigma_1 N_1^0(1 - e^{-\lambda_2 t})$$

Because of the factor $(1 - e^{-\lambda t})$, irradiation of a target for one half-life or more yields a substantial fraction of the saturation amount of the desired radionuclide (see Figure 9-2). If the half-life of the product radionuclide is more than 10 times the irradiation time, the equation above can be reduced to a simpler form by expanding the exponential term in a series to yield

$$A(t) = \phi\sigma_1 N_1^0(\lambda t)$$

with an error of less than 3%. These equations assume insignificant burnup of target atoms and no significant alteration of the projectile flux over the target, both

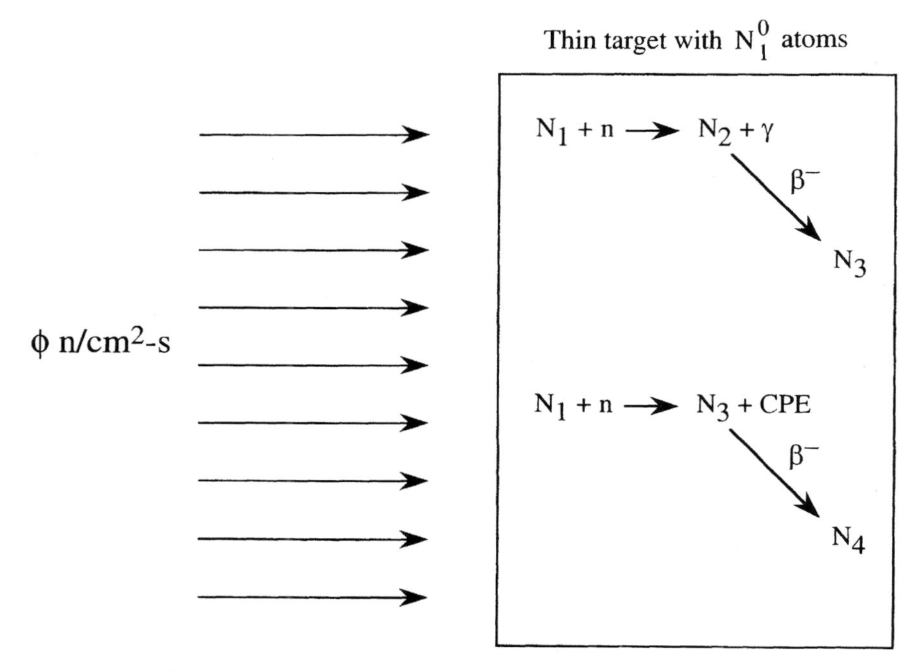

FIGURE 9-2. Schematic of neutron irradiation of a thin (greatly amplified) target material to produce new atoms of a radioactive product.

of which are conservative but realistic assumptions when thin foils are used for target materials.

Depletion of target atoms is of little concern for short irradiation of most materials because the number of target atoms can be presumed to remain constant. However, several products, such as cobalt, have long half-lives and it may require several years to produce them, which can significantly deplete the number of target atoms available. To account for target depletion, N_1 is treated as a variable, such that

$$N_1(t) = N_1^0 e^{-\phi\sigma_1 t}$$

where $N_1(t)$ is the number of target atoms after an irradiation time t, N_1^0 is the initial number of target atoms, and $\phi\sigma_1$ in the exponential term is the depletion constant. The rate of change of atoms of the product N_2 is, as before,

$$\frac{dN_2}{dt} = \phi\sigma_1 N_1(t) - \lambda_2 N_2$$

but since $N_1(t)$ is treated as a variable,

$$\frac{dN_2}{dt} = \phi\sigma_1 N_1^0 e^{-\phi\sigma_1 t} - \lambda_2 N_2$$

Co59	Co60	
100	10.47 m	5.271 y
	IT 58.6, e− β− 1.6ω, ··· γ 1332.5ω, ···	β− 0.318, ··· γ 1332.5, 1173.2, ···
σγ (21+16), (39+35)	σγ 60	σγ 2.0, 4
58.933200	E2.3E2	E 2.824

FIGURE 9-3. Excerpt from the chart of the nuclides showing the thermal neutron radiative cross section σ_γ for ^{59}Co to produce the activation products ^{60m}Co and ^{60}Co and σ_γ for the first resonance interval for higher-energy neutrons as well as radioactive properties of ^{60m}Co and ^{60}Co.

which when solved yields

$$N_2 = \frac{\phi\sigma_1 N_1^0}{\lambda_2 - \phi\sigma_1}(e^{-\phi\sigma_1 t} - e^{-\lambda_2 t})$$

Target depletion and *product activation* may both be significant in special circumstances. In this case, a similar approach can be used to obtain the number of product atoms with time, or

$$\frac{dN_2}{dt} = \phi\sigma_1 N_1(t) - \lambda_2 N_2 - \phi\sigma_2 N_2$$

which has the solution

$$N_2 = \frac{\phi\sigma_1 N_1^0}{\lambda_2 + \phi\sigma_2 - \phi\sigma_1}(e^{-\phi\sigma_1 t} - e^{-(\lambda_2 + \phi\sigma_2)t})$$

Activation product calculations thus require several parameters: the number of atoms in the target material, the neutron flux in the area where irradiation occurs, the activation cross section σ_1 for the primary reaction, the disintegration constant λ_2 of the product, and if applicable, the activation cross section σ_2 of the product.

Neutron activation product calculations, which are perhaps the most common, are facilitated by the cross section(s) for thermal neutron interaction(s) listed in the chart of the nuclides for most isotopes of stable elements and for many radioactive isotopes which may be formed and/or depleted by neutron irradiation. The chart also lists the activation cross section for the first resonance interval above thermal energy, but does not state at what energy the resonance occurs. For example, four values of σ_γ are provided for neutron activation of ^{59}Co to ^{60m}Co and ^{60}Co in the chart of the nuclides as shown in Figure 9-3: (21 + 16), (39 + 35). The first two values in parentheses are (21 + 16), which means that ^{60m}Co is formed from thermal neutron irradiation with a cross section of 21 barns, and ^{60}Co is formed directly (i.e., in its ground state) with 16 barns. The second set of values (39 + 35) have the same

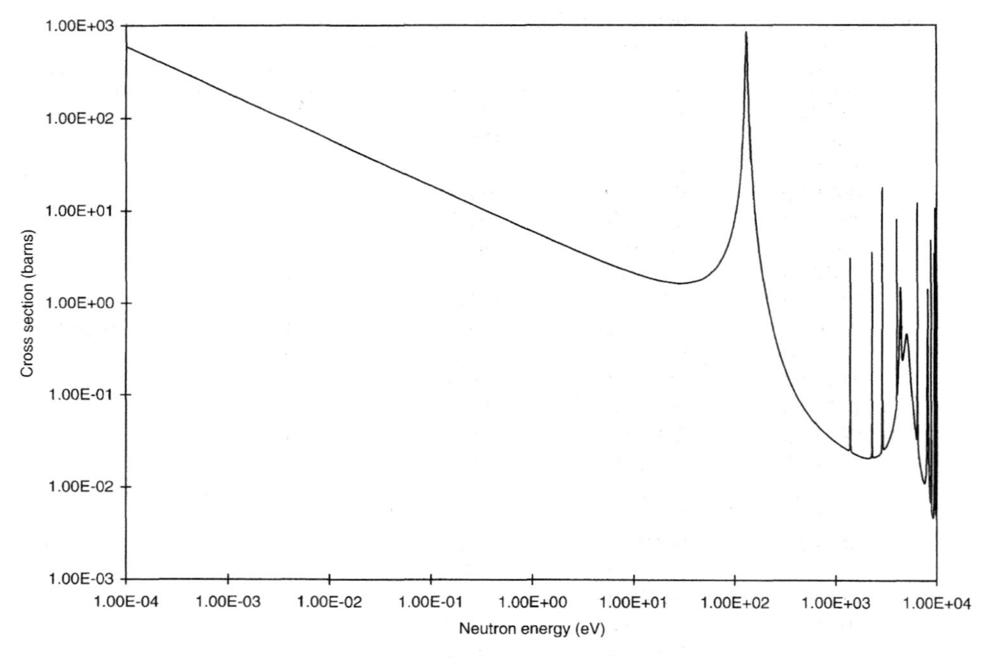

FIGURE 9-4. Neutron activation cross section versus energy for ^{59}Co.

designation except for the important distinction that these are the metastable and ground-state cross sections at the first resonance energy above thermal neutron energies. The neutron energy at which the first resonance interval occurs is not given; it is obtained by consulting a plot of σ versus neutron energy for cobalt. As shown in Figure 9-4, these cross sections occur at a neutron energy of 132 eV.

Figures 9-5 and 9-6 contain excerpts from the chart of the nuclides for several important elements that are the source of activation products. Such information can be used to calculate activation product inventories as shown in Example 9-1.

Example 9-1: A thin foil of 0.1 g of tungsten is placed in a neutron flux of 10^{12} neutrons/cm$^2 \cdot$ s. What is the activity of ^{187}W (a) after 24 h and (b) at saturation?

SOLUTION: (a) The reaction equation to produce ^{187}W is

$$^{186}\mathrm{W} + {}^{1}_{0}\mathrm{n} \rightarrow [^{187}\mathrm{W}] \rightarrow {}^{187}\mathrm{W} + \gamma + Q$$

Thus only the ^{186}W portion of the foil yields ^{187}W, and from the excerpt of the chart of the nuclides in Figure 9-6, the atom percent of ^{186}W in natural tungsten is 28.426% . The number of ^{186}W atoms in the foil is

$$N_1^0 = \frac{0.1 \text{ g} \times 6.022 \times 10^{23} \text{ atoms/mol}}{183.85 \text{ g/mol}} \times 0.28426 = 9.31 \times 10^{19} \text{ atoms}$$

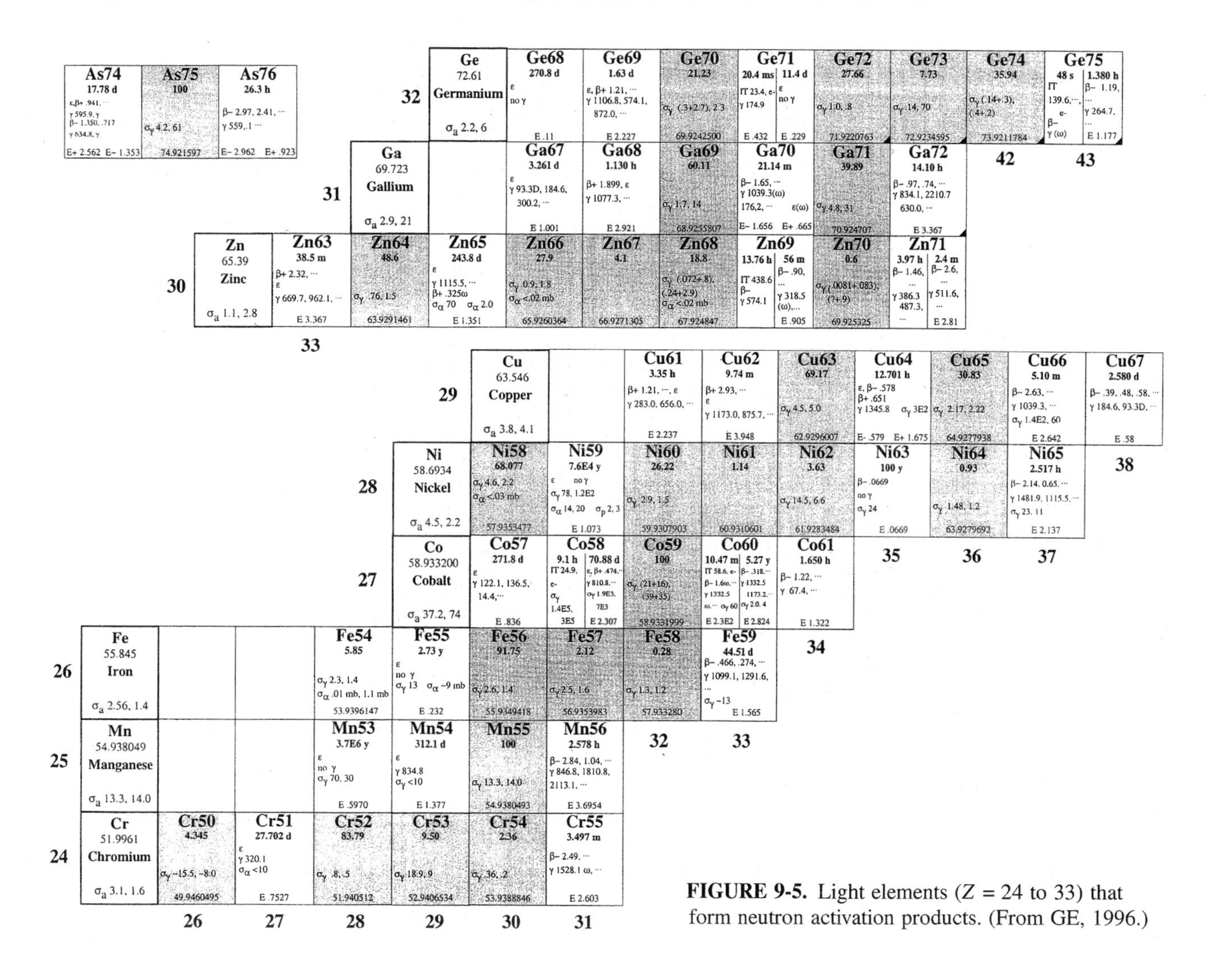

FIGURE 9-5. Light elements (Z = 24 to 33) that form neutron activation products. (From GE, 1996.)

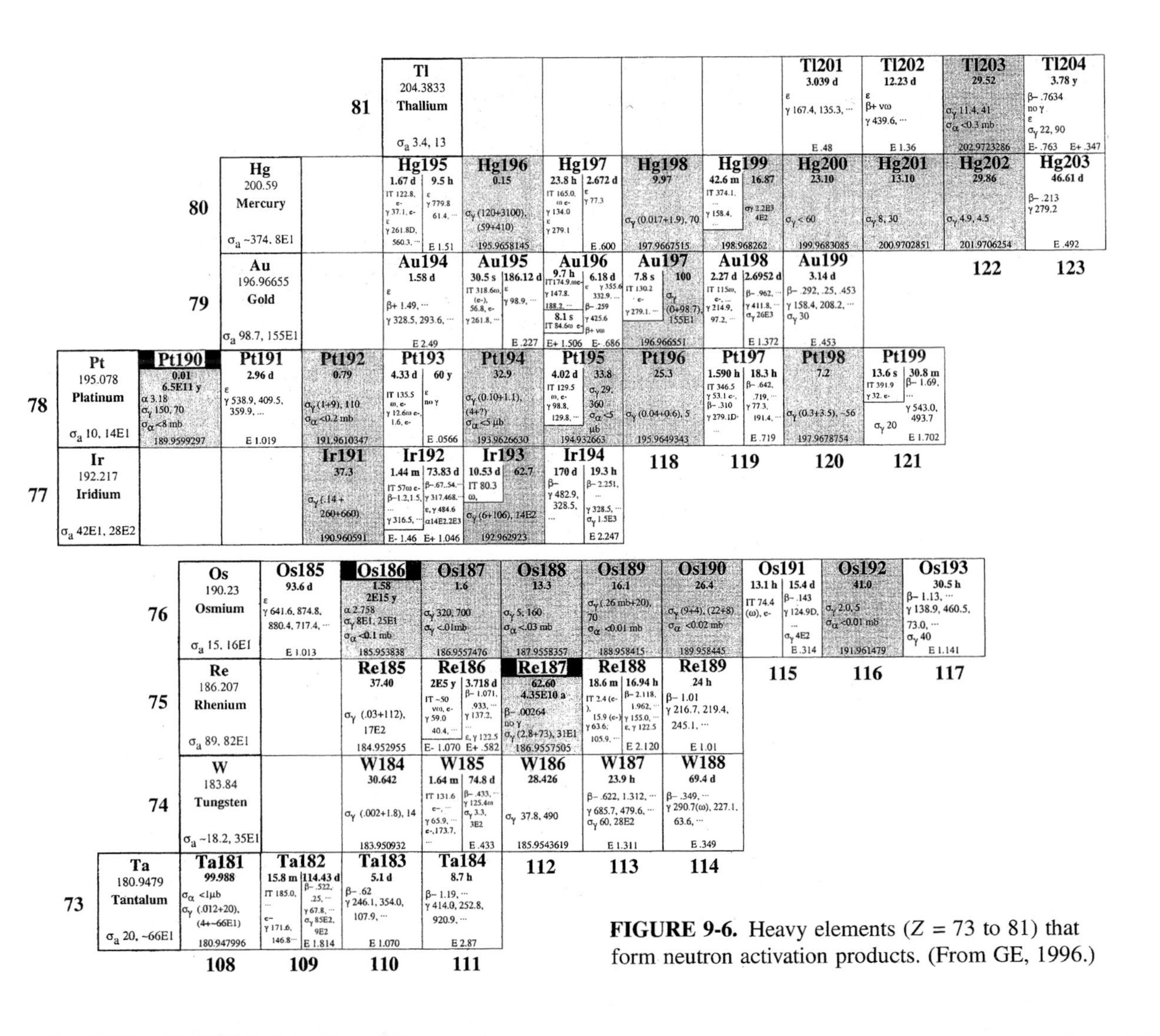

FIGURE 9-6. Heavy elements ($Z = 73$ to 81) that form neutron activation products. (From GE, 1996.)

and the activation cross section $\sigma_\gamma = 38$ b (also from the chart of the nuclides). The activity at 24 h is

$$A(24 \text{ h}) = \phi\sigma N_1(1 - e^{(-\ln 2/23.9 \text{ h})(24\text{th})})$$
$$= 10^{12} \text{ neutrons/cm}^2 \cdot \text{s} \times 38 \times 10^{-24} \text{ cm}^2 \times 9.31 \times 10^{19}(1 - 0.5)$$
$$= 1.77 \times 10^9 \text{ d/s} = 47.8 \text{ mCi}$$

(b) At saturation the exponential term will approach zero (at about 10 half-lives or 240 h), and the saturation activity is

$$A_{\text{sat}} = \phi\sigma N_1^0 = 3.54 \times 10^9 \text{ d/s} = 95.7 \text{ mCi}$$

CHARGED PARTICLE ACTIVATION

Protons and deuterons can be accelerated for various target irradiations by stripping electrons from hydrogen or deuterium gas and introducing the ionized atoms into a source region as shown in Figure 9-7 for a cyclotron (a similar input of source

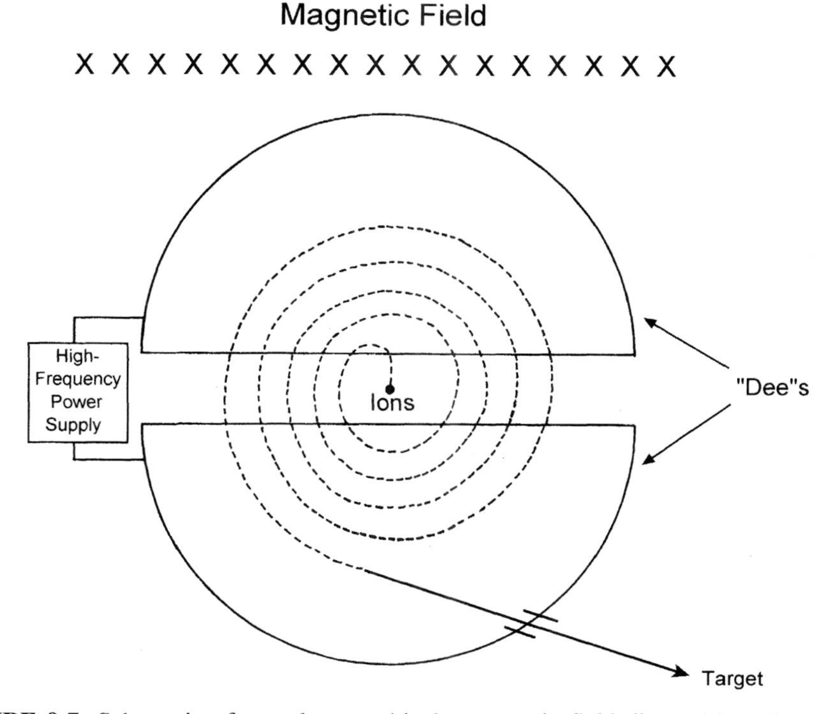

FIGURE 9-7. Schematic of a cyclotron with the magnetic field directed into the plane of the paper. Ions introduced into the gap are accelerated numerous times by the alternating voltage, causing them to travel in increasingly larger circles in the dee space before being deflected toward a target.

ions is shown in Figure 4-12 for a linear accelerator). Ionized tritium (tritons) and helium atoms can also be used as projectiles in accelerators.

Acceleration of particles in a cyclotron is based on the curvilinear deflection of a charged particle moving in a magnetic field. As shown in Figure 9-7, positively charged ions introduced into the gap will be "kicked" toward the negatively charged pole by an alternating electric field placed across the gap. The accelerated particle passes through a slit into an evacuated zone called a *dee* because of its shape. A strong magnetic field across the dee and perpendicular to the path of the particle deflects it into a circular path, bringing it back to the gap on the other side. During this passage, the electric field is reversed by the alternating voltage and when the particle reenters the gap, it receives another "kick" before it passes into the other dee, where it is deflected once again back to the gap, where by now the electric field has once again been reversed. This series of field reversals at the gap provides a series of kicks increasing the velocity (and energy) of the particle at each pass through the gap. The ever-increasing circular path is convenient because it allows the particle to travel greater and greater distances in just the time required for the alternating field to switch. When the desired energy is reached, another magnetic field is used to draw off the accelerated particles and direct them to the target.

Positron-emitting radionuclides, which can be used to diagnose various medical conditions by tagging them onto pharmaceuticals, are readily produced in cyclotrons because protons and deuterons increase the proton number of many target materials. Since many of the light-element positron emitters are short lived, they need to be produced near where they will be used. Positron emitters produce annihilation radiation, and the differential absorption patterns of these photons can provide important medical information using a technique known as *positron emission tomography* (PET). Two common PET nuclides are ^{11}C and ^{18}F, which are produced by the reactions

$$^{14}_{7}\mathrm{N} + ^{1}_{1}\mathrm{H} \rightarrow [^{15}_{8}\mathrm{O}^{*}] \rightarrow ^{11}_{6}\mathrm{C} + ^{4}_{2}\mathrm{He} + Q$$

$$^{18}_{8}\mathrm{O} + ^{1}_{1}\mathrm{H} \rightarrow [^{19}_{9}\mathrm{F}^{*}] \rightarrow ^{18}_{9}\mathrm{F} + ^{1}_{0}\mathrm{n} + Q$$

The cross sections for these reactions are shown in Figures 9-8 and 9-9. ^{18}F is also produced by

$$^{20}_{10}\mathrm{Ne} + ^{2}_{1}\mathrm{H} \rightarrow [^{22}_{11}\mathrm{Na}^{*}] \rightarrow ^{18}_{9}\mathrm{F} + ^{4}_{2}\mathrm{He} + Q$$

and ^{13}N by the reaction

$$^{16}_{8}\mathrm{O} + ^{1}_{1}\mathrm{H} \rightarrow [^{17}_{9}\mathrm{F}^{*}] \rightarrow ^{13}_{7}\mathrm{N} + ^{4}_{2}\mathrm{He} + Q$$

Each of these reactions requires MeV-range projectiles, and the reaction cross sections are relatively smaller than neutron interactions. As shown in Figure 9-8, the ^{14}N(p,α)^{11}C reaction has several resonance peaks, and the energy of the proton beam is an important factor in determining the yield of ^{11}C in a given target.

Tritium can be produced by (d,p) reactions in a deuterium target. The cross sections for this reaction are shown in Figure 9-10.

Accelerator neutron sources can also be fabricated in which various targets are bombarded with deuterons or protons to produce neutrons. Two of the most im-

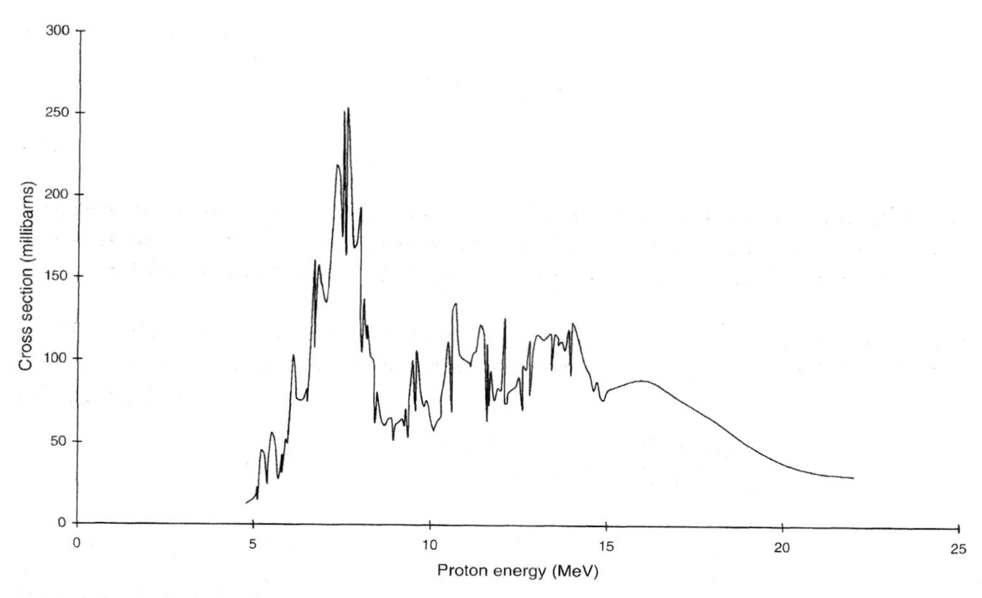

FIGURE 9-8. Cross section versus proton energy for (p, α) production of ^{11}C from ^{14}N.

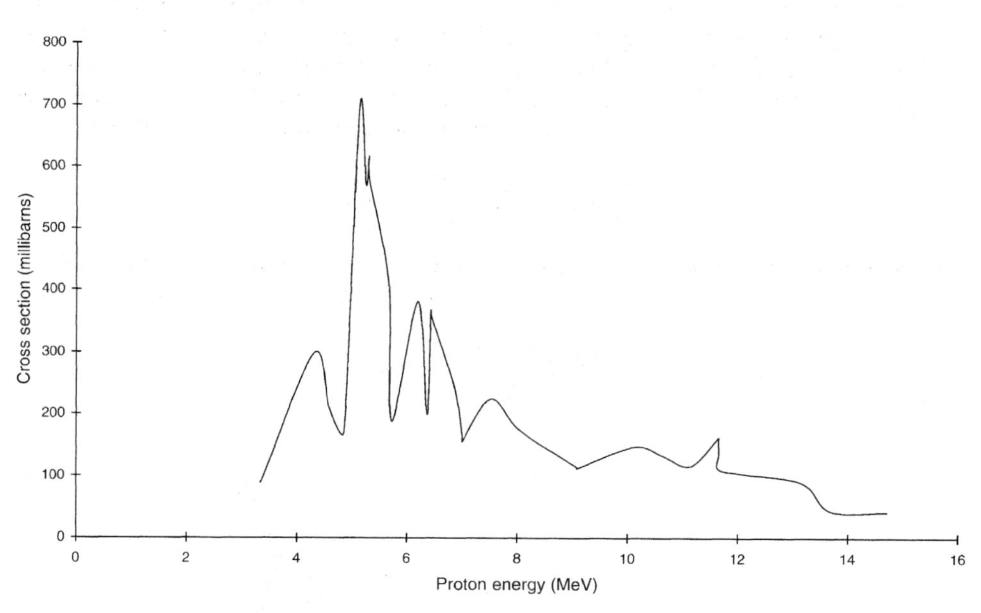

FIGURE 9-9. Cross section versus proton energy for (p, n) production of ^{18}F from ^{18}O.

portant neutron generators use deuterons on targets of deuterium (a D-D neutron generator) and deuterons on tritium (the D-T generator). The cross sections for these reactions are shown in Figures 9-11 and 9-12, respectively. Other typical (d, n) targets are ^{7}Li, ^{9}Be, ^{12}C, and ^{14}N. Protons can also be used to produce neutrons by

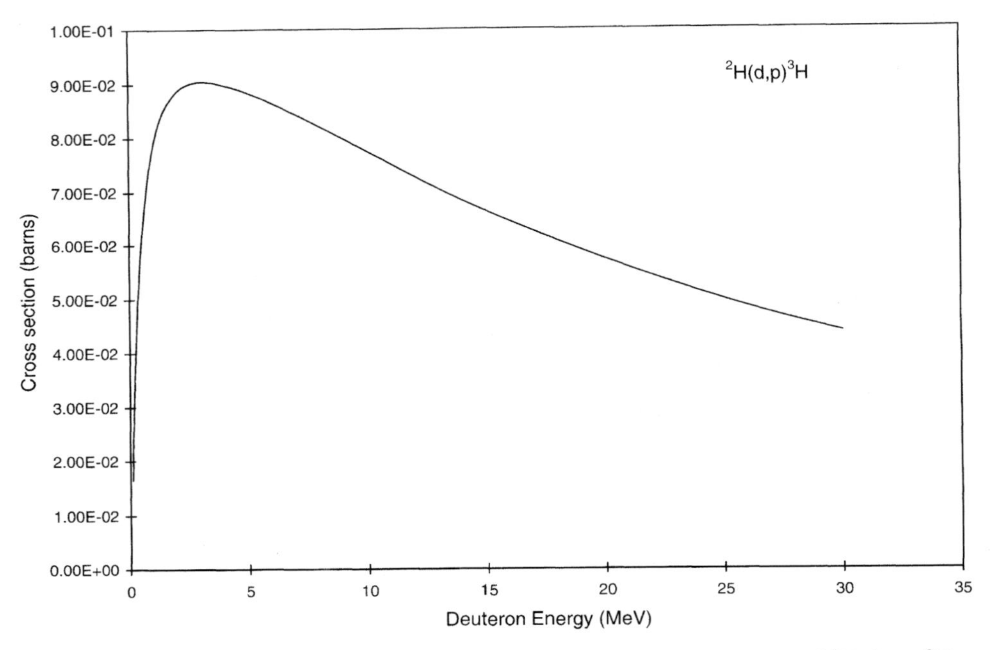

FIGURE 9-10. Cross section versus deuteron energy for (d, p) production of ^3H from ^2H.

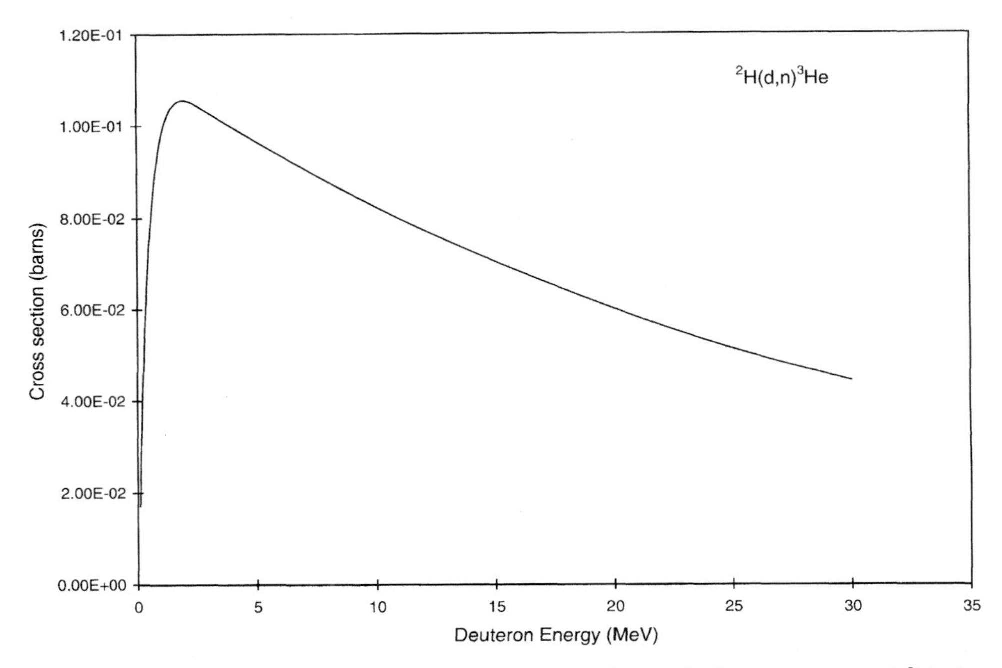

FIGURE 9-11. Cross section versus deuteron energy for producing neutrons and ^3He by (d, n) reactions with ^2H target atoms.

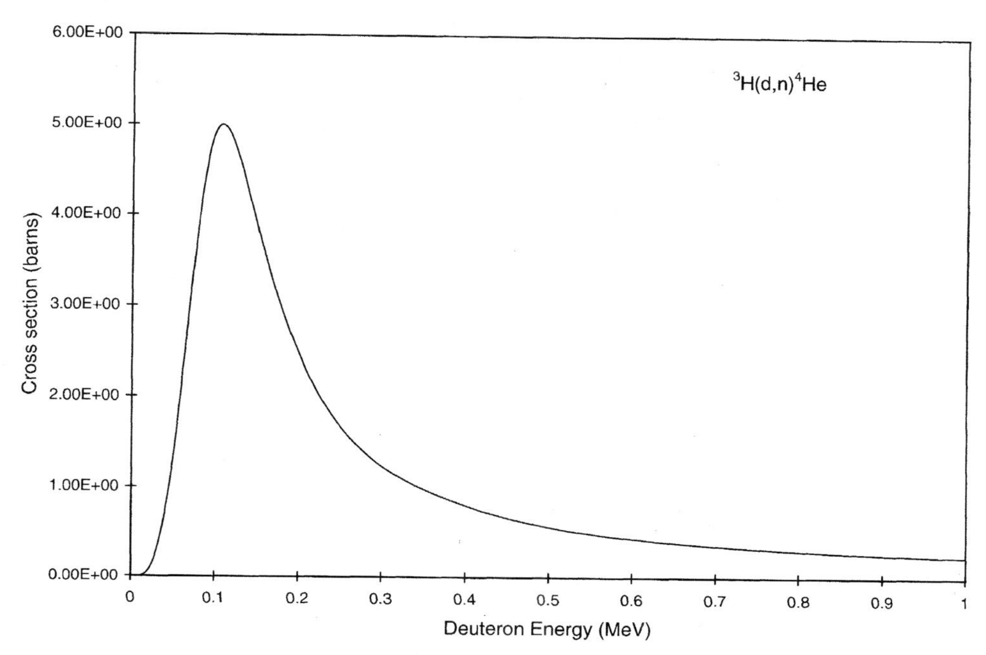

FIGURE 9-12. Cross section versus deuteron energy for producing neutrons and ^{4}He by (d, n) reactions with ^{3}H target atoms.

(p, n) reactions on foils of ^{11}B, ^{56}Fe, ^{58}Ni, and ^{65}Cu, as shown in Example 9-2 for ^{56}Fe(n, p)^{56}Co.

Example 9-2: The (p, n) reaction with iron has a cross section of 0.6 b. How many neutrons/s would be produced by a 3.0 μA proton beam incident on an iron foil of 1 cm^2 and 1.0 μm thick?

SOLUTION: The reaction is

$$^{56}\text{Fe} + {}^1_1\text{H} \rightarrow [^{57}\text{Co}] \rightarrow {}^{56}\text{Co} + {}^1_0\text{n}$$

The number of nuclei in the target (density = 7.9 g/cm^3; 91.75 atom %) is

$$N = \frac{(1.0 \times 10^{-4}\ \text{cm}^3)(7.9\ \text{g/cm}^3)(6.022 \times 10^{23}\ \text{atoms/mol})}{55.845\ \text{g/mol}} \times .9175$$

$$= 7.8 \times 10^{18}\ \text{atoms of}\ ^{56}\text{Fe}$$

The flux of protons per second in the incident beam of 3 μA (3 $\times$ 10^{-6} C/s) is

$$\phi_\text{p} = \frac{3.0 \times 10^{-6}\ \text{C/s}}{(1.6022 \times 10^{-19}\ \text{C/p})(1\ \text{cm}^2)} = 1.87 \times 10^{13}\ \text{protons/cm}^2 \cdot \text{s}$$

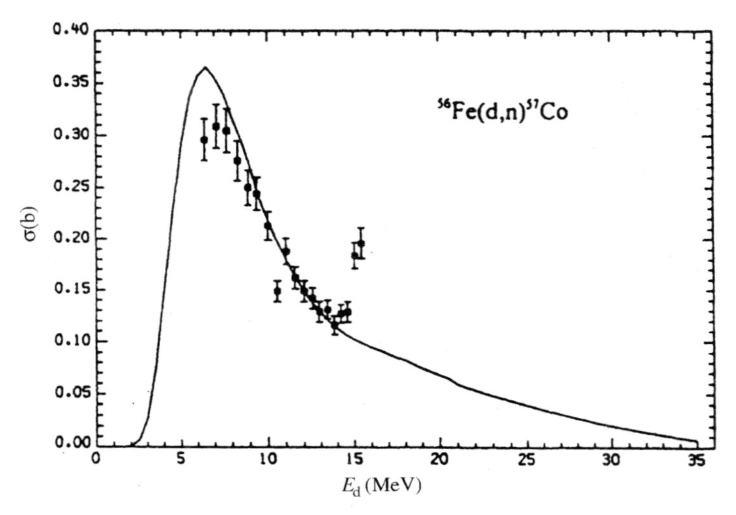

FIGURE 9-13. Cross section versus deuteron energy for (d, n) production of ^{57}Co from ^{56}Fe. (From Xiaoping et al., 1999.)

and the number of neutrons produced per second is

$$\phi_n = (7.8 \times 10^{18} \text{ atoms})(0.6 \times 10^{-24} \text{ cm}^2/\text{atom})(1.87 \times 10^{13} \text{ protons/cm}^2 \cdot \text{s})$$
$$= 8.77 \times 10^7 \text{ neutrons/cm}^2 \cdot \text{s}$$

Charged particle interactions can also produce other products. Cross sections for such reactions are typically in the millibarn range and much less available. Some data are available from the National Nuclear Data Center, and other data, such as those of Xiaoping et al. (1999), are provided in Figures 9-13 to 9-18. Particle beams from accelerators are often given in terms of beam current, usually in micro- or milliamperes. The beam current can, in turn, be converted into the number of particles per second and used to determine the production of various products, as illustrated in Examples 9-2 and 9-3.

Example 9-3: A 2 μA beam of 15 MeV deuterons with an area of 1 cm^2 is focused on an iron wall 0.2 cm thick. What activity of ^{56}Co will be induced in the wall if the beam strikes the wall for 10 min?

SOLUTION: The reaction for producing ^{56}Co is

$$^{56}_{26}\text{Fe} + ^{2}_{1}\text{H} \rightarrow [^{58}_{27}\text{Co}^*] \rightarrow ^{56}_{27}\text{Co} + 2^{1}_{0}\text{n} + Q$$

and the number of ^{56}Fe atoms (91.75 atom % of iron) in the beam is

$$\text{atoms}(^{56}\text{Fe}) = \frac{0.2 \text{ cm}^3 \times 7.9 \text{ g/cm}^3 \times 6.022 \times 10^{23} \text{ atoms/mol} \times 0.9175}{55.845 \text{ g/mol}}$$
$$= 1.563 \times 10^{22} \text{ atoms of } ^{56}\text{Fe}$$

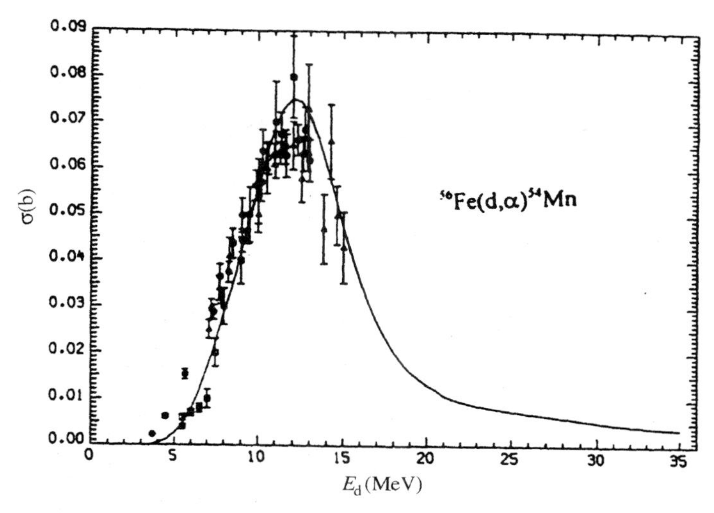

FIGURE 9-14. Cross section versus deuteron energy for (d, α) production of ^{54}Mn from ^{56}Fe. (From Xiaoping et al., 1999.)

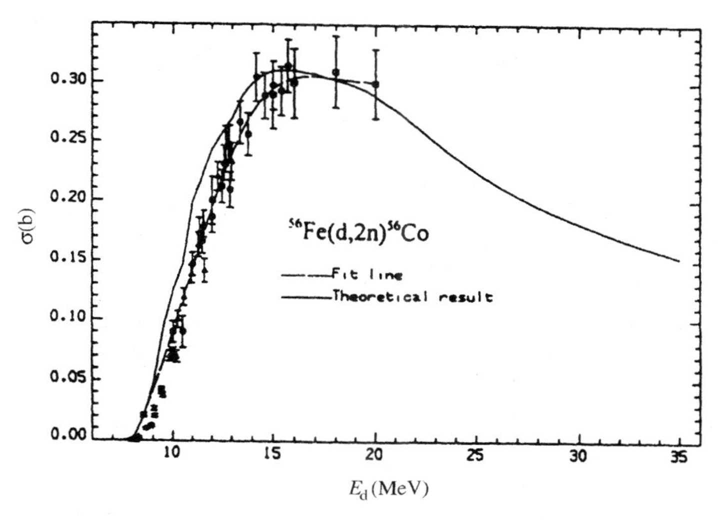

FIGURE 9-15. Cross section versus deuteron energy for (d, 2n) production of ^{56}Co from ^{56}Fe. (From Xiaoping et al., 1999.)

The number of deuterons striking the 1-cm^2 area of the wall, or the flux ϕ, is

$$\frac{2 \times 10^{-6}\ \text{A} \times 1\ \text{C/s} \cdot \text{A}}{1.6022 \times 10^{-19}\ \text{C/d}} = 1.25 \times 10^{13}\ \text{deuterons/cm}^2 \cdot \text{s}$$

From Figure 9-15, the ^{56}Fe(d, 2n) cross section is about 0.3 b; thus the production rate of atoms of ^{56}Co, ignoring radioactive transformation of the product,

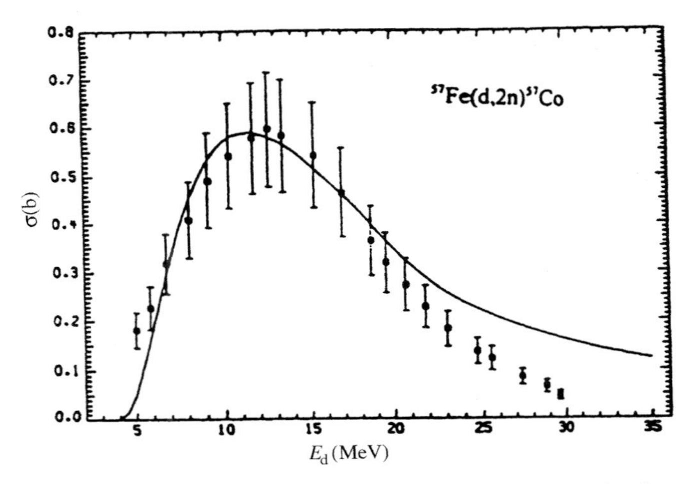

FIGURE 9-16. Cross section versus deuteron energy for (d, 2n) production of ^{57}Co from ^{57}Fe. (From Xiaoping et al., 1999.)

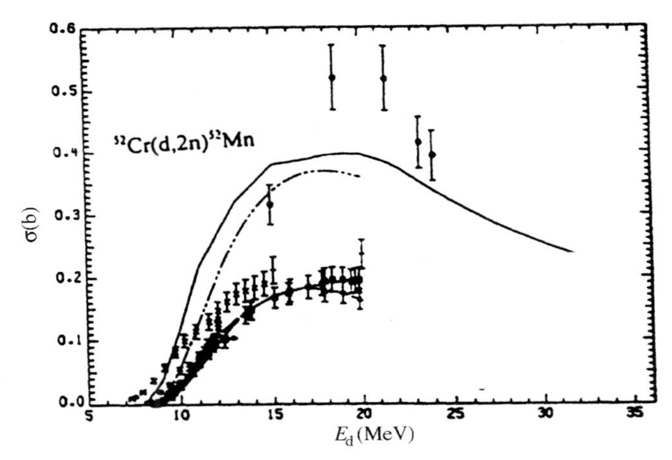

FIGURE 9-17. Cross section versus deuteron energy for (d, 2n) production of ^{52}Mn from ^{52}Cr. (From Xiaoping et al., 1999.)

is

$$\begin{aligned} N_i &= \phi\sigma N \\ &= 1.25 \times 10^{13} \text{ deuterons/cm}^2 \cdot \text{s} \times 0.3 \times 10^{-24} \text{ cm}^2/\text{atom} \times 1.563 \times 10^{22} \text{ atoms} \\ &= 5.86 \times 10^{10} \text{ atoms/s} \end{aligned}$$

or for a 10-min (600 s) irradiation time,

$$N_i = 3.52 \times 10^{13} \text{ atoms of } ^{56}\text{Co}$$

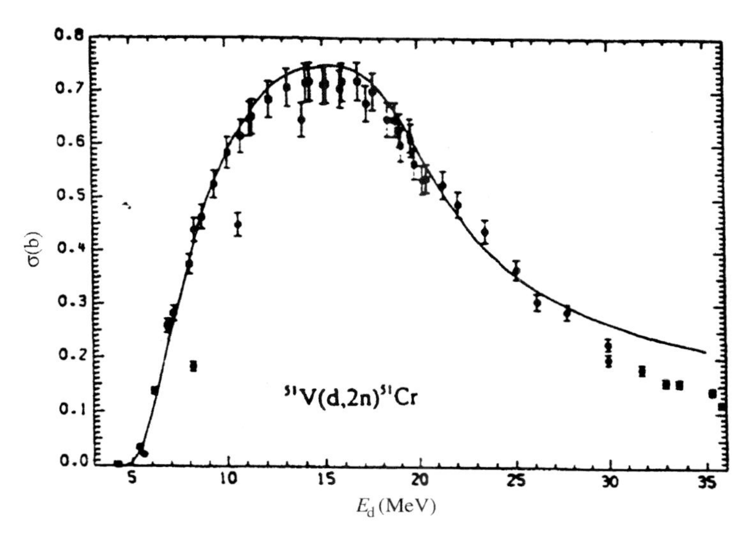

FIGURE 9-18. Cross section versus deuteron energy for (d, 2n) production of ^{51}Cr from ^{51}V. (From Xiaoping et al., 1999.)

Since the half-life of ^{56}Co is 77.3 d, the activity induced in the steel wall is

$$A_i = \lambda N_i = \frac{\ln 2}{T_{1/2}} N_i = 3.65 \times 10^6 \text{ t/s}$$

ACTIVATION BY PHOTONS

High-energy photons can be used to activate various materials, usually by (γ, n), (γ, p), and $(\gamma, 2n)$ reactions. These are threshold reactions with high Q-values, and high-energy photons (well in excess of 8 MeV) are required to induce them. Production of these high-energy photons requires high-energy acceleration of electrons to produce bremsstrahlung, which can be problematical since high energies are required and photon yields are low. Such sources, known as *synchrotron light sources*, are quite expensive, which limits their use for photonuclear activation.

Brey and collegues at Idaho State University (Brey et al., 1998) have used photonuclear activation for the analysis of ^{129}I ($T_{1/2} = 1.57 \times 10^7$ y) by a (γ, n) reaction:

$$^{129}_{53}\text{I} + ^{0}_{0}\gamma \rightarrow [^{129}_{53}\text{I}^*] \rightarrow ^{128}_{53}\text{I} + ^{1}_{0}\text{n} + Q$$

The ^{128}I product has a 25-min half-life, which allows use of a much smaller number, relatively speaking, of ^{129}I atoms for measurement than would be required for radioactivity measurement of the much longer lived ^{129}I atoms. A beam of 30 μA on a thick tungsten target produces about 10^{13} $\gamma/\text{cm}^2 \cdot \text{s}$, yielding sufficient activity of ^{128}I for measurement after a 2-h irradiation.

Photonuclear cross sections are quite small, usually in the millibarn range. As shown in Figure 9-19 for a target of ^{129}I, the (γ, n) cross section for most materials

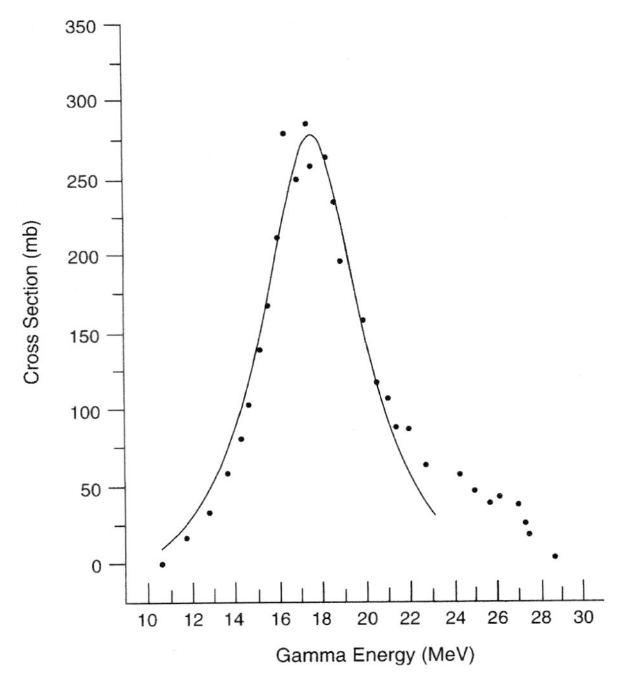

FIGURE 9-19. Photonuclear cross section versus energy for (γ,n) reaction in ^{129}I, adjusted from the well-established photonuclear cross-section spectrum for ^{127}I. (Adapted from Brey et al., 1998, minus error bars, with permission.)

is zero below a threshold energy but then rises smoothly through a peak, known as the *giant resonance*, and then decreases. For ^{129}I, the cross section peaks at about 280 mb for 18 MeV photons.

The calculated number of product atoms, N_p, is, as for other types of activation products, directly determined by the photon flux ϕ_γ, the photonuclear cross section σ_γ at the photon energy, the number of target atoms per cm^2, N, and the irradiation time, or

$$\text{product atoms} = \frac{\phi_\gamma \sigma_\gamma N}{\lambda_p}(1 - e^{-\lambda_p t})$$

where λ_p is the disintegration constant of the radioactive product.

Example 9-4: A synchrotron accelerator is used as a light source to produce 18 MeV bremsstrahlung peaked in the forward direction at a flux of 10^{13} γ/cm$^2\cdot$s. If a sample that contains 0.1 Bq of ^{129}I is irradiated uniformly in the beam for 2 h, what activity of ^{128}I will exist upon removal?

SOLUTION: The number of atoms of ^{129}I in the sample is

$$0.1 \text{ t/s} = \lambda N$$

Therefore, the number of ^{129}I target atoms is

$$N = \frac{0.1 \text{ t/s}}{\ln 2} \times 1.57 \times 10^7 \text{ y} \times 3.15576 \times 10^7 \text{ s/y}$$
$$= 7.148 \times 10^{13} \text{ atoms}$$

From Figure 9-19, σ_γ at 18 MeV is 280 mb (0.28 b), and the number of product atoms after 2 h is

$$N = \frac{7.148 \times 10^{13} \times 0.28 \text{ b} \times 10^{-24} \text{ cm}^2/\text{atom} \times 10^{13} \ \gamma/\text{cm}^2 \cdot \text{s}}{4.62 \times 10^{-4} \text{ s}^{-1}} (1 - e^{-(\ln 2/25)(120 \text{ m})})$$
$$= 4.176 \times 10^5 \text{ atoms}$$

and the activity of ^{128}I is

$$A = \lambda N$$
$$= \frac{\ln 2}{25 \text{ m} \times 60 \text{ s/m}} 4.176 \times 10^5 \text{ atoms}$$
$$= 193 \text{ t/s}$$

which can readily be measured by detection of the 0.443 MeV gamma rays emitted by ^{128}I.

SUMMARY

Radioactive activation products are of concern to radiation protection, and the particular features of their production are important in characterizing the source of these products and assessing their control. Activation products are produced when various particles are absorbed by target nuclei to form a compound nucleus which then releases energy and sometimes other particles. Such products can be produced intentionally by placing the appropriate target material in a flux of bombarding particles, or such production can occur unintentionally as the by-product of operating a source that produces emitted particles (e.g., neutrons). For example, the structural materials in a nuclear reactor contain various metals that may be activated by neutrons to produce radioactive products.

Regardless of the type of nuclear activation being considered, the cross section of the target nuclei, given in units of barns (or 10^{-24} cm^2/atom) determines the production of new material. The cross section denotes the target size or apparent area that an atom of a particular element presents to a bombarding particle; it is not specifically an area, but is better described as a measure of the probability that a reaction will take place. Cross sections are highly dependent on the type and energy of the particles being used and the atoms of the target material. Neutron activation cross sections, which are perhaps of most importance, are listed in the chart of the nuclides for most isotopes of stable elements and for many radioactive isotopes which may be transformed to other nuclei by neutron irradiation. The chart also lists the activation cross section for the first resonance interval above thermal energy, but since the resonance energy is not given, other sources need to be consulted to determine it.

Target irradiations also occur by charged particles. Protons and deuterons can be accelerated in cyclotrons and linear accelerators by stripping electrons from hydrogen or deuterium gas and introducing the ionized atoms into the source region of the accelerator and ionized tritium (tritons) and helium atoms can also be used as projectiles in accelerators. Cross sections for charged particles are typically much lower than those for neutrons and are often in the millibarn range. Cross-section data for charged particles are also much less available; however, some are available from the National Nuclear Data Center and the literature, and these are summarized herein.

REFERENCES AND ADDITIONAL RESOURCES

Brey, R., F. Harmon, D. Wells, and A. Tonchev, *The Possibility of Photon Activation Analysis of Radionuclides at Environmental-Levels*, Department of Physics, Idaho State University, Pocatello, ID, 1998.

GE, Chart of the nuclides, *Nuclides and Isotopes*, 15th ed., General Electric Company, San Jose, CA, 1996.

National Nuclear Data Center, Brookhaven National Laboratory, Upton, NY. Data resources available through the Internet at *www.nndc.bnl.gov*.

Xiaoping, X., H. Yinlu, and Z. Youxiang, Evaluation and calculation of nuclear data for deuteron-induced reaction on ^{51}V, ^{52}Cr, ^{56}Fe, and ^{57}Fe, *Health Physics* **76** (1), January 1999.

PROBLEMS

9-1. Calculate the maximum activity (d/s), assuming no source depletion, that could be induced in a copper foil of 100 mg exposed to a thermal neutron flux of 10^{12} neutrons/cm$^2 \cdot$ s. (*Note*: Natural copper consists of 69.1% ^{63}Cu and 30.9% ^{65}Cu.)

9-2. For the conditions of Problem 9-1, determine the activity (Ci) that would exist for ^{64}Cu 26 h after irradiation.

9-3. Derive the equation for activation of a product accounting for source depletion (ignore activation of the product nucleus).

9-4. A 20-g sample of cobalt is irradiated in a power reactor with a flux of 10^{14} neutrons/cm$^2 \cdot$ s for 6 y. Calculate (**a**) the activity of ^{60m}Co immediately upon removal from the reactor and (**b**) the activity of ^{60}Co 50 h after removal. (Assume no source depletion for both solutions.)

9-5. Repeat part (b) of Problem 9-4, assuming source depletion. Explain the change.

9-6. A sample of dirt (or CRUD) is irradiated for 4 h in a beam tube of 10^{12} neutrons/cm$^2 \cdot$ s. By counting the ^{56}Mn product, it is determined that the disintegration rate is 10,000 d/m. How much manganese is in the sample?

9-7. Measurement of ^{129}I can be done by neutron activation. Describe the nuclear reaction for the technique and recommend an irradiation time to optimize the sensitivity of the procedure using a constant thermal neutron flux.

9-8. Carbon-14 is produced in a swimming pool reactor by an (n, p) reaction with nitrogen that is dissolved in the pool water. If the water in and surrounding the reactor core contains 2 g of nitrogen and the neutron flux is 10^{13} neutrons/cm$^2 \cdot$ s, how much ^{14}C is produced each year?

9-9. Ten grams of HgO is irradiated uniformly in a research reactor at a flux of 10^{12} neutrons/cm$^2 \cdot$ s for 10 d to produce ^{203}Hg and ^{197}Hg. How much activity of each will be in the sample upon removal from the reactor?

9-10. A target of iridium metal is irradiated for 1 y to produce ^{192}Ir; however, the product also has a relatively important activation cross section. Determine (**a**) an equation for the product activity that accounts for activation of the product and (**b**) the activity per milligram that will be present upon removal from a neutron flux of 10^{14} neutrons/cm$^2 \cdot$ s (ignore the production of ^{192m}Ir).

9-11. Ten grams of thallium is irradiated uniformly in a research reactor at a flux of 10^{12} neutrons/cm$^2 \cdot$ s for 5 y to produce ^{204}Tl. How much activity will be in the sample upon removal from the reactor?

9-12. A cyclotron is used to produce 16 MeV protons that are directed at a transparent capsule containing 10 g of nitrogen. If the target is irradiated uniformly with a flux of 10^{10} protons/cm$^2 \cdot$ s for 50 min, what is the activity of ^{11}C produced?

9-13. A 1-g sample of common salt, NaCl, is irradiated in a reactor for 30 d at 10^{14} neutrons/cm$^2 \cdot$ s. Determine (**a**) the total activity (in curies) upon removal from the reactor, (**b**) the total activity after 1 d, and (**c**) The activity of each species after 1 d.

10

NUCLEAR FISSION AND ITS PRODUCTS

The Italian navigator has landed in the new world, the natives are very friendly.
—December 2, 1942

Many radiation protection considerations deal with the products of fission. A number of materials undergo fission, but the most important ones are ^{235}U, ^{238}U, and ^{239}Pu. Fission occurs when a fissionable atom absorbs a neutron as shown schematically in Figure 10-1.

The liquid-drop model developed by Bohr and Wheeler can be used to explain fission. Absorption of a neutron by a heavy nucleus such as uranium adds excitation energy that causes it to oscillate, producing an elongation of the nucleus much like that observed in a drop of liquid. If, as shown in Figure 10-1, the elongation causes the asymmetrical "halves" of the dumbbell-shaped drop to separate by a distance r,

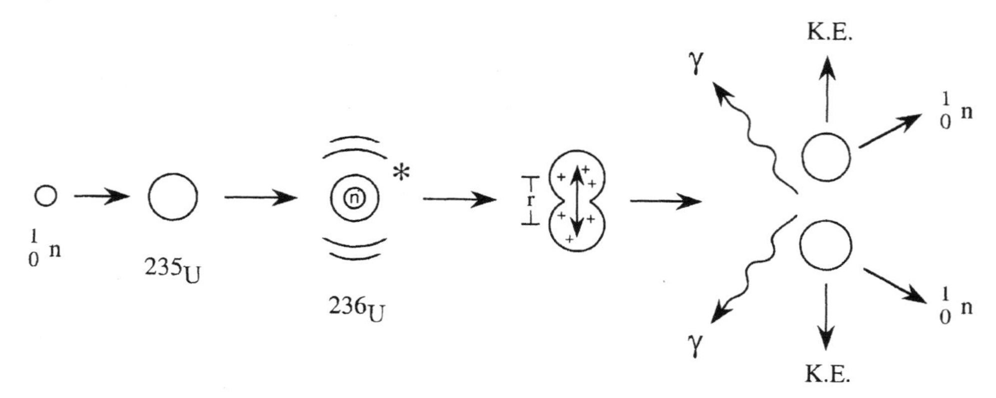

FIGURE 10-1. An absorbed neutron in uranium produces an elongated nucleus due to added excitation energy. If the centers of the "halves" reach a separation distance r, fission is likely to occur, which it does 85% of the time in ^{235}U; if not, the nucleus returns to its original shape and the excitation energy is emitted as a gamma photon, which occurs 15% of the time for ^{235}U to yield an atom of ^{236}U.

repulsive forces exerted by the positive charges in each will cause the two parts to separate, or fission.

The energy associated with the critical distance r represents a barrier energy E_b that must be overcome for separation to occur. For a neutron added to ^{235}U and ^{239}Pu, E_b is 5.2 and 4.8 MeV, respectively; for ^{238}U it is 5.7 MeV. Absorption of a thermal neutron in ^{235}U adds 6.55 MeV of excitation energy (see Example 10-1); in ^{239}Pu, 6.4 MeV; and in ^{238}U, only 4.8 MeV. Consequently, ^{235}U and ^{239}Pu fission readily when they absorb thermal neutrons. On the other hand, ^{238}U can fission only if the incoming neutron has at least 1.4 MeV of kinetic energy. The amount of excitation energy added by absorbing a neutron is just the Q-value of the reaction, as shown in Example 10-1.

Example 10-1: Calculate the amount of excitation energy added to a nucleus of ^{235}U by a thermal neutron.

SOLUTION: The excitation energy is just the Q-value of the reaction, or

$$\begin{aligned} Q &= 235.043924 \text{ u} + 1.008665 \text{ u} - 236.045563 \text{ u} \\ &= 0.007026 \text{ u}(931.5 \text{ MeV/u}) = 6.55 \text{ MeV} \end{aligned}$$

As shown in Example 10-1, adding a neutron to ^{235}U yields an excited nucleus of ^{236}U, which is 6.55 MeV above the ground state of the ^{236}U nucleus. This excitation energy must be relieved; it can occur by emission of a gamma photon, or since the added energy is more than the barrier energy (5.2 MeV) by fission. These are competing processes, but fission is much more likely and occurs 85% of the time. The fissioning nucleus yields energy, high-speed neutrons, prompt gamma rays, and highly charged fission fragments, or fission products, as they are more commonly called. All of these components are released almost instantaneously at the moment of fission.

Uranium, plutonium, and other heavy nuclei are very neutron rich because extra neutrons help to distribute the nuclear force and counteract the repulsive forces exerted by the large number of protons in the nucleus. This excess ratio of neutrons to protons carries over to the fission products, hence each product that is formed will be extremely neutron rich and will appear quite far from the line of stability. Consequently, these products are very unstable and several beta particles must be emitted to lower the neutron number (and increase the proton number) so that the fission product nucleus can become stable. Several gamma rays are also emitted during the series of beta particle transformations. Two typical mass chains of the series of radionuclide transformations that occur to yield a stable end product are shown in Figure 10-2. The stable end product eventually formed from each series of transformations will have the same mass number as the product initially formed.

FISSION ENERGY

The reaction for fissioning ^{235}U into two products plus two neutrons and energy is

$$^{235}_{92}\text{U} + ^{1}_{0}n \hookrightarrow [^{236}_{92}\text{U}^*] \rightarrow \text{X} + \text{Y} + 2^{1}_{0}\text{n} + \gamma\text{'s} + Q$$

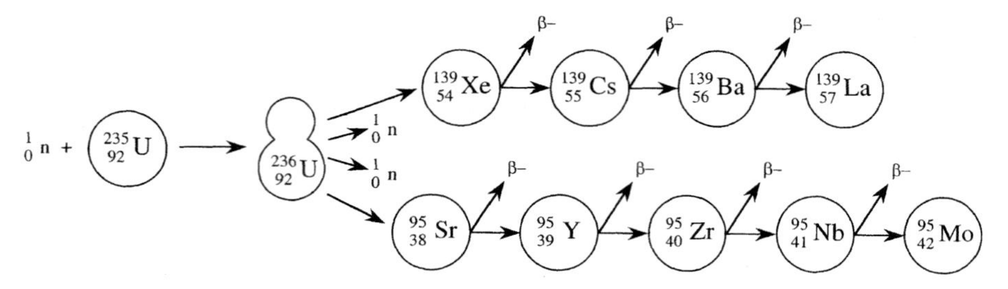

FIGURE 10-2. Uranium-235 fission and the series of radioactive transformations by beta emission of the fission products to stable end products when mass numbers 95 and 139 are produced.

TABLE 10-1. Masses Before and After Fission of a ^{235}U Atom by a Thermal Neutron

Masses Before Fission		Masses After Fission	
^{235}U	235.043923 u	^{95}Mo	94.905841 u
1 n	1.008665 u	^{139}La	138.906348 u
Total	236.052588 u	2 n	2.017330 u
		7 β^-	0.003840 u
		Total	235.837331 u

where X and Y are the fission products of the reaction. The ^{235}U nucleus can split in a great many ways, so it is impossible to say in any given fission reaction just exactly what the fission products will be. One typical split, as shown in Figure 10-2, is one in which the stable end products are ^{95}Mo and ^{139}La. This reaction can be written as

$$^{235}_{92}\text{U} + ^{1}_{0}\text{n} \rightarrow [^{236}_{92}\text{U}^*] \rightarrow ^{95}_{42}\text{Mo} + ^{139}_{57}\text{La} + 2^{1}_{0}\text{n} + 7^{0}_{-1}\beta^- + Q$$

and the energy liberated in the reaction is simply the Q-value. As shown in Table 10-1 the mass difference is 0.21526 u, and the energy released is

$$E_{\text{rel}} = 0.21526 \text{ u} \times 931.5 \text{ MeV/u} = 200.5 \text{ MeV}$$

Although this calculation was made for one particular fission of an atom of ^{235}U, it may be regarded as typical. Each fission produces its own distinct mass chain with a slightly different Q-value; however, 200 MeV of energy is released on average for each atom of ^{235}U fissioned. Most (85% or so) of this energy (see Table 10-2) appears as kinetic energy of the fission fragments, and this immediately manifests itself as heat as these heavy, highly charged fission fragments are absorbed in the fuel matrix. (The fission product nuclei are ejected with so much energy that they tend to leave their orbital electrons behind; therefore, they are highly charged particles with about 20 to 22 units of positive charge, which causes them to be absorbed in a few micrometers, producing considerable amounts of heat in the fuel.) Part of the remaining energy is released instantaneously by gamma rays and fission neutrons;

TABLE 10-2. Emitted and Recoverable Energies for Fission of ^{235}U

Form	Emitted Energy (MeV)	Recoverable Energy (MeV)
Fission fragments	168	168
Fission neutrons	5	5
Fission product decay		
Beta particles	8	8
Gamma rays	7	7
Neutrinos	12	—
Prompt gamma rays	7	7
Capture gamma rays in ^{235}U and ^{238}U	—	3–12[a]
Total	207	198–207

[a]Not associated with fission.

the rest is released gradually by radioactive transformation of the fission products as they emit beta particles, neutrinos, and gamma rays. The energy values of each of these processes are summarized in Table 10-2.

Prompt gamma rays are emitted from the fissioning nucleus at the instant of fission to relieve some of the excess energy in the highly excited compound nucleus. Their number and energy vary, but the average energy is about 7 MeV. They are referred to as prompt gamma rays to distinguish them from the delayed gamma rays emitted by the radioactive fission products, which are usually emitted with energies that are quite a bit less.

Capture gamma rays are emitted when neutrons are absorbed in ^{235}U, ^{239}Pu, and ^{238}U without producing fission (about 15% of the time in ^{235}U). Even though the capture gamma rays are due to (n,γ) activation (e.g., ^{235}U to ^{236}U) and not fission, they contribute to the energy that is released and recovered in fission reactors.

Table 10-2 indicates that not all of the energy released in fission can be recovered as heat. The energies of the fission products, the fission product beta particles and gamma rays, the prompt and delayed neutrons, the capture gamma rays, and the prompt gamma rays are usually absorbed and recovered, especially in nuclear reactors. The neutrinos emitted in beta transformations of the radioactive fission products constitute about 12 MeV per fission; however, these escape completely and their energy is lost irrevocably.

It can be shown that complete fission of 1 g of ^{235}U yields about 1 MW of thermal energy, which probably created an irresistible temptation for pioneering physicists to produce and created an abundant source of energy. The modern world has chosen to do this by building nuclear reactors to release fission energy at a controlled rate, and also in nuclear weapons that seek to have large instantaneous releases of fission energy at an uncontrolled but predictable rate.

The release of fission energy is governed by a fundamental set of physical principles that eventually dictate the design and performance of nuclear reactors and other critical assemblies. Radiation protection for these circumstances must also recognize these same principles in the various applications of nuclear fission, applications that range from sustained and controlled chain reactions in nuclear reactors for nuclear research and electricity production to nuclear criticalities for various purposes, including nuclear weapons. Radioactive fission products and activation products are by-products of these reactions, and these represent a number of radiation protection issues for workers, the public, and the environment; nuclear power reactors are particularly challenging because they are designed to operate for 30 to 40 years.

Fission energy, determined by the fission rate and reactor thermal power in watts, is an important parameter for calculating fission product inventories in reactor fuel. Since 1 W is 1 J/s, and each fission yields about 200 MeV, 1 W of reactor power corresponds to a fission rate of

$$\text{fission rate} = \frac{1\ \text{J/s} \cdot \text{W}}{(1.6022 \times 10^{-13}\ \text{J/MeV})(200\ \text{MeV/fission})}$$
$$= 3.12 \times 10^{10}\ \text{fissions/s per watt}$$

and for 1 MW of thermal energy,

$$1\ \text{MW} = 3.12 \times 10^{16}\ \text{fissions/s}$$

A common unit of energy production is the megawatt · day, which is directly related to the number of fissions that occurred; therefore,

$$1\ \text{MW} \cdot \text{d} = 3.12 \times 10^{16}\ \text{fissions/s} \cdot \text{MW} \times 86{,}400\ \text{s/d} = 2.696 \times 10^{21}\ \text{fissions}$$

PHYSICS OF SUSTAINED NUCLEAR FISSION

Release of the large quantities of energy available from fissioning uranium (or other fissile materials) depends on sustaining fission reactions once they are begun. This requires careful management of the neutrons released when fission occurs. Fission of ^{235}U yields, on average, 2.44 neutrons per fission, which can be induced to produce other fissions if certain conditions exist. Achieving and sustaining a controlled chain reaction is governed by the data shown in Figures 10-3 to 10-5 and Table 10-3.

First, as shown in Figure 10-3, most of the neutrons produced in fission have energies of 1 to 2 MeV, ranging up to about 5 MeV. Therefore, most of the neutrons available for fissioning other ^{235}U atoms are fast neutrons, and these have a fairly low fission cross section (about 1 b). Modern fission reactors incorporate features to slow them down to take advantage of the very high fission cross section for thermal neutrons in ^{235}U (584 b) as shown in Figure 10-4. Figure 10-4 also shows that fission in ^{238}U, the most abundant isotope of natural uranium at 99.3%, occurs *only* with neutrons above 1.4 MeV and is effectively a threshold phenomenon with a very low fission cross section of just 0.55 b. As shown in Table 10-3, a fast neutron striking ^{238}U is more likely to be scattered than absorbed because of significantly higher elastic and inelastic scattering cross sections (4.55 and 2.1 b versus 0.55 b for fission); thus a neutron born fast with enough energy (> 1.4 MeV) to induce fission in ^{238}U is much more likely to be scattered than absorbed to produce fission, and only one such interaction will reduce its energy below the fission threshold. Also, once a neutron in uranium is slowed down below the fission threshold of 1.4 MeV, it is very likely to be absorbed at an intermediate energy because of the

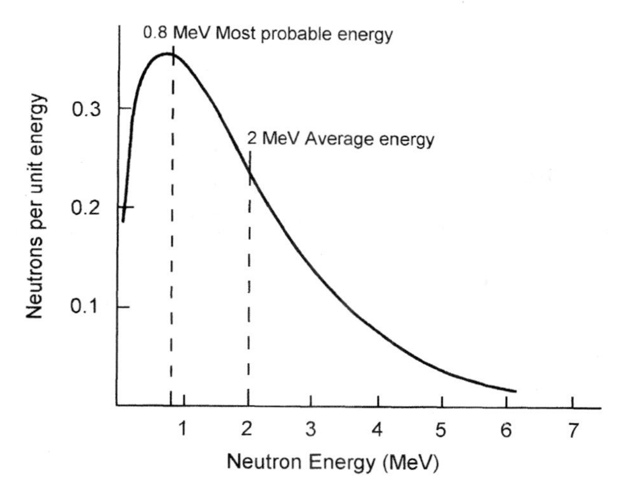

FIGURE 10-3. Spectrum of neutron energies released in uranium fission.

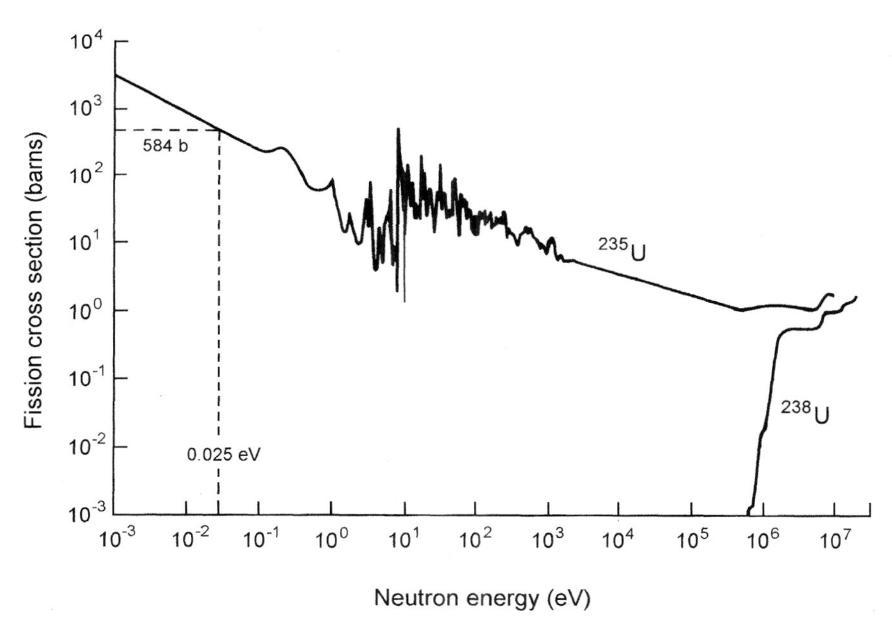

FIGURE 10-4. Neutron fission cross sections for ^{235}U and ^{238}U showing fast fission effect of ^{238}U and thermal fission effect for ^{235}U.

prominent and very high resonance absorption cross sections in ^{238}U (see Figure 10-5) and lost before it can induce fission in the less abundant (0.72%) but highly fissionable ($\sigma_f = 584$ b) ^{235}U. Because of these factors, a nuclear chain reaction cannot occur in even a large block of natural uranium. There are enough ^{235}U atoms in natural uranium to sustain a chain reaction if the neutrons can be slowed to thermal energies without losing them by absorption in ^{238}U. To have a sustained nuclear chain reaction in natural uranium, it is essential to get the neutrons outside

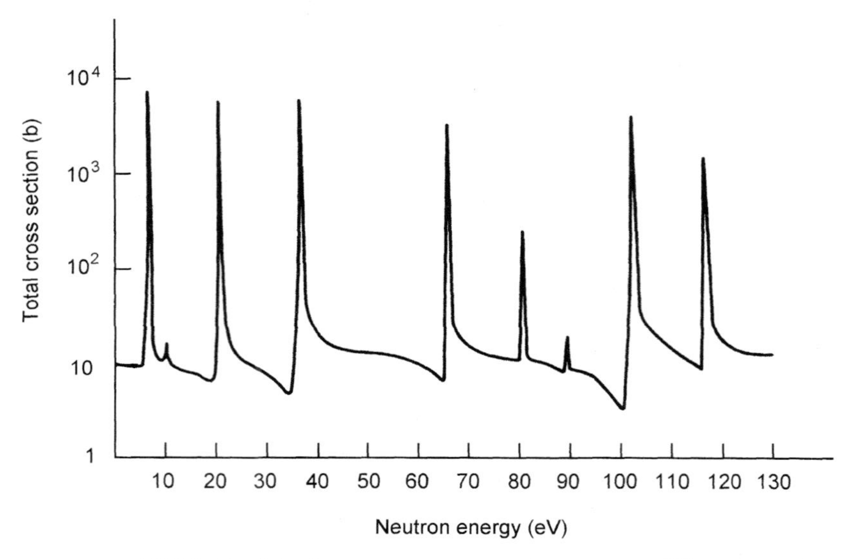

FIGURE 10-5. Resonance capture of neutrons in ^{238}U.

TABLE 10-3. Neutron Cross Sections in ^{235}U and ^{238}U

Cross Section	^{236}U (b)	^{238}U (b)
σ_f		
Thermal	584	0
Fast	~1.1	~0.55
σ_γ, Thermal	107	0.05
σ_s		
Elastic	9.0	4.55
Inelastic		2.10

the uranium matrix so they can be slowed down in a good moderator without losing them due to resonance capture in ^{238}U.

Neutrons of any energy (see Figure 10-4) can cause fission in ^{235}U and resonance is not a problem because it simply enhances σ_f since it is larger than σ_a at resonance energies (this is also true for fissionable ^{239}Pu, which is produced by neutron absorption in ^{238}U). If neutrons are moderated to thermal energies with minimal losses, the high probability of fission in ^{235}U makes it possible to use natural uranium even though the amount of ^{235}U is small. This is accomplished in a very clever way by surrounding small clusters of natural uranium with a low-Z moderator to slow neutrons down outside the uranium where resonance capture cannot occur. Fermi did so by placing lumps of uranium in a graphite pile, and the CANDU (Canadian deuterium–uranium) reactor accomplishes the same effect with small-diameter uranium rods surrounded by a deuterium oxide (heavy water) moderator. Ordinary water captures too many neutrons for it to be used as a moderator in a reactor fueled with natural uranium even though H_2O is very effective

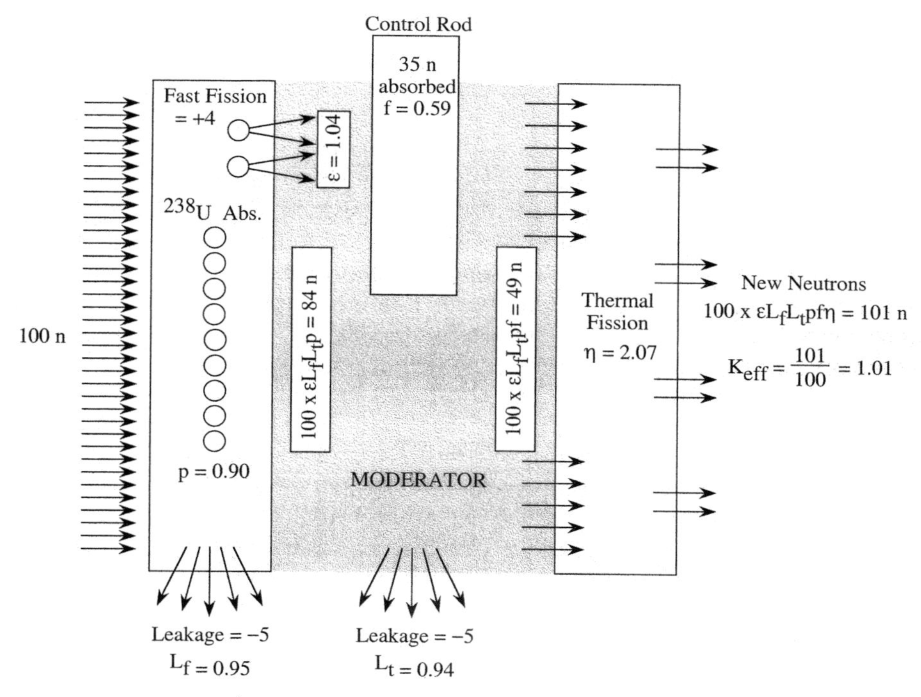

FIGURE 10-6. Fate of 100 neutrons born fast in one fuel element (on the left) and average effects of absorption, leakage, and control rods in assuring that the right number of neutrons (49) are thermalized in the moderator before they enter the next fuel element (at the right) to fission enough atoms to yield a new generation of 101 neutrons, thus sustaining the chain reaction.

in slowing neutrons to thermal energies. If H_2O is used, it is necessary to enrich the amount of ^{235}U in uranium, but even with enrichment, it is still necessary to surround the uranium fuel with a moderator to slow neutrons to thermal energies.

The physical design of a nuclear reactor is governed by those factors that influence interactions of neutrons with the fissionable material. Figure 10-6 illustrates the effects of the major scattering and absorption interactions on a population of 100 neutrons for uranium rods interspersed in a moderator material. Figure 10-6 shows the fate of 100 neutrons born fast in one fuel element (at the left) and the average effects of absorption, leakage, and control rods in determining that the right number of neutrons are thermalized in the moderator before they enter the next fuel element (on the right), where they yield a new generation of 101 neutrons, the number required to sustain the chain reaction (or, more precisely, 100.7).

It is first necessary to have a *critical mass* of fissionable material to sustain the fission chain reaction. This is the smallest mass in which fission will continue once started (i.e., where at least one neutron will be absorbed to produce fission from one generation to another). This mass is a function of the configuration involved and is dependent on the leakage fraction and the likelihood of capture without producing fission. The 100 neutrons born in a fuel element will be fast neutrons, and two or so will fission ^{238}U by the fast fission effect (ε) to increase the number to about

104; however, about 5 of these will be lost due to leakage, characterized as the fast leakage effect ($L_f = 0.95$). A fast neutron that remains in the uranium fuel matrix is much more likely to undergo a scattering interaction with ^{238}U than fission it since the fission cross section is lower. A scattering reaction will reduce its energy below the fission threshold where resonance capture is very likely. Elastic scattering (4.55 b) has little effect on fast neutrons; however, inelastic scattering ($\sigma = 2.1$ b) reduces the neutron energy below the 1.4 MeV threshold in just one collision. The probability of a neutron reaching thermal energy without capture in the abundant ^{238}U atoms is referred to as the *resonance escape probability*, p (i.e., it escapes resonance capture). A typical value of p is 0.9. Although the neutron number is momentarily increased to 104 by the fast fission effect ε, the fast leakage effect L_f and the resonance escape probability p combine to reduce the number of neutrons that enter the moderator to about 89.

The 89 neutrons that enter the moderator are slowed down to thermal energies; in doing so, another 5 or so are lost due to the thermal leakage effect (L_t), and the population of neutrons is reduced to about 84. At this point it is necessary to look ahead. The chain reaction will continue if > 100 neutrons are available after the next generation of fission; therefore, about 49 are needed to cause thermal fission in the ^{235}U in the adjacent fuel element. Control rods are used to reduce the number to 49 by soaking up about 35 of the available neutrons. About 85% of the remaining 49 neutrons can be relied on to cause fission, yielding on average 2.44 neutrons per fission with a thermal utilization factor, η, of 2.07 ($2.44 \times 0.85 = 2.07$); therefore, the 49 neutrons remaining will produce 101 new neutrons after they fission atoms of ^{235}U in the adjacent fuel rod. This number is the result of multiplying 100 initial neutrons by the six factors just considered, or

$$(\varepsilon L_f L_t p f \eta) \times (100 \text{ neutrons}) = 101 \text{ neutrons}$$

The multiple $\varepsilon L_f L_t p f \eta$ is commonly referred to as the *six-factor formula*. If systems are large and leakage of both fast and thermal neutrons is minimal, the factors L_f and L_t are essentially 1.0, and the number of neutrons is given by $\varepsilon p f \eta$, commonly called the *four-factor formula*:

$$(\varepsilon p f \eta) \times (100 \text{ neutrons}) = 101 \text{ neutrons}$$

These factors can combine to sustain a chain reaction in natural uranium only if a moderator is used that doesn't absorb neutrons; if light water is used both as moderator and coolant, it is necessary to enrich the number of ^{235}U atoms in natural uranium to overcome absorption losses.

NEUTRON ECONOMY AND REACTIVITY

The various factors illustrated in Figure 10-6 are multiplicative. The six-factor formula accounts for both fast and thermal neutron leakage; the four-factor formula is appropriate when neutron leakage can be considered minimal. In either case, the value of the multiplication formula yields the number of neutrons in a given generation, and the change in the number of neutrons from one generation to the next is defined as the effective reactivity, k_{eff},

$$k_{eff} = \frac{\text{neutrons in next generation}}{\text{neutrons in first generation}}$$

The chain reaction will be sustained as long as k_{eff} is greater than 1.00, which in turn is determined by the combined effect of the net positive or negative amount of reactivity provided by each parameter.

More than 99% of the neutrons in a reactor are released at the instant of fission. They are called *prompt neutrons*. Others are emitted by radioactive fission fragments as *delayed neutrons*, with half-lives ranging between 0.2 s and 56 s. They account for 0.65% of the total for ^{235}U (but only 0.21% for ^{239}Pu). Delayed neutrons in an otherwise critical reactor contribute reactivity of 0.0065, which is often rounded to 0.007. These are dominated by 55-s ^{87}Br and 22-s ^{127}I, both of which are fission products. Without these delayed neutrons, reactors would be very difficult to control because the necessary adjustments in reactor reactivity cannot be made in the lifetimes (about 10^{-3} s) of the prompt neutrons. A reactor with reactivity above 1.007 is said to be prompt critical; that is, it does not need to wait for the delayed neutrons to run; between 1.00 and 1.007 it is delayed critical and easily controlled because the delayed neutrons appear to sustain criticality at a much slower rate.

The amount by which k_{eff} exceeds 1.000 is termed the *excess reactivity*, Δk. If there are n neutrons in the core in one generation, there will then be $n\,\Delta k$ additional neutrons in the next generation. The growth of neutrons with time is

$$n(t) = n_0 e^{t/T}$$

where T is the reactor period. It is the time required to increase reactor power by e times (i.e., by a factor of 2.718). The reactor period is calculated as

$$T = \frac{L}{\Delta k}$$

where L is the average lifetime of the neutrons released in fission. For prompt neutrons, $L = 0.001$ s; for delayed neutrons it is about 0.082 s.

Example 10-2: If the excess reactivity in a reactor increases by 0.005, what would be the increase in reactor power in 1 s?

SOLUTION: The prompt neutrons would cause the reactor to have a period of $T = 0.001/0.005 = 0.2$ s, and in 1 s the power level (neutron flux) would increase by e^5, or a factor of 150!; thus control adjustments would need to be made within a fraction of a second, which is virtually impossible. The delayed neutrons, however, have an effective lifetime of 0.082 s, and for the same amount of excess reactivity (0.005) the reactor period would be $0.082/0.005 = 17$ s, which provides sufficient time to adjust control rods to the desired reactivity level. Clearly, the delayed neutrons are vital for controlling nuclear reactors.

When excess reactivity exceeds 0.0065, the reactor is said to be *prompt critical* and the chain reaction is sustained by the prompt neutrons regardless of any contribution by the delayed neutrons. This causes the reactor period to be very short (as shown in Example 10-2), a condition that must always be avoided. A prompt critical condition led to the explosion of the Chernobyl reactor and was brought about because it had a net positive reactivity coefficient (since redesigned and made negative by reducing the amount of graphite moderator).

Excess reactivity is commonly expressed in dollars and cents, where \$1 is exactly the amount of reactivity required to make the reactor prompt critical. (This requires a Δk value of 0.0065 for ^{235}U-fueled reactors, which is, of course, the delayed neutron fraction; for a ^{239}Pu-fueled reactor, \$1 of excess reactivity would correspond to a Δk value of 0.0021.) Thus if a control rod movement inserts \$0.10 of reactivity into a reactor running with $k_{\text{eff}} = 1.0000$, the reactor is one-tenth of the way to becoming prompt critical, which corresponds to $\Delta k = 0.0065/10$, or 0.00065, in a reactor fueled with ^{235}U.

NUCLEAR POWER REACTORS

Electricity is generated when a conductor is passed through a magnetic field, which is usually done by rotating it between the poles of a stationary magnet. The electromotive forces associated with cutting through the magnetic field lines cause the electrons in the conducting wire to flow as a current and to create a potential drop, or voltage. A rotational force is required to move the conductor (actually, multiple conductors in a generator) through the magnetic field. This rotational force can be provided by causing a high-pressure gas, usually steam, to expand against the blades of a turbine connected in turn to a shaft that rotates the conductors. In essence, electrical energy is just converted mechanical energy, which in turn is produced from heat when steam is used to induce the energy. These interconnected systems constitute the turbogenerator—it is a key component of all electric power plants, regardless of the energy source used to turn it.

A typical turbogenerator driven by superheated high-pressure steam is shown in Figure 10-7. A turbine may have high- and low-pressure stages, and the energy

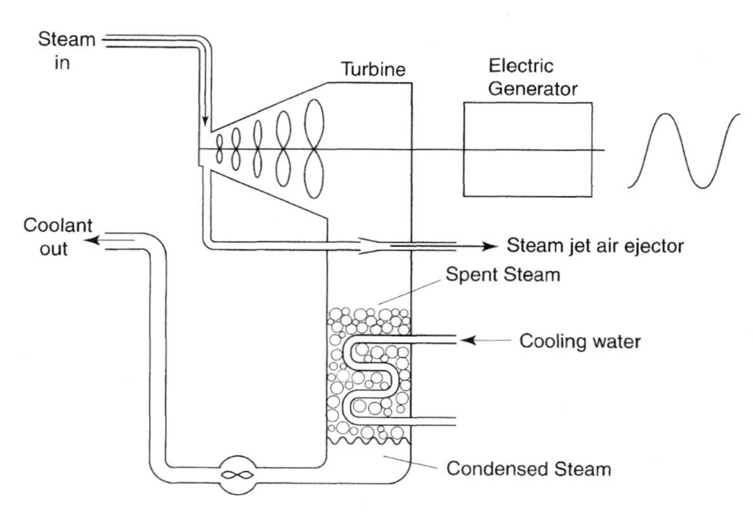

FIGURE 10-7. Schematic of a turbogenerator.

TABLE 10-4. General Parameters for Various Reactor Types

	BWR	PWR	CANDU	HTGR	LMFBR
Fuel	3–4% UO_2	3–4% UO_2	Natural U	HEU[a]	HEU
Moderator	H_2O	H_2O	D_2O	Graphite	None
Coolant	H_2O	H_2O	D_2O	He, CO_2	Na
Inlet T (°C)	260	290	266	405	~ 300
Outlet T (°C)	290	325	310	775	500
Pressure (psig)	1100	2200	1500	100–400	~ atm
Efficiency (%)	33	32–33	29	~ 40	~ 40

[a]Highly enriched uranium.

expended in turning it is directly proportional to the pressure drop across the turbine. To enhance this effect a vacuum is induced on the low-pressure side of the turbine such that the largest pressure drop possible occurs. The vacuum on the low-pressure side of the turbine is maintained by a steam-jet air ejector that entrains any gases present, including those that are radioactive, and exhausts them from the steam system.

Energy transfer is also a function of temperature drop; therefore, a condenser is provided to lower the exit temperature of the steam. A coolant loop through the condenser removes residual heat energy from the spent steam, thus condensing it back to water, which is then pumped back to the heat source.

The turbogenerator/air ejector/condenser system shown in Figure 10-7 is typical of modern power plants regardless of whether the heat energy is provided by fossil fuels or nuclear fuels. Nuclear reactors of various designs can be and are used to produce the heat energy supplied to the turbogenerator system. The major designs are light water–cooled pressurized water reactors (PWRs), boiling water reactors (BWRs), heavy water reactors, liquid metal fast breeder reactors (LMFBRs), and high-temperature gas-cooled reactors (HTGRs). Each design has particular features, listed in Table 10-4.

Reactor Design: Basic Systems

Nuclear reactors contain a number of basic components, shown in Figure 10-8. Fission energy is released as heat energy in the core of the reactor, which contains the fuel, the control rods, the moderator, and a coolant. This dynamic system in enclosed in a vessel with various components and subsystems to enhance fissioning in the fuel, to control its rate, to extract heat energy, and to provide protection of materials and persons.

Coolant is required to remove the heat produced by fission and to maintain the fuel well below its melting point. The heat extracted into the coolant is used to produce steam either directly or through a heat exchanger. Coolant is typically water but can be a gas, a liquid metal, or in rare cases another good heat transfer material.

Fuel for light water reactors is UO_2 fabricated into ceramic pellets, which have good stability and heat transfer characteristics. The pellets are loaded into zirconium alloy (zircalloy) tubes to make a fuel rod. A typical 1000-MWe plant contains about 50,000 such rods. The rods are thin, to promote heat transfer and to minimize effects

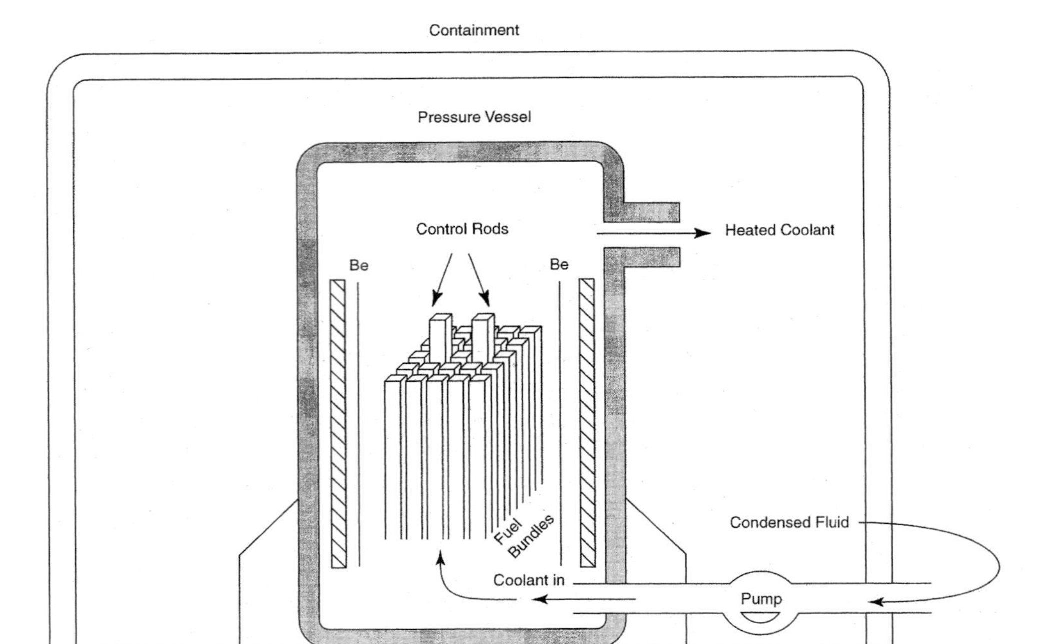

FIGURE 10-8. Schematic of major components of a nuclear reactor.

TABLE 10-5. Slowing Down Properties of Moderators

Moderator	Slowing Power (cm^{-1})	Moderating Ratio[a]
H_2O	1.28	58
D_2O	0.18	21,000
He (STP)	10^{-5}	45
Beryllium	0.16	130
Graphite	0.065	200

[a]Moderating ratio is a combination of the minimum number of collisions required to slow neutrons to thermal energies and their loss due to absorption.

on neutron flux. The large number of rods and the thin walls that each contains almost guarantees that some imperfections will exist through which volatile fission products can pass into the circulating coolant.

Moderators are usually heavy water, light water, or graphite, and are used to moderate neutrons to thermal energies. A water moderator can also serve to cool the fuel and extract the heat produced. If light water is used, the fuel is enriched to 3 to 4% to overcome neutron absorption by hydrogen in the water. The properties of typical moderators are shown in Table 10-5.

A *reflector* made of a light metal such as beryllium ($Z = 3$) is provided next to the core to reflect neutrons back into the core, where they can cause fission. The relatively light atoms in the reflector also help moderate the neutrons. Ordinary water is also a good reflector, and light water–cooled reactors consider the effect of this feature on the neutron population.

Control rods are movable pieces of cadmium or boron that are used to absorb and stabilize the neutron population. Withdrawing these rods increases the multiplication factor and hence the power level; insertion decreases it.

A *pressure vessel* encloses the entire fission reaction and must be strong enough to withstand the stresses of pressure and heat, a *thermal shield* is provided to absorb radiation and reduce embrittlement of the vessel, and a *biological shield* is placed outside the pressure vessel to reduce radiation exposure levels.

A *containment structure* encloses the entire reactor system to prevent potential releases of radioactivity to the environment, especially during incidents when fission products may escape. This may include a primary containment around the steam supply system and a secondary containment associated with the reactor building.

LIGHT WATER REACTORS

Pressurized Water Reactors

The pressurized water reactor (PWR) was one of the first reactors designed to produce power, originally for use in submarines. As indicated in Figure 10-9, water enters the pressure vessel at a temperature of about 290°C, flows down around the outside of the core, where it helps to reflect neutrons back into the core, passes upward through the core, where it is heated, and then exits from the vessel with a temperature of about 325°C. This primary coolant water is maintained at a high

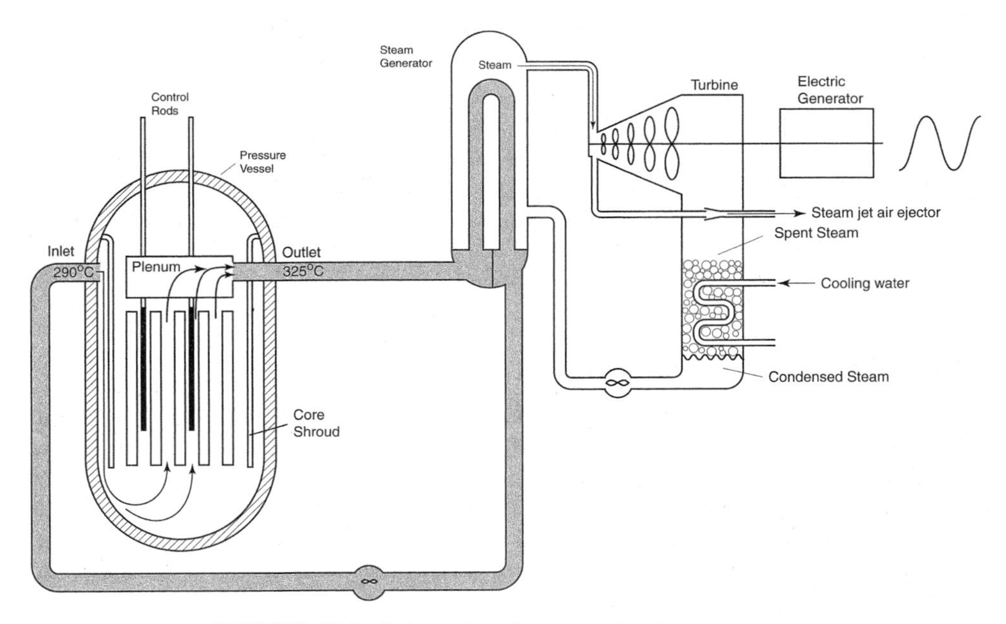

FIGURE 10-9. Schematic of a pressurized water reactor.

pressure (2200 psi) so that it will not boil. Any decrease in the coolant volume and the subsequent drop in pressure could vaporize some of the primary coolant, which in turn could lead to fuel damage. PWRs have four coolant pumps, one for each coolant loop, but only one pressurizer for the entire system.

Since the primary coolant is not allowed to boil, steam for the turbines must be produced in a secondary circuit containing steam generators (also shown in Figure 10-9). The heated coolant water from the reactor enters the steam generator at the bottom and passes upward and then downward through several thousand tubes, each in the shape of an inverted U. The outer surfaces of these tubes are in contact with lower-pressure and cooler feedwater, which extracts heat, boils, and produces steam that is routed to the turbine. The spent steam is condensed and pumped back to the steam generators to complete the secondary circuit.

Where the size of the reactor is an important consideration, as it is in a submarine, the fuel is enriched to over 90% in ^{235}U [often called highly enriched uranium (HEU)], which makes it possible to have a smaller core and pressure vessel. Highly enriched uranium is expensive, however, and for stationary power plants where size is less of an issue, the fuel is only slightly enriched (from 2 to 4%) uranium dioxide, UO_2, which is a black ceramic material with a high melting point of approximately 2800°C. The UO_2 is in the form of small cylindrical pellets, about 1 cm in diameter and 2 cm long, which are loaded into sealed tubes, usually made of zircalloy. Each tube is about 4 m long and designed to contain the fission products, especially fission product gases, that are released from the pellets during reactor operation. In this context, the fuel tubes are also known as the fuel *cladding*. Fuel of this type is nominally capable of delivering about 30,000 MW · d per metric ton of heavy metal (mostly uranium) before it is replaced, typically after about 3 y.

The fuel rods are loaded into fuel assemblies and kept apart by various spacers to prevent contact between them. Otherwise, they may overheat and cause fission products to be released. Some of the fuel pellets in the rods may contract slightly at operating temperatures and create void spaces within the fuel tubes. To prevent fuel ruptures because of the high pressure of the moderator-coolant, the fuel tubes are pressurized with helium. Fission product gases also build up pressure in the tubes, which can be as much as 14 MPa by the time they are replaced.

Control of the PWR is accomplished by the use of control rods, which normally enter the core from the top (see Figure 10-9), and by dissolving a neutron-absorbing chemical shim (usually boric acid) into the coolant water to reduce the multiplication factor. The chemical shim concentration is continually adjusted over the life of a given core to optimize the reactivity balance.

Boiling Water Reactors

The boiling water reactor (BWR) uses light water for both coolant and moderator. The water boils within the reactor, which produces steam that is routed directly to the turbines, as shown in Figure 10-10. This direct cycle design can be a major advantage because it eliminates the need for a separate heat transfer loop. Since the BWR is a direct cycle plant, more heat is absorbed to produce steam than in a PWR, in which the heat is extracted as sensible heat, which only changes the temperature of the fluid; therefore, less water is required to extract heat from a BWR for a given power output. On the other hand, the water becomes radioactive

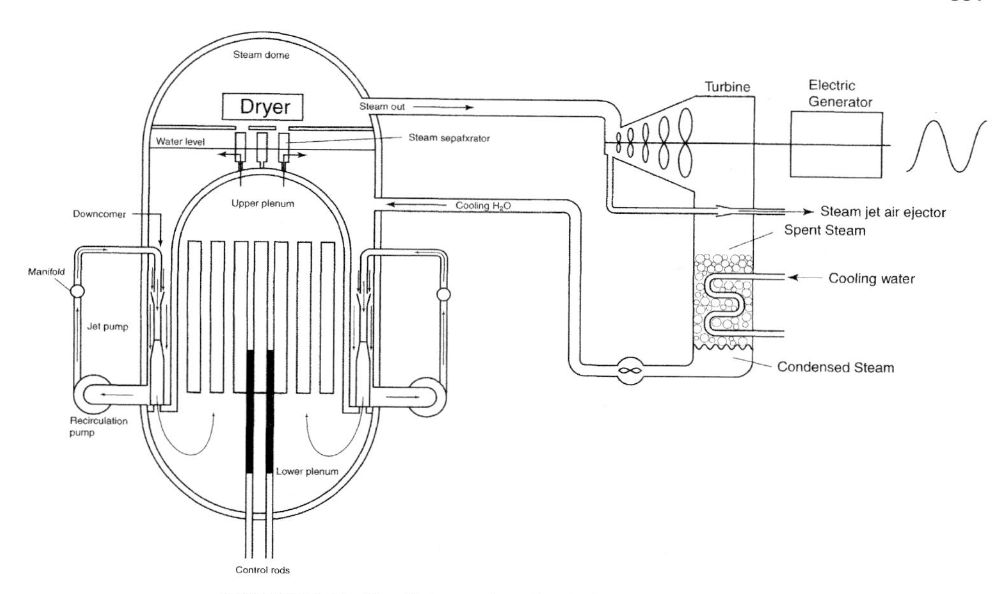

FIGURE 10-10. Schematic of a boiling water reactor.

in passing through the reactor core, which requires shielding of the steam piping, the turbines, the condenser, reheaters, pumps, pipings, and so on. The radioactivity entrained in the steam contains radioactive isotopes of xenon, krypton, iodine, and other halogens in addition to disassociated oxygen and hydrogen. These gases will be exhausted by the steam jet air ejector after they pass through the turbine and must be held for decay before release to the environment.

Cooling water in a BWR enters a chamber at the bottom and flows upward through the core. Voids are produced as the water boils, and by the time it reaches the top of the core, the coolant is a mixture of steam and liquid water. Uneven formation and behavior of the steam bubbles could make control difficult; however, the boiling is stable if done at high pressures. The turbine requires only steam, so the steam–water mixture passes through stream separators and dryers located at the top of the reactor vessel. Residual water from the separators and dryers is mixed with feedwater returning from the condenser and is routed back into the core.

Water entering the reactor passes through an annular region between the core shroud and the reactor vessel, known as the *downcomer*. Pumps withdraw water near the bottom of the downcomer and pump it at high pressure through a pipe manifold to 18 to 24 jet pumps, depending on the reactor power level. The jet pumps are located within the downcomer, as indicated in Figure 10-10. Water emerges from the nozzle of the jet pumps at high speed, which entrains water by suction flow and forces it into the lower plenum of the reactor, where it again passes up through the core.

A typical BWR produces saturated steam at about 290°C and 7 MPa and has an overall efficiency of 33 to 34%. The pressure in a BWR is approximately half that in a PWR, so that thinner pressure vessels can be used, but because the power density (W/cm^3) is smaller, the pressure vessel must be made larger, which tends to equalize the costs for each. The fuel in a BWR is slightly enriched UO_2 pellets in sealed tubes, and the core configuration is similar to that of a PWR. However, BWR control rods are always placed at the bottom (in a PWR they are at the top). Since steam voids exist in the coolant at the upper portion of the core, the value of k_{eff} for a given control rod movement will be higher in the lower part of the core.

Inherent Safety Features of LWRs

Power reactors in the United States are designed such that if k_{eff} is increased to slightly greater than 1.00 and then left alone, the power level does not continue to rise, but instead, levels out at some equilibrium value. The power rise creates effects that cause k_{eff} for the core to be held in check as a kind of "internal brake" on the reactor. This negative reactivity coefficient is due to the combined effect of changes in three quantities: moderator temperature, steam void fraction, and fuel temperature.

Moderator temperature increases affect reactivity in three major ways. First, a reduction in moderator density results in fewer hydrogen nuclei per unit volume to thermalize neutrons, so they travel farther during their lifetime and more are lost due to leakage, reducing k_{eff}. Second, the decrease in coolant density reduces the number of neutrons lost to absorption by hydrogen and oxygen, creating a positive reactivity effect (i.e., k_{eff} increases, as does reactor power from this effect). Third, the reduction in moderator density results in neutrons having a higher average energy, which in turn increases their absorption in the ^{238}U resonances because they spend more time in the resonance energy range, causing a decrease in k_{eff} due to decrease in the resonance escape probability. U.S. reactors are designed such that the negative effects of an increase in temperature outweigh the positive effects; therefore, the overall effect of an increase in moderator temperature is to introduce negative reactivity, and control rods need to be moved to overcome these negative temperature effects.

A slightly positive moderator temperature coefficient may also exist for large PWRs that use chemical shim control (boric acid dissolved in the coolant to supplement the control rods). This effect occurs because an increase in moderator temperature decreases moderator density and effectively removes chemical shim from the core. The effect, if it exists, is most pronounced early in the life of the core, but as the reactor operates, the chemical shim is burned up, and the coefficient becomes more negative. This positive temperature coefficient is small, and PWRs are overall quite stable. In reactor designs that allow temperature increases to produce a net positive effect, k_{eff} would continue to increase with temperature increases and the reactor would be very difficult to control. The Chernobyl reactor that exploded in April 1986 had a positive moderator temperature coefficient, which accelerated a change in power level to a runaway condition. (This effect was especially pronounced at low power, which was the condition established for an experimental test, as discussed in Chapter 11.)

Steam voids affect reactivity by reducing the density of the moderator, which tends to increase resonance absorption of neutrons by ^{238}U since neutrons are not slowed down. An increase in resonance absorption produces a negative reactivity effect, which tends to shut the reactor down. Steam voids also reduce neutron absorption by hydrogen and oxygen, which is a positive effect, but this is offset by the reduction in moderator density, and the overall effect of steam voids on reactivity is negative.

The effect of steam voids in boiling water reactors is pronounced because they can represent 30% or more of the total moderator volume, and at the top of the core can be as high as 70% (these effects are less in a PWR because the reactor pressure prevents boiling). Reactor power can be increased simply by adding cold water to the core to reduce the number of voids; therefore, feedwater valves and pressure control devices offer very effective supplementary means for controlling a reactor.

Fuel temperature increases cause a broadening and flattening of the resonance absorption cross-section resonance in ^{238}U. This is a Doppler effect which causes a neutron to "see" a larger target over a wider range of energies, resulting in more resonance capture and a reduction in k_{eff}. The Doppler coefficient is particularly important because it is a prompt effect that occurs in the fuel itself and immediately follows an increase in reactor power. On the other hand, moderator temperature effects and steam void effects are delayed because it takes longer for the heat liberated in fission to be conducted out of the UO_2 and the fuel rods to change the moderator temperature. Consequently, the moderator temperature and void coefficients will be delayed for several seconds should a sudden change in power occur. The fuel temperature coefficient changes instantly and is the primary mechanism for stopping the effects of fast transients.

Decay Heat in Power Reactors

A reactor can be shut down by interrupting the generation of neutrons; however, as shown in Table 10-3, a significant fraction of the energy released in a nuclear reactor is due to the beta particles and gamma radiation emitted as the highly radioactive fission products undergo radioactive transformation. This energy release occurs after fission, and because the various fission products persist well after reactor shutdown, it is still necessary to continue core cooling even after the fission reaction is safely stopped. For a 3300 MWt reactor (about 1000 MWe), about 200 MWt (or about 7% of the total) is represented by radioactive decay heat at the instant of shutdown. Although the latent heat due to fission product radioactivity falls off rapidly, as shown in Figure 10-11, it is large enough that continued cooling of the reactor core is required for several hours after reactor shutdown to prevent fuel damage.

The accident at Unit 2 of the Three Mile Island (TMI) reactor occurred because the reactor lost the ability to remove the delayed heat. The reactor had operated at 2700 MWt for about 3 months when the incident occurred (see Chapter 11). The coolant temperature was 600°F and the system pressure was 2200 psi. A pressure relief value stuck open, which caused the operators to misinterpret a number of circumstances, and they mistakenly shut down the primary coolant pumps. Without core cooling the fission products in the fuel continued to produce heat, and the fuel overheated and melted, causing fission products to be released to the primary system and the containment. Fortunately, these systems, which were designed to contain such an event, contained the fission products and releases to the environment were minimal.

Uranium Enrichment

Ordinary light water, H_2O, has desirable features for use in reactors. First, it has good heat transfer and is plentiful, so that it can be replaced easily if leakage or evaporation occurs. It is also very effective in slowing down neutrons, but this good

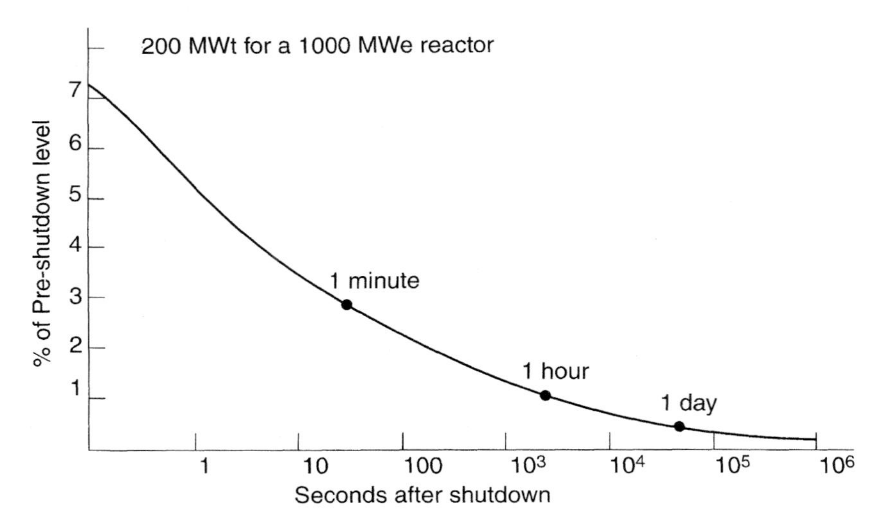

FIGURE 10-11. Decay heat in a 1000 MWe power reactor after shutdown.

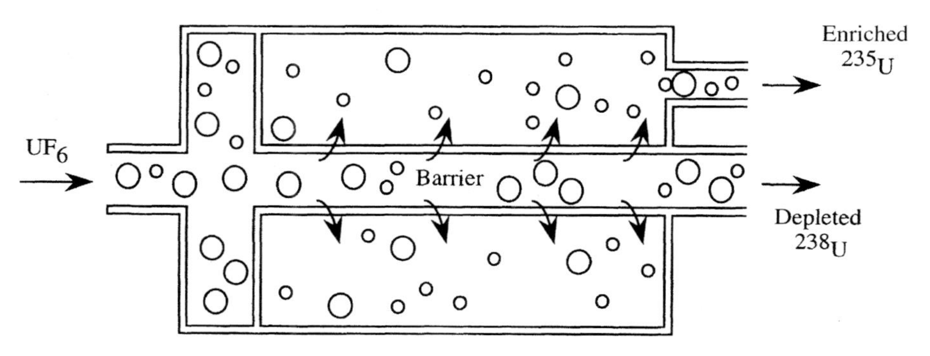

FIGURE 10-12. Uranium enrichment process.

feature is offset by a high capture cross section for neutrons, making it unsuitable as the coolant or moderator for reactors fueled with natural uranium.

Light water can be used as a coolant or moderator if the amount of ^{235}U in the fuel is enriched to about 3 to 4%. The United States could readily choose this option because it had developed gaseous diffusion plants during World War II. The Canadian reactor industry chose not to develop enrichment technology but rather, to develop the technology to produce heavy water so that natural uranium could be used as reactor fuel. Since most nuclear utilities in the United States have chosen light water–moderated and light water–cooled reactors, enrichment technology is an important adjunct of the nuclear fuel cycle.

Enrichment technology in the United States uses gaseous diffusion, in which uranium hexaflouride gas is pumped through a porous nickel diffusion barrier as shown in Figure 10-12. About 1700 such stages are required to increase the amount of ^{235}U in uranium to 3 to 5% . Since uranium is not a gas, enrichment of uranium by gaseous diffusion requires its conversion to uranium hexafluoride (UF_6), which is easily vaporized if held at an appropriate temperature. UF_6 is a stable compound but is highly reactive with water and corrosive to most common metals. As a consequence, any surface in contact with it must be fabricated from nickel or austenitic stainless steel, and the entire system must be leak-tight. Despite these requirements, UF_6 is the only compound of uranium sufficiently volatile to be used in the gaseous diffusion process.

Fortunately, natural fluorine consists of only one isotope, ^{19}F, so that the diffusion rates of UF_6 are due only to the difference in weights of the uranium isotopes. On the other hand, the molecular weights of $^{235}UF_6$ and $^{238}UF_6$ are very nearly equal, and little separation of the ^{235}U and ^{238}U occurs in passing through a barrier, thus a sequence of stages is required in which the outputs of each stage become the inputs for two adjoining stages. The gas must be compressed at each stage, which produces compression heating, which then must be cooled; the requirements for pumping and cooling UF_6 through hundreds of stages make diffusion plants enormous consumers of electricity.

HEAVY WATER REACTORS

An alternative to using ordinary water (H_2O) as the moderator or coolant in a thermal reactor is to choose D_2O, or heavy water, for one or both of these purposes. Because heavy water absorbs many fewer neutrons than does ordinary water, heavy water–moderated reactors do not require enriched uranium fuel but can be designed to use natural uranium (0.72% ^{235}U). Neutrons must travel farther in heavy water to reach thermal energies, so the separation distance between fuel bundles is greater than in LWRs. This increased separation distance has been used to advantage in the Canadian deuterium–uranium (CANDU) reactor, which uses a series of individually cooled fuel channels one bundle thick with the fuel bathed in D_2O moderator. The use of fuel channels allows a CANDU reactor to be refueled on-line without shutting down the reactor.

A typical CANDU reactor and coolant system is shown schematically in Figure 10-13. It has many similarities to a pressurized water reactor (see Figure 10-9) except that the primary coolant is usually heavy water (each channel could be cooled with light water as long as the moderator between the fuel channels is heavy water; the Japanese Fugen reactor is designed this way). Coolant from a primary pump passes through a distribution header to the individual tubes, goes once through the reactor, through another header to the steam generator, through the U-tube steam generator, and finally back to the primary coolant pump. The heavy water coolant in each fuel tube is maintained at a pressure of about 1500 psi (10 MPa) and in passing through the pressure tubes reaches a temperature of 310°C (590°F), which is below the boiling point at that pressure. It is this characteristic that makes a CANDU reactor very similar to a PWR, at least in terms of how the primary coolant is managed.

The secondary coolant is light water, which is converted to steam in the steam generators, routed to a turbine, and then condensed and returned to the steam generators as feedwater. There are four steam generators and pumps, a single pressurizer, and pumps that induce a flow rate through the pressurized tubes, which is about 60 million pounds per hour for a 600 MWe CANDU. The overall electrical efficiency

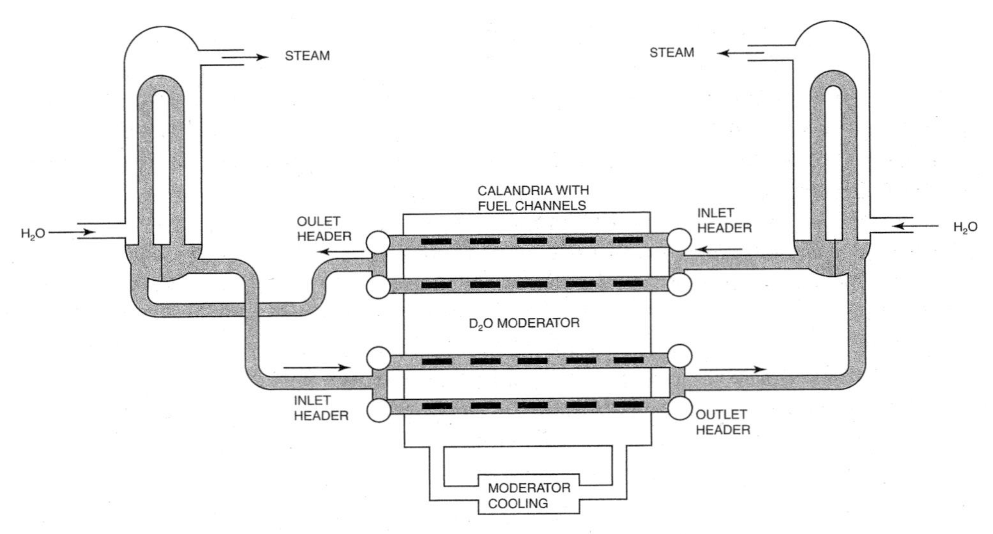

FIGURE 10-13. Major components of a CANDU heavy water reactor. Heavy water flows through zircalloy fuel channels immersed in a calandria of heavy water moderator, where it is heated under pressure before flowing through a header to steam generators for extraction of the heat energy to produce steam in the light water secondary coolant system.

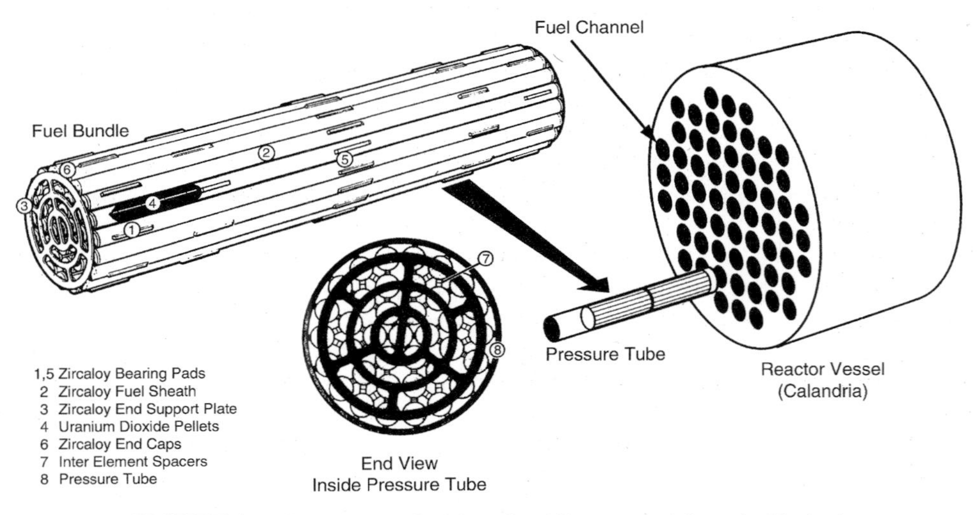

FIGURE 10-14. CANDU fuel bundle. (Courtesy of Ontario Hydro.)

of a CANDU system is about 29%, which is significantly lower than that of most LWR commercial nuclear power plants.

Figure 10-13 shows only four of the fuel channels; however, a typical CANDU has hundreds of channels, containing some 4500 fuel bundles arranged end to end. Each fuel bundle is made up of 37 zircalloy-clad fuel pins containing 0.72% $^{235}UO_2$, as shown in Figure 10-14. The fuel channels, which are made of zircalloy and are pressurized, pass horizontally through a lattice of tubes placed in a large *calandria* containing the heavy water moderator, which is maintained near atmospheric pressure; therefore, a large pressure vessel is not required. The calandria is a stainless steel cylinder about 25 ft (7.6 m) in diameter and 25 ft (7.6 m) long, with walls that are about 1 in. (2.5 cm) thick and end pieces that are about 2 in. (5 cm) thick. The moderator picks up some of the reactor heat since it envelopes the pressure tubes where fission occurs. This heat is removed by two pumps and two heat exchangers such that the moderator temperature is about 160°F (70°C) when the reactor is running. Representative values of these various parameters are listed in Table 10-4.

Heavy water reactors have two significant advantages over LWRs: Very few neutrons are lost due to absorption in the moderator, and the design permits on-line refueling. The heavy water moderator is very efficient since less than 0.1 of the 2.44 neutrons produced per fission are lost to absorption in D_2O, versus 0.3 for H_2O. This efficiency allows burnup of ^{235}U in natural uranium fuel (originally 0.72%) down to about 0.25% . When the fuel is removed from the reactor, the ratio of fissile material produced to fissile material destroyed is 0.75 to 0.80. This factor is called the *conversion ratio*, and since a significant amount of new fissile material is produced, CANDUs are often called converter reactors. The fissile content of the discharged fuel is about 0.5% , about half of which is fissile plutonium; therefore, if the plutonium produced were recycled, the overall amount of uranium used could be reduced by about half. Use of natural uranium reduces fuel costs; however, the reactor requires a million pounds of very expensive heavy water which tends to offset these savings.

Reactivity control in a heavy water reactor is achieved by several systems, including light water zone absorbers, solid control rods, and neutron poisons added to the moderator. Routine on-line control is accomplished by the zone absorbers, which consist of compartments in the core into which light water, a neutron absorber, can be introduced. Cadmium control rods can be dropped under gravity for quick power reduction, but long-term reactivity control and startup reactivity control are usually provided by mixing neutron-absorbing compounds of boron and gadolinium into the moderator. Power distribution across the core can be controlled effectively by the refueling sequence, since each pressure tube can be serviced individually, and optimal distributions of fuel can be made.

Continuous refueling can reduce outage times, but its main advantage is in fuel management since refueling can take place as needed and the maximum energy may be extracted from each fuel bundle. Refueling also serves as a reactivity control, increasing the fissile content precisely when and where it is required in the core. On the average, about 15 bundles are replaced per day of operation. Relatively little neutron absorber is necessary during reactor operation because there are no large swings in fissile content or fission product poisons during the fuel cycle. This leads to a higher conversion ratio and, under some conditions, to significantly improved use of the uranium resource.

Since heavy water is expensive (about $100 per kilogram), the reactor building contains systems for the collection, purification, and upgrading of heavy water. A moderator cleanup system uses filters and ion-exchange resins to control impurities, including adjustments for boron and gadolinium neutron poisons. Two shutdown cooling systems, each consisting of a pump and heat exchanger, are also provided to remove decay heat during reactor shutdown because the primary pumps are isolated along with the steam generators.

HWR Safety Systems

Under abnormal conditions, HWRs can be shut down by gravity drop of the shutdown control rods. Earlier CANDUs were designed to dump the moderator out of the calandria into a large tank to stop the chain reaction, which is a very effective procedure because the reactor cannot run unless the neutrons are slowed down. This capability is no longer used in CANDUs because it is burdensome to replace the moderator in an optimal condition, and fast injection of gadolinium into the moderator has proved to be just as effective and reliable.

An emergency core cooling system is provided for controlling loss-of-coolant accidents. Should a reactor coolant system rupture, valves close to isolate the intact system, and light water from a storage tank (dousing tank) built into the roof of the containment system is injected into the ruptured system. Heat is initially rejected through the steam generators. As the dousing tank is emptied, water is recovered from the bottom of the reactor building, passed through a heat exchanger, and reinjected into the ruptured system. The moderator in the calandria provides some independent heat capacity, and the moderator heat exchangers remove this heat.

Since CANDU HWRs have many pressure tubes, gross failure of the pressure vessel is highly unlikely. Even if one of the headers is ruptured, the other independent coolant loop would presumably still be intact, and it and the moderator could carry off enough heat to prevent gross melting. Consequently, CANDU HWR reactors are inherently safe designs; however, as a precaution, a prestressed concrete containment structure is provided and operated at negative pressure. A spray system and air coolers are also provided to reduce the building pressure from a loss of coolant incident.

It is perhaps significant and reflective of the CANDU design that no major accidents have occurred with CANDU reactors. Since D_2O is used as a coolant as well as a moderator, considerable amounts of tritium are produced and require management during operation and water handling, and precautions against its environmental release are required. Routine release of radioactive fission and activation products to the environment are, like their U.S. counterparts, minimal to nonexistent.

BREEDER REACTORS

All reactors that contain natural uranium can be thought of as converter reactors because neutron absorption in ^{238}U produces ^{239}Pu, another fissile material. Practically all of the fuel in a light water reactor is ^{238}U, and the conversion of ^{238}U to ^{239}Pu takes place as a matter of normal operation. Plutonium, being a different element, can be chemically separated from the uranium, but this is generally limited to production reactors. After ^{239}Pu has been formed in a reactor, it may absorb a neutron and undergo fission or be transformed into ^{240}Pu. The ^{240}Pu, which is not fissile, may in turn capture another neutron to produce ^{241}Pu, which is. Finally, the ^{241}Pu may undergo fission or be transformed into ^{242}Pu. Thus a reactor contains, in

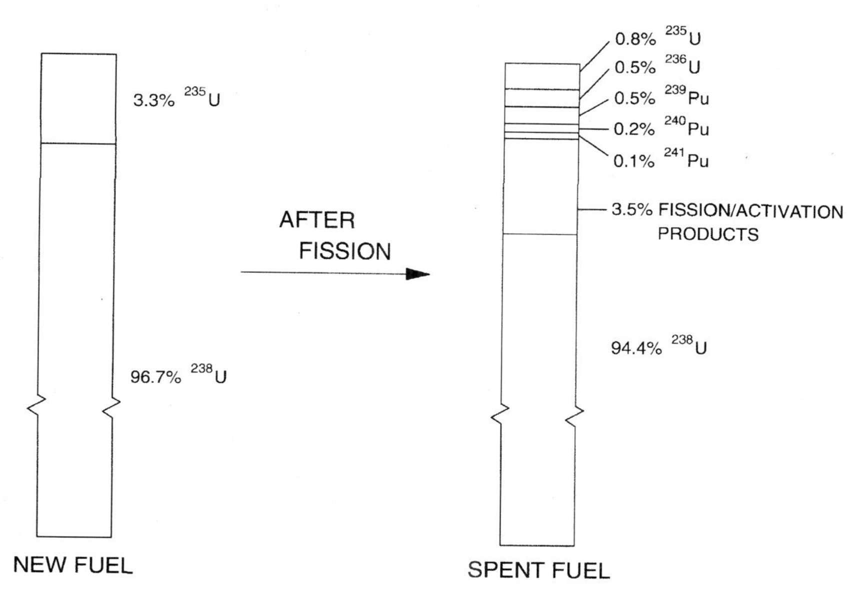

FIGURE 10-15. Typical composition (in percent) of ^{235}U and ^{238}U in new fuel and typical composition (in percent) of fission and activation products and ^{236}U from neutron absorption in ^{235}U and plutonium isotopes from neutron absorption in ^{238}U (fuel burnup of 33,000 MW · d/MT).

decreasing amounts, the isotopes ^{239}Pu, ^{240}Pu, ^{241}Pu, and ^{242}Pu, depending on the burnup of the fuel (see Figure 10-15).

Light water–cooled and light water–moderated reactors strive to minimize the production of ^{239}Pu and other fissile materials because it robs the system of neutrons that need to be made up by enrichment of ^{235}U in the fuel. The phenomenon can, however, be used to advantage to create a *breeder reactor*, one that produces more fissionable material than it consumes. To do so, it is necessary to maintain a population of fast neutrons to enhance resonance capture of neutrons in ^{238}U (see Figure 10-4) to produce (or breed) more atoms of ^{239}Pu. The major design feature of such reactors is to keep neutrons from slowing down appreciably (hence they are called *fast reactors*), and for these reasons the coolant must have minimal moderating properties and at the same time provide effective heat removal. A gas can be used or a liquid metal such as sodium, and both designs have been developed for breeder reactors. Each has advantages and disadvantages, but the sodium-cooled liquid metal fast breeder reactor has received the most attention; it is based on the following reaction/radioactive transition sequence

$$^{238}\mathrm{U}(\mathrm{n},\gamma)^{239}\mathrm{U} \underset{23.5\ \mathrm{m}}{\xrightarrow{\beta^-}} {}^{239}\mathrm{Np} \underset{2.35\ \mathrm{d}}{\xrightarrow{\beta^-}} {}^{239}\mathrm{Pu}$$

It is more difficult to design a reactor that will breed more fissionable material than it consumes than to design one that merely converts ^{238}U to ^{239}Pu because η must be greater than 2.00 (i.e., one fission neutron must be absorbed to keep the chain reaction going, and more than one neutron must be absorbed in fertile material to produce a new fissile atom). In actual fact, η must be substantially greater than 2.00 because some neutrons inevitably are absorbed by nonfuel atoms or lost by leakage. The value of η is 2.07 for ^{235}U and 2.14 for ^{239}Pu if neutron energies can be kept above about 100 keV, and it is possible to breed fissile material with these fuels if the bulk of the fissions are induced by fast neutrons. If, on the other hand, neutrons are in the thermal energy range, ^{233}U, which has a value of η of 2.3, is the only fuel that can be used for a thermal breeder reactor. ^{233}U is produced by

$$^{232}\text{Th}(\text{n},\gamma)^{233}\text{Th} \xrightarrow[22.3\text{ m}]{\beta^-} {}^{233}\text{Pa} \xrightarrow[27.0\text{ d}]{\beta^-} {}^{233}\text{U}$$

Naturally occurring thorium is entirely ^{232}Th, which is not fissionable; therefore, it is used as a target material in a reactor to produce ^{233}U, which in turn can be chemically separated from thorium because it is a different element. Production of ^{233}U also occurs readily with fast neutrons.

An experimental breeder reactor was built at Los Alamos in 1946, and its potential impact on future energy supplies was immediately recognized. It was a small plutonium-fueled, mercury-cooled device that operated at a power level of 25 kW. This was followed in 1951 by the Experimental Breeder Reactor-I (EBR-I) located in Idaho and EBR-II a few years later. The EBR-I was cooled with a mixture of sodium and potassium (NaK); a secondary loop and a turbine-generator were provided and it produced 200 kW of electricity, the world's first nuclear-generated electricity. The EBR-II was designed to recycle fuel, which was demonstrated successfully, but it, too, has been discontinued.

Liquid Metal Fast Breeder Reactors

The liquid metal fast breeder reactor (LMFBR) uses uranium–plutonium fuel and has a blanket of natural or depleted uranium that absorbs neutrons to produce ^{239}Pu. Sodium is used as a coolant and is excellent for this purpose because it has an atomic weight of 23 and does not appreciably slow down neutrons by elastic scattering, although it does moderate neutrons to some extent by inelastic scattering. It is also an excellent heat transfer agent, allowing the LMFBR to be operated at high power density. Sodium is not corrosive, so it does not deteriorate structural materials, and its high boiling point (882°C at 1 atm) allows high-temperature operation at essentially atmospheric pressure, which eliminates the need for a heavy pressure vessel. Steam from a LMFBR plant is delivered superheated to the turbines at about 500°C and between 16 and 18 MPa. The overall plant efficiency is in the neighborhood of 40%.

Sodium also has some undesirable characteristics. It solidifies at room temperature, thus the entire coolant system must be kept above 98°C at all times. Sodium reacts violently with water and catches fire when it comes in contact with air; therefore, a cover gas of nitrogen and/or argon is maintained in empty regions above the sodium to reduce the likelihood of fires. Sodium also absorbs neutrons to form 15-h ^{24}Na, which emits high-energy gamma rays. Two sodium loops are used to mitigate these potential effects: A primary reactor loop contains the radioactive sodium, and an intermediate sodium loop containing nonradioactive sodium extracts the heat from the primary loop through an intermediate heat exchanger. A third loop

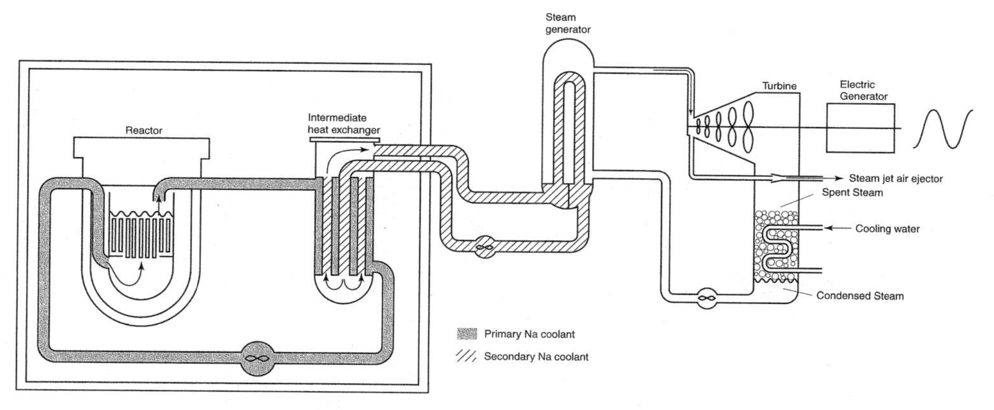

FIGURE 10-16. Schematic of a loop-type liquid metal fast breeder reactor.

containing ordinary water and a steam generator extracts the heat from the secondary loop, producing very hot steam that is routed to a conventional turbine-generator. Since the sodium in the primary loop is radioactive, heavy shielding is provided. The sodium in the secondary loop is not radioactive, and neither is the water or steam, so these loops are not shielded. Because of these characteristics, LMFBRs are designed with very tight systems, and they emit far less radiation to the environment than do comparable LWRs.

There are two principal LMFBR designs: loop type and pool type. Loop-type LMFBRs generally incorporate the features shown in Figure 10-16. The core and blanket are enclosed in a reactor vessel, not unlike that of an LWR, except that the vessel need not withstand high pressure. The intermediate heat exchanger and all other components of the heat transfer system are located external to the reactor vessel; therefore, those portions of the primary loop outside the reactor vessel must be shielded. A typical 1000 MWe LMFBR plant has three or four primary loops, each of which is connected to a separate intermediate water-steam loop, and since these separate systems are external to the reactor vessel, inspection, maintenance, and repairs are straightforward. Substantial amounts of shielding are required around the primary loops, which makes these plants rather large and heavily built.

In the pool-type LMFBR, all of these components are immersed in hot, radioactive, and opaque sodium. Pool-type reactor vessels are located at least partially underground, so that only the uppermost portion of the vessel requires heavy shielding. No radioactivity leaves the reactor vessel, so no other component of the plant must be shielded; therefore, one can walk into the reactor room and even across the top of the reactor while it is operating without receiving a significant radiation dose.

Fission heat is extracted by the liquid sodium as it passes upward around each fuel assembly, which contain stainless steel fuel pins 6 or 7 mm in diameter that are separated from each other by spacers. The fuel pins contain pellets of a mixture of oxides of plutonium (PuO_2) and uranium (UO_2) with an equivalent enrichment (the percent of the fuel that is plutonium) of 15 and 35%, respectively. The pins in the blanket, which contain only natural or depleted UO_2, are larger in diameter (about 1.5 cm) because they require less cooling than the fuel pins. The fuel and blanket

TABLE 10-6. Breeding Ratios for LMFBRs with Oxide or Carbide Fuel

	U–^{239}Pu fuel	Th–^{233}U fuel
Oxide	1.277	1.041
Carbide	1.421	1.044

pins can be more tightly packed than in LWRs or HWRs because of the excellent heat transfer properties of sodium.

The control rods for LMFBRs usually contain boron carbide, although other materials have also been used. At one time it was thought that LMFBRs would be more difficult to control than most thermal reactors, because the fission neutrons do not spend as much time slowing down before inducing further fissions. However, years of operating experience have shown that fast reactors, in fact, are highly stable and easily controlled.

Present-day LMFBRs operate with uranium–plutonium oxide fuel; however, there is considerable interest in uranium–plutonium carbide fuel because with only one atom of carbon per uranium atom, it produces less moderation of neutrons than UO_2, which contains two atoms of oxygen. The spectrum of neutrons in a carbide-fueled LMFBR would contain neutrons of somewhat higher energies than an oxide-fueled LMFBR, and as shown in Table 10-6, a higher breeding ratio is possible because η increases with neutron energy.

GAS-COOLED REACTORS

Carbon, usually in the form of high-purity graphite, is another good moderator because it absorbs few neutrons. With an atomic mass of 12, more collisions are required in graphite than in water to slow down neutrons, but like D_2O, it has a better moderator ratio than light water (see Table 10-5), which allows natural uranium to be used as fuel. The coolant in a carbon-moderated reactor is usually a gas such as air, helium, or carbon dioxide, but water can be used if it is confined to tubes containing the fuel bundles.

A number of carbon dioxide–cooled graphite-moderated reactors have been operated in the United Kingdom for electricity production. A 330 MWe high-temperature gas-cooled reactor (HTGR) was built in the United States but is now shut down, and the larger commercial versions of the design were withdrawn from the market in 1976. A pebble-bed graphite-moderated gas-cooled reactor is being developed in Germany.

High-Temperature Gas Reactors

The high-temperature gas reactor (HTGR) design is shown schematically in Figure 10-17. The reactor is designed to use helium as the coolant since it absorbs essentially no neutrons and can be operated at high temperatures to produce electricity at an efficiency of about 40%, which is unusually high for nuclear power plants. Steam is produced in the HTGR by heating ordinary water through a heat exchanger

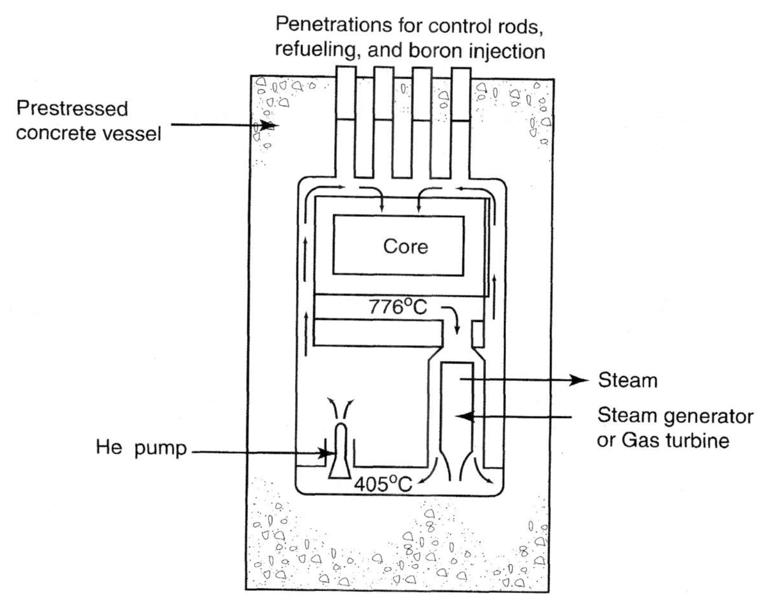

FIGURE 10-17. Schematic of a high-temperature gas reactor.

similar to those used in PWRs, or the HTGR can be operated in a direct cycle with a gas turbine to further improve plant efficiency.

The core of the HTGR does not use metal fuel rods but consists of stacked carbon blocks that enclose small uranium–thorium fuel regions. The fuel–moderator system is also quite distinct, since it consists of uranium and thorium pellets fabricated inside carbon moderator blocks. The fuel pellets contain 93% ^{235}U and are coated with pyrolytic carbons and silicon carbide, and the fertile pellets (^{232}Th) are coated with carbon to moderate neutrons to thermal energies so they will be absorbed in thorium to produce fissile ^{233}U. Because it does not burn, the silicon carbide aids in separating the two particle types at reprocessing, which is performed by burning away the carbon. All the fuel elements have holes through which the coolant flows. Refueling occurs around a central stack, which has two vertical control rod penetrations, and six adjacent stacks, which do not have fuel rod channels. The central stack also has a safety channel designed for rapid injection of boron carbide balls to provide quick shutdown.

The helium coolant in the HTGR is pumped at a pressure of 700 psi (5 MPa) downward through the core, where it exits with a temperature of about 1430°F (776°C), and is routed to a steam generator. The cooled helium exiting the steam generators is purified by filtration, adsorption, and a hydrogen getter to remove particulates and contaminant gases, and is pumped back into the core. Two purification systems are provided; one operates while the other is shut down for decay and regeneration. Regeneration of the purification system produces radioactive gases, which are separated and returned to the reactor; it also produces a stable component, which is released to the atmosphere. Liquid wastes are produced only from decontamination operations, and the principal solid wastes are the tritium-contaminated getters from the helium purification systems.

The core, the entire primary coolant system, and other components of the nuclear steam supply system are contained in a prestressed concrete reactor vessel (PCRV), which is a unique feature of the HTGR. The reactor vessel has penetrations for refueling, control rods, and steam pipes. Removal plugs are provided for servicing of steam generators, helium circulators, and so on.

Inherent safety features for the HTGR are numerous and substantially different from those of water-cooled reactors. The core provides a massive heat sink, and the HTGR fuel particles, with their ceramic coatings, can survive without coolant flow for as much as an hour (decay heat will melt LWR fuel in a minute or so if cooling is lost). The structural strength of the core is provided by the graphite, and this strength increases with temperature. A complete loss of helium cooling is extremely improbable because there are several primary cooling loops which are largely independent; however, just in case, auxiliary cooling loops are provided to handle the decay heat. Furthermore, helium is always a gas and is nonreactive; therefore, metal water reactions that could produce explosive hydrogen gas are eliminated.

REACTOR RADIOACTIVITY

Reactors are sources of many radioactive materials, many of which exist in off-gas systems, effluents, or in waste materials. Large amounts of radioactive fission products and activation products are produced because of the large population of neutrons. These products accumulate in some reactor systems where they can cause radiation exposures of workers, and some of the products are released during routine operations, incidents, or in the form of radioactive wastes. Releases to the environment can, of course, produce radiation exposures of the public depending on various pathways.

Release of radioactivity during routine operations is a complex process governed by reactor type, power level and operating history, fuel performance and overall system cleanliness, and the particular waste processing and cleanup systems used. The direct cycle design of the BWR introduces any entrained radionuclides into the various cleanup systems, whereas these materials would remain in the primary coolant of the PWR and would not be released unless steam generator leaks occur. These features are reflected in the capacities and designs of auxiliary systems for the two reactor types. The general patterns of environmental releases for BWRs and PWRs and the influence of the various elements of each design are shown schematically in Figure 10-18.

Fuel Cladding

The fuel in light water reactors is the first and perhaps the most important barrier to release of fission products. Enriched UO_2 is fabricated into ceramic pellets, which have good stability and heat transfer characteristics. These ceramic pellets are loaded into fuel rods of zircalloy tubing that are welded at each end, as shown in Figure 10-19. As fissioning proceeds, gaseous and volatile fission products migrate slowly through the ceramic fuel matrix into the gap between the pellets and the zircalloy tube walls. As more and more atoms of these products are produced, they build up

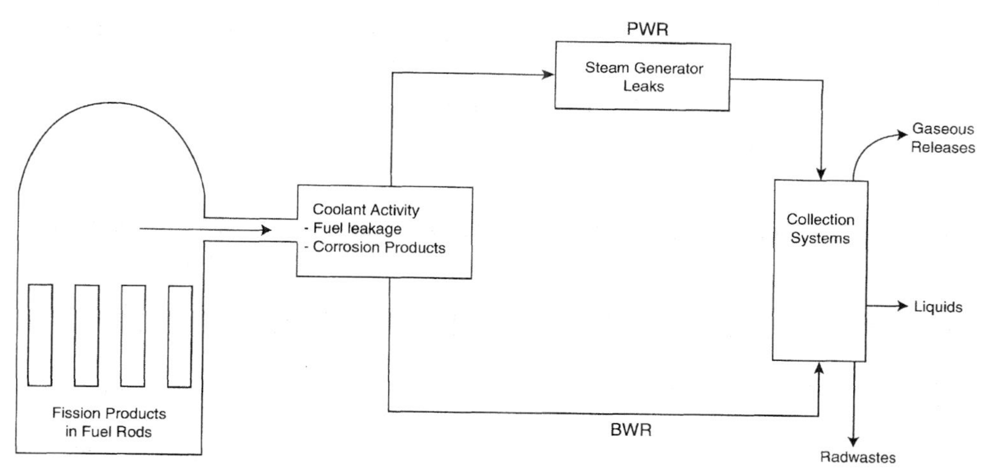

FIGURE 10-18. Environmental release pathways for radioactive fission and activation products produced in pressurized water and boiling water reactors.

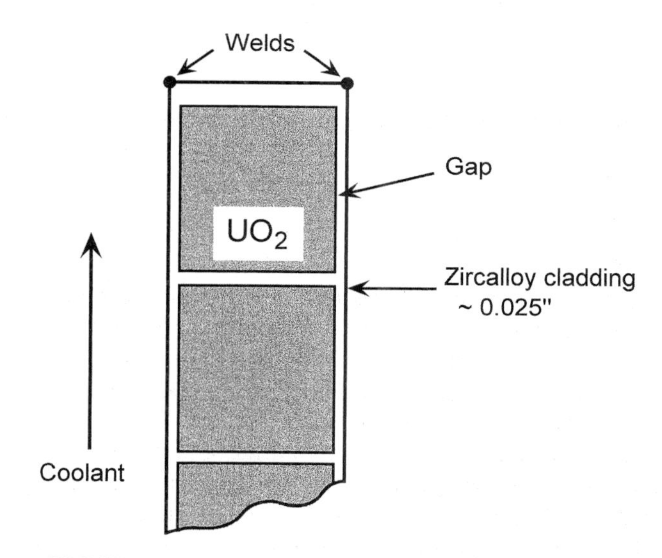

FIGURE 10-19. UO_2 fuel pellet in zircalloy cladding.

pressure, which may drive them through any imperfections in the zircalloy cladding. The rods are thin to promote heat transfer and minimize effects on neutron flux, and because of this and the large number, some imperfections are to be expected that may result in leakage of volatile fission products into the passing coolant. Fuel manufacture is subjected to intense quality control to minimize such leakage.

Fuel cladding is the major barrier to eventual release of fission products. Zircalloy cladding thickness is quite thin at 0.024 to 0.034 in. (0.61 to 0.81 mm) and receives considerable stress, especially from gaseous products. Most reactors have cleanup systems designed to tolerate releases of 0.25 to 1% of the volatile and gaseous radioactivity into the coolant from such flaws. If fuel flaws exceed these design considerations, it is necessary to reduce power or refuel to deal with the fission products that may be circulating in the reactor or perhaps released in various effluents. This performance specification provides a good starting place for estimating fission product release.

Radioactive Products of Fission

Thermal neutron fission of ^{235}U yields over 88 primary fission products with mass numbers from 72 to 160. Being neutron rich, these products undergo radioactive transformation by emission of negatively charged beta particles, and several such transformations can occur for each atom formed before a product of fission reaches a stable end product. As shown in Figure 10-20, an excerpt from the chart of the nuclides for mass numbers 94, 95, 139, and 140, the products of fission are formed far from the line of their stable endpoints and several beta transformations are necessary for them to become stable. The fission product sequences in Figure 10-20 represent chains that can be represented by the reactions

$$^{235}_{92}\mathrm{U} + ^{1}_{0}\mathrm{n} \rightarrow [^{236}_{92}\mathrm{U}^*] \rightarrow ^{94}_{40}\mathrm{Zr} + ^{140}_{58}\mathrm{Ce} + 2^{1}_{0}\mathrm{n} + 6\,^{0}_{-1}\beta^- + Q$$

$$^{235}_{92}\mathrm{U} + ^{1}_{0}\mathrm{n} \rightarrow [^{236}_{92}\mathrm{U}^*] \rightarrow ^{95}_{42}\mathrm{Mo} + ^{139}_{57}\mathrm{La} + 2^{1}_{0}\mathrm{n} + 7\,^{0}_{-1}\beta^- + Q$$

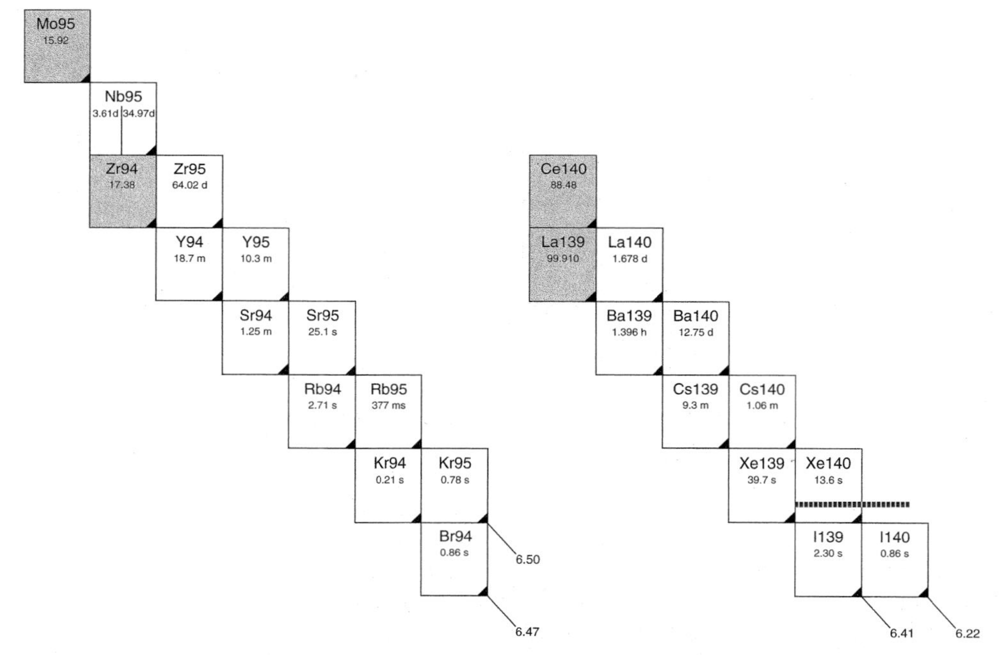

FIGURE 10-20. Excerpt from the chart of the nuclides for formation of fission products with masses 94, 95, 139, and 140 through a series of elements that transform by negative beta-particle emission. Percent cumulative fission yields for the stable endpoints of the mass chain are shown by the numbers at each diagonal (e.g., 6.47% for mass number 94).

The total fission product yields for each of the stable end products are shown at the beginning of each diagonal and represent the sum of the independent fission yields of each element in the fission product chain. Since there are some 88 fission product chains and since each has numerous beta transformations, there are well over 1200 different products formed from fission of ^{235}U, ^{239}Pu, and other fissionable elements.

The fission yield of a given mass number is expressed as a percentage of the fissions that occur, as shown in Figure 10-21 for ^{235}U and ^{239}Pu. Figure 10-21 also shows that most fission is asymmetrical and that the masses of the major fission products fall into two broad groups, a light group, dominated by mass numbers from 80 to 110, and a heavy group, with mass numbers from 130 to 150. The most likely mass numbers produced in fission are products with mass numbers of 95 and 134, which occur with a percentage yield of about 6.5 and 7.87% respectively. The highest fission yield is for mass number 134 at 7.87%.

The fission yield data in Figure 10-21 are total yields for each mass number, but no information is gained on how a given fission product is formed, nor its independent or cumulative yield. Such information is, however, provided in a detailed fission product mass chain, as shown in Figure 10-22 for mass numbers 85 and

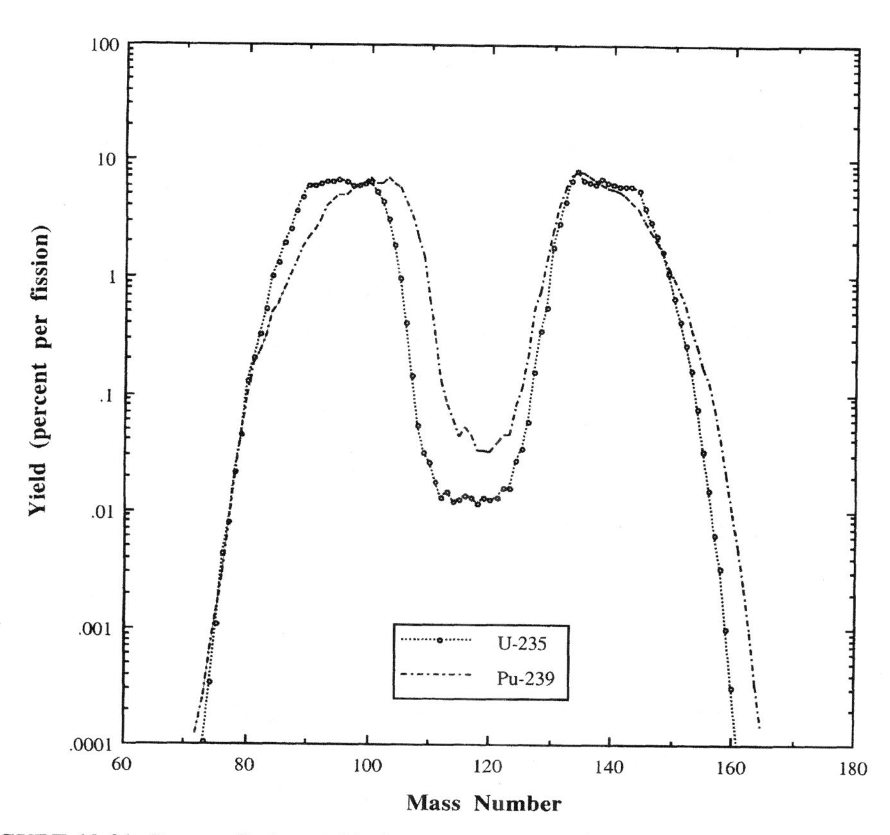

FIGURE 10-21. Percent fission yields for each fission product mass number due to thermal fission of ^{235}U and ^{239}Pu.

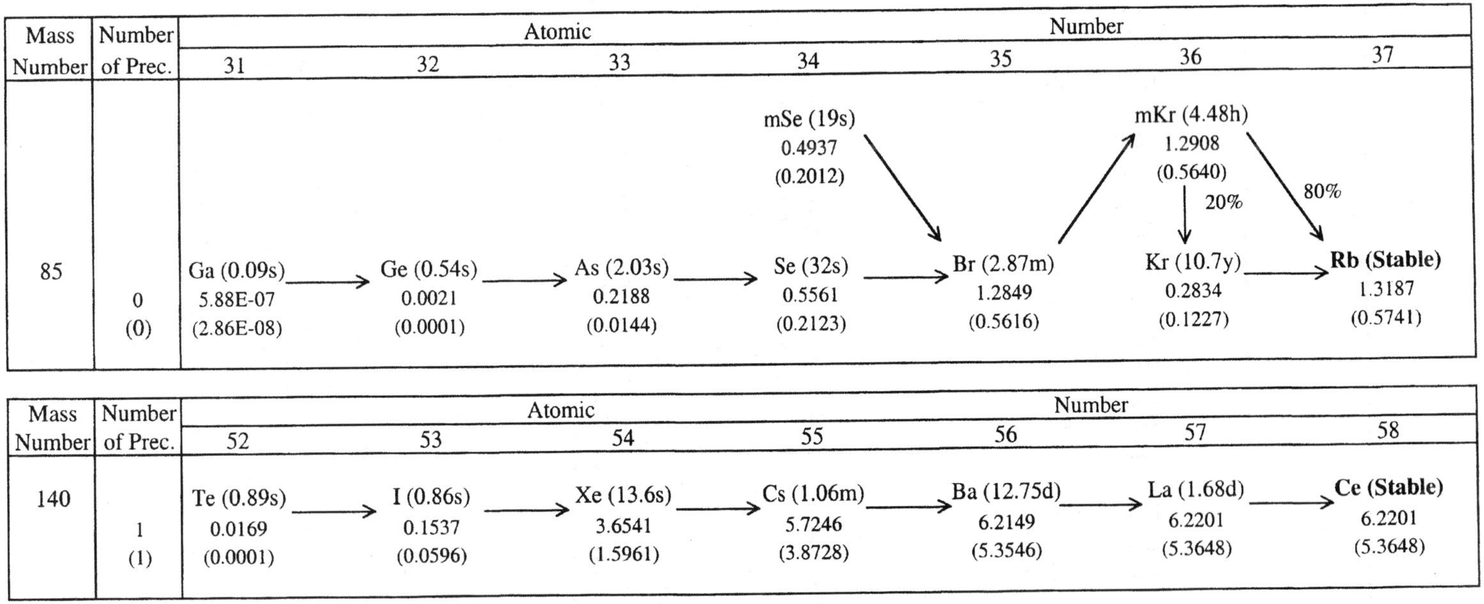

Mass Number	Number of Prec.	Atomic Number 31	32	33	34	35	36	37
					mSe (19s) 0.4937 (0.2012) → Br		mKr (4.48h) 1.2908 (0.5640) (20% → Kr; 80% → Rb)	
85	0 (0)	Ga (0.09s) 5.88E-07 (2.86E-08) →	Ge (0.54s) 0.0021 (0.0001) →	As (2.03s) 0.2188 (0.0144) →	Se (32s) 0.5561 (0.2123) →	Br (2.87m) 1.2849 (0.5616) → mKr	Kr (10.7y) 0.2834 (0.1227) →	**Rb (Stable)** 1.3187 (0.5741)

Mass Number	Number of Prec.	Atomic Number 52	53	54	55	56	57	58
140	1 (1)	Te (0.89s) 0.0169 (0.0001) →	I (0.86s) 0.1537 (0.0596) →	Xe (13.6s) 3.6541 (1.5961) →	Cs (1.06m) 5.7246 (3.8728) →	Ba (12.75d) 6.2149 (5.3546) →	La (1.68d) 6.2201 (5.3648) →	**Ce (Stable)** 6.2201 (5.3648)

FIGURE 10-22. Fission product chains for masses 85 and 140.

140. Similar data are provided in Appendix E for all the fission product chains produced by thermal fission of ^{235}U and ^{239}Pu (cumulative fission yields shown in parentheses). Such information is quite useful for determining fission product inventories.

The fission yields for each element in a mass chain are given in cumulative percent yield. It is important to note that each yield value (given in percent) may contain a portion due to formation by its own independent yield, a value that can be obtained by subtracting the cumulative yield of the most immediate precursor. If precursors are short lived, as is often the case, the cumulative yield data for the isotope of interest can be used directly in calculations of fission product inventories, but if they are not, it is necessary to consider the independent formation of the element of interest and any precursors that are important in its formation and then to account for their buildup and removal by radioactive transformation or other processes (see Example 10-3).

The mass number for any given fission product chain remains the same even though the element changes as shown in Figure 10-22 for mass numbers 85 and 140. The formation of ^{85}Kr, which is perhaps of most importance in the chain for mass number 85, is typical of many of the important fission product radionuclides encountered in radiation protection in that it is not formed directly, but is the product of several precursors. A similar situation exists for mass number 140 in which ^{140}Ba and ^{140}La are of most interest. Most fission products of interest to radiation protection are the products of precursor radionuclides that are formed directly with their own independent fission yield or are the transformation products of products that are (e.g., ^{90}Sr, ^{131}I, ^{137}Cs, etc.).

Production of Individual Fission Products

The number of atoms of any given fission product, N_i, formed from fission of fissionable nuclei (e.g., ^{235}U, ^{239}Pu, ^{233}U, etc.) can be determined from the total number of fissions (or fission rate) and the fission yield Y_i of the particular product, or

$$N_i = \text{number of fissions } \times Y_i$$

where Y_i is the fractional yield of atoms of species i that are formed. Tabulations of Y_i account for the fact that two new atoms are formed from each atom that fissions [i.e., the total of all possible modes of fission is 200% (fission of 100 ^{235}U atoms yields 200 fission products)]. Since Y_i is usually listed as a percent, it is necessary to convert it to a fractional value when calculating the number of fission product atoms. It is also necessary to account for removal of atoms of N_i by radioactive transformation or activation in determining the inventory of atoms of N_i at any given time.

The rate of change of the number of atoms of fission product, i, in a nuclear reactor of power level P (in thermal megawatts) is

$$\frac{dN_i}{dt} = (\text{amount produced}) - (\text{amount removed})$$

$$= 3.12 \times 10^{16} \text{ fissions/s} \cdot \text{MW} \times P(\text{MW}) \times Y_i - \lambda_i N_i - \sigma_i \phi N_i$$

For the case where $\sigma_i \approx 0$, and assuming that $N_i = 0$ at $t = 0$, the number of atoms N_i at any time t after operation begins for a sustained power level P (in MWt) is

$$N_i(t) = \frac{3.12 \times 10^{16} PY_i}{\lambda_i}(1 - e^{-\lambda_i t}) \qquad \text{atoms}$$

and since activity $= \lambda_i N_i$, the activity (or transformation rate) of fission product i is

$$A_i(t) = 3.12 \times 10^{16} PY_i(1 - e^{-\lambda_i t}) \qquad \text{t/s}$$

or

$$A_i(t) = 8.43 \times 10^{5} PY_i(1 - e^{-\lambda_i t}) \qquad \text{Ci}$$

where

$A_i(t)$ = activity of fission product i at t
P = power level in megawatts (thermal)
Y_i = fractional yield of fission product i
λ_i = disintegration constant of fission product i
t = time after reactor startup (operating time)

Example 10-3: Calculate the number of curies of ^{131}I in a 3000 MW reactor (thermal) at 8 d after startup.

SOLUTION: From Appendix E, the cumulative fission yield of ^{131}I is 0.029 (2.9%). By assuming that all precursors have reached equilibrium with ^{131}I, the activity of ^{131}I at 8 d is

$$\begin{aligned} A &= 8.43 \times 10^5 (3000) 0.029 (1 - e^{-(0.693 \cdot 8)/8}) \\ &= 3.67 \times 10^7 \ \text{Ci}\,^{131}\text{I} \end{aligned}$$

As shown in Example 10-3, it is relatively straightforward to determine the amount of a fission product if it can be assumed that all the precursors are short compared to the reactor operating time. When this is the case, the cumulative fission yield for the nuclide of interest can be used; however, this is not always the case, as illustrated for ^{140}La in the chain for mass number 140. Lanthanum-140 ($T_{1/2} = 40.28$ h) is produced solely by ingrowth due to radioactive transformation of ^{140}Ba ($T_{1/2} = 12.76$ d), which has a cumulative fission yield of 6.3%. In most cases, the short-lived precursors of ^{140}Ba can be ignored and the ^{140}Ba inventory calculated from the cumulative fission yield; however, the formation of ^{140}La must consider not only its formation from ^{140}Ba but its removal by radioactive transformation as well. The general case for formation of N_2 from the transformation of N_1 which is also being formed and removed with time is

$$\frac{dN_2}{dt} = \lambda_1 N_1 - \lambda_2 N_2$$

and since N_1 atoms are produced at a rate of

$$\frac{3.12 \times 10^{16} P Y_1}{\lambda_1}(1 - e^{-\lambda_1 t})$$

we have

$$\frac{dN_2}{dt} + \lambda_2 N_2 = 3.12 \times 10^{16} P Y_1 (1 - e^{-\lambda_1 t})$$

which can be converted to a form for direct integration by the integrating factor $e^{\lambda_2 t}$, and solved to yield

$$N_2 = \frac{3.12 \times 10^{16} P Y_1}{\lambda_2}(1 - e^{-\lambda_2 t}) + \frac{3.12 \times 10^{16} P Y_1}{\lambda_2 - \lambda_1}(e^{-\lambda_2 t} - e^{-\lambda_1 t})$$

An almost exact solution for fission-product formation from a relatively long lived fission-product parent is to assume that the product and all of its precursor radionuclides are in equilibrium such that the production of the product is at a steady state. In this case, the activity of the shorter-lived product will be the same as the activity of the parent while both are in the reactor and the reactor is still running. If both are removed from the reactor at the same time (e.g., in a waste product or spent fuel), the number of product atoms at a time t_r after removal is

$$N_2 = \frac{N_1^0}{\lambda_2}(e^{-\lambda_1 t_r} - e^{-\lambda_2 t_r}) + N_2^0 e^{-\lambda_2 t_r}$$

where $N_1^0 = (3.12 \times 10^{16} P Y_1)/\lambda$ the saturation value of the number of atoms of N_i, and N_2^0 are stated at the time of shutdown. The activity corresponding to atoms of N_2 is obtained by multiplying by λ_2.

Fission Products in Spent Fuel

Most nuclear power plants manage the fuel such that it lasts about 3 y and produces from 25,000 to 35,000 MW · d of energy per ton of fuel. This is referred to as *fuel burnup*. Other reactors can have shorter or longer burnup periods and different burnup values, and future research is expected to develop fuels with burnups well above 50,000 MW · d per ton of fuel.

Since the total energy produced is related directly to the number of fissions that occurred, there is a direct relationship between fuel burnup and the inventory of radionuclides in the fuel elements. Since 1 MW · d of energy corresponds to 2.7×10^{21} fissions, the activity of relatively long-lived products in fuel can readily be determined when the burnup is known, as shown in Example 10-4.

Example 10-4: A fuel element is removed from a university research reactor when it reaches a burnup of 10 MW · d. Determine the inventory of ^{90}Sr in such a fuel element if it has been in the reactor for about 1 y by (a) fuel burnup and (b) average power level.

SOLUTION: (a) At 10 MW · d of burnup, 2.7×10^{22} fissions will have occurred. The approximate number of atoms of ^{90}Sr (ignoring all precursors) can be determined based on the number of fissions since the 1-y period of operation is small compared to the half-life of ^{90}Sr ($T_{1/2} = 28.78$ y) (i.e., the removal of ^{90}Sr atoms during the year is relatively small). On this basis, then

$$\text{atoms of } ^{90}\text{Sr} = 2.7 \times 10^{22} \text{ fissions} \times 0.058 \text{ atoms of } ^{90}\text{Sr/fission}$$
$$= 1.57 \times 10^{21} \text{ atoms}$$
$$\text{activity} = \lambda N = (\ln 2/28.78 \text{ y} \times 3.154 \times 10^7 \text{ s/y})(1.57 \times 10^{21} \text{ atoms})$$
$$= 1.182 \times 10^{12} \text{ d/s} = 31.94 \text{ Ci}$$

(b) The effect of both buildup and radioactive removal on the ^{90}Sr activity can be considered by determining the average power level over the 1-y period and calculating the activity in the usual way based on power level. In this case the average power level would be 10 MW · d/365 d = 0.0274 MW, and the activity of ^{90}Sr is

$$A(^{90}\text{Sr}) = 8.43 \times 10^5 P Y_i (1 - e^{-\lambda_i})$$
$$= 8.43 \times 10^5 (0.0274)(0.058)(1 - e^{-(\ln 2 \times 1)/28.3})$$
$$= 31.88 \text{ Ci}$$

which is essentially the same because radioactive transformation of long-lived ^{90}Sr is minor over the period of operation.

For shorter-lived fission products in which removal by radioactive transformation is significant, it is necessary to determine the average power level over the period of fuel burnup [as shown in part (b) of Example 10-4] to calculate the fission product inventory. This approach is somewhat more exact even for fission products with half-lives considerably longer than the burnup time, but the approximation is more straightforward for most long-lived products ($T_{1/2} > 10$ y or so) and provides satisfactory accuracy for most circumstances (see Problems 10-7 and 10-8).

Fission Product Poisons

Some of the fission fragments produced in fission and their progeny have substantial neutron absorption cross sections, which can have a significant effect on the multiplication factor k_{eff}. Xenon-135 ($\sigma_a = 2.6 \times 10^6$ b) and samarium-149 ($\sigma_a = 4.0 \times 10^4$ b) have fission yields and cross sections large enough to decrease the thermal utilization factor, f. Both change with power level, since their formation is a direct function of the fission rate, and it is necessary to adjust control rods and chemical shim to compensate for the negative reactivity effects induced by their ingrowth and burnup as fissioning proceeds.

Xenon-135 is the most important fission product poison. It is formed directly as a fission product and also by decay of its ^{135}I precursor ($Y_i = 6.28\%$). As shown in

Appendix E for mass chain 135, ^{135}Xe is formed as follows:

$$\begin{array}{ccccccc} ^{135}Sn & \rightarrow & ^{135}Sb & \rightarrow & ^{135}Te & \rightarrow & ^{135}I & \rightarrow & ^{135}Xe & \rightarrow & ^{135}Cs & \rightarrow & ^{135}Ba \\ (0.0006) & & (0.1458) & & (3.3395) & & (6.2823) & & (6.5390) & & (6.5395) & & (6.5395) \end{array}$$

The inventory of ^{135}Xe is affected by radioactive transformation of ^{135}I, which produces it, its own transformation, which removes it, and its very large cross section, which results in neutron absorption, or burnout. For all practical purposes it can be assumed that all of the ^{135}I present is formed directly by fission even though it is produced by short-lived ^{135}Te and its precursors. Since the half-life of ^{135}I is 6.57 h, it and its ^{135}Xe product will have exceeded 99% of their final equilibrium values in 46 h, or about 2 d. After shutdown, xenon continues to be produced in the reactor by the transformation of ^{135}I to ^{135}Xe, and since it too is radioactive, with a half-life of 9.1 h, it in turn is removed by its own radioactive transformation. After about 2 d of steady-state reactor operation, the amount of ^{135}I (and ^{135}Xe) present in the reactor is in equilibrium; however, after reactor shutdown, ^{135}Xe will be produced more rapidly than it is removed and the amount will peak about 7 h after shutdown and then decrease due to iodine transformation, as shown in Figure 10-23. Some 18 h after shutdown, ^{135}Xe will have peaked and returned to

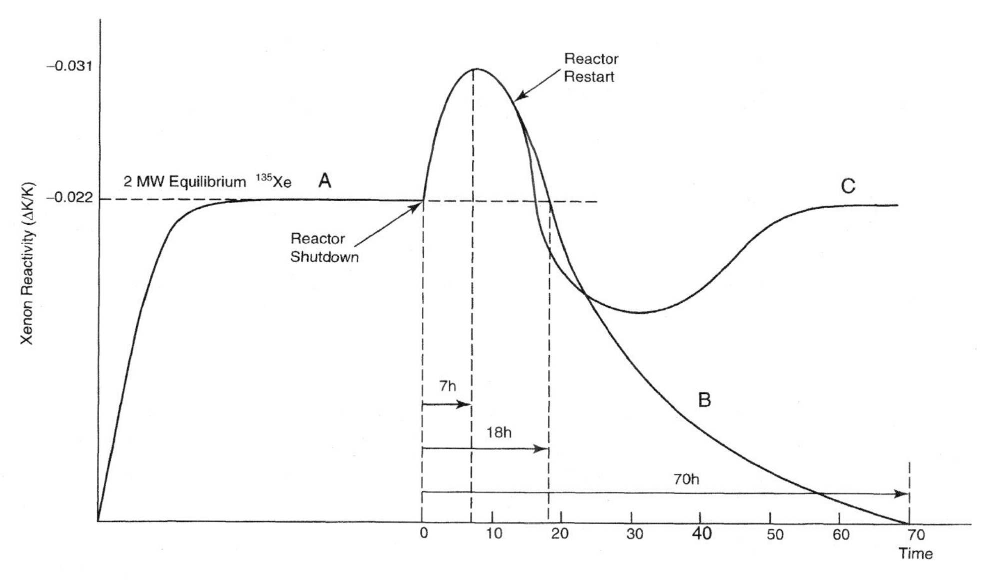

FIGURE 10-23. Effect of ^{135}Xe on reactor reactivity for steady-state operation (curve A), after shutdown (curve B), and shutdown followed by restart (curve C). The ^{135}Xe at steady state introduces negative reactivity of $\Delta k/k$ of -0.022, which becomes even more negative due to ingrowth by radioactive transformation reaching a maximum ($\Delta k/k = -0.031$) 7 h later, at which point removal by radioactive transformation overcomes its production, decreasing it to zero if the reactor remains shutdown for a period of about 70 h (curve B). If the reactor is restarted before the ^{135}Xe decays, the population of neutrons further accelerates the removal of ^{135}Xe by neutron activation; however, new ^{135}Xe is produced by fission and the negative reactivity eventually returns to an equilibrium condition ($\Delta k/k = -0.022$) some 48 h later.

the starting equilibrium level, and by about 72 h, essentially all of the ^{135}Xe atoms will have transformed to provide a clean core free of ^{135}Xe poisons.

If sufficient excess core reactivity is available to overcome the negative reactivity induced by xenon buildup following a shutdown, the reactor can be restarted. Returning the power level reestablishes the full neutron flux, and any xenon present is first depleted rapidly by neutron capture and its associated negative reactivity will drop rapidly as shown by curve C in Figure 10-23. The return to power also results in fissions that produce new atoms of iodine and xenon, which balances the removal of xenon by transformation and burnout, and equilibrium will be reestablished. If the power level were to be returned to a value different from the pre-shutdown level, a different equilibrium level would eventually be established, but at essentially the same time (i.e., 46 h) due to the ingrowth of 6.7-h ^{135}I.

Samarium-149 also affects reactor reactivity in a similar way because of its large cross section. It is a stable nuclide produced by radioactive transformation of ^{149}Pm, which is formed in the fission process by the decay of ^{149}Nd. Samarium-149 is stable, but it is depleted by burnout. Because the half-life of ^{149}Nd is short in comparison to that of ^{149}Pm, the two are treated as if ^{149}Pm were formed directly from fission with a fission yield of 1.082% (see Appendix E). The half-life of ^{149}Pm (2.212 d) is the controlling time constant; therefore, promethium and samarium will reach equilibrium in about 16 d.

Increases and decreases in reactor power represent transient conditions as illustrated in Figure 10-24. If reactor power is increased, the samarium concentration will decrease initially due to higher burnout, but ^{149}Pm production will eventually bring the samarium concentration back up to its equilibrium value. If power is decreased, burnout decreases, production of promethium continues, and the samarium concentration increases due to replenishment by radioactive transformation of new atoms of promethium produced by fission. A power decrease eventually decreases the concentration of promethium due to the lower neutron flux, and the samar-

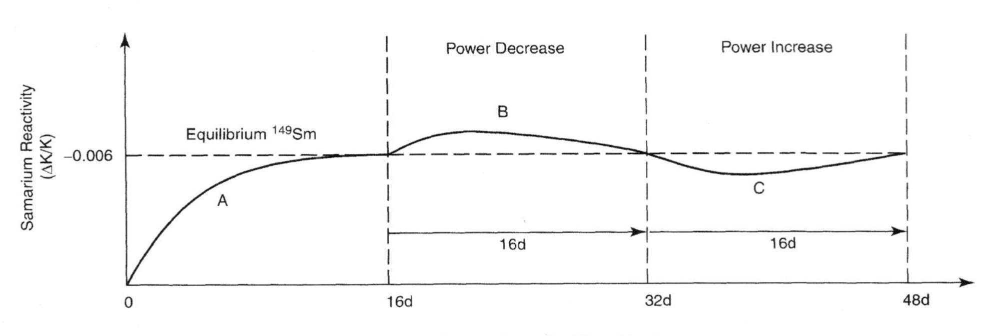

FIGURE 10-24. Effect of stable ^{149}Sm, a product of fission, on reactor reactivity. An equilibrium condition of negative reactivity ($\Delta k/k = 0.006$) is achieved in about 16 d (curve A). If the power level is decreased, $\Delta k/k$ becomes even more negative due to ingrowth of ^{149}Sm from the decay of ^{149}Pm (curve B), which returns to equilibrium some 16 d later. If the power level is increased back to its original level (curve C), ^{149}Sm is at first decreased due to neutron absorption, but the increased fission rate produces more ^{149}Pm and ^{149}Sm and the negative reactivity coefficient returns to the equilibrium value ($\Delta k/k = -0.006$).

ium concentration reaches an equilibrium value. In both cases, the transient time is approximately 16 d.

RADIOACTIVE WASTES FROM REACTORS

Fission reactions in reactor fuel produce numerous new products as the fissile material is consumed. As shown in Figure 10-15, fuel that initially contains 3.3% ^{235}U and 96.7% ^{238}U will, after about 3 y in the reactor, still contain mostly ^{238}U but substantial amounts of fission and activation products and isotopes of plutonium. These spent fuel rods or any by-products of processing them constitute high-level wastes.

There are over 1200 individual fission products produced in fission; however, the major ones fall into two groups, with mass numbers of 82 to 106 and 121 to 144 or so. The features of key fission products in these groupings are provided in Figures 10-25 and 10-26.

Low-Level Radioactive Waste

Even with good fuel performance, some radioactive materials will be present in various parts of a nuclear reactor that must be removed as wastes. Although the sources, amounts, and types vary from plant to plant, most radioactive wastes contain fission and activation products that escaped the fuel matrix or were entrained in coolant and removed. These are characterized as low-level radioactive wastes, and for reactors generally fall into one of the following categories:

- *Condensate filter sludges and resins*, which are used to clean up condensate from the turbine before it is returned to the reactor. Deep-bed or powdered resins are typically used, including precoat filter/demineralizers used in many plants.
- *Filter sludges and resins* and associated prefilters, used to clean up reactor water. Cleanup demineralizer resins are used in BWRs and in the primary coolant let-down system in PWRs; both can become very radioactive.
- *Evaporator concentrates (or bottoms)* remaining from the distillation of liquids in both BWR and PWR plants. In BWRs, most such liquids are produced during regeneration of demineralizers used to clean up condensate and various waste streams. In PWRs, evaporator concentrates are produced from primary equipment drain sumps, radwaste sumps, and liquids from regeneration of demineralizer resins.
- *Resins and filters* used to cleanse water in the fuel storage pool and various other liquids before they are evaporated.
- *Crud*, which is a scale or deposit on surfaces of vessels, pipes, and equipment. Crud builds up radioactivity, some of which may break off and reach a waste processing system. Crud is normally not discharged from the plant, but these deposits can be important in maintenance and decommissioning.
- *Trash*, which includes paper used on floors, plastic bags and sheeting, disposable shoe covers, and contaminated clothing.

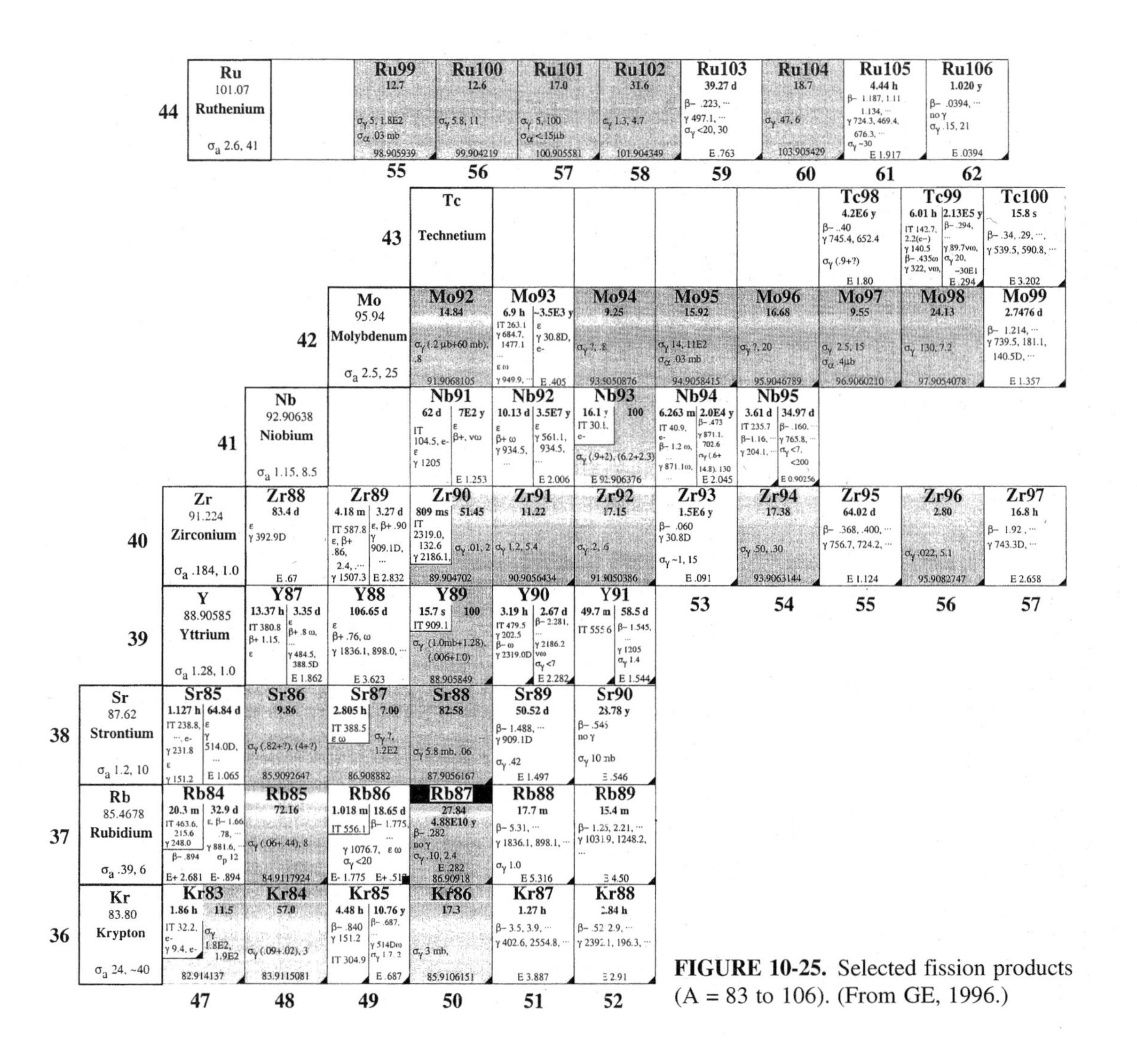

FIGURE 10-25. Selected fission products (A = 83 to 106). (From GE, 1996.)

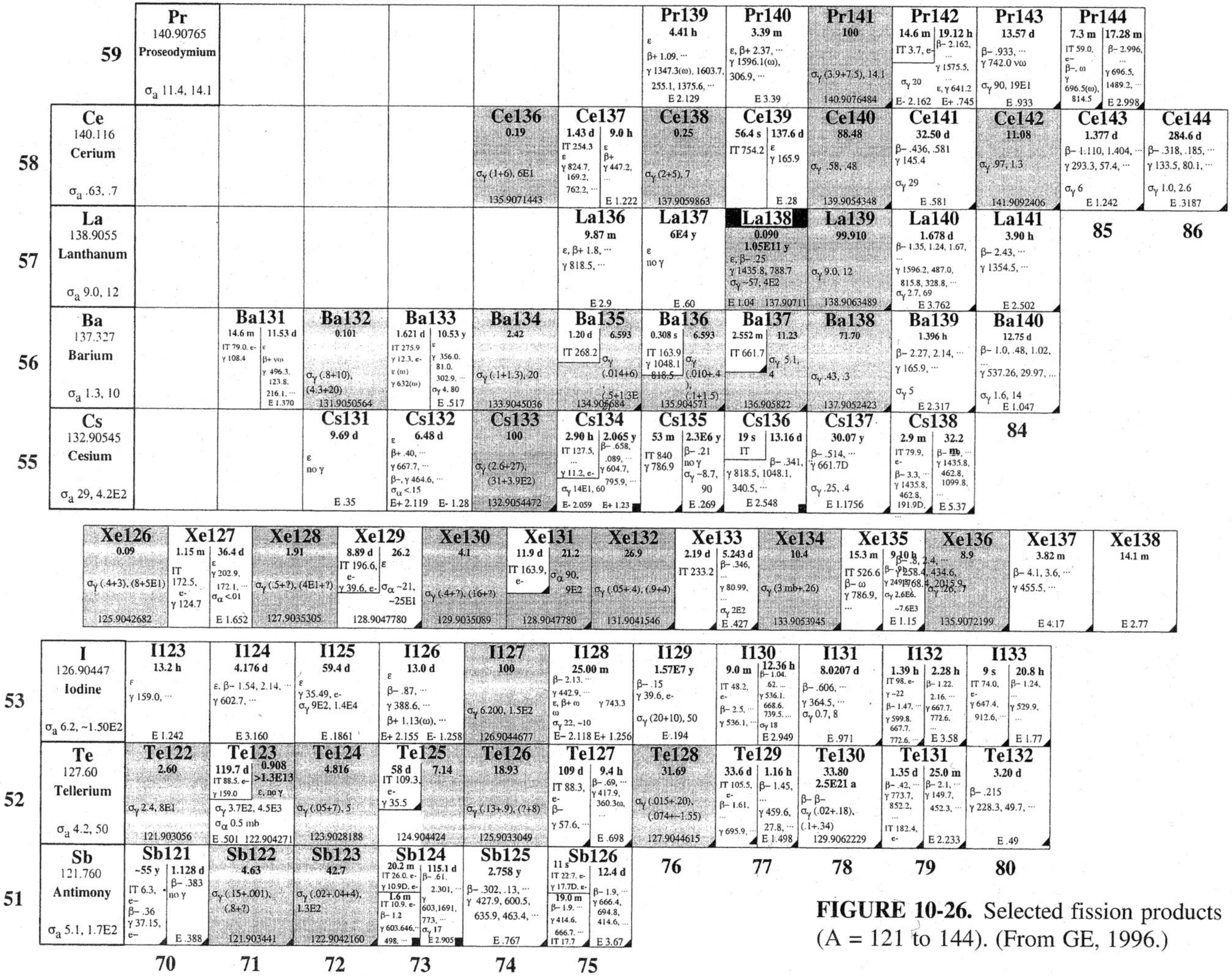

Z = 59 — Pr, 140.90765, Proseodymium, σ_a 11.4, 14.1

- **Pr139** — 4.41 h; ε; β+ 1.09, ···; γ 1347.3(ω), 1603.7, 255.1, 1375.6, ···; E 2.129
- **Pr140** — 3.39 m; ε, β+ 2.37, ···; γ 1596.1(ω), 306.9, ···; E 3.39
- **Pr141** — 100; σ_γ (3.9+7.5), 14.1; 140.9076484
- **Pr142** — 14.6 m | 19.12 h; IT 3.7, e-; β− 2.162, ···; γ 1575.5, ···; σ_γ 20; ε, γ 641.2; E- 2.162 E+ .745
- **Pr143** — 13.57 d; β− .933, ···; γ 742.0 vω; σ_γ 90, 19E1; E .933
- **Pr144** — 7.3 m | 17.28 m; IT 59.0, e-, β−, ω, γ 696.5(ω), 814.5; β− 2.996, ···; γ 696.5, 1489.2, ···; E 2.998

Z = 58 — Ce, 140.116, Cerium, σ_a .63, .7

- **Ce136** — 0.19; σ_γ (1+6), 6E1; 135.9071443
- **Ce137** — 1.43 d | 9.0 h; IT 254.3, ε, γ 824.7, 169.2, 762.2, ···; ε, β+, γ 447.2, ···; E 1.222
- **Ce138** — 0.25; σ_γ (2+5), 7; 137.9059863
- **Ce139** — 56.4 s | 137.6 d; IT 754.2; ε, γ 165.9; E .28
- **Ce140** — 88.48; σ_γ .58, .48; 139.9054348
- **Ce141** — 32.50 d; β− .436, .581; γ 145.4; σ_γ 29; E .581
- **Ce142** — 11.08; σ_γ .97, 1.3; 141.9092406
- **Ce143** — 1.377 d; β− 1.110, 1.404, ···; γ 293.3, 57.4, ···; σ_γ 6; E 1.242
- **Ce144** — 284.6 d; β− .318, .185, ···; γ 133.5, 80.1, ···; σ_γ 1.0, 2.6; E .3187

(N = 85, 86)

Z = 57 — La, 138.9055, Lanthanum, σ_a 9.0, 12

- **La136** — 9.87 m; ε, β+ 1.8, ···; γ 818.5, ···; E 2.9
- **La137** — 6E4 y; ε; no γ; E .60
- **La138** — 0.090; 1.05E11 y; ε, β− .25; γ 1435.8, 788.7; σ_γ ~57, 4E2; E 1.04; 137.90711
- **La139** — 99.910; σ_γ 9.0, 12; 138.9063489
- **La140** — 1.678 d; β− 1.35, 1.24, 1.67, ···; γ 1596.2, 487.0, 815.8, 328.8, ···; σ_γ 2.7, 69; E 3.762
- **La141** — 3.90 h; β− 2.43, ···; γ 1354.5, ···; E 2.502

Z = 56 — Ba, 137.327, Barium, σ_a 1.3, 10

- **Ba131** — 14.6 m | 11.53 d; IT 79.0, e-, γ 108.4; ε, β+ vω, γ 496.3, 123.8, 216.1, ···; E 1.370
- **Ba132** — 0.101; σ_γ (.8+10), (4.3+20); 131.9050564
- **Ba133** — 1.621 d | 10.53 y; IT 275.9, γ 12.3, e-, ε (m), γ 632(ω); ε, γ 356.0, 81.0, 302.9, ···; σ_γ 4, 80; E .517
- **Ba134** — 2.42; σ_γ (.1+1.3), 20; 133.9045036
- **Ba135** — 1.20 d | 6.593; IT 268.2; σ_γ (.014+6), (.5+1.3E); 134.905684
- **Ba136** — 0.308 s | 6.593; IT 163.9, γ 1048.1, 818.5; σ_γ (.010+.4), (1+1.5); 135.904571
- **Ba137** — 2.552 m | 11.23; IT 661.7; σ_γ 5.1, 4; 136.905822
- **Ba138** — 71.70; σ_γ .43, .3; 137.9052423
- **Ba139** — 1.396 h; β− 2.27, 2.14, ···; γ 165.9, ···; σ_γ 5; E 2.317
- **Ba140** — 12.75 d; β− 1.0, .48, 1.02, ···; γ 537.26, 29.97, ···; σ_γ 1.6, 14; E 1.047

(N = 84)

Z = 55 — Cs, 132.90545, Cesium, σ_a 29, 4.2E2

- **Cs131** — 9.69 d; ε; no γ; E .35
- **Cs132** — 6.48 d; ε; β+ .40, ···; γ 667.7, ···; β−, γ 464.6, ···; σ_α <.15; E+ 2.119 E- 1.28
- **Cs133** — 100; σ_γ (2.6+27), (31+3.9E2); 132.9054472
- **Cs134** — 2.90 h | 2.065 y; IT 127.5, ···, γ 11.2, e-; β− .658, .089, ···, γ 604.7, 795.9, ···; σ_γ 14E1, 60; E- 2.059 E+ 1.23
- **Cs135** — 53 m | 2.3E6 y; IT 840, γ 786.9; β− .21, no γ; σ_γ ~8.7, 90; E .269
- **Cs136** — 19 s | 13.16 d; IT; β− .341, ···; γ 818.5, 1048.1, 340.5, ···; E 2.548
- **Cs137** — 30.07 y; β− .514, ···; γ 661.7D; σ_γ .25, .4; E 1.1756
- **Cs138** — 2.9 m | 32.2 m; IT 79.9, e-, β− 3.3, ···, γ 1435.8, 462.8, 191.9D, ···; β− ···, γ 1435.8, 462.8, 1099.8, ···; E 5.37

Z = 54 — Xe

- **Xe126** — 0.09; σ_γ (.4+3), (8+5E1); 125.9042682
- **Xe127** — 1.15 m | 36.4 d; IT 172.5, e-, γ 124.7; ε, γ 202.9, 172.1, ···; σ_α <.01; E 1.652
- **Xe128** — 1.91; σ_γ (.5+?), (4E1+?); 127.9035305
- **Xe129** — 8.89 d | 26.2; IT 196.6, e-, γ 39.6, e-; ε; σ_α ~21, ~25E1; 128.9047780
- **Xe130** — 4.1; σ_γ (.4+?), (16+?); 129.9035089
- **Xe131** — 11.9 d | 21.2; IT 163.9, e-; σ_α 90, 9E2; 128.9047780
- **Xe132** — 26.9; σ_γ (.05+.4), (.9+4); 131.9041546
- **Xe133** — 2.19 d | 5.243 d; IT 233.2; β− .346, ···, γ 80.99, ···; σ_γ 2E2; E .427
- **Xe134** — 10.4; σ_γ (3 mb+.26); 133.9053945
- **Xe135** — 15.3 m | 9.10 h; IT 526.6, β− ω, γ 786.9, ···; β− .91, γ 249.8, 608.2, ···; σ_γ 2.6E6, ~7.6E3; E 1.15
- **Xe136** — 8.9; β− .8, 2.4, ···; 258.4, 434.6, 1768.4, 2015.9; σ_γ .26, .7; 135.9072199
- **Xe137** — 3.82 m; β− 4.1, 3.6, ···; γ 455.5, ···; E 4.17
- **Xe138** — 14.1 m; E 2.77

Z = 53 — I, 126.90447, Iodine, σ_a 6.2, ~1.50E2

- **I123** — 13.2 h; ε; γ 159.0, ···; E 1.242
- **I124** — 4.176 d; ε, β− 1.54, 2.14, ···; γ 602.7, ···; E 3.160
- **I125** — 59.4 d; ε; γ 35.49, e-; σ_γ 9E2, 1.4E4; E .1861
- **I126** — 13.0 d; ε; β− .87, ···; γ 388.6, ···; β+ 1.13(ω), ···; E+ 2.155 E- 1.258
- **I127** — 100; σ_γ 6.200, 1.5E2; 126.9044677
- **I128** — 25.00 m; β− 2.13, ···; γ 442.9, ···; ε, β+ ω; γ 743.3; σ_γ 22, ~10; E− 2.118 E+ 1.256
- **I129** — 1.57E7 y; β− .15; γ 39.6, e-; σ_γ (20+10), 50; E .194
- **I130** — 9.0 m | 12.36 h; IT 48.2, e-, β− 2.5, ···, γ 536.1, ···; β− 1.04, .62, ···, γ 536.1, 668.6, 739.5, ···; σ_γ 18; E 2.949
- **I131** — 8.0207 d; β− .606, ···; γ 364.5, ···; σ_γ 0.7, 8; E .971
- **I132** — 1.39 h | 2.28 h; IT 98, e-, γ ~22, β− 1.47, ···, γ 599.8, 667.7, 772.6, ···; β− 1.22, 2.16, ···, γ 667.7, 772.6, ···; E 3.58
- **I133** — 9 s | 20.8 h; IT 74.0, e-, γ 647.4, 912.6, ···; β− 1.24, ···, γ 529.9, ···; E 1.77

Z = 52 — Te, 127.60, Tellerium, σ_a 4.2, 50

- **Te122** — 2.60; σ_γ 2.4, 8E1; 121.903056
- **Te123** — 119.7 d | 0.908, >1.3E13; IT 88.5, e-, γ 159.0; ε, no γ; σ_γ 3.7E2, 4.5E3; σ_α 0.5 mb; E .501 122.904271
- **Te124** — 4.816; σ_γ (.05+7), 5; 123.9028188
- **Te125** — 58 d | 7.14; IT 109.3, e-, γ 35.5; 124.904424
- **Te126** — 18.93; σ_γ (.13+.9), (?+8); 125.9033049
- **Te127** — 109 d | 9.4 h; IT 88.3, e-, β−, γ 57.6, ···; β− .69, ···, γ 417.9, 360.3ω, ···; E .698
- **Te128** — 31.69; σ_γ (.015+.20), (.074+.1.55); 127.9044615
- **Te129** — 33.6 d | 1.16 h; IT 105.5, e-, β− 1.61, ···, γ 695.9, ···; β− 1.45, ···, γ 459.6, 27.8, ···; E 1.498
- **Te130** — 33.80; 2.5E21 a; β− β−; σ_γ (.02+.18), (.1+.34); 129.9062229
- **Te131** — 1.35 d | 25.0 m; β− .42, ···, γ 773.7, 852.2, ···, IT 182.4, e-; β− 2.1, ···, γ 149.7, 452.3, ···; E 2.233
- **Te132** — 3.20 d; β− .215; γ 228.3, 49.7, ···; E .49

Z = 51 — Sb, 121.760, Antimony, σ_a 5.1, 1.7E2

- **Sb121** — ~55 y | 1.128 d; IT 6.3, e-, β− .36, γ 37.15, e-; β− .383, no γ; E .388
- **Sb122** — 4.63; σ_γ (.15+.001), (.8+?); 121.903441
- **Sb123** — 42.7; σ_γ (.02+.04+4), 1.3E2; 122.9042160
- **Sb124** — 20.2 m | 1.6 m | 115.1 d; IT 26.0, e-, γ 10.9D, e-; IT 10.9, e-, β− 1.2, γ 603.646, ···, 498, ···; β− .61, 2.301, ···, γ 603.1691, 773, ···; σ_γ 17; E 2.905
- **Sb125** — 2.758 y; β− .302, .13, ···; γ 427.9, 600.5, 635.9, 463.4, ···; E .767
- **Sb126** — 11 s | 19.0 m | 12.4 d; IT 22.7, e-, γ 17.7D, e-; β− 1.9, ···, γ 414.6, 666.7, ···, IT 17.7; β− 1.9, ···, γ 666.4, 694.8, 414.6, ···; E 3.67

Neutron numbers: 70, 71, 72, 73, 74, 75, 76, 77, 78, 79, 80, 84, 85, 86

FIGURE 10-26. Selected fission products (A = 121 to 144). (From GE, 1996.)

- *Scrap*, which consists of structural materials, piping, small valves, and various other components from maintenance and plant modifications.

SUMMARY

The principal fissionable materials are ^{235}U, ^{238}U, and ^{239}Pu. Absorption of a neutron by a fissionable nucleus produces a dumbbell-shaped deformation in which repulsive forces exerted by the positive charges in each of the two "halves" can cause the two parts to separate, or fission. The excitation energy added by the neutron can also be relieved by emission of a gamma photon, but fission, which occurs 85% of the time, is more likely. Radioactive fission products and activation products are by-products of these reactions, and these represent a number of radiation protection issues for workers, the public, and the environment.

Uranium, plutonium, and other fissionable nuclei are very neutron rich because extra neutrons help to distribute the nuclear force and counteract the repulsive forces exerted by the large number of protons in the nucleus. This excess ratio of neutrons to protons carries over to the fission products; hence each product that is formed will also be extremely neutron rich and very unstable, and several beta particles and gamma rays are emitted to lower the neutron number (and increase the proton number) in order for the fission product to become stable.

Each fission produces its own distinct mass chain with a slightly different *Q*-value; however, about 200 MeV of energy is released per fission and most (85% or so) of this energy appears as heat as the heavy, highly charged fission fragments are absorbed in the fuel matrix. Part of the remaining energy is released instantaneously by gamma rays and fission neutrons, and the rest is released gradually by radioactive transformation of the fission products as they emit beta particles, neutrinos, and gamma rays. The release of fission energy is governed by a fundamental set of physical principles which eventually dictate the design and performance of nuclear reactors and other critical assemblies. Radiation protection for these circumstances must also recognize these principles.

The central feature of a nuclear reactor is the core, which contains the fuel, the control rods, the moderator, and a coolant; it is enclosed in a vessel with various components and subsystems to enhance fissioning in the fuel, to control its rate, to extract heat energy, and provide protection of materials and persons. The two principal designs used in the United States and many other countries are the pressurized water reactor (PWR) and the boiling water reactor (BWR).

The primary coolant water in a PWR is maintained at a high pressure (2200 psi) so that it will not boil; therefore, steam for the turbines must be produced in a secondary circuit containing steam generators. The heated coolant water from the reactor enters the steam generator at the bottom and passes upward and then downward through several thousand tubes each in the shape of an inverted U. The outer surfaces of these tubes are in contact with lower-pressure and cooler feedwater, which extracts heat, boils, and produces steam that is routed to the turbine. The spent steam is condensed and pumped back to the steam generators to complete the secondary circuit.

The BWR uses light water for both coolant and moderator, and the water is allowed to boil to produce steam that is routed directly to the turbines. This direct cycle design can be a major advantage because it eliminates the need for a separate heat transfer loop; however, the water becomes radioactive in passing through the reactor core, which requires shielding of the steam piping, the turbines, the condenser, reheaters, pumps, pipings, and so on.

Ordinary light water, H_2O, has desirable features for use in reactors. First it has good heat transfer and is plentiful, so that it can easily be replaced if leakage or evaporation occurs. It is also very effective in slowing down neutrons, but this good feature is offset by a high capture cross section for neutrons, making it unsuitable as the coolant or moderator unless the fuel is enriched to about 3 to 4% ^{235}U. Alternatively, D_2O, or heavy water, which absorbs very few neutrons, can be used as the moderator and coolant, allowing the use of natural uranium (0.72% ^{235}U) as fuel; heavy water reactors (e.g., the CANDU) use such features.

Reactors are sources of many radioactive materials, many of which exist in offgas systems, effluents, or in waste materials to cause radiation exposures of workers, or to be released during routine operations, incidents, or in the form of radioactive wastes. Thermal neutron fission of ^{235}U yields over 88 primary fission products with mass numbers from 72 to 160 and well over 1200 different fission products. Inventories of these products can be calculated from the fission rate (or reactor power) and the fission yield of the product.

ACKNOWLEDGMENTS

This chapter was compiled with major help from Chul Lee and Ihab R. Kamel. Dr. Kamel compiled the table of fission product chains and yields from recent data graciously provided by Dr. Tal England, who made it available before it was published.

REFERENCES AND ADDITIONAL RESOURCES

England, T. R., and B. F. Rider, *Evaluation and Compilation of Fission Product Yields*, Report LA-UR-94-3106 (ENDF-349), Los Alamos National Laboratory, Los Alamos, NM, October 1994.

GE, Chart of the nuclides, *Nuclides and Isotopes*, 15th ed., General Electric Company, San Jose, CA, 1996.

Lamarsh, J. R., *Introduction to Nuclear Engineering*, 2nd ed., Addison-Wesley, Reading, MA, 1983.

PROBLEMS

10-1. In 1000 atoms of present-day natural uranium, 993 are ^{238}U and 7 are ^{235}U. (**a**) What would have been the percentage of ^{235}U in natural uranium 2 billion years ago? (**b**) Could a natural reactor have existed 2 billion years ago? Why?

10-2. A nuclear reactor of 1000 MWe and a generation efficiency of 32% starts up and runs for 1 d. (**a**) What is the inventory of ^{133m}Xe and ^{133}Xe in the core? (**b**) How long would it have to run to reach an equilibrium level of ^{133}Xe, and what would be the inventory at equilibrium?

10-3. A nuclear reactor operates for 3 y at a power level of 1000 MW (3300 MW), after which all the fuel is removed (a problem core) and processed soon after releasing all of the noble gases. How many curies of ^{85}Kr would be released to the world's atmosphere?

10-4. A nuclear reactor of 1000 MWe and a generation efficiency of 32% starts up and runs for 10 d. (**a**) What is the inventory of ^{140}Ba in the core? (**b**) How long would it have to run to reach an equilibrium level of ^{140}Ba, and what would be the inventory at equilibrium? (**c**) At 10 d after startup, what would be the inventory of ^{140}La?

10-5. (**a**) Estimate the ^{14}C inventory in a 1000 MWe PWR after 350 d of operation if the coolant volume is 80,000 gal and the coolant contains dissolved air of 50 ppm in the primary coolant. (**b**) If this is released continuously as CO_2 with other stack gases, estimate the dilution needed to reduce the ^{14}C concentration to 10^{-6} μCi/cm^3.

10-6. Estimate the ^{85}Kr content in a reactor core containing 100 tons of UO_2 after 2 y of operation at a burnup of 33,000 MW · d per metric ton. (Assume 3% enrichment and no leakage of ^{85}Kr from the fuel.)

10-7. High-level radioactive waste standards for spent nuclear fuel are based on inventories of various radionuclides. (**a**) What is the inventory of ^{137}Cs in 1000 metric tons of spent fuel based on a burnup of 33,000 MW · d/ton over 3 y? (**b**) What would be the ^{137}Cs inventory based on an average power level for the 3-y period?

10-8. What is the inventory of ^{99}Tc for the fuel in Problem 10-7? Show whether it is necessary to use average power level in this calculation, and explain why.

10-9. Find the energy difference between ^{235}U + n and ^{236}U, which can be regarded as the excitation energy of a compound nucleus of ^{236}U, and repeat for ^{238}U + n and ^{239}U. Use these data to explain why ^{235}U will fission with very low energy neutrons, while fission of ^{238}U requires fast neutrons of 1 · 4 MeV or greater. From a similar calculation, predict whether ^{239}Pu requires low- or higher-energy neutrons to fission.

11

NUCLEAR CRITICALITY

If the radiance of a thousand suns were to burst into the sky, that would be the splendor of the Mighty One.

—Bhagavad-Gita, ~ 1 AD

Anyone who has witnessed the explosion of a nuclear device will attest that the emitted light is indeed brighter than "a thousand suns," a connection easily made by many of the physicists who saw the first nuclear explosion on July 16, 1945. J. Robert Oppenheimer, who led the development of the atomic bomb, learned Sanskrit so that he could read the ancient 700-stanza Bhagavad-Gita in the original. His recollection of the phrase, "Now I have become death, destroyer of worlds" from the ancient scripture is frequently cited to characterize the new age the world entered that day; however, the Gita's reference to "the radiance of a thousand suns" also captured the awe-inspiring splendor, and the physics, of the event.

This author, after shivering with Morgan Seal and R. H. Neill in the early morning darkness on a plateau above Yucca Flats in Nevada, was struck by the blinding brilliance, felt the heat, and saw the changing colors of the nuclear fireball as it spread and rose into the dawn of a September day in 1957; experienced several months later the persistent energy of "Oak," the largest of several multimegaton devices exploded in the Marshall Islands in 1958; and saw the brightness of day burst upon and persist for tens of seconds in the night sky above Fiji from a high-altitude detonation some 2500 miles away at Johnston Island in the Pacific. The light of nuclear criticality, once seen, is like no other. It is unfortunate that it has also been seen, as detailed in several accounts that follow, by those exposed to the burst of neutrons and gamma rays that also accompany the light of nuclear excursions.

Prevention of inadvertent criticality of fissionable materials events requires consideration of the same factors that must be met to obtain an explosive release of fission energy in nuclear devices or in a controlled fashion in nuclear reactors. The purpose of this chapter is to describe the features of criticality events, the factors related to their occurrence, and the relevant radiation protection issues associated with each.

NUCLEAR REACTORS AND CRITICALITY

"Natural Reactor" Criticality Event

A *"natural reactor" criticality event* occurred in Gabon, Africa some 2 billion years ago. It was discovered when assays of uranium ore from the region found that the ^{235}U content was well below 0.72% found in most current-day natural uranium, which suggested that the ^{235}U had been depleted by fission. The Gabon natural reactor provides a good example of the interplay of various parameters essential for nuclear criticality to occur. First, the ore body was more than large enough to provide a critical mass. Second, with arid conditions the surface layer would dry out and crack so that rainwater could settle in the cracks and provide a zone of moderator to slow any neutrons to thermal energies for enhanced fission of ^{235}U in the ore. Third, the ^{235}U content was sufficient to overcome neutron absorption losses in hydrogen atoms present in light water. Fourth, spontaneous fission in uranium would provide neutrons to begin the reaction. Fifth, the heat of the reaction would boil the water to steam, forcing it out of the matrix to stop the reaction, which would start up again when more water accumulated in the cracks and a spontaneous fission event occurred to begin a new chain reaction. It appears that this sequence was repeated many times over hundreds to thousands of years until the semi-ideal conditions changed, perhaps by rearrangement of the ore body due to the energy of the reactions, changes in climate, and/or reduction of the percent abundance of ^{235}U below that necessary to sustain a chain reaction with a light water moderator.

The first human-induced sustained chain reaction was achieved on December 2, 1942 by Enrico Fermi, who duplicated many of the conditions that occurred naturally some 2 billion years ago. Since the ^{235}U content of present-day uranium is much less, Fermi chose to intersperse natural uranium metal slugs in a graphite moderator (ordinary water absorbs too many neutrons) in a size and configuration to keep the reaction going once started (in this case with a Ra–Be neutron source). The results of Fermi's genius led to the design and operation of nuclear reactors, first to generate plutonium for nuclear weapons and later for electricity production.

Three Mile Island Accident

On March 28, 1979, a defining event in the history of nuclear power occurred at the Three Mile Island (TMI) Unit 2 reactor near Harrisburg, Pennsylvania. The reactor was operating at near its full power level of 2700 MWt (850 MWe), having reached that level about three months earlier. It had gone critical for the first time exactly one year before the incident. The coolant temperature was 600°F and the system pressure was 2200 psig, both normal conditions for a pressurized water reactor.

The incident that occurred was not a criticality event, but rather, involved overheated fuel, which then led to release of fission products within the reactor, and some of these escaped to the environment. The incident was defining in nature because it highlighted design deficiencies, but more important, it demonstrated that operator training, procedures, and communication of information are essential elements to offset human frailty.

At approximately 4:00 A.M. on March 28, a condensate pump on the secondary loop stopped, which is not unusual. When this occurred, there was a loss of flow to the feedwater pumps, which circulate water to the steam generators. Without suction, they too stopped, as designed. This produced an automatic shutdown of the turbine—with no feedwater, no steam is produced, and the turbine shuts down to protect itself. PWRs are designed with auxiliary feedwater pumps to supply feedwater to the steam generators in such events. These, in fact, came on automatically to supply water to the steam generators, but the valves had been left closed by the last maintenance crew, a circumstance that could be called the first human frailty in the event. Without coolant water, the secondary side began to boil dry. Eight minutes elapsed before the closed valves were discovered and opened, but by this time the steam generators had boiled dry, and trouble had begun. Safety systems designed into the reactor would have righted the situation; however, more confounding events, many of which were induced by the operators, followed.

Without water in the secondary system the reactor heat could not be removed, and since the reactor was still running, the primary loop pressure began to build up. This caused the pressure relief value on top of the pressurizer to open, as designed. This allowed primary coolant water to flow out the valve and into a discharge tank, put there for just such occurrences. But without feedwater to the steam generators to remove heat from the primary loop, the pressure started to increase again, and the reactor shut down automatically. At this point, a malfunction exists but the reactor is still safe. But unfortunately, the pressure relief valve had stuck in the open position. If the valve had closed, the system pressure would have leveled off, the automatic shutdown system would have stabilized, and the operator could have proceeded to a stable shutdown (and power could have been restored after the feedwater pumps and other systems were returned to normal function). But with the pressure valve stuck open, the system pressure continued to drop, and per design the emergency core cooling system (ECCS) came on to remove the fission product decay heat (about 200 MWt) which was still being produced in the fuel.

A second human frailty occurred when the operators didn't notice the stuck valve (and may have had no telemetry to indicate its position). Fearing a flooding of the loop with soiled water, they shut down the ECCS, a grave mistake. The primary system pressure dropped even further with the ECCS off, and steam bubbles began to form, which in turn caused a troubling vibration in the coolant pumps. The operators then further exacerbated the situation when they decided to shut the pumps off (the third human frailty), with the idea that cooling could be established with natural convection. However, the steam voids prevented convection flow and the core began to overheat since all means of removing the fission product decay heat (which, as shown in Figure 10-11, is about 7% of total energy production) were now out of commission. The water level dropped significantly, exposing the fuel. Without cooling the fuel cladding began to fail, and fission products were released. At the same time, water began to react with the overheated metal (a dreaded metal–water reaction), releasing hydrogen. This actually led to an explosion later, but it was contained so well by the containment structure that it went unnoticed.

The coolant containing fission products continued to flow out the stuck pressure relief valve, first overflowing the discharge tank and then onto the containment floor and into a sump. The sump pump came on, and contaminated water was pumped to the auxiliary building outside the containment structure, where volatile fission products could vent directly to the atmosphere. This pathway released about 10 to 20 Ci of ^{131}I and about 10^6 Ci of ^{133}Xe. The highest dose to the public was estimated to be 80 mrem for a person assumed to be outdoors at the site boundary for 11 d. The rest of the fission products, and surprisingly most of the ^{131}I, was contained in the primary coolant water that remained in the containment building. At 7:04 A.M. (3 h later) the Pennsylvania Health Department was notified and emergency crews from everywhere began to descend on the scene.

The TMI accident was the result of operator error, which was compounded by design deficiencies and equipment failure, especially the stuck pressurizer valve. Operator errors were numerous, including the closed valves in the feedwater line to the steam generators, misreading the condition of the pressurizer, and shutting off both the emergency core cooling pumps and the reactor cooling pumps.

The TMI accident led to significant actions that have improved the safety and performance of all nuclear power reactors. Among these are (1) an increase in the number of qualified operating personnel; (2) upgrading of training and operator licensing practices; (3) reviews of control room design to take account of human factors; (4) new detectors and instruments that would permit operators to know the status of the reactor at all times; (5) hydrogen-detecting equipment; (6) improvement in monitoring of accident conditions, including inadequate core cooling; (7) improved intercommunications between regulators and the plants; and (8) better emergency preparedness. These actions were taken by the nuclear industry as a whole primarily through the formation of the Institute of Nuclear Power Operations, in which senior managers share experiences and knowledge. If the industry had not imposed these improvements, the U.S. Nuclear Regulatory Commission certainly would have.

Chernobyl Accident

A very serious reactor accident occurred at Unit 4 of the Chernobyl reactor station near Kiev in Ukraine on April 26, 1986. The Chernobyl-4 reactor was an RBMK design with a graphite moderator pierced by over 1600 pressure-tube fuel channels spaced 25 cm apart in the graphite. Clusters of 18 slightly enriched (2% ^{235}U) uranium oxide fuel assemblies were placed inside each pressure tube and the heat of fission was removed by ordinary light water. Separate channels contained 222 control and shutdown rods, and the reactor could be refueled on line by a refueling machine above the core.

The sequence of events leading to the accident was begun with an experiment conducted by a group not responsible for operations of the reactor. The test sought to determine if electricity produced during coastdown of the turbine could be used to run emergency core cooling systems until diesel generators could be started to produce emergency power. This separate organization was not familiar with the reactor and apparently placed more attention on the test than on safety measures. The operators were under some pressure to complete the test because the next maintenance period was over a year away, a situation made even more stressful when the local dispatcher requested a delay of 8 h. The emergency core cooling system was deactivated for the test.

The first step in the test was to reduce the power from 3200 MWt to about 800 MWt, but it was inadvertently reduced to 30 MWt, which was too low to offset the ingrowth of ^{135}Xe fission product poison. The rapid ingrowth of ^{135}Xe made it very difficult to bring the power level back up. In violation of all rules the operators

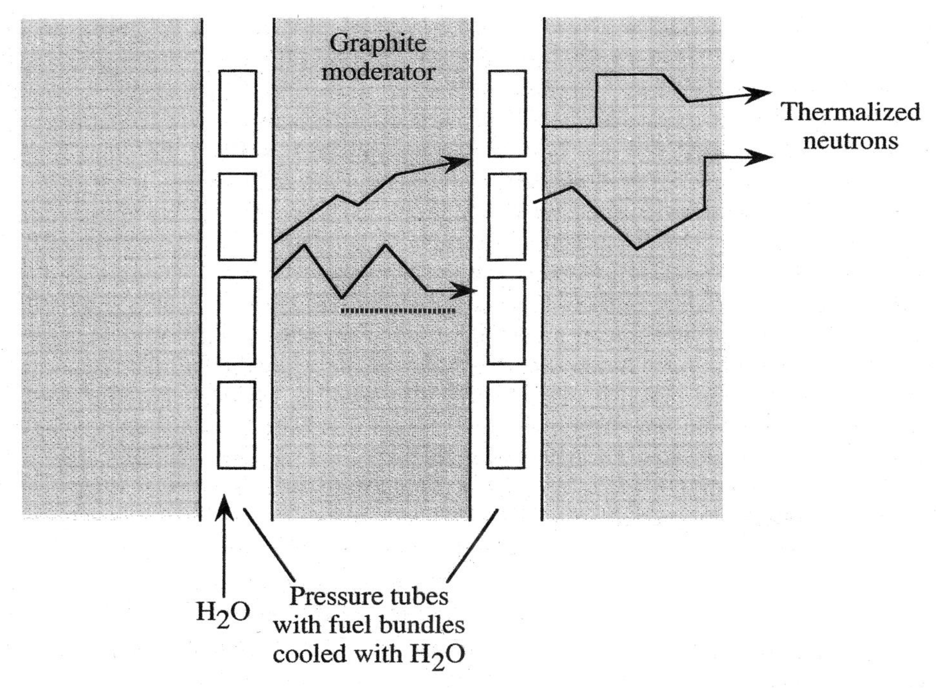

FIGURE 11-1. Schematic of fuel channels in a graphite-moderated, light-water-cooled RMBK-type reactor in which an increase in fuel temperature causes a reduction of neutron absorption by coolant, thus inducing positive reactivity.

pulled out most of the control rods in an attempt to override the ^{135}Xe poison and overcome their earlier error; however, they were unable to bring the power level back to 800 MWt, and in fact could not get it above 200 MWt. The RBMK reactor is unstable at 200 MWt, a situation made even worse because various safety systems had been disabled to prevent circuit trips while the experiment was in progress. Under these conditions steam voids began to form in the coolant flowing in the pressure tubes, which adds positive reactivity because fewer hydrogen atoms exist to absorb neutrons (i.e., the RBMK has an inherent positive void coefficient, in sharp contrast to LWRs, which have a negative void coefficient).

The positive reactivity coefficient was a fatal design flaw. As shown in Figure 11-1, an increase in temperature reduces the density of the cooling water in the pressure tubes, which results in fewer neutrons being absorbed in hydrogen atoms. But these extra neutrons would still be thermalized in the graphite moderator; thus any increase in fuel temperature produces a net increase in fissioning reactions (and an increase in power level), an effect that multiplies rapidly unless neutron absorption is added quickly. Even in normal conditions an elaborate system of detectors, circuits, and control rods was required to control the power level in the RMBK reactor.

When steam voids began to form, the result was a sudden insertion of positive reactivity. This caused a power surge to around 30,000 MWt, which was 10 times normal, and it could not be reduced quickly because too many rods were too far out to have any effect. The power surge pulverized the fuel and the increased steam pressure ruptured the coolant tubes. Chemical reactions between steam, graphite, and zirconium produced large amounts of carbon monoxide and hydrogen, which exploded through the roof (designed to provide confinement but not containment). The hot graphite caught fire and burned for several days, melting the core. Large amounts of fission products were carried high into the atmosphere to drift over Scandinavia and Eastern Europe and eventually worldwide. All the noble gases were released, 3% of the transuranic elements, 13% of the ^{137}Cs, and 20% of the ^{131}I. Exposures of emergency workers ranged between 100 and 1500 rem, and 31 persons, mainly firefighters, died. Thousands of people were evacuated, including 45,000 from the town of Pripyat, many of whom were permanently relocated with great cost and undoubtedly much distress. Most of these people received less than 25 rem. Although the amount of radiation exposure to workers and the public (1.6 million person-rem has been estimated) is not precisely known, the collective dose to the public has been projected to increase the cancer risk significantly. An International Chernobyl Project, which involves more than 200 specialists from around the globe, continues to follow the accident.

Many factors contributed to the Chernobyl accident, but the fatal flaw was the positive void coefficient, which has since been changed in the other Chernobyl-type reactors in the region. The hard lesson learned at Chernobyl was the consequence of urgent actions by well-intentioned persons who were so focused on a different agenda that considerations by experts in criticality and nuclear safety were either not sought or were ignored or overridden. Similar human frailties have contributed to most criticality incidents, particularly those that have occurred with aqueous solutions and critical assembly tests.

NRX Reactor Accident

The NRX reactor in Chalk River, Ontario, was developed in Canada during World War II to demonstrate the use of D_2O, or heavy water, as a moderator for a reactor fueled with natural uranium. It has the distinction of being the forerunner of a very useful nuclear power reactor, the CANDU (see Chapter 10). The uranium rods in the NRX were contained in a pressure tube and were cooled by a thin sheath of light water flowing between the aluminum-clad fuel rods and the pressure tube wall, a slightly larger concentric aluminum cylinder. The heavy water moderator, which flowed in the space between the pressure tubes, was sufficient to moderate all neutrons to thermal energies; therefore, the light water coolant was effectively a poison in that it slightly depleted the number of neutrons available for fissioning uranium atoms.

The NRX accident occurred while experiments were under way that required a reduced flow of the light water coolant for several of the fuel rods. Operator errors and electrical and mechanical safety circuit failures also contributed to the incident. With the experiments in process, a rapid power increase began on December 12, 1952, but it appeared that it would level off at about 20 MW as a slowly moving control rod took effect. Normally, this would have been a high, but tolerable power level and the situation probably would have been controllable if low-flow conditions had not been imposed. At a power of about 17 MW, the cooling water began to boil in the channels with reduced coolant flow, which effectively removed some of the H_2O poison due to the decreased density, causing the power to rise once more. The power increased to 60 to 90 MW, at which point the heavy water moderator was dumped automatically and the reaction stopped. About 1.2×10^{20} fissions had occurred and the core and calandria (fuel element support structure) were damaged beyond repair. Some 104 Ci of long-lived fission products were flushed to the basement in about 106 gal of cooling water, which had to be cleaned up. Personnel exposures were low, and the reactor was restarted about a year later.

The NRX accident also demonstrated the positive reactivity effect when light water is used to cool pressure-tube fuel channels, a situation that was similar in some respects to the RBMK Chernobyl reactor. It is also another example of the violence of the steam explosion that accompanies a rapid power surge. Also noteworthy is the ability to stop the fission reaction by dumping the D_2O moderator, which is one advantage a heavy water reactor has that a graphite-moderated reactor does not.

SL-1 Accident

A severe nuclear criticality accident, with three deaths, occurred at the stationary low-power unit one (SL-1) reactor in Idaho on January 3, 1961. The reactor had been designed for the U.S. Army to provide power in remote regions and had minimal containment and fixed systems to allow portability. The reactor had been shut down for maintenance, and three operators (the only three people at the plant) came in on the night shift to reconnect the control rods to the drive mechanisms so that the reactor could be restarted the next morning. It was necessary to physically lift the control rods to reconnect them. No one knows exactly what happened, but it is believed that the central rod was stuck and when they pulled on it, it let go with a large and sudden injection of reactivity. The steam explosion from the excursion blew out all of the control rods and lifted the heavy reactor vessel some 9 ft. All three operators were killed.

Alarms sounded in the fire station some miles away, and the responding firemen found the plant deserted, but followed procedures and took radiation readings before entering the building. The radiation levels pegged their survey meters, and they called for backup and initiated emergency response procedures. The plant health physicist responded to the site and entered the building to try and save the one person still alive. He and the medical personnel who attempted to treat the person received doses of a few rad up to 27 rad. The reactor was so badly damaged that it was dismantled and buried on site, a recovery process that took several weeks.

K-Reactor Event

K-reactor is one of several heavy water–moderated reactors built in the early 1950s to produce tritium by neutron irradiation of ^6Li targets. The reactor had operated successfully at the Savannah River site for about 25 years when it was shutdown in midcycle for various safety upgrades, primarily for improved seismic stability. These upgrades took about a year, which was longer than expected, and during this period tritium in the targets decayed to ^3He, a significant neutron absorber with a cross section of 5330 b. When the reactor was restarted in 1988, the operators had to withdraw the control rods considerably more than normal. When this was noted, the Department of Energy official ordered the reactor to be shut down until this unusual condition was formally evaluated. It was at this point that the poisoning effect of the ingrown ^3He was determined, but further questions on the formality of procedures for identifying and reviewing off-normal circumstances (even self imposed ones without prior consideration of effect on reactor criticality) were raised.

The event is yet another example of poisoning materials on reactor criticality and the need to consider these when circumstances change. Even though the operators were very familiar with their reactor and brought it to power safely, it was done so without full understanding of the conditions that existed. The K-reactor remained shut down for several years after 1988, was restarted after costly upgrades in equipment and formality of operations, and was then placed in a standby condition. It is questionable whether further operations at this relatively old reactor will ever resume; other means of producing tritium for nuclear weapons are likely to be developed.

Special Reactor Experiments

Several criticality experiments have been associated with reactors designed specifically to test various criticality parameters. Among these are the BORAX (boiling reactor experiment), SPERT (special power excursion reactor test), TREAT (transient reactor test), and LOFT (loss of fluid test) reactors built and operated in Idaho in the 1950s and 1960s. The criticality events at these special test facilities are notable because some of them produced more energy release and consequences than expected as criticality was induced, especially the BORAX experiment. The BORAX, SPERT, and TREAT tests were accompanied with severe steam explosions following insertion of significant amounts of reactivity. The LOFT experiment yielded valuable data on the need to remove fission product decay heat and the design of emergency core cooling systems (ECCS) to limit fuel damage from nuclear power excursions.

The VENUS facility in Mol, Belgium and the RA-2 facility in Buenos Aires, Argentina are two experimental facilities, among others, that substantiate the value of clear and rigorous procedures with respect to criticality events. VENUS was a tank-type, heavy water–moderated facility used for criticality experiments. The D_2O moderator could be diluted with H_2O to soften the neutron spectrum and maintain reactivity as fuel was consumed. A criticality incident occurred at VENUS in an experiment to test a new rod pattern. The moderator–H_2O composition was 30% D_2O and 70% H_2O and the fuel was UO_2 enriched in ^{235}U to 7%. Under these conditions the reactor should have been subcritical by one safety rod, two control rods, and one manual rod, which would allow a different manual rod to be pulled out of the core and the reactor made critical again by lifting two safety rods. The test required a person first to insert one manual rod and then extract another. Operating rules required that the reactor vessel be emptied before manipulation of a manual rod in the core, but the operator did not adhere to the rule and gave a written order to a technician to load a manual rod and then to unload a different one. The technician started the manipulation in the wrong order: He extracted a manual rod without first inserting the other, and the reactor became critical. He noticed a flash of light in the bottom of the reactor, immediately dropped the control rod, and the excursion (estimated at about 4.3×10^{17} fissions) was apparently stopped by the falling rod. No steam was created, no damage was done to the fuel, and there was no contamination. The technician received severe radiation doses to his head and upper body of 300 to 500 rem and 4000 rem to his left foot, which rested on a grating above the vessel. He survived, but the left foot had to be amputated.

The RA-2 facility in Argentina was an essentially zero-power experimental reactor facility with a graphite reflector and a large vessel filled with demineralized water for test operations. A criticality occurred while a qualified operator with 14 years of experience was making a change in the fuel configuration. He was alone in the reactor room, and in violation of procedures, had not drained the moderator from the reactor vessel. Two fuel elements, which were supposed to have been removed completely from the tank, remained in the tank just outside the graphite reflector. A nuclear criticality of 4.5×10^{17} fissions occurred, apparently as the second of two control elements was being installed since it was found only partially inserted after the accident. The operator received an absorbed dose of about 2000 rad of prompt gamma radiation and 1700 rad due to neutrons. He survived for only 2 d.

NUCLEAR EXPLOSIONS

The detailed information on nuclear weapons is highly classified, but the general physical concepts of the two major types are fairly well known. The first, based on fission, uses plutonium or highly enriched uranium, and the second depends on thermonuclear fusion to produce a hydrogen bomb. The physical principles of modern fission weapons are shown schematically in Figure 11-2. The predetonation configuration arrays a critical mass of plutonium or uranium in a thin spherical geometry which prevents criticality by providing a large surface area to promote neutron leakage. A supercritical configuration is produced by the use of high explosives to rapidly compress the spherical shell with pressures of 10 million pounds per square inch or more. Compression reduces the core radius, which increases the surface/volume ratio, and even though this increases neutron leakage, the increased density decreases the mean free path of the neutrons, which is the dominant effect. Nuclear weapons are designed to hold the fissionable material together long enough to fission the maximum possible number of uranium or plutonium atoms. Such designs depend on fast neutrons; there is not enough time to moderate neutrons to thermal energies, and moderating material is kept out of these devices intentionally. As postulated in Figure 11-2, a burst of neutrons is required at the moment of compression. This could be provided by an (α,n) source of neutrons (called an *initiator*). Under these conditions the neutron multiplication accelerates, power increases rapidly as the fission energy is released, and the rapid increase in power eventually blows the material apart.

An unreflected plutonium assembly has a critical mass of about 16 kg, which can be reduced by adding a 1-in. reflecting layer of natural U to about 10 kg. Devices composed of unreflected ^{235}U require about 23 kg for 93% enriched metal; below 20% enrichment they are impractically large.

Fusion Weapons

Hydrogen bombs are possible because a fission trigger (or primary) can be used to produce the intense heat necessary for fusion reactions (as described in Chapter

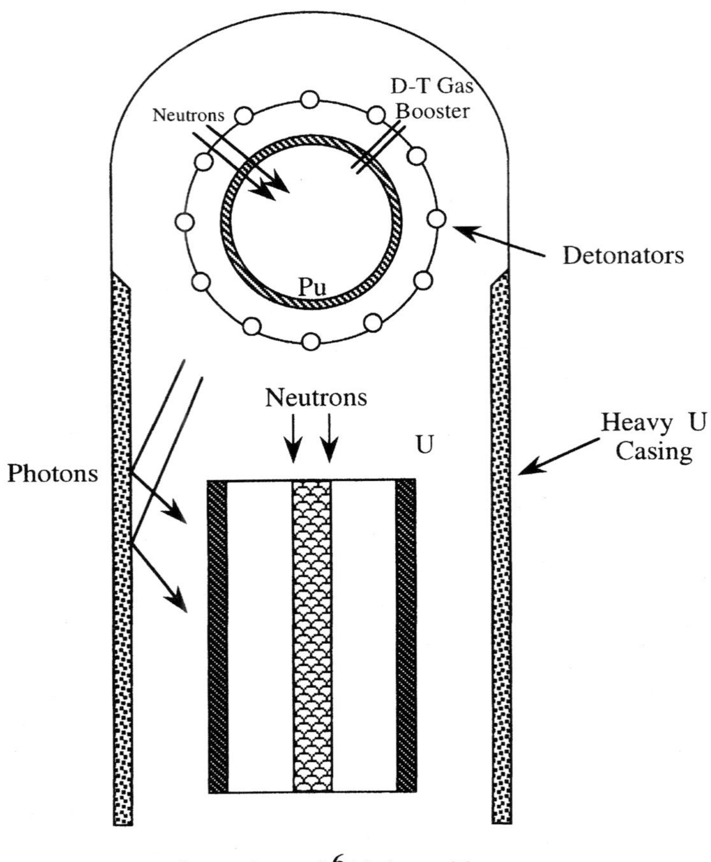

FIGURE 11-2. Hypothetical configuration of a nuclear weapon containing a tritium-boosted primary that fissions when imploded and a secondary of $^{6}Li^{2}H$ around a fission "spark plug" that is compressed by radiation to fusion temperature.

4)—hence the term *thermonuclear weapons*. Some of the fission neutrons interact with the ^{6}Li in an assembly of lithium deuteride ($^{6}Li^{2}H$) to produce ^{3}H via (n, α) reactions. This tritium in turn fuses with the deuterium in the lithium deuteride by a D-T fusion reaction to yield 17.6 MeV/fusion. The high temperature created by the fission trigger gives the charged deuterons and tritons enough energy to overcome the electrostatic repulsion between the nuclei and fuse via the D-D and D-T fusion reactions, thus yielding energy, additional tritium, and more neutrons. The neutrons from D-D and D-T reactions can be used to induce more fissions or produce more tritium in ^{6}Li, which in turn yields additional fusions with deuterium to produce yet more neutrons and energy. The neutrons produced in the D-T reactions have an energy of 14 MeV, which can induce fast neutron fission in ^{238}U, which is sometimes provided expressly for this purpose, thus creating a fission–fusion–fission weapon. The explosive yields of fusion weapons range between 100 kilotons (kT) and a few megatons (MT) of TNT equivalent; fission–fusion–fission weapons can yield explosive energies of several megatons or more, apparently limited only by the amount of material added to the design.

Fission Products from Nuclear Explosions

The energy released in nuclear explosions is commonly expressed in kilotons or megatons of TNT equivalent. Since each fission represents about 200 MeV of energy release, the explosive energy yield of a weapon is related directly to the number of fissions that occurr. One kiloton of TNT is equivalent to 9.1×10^{11} cal, and the number of fissions per kiloton can be calculated as

$$\text{no. fissions} = \frac{9.1 \times 10^{11}\ \text{cal/kT}}{(3.83 \times 10^{-14}\ \text{cal/MeV})(200\ \text{MeV/fission})}$$

$$= 1.19 \times 10^{23}\ \text{fissions/kT}$$

The number of atoms of a given fission product is determined by its fission yield Y_i, or

$$N_i = (1.19 \times 10^{23}\ \text{fissions/kT of fission yield})\ Y_i$$

This relationship holds only for the fission yield of nuclear weapons. For weapons in which a substantial fraction of the explosive yield is due to fusion, it is necessary to know the fission yield to compute the inventory of fission products. The fission yield Y_i of an element is directly related to the material fissioned and the energy of the fissioning neutrons; therefore, these also need to be known if precise calculations are to be made. Reasonable values can be calculated with the thermal neutron fission yield data given in Appendix E for ^{235}U and ^{239}Pu.

Example 11-1: The Hiroshima explosion was estimated at 14 kT. If all the ^{137}Cs produced fell out over Japan, how much ^{137}Cs (in Ci) was deposited on the ground?

SOLUTION: The number of fissions that occurred is

$$\text{fissions} = 1.19 \times 10^{23}\ \text{fissions/kT} \times 14\ \text{kT}$$

$$= 1.663 \times 10^{24}\ \text{fissions}$$

The nuclear device had a core of ^{235}U, and since the fission yield of ^{137}Cs in ^{235}U is 6.189%, the number of atoms N of ^{137}Cs formed is

$$N(^{137}\text{Cs}) = 1.663 \times 10^{24} \times 0.06189 = 1.03 \times 10^{23}\ \text{atoms}$$

$$\text{activity} = \lambda N = 7.53 \times 10^{13}\ \text{t/s} = 2035\ \text{Ci}$$

Fission Product Activity and Exposure

The beta-emission rate from fission can be used to determine the activity associated with a single fission, which is

$$A_{\mathrm{FP}} = 1.03 \times 10^{-16} t^{-1.2} \text{ Ci/fission}$$

The variation of fission product activity as a function of $t^{-1.2}$ is very useful in determining radiation exposure rates from fission products. The exposure rate from deposition of fresh fission products in an area following a criticality or reactor incident, or perhaps a nuclear detonation, changes with time t according to the relationship

$$R_t = R_1 t^{-1.2}$$

where R_t is the exposure rate at a time t after fission products are formed; R_1 the exposure rate at 1 h, 1 d, and so on; and t the time after formation of fission products. This relationship, which is plotted in Figure 11-3, holds reasonably well for fission products with ages between 1 min and about 200 d. Exposure rates at

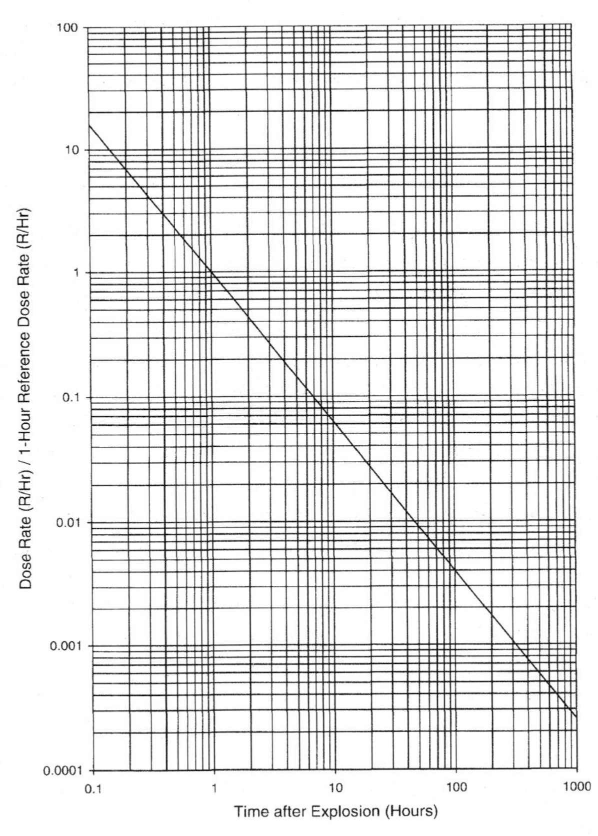

FIGURE 11-3. Decrease of dose rate from fission products with time.

various times after the production of fresh fission products can be determined from Figure 11-3; however, direct calculations are relatively straightforward, as shown in Example 11-2.

Example 11-2: The exposure rate downwind from an atmospheric nuclear explosion is measured to be 100 mrem/h 4 h after detonation. What will be the exposure rate 24 h later?

SOLUTION: First, the exposure rate at 1 h is calculated:

$$100 \text{ mrem} = R_1(4)^{-1.2}$$
$$R_1 = 528 \text{ mrem/h}$$

and the exposure rate at 4 h + 24 h is

$$R_t = 528 \text{ mrem/h } (28 \text{ h})^{-1.2}$$
$$= 9.68 \text{ mrem/h}$$

Although Example 11-2 yields the exposure rate for fission products at a given time t after their formation, the actual radiation exposure received depends on the exposure time. Since the exposure rate changes with time, it is necessary to integrate the instantaneous exposure rate over the period of actual exposure to obtain the total exposure, E_{TOT}, which is

$$E_{TOT} = 5R_1(t_1^{-0.2} - t_2^{-0.2})$$

If R_1 is per hour, t_1 and t_2 must also be in hours. It has been demonstrated (in Glasstone, 1957) that total exposures from a few seconds after formation of fresh fission products to infinity is 11.3 times the 1-h exposure rate.

Example 11-3: For the conditions of Example 11-2, calculate the total exposure that a person would receive between 4 h and 28 h.

SOLUTION: Since the exposure rate at 1 h is 528 mrem/h,

$$E_{TOT} = 5 \times 528[(4)^{-0.2} - (28)^{-0.2}]$$
$$= 645 \text{ mrem}$$

CHECKPOINTS

There are two rules of thumb that are very useful in radiation protection considerations for fresh fission products:

- *Rule of Thumb*: For every sevenfold increase in time, the exposure rate will decrease by a factor of 10 for times between 1 min and about 200 d. (i.e., R_1 at 1 h will equal $0.1R_1$ at 7 h, R_1 at 1 d will be $0.1R_1$ in 7 d, etc.).

- *Rule of Thumb*: The total dose to infinity will be about 11.3 times the 1-h dose rate.

The first rule of thumb ($7t = 0.1R_1$) is illustrated by Example 11-3, where a sevenfold increase in time from 4 h to 28 h reduced the 100 mrem/h exposure rate to about 10 mrem/h (actually, 9.7). Application of the second rule of thumb for total exposure from a few seconds after detonation to infinity would yield an infinity dose of 5966 mrem.

It is important to remember that the exposure rate and decay relations for fission products are empirical and are based on a uniform mixture of fresh fission products such as that produced in a criticality event or nuclear detonation. Any given fission product mixture is affected by many factors, such as plateout on surfaces before being released and alteration or selective depletion of reactive components during transport to the location where they are deposited. Even with such caveats, these relationships can be quite useful for assessing radiation exposures due to fission products and how they vary over time.

CRITICALITY ACCIDENTS

There have been numerous supercritical accidents worldwide, many of which involved mixing of ^{235}U or ^{239}Pu with liquid, which provided a moderator–reflector for neutrons, thus promoting thermal neutron fission because of the higher fission cross section at thermal energies. Eight of these, seven of which occurred in the United States, involved aqueous solutions in chemical processing operations: five with highly enriched uranium and three with plutonium. These caused two deaths and 19 significant overexposures of personnel to radiation, but negligible loss of fissile material, equipment damage, or impact on the general public. Three of the excursions took place in shielded areas designed for processing irradiated fuel; consequently, personnel were protected from the direct radiation released. Each accident was related to misuse of equipment, procedural inadequacies or violations, or various combinations of each. Most of these accidents resulted in prompt criticality in which there was a "fission spike" with an exponential increase in energy generation, and some were followed by secondary spikes or pulses. Such spikes were terminated when bubbles formed or the energy release and radiolytic dissociation of water resulted in subcritical concentrations and/or configurations.

Y-12 Plant, Oak Ridge National Laboratory

At the Y-12 plant at the Oak Ridge National Laboratory in Tennessee, a bank of geometrically subcritical storage vessels for highly enriched uranium recovered from scrap had been disassembled, cleaned, and reassembled for leak testing with water. The leak test was to be done some time after reassembly; however, during this interval, and unbeknownst to the operators, an isolation valve between the vessels and process equipment upstream leaked, which allowed highly enriched uranium to drain into the vessels. Unfortunately, the presence of the HEU did not affect the leak test and it went unnoticed. After the leak tests were done, the water was drained into a 55-gal drum, and the HEU solution containing about 2.1 kg of ^{235}U was suddenly present in an unsafe geometry, and criticality occurred (June 16, 1958). A succession of pulses then produced a total of 1.3×10^{18} fissions (in about 2.8 min). The reaction was terminated after about 20 min by additional water, which continued to flow into the drum, diluting the uranium density to a subcritical concentration.

An initial "blue flash" was observed from the first pulse (estimated at 10^{16} fissions), but the solution stayed within the open container. One person who was about 2 m away received a whole-body dose of about 461 rem; other exposures were 428 rem at about 5.5 m, 413 rem at about 4.9 m, 341 rem at about 4.6 m, 298 rem at 6.7 m, 86.5 rem at 9.4 m, 86.5 rem at 11 m, and 28.8 rem at 15.2 m. All responded to the evacuation alarm within 5 to 15 s, which is roughly the interval between the first two pulses, which apparently limited exposures considerably. After the accident it became standard practice to disconnect transfer lines containing fissile material instead of relying on valves, and "always safe" containers were required for use with ^{235}U-enriched solutions.

Los Alamos Scientific Laboratory

On December 30, 1958, an accident involving plutonium occurred at the Los Alamos Scientific Laboratory in New Mexico under conditions that were very similar to those at the Y-12 plant just 6 months earlier. Four vessels were to be emptied and cleaned individually, but somehow the residues and acidic wash solutions from the four vessels were combined into a single 850-L, 96.5-cm-diameter tank. The wash solutions contained 3.10 kg of plutonium that had accumulated gradually in the process vessels and transfer lines over a period of about 7.5 y. The solution in the tank was subcritical at first because the plutonium had stratified into an organic layer that was about 20 cm thick and represented about 160 L of the total volume in the tank.

A stirrer in the tank was started to mix the solution for sampling before further transfer. When the stirrer blades, which were near the bottom of the tank, began to rotate, solution was forced up the walls of the tank, which squeezed the aqueous plutonium layer toward the center, increasing its thickness and decreasing its lateral dimension. This created a cubelike or oblong spherical geometry, which was more favorable for criticality and an excursion estimated at 1.5×10^{17} fissions occurred. This supercritical condition lasted only a few seconds before the stirrer created a vortex in the liquid and the configuration quickly became subcritical, perhaps aided by the energy of the excursion. The excursion lasted only a few seconds because further stirring rapidly mixed the two phases to a subcritical concentration. The operator who started the stirrer received an exposure of about 12,000 rem and died 36 h later. Two men who went to his aid received doses of 134 and 53 rem.

The entire recovery plant, which had been scheduled for rebuilding after another 6 months of operation, was retired immediately. The replacement facility used geometrically subcritical equipment, eliminated unnecessary solution-transfer lines, auxiliary vessels were "poisoned" with borosilicate glass Rasching rings, and written procedures and nuclear safety training were improved.

Idaho Chemical Processing Plant

Three incidents involving highly enriched uranium (HEU) solutions occurred at the Idaho chemical processing plant, all of them behind heavy shielding. In the first (October 16, 1959), a bank of geometrically subcritical storage cylinders was inadvertently siphoned into a large waste tank. About 200 L of solution containing 34 kg of ^{235}U were transferred into the tank with about 600 L of water. The large geometry with H_2O moderator and reflector went critical and produced about 4×10^{19} fissions over a period of about 20 min. An initial spike of about 10^{17} fissions was followed by smaller pulses, then by more-or-less stable boiling that distilled 400 L of water into another tank. The exceptionally large yield was the result of the large solution volume and long duration of the reaction. Fission products were vented into working areas, which resulted in doses of 50 and 32 rem to two people, mostly in the form of beta radiation to the skin.

The second event occurred some 15 months later (January 25, 1961), again inside shielding, when a large air bubble forced 40 L of solution containing 8 kg of ^{235}U out the top of a "geometrically safe" 12.7-cm-diameter section of an evaporator into a large (61-cm-diameter) cylinder. The resulting excursion, probably a single pulse, had a magnitude of 6×10^{17} fissions. Personnel were protected from direct radiation and the ventilation system prevented airborne activity from entering work areas. The 61-cm-diameter cylinder ultimately was "poisoned" by a grid of stainless steel plates containing 1.0 wt % natural boron. The incident provided valuable, if unintended, information that led to steps to prevent the introduction of air into solution lines that could contain fissionable material.

In the third criticality incident (October 17, 1978) a dilute aqueous solution of dissolved reactor fuel was introduced into the first of a series of pulsed columns for extracting and purifying highly enriched uranium. Uranium was to be extracted into an organic stream, which then flowed to a second column for removal of fission products into a stream of water. Under normal operations, the water in the second column is buffered with aluminum nitrate to prevent significant entrainment of uranium, and after fission products were removed, the solution was then reintroduced along with feedstock into the first column to remove traces of uranium. The uranium was supposed to remain in the organic stream flushed from the second column. But water had leaked into the aluminum nitrate makeup tank, and it was too dilute to prevent appreciable uptake of uranium. Instead of leaving with the organic, as it was supposed to, the uranium was entrained into the water and recycled successively through the first and second columns, building up to an estimated 10 kg in the second column, and with H_2O moderation and reflection, a criticality condition was eventually established. About 2.7×10^{18} fissions occurred over a 30-min period. The flow of HEU feed material was stopped and the reaction terminated due to improved mixing.

Hanford Recuplex Plant

On April 7, 1962 at the Hanford recuplex plant, a liquid, unidentified at the time and containing some 1400 to 1500 g of plutonium, was collected from a large glove-box sump into a 45.7-cm-diameter vessel. Apparently, 46 L of the concentrated liquid had overflowed from a favorable geometry tank and was sucked into the vessel through a temporary line used for cleanup operations that were in progress. This change in geometry produced an excursion of 8.2×10^{17} fissions that occurred over 37 h, 20% of which occurred in the first half-hour. The initial pulse (estimated at 10^{16} fissions) was followed by smaller pulses for about 20 min, after which boiling ultimately distilled off enough water to stop the reaction. The initial pulse, accompanied by the usual blue flash, triggered a criticality accident alarm, and the area was evacuated promptly, presumably before a second pulse. Exposures of 110, 43, and 19 rem were received by personnel at distances of about 2.1, 3.2, and 7 m, respectively.

Wood River Junction

The plant at Wood River Junction, Rhode Island was designed to recover highly enriched uranium from scrap. Startup difficulties on July 24, 1964 led to an unusual accumulation of trichloreoethane (TCE) solution containing low concentrations of uranium, which was recovered by tedious hand agitation with sodium carbonate solution. An easier process was improvised in which the TCE was treated in a 46-cm-diameter tank which was normally used only for making up sodium carbonate solution. Neither the plant superintendent nor one of three shift supervisors was aware of this improvisation. Unbeknownst to these operators, a plug of uranium nitrate crystals containing 240 g of ^{235}U per liter had been drained the previous day from a connecting line into polyethylene bottles identical to those they were using for the low-concentration TCE solution. Thinking it was TCE solution, an operator poured a bottle of the concentrated solution into the large-diameter makeup tank, which was being agitated by an electric stirrer. This configuration of highly concentrated ^{235}U in solution produced a criticality excursion of 10^{17} fissions with the usual flash of light, splashed about 8 L (20% of the total) out of the makeup tank, and knocked the operator to the floor. He received about 10,000 rad and died 49 h later.

The loss of solution and the vortex created by the stirrer, which continued to run, was sufficient to maintain subcriticality. Two men entered the area 90 min later to drain the solution into safe containers. As they were leaving, they turned off the stirrer, and the change in geometry created by the collapse of the stirrer-induced vortex apparently caused at least one more excursion of 2 to 3×10^{16} fissions. They received radiation doses estimated at between 60 and 100 rad. Modifications were made in operating and emergency procedures, criticality limits and controls, uranium accountability and material balance practices, and training. Favorable geometry equipment was installed for recovering uranium from TCE.

UKAEA Windscale Works

An unsuspected buildup of plutonium occurred in an organic solvent layer at the UKAEA Windscale Works in Great Britain on August 24, 1970. The excursion, which was similar to the Los Alamos accident, took place at the head end of a process for recovering plutonium by solvent extraction from an aqueous solution of about 6 g of Pu per liter. Forty liters of organic solvent from an unknown source entered a vacuum transfer vessel and was then transferred to a trap that isolated the floating layer of solvent instead of permitting it to drain. Although the trap had been provided for safety, it allowed the solvent to accumulate plutonium gradually, and aqueous batches pouring through it eventually built up a concentration of 55 g of Pu per liter in the solvent. It appears that an emulsion layer formed between the solvent and the aqueous solutions, which led to a criticality of about 10^{15} fissions during the brief period after the flow stopped and before the two phases of emulsion separated. Exposures were less than 2 rad for the two closest workers, who were protected somewhat by shielding.

Bare and Reflected Metal Assemblies

Two fatal accidents occurred at Los Alamos in 1945 and 1946 to scientists working with a bare assembly of two hemispheres of delta-phase plutonium (density of about 15.7 g/cm^3). In the first accident, a critical assembly was being created by hand stacking 4.4-kg tungsten carbide reflector bricks around the plutonium hemispheres. The experimenter was working alone, in violation of procedures. As he was moving the final brick over the assembly (he was supposed to add it from below), he noticed, from the nearby neutron counters, that the addition of this brick would make the assembly supercritical, and he began to withdraw it. But he dropped it onto the center of the assembly, which added sufficient reflection to make the system super-prompt critical. He quickly pushed off the final brick and proceeded to unstack the assembly, but an excursion estimated at 10^{16} fissions had already occurred. The experimenter received an estimated exposure of 510 rem and died 28 d later.

In the second accident, an experimenter was demonstrating a metal-critical assembly using the same plutonium sphere, which in this case was reflected by beryllium, a better neutron reflector. The top and final hemispherical beryllium shell was slowly being lowered into place; one edge was touching the lower beryllium hemisphere while the edge 180° away was resting on the tip of a screwdriver. The person conducting the demonstration was holding the top shell with his left thumb inserted in an opening at the polar point while slowly working the screwdriver out with his right hand, when the screwdriver slipped and the shell seated on the lower hemisphere, creating a supercritical assembly and a power excursion of 3×10^{15} fissions, which blew the shell off the assembly and onto the floor. The eight people in the room received doses of about 2100, 360, 250, 160, 110, 65, 47, and 37 rem. The man who performed the experiment died 9 d later.

These types of experiments were characterized by Richard Feynman, the Nobel prize–winning physicist, as "tickling the dragon's tail"—get too close and the beast will breathe fire on you.

RADIATION EXPOSURES IN CRITICALITY EVENTS

Figure 11-4 shows a general picture of personnel dose versus distance from a criticality event; the exposures are normalized to 10^{17} fissions and exposure times of about 15 s. The relationships are based on empirical data rather than calculable parameters, which can vary considerably from one event to another; hence the exposures should be regarded as representing typical excursions. The short exposure times presume rapid evacuation, a major factor in the dose received. Good alarms and personnel training to assure rapid response are essential elements of criticality safety programs. Our first knowledge that the $LD_{50,30}$ for humans was 450 to 500 rad came from the fatality associated with the Los Alamos criticality accident in 1946.

The criticality accidents involving aqueous solutions, six of which occurred in a 6-y period between 1958 and 1964, appear to be related to a wider variety of

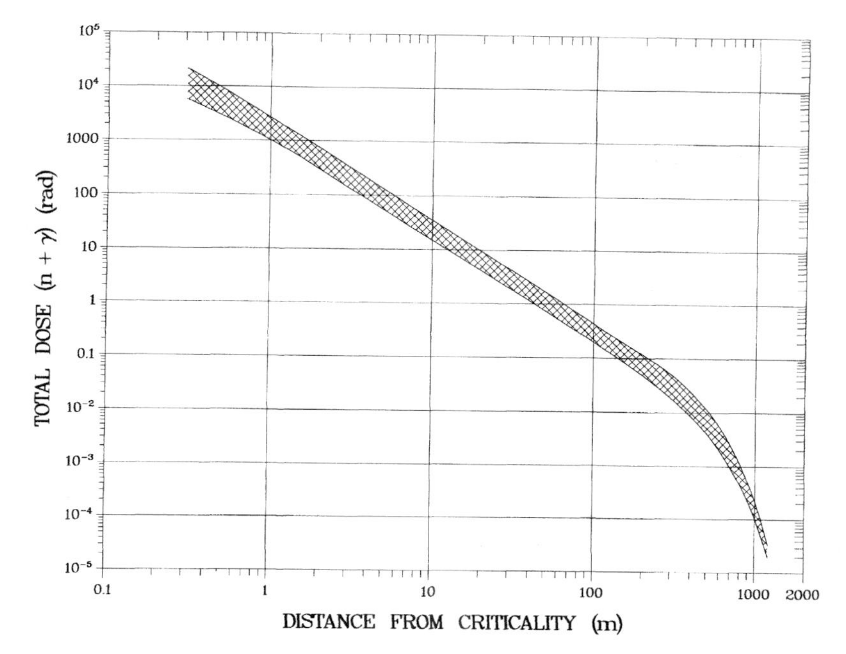

FIGURE 11-4. Radiation dose versus distance from the criticality event.

tasks associated with an increased demand for plutonium and enriched-uranium production. They are also dramatic examples of the consequences of major changes in operations without a corresponding reassessment of criticality control. Facilities left over from the Cold War have interesting inventories and arrays of fissionable material and the potential for criticality conditions. Particular attention needs to be paid to anticipate and prevent criticality events when solutions of plutonium or highly enriched uranium are processed by the nonroutine circumstances that are associated with stabilizing, dismantling, and recovering old sites and facilities.

CRITICALITY SAFETY

Preventing a criticality of fissile material is based on a consideration of the same factors that must be overcome if a supercritical condition is to be achieved in controlled nuclear reactors or nuclear weapons. The mass, volume, and density/concentration of fissionable material, its geometrical shape and configuration, and the moderation/reflection, absorption, and leakage of neutrons are all important factors in achieving or preventing a supercritical system. Criticality safety depends on the following:

- Limiting the quantity of fissile material to less than the critical mass under any conceivable configuration
- Restricting the geometry of the fissile material, preferably to one that is "always safe" (i.e., one where the surface/volume ratio is such that excessive neutron leakage makes it impossible to attain a multiplication factor of 1.00 or more)

- Limiting the concentration of any fissile material that may be in a solution of moderating material, such as H_2O, D_2O, or other liquids, and considering the presence or absence of neutron-reflecting material(s)
- Preventing the interaction between two or more subcritical arrays of fissile materials, especially in transport and storage

Mass/density effects influence the critical mass of a fissile material. The critical mass of a sphere of ^{239}Pu metal is much less than that of a sphere containing unmoderated ^{239}Pu filings or chips. Nuclear weapon designs take advantage of this fact by compressing a spherical mass to high density.

Geometry effects can be used to establish an "always safe" configuration to prevent inadvertent criticality by promoting neutron loss even if sufficient mass and/or moderation and reflection are available. If a given volume of fissile solution departs from a spherical shape, there is an increase of surface area through which neutrons can escape; therefore, two practical "always safe" geometries are an elongated cylinder of sufficiently small diameter or an extensive slab of restricted thickness; parameters for these shapes, reflected by a thick layer of water, are provided in Table 11-1. Another very important geometric factor is separation of subcritical units to preclude mutual exchange of neutrons between them. Such separation is especially important in facility design and in the storage and transport of subcritical units.

Moderator/reflector effects of light materials, especially water, can greatly influence the critical configuration of fissile material, reducing its critical mass to as little as a few hundred grams. On the other hand, very dilute solutions in which neutron absorption by hydrogen predominates can contain unlimited quantities of fissionable material without criticality. A close-packed subcritical array may become critical if flooded; conversely, a flooded subcritical array of large, less closely packed units may become critical if the water is removed, since the water, as a neutron absorber, may diminish neutron coupling of the units.

Subcritical systems may also become critical if a reflector such as beryllium, H_2O, or tissue is added that returns escaping neutrons into the array. For example, Yuli Khariton recalled that as Vannikov, who was a rather large fat man, "went back and forth to read the gauges...during [a test of critical assembly for the first Soviet nuclear weapon] we understood: the bomb would definitely work" (described by Rhodes, 1995).

Criticality Safety Parameters

The interrelationships of mass, moderator, and reflector are shown in Figures 11-5 and 11-6 for spherical shapes containing ^{235}U and ^{239}Pu and in Figure 11-7 and 11-8 for slabs of various concentrations of the two materials. These two geometries are conservative ones, and the relationships determined for them provide indices for preventing criticality conditions. For example, spherical geometries of fully reflected pure uranium or plutonium metal will be safe (Figures 11-5 and 11-6) as long as the total mass of each is less than 1.5 and 0.9 kg, respectively; any other shape will also be safe since this is an optimal geometrical configuration requiring the least amount of fissionable material. Without reflection, an assembly can contain more mass without becoming critical. If, however, the ^{235}U or ^{239}Pu is in a solution that

TABLE 11-1. Minimum Critical Safety Parameters for Fissionable Materials in Moderated and Reflected Conditions and as Unmoderated Metal

Fissile Material	Parameter	Minimum Critical Value: Moderated and Reflected[a]	Minimum Critical Value: Unmoderated[b]
Always safe	^{235}U enrichment	≤ 0.93 wt %	5.0 wt %
	Aqueous ^{235}U	≤ 11.94 g/L	—
	$^{235}UO_2(NO_3)_2$ at 2.88 wt %	≤ 595 g/L	—
	$^{235}UO_2(NO_3)_2$	≤ 1.96 wt %	—
	Dry oxides Pu + U	≤ 4.4 wt % Pu	—
	Damp oxides (H/Pu + U) ≤ 0.45	≤ 1.8 wt % Pu	—
^{235}U, 93%	Mass	830 g	22.8 kg
	Diameter of infinite cylinder	8.0 cm	—
	Thickness of infinite slab	1.80 cm	—
^{235}U, 100%[c]	Mass ($\rho = 17.484$)	760 g	20.1 kg
	Diameter of infinite cylinder	13.7 cm	7.3 cm
	Thickness of infinite slab	4.4 cm	1.3 cm
	Volume of solution	5.5 L	—
	Concentration (aqueous)	11.6 g/L	—
^{233}U, 100%	Mass ($\rho = 18.05$)	540 g	7.5 kg
	Diameter of infinite cylinder	10.5 cm	5.0 cm
	Thickness of infinite slab	2.5 cm	0.6 cm
	Volume of solution	2.8 L	—
	Concentration (aqueous)	10.8 g/L	—
^{239}Pu, 100%[d]	Mass ($\rho = 19.74$)	480 g	5.0 kg
	Diameter of infinite cylinder	15.4 cm	4.40 cm
	Thickness of infinite slab	5.5 cm	0.65 cm
	Volume of solution	7.3 L	—
	Concentration (aqueous)	7.3 g/L	—

Source: Adapted from Provost and Paxton, 1996.
[a]Moderation and reflection by H_2O is assumed.
[b]Unmoderated values assume water reflection.
[c]Values for moderated and reflected solutions are based on $^{235}UO_2F_2$; for $^{235}UO_2(NO_3)_2$, these may be increased to 780 g, 14.4 cm, 4.9 cm, 6.2 L, and 11.6 g U/L, respectively.
[d]Values based on moderated and reflected pure ^{239}Pu metal or $^{239}Pu(NO_3)_4$ solutions; if ≥ 5 wt % of ^{240}Pu is present, the values increase to 570 g, 17.4 cm, 6.7 cm, 10.0 L, and 7.8 g Pu/L.

provides moderation and reflection, the critical mass is reduced considerably and can be as low as 0.76 kg of ^{235}U or 0.48 kg of ^{239}Pu in solution. On the other hand, if the mass of either is kept below these quantities, they will always be subcritical, regardless of how they are mixed; therefore, a minimum critical mass of material can be stated that will always be safe. These various limiting parameters and several other useful data are also listed in Table 11-1.

Similar interpretations can be made for slab geometries in which layered material containing ^{235}U or ^{239}Pu will always be subcritical (Figure 11-7 and 11-8) if the thickness of a fully reflected slab of a very large ("infinite") area is less than 1.9 or 0.8 cm, respectively. If a reflector is absent, the layer of fissionable material could

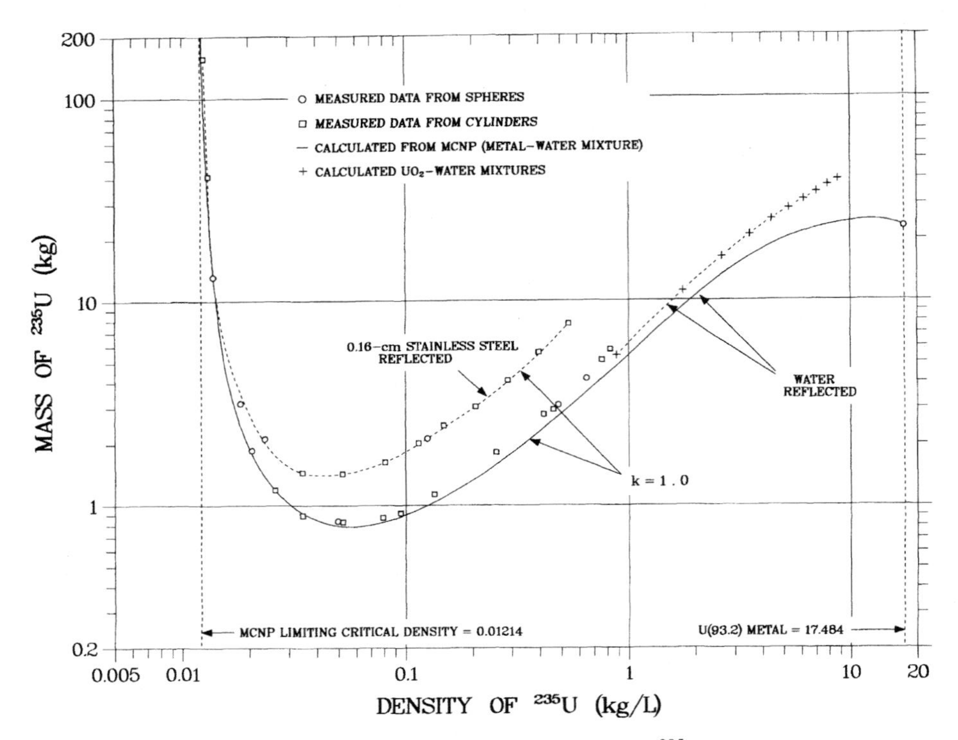

FIGURE 11-5. Critical masses of spherical shapes of 93.2% ^{235}U in homogeneous water-moderated solution as function of ^{235}U density and reflection by thick water and stainless steel. (Adapted from Provost and Paxton, 1996, with permission.)

be thicker, but most criticality assessments of layered material would want it to be safe even if flooded with water.

A cylindrical geometry of ^{235}U will be "always safe" regardless of the concentration or reflection if the diameter is less than 7.3 cm (100% ^{235}U, fully reflected and unmoderated); a cylindrical geometry of fully reflected pure ^{239}Pu will be "always safe" if the diameter is less than 4.4 cm (100% ^{239}Pu, fully reflected and unmoderated). Most processes developed for solutions of fissionable material use these dimensions because as the accidents described above demonstrate, fissionable materials have a way of accumulating in concentrations that would be undesirable if "always safe" geometries are not assured.

Most real-world situations are not represented by spherical or infinite slab geometries, nor are they fully reflected or moderated. For such configurations a larger mass of fissionable material could be present without criticality, but allowing these conditions to exist should be done with the utmost care.

Separation of subcritical units is also an important consideration in assessing criticality safety, especially if either could contribute significant neutrons to the other. Such separation is essential between "always safe" vessels, transfer lines, and pipes, lest they assume a critical configuration, and needs to be exercised in placing either near materials that are good reflectors of neutrons. Densely formed concrete can be a good reflector; therefore, placement of vessels containing concentrated

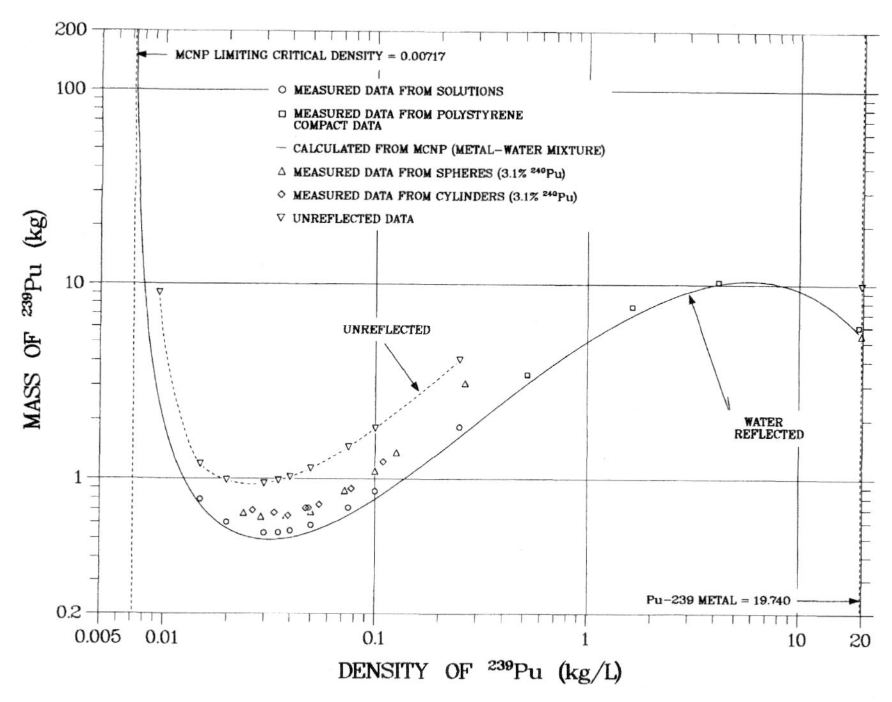

FIGURE 11-6. Critical masses of spherical shapes of ^{239}Pu in a homogeneous water-moderated solution as a function of ^{239}Pu density with and without water reflection. (Adapted from Provost and Paxton, 1996, with permission.)

fissionable material next to a wall or floor of such concrete could cause subcritical concentrations to become critical.

Containers that assure "always safe" spacing are commonly called *birdcages*. A 55-gal drum with an "always safe" unit container welded into the center by appropriate spacers provides an excellent birdcage. The drum becomes the spacing container, assuring that the closest distance between the surfaces of fissile materials is always less than a safe separation distance even if the containers are immersed in water.

Criticality safety practices limit at least one, and preferably two, of the factors that determine criticality below specified minimum values as listed in Table 11-1. The limiting values in Table 11-1 are independent of each other, and restriction of any one of the parameters is protective; however, processes and equipment are generally designed such that at least two of these independent parameters are met to provide additional safety, the *double-contingency principle*. It is also good practice to rely on equipment designs rather than principally on a process to assure safety. A process designed to meet a concentration or enrichment criterion could well fail if the wrong material inadvertently appears in the process, circumstances that can be precluded by equipment designs that provide an "always safe" geometry.

Limiting the degree of enrichment of fissionable material (^{235}U, ^{239}Pu, ^{233}U) in a concentration of liquid is also an effective practice. It is not possible for an

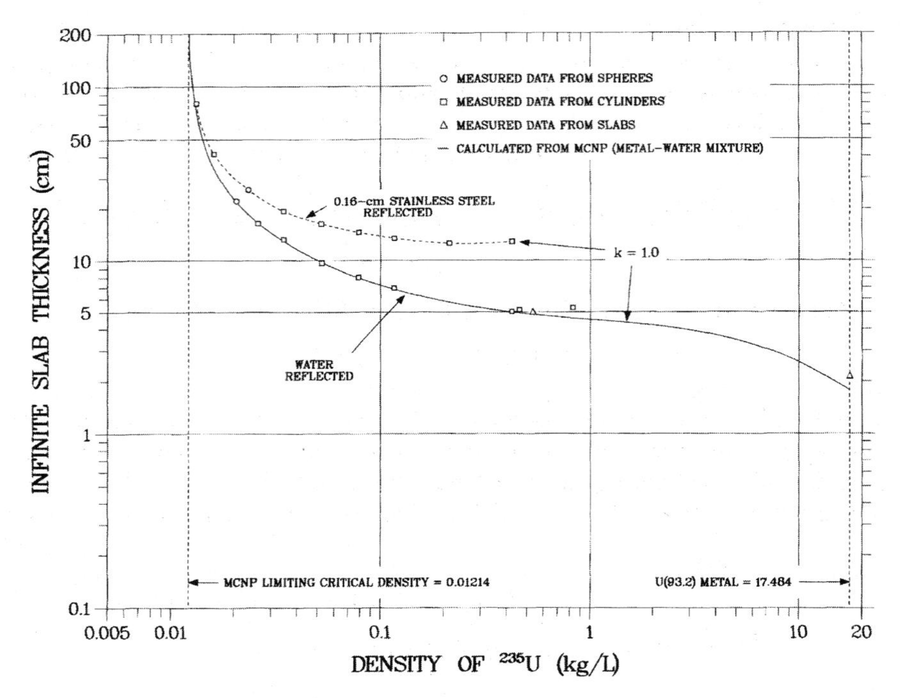

FIGURE 11-7. Thicknesses of infinite slabs of 93.2% ^{235}U in a homogeneous water moderator as a function of ^{235}U density and reflection by water and stainless steel. (Adapted from Provost and Paxton, 1996, with permission.)

aqueous mixture or solution of uranium to attain criticality unless the ^{235}U is enriched to above 0.95% or more; therefore, no criticality restrictions need be made for transportation or storage of natural uranium or homogeneous aqueous mixtures containing less than 0.95% ^{235}U. Similarly, uranium metal with an enrichment of less than 5% ^{235}U cannot achieve criticality provided that the metal is not interspersed as chunks or rods in hydrogenous material.

Example 11-4: Vessels containing highly enriched (> 93%) $UO_2(NO_3)_2$ solutions are located above a bermed concrete floor area of 9 m^2. What conditions should be required for an overflow line above the floor to limit the thickness of solution should the vessels leak?

SOLUTION: The configuration of the solution can be conservatively approximated by an effectively infinite uniform slab with a thick concrete reflector on one side and incidental reflection on the other side. From Table 11-1 the subcritical thickness of an infinite slab of $UO_2(NO_3)_2$ fully reflected by water is 4.9 cm. Assumption of a thick water reflector on both surfaces is more conservative than concrete reflection by the floor and the berm; therefore, an overflow pipe placed 4.9 cm above the lowest portion of the floor would assure that a critical configuration of $^{235}UO_2(NO_3)_2$ solution could not occur in the bermed area regardless of the areal extent.

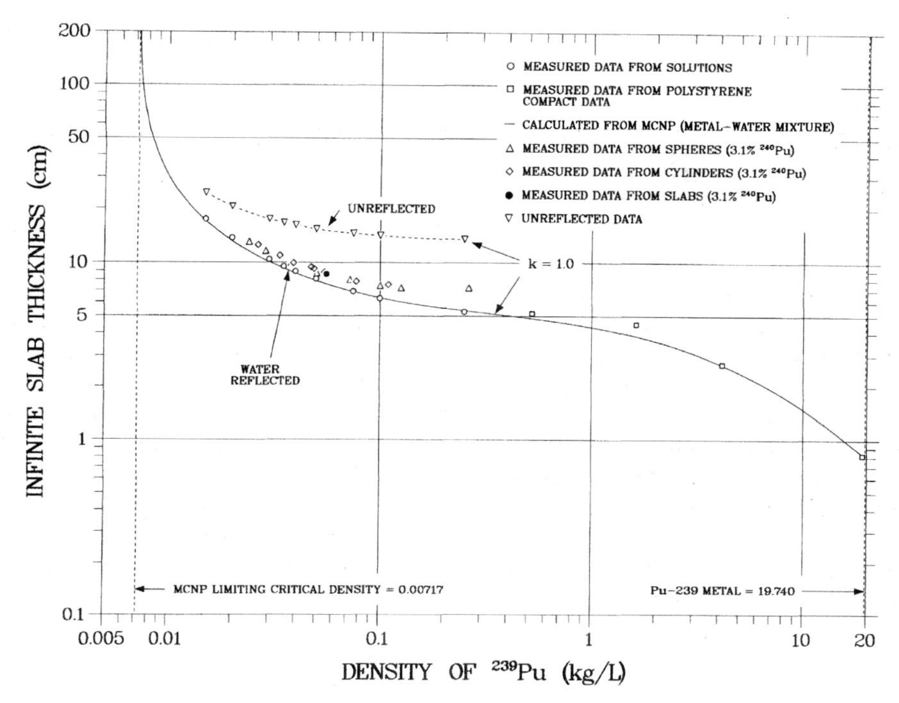

FIGURE 11-8. Thicknesses of infinite slabs of 93.2% ^{239}Pu in a homogeneous water moderator as a function of ^{239}Pu density with and without water reflection. (Adapted from Provost and Paxton, 1996, with permission.)

FISSION PRODUCT RELEASE IN CRITICALITY EVENTS

Criticality events yield fission products in addition to the high exposure field of neutrons and gamma rays. Volatile fission products such as iodines, bromines, and noble gases usually escape because of the heat generated, and these can produce airborne levels in buildings or be released to the environment. Others remain in the medium where the criticality occurred and can be used to characterize the magnitude of the event, usually in terms of the number of fissions that occurred.

A calculation of fission product releases from criticality events must first recognize that such events occur instantaneously, which means that any gaseous and volatile fission products available for release will be those formed instantaneously, which is determined by their independent fission yields. The inventory will of course change with time as the fission products undergo transformation and transition down the fission product mass chain, in some cases to produce other gaseous and volatile radionuclides. Alternatively, those volatiles that were formed directly and released in the excursion will not yield their usual products from radioactive transformation associated with fission products chains.

Example 11-5: A criticality event occurs in a container that overpressurizes, causing all the ^{138}Xe produced to be vented. The container is sampled 3 h later and found to contain 10 Ci of ^{138}Cs. How many fissions occurred in the event?

SOLUTION: The formation of ^{138}Cs (see Appendix E) occurs by radioactive transformation of ^{138}Xe, its precursor, and by independent yield. Since all the ^{138}Xe formed at the instant of the excursion can be assumed to have been vented from the container due to the overpressurization, the only ^{138}Cs in the container would be formed by direct fission yield (6.7080 – 6.2973).

The amount of ^{138}Cs that existed when the excursion occurred (i.e., 3 h earlier) is

$$10 \text{ Ci} = A_0 e^{-(\ln 2/32.2 \text{ m})(180 \text{ m})} \quad \text{or} \quad A_0 = 481.3 \text{ Ci}$$

The number of atoms in 481.3 Ci is

$$N_{\text{Cs}} = \frac{1.78 \times 10^{13} \text{ t/s}}{3.588 \times 10^{-4} \text{ s}^{-1}} = 4.96 \times 10^{15} \text{ atoms of } ^{138}\text{Cs}$$

and the number of fissions is

$$\frac{4.96 \times 10^{15} \text{ atoms}}{0.00411} = 1.21 \times 10^{18} \text{ fissions}$$

Fast Fission in Criticality Events

Many criticality events occur with fast neutrons, especially nuclear explosions. The fission product yield spectrum for fast neutron fission is somewhat different from that due to thermal fission, as shown in Figure 10-21; therefore, the most accurate calculations of the yield of a given fission product for such events is obtained if the fission yield spectrum for fast neutrons is used. Fast fission yields for ^{239}Pu and ^{235}U for some of the more important mass chains are contained in Figure 11-9, using the same nomenclature given in Appendix E. If fast fission yields are not otherwise available, the fission yields in Appendix E for thermal neutron fission can be used for fission with an accuracy within 10 to 20%. Fast fission yields are available for other mass chains from the National Nuclear Data Center Web site, as referenced in Appendix E.

Example 11-6: Estimate the amount of ^{131}I produced from a 1 kT ^{239}Pu-fueled weapon based on fast fission. How does this compare with a calculation based on the thermal neutron fission yield in ^{239}Pu?

SOLUTION: From Figure 11-9 the cumulative fast fission yield in ^{239}Pu for ^{131}I atoms is 4.355%, and since 1 kT of weapon yield corresponds to 1.19×10^{23} fissions, the number of atoms of ^{131}I formed is

$$^{131}\text{I}_f \text{ atoms} = 1.19 \times 10^{23} \text{ fissions} \times 0.04355 \text{ atom/fission}$$
$$= 5.18 \times 10^{21} \text{ atoms} \quad \text{or} \quad 140{,}000 \text{ Ci}$$

If the thermal fission yield for ^{239}Pu were assumed (Appendix E, 3.856%), the number or atoms of ^{131}I would be

$$^{131}\text{I}_t \text{ atoms} = 1.19 \times 10^{23} \text{ fissions} \times 0.03856 \text{ atom/fission}$$
$$= 4.59 \times 10^{21} \text{ atoms} \quad \text{or} \quad 124{,}000 \text{ Ci}$$

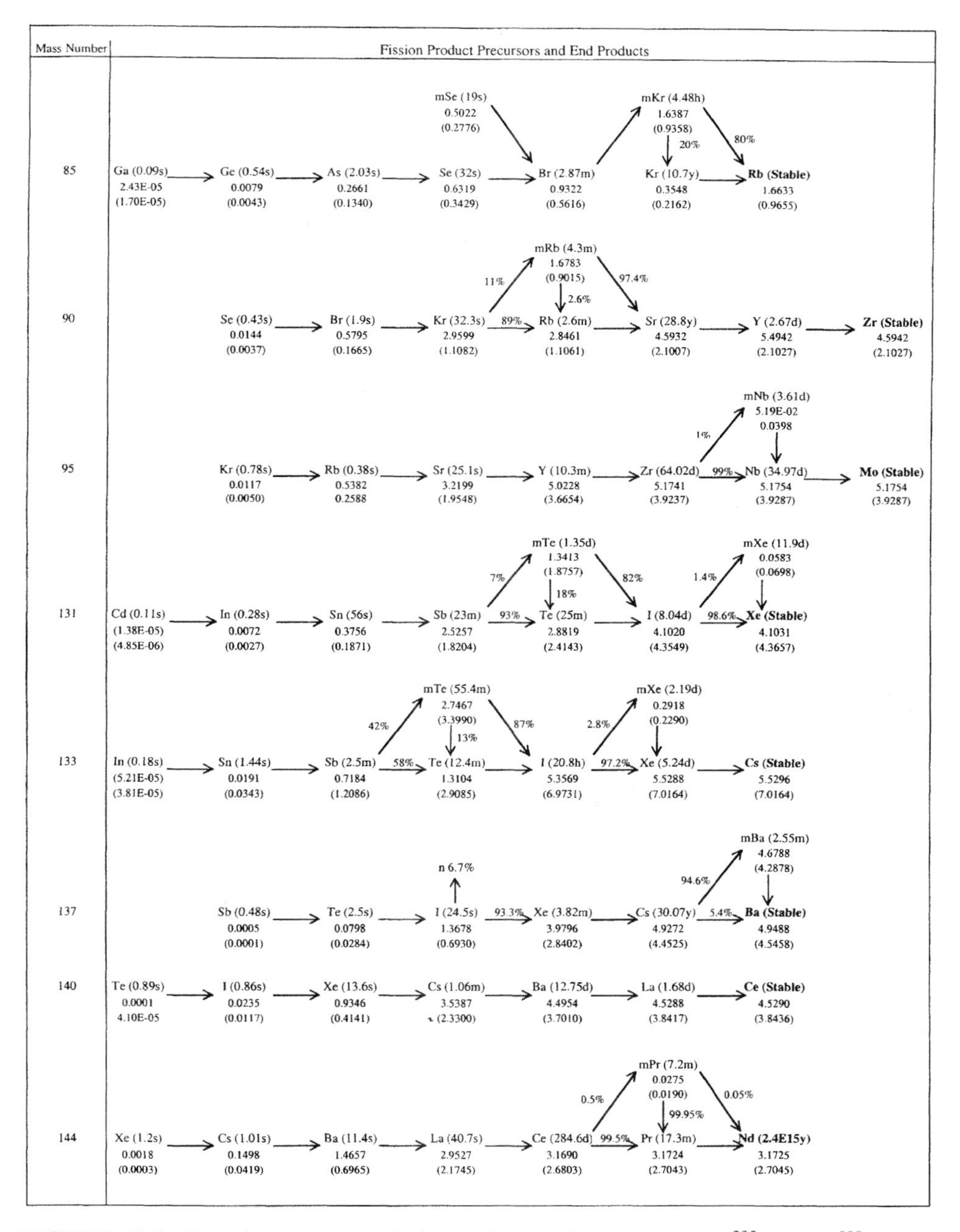

FIGURE 11-9. Cumulative percent fission yields for fast neutrons in ^{235}U and (^{239}Pu) for each isotope in selected mass chains.

Example 11-6 shows the importance of the type of fission event in determining fission product inventories from criticality events. The calculated ^{131}I activity is a factor of about 1.13 larger (or about 13%) when the fast fission yield is used, which is just the ratio of the fast and thermal fission yields.

SUMMARY

Criticality safety is an important consideration in radiation protection because exposures and consequences can be very high. Criticality events are generally of three types: (1) the intentional assembly of fissionable material to release energy, such as in nuclear weapons; (2) supercritical conditions in reactors; and (3) unintended accumulations or configurations of fissionable materials such as enriched ^{235}U or ^{239}Pu in solution. Preventing a criticality of fissile material is based on a consideration of the same factors that must be overcome to achieve a sustained supercritical condition in a nuclear reactor or the explosive release of energy in nuclear weapons. The mass, volume, and density/concentration of fissionable material; its geometrical shape and configuration, and the moderation/ reflection, absorption, and leakage of neutrons are all important factors in achieving or preventing a supercritical system.

Processes and equipment are generally designed such that at least two of the independent parameters that determine criticality are met. Equipment designs are more reliable for assuring safety than is a process or a procedure since these could fail if the wrong material appears inadvertently in the process. Equipment designs that provide an "always safe" geometry preclude the occurrence of an accidental critically even if concentration limits are exceeded and a good reflector is present. Whenever possible, fissionable material should be processed and/or stored in cylindrical containers with safe diameters, and these should be physically spaced with birdcages or other fixed arrays far enough apart that they could not go critical even if flooded with water.

Criticality events produce high exposure fields of neutrons and γ-rays and the release of volatile fission products such as iodines, bromines, and noble gases. The fission products available for release will be those formed instantaneously, which is determined by their independent fast fission yields; however, thermal neutron fission yields can be used for such determinations with an accuracy within 10 to 20% if fast fission yields are not otherwise available.

ACKNOWLEDGMENTS

N. L. Provost and H. C. Paxton of the Criticality Safety Group at Los Alamos National Laboratory were especially helpful in providing comprehensive information on criticality safety, which the author hopes he did justice to in the summary of a very complex subject.

REFERENCES AND ADDITIONAL RESOURCES

Glasstone, S., *Effects of Nuclear Weapons*, Atomic Energy Commission, U.S. Department of Defense, Washington, DC, 1957.

Provost, N. L., and H. C. Paxton, *Nuclear Criticality Safety Guide*, Report LA-12808, Los Alamos National Laboratory, Los Alamos, NM, September 1996.

Rhodes, R., *Dark Sun*, Simon & Schuster, New York, 1995.

Stratton, W. R., *A Review of Criticality Accidents*, Report DOE/NCT-04, U.S. Department of Energy, Washington, DC, March 1989.

PROBLEMS

11-1. In 1000 atoms of present-day natural uranium, 993 are ^{238}U and 7 are ^{235}U. What would have been the percentage of ^{235}U in natural uranium 2 billion years ago, and could it have yielded a natural reactor?

11-2. If the earth is 4.5 billion years old, what would have been the percentage of ^{235}U in uranium at that time, and how might this have contributed to early changes in the earth's crust?

11-3. Careful measurement of the soil around the Gabon natural reactor site established the current inventory of ^{129}I formed in the soils at 10 mCi. If it can be assumed that none of the ^{129}I initially has left the site, estimate the number of fissions that occurred some 2 billion years ago. If humans existed and could recover the energy, how much thermal energy (kW) was available for their use?

11-4. A 100 kT nuclear fission device is exploded underground in Nevada and the iodines produced vent through a fissure. How many curies of ^{131}I would be available for release?

11-5. A foreign country tests a nuclear device in the atmosphere. The exposure rate measured downwind was 10 mR/h 3 h later. (**a**) What would be the exposure rate 24 h after the test? (**b**) What would be the total dose from 3 h onward if people remained in the area indefinitely?

11-6. It has been estimated that 20 Ci of ^{131}I was released to the environment from the Three Mile Island incident, which occurred when the reactor had been running at full power (2700 MWt) for 3 months. How many curies of ^{131}I were released in the reactor system/containment structure?

11-7. The Chernobyl reactor had operated at 3200 MWt for about a year before the explosion. The inventory of ^{133m}Xe and ^{133}Xe was released as a large puff that drifted over the nearby area to produce external exposure by sky shine and submersion. How many curies of ^{133m}Xe and ^{133}Xe were released in this puff?

11-8. A nuclear criticality incident occurred instantaneously in a tank containing enriched uranyl nitrate and water when a mixer was turned on. It was determined that 10^{20} fissions occurred. (**a**) What would be the activity (in curies) of ^{138}Xe 10 s after the incident? (**b**) What would be the activity of ^{138}Xe 1 h later? (Use exact solution.)

11-9. A criticality occurred in a tank containing a solution of fissionable material producing 2×10^{17} fissions. Estimate the radiation exposure to a worker standing 2 m away who responded quickly to the criticality alarm and exited the area within 15 s.

11-10. A solution containing highly enriched uranium was processed after a criticality of 10^{18} fissions had occurred to recover the uranium. How many curies of ^{137}Cs would be in the residual radioactive waste process solution?

12

RADIATION DETECTION AND MEASUREMENT

When you can measure what you are speaking about and express it in numbers, you know something about it; when you cannot...your knowledge is of a meager and unsatisfactory kind.

—Lord Kelvin, 1889

The interactions of various types of radiation in matter provide mechanisms for measuring the amount of radiation emitted and absorbed by a source(s), and with careful techniques the identity of the radiation source or radionuclide producing it. Radiation instruments include portable survey instruments that are designed to detect radiation and measure exposure or absorbed dose and laboratory instruments that allow precise quantitation and identification of the radiation source. Various detectors are used in them, and these can be roughly divided into two categories: gas-filled chambers and crystalline materials. Each of the various devices is based on the liberation of electrons in a medium and the collection and processing of the ions by electronic means. Although many of these same considerations apply to neutrons, their detection and measurement require special considerations which are described in Chapter 14.

TYPES OF DETECTORS

Gas-Filled Detectors

Absorption of radiation in a gas-filled chamber is accomplished by establishing an electrostatic field between the wall of the chamber and a positive electrode located on the axis of the chamber and insulated from it as illustrated in Figure 12-1. The gas is usually at a pressure of 1 atm or less, but with appropriate care can be pressurized to enhance interactions. When radiation is absorbed in the gas contained by the chamber, ion pairs (a positive ion and an electron) are produced, which are collected and amplified for recording the signal produced.

The collection of ion pairs in a gas-filled detector is a function of the applied voltage as shown in Figure 12-2, which characterizes the operation of such detectors for beta particles (lower curve) and for alpha particles (upper curve). If there is no

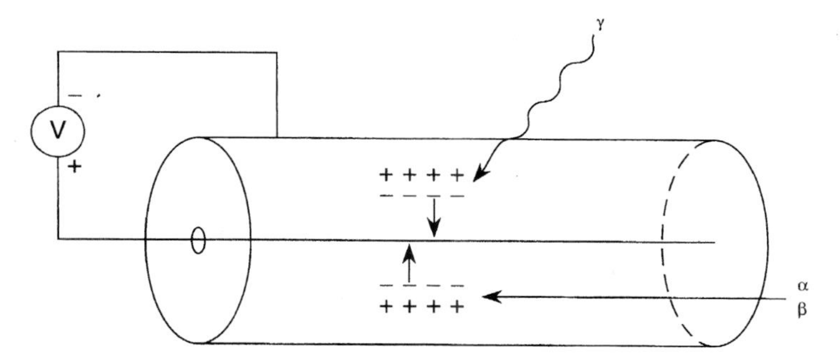

FIGURE 12-1. Schematic of gas-filled detector operated with a varying voltage applied between the chamber wall (cathode) and a central collecting electrode (the anode).

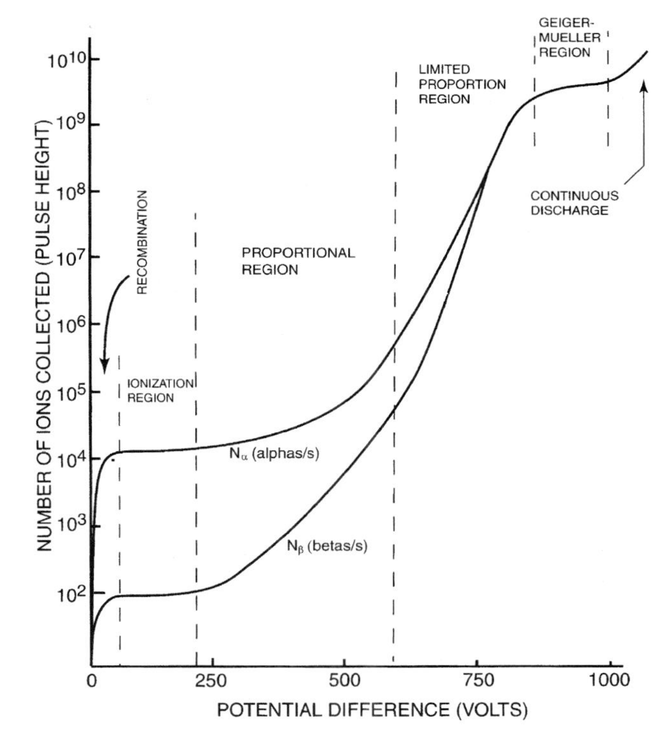

FIGURE 12-2. Ionization rates for constant source strengths of a beta emitter and an alpha emitter in a gas-filled detector operated in different voltage ranges.

voltage across the chamber, the ion pairs will recombine and no charge will flow in the external circuit for either type of radiation. As the voltage is increased, say to a few volts, some ion pairs will still recombine, but others will flow to the electrodes and be collected. At a voltage of perhaps 10 V or more, recombination becomes negligible and all of the electrons produced by ionization will reach the central electrode. As the voltage is increased to several tens of volts, the number of ion pairs collected is independent of the applied voltage, and the curve will remain horizontal as long as the radiation source produces ionizing radiation at a steady rate. If 10 ion pairs are formed initially, the response will be steady, as shown by the lower curve; if the source strength or ionizing rate (e.g., an alpha emitter) is a factor of 10 higher, 100 ion pairs will be formed and the detector response will also be a factor of 10 higher, as shown in the upper curve. These curves are parallel to each other (i.e., the current collected is in direct proportion to the ionization produced by the incoming radiation). This region is called the *ionization region*; its magnitude will be quite different for equal source strengths of alpha particles, beta particles, and x or gamma rays due to differences in the number of ion pairs produced by each as they interact in the medium.

Increasing the voltage on a gas-filled detector above the ionization region causes the electrons released by the primary ionizations to acquire enough energy to produce additional ionizations as they collide with the gas molecules in the chamber. The number of electrons collected increases roughly exponentially with the applied voltage because each initial electron is then accelerated to produce a small "avalanche" of electrons, most of which are liberated close to the central electrode. Each electron liberated by the incoming radiation produces its own independent avalanche such that at a given voltage on the detector the ionization produced is amplified by a constant amount (i.e., the number of ion pairs collected is proportional to the initial ionization). Gas-filled detectors operated at these voltages are characterized as being operated in the *proportional region*.

When the voltage across the chamber reaches several hundred volts, the gas multiplication effect increases very rapidly, and as more electrons produce avalanches, the latter begin to interact with one another, creating a region of limited proportionality. As voltage is increased further the charge collected becomes independent of the ionization initiating it, and the two curves not only become identical but form a plateau as voltage is increased. This is the *Geiger–Mueller region*, and it is characterized by a plateau. At voltages above the plateau, the detector produces a region of continuous discharge.

Three radiation detector types have been developed based on the regions of applied voltage illustrated in Figure 12-2: the ionization chamber (based on the ionization region), the proportional counter (based on the proportional region), and the Geiger–Mueller counter (based on the G-M region). The *ionization chamber*, operated at voltages in the ionization region, is characterized by complete collection, without gas amplification, of all the electrons liberated by the passage of the ionizing radiation, whether it is due to particles or photons. The electrodes of an ionization chamber are typically cylindrical with a center electrode, but it may also be a parallel-plate design. The applied voltage is selected to assure collection of all the ions formed (i.e., it is high enough to prevent recombination of the ions produced but still on the plateau where current amplification doesn't occur). The current pulse is proportional to the number of ionizing events produced, and the collected charge can be used directly to establish radiation exposure.

The *proportional counter*, which is operated at voltages in the proportional region, is characterized by gas multiplication, which produces a pulse proportional to the initial ionization. Since alpha particles are highly ionizing relative to beta particles, proportional counters are useful both to count the particles and to discriminate between them based on the size of the pulse each produces. The proportional counter thus offers a particular advantage for pulse-type measurements of beta radiation and can be used in conjunction with electronics to sort the smaller beta pulses from the larger ones produced by alpha particles. With modern electronics, samples that contain both alpha and beta particles can easily be sorted and measured at the same time.

The *Geiger–Mueller* (GM) *counter* is operated in the Geiger region and is characterized by a plateau voltage that produces an avalanche of discharge throughout the counter for each ionizing radiation that enters the chamber. This avalanche of charge produces a pulse whose size is independent of the initial ionization; therefore, the GM counter is especially useful for counting lightly ionizing radiations such as beta particles or gamma rays and is specially designed to take advantage of this effect. Since it is difficult to make tubes with windows thin enough for alpha particles to penetrate the gas chamber, GM counters are used mainly for beta and gamma radiations, which are more penetrating. A thin end window is often incorporated in GM tubes to enhance detection of beta particles. A GM tube usually consists of a fine wire electrode (e.g., tungsten) mounted along the axis of a tube containing a mixture of 90% argon and 10% ethyl alcohol (for quenching) at a fraction of atmospheric pressure. A potential difference of 800 to 2000 V (nominally, 900 V) is applied to make the tube negative with respect to the wire. Because of their mode of operation, they cannot be used to identify the type of radiation being detected, nor its energy.

In sum, gas-filled radiation detectors respond to incoming radiation by the ionization it produces. The ions, either multiplied in number or not, are collected to produce a voltage pulse, which may be as small as 10 μV. These small pulses are amplified to 5 to 10 V and fed to a galvanometer for a direct reading or to a pulse counter (or scaler) so that their rates can be measured. The magnitude of the pulses produced can also be measured in the proportional region and sorted according to their energy (e.g., alpha particles from beta particles).

Crystalline Detectors and Spectrometers

Various crystals and solid-state detectors can be used with electronic instrumentation to quantitate the number of radiations of a given energy that activate the detector and the rate at which such events occur. A scintillating crystal such as sodium iodide with a thallium additive [NaI(Tl)] will produce a "flash" of light proportional to the energy deposited by a photon that interacts in the crystal. The light (also a photon with a wavelength in the visible region of the electromagnetic spectrum) is reflected onto a photocathode connected to a photomultiplier (PM) tube. When these photons strike the light-sensitive cathode, a small number of electrons (Figure 12-3), proportional to the energy of the absorbed photon, are emitted. These are accelerated by a potential difference across the first dynode, which emits approximately four secondary electrons for each incident electron. A series of 10 dynodes gives an amplification factor of 4^{10}, or approximately a million. Each pulse leaving the tenth dynode is proportional to the amount of energy absorbed, which in turn is proportional to the energy of the photon striking the crystal. These pulses are sorted according to their size and stored in corresponding channels of a multichannel analyzer as counts, allowing the source to be identified by its unique gamma energy.

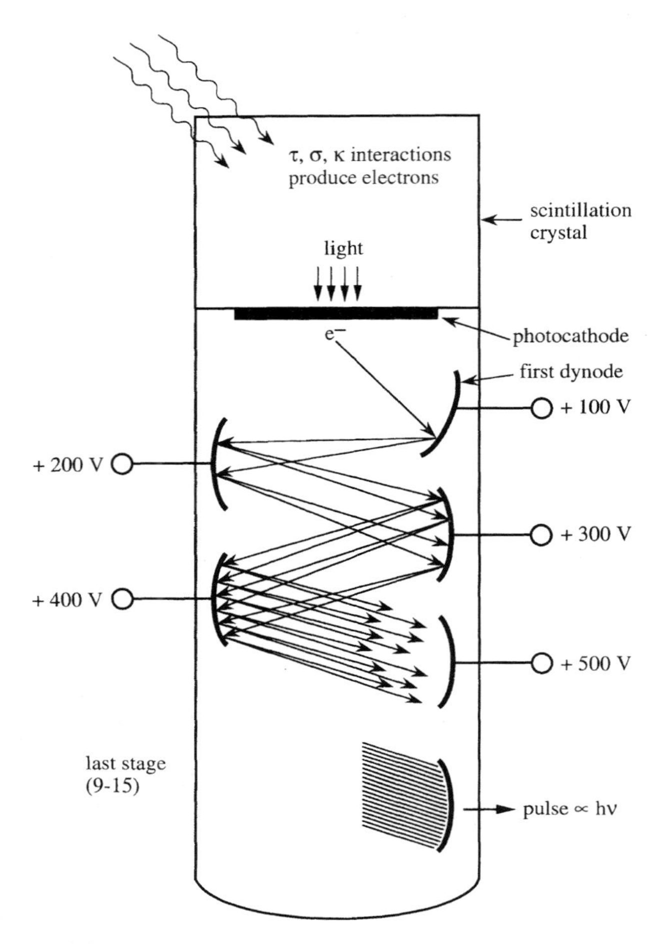

FIGURE 12-3. Scintillation detector with associated photomultiplier tube.

Sodium iodide crystals have a high efficiency for detecting gamma rays (because of their density and the high Z value of iodine), and their pulse resolving time of about 0.25 μs permits their use for high count rates. Plastic scintillators have even shorter resolving times (several nanoseconds), but their detection efficiency for photons is low and the light output with energy is not linear, resulting in spectra with poor resolution.

Semiconducting Detectors

The deposition of energy in a semiconducting material such as intrinsically pure germanium excites electrons from filled valence bands to conduction bands, producing pairs of conduction electrons and electron vacancies, or *holes*. A bias voltage is applied across the semiconductor, which causes these charge carriers to move, producing a current pulse. The energy needed to produce an electron–hole pair in a semiconductor is typically about 1 eV, which is considerably less than that required to produce ionizations in a scintillator; therefore, a relatively large number of charge carriers is produced for each photon absorbed. Consequently, the statistical fluctuations in the number of atoms excited or ionized is much less for a semiconductor detector, and when used with a multichannel analyzer, very sharp peaks (high resolution) are obtained, which allows energies to be determined very accurately. This technology has advanced to a stage that high-efficiency germanium detectors are routinely produced that provide excellent photon resolution (an example of a spectrum obtained with such a detector is shown in Figure 12-7).

Semiconductor detectors have several advantages over scintillators. Although they are not as efficient as NaI(Tl) they readily detect photons and produce fast pulses (typically, a few nanoseconds) with excellent resolution. Germanium and lithium-drifted germanium Ge(Li) detectors must be operated at liquid-nitrogen temperatures to reduce thermal noise, which can be a disadvantage. Ge(Li) (sometimes called "jelly") detectors will quickly deteriorate if not stored at liquid-nitrogen temperature, due to unwanted drifting of the Li into the germanium matrix, and for this reason Ge(Li) detectors have been supplanted by intrinsically pure germanium detectors, which can be stored at room temperature between uses; however, they must be cooled during use.

Other semiconductor detectors are made of silicon with lithium additives to detect electrons and alpha particles, and when coupled with pulse height analyzers are used for spectral analysis. These detectors also provide excellent resolution and can be stored at room temperatures because the mobility of Li is less in Si then it is in Ge.

GAMMA SPECTROSCOPY

Gamma ray spectrometers use photon detectors designed to take advantage of the various absorptive processes in a high-Z medium to maximize sensitivity and provide spectral information about the photons being detected. The complexities of photon interactions are evident in the energy spectra of gamma emitters. The most distinctive features can be roughly divided as those that occur below the threshold for pair production ($h\nu \leq 1.022$ MeV) and those that occur above it.

Gamma-Ray Spectra: $h\nu \leq 1.022$ MeV

A crystal of NaI(Tl) is a commonly used scintillation detector that produces light scintillations that are proportional to absorbed photon energy. The light scintillations are due solely to ionizations and excitations produced by the electrons that are released when photons interact in the crystal. Being charged particles, virtually all of the electrons so released are absorbed in the crystal to produce an output pulse that corresponds to the amount of photon energy absorbed, thus counting the photons that interact and their energy. A typical NaI(Tl) detector/spectrometer is shown in Figure 12-4.

The interaction probability of photons in NaI(Tl) is determined by the attenuation coefficient, which is dominated by the presence of high-Z iodine atoms. As shown in Figure 12-5, the photoelectric effect is the dominant interaction with the iodine atoms ($Z = 53$) in NaI at all energies up to about 250 keV. Some interactions occur in sodium, but with $Z = 11$ these are considerably less probable.

Electrons ejected by photoelectric absorption will contain the photon energy minus 33.17 keV supplied to overcome the binding energy of K-shell electrons in

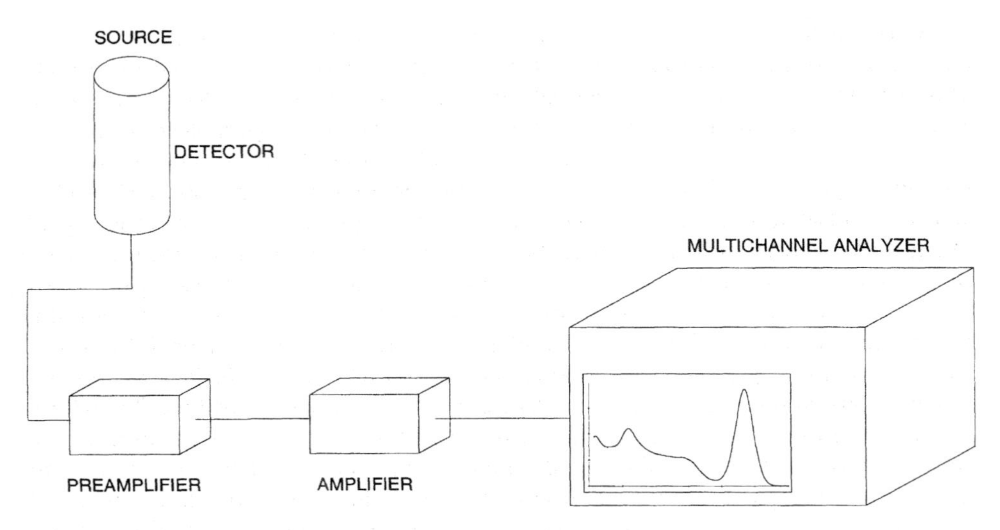

FIGURE 12-4. Schematic of a NaI(Tl) photon detector/spectrometer.

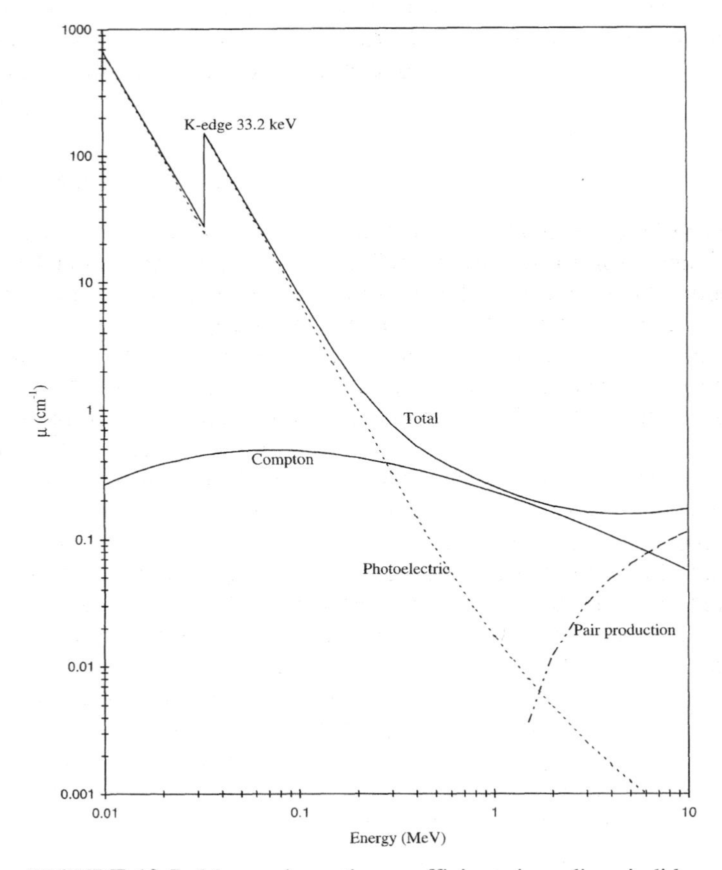

FIGURE 12-5. Linear absorption coefficients in sodium iodide.

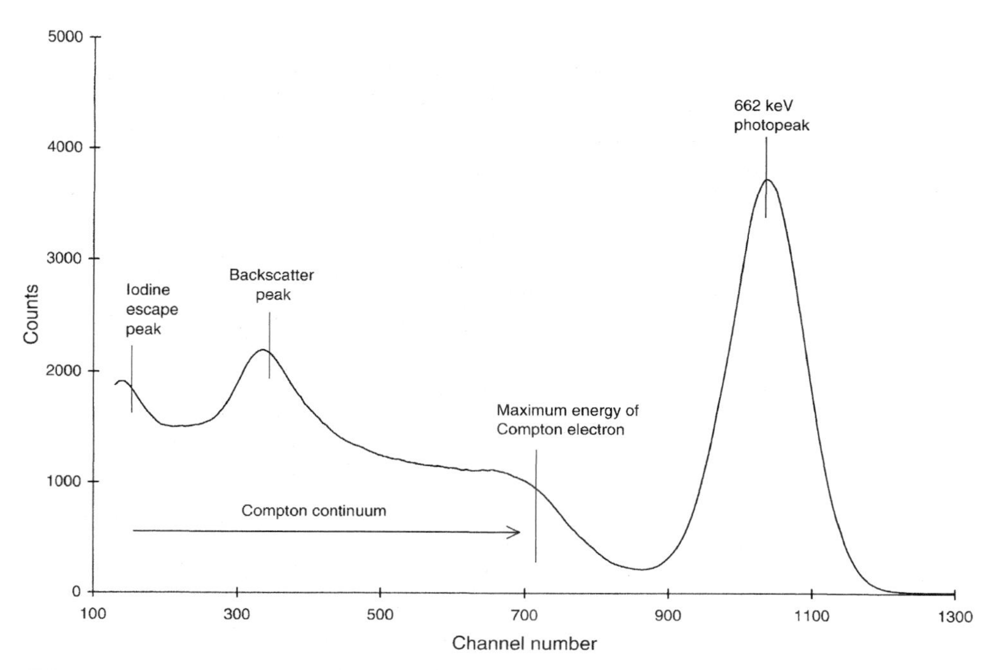

FIGURE 12-6. Gamma-ray spectrum obtained with a sodium iodide crystal exposed to 662 keV gamma rays of ^{137}Cs.

iodine atoms. The orbital vacancy thus produced will be filled promptly, followed by the emission of characteristic iodine x rays. Since these are very soft, they are highly likely to be absorbed in the crystal by other photoelectric interactions. If all of these processes (absorption of the photoelectron and the subsequent photoelectron liberated by the K_α x ray) take place within the luminous lifetime of the phosphor, the visible light produced will correspond to that of the original photon energy, $h\nu$. This in fact occurs for a large fraction of absorbed photons (determined by crystal size and efficiency) and yields a full-energy photopeak, as illustrated in Figure 12-6 for ^{137}Cs.

The full-energy photopeak is the most prominent feature of the gamma-ray spectrum. It is spread over several channels because the number of electrons produced in the photocathode of the photomultiplier tube is relatively small and is subject to statistical fluctuations which are nearly normally distributed and serve to broaden the peak (Figure 12-6). About 12% of the energy of an incident photon will be deposited in NaI, and since each light photon produced will have an energy of about 3 eV, the number of photoelectrons produced in the photocathode of the photomultiplier tube can be determined as well as the statistical fluctuation. As shown in Example 12-1, the number of photoelectrons produced will be relatively low, on the order of a few thousand.

Example 12-1: For 0.364 MeV gamma rays from ^{131}I, determine the number of photoelectrons in NaI(Tl) for each gamma ray at the photocathode of the photomultiplier tube (see Figure 12-3). Assume a 12% scintillation efficiency, an average light photon energy of 3 eV, 25% loss of light photons in the crystal/photomultiplier assembly, and a conversion efficiency in the photocathode of 20%.

SOLUTION: At 12% efficiency, a 0.364 MeV gamma ray will deposit 43,800 eV of energy in NaI, yielding 14,600 light photons of 3 eV each, 11,000 of which strike the photocathode. Twenty percent (or 0.2) of these will produce photoelectrons in the photocathode, or

$$\begin{aligned}\text{no. photoelectrons} &= 11{,}000 \times 0.2 \\ &= 2200 \text{ photoelectrons}\end{aligned}$$

Energy resolution of a gamma spectrometer is an important parameter. If the spread of the data points that make up a measured peak is small, energy resolution is good and peak identification is reasonably accurate; if the spread is wide, energy resolution is much less, perhaps even poor and identification of the peak energy suffers. Qualitatively, sharp peaks have good resolution; wide peaks do not.

Energy resolution is defined as the ratio of the full width of the spectral energy peak measured at half the height of the peak center (both expressed as counts or a count rate), which is known as *full width at half maximum* (FWHM), or

$$\text{resolution} = \frac{\text{FWHM}}{E_0}$$

where E_0 is the energy value at the center of the peak and FWHM is the number of energy units encompassed by the peak measured at one-half the peak height. It is important to note that the peak height is determined as the net count (or count rate) above any background spectrum or Compton continuum upon which it may be superimposed.

Example 12-2: The centroid of a gamma peak obtained with a NaI(Tl) detector is observed at 0.662 MeV to have a height of 10,000 counts above the spectrum background at the peak. At a half-height of 5000 counts (FWHM), the peak width varies from 0.630 to 0.694 MeV. What is the energy resolution?

SOLUTION:

$$\begin{aligned}\text{Resolution} = \frac{\text{FWHM}}{E_0} &= \frac{0.694 - 0.630}{0.662} \\ &= 0.097\end{aligned}$$

or 9.7%. The energy resolution for NaI(Tl) detectors is usually in the range of a few percent because of the statistical variability due to the relatively small number of photoelectrons produced in the photocathode of the photomultiplier tube (the energy resolution of the peak in Figure 12-6 is about 9%).

By contrast, deposition of photon energy in a semiconductor produces ion pairs that are collected directly (i.e., without the need for producing light photons, which in turn produce photoelectrons with inherent losses along the way). If the 0.364 MeV gamma ray in Example 12-1 were to be absorbed in germanium at about 1 eV per ion pair, about 360,000 electrons would be produced and collected. Statistical fluctuations in the energy peaks thus produced are much less in a semiconductor such as germanium, and excellent resolution is obtained.

Other prominent features of spectra for photons with $h\nu \leq 1.022$ MeV in NaI(Tl) are the *iodine escape peak*, the Compton continuum, the Compton edge, a lead x-ray peak, and a backscatter peak. All are denoted in Figure 12-6. An iodine x-ray peak could possibly occur because some of the characteristic x rays produced by filling the empty K shell of the ionized iodine atoms could escape capture during the luminous period of the phosphor and instead be absorbed as soft x rays and as independent events, producing a separate iodine x-ray peak in the gamma spectrum. This iodine x-ray peak is rarely absorbed; however, an *iodine escape peak* occurs at an energy of the photon minus the energy of the iodine x ray that "escaped" detection (see below). The process also causes a slight depression on the low-energy side of the photopeak.

Compton interactions, which are prominent in NaI(Tl) above about 250 keV (Figure 12-6), yield multistage processes, each of which can produce light scintillations that enter the photomultplier tube either as separate events or summed in various combinations to produce features of the gamma spectrum. The Compton electron will be completely absorbed, and if the Compton-scattered photon is also absorbed in the crystal along with it, the light output will be due to the energy dissipated in both events and will correspond to the total energy of the photon, thus producing a pulse that appears in the full-energy photopeak. This is quite likely if the crystal is fairly large, since many of the scattered photons will be of lower energy and photoelectric absorption is favorable. If, however, the Compton-scattered photon escapes from the crystal, the light output will correspond only to the energy transferred to the Compton electron. These electrons form a continuum of light scintillations and output pulses that are registered in the spectrometer (labeled as the Compton continuum in Figure 12-6). The energies range from zero up to the maximum that the photon in question can transfer to an electron in a Compton interaction, which occurs when the photon is scattered backward toward the source or when $\theta = 180°$ in the equation for the energy of a Compton-ejected electron,

$$E_{\text{ce}} = h\nu \frac{\alpha(1 - \cos\theta)}{1 + \alpha(1 - \cos\theta)}$$

where $\alpha = h\nu/m_0c^2$. When this happens the electron recoils along the original direction of the photon and will receive a maximum energy

$$E_{\text{ce,max}} = h\nu \frac{2\alpha}{1 + 2\alpha}$$

The Compton continuum drops off sharply at this maximum energy to form the *Compton edge*. The ratio of the Compton continuum to the photopeak expresses this likelihood, which varies with each detector.

Compton interactions may also cause a *backscatter peak* to appear in the spectrum at an energy of about 0.25 MeV. This peak, which is also shown in Figure 12-6, occurs because some of the primary photons from a source miss the detector and produce Compton-scattered photons from interactions in the shield and other materials. Because they must undergo large-angle scattering to reach the detector, the energy of these Compton-backscattered photons approaches $m_0c^2/2$ or about 0.25 MeV, or slightly less. These lower-energy backscattered photons are very likely to be detected separately by photoelectric interactions to produce a small backscatter peak near the low-energy end of the continuum.

A *lead x-ray peak* at about 88 keV is often observed in gamma spectra because most gamma detectors are shielded with lead. Primary photons from the source interact in the lead shield, as do cosmic rays and other background radiation. Since lead is a high-Z material, many of these photons are absorbed in lead atoms by photoelectric interactions, which are followed by the emission of 88 keV K_α x rays, which can exit the shield and be absorbed in the detector, thus producing a separate peak. Low background shields are often lined with copper and aluminum to absorb lead K_α x rays; the K_α x rays from copper and aluminum are low energy, which generally do not penetrate the detector cover, or if they do so, they appear very near the low-energy end of the spectrum.

The various interactions and the spectral results produced will be random events for any individual photon; however, the average effect of a large number of photon events for a given source/detector produce predictable and reproducible spectra. Such results allow photon sources to be identified and quantified with reasonable accuracy.

Gamma-Ray Spectra: $h\nu \geq 1.022$ MeV

More complicated spectra occur when the energy of the primary photon is sufficient to undergo absorption by pair production (i.e., $h\nu \geq 1.022$ MeV). If the detector is relatively large, the full-energy photopeak will be seen as usual, but other features will also exist, as illustrated in the gamma spectrum of ^{24}Na (see Figure 12-7), which was measured with an intrinsically pure germanium crystal. Sodium-24 emits two high-energy gamma rays at 1.369 and 2.754 MeV, both of which produce photopeaks at these energies in the gamma spectrum. Figure 12-7 also shows the excellent resolution provided by an intrinsically pure germanium detector, which is much larger than that obtained with a NaI(Tl) detector (as shown in Figure 12-8 for ^{24}Na).

All three of the primary photon interactions can occur for ^{24}Na gamma rays, but most interactions occur as Compton scattering or pair production since the photoelectric absorption coefficient in germanium is low for these high energies. When Compton interactions occur and all the processes following such interactions are completed within the luminous lifetime of the crystal, these will yield two photopeaks, at 1.369 and 2.754 MeV. These peaks are produced by pulses corresponding to deposition of all of the energy of the photons by the various processes that can occur in the crystal. A Compton continuum of pulses produced by the absorption of the Compton electrons exists below each photopeak because some of the respective Compton-scattered photons are lost from the crystal, and only the energy transferred to each ejected electron is absorbed and registered.

Pair production interactions are also prominent for the 1.369 and 2.754 MeV photons of ^{24}Na since they are well above the pair production threshold. When

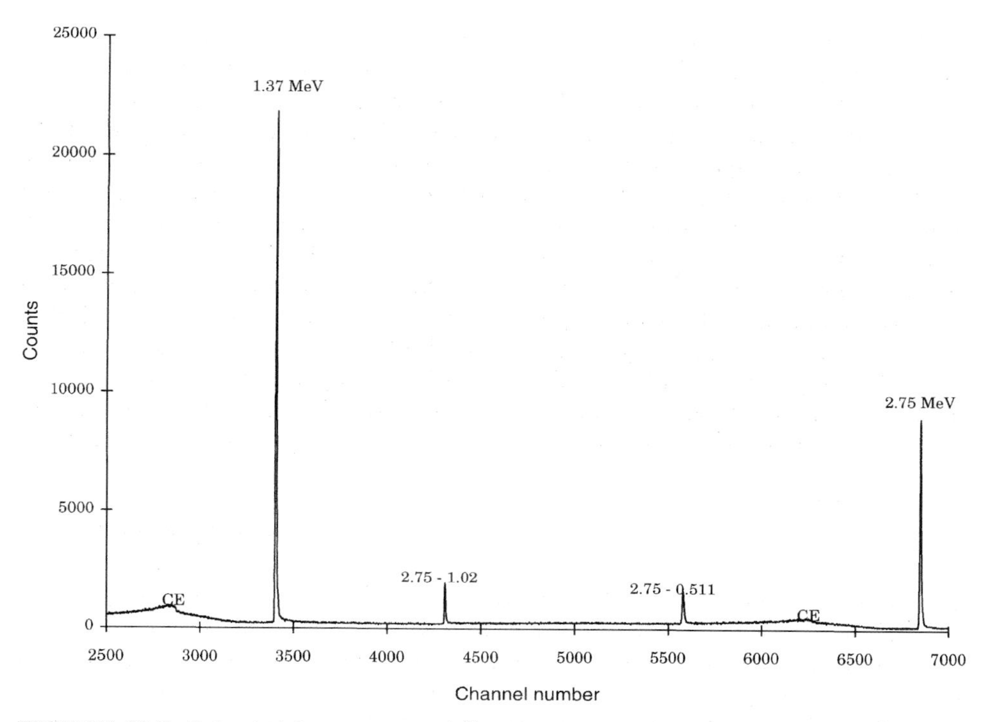

FIGURE 12-7. Pulse height spectrum of ^{24}Na measured with a germanium detector with excellent resolution.

these occur, the total kinetic energy shared between the positron and electron pair ($h\nu$ – 1.022 MeV) will be absorbed in the crystal followed by positron annihilation. If both of the annihilation photons are absorbed in the crystal, their energy adds to that of the absorbed electron pair, and the pulse from the detector will correspond to the full energy of the original photon; the summation of these events will be registered in the full photopeak. Since the annihilation photons are fairly energetic at 0.511 MeV, one or both may escape the crystal, such that this energy is absent in forming the photopeak.

Gamma emitters that produce pair production interactions may also produce a peak at 0.511 MeV. This peak occurs because many of the photons emitted by the source will miss the crystal entirely and be absorbed in the shield around the crystal, and due to the 180° separation between the two annihilation photons thus produced, only one will strike the crystal and be detected.

Escape Peaks and Sum Peaks

Escape peaks are those peaks in a gamma spectrum that occur separate and distinct from the full-energy photopeak and are due to the net photon energy absorbed [i.e., the escape peak energy is that of the interacting photon minus that carried off by the "escaping" photon(s)]. The escape of annihilation photons following pair production interactions in a crystal causes the production of either a single escape peak or a double escape peak. If both of the annihilation photons escape from the

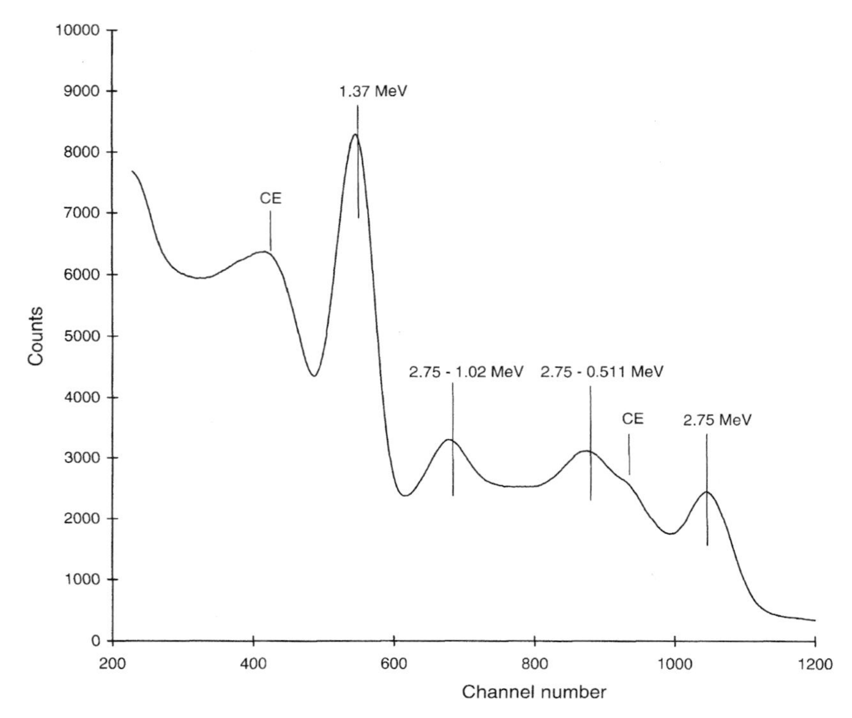

FIGURE 12-8. Pulse height spectrum of ^{24}Na measured with a NaI(Tl) detector.

crystal, a peak will be seen at an energy of $h\nu - 1.022$ MeV. If only one of the 0.511-MeV photons escapes and the other is absorbed, a peak will occur at an energy of $h\nu - 0.511$ MeV.

The peak shown in Figure 12-7 at $2.75 - 0.511$ MeV is labeled as a single escape peak, due to escape of one of the annihilation photons, and the one at $2.75 - 1.022$ MeV is the double escape peak because both annihilation photons escaped the detector without depositing energy. The Compton continuum in a gamma spectrum is also due to escape photons, but since Compton-scattered photons represent many energies, an escape continuum is produced rather that an escape peak, even though each point in the continuum corresponds to $h\nu$ minus the energy of the escaping Compton-scattered photon.

Escape peaks are distinct, but they do not represent primary photon energies and do not enter into the primary identification of the gamma emitter; they are only observable consequences of the patterns of photon interactions and losses in the detector itself. Consequently, their size and shape are a unique feature of an individual detector system because their patterns of interaction and detection are determined by the size and configuration of the detector system, which can be quite variable. Detector size has perhaps the greatest influence on whether secondary photons escape detection.

Charateristic x-ray escape peaks may be observed when detectors are quite small and photon energies are low, say < 0.5 MeV or so, where photoelectric interactions are dominant. For low-energy photons and small NaI(Tl) detectors, the 33.17 keV charateristic x rays produced in iodine atoms may well leave the detector, such that the actual photon energy deposited is at $h\nu - 0.033$ MeV; thus an iodine escape peak is produced below the full-energy photopeak at $h\nu - 0.033$ MeV. By a similar process a germanium escape peak can occur for low-energy photons in a small germanium detector at $h\nu - 0.011$ MeV.

Sum peaks are another feature of gamma-ray spectra. These are produced when two or more photons are emitted in quick succession such that they are absorbed within the luminous lifetime of the crystal (i.e., simultaneously, or very nearly so). When this happens they are recorded as a single event but at an energy equal to the sum of the two energies. Sum peaks are often observed for high-activity sources that emit two or more photons per transformation, or for photon sources with $h\nu \geq 1.022$ MeV which yield annihilation photons; these peaks will be much smaller than the photopeaks since they depend on simultaneous detection in coincidence of two emission events, which is much less probable.

Gamma Spectroscopy of Positron Emitters

Positron emitters that also emit gamma rays produce gamma spectra with distinct features and other complexities. These additional features are illustrated in Figure 12-9 in the spectrum for ^{22}Na, which emits positrons followed by emission of a

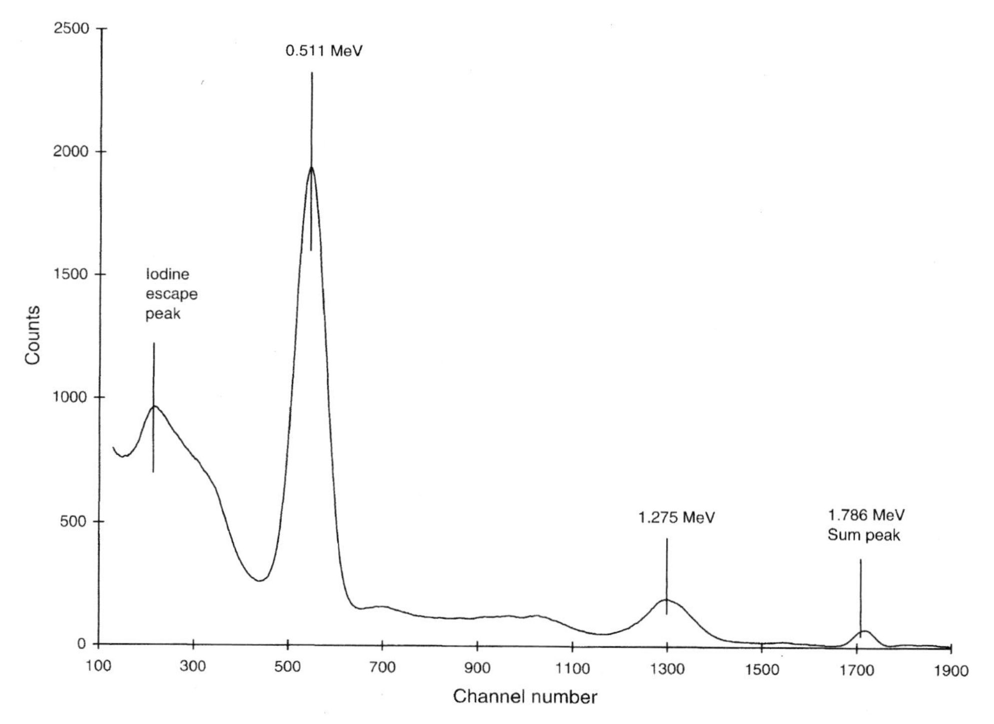

FIGURE 12-9. Gamma-ray spectrum of ^{22}Na measured with a NaI(Tl) detector. The spectrum shows the typical 0.511 MeV annihilation peak of a positron emitter, the photopeak of the 1.275 MeV gamma ray emitted, two sum peaks, the Compton continuum, and the backscatter peak.

1.275 MeV gamma ray. A positron emitter will always show a peak at 0.511 MeV due to the annihilation of the positron. The strong peak at 0.511 MeV occurs because most of the annihilation events occur within the source itself, outside the crystal, or at the surface of the crystal where the positrons are absorbed. Since the annihilation photons are emitted 180° from each other, only one of them is likely to be absorbed in the crystal, and thus it will be detected as a separate event with an energy of 0.511 MeV. The Compton continuum also occurs below the 0.511 MeV photopeak (and the 1.275 MeV photopeak as well), as does the usual backscatter peak, and if two annihilation peaks are detected simultaneously, a small sum peak at 1.022 MeV will be produced.

The 1.275 MeV gamma ray of ^{22}Na will also be detected by photoelectric, Compton, and pair production interactions. If the sequence of electron-producing events for each interaction all occur within the luminous lifetime of the crystal, the pulses will produce a photopeak at 1.275 MeV, as shown in Figure 12-9. If these events are detected simultaneously with an annihilation photon, a small sum peak will be produced at 1.786 MeV due to the reduced probability of both sequences occurring simultaneously. This sum peak is evident in the spectrum shown in Figure 12-9, in addition to the strong peak at 0.511 MeV produced when ^{22}Na positrons are annihilated and the full-energy photopeak produced by the 1.275 MeV gamma rays.

PORTABLE FIELD INSTRUMENTS

The general principles of radiation detection discussed above can be incorporated into portable field instruments, and instruments based on G-M detectors, proportional counting, and ionization are available. The portability of such instruments makes field measurements convenient and reliable, although with differing degrees of sensitivity compared to laboratory instruments, where greater shielding of detectors is practicable.

Geiger Counters

Geiger counters, especially with pancake probes, are very useful for general surveys of personnel contamination, area contamination, and the presence of external radiation fields (see Figure 12-10). In making personnel surveys, the GM probe should normally be moved slowly to allow time for the instrument to respond. An energy correction factor may need to be applied to GM-measured readings if the photon field contains a large fraction of low-energy gamma rays because it will overrespond in this region because the photoelectric cross section is high for low-energy photons. The shield assembly provided by the manufacturer flattens out this overresponse by selectively filtering out lower-energy photons more than medium-energy photons, and the instrument should be used with the shield closed when low-energy photons are dominant. GM detectors should be used with care in high-exposure fields because they can saturate, yielding a reading near zero when in fact the exposure rate is well above the highest range on the survey meter. In such circumstances, an ion chamber should be used.

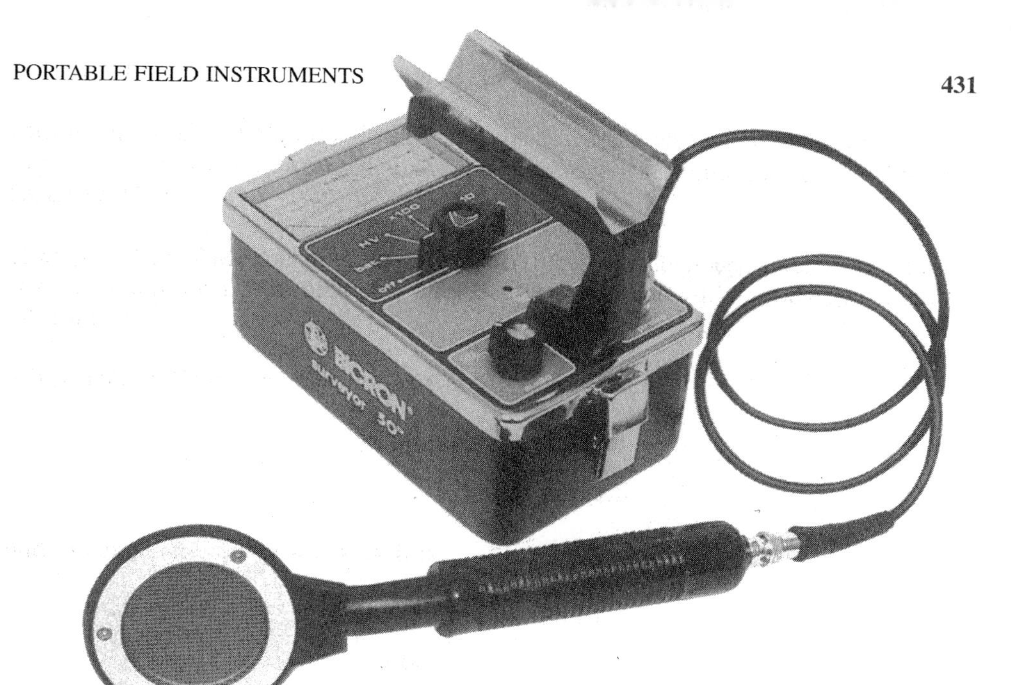

FIGURE 12-10. Geiger counter with a thin beta window.

Ion Chambers

Ion chambers have a very flat energy response for gamma-ray measurements, and although they are less sensitive than the Geiger counter, they respond correctly in much higher fields. They are, however, subject to *zero drift*, especially on the more sensitive ranges; therefore, they are provided with a control to set the meter scale to zero before an exposure is read.

A typical ion chamber survey instrument contains a chamber of about 200 cm^3 filled with ambient air, although other gases can be used. The chamber is sealed at one end with aluminized Mylar with a density thickness of 7 mg/cm^2, and a removable plastic shield is provided to protect the Mylar window. The sliding shield is commonly made of phenolic plastic ($\rho = 1.25$) about 0.34 cm thick, which has a density thickness of about 0.439 g/cm^2, which stops all beta particles of energy below about 1 MeV (see Figure 12-11). When operated with the shield closed, only gamma radiation and high-energy beta particles are detected; when the shield is removed (or opened in some designs), beta particles of sufficient energy to penetrate the dead skin layer are detected as well as gamma radiation. These features allow detection of mixed fields of gamma and beta emitters when both are present, which they often are, especially in various nuclear facilities. As a practical matter, the gamma reading should be significantly smaller than the beta + gamma reading, or else large uncertainties are introduced.

The basic survey procedure is to take two readings, one with the window open and the other with the window closed. The beta component can then be determined by subtraction of the gamma reading (taken with the cap on) from the open-window reading. The open-window reading may need to be adjusted to account for the energies of the beta particles and the density of air in the chamber due to atmospheric

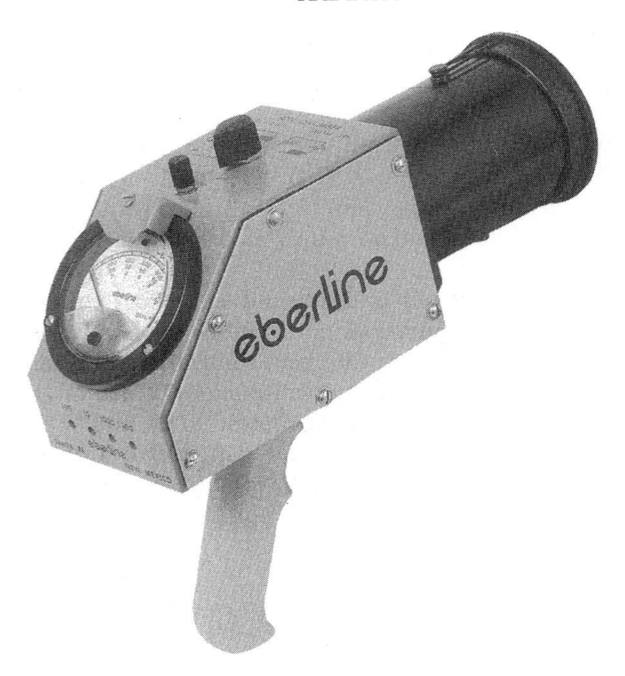

FIGURE 12-11. Portable ion chamber instrument with a removable beta shield cap.

pressure changes, but this is usually a minor effect, except perhaps at high altitudes. If the ion chamber is exposed to a highly collimated beam of radiation so that only part of the chamber volume is irradiated, the instrument will read low by an amount equal to the ratio of the volume exposed to the total chamber volume.

Microrem Meters

Microrem meters are used to obtain accurate measurements of low-level gamma radiation (e.g., near background level). Commercial microrem survey meters use a scintillation crystal as the detector to increase sensitivity, and these are typically about 10 times more sensitive than commercial Geiger counters. The microrem meter consists of a suitable phosphor, which is optically coupled to a photomultiplier tube, which in turn is connected to an electronic circuit. If the counter is used simply to detect radiation, the output circuit often consists of a battery-operated power supply, an amplifier and pulse shaper, and a rate meter. If the device is used for energy analysis, the output circuit includes a pulse height analyzer and a scaler and can be operated as a single- or multichannel analyzer.

Alpha Radiation Monitors

Generally, alpha radiation is encountered only as a result of surface contamination or in the form of airborne particulates, both of which may be taken internally by workers or others. Two types of instruments are in practical use for locating surface contamination: the portable proportional counter and a portable scintillation counter. Ion chamber survey meters with thin windows can also be used, but since they also

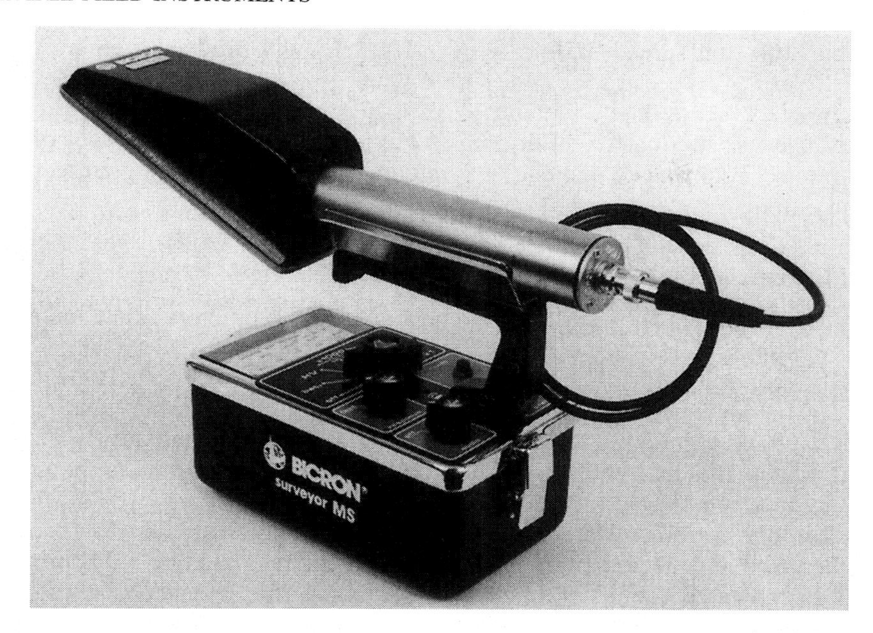

FIGURE 12-12. Portable alpha scintillation counter.

respond to both photons and particles, they are best used for monitoring sources of pure alpha emitters.

Portable proportional counters for alpha-particle measurements are of two types: One uses air at ambient pressure as the counting gas, while the other is supplied by propane gas, usually by a small cylinder attached to the counter. Both must have an extremely thin window for the alpha particles to penetrate to the sensitive region of the detector. They also require special discriminator circuits, very stable high-voltage supplies, and in some cases, very sensitive amplifiers. Alpha particles produce a rather large electrical pulse in a proportional counter, and good discrimination is possible against beta- and gamma-ray interference or fast neutrons, if present. Such interference can be detected simply by moving the alpha probe about 10 cm away from the surface being monitored; if the counts cease, they are due to true alpha contamination.

An air proportional counter does not require the gas cylinder and associated plumbing necessary for a propane proportional counter; consequently, it is lighter and less cumbersome to set up and use, although it is generally less efficient and sensitive to changes in humidity, a condition that is avoided with the propane counter. The propane regulator and needle valve tend to plug up after long use, and the counter is not suitable for monitoring large quantities of plutonium because the propane released can be a hazard since plutonium is pyrophoric.

Scintillation counters for alpha-particle measurement use a silver-activated zinc sulfide [ZnS(Ag)] phosphor that is quite sensitive to alpha particles; however, it is less rugged than either the air or propane proportional counters because of the fragile photomultiplier assembly (see Figure 12-12). It does have good detection efficiency, ranking between the air and propane proportional counters. A ZnS alpha probe has a sensitive area of about 50 cm^2 and a total efficiency [(c/m)/(d/m)] of

25%; thus 100 d/m spread uniformly over 100 cm^2 will produce a count rate of about 12 c/m.

In a practical sense, alpha monitoring of a large area must be done slowly and carefully because the meter will not have time to respond to a small spot of contamination, a circumstance that can be overcome somewhat by using an earphone or audio output.

Survey Instruments for Beta Particles

The proper instrument for measuring the beta dose rate is a thin-window ion chamber because it responds to actual energy deposited, and the electrical signal produced is directly proportional to the energy deposited in the chamber gas. These instruments can be converted for use in gamma surveys by covering the window with a shield cap (see Figure 12-11). The Geiger counter cannot quantitatively measure the dose rate or the dose equivalent rate in a beta field, although it can be used to detect the presence of beta radiation.

Ion chamber instruments are usually calibrated for photons rather than beta particles; therefore, it is necessary to multiply the ion chamber reading by a correction factor, CF (expressed in units of mrad/mR), to obtain surface contact beta dose rates. The actual CF values for beta surface contamination depend on the chamber geometry (size and shape), the chamber wall thickness, and the size of the beta source; small spots of contamination require CF values that are substantially greater than unity.

Removable Radioactive Surface Contamination

The smear, swipe, or wipe test is used universally to measure removable surface contamination, whereas survey instrument measurements of alpha-, beta- and/or gamma-surface contamination include the total contamination, whether fixed or removable. In the usual technique a surface area of 100 cm^2 is swiped by a cloth, paper, plastic, foam, or fiberglass disk. These smears are in turn counted for alpha contamination or higher-energy beta emitters in a gas flow proportional counter, or for gamma contamination with a scintillation or semiconductor detector. Low-energy beta emitters such as tritium or ^{14}C are counted in liquid scintillation counters, which can be used to count other beta emitters, and with appropriate adjustment, alpha emitters as well.

Instrument Calibration

Calibration of an instrument involves a determination of its response or reading relative to a series of known radiation values covering the range of the instrument, and adjusting the instrument to provide a correct response. Three levels of calibration are generally recognized: a full characterization (usually done by the instrument manufacturer); a calibration for specific, perhaps unusual, conditions; and a routine calibration for normal working conditions using an appropriate source.

In general, two approaches are used in the calibration of portable instruments. A radiation field produced by a standard source calibrated by the National Institute of Standards and Technology (NIST) can be used, or the field can be measured with an instrument that is a secondary standard. A secondary (or transfer) standard instrument is an instrument that has been calibrated by comparison with a national standard by placing the instrument in the radiation field and noting the response. The instrument to be calibrated is then substituted for the original instrument, and calibration is based on a comparison of the two readings.

Ion chambers and GM counters are calibrated with ^{60}Co, ^{137}Cs, and ^{226}Ra to provide a known exposure rate at distances that allow the source to be treated as a point source. Calibration of ion chambers requires uniform irradiation of the detector volume; therefore, the distance between the source and the calibration point should be at least three times the chamber's longest dimension, which should be perpendicular to the source.

Personnel Dosimeters

Personnel dosimeters consist of film badges that contain photographic film or thermoluminescent dosimeters (TLDs) to measure the radiation dose received by persons over a period of time, usually a month or a quarter. For short-term monitoring of work, a pocket ion-chamber dosimeter or an electronic dosimeter is worn. Electronic dosimeters can be useful for this purpose because they can be set to indicate an alert or warning level of exposure.

Film Badges Film badges continue to be used in personnel monitoring because they are relatively inexpensive, the processed film provides a permanent record, and they provide sufficient accuracy to meet accreditation requirements, especially for x and gamma radiation. Film badges use dental x-ray film shielded in places (called *windows*) by thin absorbers to distinguish beta rays from gamma radiation. An unshielded area (the open window) gives the total dose from beta and gamma rays, and this can be apportioned by the amount of film density behind the shielded strip. Double emulsion film is used for high sensitivity (in the mR range); conversely, high-range film is made with a single emulsion. Film can also be used to measure neutrons if the film is impregnated or covered with a material such as lithium or boron, which readily absorbs neutrons.

The response of film is independent of energy above 200 keV or so; however, at low photon energies, the film may overrespond by as much as a factor of 20 to 40. With proper calibration and appropriate corrections for low-energy response, the film badge can report doses from about 0.1 mSv (10 mrem) to about 10 Sv (1000 rem).

Thermoluminescence Dosimeters (TLDs) The phenomenon of thermoluminescence (TL) was rediscovered in the 1960s and was perfected as a dosimetry technique in the 1970s when it was found that the chemical makeup of lithium fluoride (LiF) or calcium fluoride (CaF_2) phosphors could be made uniform by good quality control. As indicated in Figure 12-13, exposure of a TLD to radiation causes electrons to be excited to the conduction band, from which they can fall into one of the isolated levels provided by impurities in the crystal and be "trapped"; they will remain so until energy is supplied (usually by heat) to free it. Heating the crystal elevates the trapped electrons back to the conduction band, and when they return to a valence hole, a photon of visible light is emitted. The total light emitted is a measure of the number of trapped electrons and therefore of the total absorbed

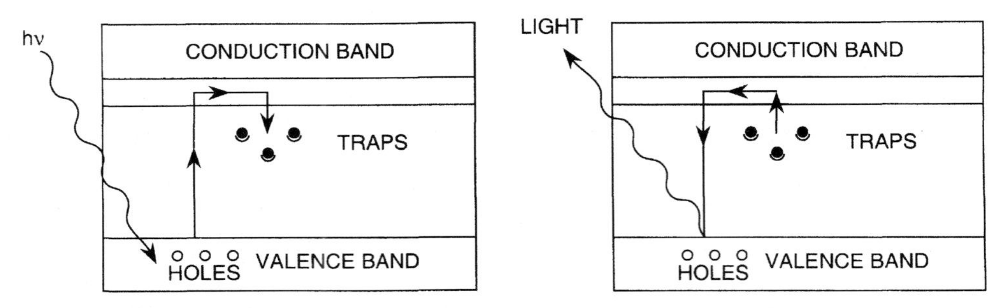

FIGURE 12-13. Energy-level diagram of a TLD crystal. Absorption of radiation excites valence electrons up to electron "traps," thus storing information on the amount of radiation absorbed; heating the TLD deexcites the trapped electrons and the emission of light photons with intensity that is proportional to the amount of radiation absorbed.

radiation even after months of storage. Thermoluminescent dosimters are aptly named because thermal heating (thermo-) produces luminescence emitted by the crystal. TLDs are generally more accurate than film badges, are less subject to fading, are usable over a much larger range of radiation levels, can be used over and over, and processing does not require film development but can be done by a one-step electronic reader.

The TLD phosphor can be reset by heating it to a high temperature to release all trapped electrons, a process known as *annealing*. After proper annealing, the phosphor has the same sensitivity as previously, so it can be reused as a dosimeter, which offers some advantage versus film. This feature can also be a disadvantage, because once the traps release their electrons, the information is lost; however, the glow curve from the TLD reader can be stored electronically or run out on a strip-chart recorder to provide a permanent record if such is required or desirable.

Pocket Dosimeters The pocket dosimeter, often used in conjunction with a TLD dosimeter or film badge, is a small electroscope, about the size and shape of a fountain pen. The dosimeter is usually fitted with an eyepiece and a calibrated scale so that an individual user can read the amount of exposure received. The dosimeter is charged before use, and exposure to radiation causes it to lose charge proportional to the radiation exposure received. Since any leakage of charge produces a reading, good insulation of the electrode is needed. The response of the dosimeter is seldom linear, except in the region of the calibrated scale; therefore, exposures should not be estimated if the device reads slightly above the full-scale reading. When reading the dosimeter the fiber image should be vertical to reduce the geotropic effect, the tendency is to give a reading that depends on the orientation of the device.

LABORATORY INSTRUMENTS

Liquid Scintillation Analyzers

Liquid scintillation analysis (LSA) of samples is achieved by mixing the radioactive sample into a liquid scintillant made up of chemicals that produce visible light when radiation is absorbed. Liquid scintillation solutions contain low-Z materials

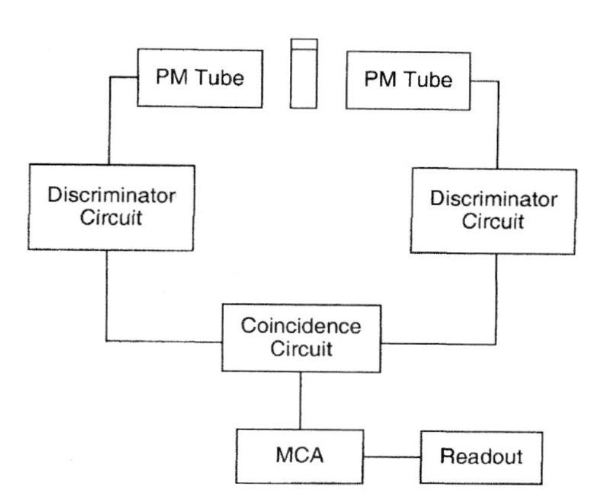

FIGURE 12-14. Schematic of a liquid scintillation counter in which two photomultiplier tubes measure light produced by absorption of emitted radiations in the liquid scintillation sample. The signals from each of the PM tubes pass through discriminators and are registered as a function of energy if they arrive in coincidence; if not, the signal is presumed to be noise and is rejected.

(typically, $Z = 6$ to 8), and consequently have relatively low counting efficiency for x and gamma rays above 40 keV or so. On the other hand, good efficiency is obtained for low-energy x and gamma rays, especially so for most beta particles because of their short range in liquids. The liquid scintillation counter is often the best, and perhaps the only practical detector for measuring low-energy beta emitters such as ^{3}H and ^{14}C. It is also useful for measuring beta-emitting radionuclides on wipe and leak test samples.

Liquid scintillation counting is done by placing a sample and the scintillating chemicals into vials that transmit light. The light flashes, which are proportional to the energy of the radiation source, are measured by one or more photomultiplier (PM) tubes, as shown in Figure 12-14. These are placed in a darkened chamber with a tight lid made of a material that excludes ultraviolet radiation so that only the light produced by the sample is observed. Electronic noise due to thermal emission of electrons from the photocathode of the PM tube can be significant, thus it also is essential to use low-noise PM tubes with coincidence circuitry to eliminate extraneous pulses electronically. This is done by coincidence detection using two PM tubes placed 180° apart, as shown in Figure 12-14. Optical reflectors around the counting vials direct the emitted light produced in the liquid scintillant to the two PM tubes with outputs routed to a coincidence circuit that accepts only those pulses that arrive simultaneously (Figure 12-14). Since the light emitted from a radiation event will be reflected into both PM tubes, two pulses are produced in coincidence and the event will be detected and registered; thermal noise that occurs randomly in the PM tube photocathode will produce only a single pulse and will be rejected.

The signal from the PM tubes is proportional to the light energy collected, which in turn is directly proportional to the radiation energy absorbed, thus the pulses that are produced can be sorted to produce a spectrum. Most modern LSCs have several thousand channels and the data can be accumulated and displayed electronically as well as plotted. With appropriate calibration, the endpoint beta energy, or $E_{\beta,\max}$ can

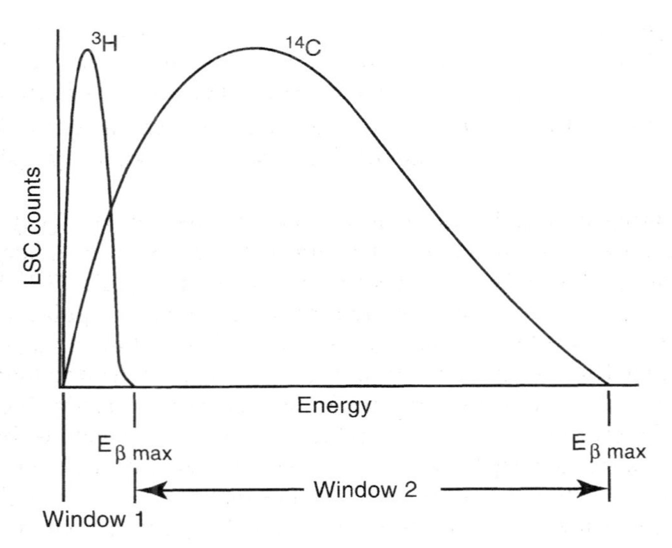

FIGURE 12-15. Typical beta spectra of ^{3}H and ^{14}C obtained by liquid scintillation counting.

be determined, and this is often a good determinant of the radionuclide. Another technique is to compare the spectrum obtained with reference spectra obtained with pure standards, many of which have unique shapes. For example, a full spectrum of ^{137}Cs will show the sharp conversion electron peak above the continuous spectrum of beta particles (see Figure 5-22) and a ^{60}Co spectrum will show a unique feature at the upper energies due to absorption of its gamma rays.

Two or more beta emitters in a sample can be identified and quantified in the presence of each other. The spectrum obtained is the sum of both, as shown in Figure 12-15 for tritium and ^{14}C, and the counts from each beta emitter can be accumulated by setting appropriate energy windows. This approach requires counting of a standard of each radionuclide to set the energy window, determining the counting efficiency in each set window, and determining the fractional overlap of each into the other window so that it may be subtracted as shown in Example 12-3.

Example 12-3: A mixed sample containing ^{3}H and ^{14}C is counted in an LSC system with two window settings with upper-level discriminator settings just above 18.6 keV for ^{3}H beta particles and 156 keV for ^{14}C beta particles. The measured counting efficiencies are 36% for the ^{3}H window and 80% for ^{14}C, and it is determined that 8% of the counts observed from a ^{14}C standard will be registered in the ^{3}H window. A sample that contains both ^{3}H and ^{14}C is counted with a net (minus background) count rate of 2000 c/m in the ^{3}H window and 6000 c/m in the ^{14}C window. Determine the activity of each in the sample.

SOLUTION: First, subtract the number of counts in the ^{3}H window due to the spectral distribution of ^{14}C counts:

$$^{3}\text{H(net c/m)} = 2000 - (0.08)6000 = 1520 \text{ c/m}$$

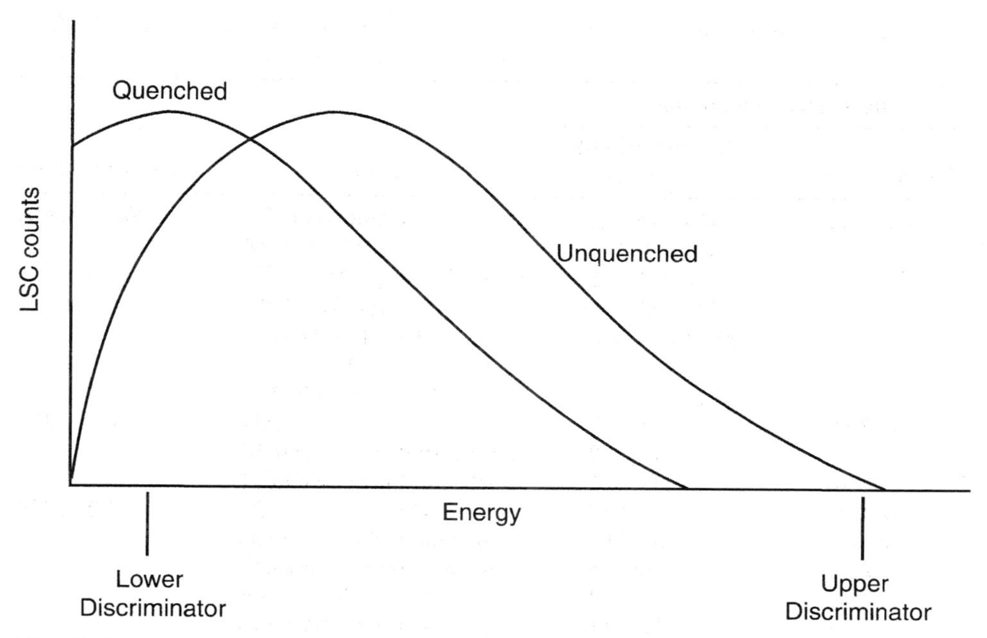

FIGURE 12-16. The effect of sample quench in liquid scintillation counting is to lower the number of counts observed in a set window.

Since ^3H beta particles are not energetic enough to exceed the lower setting of the ^{14}C window, the net count rate for ^{14}C is 6000 c/m. The activity of each is

$$^3\text{H(t/m)} = \frac{1520}{0.36} = 4222 \text{ t/m}$$

$$^{14}\text{C(t/m)} = \frac{6000}{0.80} = 7500 \text{ t/m}$$

The spectral analysis method shown in Example 12-3 can be developed and used, with appropriate standards and technique, for various other beta-emitting radionuclides.

Sample quenching, which is the term applied to any process that reduces the emitted light output, is a major consideration in liquid scintillation counting. There are two general subclasses: chemical and optical quenching. The effect of sample quench, as shown in Figure 12-16, is to shift the beta spectrum toward lower energies and thus reduce the overall number of counts observed. This problem is dealt with by measuring the amount of quench and applying a quench correction to the counter results. Sample quench can be minimized by assuring good mixing of the sample in the liquid scintillant. If incompatible liquids are used, the sample may phase-separate and accurate quantitation will be difficult. Use of a small sample volume can reduce this phenomenon but may not eliminate it; therefore, each sample should be examined before counting.

The amount of quench in each sample is usually measured by exposing it briefly to a high-activity radiation standard (e.g., ^{133}Ba) located next to the sample chamber. The measured light output produced by such exposure is calculated as a *quench number*, which has been given various designations by equipment manufacturers. A plot of a series of such measurements made on a known quantity of sample yields a quench curve of counter efficiency versus quench number, which in turn is used to adjust the count rate for each sample to the true count rate. The known sample is mixed with distilled water, added to the LSC cocktail, and counted. A chemical contaminant such as carbon tetrachloride is then added to the "known" in the LSC cocktail in increasing amounts, and the quench number and count rate are recorded and plotted to produce the quench curve. Modern LSC instruments can store quench correction curves, and such corrections can be made electronically to increase the number of recorded counts to those that would have been recorded if quench losses had not occurred.

Proportional Counters

Gas proportional counters take advantage of the amplification obtained when radiation is absorbed in the counting gas to provide good sensitivity as well as to discriminate charged particles on the basis of the proportional pulse sizes produced (e.g., beta emitters from alpha emitters). Because of these factors and their relatively poor sensitivity to x and gamma radiation, proportional counters are used primarily for counting charged particles, principally beta particles, alpha particles, or one in the presence of the other. A hemispherical counting chamber (Figure 12-17) with a central electrode, or anode, usually in the form of a small wire loop, is used that encloses an optimal counting gas in intimate contact with the sample to be counted. The electric field changes rapidly with distance in the immediate vicinity of the an-

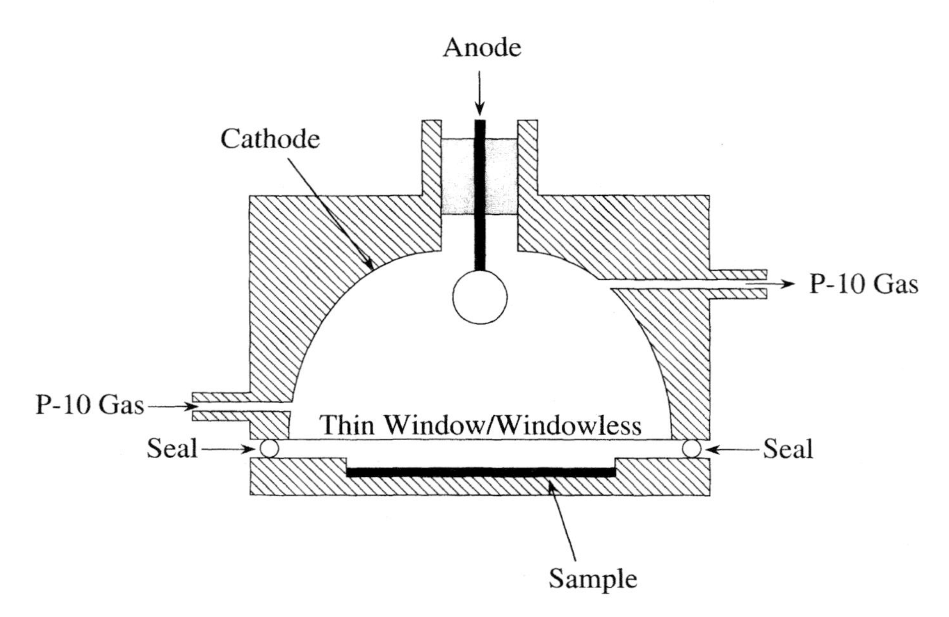

FIGURE 12-17. A 2π proportional counter with a window; if used without a window in either 2π or 4π geometry, it is necessary to provide a mechanism to reseal the chamber after samples are inserted and to provide rapid flushing with the counting gas to replace any outside air.

ode, and most of the amplification occurs in the vicinity of the anode wire. In this configuration (typically, referred to as a 2π chamber or 2π geometry), the counter may be windowless, or more typically, it will have a thin window such that the sample to be counted is placed just outside the window. Both designs have advantages and disadvantages in terms of sensitivity and control of contamination.

Proportional counters have good counting efficiency and can be used at very high counting rates because the negative ions, which are moving in the most intense field in the counter, have to move only a few mean free paths to be collected. The movement of the negative ions is little influenced by the presence of the positive ions. When operated in the proportional region, secondary electrons are formed in the immediate vicinity of the primaries produced by the incoming radiation as it is absorbed, and with a typical gas amplification of 10^3, the absorption of an alpha particle might lead to a pulse containing 10^8 ions, while a pulse initiated by a beta particle or a gamma ray would contain more nearly 10^5 ions. These pulses are in turn amplified to a desired output level, but the relative pulse sizes remain the same and a discriminator can be set to reject all pulses below some desired level, allowing alpha particles to be counted in the presence of beta and gamma radiation. These characteristics lead to different degrees of amplification versus detector voltage for alpha and beta particles, producing two distinct plateaus for counter operation, as shown in Figure 12-18. The counter can be operated to just detect alpha particles or at a higher voltage to detect both alpha and beta particles. Electronic circuitry can be used to register one or the other or both.

Proportional counting is usually done at atmospheric pressure because the voltages, which range from 1000 to 5000 V, present no technical problems and there are advantages to a counter that can be opened readily for the insertion of samples, a

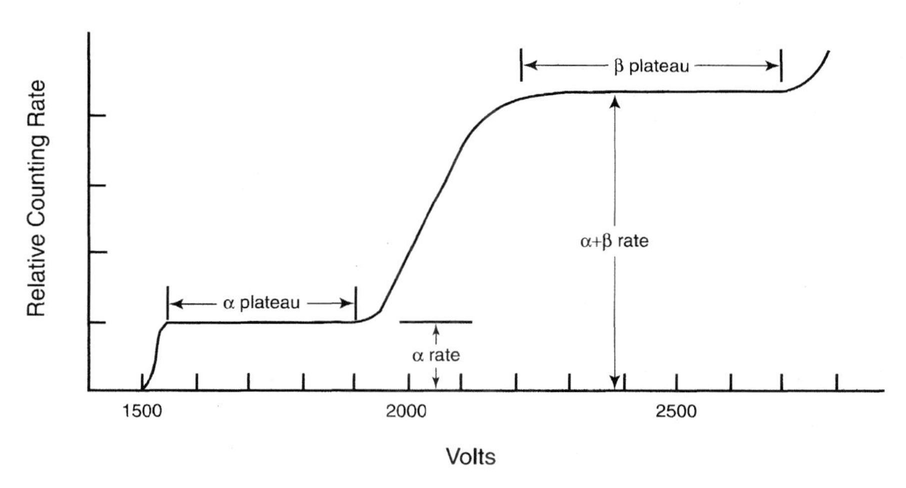

FIGURE 12-18. Response of a proportional counter versus applied voltage showing a voltage plateau for counting energetic alpha particles and a much higher voltage plateau for beta particles. Whereas beta particles, because of insufficient energy, are not registered in the alpha-counter region, both are registered when the counter is operated in the beta-particle region; however, the sizes of the pulses are significantly different and can be sorted electronically for a separate alpha count in the alpha region, where beta particles are excluded and subtracted from the count which includes both particles.

necessity for windowless counters. If a windowless chamber is used, it is necessary to provide a mechanism to reseal the chamber after samples are inserted. This is usually done by causing the sample tray to press against an O-ring or similar sealing device when the sample holder brings the sample into the detector volume (or under it for a thin-window detector). It is also necessary to flush the windowless chamber with counting gas to assure that no outside air remains in the chamber. A detector with a window does not require flushing since a steady gas flow is maintained; however, the window must be quite thin to minimize absorption of alpha and beta particles.

Various gases and mixtures are suitable for proportional counting, but a mixture of 90% argon and 10% methane is often used. Argon has good performance because of its high density, but it has long-lived excited states which may trigger spurious discharges; therefore, methane is added to counteract these. The 90% argon–10% methane mixture is known as P-10 gas.

Windowless proportional counters can be used to eliminate absorption of particles to be counted, and they are essential for proportional counting of very low energy beta emitters. If 4π geometry is used, the chamber volume is a complete sphere, typically with two anode loops. The sample is placed on a thin suspension in the center, and in a double chamber with a suitable sample mount essentially all the particles emitted will be counted. Such counters are used in absolute counting where it is necessary to determine the absolute disintegration rate of a radioactive sample. Consequently, 4π measurements are a special technique and are generally used primarily for standardization of sources.

Contamination of the windowless chamber surfaces can be problematic because the surfaces are much more difficult to clean should they become contaminated. This can be controlled to some extent by assuring that the sample remains fixed, perhaps by a fixative of some type. After a sample is in position in the chamber volume, the counter is flushed rapidly with the counting gas until any room air inside the chamber has been displaced. After stable counting conditions have been established, the gas flow can be reduced to provide a slight positive pressure in the chamber to prevent the inward diffusion of room air.

End-Window GM Counters

A laboratory version of the Geiger–Mueller counter is used for counting various samples, such as smears or determining beta-particle absorption. The GM tube has been replaced in many applications by newer devices because of its inability to distinguish between radiations. It was, however, the first practical detector used in the development of radiation physics and is still widely used. The GM counter can be made quite sensitive for laboratory conditions by providing a stable mount for a high-quality thin-window GM tube and enclosing it in a shield to both reduce background radiation and provide physical protection of the GM tube itself. Samples are placed just below the end window and the output routed to a scaler unit, often with a timer, to record the counts. The operating voltage is set in the GM plateau region (see Figure 12-2) and the counter is usually calibrated for efficiency with a medium-energy beta source, or a standard of the radionuclide that is routinely counted. Such systems provide good sensitivity for routine counting of many types of samples.

Surface Barrier Detectors

Specially constructed silicon surface barrier semiconductors are used for alpha and beta spectroscopy. A pure silicon crystal will normally have an equal number of electrons and holes, but impurities can be added to create an excess number of electrons (an n-region) or an excess number of holes (a p-region). Silicon (and germanium) is in group IV of the periodic table. If atoms of group V, each of which has five valence electrons, are added, four of the five electrons in each of the added atoms will be shared by silicon atoms to form a covalent bond. The fifth electron from the impurity is an excess electron and is free to move in this n-type crystal and to increase the flow of electric current. Similarly, adding an impurity from group III with three valence electrons creates bonds with a missing electron, or hole, or a p-type crystal.

Surface barrier detectors are constructed to provide a junction between an n-type and a p-type crystal. When a voltage bias is applied to a silicon-layered crystal with an n-p junction, the excess electrons are swept in one direction and the holes in the opposite direction. This creates a depletion layer between the two that is nonconducting. If, however, an alpha (or beta) particle is absorbed in the depletion layer, it creates numerous charge pairs by ionizing silicon atoms, and these migrate quickly to create a pulse. The size of the pulse is directly proportional to the energy deposited and can be processed electronically to provide a spectrum for identification and quantification of the source. A large number of charge units is produced because only 1 eV or so is required to ionize silicon; thus excellent resolution is obtained because the statistical fluctuation of the collected ions is minimal.

The depletion layer can be constructed just thick enough to equal the maximum range of the particles to be detected and thus minimize interference by other types of radiation. A detector with a very thin depletion layer is used for alpha spectroscopy, and these detectors have essentially zero background since the probability of photon interactions is minimal. Beta spectroscopy is achieved by increasing the depletion layer, and both alpha and beta particles will be detected with such a detector. When used for alpha spectroscopy, silicon surface barrier detectors provide good energy resolution because alpha particles are emitted monoenergetically. Any self-absorption of the alpha particles in the source will degrade their energy before they reach the detector, and only the energy actually deposited will be recorded; therefore, sharp peaks will be attained only if the alpha source is deposited in a thin layer to minimize self-absorption. Beta particles, which are less subject to self-absorption, will be detected and displayed as a continuous energy spectrum up to $E_{\beta,\max}$ because of the mode of beta transformation. Since no peaks will be formed except perhaps for conversion elections associated with gamma emission (e.g., ^{137}Cs), beta spectroscopy with silicon detectors must be done with the entire spectrum as in liquid scintillation analysis.

Range versus Energy of Beta Particles

Proportional counters and laboratory-type end-window GM counters are used to determine the range of beta particles. Since the amount of beta absorption is related to its energy, the amount of absorber required to just stop all of the particles emitted

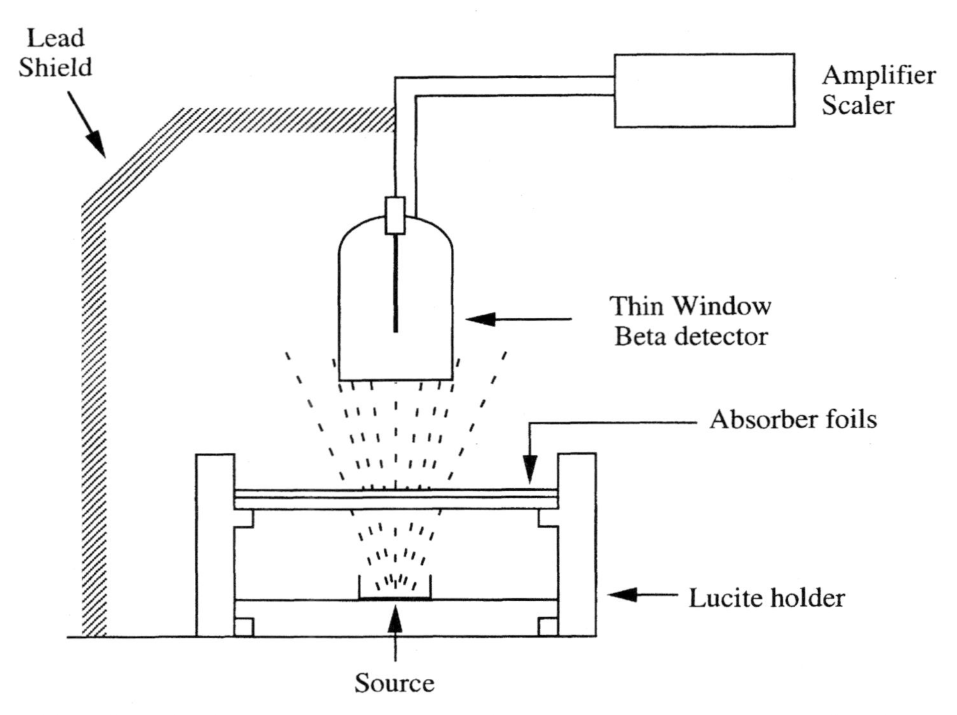

FIGURE 12-19. Experimental arrangement for measuring the absorption of β-particles.

by a source is also a measure of the beta-particle energy. A range–energy plot for a given beta source is obtained by placing different thicknesses of an absorber such as aluminum between the source and a thin-window detector as shown in Figure 12-19 and measuring the change in activity for each absorber thickness. A plot of these measurements versus absorber thickness (usually in units of g/cm^2 or mg/cm^2) produces a curve of the type shown in Figure 12-20.

As shown in Figure 12-20, the counting rate when plotted on a logarithmic scale decreases as a straight line, or very nearly so, over a large fraction of the absorber thickness, eventually tailing off into another straight-line region represented by the background, which is always present, and any gamma rays that may be emitted by the source (many beta-emitting radionuclides also emit photons). It is very important to subtract the background curve from the measured curve to create another curve (shown in Figure 12-20) that represents absorption due to beta particles alone. The point where the corrected beta-absorption curve meets the residual radiation level is the range, R_β, traversed by the most energetic beta particles emitted by the source.

The measured range in mg/cm^2 in aluminum or similar absorber is then used to determine the energy by an empirical range–energy curve as shown in Figure 12-21, or alternatively, from a listing (usually provided in mg/cm^2) for different beta-emitting sources. The energy–range relationship in Figure 12-21 can be considered generic for most light absorbers as long as the equivalent mg/cm^2 of absorber is known because most absorber materials will have essentially the same electron

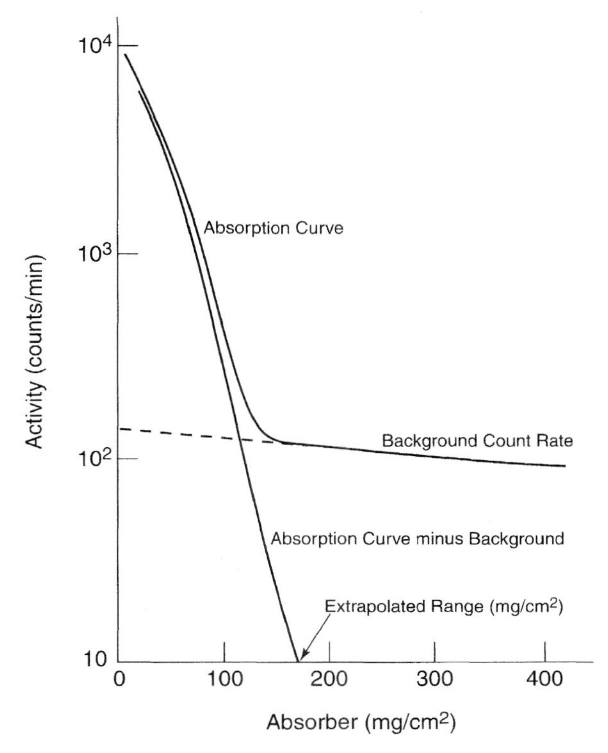

FIGURE 12-20. Decrease in strength of a beta-particle source versus mg/cm^2 of absorber thickness and subtraction of the extrapolated background to produce a new curve, with background subtracted, of activity versus absorption for beta particles emitted by the source.

density, since with the exception of hydrogen, the Z/A ratio varies only slowly with Z.

Three techniques are generally used to obtain the range, R_β, from the net beta-absorption curve: (1) extrapolation of the beta curve to intersect the residual radiation background; (2) a modified Feather analysis; and (3) a complete Feather analysis. The extrapolation technique, which is easiest, is only approximate but is often of sufficient accuracy to determine the energy, and hence the identity, of the beta emitter. The modified Feather analysis is, like the complete Feather analysis, based on an absorption curve obtained for a known standard such as ^{210}Bi, which has a well-established value of R_β. In this technique the amount of absorber required to reduce the count rate of the standard by one-half is compared to that required to reduce the unknown by one-half, and the two are compared as illustrated in Example 12-4.

Example 12-4: A beta standard with a known range of 200 mg/cm^2 is counted with different thicknesses of absorber, and from a data plot it is found that the count rate is reduced to one-half by 50 mg/cm^2 of aluminum. An unknown that is measured the same way requires 100 mg/cm^2 to reduce the count rate by one-half. What is the approximate beta range of the unknown and its energy?

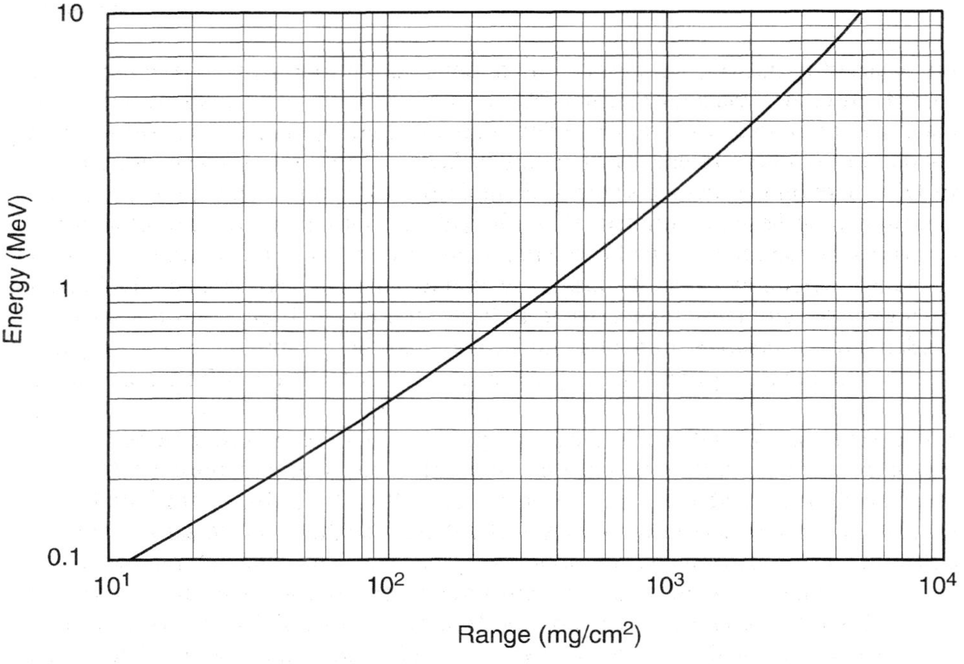

FIGURE 12-21. Equivalent range of electrons in mg/cm^2 of low-Z absorbers.

SOLUTION: For the standard, 50 mg/cm^2 corresponds to $0.25R_\beta$; therefore, since 100 mg/cm^2 is required to reduce the count rate of the unknown by one-half,

$$R_\beta = \frac{100 \text{ mg/cm}^2}{0.25} = 400 \text{ mg/cm}^2$$

From Figure 12-19, 400 mg/cm^2 corresponds to $E_{\beta,\max} \simeq 1.0$ MeV.

It is common practice to extrapolate the straight-line portion of the curve in Figure 12-20 to the background and to use the extrapolated value as the maximum range. This procedure is often, but erroneously, called a *Feather analysis*. A true Feather analysis, named after its originator, is described by Lapp and Andrews as a more deliberate and accurate process which compares the absorption data from an unknown to a known beta standard, usually ^{210}Bi ($E_{\beta,\max} = 1.162$ MeV). The maximum range of ^{210}Bi has been carefully established as 510 mg/cm^2 and is thus used as a reference for calculating unknown range.

ACKNOWLEDGMENTS

Contributors to this chapter are Wendy Drake, who drafted the section on portable instruments, and Chul Lee, who measured the various spectra and patiently and expertly compiled the various sections.

REFERENCES AND ADDITIONAL RESOURCES

Knoll, G. F., *Radiation Detection and Measurement*, 2nd ed., Wiley-Interscience, New York, 1989.

Lapp, R. E., and H. L. Andrews, *Nuclear Radiation Physics*, 4th ed., Prentice Hall, Upper Saddle River, NJ, 1972.

Moe, H. G., *Operational Health Physics Training*, Report ANL-88-26, Argonne National Laboratory, Argonne, IL, 1988.

PROBLEMS

12-1. K- and L-shell binding energies for cesium are 28 and 5 keV, respectively. What are the kinetic energies of photoelectrons released from the K and L shells when 40 keV photons interact in cesium?

12-2. Calculate the number of electrons entering the first stage of a photonmultiplier tube from the interaction of a 0.46 MeV photon in a 12% efficient NaI(Tl) crystal.

12-3. Calculate the energy of the Compton edge for 1 MeV photons.

12-4. For a small NaI(Tl) detector exposed to 0.2 MeV photons, calculate the energy of the iodine escape peak and provide an approximate plot of the energy spectrum.

12-5. For ^{88}Y photons incident on an intrinsically pure germanium detector/spectrometer, calculate the energies of the single and double escape peaks.

12-6. The absorption of β^- particles from a source was measured with aluminum and the results shown in Table 12-1 were obtained. (**a**) Plot the data and estimate the maximum range of the beta particles in the source. (**b**) From the range versus energy plot, determine $E_{\beta,\max}$. (**c**) Determine the maximum energy from Feather's empirical formula: $R = 543E - 150$, where R is in mg/cm^2 and E is in MeV.

TABLE 12-1. Data for Problem 12-6

Al (mg/cm^2)	c/m	Al (mg/cm^2)	c/m
0	1000	550	8
100	600	600	3.5
200	375	650	1.5
250	250	700	0.75
300	165	750	0.5
350	110	800	0.4
400	65	850	0.35
450	37	900	0.33
500	18	950	0.32

12-7. From the data in Problem 12-6, determine the range using the approximate Feather analysis method if the count rate of a ^{210}Bi standard (R_β = 510 mg/cm^2) is reduced to one-half by 115 mg/cm^2 of absorber.

12-8. A biomedical research laboratory uses ^{14}C and ^{32}P in tracer experiments. Devise a method for setting up a liquid scintillation counter to quantitate each radionuclide separately and in the presence of each other.

12-9. An LSC is calibrated such that a low-energy window has a counting efficiency of 30% for ^{3}H and a higher-energy window has a counting efficiency of 72% for ^{32}P. When a ^{32}P standard is counted, it is noted that 12% of the counts in window 2 are recorded in window 1. A mixed sample containing ^{3}H and ^{32}P yields 3800 counts in window 1 and 5800 counts in window 2. Determine the activity of each radionuclide.

12-10. What count rate will be observed by a portable air proportional alpha counter above a surface uniformly contaminated with an alpha emitter that emits 200 5.16 MeV alpha particles per 100 cm^2 if the probe area is 60 cm^2 and the overall counting efficiency is 22%?

12-11. An ionization chamber detector has been calibrated at STP to provide accurate measurements of dose from photons ranging from about 1 to 3 MeV. (**a**) If the counter is taken to an elevation of 6000 ft what modifications would need to be made in interpreting the dose measurements to obtain accurate results? (**b**) If a proportional counter calibrated at STP is substituted, what modifications would need to be made to interpret the results properly?

12-12. What relative meaning does the term "quench" *have* regarding a proportional counter, a GM counter, and a liquid scintillation counter?

13

STATISTICS IN RADIATION PHYSICS

A statistic is worthless unless you know its use.
—John Thompson, Georgetown University, 1998

Many nuclear processes are random (i.e., statistical) in nature. The transformation of any given radioactive atom is subject to the laws of chance and thus has the same probability for transformation in an interval of time regardless of the past history. The science of statistics is used to describe radioactive transformation in several important ways:

- Specifying the amount of uncertainty at a given level of confidence in a measurement of radioactivity
- Determining whether a sample actually contains radioactivity, especially if the measured activity level in the sample is very close to the natural background
- Checking whether a counting instrument is functioning properly by comparing the statistically predicted variance of the sample counts to that obtained experimentally

Practical circumstances dominate these determinations, and much of the presentation that follows emphasizes applications; however, it is important to recognize that all such applications are due to the random statistical behavior of radioactive atoms and to focus to some degree the fundamental statistical principles that support the rather straightforward applications that can be carried out once the statistical basis of the phenomena and models used are understood.

NATURE OF COUNTING DISTRIBUTIONS

If one makes successive measurements of a steady radiation source in a fixed geometry with a stable counting system, it is found that the source does not consistently produce the same count, but that the counts vary, sometimes considerably, due to the random nature of radioactive transformations. Figure 13-1, which is a scatter

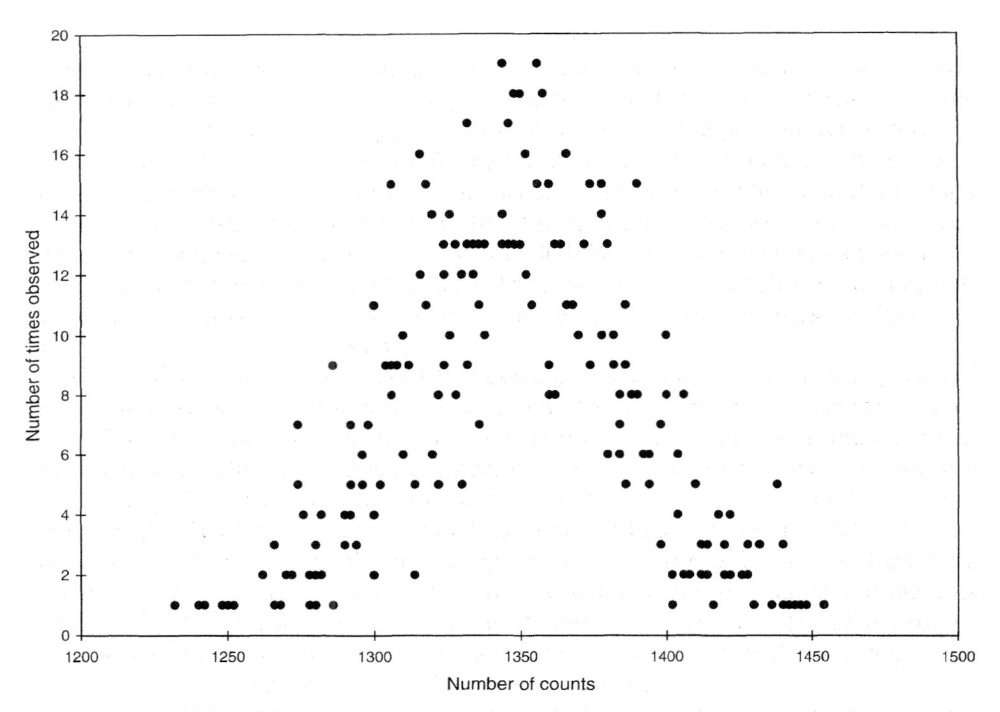

FIGURE 13-1. Plot of the number of times (i.e., the frequency) a given count was observed when a sample was counted repeatedly for a 1-min period.

plot of one thousand 1-min counts of a sample containing a long-lived radionuclide, demonstrates two important features of such data sets: (1) the 1-min counts cluster around a central value that is likely to be the true count rate of the sample; and (2) the values around the central value form a fairly symmetrical distribution pattern. If one additional count is made on the sample, it is highly likely that it would be one of the values obtained before; however, it is also possible, but not as likely, that the new value would be outside the distribution of the data.

When a radioactive sample is measured, it is not generally practical to make more than one or at most a few repeat measurements. Obviously, the objective is to arrive at a value as close to the true transformation rate, or activity, of the sample as possible. It is a fact of nature that the true value cannot be determined exactly unless an infinite number of measurements is made, which is not practical. It is possible, however, due to the random nature of radioactive transformation to use one or several measurements to describe the measurement and its reliability through the use of statistical models. The features of the binomial, Poisson, and normal (or Gaussian) distributions, each of which is summarized briefly, are appropriate for this purpose.

Binomial Distribution

The binomial distribution is the most general statistical model and is widely applicable to processes that occur with a constant probability p: for example, the flip of a coin, choosing a card, or an atom that may or may not disintegrate in a set time. If n is the number of events, each of which has a probability p, the predicted probability of observing exactly x events (e.g., transformed atoms) in an interval (or a number of trials) is

$$P(x) = \frac{n!}{(n-x)!x!} p^x (1-p)^{n-x}$$

where $P(x)$ is the predicted binomial probability distribution function. It is important to recognize that $P(x)$ is defined only for integer values of n and x.

Poisson Distribution

The binomial distribution, despite its fundamental applicability to random events, is too complex to use for describing the distribution of radioactive events. Most observations of radioactive transformations of interest in radiological protection involve relatively large numbers of events that are produced by a large number of atoms each of which has a very small probability (i.e., $p \ll 1.0$) of transformation in a practical time interval, say minutes or hours. When $p \ll 1.0$ the binomial distribution can be simplified mathematically into a quite useful function, the Poisson distibution:

$$P_n = \frac{x^n e^{-x}}{n!}$$

where, for radioactivity determinations, P_n is the probability of obtaining a count n, and x is the "true" average count for the sample. Since the true average count or true mean for the sample cannot be measured, the average measured count (properly termed the estimated or *sample mean*, $\overline{x}$) is used for x. When the sample mean $\overline{x}$ is substituted in the general expression for the Poisson distribution of a set of events,

$$P_n = \frac{\overline{x}^n e^{-\overline{x}}}{n!}$$

This expression can be used to characterize the distribution of measured data for a radioactive source as shown in Example 13-1.

Example 13-1: What is the probability of obtaining a count of 12 for a radioactive source when the true average count is 15?

SOLUTION: Since $p \ll 1.0$ for radioactive transformation of a large number of radioactive atoms, the Poisson distribution can be used. This yields

$$P_n = \frac{\overline{x}^n e^{-\overline{x}}}{n!}$$

$$P_{12} = \frac{(15)^{12} e^{-15}}{12!}$$

$$= \frac{(129.7 \times 10^{12})(30.6 \times 10^{-8})}{4.79 \times 10^8}$$

$$= 0.0829$$

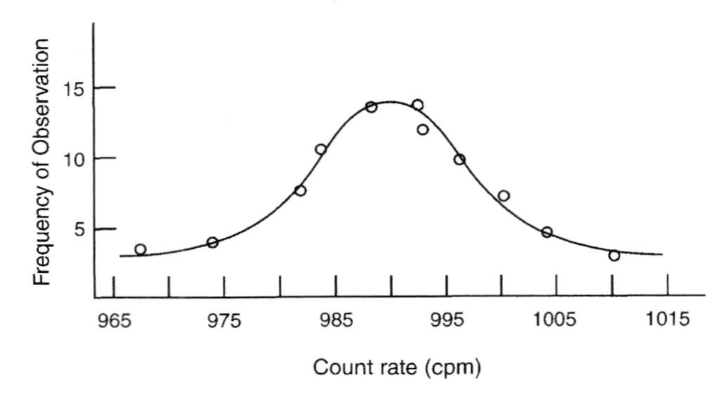

FIGURE 13-2. Data from 100 ten-minute measurements of a ^{137}Cs source with a mean count rate of about 990 c/m plotted as a function of the number of times that the count rate fell within successive increments of 5 (i.e., between 975 and 980, 980 to 985, etc.). The Poisson curve of best fit is drawn through the central values of the data increments.

The probability is 0.0829 (or 8.29%) that a count of 12 will be obtained when the true average count is 15.

Use of the Poisson distribution law is fairly straightforward when n is small (as in Example 13-1); however, it too can be quite cumbersome for determining the probability of observing a given count from a source when n is large. Obviously, a simpler function that accurately represents the random statistical events that occur in radioactive transformation is desirable. Such a function is the normal distribution, which can be used to represent the Poisson-distributed events if n is ≥ 17. Figure 13-2 illustrates the application of the Poisson distribution for intervals of counts from a ^{137}Cs sample.

Normal Distribution

If a large number of measured counts produced by a radioactive sample are grouped into smaller and smaller intervals of incremental counts, a plot of the frequency of occurrence of a given counting interval becomes very symmetrical, as illustrated in Figure 13-3. The symmetry of such a plot for events where $n \geq 17$ allows statement of a very fortunate principle: Poisson-distributed events such as radioactive transformation can be represented by the normal distribution, which is completely described by two parameters, the true mean μ and the standard deviation σ, which in turn is derived from the estimated mean by $\sqrt{\overline{x}}$.

The variation in a large number of repeated measurements (e.g., in Figure 13-3) can be described by plotting a histogram of the number of measurements observed for a given interval (frequency) versus that interval. The intervals still represent discrete events (i.e., an atom transforms or it doesn't); however, if the counting intervals are made very small, the discrete values can be replaced by a continuous function that fits the data points. The equation for the function (or curve) that fits the data points is the normal (or Gaussian) distribution, or

$$P_N(n) = \frac{1}{\sqrt{2\pi\overline{x}}} e^{-(x_i-\overline{x})^2/2\overline{x}}$$

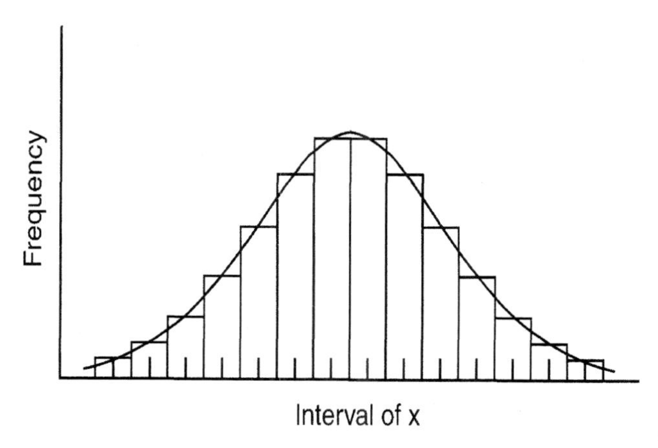

FIGURE 13-3. Histogram illustrating the conceptual basis of a normal curve.

where $P_N(n)$ is the probability of observing a count n when the true count (estimated as the average count) is $\bar{x}$ ($\bar{x} = \sigma_{\bar{x}}^2$, which can be substituted). This, like the binomial and Poisson distributions, is a distribution function for integer values of x; however, the continuous function of the normal distribution makes the interpretation and description of probabilistic events considerably easier.

Example 13-2: Apply the normal distribution function to Example 13-1.

SOLUTION: The probability $P_N(n)$ of obtaining a count of 12 when the true average count is 15 is

$$P_N(n) = \frac{1}{\sqrt{2\pi\bar{x}}} e^{-(x_i-\bar{x})^2/2\bar{x}}$$

$$= \frac{1}{\sqrt{2\pi(15)}} e^{-(12-15)^2/2(15)}$$

$$= 0.103e^{-0.3} = 0.0763$$

Example 13-2 shows that the normal distribution yields a probability close to that calculated by the more accurate Poisson distribution in Example 13-1 for a mean of 15 events. The results for 17 to 20 or more are essentially the same.

A data set that is governed by the normal distribution is fully defined by two parameters, the true mean μ and the standard deviation σ, which indicates the degree of spread among values of the individual observations x_i, as illustrated in Figure 13-4 for two data sets with the same true mean but different values of σ.

A normal distribution curve, as shown in Figures 13-4 and 13-5, has two distinct features. First, it is symmetrical about a vertical line drawn through the value for the true mean, which is at the maximum of the curve; and second, regardless of the relative magnitude of σ and μ, zero probability only occurs at $+\infty$ and $-\infty$. Integration of the normal distribution function between the limits $-\infty$ and $+\infty$ yields the total area (i.e., $p = 1.0$), and all normal curves, regardless of their values of μ and σ, have the property that the area between $\mu - \sigma$ and $\mu + \sigma$ is 68.3% of the

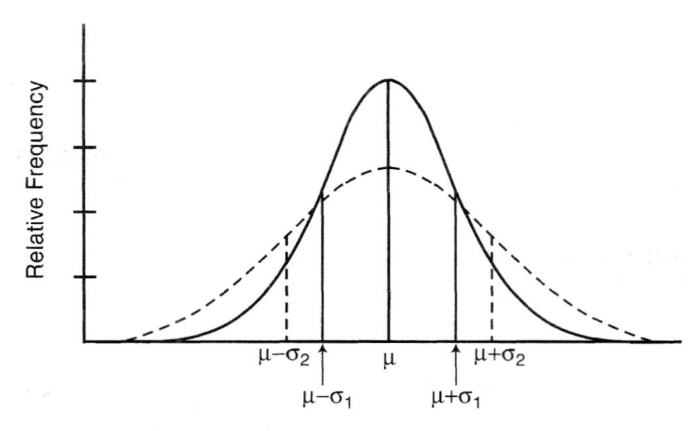

FIGURE 13-4. Normal distribution curves with the same true mean but different standard deviations.

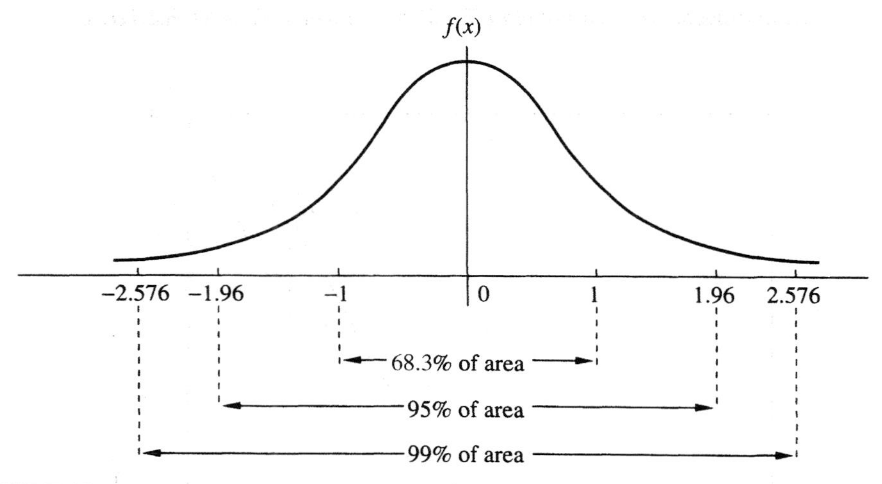

FIGURE 13-5. Percent area between intervals of the standard deviation for the normal distribution curve.

total area, as shown in Figure 13-5. Extending the interval to $\mu \pm 2\sigma$ includes 95.5% of the total area, and between $\mu \pm 3\sigma$ it encompasses 99.7% of the area under the normal curve. In practical terms, when an estimated mean is assigned an uncertainty of 1σ, 2σ, or 3σ, one can be 68.3%, 95.5% , and 99.7% confident that the true mean lies somewhere in the assigned interval. These degrees of confidence, referred to as *confidence intervals*, are generally stated as percent confidence or as multiples of σ; Table 13-1 summarizes and relates the two.

A normal curve can be fitted to a data set of 17 or more events if only one parameter, the arithmetic mean $\bar{x}$, is given. This fit can be made by determining the standard deviation of the mean, which is obtained by $\sqrt{\bar{x}}$. For Poisson-distributed events numbering 17 or more, all the properties of the normal distribution can be applied to the data obtained, and the standard deviation can be readily obtained once

TABLE 13-1. Confidence Intervals Based on the Normal Distribution for Multiples of the Standard Deviation (σ) of a Measurement

Interval	Confidence Level (%)
$\bar{x} \pm 1\sigma$	68.3
$\bar{x} \pm 1.96\sigma$	95.0
$\bar{x} \pm 2\sigma$	95.4
$\bar{x} \pm 2.58\sigma$	99.0
$\bar{x} \pm 3\sigma$	99.7

TABLE 13-2. Poisson-Distributed 95% Confidence Limits of the True Mean for Low Numbers of Counts

Observed Count	Lower Limit	Upper Limit
1	0.03	5.57
2	0.24	7.22
3	0.62	8.77
4	1.09	10.24
5	1.62	11.67
6	2.20	13.06
7	2.81	14.42
8	3.45	15.76
9	4.12	17.08
10	4.80	18.39
11	5.49	19.68
12	6.20	20.96
13	6.92	22.23
14	7.65	23.49
15	8.40	24.74
16	9.15	25.98

the best estimate of the mean is made. If, however, the number of counts falls to a very small value (less than 17), the distribution no longer follows the normal curve and the statement of confidence limits must be based on the Poisson distribution. The confidence limits for small numbers of counts at the 95% confidence level are shown in Table 13-2.

Mean and Standard Deviation of a Set of Measurements

The *arithmetic mean value* $\bar{x}$ of a set of measurements is the best estimate of the transformation rate of the source; it is calculated as

$$\bar{x} = \frac{1}{n} \sum_{i=1}^{i=n} x_i$$

where x_i is the value of each measurement and n is the number of measurements. If only one measurement is made, $\overline{x}$ is obviously that value.

The *standard deviation*, which describes the degree of fluctuation of the data set about the mean, is the next most useful descriptor of a statistical data set. The standard deviation is denoted as σ and is calculated for a single radioactivity measurement as $\sqrt{N}$, where N is the number of observed counts. If several measurements are made, σ is calculated from the arithmetic mean value and each measured value x_i in the data set as

$$\sigma = \left[\frac{\sum_{i=1}^{n}(x_i - \overline{x})^2}{n-1}\right]^{1/2}$$

where the term $(n - 1)$ is associated with the number of *degrees of freedom*. When n independent observations of x are made, n values are obtained; however, the calculation of σ (the sample standard deviation) includes $\overline{x}$, which is computed from the data for n observations; thus there are only $(n - 1)$ independent data left. This is consistent with only one observation in which $\overline{x} = x$ and σ is indeterminate because $(n - 1)$ goes to zero.

Uncertainty in the Activity of a Radioactive Source

For any radioactive source the challenge is to estimate from a finite number of observations a result that can only be known exactly by making an infinite number of observations, which of course is impossible. Successive counts of a long-lived radioactive source will, as shown in Example 13-3, yield different results, which raises the question of which count represents the true activity of the source. The best estimate of the true count is the arithmetic mean (or the average value), which for the data in Example 13-3 is $\overline{x} = 328$ counts. The value of $\overline{x}$ itself is only an estimate, but since it is a calculated average of five observations, there is added confidence that it represents the true activity of the source. Having made this estimate, however, the statistical uncertainty in the estimate is determined by the standard deviation of the data.

Two approaches can be taken to determine the standard deviation (i.e., uncertainty) of the mean of several measurements. If it is not known that the data are radioactivity counts, it is common practice to determine the mean and the standard deviation in the usual way using the specific definitions of the mean and standard deviation of a set of observed counts. But this approach unnecessarily constrains the precision of the estimated mean. Since the data in Example 13-3 are radioactive transformations, which are independent random events, and each measurement was taken the same way, the individual counts can be summed and used directly to compute σ. Examples 13-3 and 13-4 show the relative features of treating observed counts as individual versus separate independent random events.

Example 13-3: A sample is counted five times for 10 min each and 315, 342, 335, 321, and 327 counts recorded. What are the mean count and the standard deviation of the data set, and the mean count rate and its uncertainty based on the data set?

TABLE 13-3. Calculations for Example 13-3

Counts, x_i	$x_i - \overline{x}$	$(x_i - \overline{x})^2$
315	−13	169
342	+14	196
335	+7	49
321	−7	49
327	−1	1
$\sum \overline{x}_i = 1640$		$\sum (x_i - \overline{x})^2 = 464$

SOLUTION: See the calculations in Table 13-3. The mean is

$$\overline{x} = \frac{1640}{5} = 328 \text{ counts}$$

The standard deviation of the data set

$$\sigma = \sqrt{\frac{464}{4}} = 11 \text{ counts}$$

$$\text{count rate} = \frac{328 \pm 11 \text{ counts}}{10 \text{ min}} = 32.8 \pm 1.1 \text{ c/m}$$

Example 13-4: For the data in Example 13-3, determine the mean, the count rate, and the standard deviation of each based on the fact that these are counts of radioactive transformations, all of which are measured the same way.

SOLUTION: Since radioactivity counts are discrete independent random events, the five 10-min counts can be examined as a group of 1640 counts. The mean is

$$\overline{x} = \frac{1640}{5} = 328 \text{ counts}$$

The standard deviation is

$$\sigma = \sqrt{1640} = 40.5 \text{ counts}$$

and the mean count rate and its uncertainty are

$$\overline{x} = \frac{1640 \pm 40.5}{50 \text{ min}} = 32.8 \pm 0.81 \text{ c/m}$$

It is clear from Examples 13-3 and 13-4 that a certain amount of precision is given up by the statistical treatment of the data as a group of numbers. If it is known that the detector system and source are steady over the period that the measurements are made, the measurement can be treated as a set of cumulative counts that can be used directly to estimate the count rate and its uncertainty with a somewhat smaller value for the uncertainty of the estimated count rate.

Uncertainty in a Single Measurement

A common procedure is to make a single observation of a radioactive sample. Consequently, it is the best estimate of the true transformation rate of the sample since no other value is in hand. The true mean is probably different from this measurement, but limits can be specified within which the true mean probably lies by use of the normal distribution. When a single measurement is used to estimate the mean and a standard deviation (or two or three) is assigned to it, in effect one is drawing a normal curve around the estimated mean with a data spread equal to the square root of the measured counts. This can be shown to be a reasonable practice using the data in Example 13-3. If only one count is selected randomly from the table, say 321, the standard deviation is $\sqrt{321} = 18$. This single measurement would be reported as 321 ± 18 counts, which is equivalent to stating that the "true" activity of the source has a 68.3% probability (or 1σ) of being between 303 and 339 counts in 10 min. The mean count for the five separate observations listed in Example 13-3 is 328 counts in 10 min, which falls within the range.

CHECKPOINTS

Before proceeding on to practical applications of statistics to common radiation protection circumstances, the following points need to be emphasized:

- Radioactivity measurements simply record the number of random events (transformations) that occur in a given time interval, and these events are truly random, with very small probabilities of occurrence.
- Since radioactivity events are binary (they either occur or they don't), their predictability is governed by the broadly applicable binomial probability distribution function, but this is generally too cumbersome to use for more than a few counts.
- The Poisson distribution can be used instead of the binomial distribution when the probability of any given event is $\ll 1$, which is the case for radioactive transformation.
- The normal (or Gaussian) distribution is a continuous function that provides statistical results equivalent to the Poissson distribution if the number of events (counts) is greater than 17 to 20. It is symmetrical about the mean and is easier to use, plot, and describe than either the binomial or Poisson distributions, which is the main reason it is used to describe radioactivity measurements.
- Both the Poisson and normal distributions are characterized by the mean (estimated as $\bar{x}$ from a single measurement or from n measurements), from which the standard deviation can be calculated as $\sqrt{\bar{x}}$.

PROPAGATION OF ERROR

It is important to remember that the statistical model assumptions apply only to an observed number of counts. For radioactivity measurements, the standard deviation or any multiples of it cannot be associated with the square root of any quantity that is not a directly measured number of counts. Two values, $A \pm \sigma_A$ and $B \pm \sigma_B$, can be combined arithmetically in such a way as to aggregate their respective uncertainties. The appropriate term for this process is propagation of error. The aggregate uncertainty for arithmetic operations involving A and B and their computed standard deviations can be propagated as follows:

- For addition or subtraction of $(A \pm \sigma_A)$ and $(B \pm \sigma_B)$, the results and the aggregate standard deviations are

$$(A \pm B) \pm \sqrt{\sigma_A^2 + \sigma_B^2}$$

- For multiplication or division of $(A \pm \sigma_A)$ and $(B \pm \sigma_B)$, the results and the aggregate standard deviations are

$$\frac{A}{B} \pm \sqrt{\left(\frac{\sigma_A}{A}\right)^2 + \left(\frac{\sigma_B}{B}\right)^2}$$

or

$$AB \pm \sqrt{\left(\frac{\sigma_A}{A}\right)^2 + \left(\frac{\sigma_B}{B}\right)^2}$$

Multiplying or dividing total counts or a count rate by a constant is not propagated by a statistical operation but is simply altered by the constant applied. A common example of this is conversion of an observed number of counts to a count rate r by dividing by the count time t, or

$$r = \frac{C_{s+b}}{t}$$

The standard deviation for a counting rate r can be calculated in one of two ways:

$$\sigma = \frac{\sqrt{C_{s+b}}}{t}$$

or

$$\sigma = \sqrt{\frac{r}{t}}$$

where C_{s+b} is the gross sample counts including background, r the counting rate, and t the sample counting time.

Example 13-5: A long-lived sample was counted for 10 min and produced a total of 87,775 counts. What are the counting rate and its associated standard deviation?

SOLUTION: The counting rate r is

$$r = \frac{87{,}775}{10} = 8778 \text{ c/m}$$

and the standard deviation of the count rate is

$$\sigma = \frac{\sqrt{87{,}775}}{10} = 30 \text{ c/m}$$

or alternatively, the standard deviation can be determined from the count rate r:

$$\sigma = \sqrt{\frac{8778}{10}} = 30 \text{ c/m}$$

and the counting rate $\pm 1\sigma$ would be expressed as 8778 ± 30 c/m.

Statistical Subtraction of a Background Count or Count Rate

When a measurement includes a significant contribution from the radiation background, the background value must be subtracted and the total uncertainty in the net result is obtained by propagation of the error associated with each. The resulting net count has a standard deviation associated with it that is greater than that of either the sample or the background alone.

Example 13-6: What are the net count and its standard deviation for a sample if the sample count is 400 ± 20 and the background count is 64 ± 18?

SOLUTION: Since the uncertainty of each measurement is given, the net count and its uncertainty are determined by subtracting the background count from the sample count and propagating the error of each measurement; therefore, the net sample count $\pm 1\sigma$ is

$$\begin{aligned} \text{sample net count} &= (A - B) \pm \sqrt{\sigma_A^2 + \sigma_B^2} \\ &= (400 - 64) \pm \sqrt{(20)^2 + (18)^2} \\ &= 336 \pm 27 \text{ counts} \end{aligned}$$

Example 13-6 is computed with a 1σ uncertainty or a 68.3% confidence interval; if a 95% confidence interval is desired, the net count would be computed as $A - B$ with an uncertainty of 1.96σ for each value and reported as 400 ± 39 and 64 ± 35. The result and the aggregate uncertainty at the 95% confidence level would be $(400 - 64) \pm 53$ counts, or 336 ± 53 counts. The uncertainty of 1.96σ can be calculated for A and B separately and aggregated as shown in Example 13-6 or applied to the result (i.e., $1.96 \times 27 = 53$ counts).

The *uncertainty for a net count rate* is often determined directly from the counting data for the sample and the background associated with its measurement. The net count rate R is calculated as

$$R = \frac{C_{s+b}}{t_{s+b}} - \frac{B}{t_b}$$

where

$$R = \text{net count rate (c/m)}$$

$$C_{s+b} = \text{gross count of sample plus background}$$

$$B = \text{background count}$$

$$t_{s+b} = \text{gross sample counting time}$$

$$t_b = \text{background counting time}$$

The standard deviation of the net count rate is

$$\sigma_{\text{net}} = \sqrt{\frac{C_{s+b}}{t_{s+b}^2} + \frac{B}{t_b^2}}$$

This general expression for the standard deviation of a net count rate is applicable to any circumstance when the background count (measured for a count time t_b) is subtracted from a sample count measured for a time t_{s+b} regardless of whether t_b and t_{s+b} are equal or not, as shown in Example 13-7. If the counting time is the same for the sample and the background, this simplifies to

$$R = \frac{C_{s+b} - B}{t} \qquad \text{and} \qquad \sigma = \frac{\sqrt{C_{s+b} + B}}{t}$$

Example 13-7: A sample counted for 10 min yields 3300 counts. A 1-min background measurement yields 45 counts. Find the net counting rate and the standard deviation.

SOLUTION:

$$\text{Net count rate} = \frac{3300}{10} - \frac{45}{1} = 285 \text{ c/m}$$

$$\sigma = \sqrt{\left(\frac{3300}{10^2} + \frac{45}{1^2}\right)} = 8.83 \text{ c/m}$$

The net count rate is thus

$$\text{net c/m} = 285 \pm 8.83 \text{ c/m}$$

with a 1σ (or 68.3%) confidence level; for a 95% confidence level the net count rate is

$$\text{net c/m (95\%)} = 285 \pm 1.96 \times 8.83 \text{ c/m}$$

$$= 285 \pm 17.3 \text{ c/m}$$

Error Propagation of Several Uncertain Parameters

Laboratory results are often calculated from a number of parameters, more than one of which may have a specified uncertainty. For example, a common measured result is a sample count divided by the detector efficiency and the chemical/physical recovery of the procedure to determine the activity in a collected sample (i.e., before it is processed). Detector efficiency is determined by counting a known source and dividing the measured count rate by the disintegration rate, which requires a statement of uncertainty. Similarly, the chemical yield is usually determined by processing a tracer or several known spikes and taking the mean of the recorded counts divided by the known activity, calculations that combine measured values and their uncertainties. Count times can be reasonably assumed to be constant because electronic timers are very accurate, and geometry factors can also be considered constant if samples are held in a stable configuration.

The reported uncertainty in a measured value is obtained by propagating the relative standard errors (i.e., 1σ) of each of the uncertainties according to the rule for dividing or multiplying values with attendant uncertainties. The uncertainty in a calculated value u is

$$\sigma_u = u\left[\left(\frac{\sigma_a}{a}\right)^2 + \left(\frac{\sigma_b}{b}\right)^2 + \left(\frac{\sigma_c}{c}\right)^2 + \cdots\right]^{1/2}$$

where u is the calculated value and $\sigma_a, \sigma_b, \ldots$ are the relative standard errors of each of the parameters used to calculate u. Typical cases are shown in Examples 13-8 and 13-9.

Example 13-8: A counting standard with a transformation rate of 1000 ± 30 d/m is used to determine the efficiency of a counting system. The measured count rate is 200 ± 10 c/m. Determine the efficiency of the counting system and its uncertainty.

SOLUTION: The efficiency is

$$\varepsilon = \frac{200 \text{ min}^{-1}}{1000 \text{ min}^{-1}} = 0.2 \quad \text{or} \quad 20\%$$

and the standard deviation of the efficiency is calculated as

$$\sigma_\varepsilon = 0.2\sqrt{\left(\frac{30}{1000}\right)^2 + \left(\frac{10}{200}\right)^2} = 0.058 = 5.8\%$$

thus, the recorded efficiency with a 1σ confidence interval would be $20 \pm 5.8\%$ (or 0.2 ± 0.058).

Example 13-9: A 1-g sample is processed in the laboratory and counted in a measurement system with a precise timer and a fixed geometry that has a counting efficiency of $20 \pm 2\%$. The chemical yield was determined by processing several known spikes and was recorded as 70 ± 5%. The recorded count rate was 30 ± 5 c/m. Determine the concentration of radioactivity in the sample and its associated uncertainty.

SOLUTION:

$$\text{concentration}[(\text{d/m})/\text{g}] = \frac{30 \text{ c/m}}{0.2 \text{ c/m/(d/m)} \times 0.7 \times 1 \text{ g}}$$
$$= 214 \text{ (d/m)/g}$$

The uncertainty of the measurement is

$$\sigma_{\text{conc}} = 214\sqrt{\left(\frac{5}{30}\right)^2 + \left(\frac{0.02}{0.2}\right)^2 + \left(\frac{0.05}{0.7}\right)^2}$$
$$= 214\sqrt{0.0278 + 0.01 + 0.0051}$$
$$= 214\sqrt{0.0429} = 44 \text{ (d/m)/g}$$

The result would be reported as 214 ± 44 (d/m)/g at a 1σ confidence interval [i.e., the true concentration has a 68.3% probability of being between 170 and 258 (d/m)/g]. The 95% confidence interval would be 1.96 × 44 (d/m)/g and the result would be reported as 214 ± 86 (d/m)/g.

COMPARISON OF DATA SETS

A typical problem in radiation protection is whether two measured values of radioactivity differ from each other with statistical significance. Perhaps the most common example of radiation measurement is whether a measured value is significantly above that due to the surrounding background.

Two single measurements can be compared by Student's t test. The test is performed by calculating a value τ_{calc} and comparing it with a true probability value τ_{table}, obtained from the normal distribution as listed in abbreviated form in Table 13-4. The calculated value τ_{calc} is determined from the absolute value $|r_1 - r_2|$ of the difference between two counting rates, r_1 and r_2:

$$\tau_{\text{calc}} = \frac{|r_1 - r_2|}{\sqrt{r_1/t_1 + r_2/t_2}}$$

or since $\sigma_i^2 = r_i/t_i$,

$$\tau_{\text{calc}} = \frac{|r_1 - r_2|}{\sqrt{\sigma_1^2 + \sigma_2^2}}$$

where τ_{calc} is the relative error for a given confidence interval; r_1 and r_2 are the count rate (c/m) for the sample and the background "blank"; t_1 and t_2 the counting time (minutes) for r_1 and r_2, respectively; and σ_1 and σ_2 the standard deviations of count rates r_1 and r_2, repectively.

TABLE 13-4. Student *t*-Test Values (or τ_{table}) for Selected *p* Values Where $(1-p)$ Corresponds to the Confidence Interval of the Two-Sided Normal Distribution

τ_{table}	p	τ_{table}	p
0.0	1.000	1.5	0.134
0.1	0.920	1.6	0.110
0.2	0.841	1.7	0.090
0.3	0.764	1.8	0.072
0.4	0.689	1.9	0.060
0.5	0.617	2.0	0.046
0.6	0.548	2.1	0.036
0.7	0.483	2.2	0.028
0.8	0.423	2.3	0.022
0.9	0.368	2.4	0.016
1.0	0.317	2.5	0.0124
1.1	0.272	2.6	0.0093
1.2	0.230	2.7	0.0069
1.3	0.194	2.8	0.0051
1.4	0.162	2.9	0.0037

TABLE 13-5. Common Test Values and Probabilities for Radiation Measurements

τ_{table}	p
2.580	0.010
1.960	0.050
1.645	0.100
1.000	0.317

The most-used values of τ_{table} and the corresponding probability values for radiation measurements are listed in Table 13-5. Values of p for 0.010, 0.050, 0.100, and 0.317 correspond to confidence intervals of 99%, 95%, 90%, and 68.3%, respectively.

Example 13-10: A sample measurement yields 735 counts in 50 min. A background measurement with the same detector system and geometry yields 1320 counts in 100 min. Can it be said at the 95% confidence level that this sample contains radioactivity above the background activity?

SOLUTION: First, the hypothesis is made that the sample contains no radioactivity with at least 95% certainty (i.e., $\tau_{\text{calc}} \leq \tau_{\text{table}} = 1.96$). The value of τ_{calc} is determined from the values of r_1 and r_2 based on the observed counts and the recorded count times, t_1 and t_2, or

$$r_1 = 14.7 \text{ c/m} \qquad t_1 = 50 \text{ min}$$

$$r_2 = 13.2 \text{ c/m} \qquad t_2 = 100 \text{ min}$$

Thus the value of τ_{calc} is

$$\tau_{calc} = \frac{|14.7 - 13.2|}{\sqrt{14.7/50 + 13.2/100}} = 2.3$$

From Table 13-4 for $p = 0.05$, $\tau_{table} = 1.96$, which is less then τ_{calc}; therefore, the hypothesis is rejected (τ_{calc} is not $\leq \tau_{table}$), and there is at least a 95% chance (i.e., $1 - p = 0.95 = 95\%$) that the sample contains radioactivity above background.

With the computed τ_{calc} value and values of τ_{table} from Table 13-4, it can be determined whether observed differences are caused by the random nature of radioactive transformation or whether there is a "real" difference between dissimilar samples.

Example 13-11: Two different samples from an area were counted for 1 min. The net count rate was 1524 ± 49 c/m for the first and 1601 ± 49 for the second. Is the difference significant?

SOLUTION:

$$\tau_{calc} = \frac{|x_1 - x_2|}{\sqrt{\sigma_1^2 + \sigma_2^2}}$$

$$= \frac{|1524 - 1601|}{\sqrt{(47)^2 + (49)^2}}$$

$$= 1.13$$

From Table 13-4, the probability p that corresponds to $\tau_{calc} = 1.13$ is 0.259 (or 25.9%) that measured values of two identical samples would differ due only to the random variation of the count rate; thus the probability that the difference between the samples is significant is $1 - 0.259 = 0.741$, or 74%.

STATISTICS FOR THE COUNTING LABORATORY

Statistics can be used in the counting laboratory for several circumstances: for example, assigning an uncertainty to a total count or count rate, determining an optimal count time, for an optimal distribution of available counting time between sample and background, doing γ-ray spectroscopy, checking whether equipment is working properly, determining a weighted average of several measurements, and determining whether and when it is acceptable to reject data.

Uncertainty of a Radioactivity Measurement

Perhaps the most common radioactivity measurement is to place a prepared sample in a counter for a set count time; examples of such samples are a routine air filter and an evaporated volume of water (e.g., 1 L of a stream sample). The size of the prepared sample, the count time, the background count time, and the sensitivity (efficiency) of the detector system are usually chosen to provide an optimal balance between sample workload, desired accuracy, and cost. For many practical reasons, samples are often counted just once for a set time, and typically the background will be counted for a different time, usually much longer than the sample count time.

Sample counting data are reported as the best estimate of the true activity (based on the true count rate) with a stated level of uncertainty or confidence interval. Typical reported uncertainties are 1 standard deviation (known as probable error, the 1σ error or the 68.3% confidence interval), a 90% confidence interval (1.645σ), or the 95% confidence interval (1.96σ, often rounded to 2σ), which is perhaps the value most used for low-level radioactivity measurements.

Example 13-12: An overnight background count is made for a gas-flow proportional counter, and this value is used throughout the next day for 10-min measurements of routine air filters. If the background count is 7600 counts in 800 min and the sample count plus the background is 240 counts in 10 min, determine (a) the net count rate for the sample and its uncertainty with a 95% confidence level and (b) the sample count rate and the 95% confidence interval if it were to be counted for 30 min to yield 720 counts.

SOLUTION: (a) The sample count rate with a 1σ uncertainty is $240/10 = 24 \pm 1.55$ c/m and the background count rate is 9.5 ± 0.11 c/m. The net sample count rate at the 95% confidence level (1.96σ) is

$$\text{net c/m} = 24 - 9.5 \pm 1.96\sqrt{(1.55)^2 + (0.11)^2}$$

$$= 14.5 \pm 1.96(1.554) \text{ c/m}$$

$$= 14.5 \pm 3.0 \text{ c/m}$$

(b) For a 30-min sample count the sample count rate remains at 24 c/m; however, the uncertainty is $= \pm 0.894$ c/m. At the 95% confidence level (1.96σ), the net count rate is

$$\text{net c/m} = 24 - 9.5 \pm 1.96\sqrt{(0.894)^2 + (0.11)^2}$$

$$= 14.5 \pm 1.96(0.812) \text{ c/m}$$

$$= 14.5 \pm 1.6 \text{ c/m}$$

As shown in Example 13-12, increasing the sample count time narrows the 95% confidence interval, with a corresponding reduction in the uncertainty of the measurement. The long background count is also a significant factor in reducing the uncertainty if it can be presumed that the same conditions exist during overnight and daytime hours, which may not always be the case. Measurement uncertainty can also be reduced by using larger samples to increase the total counts and/or reducing background radiation by increased shielding of the detector or by other means.

Determining a Count Time

Statistical comparisons of data can be used to optimize the time a sample is counted. Many measurements of radioactivity are simply made and the data are then used to calculate the net counting rate and its uncertainty; however, once a desired confidence level has been selected, counting the sample beyond an optimized time interval yields no improvement in the reported result. The counting time can also be calculated by the t test by making the following a priori determinations:

- The confidence level to be used in reporting the data
- The background count rate and its counting time
- The minimum detectable count rate to be detected above background

Example 13-13: A detector system in a laboratory is routinely set to accumulate a 1000-min background count overnight. If the background count rate from the 1000-min count is 11 c/m, how long must a sample be counted the next morning in order to measure 1 c/m above background with 95% confidence?

SOLUTION: From Table 13-4, $\tau_{\text{calc}} = 1.96$ for the 95% confidence interval ($p =$ 0.05). This value is used as follows:

$$1.96 = \frac{|12.0 - 11.0|}{\sqrt{12.0/t_1 + 11.0/1000 \text{ m}}}$$

Squaring both sides and solving for t_1 gives

$$\frac{12.0}{t_1} + 0.011 = \frac{1}{(1.96)^2}$$

$$t_1 = \frac{12}{0.26 - 0.011}$$

$$= 48.1 \text{ min}$$

Example 13-14: How long should a sample that has an approximate gross counting rate of 800 counts in 2 min be counted to obtain the net counting rate with an accuracy of 1% if the background for the counting system is 100 ± 2 c/m?

SOLUTION: The approximate net count rate is $400 - 100 = 300$ c/m and the desired uncertainty of 1% is thus 3 c/m. The general expression for the standard deviation of the difference in two count rates is

$$\sigma_r = \sqrt{\frac{R_{s+b}}{t_{s+b}} + \sigma_b^2}$$

Solving for t_{s+b} where $\sigma_r = 3$ c/m and the approximate count rate $R_{s+b} = 400$ c/m yields

$$t_{s+b} = \frac{400}{3^2 - 2^2} = 80 \text{ min}$$

Efficient Distribution of Counting Time

An optional distribution of available counting time can be made between the sample count and the background to reduce the estimated standard deviation of the net count rate to a minimum. The most efficient distribution of counting time between the sample count time t_{s+b} and the background count time t_b is determined when their ratio is

$$\frac{t_{s+b}}{t_b} = \sqrt{\frac{C_{s+b}}{C_b}}$$

where t_{s+b} is the counting time for sample plus background, t_b the counting time for background, C_{s+b} the estimated count rate for sample plus background, and C_b the estimated count rate for background.

Example 13-15: The count rate for a radioactive sample, including background, is estimated from a quick count to be about 3500 c/m, and the background count rate to be about 50 c/m. What is the optimal allocation of 10 min of counting time between the sample and the background?

SOLUTION: The optimal ratio for the two count rates is

$$\frac{t_{s+b}}{t_b} = \sqrt{\frac{r_{s+b}}{r_b}}$$

$$= \sqrt{\frac{3500}{50}} = 8.37$$

or

$$t_{s+b} = 8.37 t_b$$

and since $t_{s+b} + t_b = 10$ min, this can be simplified as

$$8.37 t_b + t_b = 10 \text{ min}$$

Thus the optimal counting time is

$$t_b = 1.07 \text{ min}$$

and since 10 min are available, the sample counting time is

$$t_{s+b} = 10.00 - 1.07 \text{ min}$$

$$= 8.93 \text{ min}$$

As a practical matter, the sample would be counted for 9 min and the background for 1 min.

Detection and Uncertainty for Gamma Spectroscopy

Many radiation measurements are done on gamma spectrometers in which a spectrum of counts versus photon energy is obtained as illustrated in Figure 13-6. Each

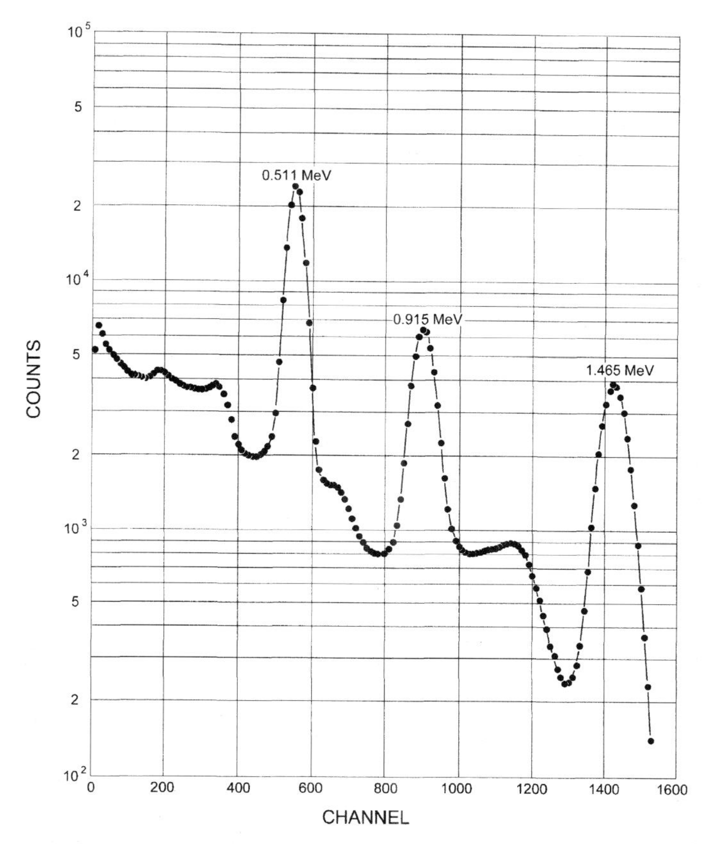

FIGURE 13-6. Gamma spectrum of ^{148}Pm ($T_{1/2}$ = 5.37 d) with an energy peak of 0.915 MeV superimposed on a detector background and Compton continuum of 800 counts and a 0.551 MeV peak on a background/continuum of 1800 average counts per channel.

peak in such a spectrum will be superimposed on a background spectrum that is due to natural background radiation striking the detector, background radiation from the shield and surrounding components, and Compton-scattered photons produced by interactions of gamma rays with energies above the peak energy. The net count at a given energy is used to quantitate the activity at the peak, its uncertainty or standard deviation, and whether or not it is considered detected, either through the decision tools of L_c and/or LLD (as discussed below).

The net count determination for the peak is made by subtracting the background at the peak energy, which is actually incorporated in the measurement of the sample. Background subtraction involves selecting a region of interest of several channels on each side of the peak, taking care to select a similar number on each side. The average number of counts per channel due to background in each channel under the peak is determined, and this value is then multiplied by the number of channels covered by the peak area to obtain the total background that should by subtracted from the total peak counts. Selecting the peak region is somewhat of an art to maximize peak counts relative to background counts. Once this is done, calculations of net counts, uncertainty, L_c, and LLD are performed as usual and any determinations of activity per unit volume, mass, and so on, are obtained by incorporating the relevant factors of the system as shown in Example 13-16. Modern gamma spectroscopy systems usually incorporate algorithms based on this approach to perform background subtraction and report net peak values in counts or activity.

Example 13-16: (a) For the 0.915 MeV peak in the gamma spectrum in Figure 13-6 determine the net count and its uncertainty at the 95% confidence level. (b) If the count time is 10 min, the detector efficiency for a 1.5-L geometry is $8 \pm 1\%$ at the 0.915 MeV peak, the chemical recovery of the processed sample is $90 \pm 4\%$, and the yield for the 0.915 MeV gamma ray is 12.5%, what is the concentration of ^{148}Pm in the measured sample?

SOLUTION: (a) From Figure 13-6, for the 200 channels encompassing the 0.915 MeV peak, the peak total is about 665,000 counts and the background under the peak is 200 × 800 counts/channel or 160,000 counts; therefore, the net count and its uncertainty at 95% confidence (1.96σ) is

$$\begin{aligned}\text{net count} &= 665{,}000 - 160{,}000 \pm 1.96\sqrt{665{,}000}\\ &= 505{,}000 \pm 1600 \text{ counts}\end{aligned}$$

(b) The sample concentration, C_s, is

$$\begin{aligned}C_s &= \frac{505{,}000 \text{ counts}}{10 \text{ min} \times 0.125\gamma/\text{d} \times 0.08 \text{ c/d} \times 1.5 \text{ L} \times 0.9 \times 2.22 \times 10^6 \text{ (d/m)}/\mu\text{Ci}}\\ &= 1.68\ \mu\text{Ci/L}\end{aligned}$$

The uncertainty for this measurement is determined by propagation of individual relative errors (σ/x_i) as illustrated in Example 13-9. The standard deviation of the concentration measurement is

$$\begin{aligned}\sigma_{\text{conc}} &= 1.68\ \mu\text{Ci/L}\sqrt{\left(\frac{1600}{505{,}000}\right)^2 + \left(\frac{0.010}{0.08}\right)^2 + \left(\frac{0.04}{0.90}\right)^2}\\ &= 1.68\ \mu\text{Ci/L}\sqrt{9.7 \times 10^{-6} + 0.0156 + 0.00197}\\ &= 1.68\ \mu\text{Ci/L}\sqrt{1.758 \times 10^{-2}}\\ &= 0.22\ \mu\text{Ci/L}\end{aligned}$$

at the 1σ level. For 95% confidence (or 1.96σ), the uncertainty is 0.44 μCi/L and the measured concentration is reported as 1.68 ± 0.44 μCi/L. It is noteworthy that the largest contribution to the uncertainty is not due to the statistics of radiation counting but to the uncertainty in the determination of detector efficiency, which is probably high for this type of measurement.

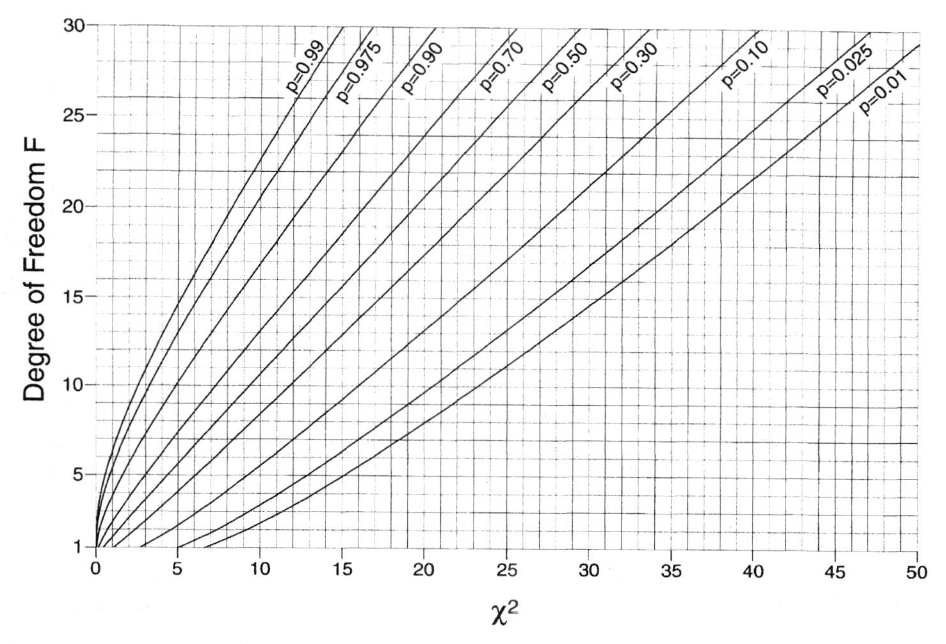

FIGURE 13-7. Integrals of the chi-square distribution.

Testing the Distribution of a Series of Counts: Chi-Square Statistic

The operation of a counting system can be tested by comparing the distribution of a series of repeat measurements with the expected distribution. One measure of performance is the Poisson index of dispersion, or the chi-square statistic, defined as

$$\chi^2 = \frac{\sum_{i=1}^{n}(x_i - \bar{x})^2}{\bar{x}}$$

where $\bar{x}$ is the arithmetic mean of the set of measurements and x_i is the value of each measurement.

The chi-square statistic is plotted as a function of the degrees of freedom F, where $F = n - 1$ (Figure 13-7). A system is expected to work normally if the probability of χ^2 lies between 0.1 and 0.9. The system performance is suspect if the p-value of χ^2 lies outside this range. If the measured value of χ^2 is < 0.1 (e.g., $p = 0.01$), the replicate counts are fluctuating more than would be expected for a Poisson distribution, and a stability problem in the electronics of the system is likely. If the measured value of χ^2 is > 0.9 (e.g., $p = 0.99$), the replicate counts are not fluctuating as would be expected in a Poisson distribution, and it should be suspected that a strong bias problem exists in the counting system.

Example 13-17: Five 1-min counts of a given sample had values of 11, 9, 11, 13, and 9. Does the system follow the Poisson distribution?

SOLUTION: The value of χ^2 for the five counts, which have an estimated mean $\bar{x} = 10.6$, is

$$\chi^2 = \frac{\sum (x_i - \bar{x})^2}{\bar{x}} = \frac{11.2}{10.6}$$

$$= 1.06$$

and from Figure 13-7, for $F = n - 1$ or 4 degrees of freedom, $p = 0.90$. The data look somewhat more uniform than expected, but they follow the Poisson distribution and the counting system can be judged to be functioning properly.

The chi-square test in Example 13-17 is marginally conclusive because only 5 measurements were made of a low-activity source. Increased confidence can be gained by counting a relatively "hot" source for 10 times or more.

Weighted Sample Mean

When measured values are to be averaged and some have better precision than others, a weighted sample mean and a weighted sample standard deviation are desirable. The weighting factor, w_i, to be assigned to each measurement is

$$w_i = \frac{1}{\sigma_i^2}$$

The weighted mean is given by

$$\bar{x}_w = \frac{\sum_{i=1}^{n} w_i \bar{x}_i}{\sum_{i=1}^{n} w_i}$$

and the weighted standard deviation is

$$\sigma_w = \sqrt{\frac{n}{\sum_{i=1}^{n} w_i}}$$

where n is the total number of measurements; $\bar{x}_i$ is the sample mean of a given measurement; and σ_i is the standard deviation of a given measurement.

Example 13-18: Five persons measured the radioactivity in a liquid waste sample. Each took an aliquot, prepared it for counting, and reported the mean of several 10-min counts. The mean of each person's measurements, the weighting factor w_i, and the values of $w_i \bar{x}_i$ are listed in Table 13.6. Determine the weighted mean and the weighted standard deviation.

SOLUTION: The weighted mean $\bar{x}_w$ is

$$\bar{x}_w = \frac{4.774}{0.01471} = 324 \text{ counts}$$

TABLE 13-6. Data for Example 13-18

Counts, $x_i \pm \sigma_i$	w_i	$w_i\bar{x}_i$
315 ± 18	0.00309	0.973
342 ± 20	0.00250	0.885
335 ± 42	0.00057	0.191
321 ± 12	0.00695	2.232
327 ± 25	0.00160	0.523
	$\sum w_i = 0.01471$	$\sum w_i\bar{x}_i = 4.774$

The weighted standard deviation is

$$\sigma_w = \sqrt{\frac{5}{0.01471}} = 18 \text{ counts}$$

and the weighted count rate is

$$\frac{324 \pm 18 \text{ counts}}{10 \text{ min}} = 32.4 \pm 1.8 \text{ c/m}$$

Rejection of Data

Occasionally, a set of measurements will contain one or more values that appear to be true outliers, and the temptation is to just cast them aside. This is not scientifically correct, nor should it be a standard practice. The appropriate procedure is first to determine if the counting system is performing satisfactorily. If it is, a sample count can be rejected (or kept) based on *Chauvenet's criterion*, which states that any count of a series of n counts (also the number of observations) shall be rejected when the magnitude of its deviation from the experimental mean is such that the probability of occurrence of all deviations that large or larger does not exceed the value $1/2n$ (assuming that the data represent a normal distribution). Chauvenet's criterion is applied to a set of measurements by determining Chauvenet's ratio (CR), which is

$$\text{CR} = \frac{|x - \bar{x}|}{\sqrt{\bar{x}}}$$

where x is the suspect value in the data set and $\bar{x}$ is the mean of the set. If the calculated CR for a given value is greater than that listed in Table 13-7 for the specified number of observations, it is appropriate to reject the datum; otherwise, the datum has an acceptable value.

Chauvenet's criterion is used to reject a suspect count when it deviates so much from the statistical mean that it adversely affects the experimental mean. If counts from a data set are rejected, the experimental mean is recalculated to obtain a new value that more closely approximates the statistical mean.

TABLE 13-7. Limiting Values of Chauvenet's Ratio

Number of Observations	Limiting Ratio	Number of Observations	Limiting Ratio
2	1.15	15	2.13
3	1.38	19	2.22
4	1.54	20	2.24
5	1.68	25	2.33
6	1.73	30	2.39
7	1.79	35	2.45
8	1.86	40	2.50
9	1.92	50	2.58
10	1.96	75	2.71
12	2.03	100	2.80

Example 13-19: The following 25 observations were made on a sample. Should any of the counts be discarded?

15	19	19	28	18
24	11	20	23	19
20	13	29	23	14
17	22	22	32	30
26	17	18	20	24

SOLUTION: First, a chi-square analysis of the data is performed to determine if the measurement system is performing satisfactorily

$$\bar{x} = \frac{\sum x_i}{25} = 21$$

and

$$\chi^2 = \frac{\sum_1^{25} (x_i - \bar{x})}{21} = 32$$

which is in the satisfactory region of Figure 13-7. Since the system that recorded the counts is functioning properly, Chauvenet's ratio can be used to test the highest and lowest values, 32 and 11, for possible rejection. The CRs for 32 and 11 counts are

$$\text{CR}_{32} = \frac{32 - 21}{\sqrt{21}} = 2.4$$

$$\text{CR}_{11} = \frac{21 - 11}{\sqrt{21}} = 2.18$$

From Table 13-7, CR = 2.33 for 25 observations; therefore, the gross count of 32 can be rejected since $\text{CR}_{32} = 2.4 > 2.33$. On the other hand, $\text{CR}_{11} = 2.18 < 2.33$, and the gross count of 11 should be kept. The chi-square statistic and the mean for the data should then be recalculated without the value of 32.

Caution must be used when rejecting data. Even if data don't meet expectations, they may in fact be quite meaningful (i.e., the data are correct; it is the expectation that is wrong). For example, if Michelson and Morley had rejected their data, we might still be discussing the "omniferous ether," or if Geiger and Marsden (graduate students of Lord Rutherford) had rejected data from their experiments on α-particle scattering, discovery of the atomic nucleus would have been missed.

LEVELS OF DETECTION

Applications of statistics have been discussed in terms of measured values in which counts, count rate, or activity levels are determined and an uncertainty is assigned to the measurement. If is often desirable to determine just how sensitive a given radioanalytical procedure, which will be influenced by statistical variations in observed counts, can be. This sensitivity is characterized as a detection level, and many regulatory programs require its determination. The detection level is also of considerable interest for measuring low activity samples and is influenced considerably by the level of background radiation.

The sensitivity of a measurement system is specified as a detection limit at or near the background level, which at best is only an estimate; it is useful as a guidepost, not an absolute level of activity that can or cannot be detected by a counting system. Three such guideposts are the critical level (L_c), the lower limit of detection (LLD), and the minimum detectable activity (MDA), each of which has a specific definition and application that should not be interchanged in dealing with measurements of radioactivity.

Critical Level

The critical level (L_c) is defined as the net count rate that must be exceeded before the sample can be said, with a certain level of confidence, to contain measurable radioactivity above background. The critical level is derived from the statistical variation that is associated with a true zero net counting rate; when the background counting rate is subtracted from the gross counting rate, the result is zero and the gross counting rate is identical to the background counting rate. A net counting rate of zero has a normal distribution of counts about a mean of zero, as shown in Figure 13-8, for which various probabilities exist for observing negative or positive net counting rates. A negative net counting rate occurs when the gross counting rate is less than the background counting rate, and it is understandable that the samples with negative net count rates would be considered as having no activity above background. A positive net counting rate may either be due to activity being present, or it may be due to the statistical probability of observing a net count rate above zero even though the true count rate is zero. Obviously, the more positive a net counting rate deviates from the zero net counting rate, the more likely it is that radioactivity is present, and conversely, the closer the net counting rate approaches the zero net counting rate, the greater the probability that a true net zero counting

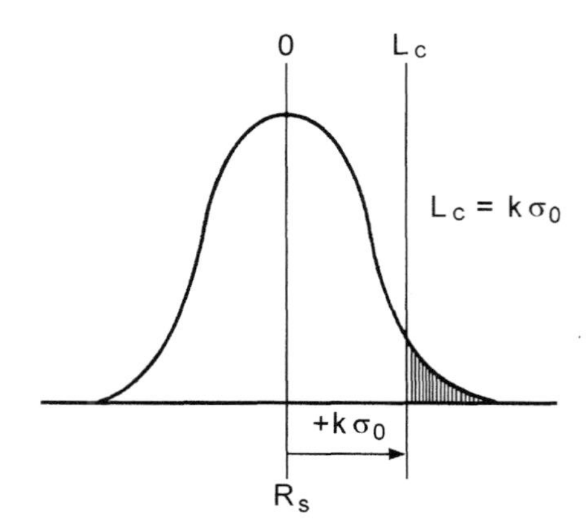

FIGURE 13-8. The critical level L_c is that point on the normal distribution curve of counts for a true zero net count above which there is a specified level of confidence that a true mean count of zero would be falsely recorded as positive (i.e., activity is recorded as present when in fact it is not).

rate exists. A net positive counting rate that is high enough to be concluded as positive with statistical confidence can be referred to as the critical level, L_c. It is expressed as

$$L_c = k_\alpha \sigma_0$$

where k_α is the one-sided (i.e., we are interested only in values that are statistically positive) confidence factor of the normal distribution and σ_0 is the standard deviation of a zero net count rate. If k_α is chosen so that 95% of the measurements of a true zero net count rate are less than L_c, that is, $k_\alpha = 1.65$, there is only a 5% chance that a true mean count of zero will be falsely recorded as a positive value.

If the net count rate of a sample exceeds L_c, it is reported as positive, and the two-sided confidence interval is reported in the usual way. If the net count is less than L_c, it can only be said that the total count rate (R_{s+b}) is not statistically different from the background at the 95% confidence level (or when the false probability = 0.05). A value of L_c based on a different confidence level can be determined from the value of k_α that corresponds to the desired one-sided confidence probability from Table 13-8. For example, if it is desired that a false positive occurs only 1% of the time, a k_α value of 2.326 would be used in determining L_c.

Because the standard deviation of the zero net count rate is a theoretical number, it is useful to express the critical level in terms of the standard deviation σ_b of the background count rate, R_b, or

$$L_c = k_\alpha \sigma_b \sqrt{1 + \frac{t_b}{t_{s+b}}}$$

where k_α is a constant corresponding to a stated false-positive probability, t_b is the background counting time, and t_{s+b} the counting time of the sample plus back-

TABLE 13-8. Probability of a False-Positive Observation of a True Net Count of Zero and the Corresponding k_α Constant Based on the Normal Distribution

False-Positive Probability	k_α Constant
0.010	2.326
0.025	1.96
0.050	1.645
0.100	1.282

ground, and where

$$\sigma_b = \sqrt{\frac{R_b}{t_b}}$$

If the count times for the sample and the background are equal,

$$L_c = k_\alpha \sigma_b \sqrt{2}$$
$$= 2.33 k_\alpha \sigma_b$$

or in terms of the background count rate and the background count time,

$$L_c = 2.33 k_\alpha \sqrt{\frac{R_b}{t_b}}$$

where R_b is the background count rate; thus L_c is directly related to the background of the detector system and inversely to the background count time. This relationship and equations for calculating L_c for various other conditions are shown in Table 13-9.

The critical level should only be used to determine if a measurement is statistically different from background. It is an a priori determination and should not be reported as a less-than value.

TABLE 13-9. Equations for Calculating the Critical Level L_c and Conditions for Their Use

Condition	Equation
General equation	$L_c = k_\alpha \sigma_b \sqrt{1 + \frac{t_b}{t_{s+b}}}$
Background counting rate	
Unequal count times for background and sample ($k_\alpha = 1.645$ for 95% confidence)	$L_c = 1.645 \left[\frac{R_b}{t_b} \left(1 + \frac{t_b}{t_{s+b}} \right) \right]^{1/2}$
Equal count times for background and sample	$L_c = 2.33 \sigma_b = 2.33 \sqrt{\frac{R_b}{t_b}}$

Detection Limit or Lower Level of Detection

The concept of *minimum detectable activity* (MDA) for a measurement procedure is often stated in terms of the sensitivity of a technique for observing the presence of radioactivity. The MDA can be thought of as a claim on how good the laboratory is, or it may be used to judge a procedure. Some regulatory programs require that the MDA be stated for routine procedures in a radiation protection program and that action be taken for measured activity above the MDA (e.g., waste material that contains measured radioactivity above the MDA is to be considered as radioactive). Such a regulatory requirement can be quite different from the laboratory perspective in which the MDA is designed to answer the question: What sample count is required to be sure of detecting it on a routine basis? It is not desirable to specify the MDA in terms of L_c because a single measurement of a true mean net count rate of L_c has an equal chance of being above or below L_c because the counts are normally distributed. This appears contradictory because the L_c itself is determined at the 95% confidence level, but it is a decision tool for assuring that a true zero net count rate is not falsely interpreted due to its statistical variability. Obviously, this is too restrictive and a more reliable value is desirable.

The detection limit (L_d) is the value that should be specified for determining a minimum activity that can be reported as detected with a degree of assurance that it in fact is present; it is defined as the smallest quantity of radioactive material that can be detected with some specified degree of confidence, preferably (and often) as 95% . The detection limit (or L_d) for a measurement is commonly referred to as the lower limit of detection (LLD), and even though LLD and L_d are synonymous, LLD will be used since it connotes the most common use of the statistical concept.

The LLD should be defined above L_c, and as shown in Figure 13-9, it is the smallest amount of sample activity that will yield a net count above background such that the probability of concluding that activity is present when in fact it is not is limited to a predetermined value, typically $p = 0.05$ [also referred to as a

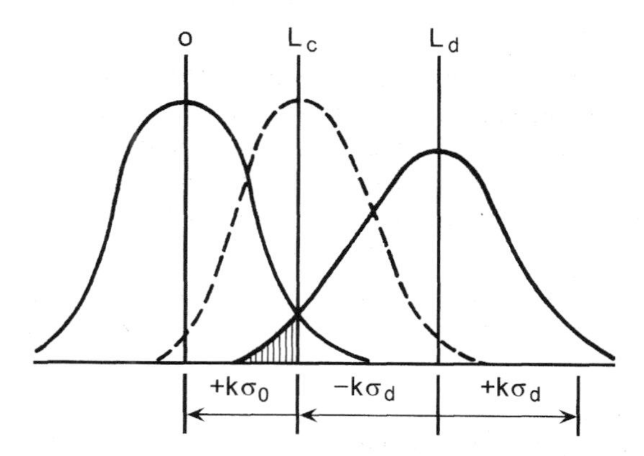

FIGURE 13-9. The lower limit of detection (LLD) is the smallest amount of net activity above background that will be registered as positive with a given level of confidence, typically with a false-positive level of $p = 0.05$, which corresponds to $k_\beta = 1.645$ for the one-sided normal distribution.

type II (or β) error]. A value of k_β that corresponds to the accepted false-positive probability (e.g., $k_\alpha = 1.645$ for $p = 0.05$) is chosen for a one-sided confidence interval since wrong determinations relative to the distribution of the net mean of zero are the only ones of interest. Defining LLD in terms of $k_\beta = 1.645$ where $p = 0.05$ states a willingness to accept being wrong only 5% of the time (i.e., when activity is measured at or above the LLD there is a 95% level of confidence that it is actually present, or similarly, a count at LLD will be erroneously concluded to contain activity only 5% of the time). Thus the LLD is a practical detection limit that considers the statistical variation in measured counts of both sample and background; it is related to L_c as follows:

$$\text{LLD} = L_c + k_\beta \sigma_d$$

where k_β is, again, the one-sided confidence factor and σ_d is the standard deviation of a net count rate equal to LLD.

The LLD can be determined in terms of counts per unit time by substituting the expression for L_c and assuming that $k_\alpha = k_\beta = k$, or

$$\text{LLD} = \frac{k^2}{t_{s+b}} + 2k\left[\frac{R_b}{t_b}\left(1 + \frac{t_b}{t_{s+b}}\right)\right]^{1/2}$$

This general relationship, the specific generalizations derivable from it, and the conditions for the use of each are summarized in Table 13-10. The latter two expressions, for which it is assumed that $k_\alpha = k_\beta = 1.645$, are frequently used for measurements of radioactivity when it is desirable that the LLD for the observed counts be at the 95% confidence interval. If, however, a different level of confidence is desired, it is necessary to use the more general equation with the value of k that correponds to the confidence interval selected (Table 13-8 and Example 13-20).

TABLE 13-10. Equations for Calculating the Lower Limit of Detection (LLD) for Various Conditions and $k = k_\alpha = k_\beta$

Prerequisites for Use	Equation[a]
General equation for LLD as a count rate	$\text{LLD} = \frac{k^2}{t_{s+b}} + 2k\left[\frac{R_b}{t_b}\left(1 + \frac{t_b}{t_{s+b}}\right)\right]^{1/2}$
LLD as a count rate for equal count times for sample and background and for any value of k	$\text{LLD} = \frac{k^2}{t} + 2k\left(2\frac{R_b}{t}\right)^{1/2}$
LLD as a count rate for equal count times for background and sample where $k = 1.645$ (95% confidence level)	$\text{LLD} = \frac{2.706}{t} + 4.653\sqrt{\frac{R_b}{t}}$
LLD in counts for a total background B, equal count times for background and sample, $k = 1.645$ (95% confidence level)	$\text{LLD} = 2.706 + 4.653\sqrt{B}$
LLD in counts for a background B and $t \gg 2.706$	$\text{LLD} = 4.653\sqrt{B}$

[a] R_b in these equations is the background count rate in counts per unit time t and $\sqrt{B} = \sigma_b$.

The LLD is related only to the observed counts produced by the detector system (i.e., it is not dependent on other factors involved in the measurement method or on the sample characteristics). These other factors, some of which have their own uncertainty, should of course be included in the calculation of the measured activity of the sample, and the individual uncertainties should be propagated through the calculation. A typical circumstance is measurement of a sample that includes detector efficiency, chemical recovery, and a measured volume or mass.

Example 13-20: What is the LLD for a detection system that records 50 background counts in 2 h at (a) a 95% confidence level and (b) a 90% confidence level?

SOLUTION: (a) For 95% confidence $k = 1.645$, and

$$\begin{aligned}\text{LLD} &= \frac{(1.645)^2}{120\ \text{min}} + 2(1.645)\left[2\left(\frac{0.42\ \text{c/m}}{120\ \text{min}}\right)\right]^{1/2} \\ &= 0.02 + 0.27 \\ &= 0.29\ \text{c/m}\end{aligned}$$

(b) For a 90% confidence level, $k = 1.282$ (Table 13-8), and

$$\begin{aligned}\text{LLD} &= \frac{(1.282)^2}{120\ \text{min}} + 2(1.282)\left[2\left(\frac{0.42\ \text{c/m}}{120\ \text{min}}\right)\right]^{1/2} \\ &= 0.014 + 0.215 = 0.23\ \text{c/m}\end{aligned}$$

Example 13-21: A single measurement of a sample yields a gross count of 465 c/m. If the counting system has a background of 400 ± 10 c/m recorded for the same time interval and for which $t \gg 2.706$, should the measurement be reported as being radioactive?

SOLUTION: The net count rate is $465 - 400 = 65$ c/m; and since $\sqrt{B} = \sigma_b = 10$ c/m, the LLD for 95% confidence is

$$\text{LLD} = 4.653 \times 10 = 46.53\ \text{c/m}$$

Since the net count of 65 exceeds the calculated value of LLD for the system, it can be concluded with 95% confidence that the sample is radioactive.

Example 13-22: A gamma peak at 0.915 MeV has a total peak activity of 665,000 counts and the peak background is 160,000 counts (see Example 13-16). Determine the critical level, the LLD, and whether it is appropriate to quantitate the 0.915 MeV peak activity.

SOLUTION: The critical level is

$$\begin{aligned}L_c &= 2.33\sqrt{B} = 2.33\sqrt{160{,}000\ \text{counts}} \\ &= 400\ \text{counts}\end{aligned}$$

and the LLD at the 95% confidence level if $t \gg 2.706$ is

$$\text{LLD} = 4.653\sqrt{160{,}000 \text{ counts}}$$
$$= 1861 \text{ counts}$$

or in terms of count rate, 186 c/m. Since the net sample count is well above L_c and LLD, the concentration can be calculated directly and the associated uncertainty assigned.

Minimum Detectable Concentration or Contamination

The LLD, or lower limit of detection, can be used to derive a *minimum detectable concentration* (MDConc.) or *minimum detectable contamination* (MDCont.) that is measurable with a given detector system and protocol. This quantity is often denoted as miminum detectable activity (MDA), but MDConc. and MDCont. are more precise terms. The MDC is the a priori activity level that a measurement system can be expected to detect and report correctly 95% of the time. Once a confident activity level is determined, it is then adjusted by various conversion factors to give units of activity per unit volume, weight, area, and so on. The value of MDC thus determined is the estimated level of activity per unit volume, weight, area, and so on, that can be detected using a given protocol. It should be used before any measurements are made, not after.

Minimum Detectable Concentration The minimum detectable concentration (MDConc.) is a level (not a limit) derived from the LLD (also the detection level, L_d). It is a concentration of radioactivity in a sample that is practically achievable by the overall method of measurement. The MDConc. considers not only the instrument characteristics (background and efficiency), but all other factors and conditions that influence the measurement. These include sample size, counting time, self-absorption and decay corrections, chemical yield, and any other factors that influence the determination of the amount of radioactivity in a sample. It cannot serve as a detection limit per se, because any change in measurement conditions or factors will influence its value. It establishes that some minimum overall measurement conditions are met, and is often used for regulatory purposes; when so used it is commonly referred to as the *minimum detectable activity* (MDA).

The MDConc. for a measurement system is derived from the LLD and expressed as pCi per unit volume or weight of the sample:

$$\text{MDConc.(pCi/unit)} = \frac{\text{LLD}}{(E \cdot Y)(\text{volume or weight})(2.22)t}$$

where

MDConc. = minimum detectable concentration of activity in a sample

$\text{LLD} = 2.706 + 4.653\sqrt{B}$

E = counting efficiency for a given detector and geometry

Y = chemical recovery

$$2.22 = (\text{d/m})/\text{pCi}$$

$$t = \text{count time for } B \text{ (min)}$$

$$B = \text{counts due to background}$$

A *regulatory MDA* is one use of the LLD, such as determining whether material is a radioactive waste. For example, various plants collect floor drains in a sump to determine whether it can be disposed without regard to its radioactivity or whether it must be considered a radioactive waste that is required to be held for decay, treated, or disposed according to radioactive waste regulations. The regulatory basis for this decision is the LLD of the detector–sampling system, as long as the activity meets other environmental and public health regulations. This situation is illustrated in Example 13-23.

Example 13-23: Floor drains are consolidated at a nuclear power plant. A two-liter sample of water from the uncontaminated plant water supply (i.e., background) is counted in a Marinelli beaker on a germanium detector that has an efficiency of 8% for ^{137}Cs γ-rays and produces 2000 counts in 2 h. (a) What concentration of ^{137}Cs in the floor drain sump could be disposed without regard to its radioactivity? (b) What concentration would be allowed if the background for the plant water system measured 500 counts in 30 min under the same conditions?

SOLUTION: (a) Although the critical level for this measurement could provide a determination of positive activity above background, it is too restrictive for this purpose because the confidence level is too low; therefore, the limiting concentration, or the regulatory MDA, should be based on the LLD, or

$$\text{LLD} = 2.706 + 4.653\sqrt{2000} = 211 \text{ counts}$$

Therefore, any 2-L sump sample that had a net number of counts above 2211 (2000 + LLD) for a 2-h measurement would be declared as containing radioactivity and would require additional consideration. An activity of 2000 counts + LLD would correspond to a measured MDConc. in sump water of

$$\text{MDConc.} = \frac{(2000 + 211) \text{ counts}}{0.08 \text{ c/d} \times 120 \text{ min} \times 2 \text{ L} \times 2.22 \text{ (d/m)/pCi}}$$
$$= 51.9 \text{ pCi/L}$$

(b) Similarly, for a 30-min measurement of the same background rate, or 500 counts in 30 min,

$$\text{MDConc.} = 56.9 \text{ pCi/L}$$

These two results indicate that a 10% greater concentration could escape regulation if the routine method of measuring sump water were based on a 30-min count of sample and background rather than 120 min. Conversely, if increased sensitivity were the main goal, the longer count would be chosen.

Minimum Detectable Contamination For an integrated measurement over a preset time, the MDCont. value for a surface activity measurement is based first on the LLD, which is then adjusted for measurement time, detector efficiency, and the active area of the detector,

$$\text{MDCont.} = \frac{2.706 + 4.653\sqrt{B}}{t\varepsilon A}$$

where

MDCont. = minimum detectable contamination [(d/m)/cm^2]
B = background counts
t = counting time (min)
ε = total detector efficiency (c/d)
A = physical probe area in (cm^2)

MDCont. values for other measurement conditions may be derived from this equation, depending on the efficiency and area of the detector and contaminants of concern.

Example 13-24: Determine the minimum detectable contamination in units of (d/m)/100 cm^2 at a 95% confidence level for a detector with a 20% detection efficiency and an active area of 15 cm^2 if the background level under identical conditions measures 40 counts in 1 min?

SOLUTION:

$$\text{MDCont.} = \frac{2.706 + 4.653\sqrt{40}}{(1)(0.2)(15)} = 10.7\ \text{(d/m)/cm}^2$$

or when adjusted to a reference area of 100 cm^2,

$$\text{MDCont.} = 1070\ \text{(d/m)/100 cm}^2$$

These results should be interpreted as follows. The level of contamination that would be reported as detected 95% of the time is 1070 (d/m)/100 cm^2, which takes into acount all the factors of the measurement, detection efficiency, probe area, scan time, and so on. It is also possible to interpret the data in terms of the critical level L_c, which is calculated from the background counts B as

$$L_c = 2.33\sqrt{B} = 15\ \text{counts}$$

This result means that any count yielding 15 counts above the background count of 40 (or 55 total counts) during a 1-min period would be regarded as greater than background. However, any net counts of 15 or more recorded in a 1-min period would represent a positive measurement of activity above background only 50% of the time, due to the normal distribution of counts around L_c. Even though it appears that use of L_c increases the sensitivity of the measurement technique, it occurs with considerable loss of confidence in the result. As the net count increases above L_c, the confidence that activity is present also increases.

Less-Than Level

It is often desirable to record a quantity of radioactive material that could be present but is below detectable levels, or a less-than level, L_t. The requirements for L_t are not met by either L_c, which is used for determining if a measured count rate is statistically greater than background, nor by the LLD, which is the true net count that a sample must have above background to be sure at a stated level of confidence (e.g., 95%) that a single determination will measure the activity. L_t is therefore defined as the maximum true count rate that a sample could have based on the measured count rate, and the standard deviation σ_s of R_s, or

$$L_t = R_s + k\sigma_s$$

where k is the one-sided confidence factor, typically selected for a 95% confidence level, or $k = 1.645$. When so used the calculation of L_t states that there is only a 5% chance that the activity actually present exceeds the stated value of L_t.

In the special case where the sample net count rate is exactly zero,

$$L_t = k\sigma_0 = L_c$$

and it is possible to have $R_s < 0$, due to the statistical nature of low-level counting but in practical terms calculated values of L_t lie somewhere between L_c and L_d. In terms of count rate,

$$L_t = R_s + k\left(\frac{R_s}{t_s} + \frac{R_b}{t_b}\right)^{1/2}$$

and if the sample plus background count time, t_{s+b}, and the background count time, t_b, are equal,

$$L_t = R_s + k\left(\frac{R_{s+b} + R_b}{t}\right)^{1/2}$$

Interpretations and Restrictions

Perhaps the most onerous aspect of nuclear counting determinations and their uncertainties is the meaning of the reported values. First and foremost, neither the statistically rigorous lower level of detection (LLD) nor quantities derived from it (e.g., MDA, MDCont., or MDConc.) should be used for routine counting and reporting of measured data. They can, however, be used when the minimum detectable values from a measurement system must be specified (e.g., to a regulatory agency). The LLD, MDCont., or MDConc. (commonly referred to as MDA) are only estimates for a system; they are not absolute levels of activity or activity concentration that can or cannot be detected in any given sample. Their practical significance is to serve as guideposts or criteria for experimental design, comparison and optimization purposes, and to serve, for regulatory purposes, as minimally acceptable levels that can practically be achieved and reported.

All measurement results should be reported directly as obtained, and any estimated MDCs should be specified separately and used only to indicate the ability of a sampling or measurement system to detect radioactivity. An MDC derived from the LLD for a detector system should not be calculated for each individual measurement. For routine low-level counting, measurements of the net count rate R_s of a source (or sample) above background should be determined and used as follows:

- If $R_s > L_c$, the result can be determined to be positive; it is reported as $R_s \pm k\sigma_s$, where k corresponds to the chosen two-sided confidence level [e.g., $k = 1.96$ at the 95% confidence ($p = 0.05$) level].
- If $R_s \leq L_c$, L_t is calculated using the one-sided confidence interval and reported as less than $R_s + k\sigma_s$, where $k = 1.645$ at the 95% confidence ($p = 0.05$) level.
- If the data are to be averaged, L_t may be reported, but the *absolute* activities (whether positive, negative, or zero) should be recorded and averaged (if the L_t values are used for the average value, the result will be biased).

Uncertainties Due to Response Time/Dead Time

The standard deviation σ_{crm} in a count rate measured by a count rate meter is a function of the time constant or response time of the instrument, or

$$\sigma_{\text{crm}} = \sqrt{\frac{\text{count rate}}{2 \times \text{time constant}}}$$

For example, a survey meter with a time constant of 1.5 s (or 0.025 min) that registers 4000 c/m will have a 1σ standard deviation of ± 283 c/m.

The *resolving time* (often called the "dead time") of a detector is the minimum time that can elapse between the arrival and the recording of two distinct pulses; it is the sum of the dead time and the recovery time and is the interval during which the detector is "busy" and unable to register counts so that some radiations will not be recorded. These losses can be significant for high counting rates, but fortunately corrections can be made for them as follows: If τ is the resolving ("dead") time of a detector system and r is the observed counting rate, the fraction of time during which the system is insensitive is $r\tau$, and n, the true counting rate, is $n = r/1 - r\tau$; e.g., the true counting rate of a detector system with $\tau = 200$ μs that registers 30,000 c/m is

$$n = \frac{r}{1 - r\tau} = \frac{30{,}000/60}{1 - (30{,}000/60)(200 \times 10^{-6})} = 555 \text{ c/s}$$

Thus 55/500 or about 10% of the counts (approximated by $r\tau = 0.10$) are lost.

The standard deviation of the net count (C_{s+b} minus background) for σ_b determined separately is

$$\sigma_{\text{net}} = \sqrt{\left(\frac{1}{1-(C_{s+b}/t_{s+b})\tau}\right)\frac{C_{s+b}}{t_{s+b}^2} + \sigma_b^2}$$

ACKNOWLEDGMENTS

Much of the information on statistical tests and examples was compiled by Ihab R. Kamel and Chul Lee, both graduates of the University of Michigan Radiological Health Program; the presentation of detection levels relied considerably on a paper by Lochamy (1981), who brought clarity to an otherwise complex concept.

REFERENCES AND ADDITIONAL RESOURCES

Altshuler, B., and B. Pasternak, Statistical measurements of the lower limit of detection of a radioactivity counter, *Health Physics* **9** (1963), 293–298.

Currie, L. A., Limits for qualitative detection and quantitative determination, *Analytical Chemistry* **40**, 3 (1968), 586–693.

Currie, L. A., *Lower Limit of Detection: Definition and Elaboration of a Proposed Position for Radiological Effluent and Environmental Measurements*, Report NUREG/CR-4007, U.S. Nuclear Regulatory Commission, Washington, DC, 1984.

Lochamy, J. C., *The Minimum Detectable Activity Concept*, Report PSD 17, EG&G ORTEC, Oak Ridge, TN, September 1981.

U.S. NRC, *Multi-agency Radiation Survey and Site Investigation Manual* (MARSSIM), Report NUREG-1575, Nuclear Regulatory Commission, Washington, DC, 1997.

Watson, J. E., *Upgrading Environmental Radiation Data*, Report HPSR-1, Health Physics Society, McLean, VA, 1980.

PROBLEMS

13-1. A single count of a radioactive sample yields 100 counts in 2 min. What is the uncertainty in the count and the count rate for (**a**) a 1σ confidence interval and (**b**) a 95% confidence interval?

13-2. Compare the 95% uncertainties of a 2-min measurement yielding 100 counts (see Problem 13-1) and another 2-min count that gives 1000 counts.

13-3. A counter registers 400 counts in 2 min. (**a**) What is the count rate and its standard deviation? (**b**) How long would the sample have to be counted to show a percent error of 5% with 90% confidence?

13-4. Calculate the total number of counts that would have to be recorded in a single counting measurement for the 1σ uncertainty to equal exactly 1%.

13-5. A single 1-min count of a radioactive ^{137}Cs source is 2000 counts. What is the expected range of values within which another count could be expected to appear with the same counting conditions at (**a**) a 68.3% confidence level and (**b**) a 95% confidence level?

13-6. If the counter setup in Problem 13-5 has a measured background rate of 100 c/m (measured during a counting time of 1 min), calculate the net count rate and its uncertainty at (**a**) a 68% confidence level and (**b**) a 95% confidence level. (**c**) What percent errors do these results have?

13-7. A bench model GM counter records 200,000 c/m on a wipe test sample. The counter has a dead time of 125 μs. (**a**) What is the true count rate? (**b**) If the counter efficiency is 16%, what is the sample activity?

13-8. A beta-sensitive probe with an area of 20 cm^2 is used with a scaler readout to measure total surface contamination. The readout test time is 1 min, and the background was measured for 10 min. If the efficiency is 32%, what is the MDCont. level in (d/m)/100 cm^2 for the detector?

13-9. A long-lived radioactive sample produces 1100 counts in 20 min on a detector system with measured background of 900 counts in 30 min. Determine the net sample count rate and its standard deviation?

13-10. A sample was measured and reported at the 95% confidence level as 65.2 ± 3.5 c/m. The background was measured at 14.7 ± 4.2 c/m. Determine the net count rate for the sample and its uncertainty at the 95% confidence level.

13-11. The gross count rate of a weak ^{60}Co source is estimated to be 65 c/m in a system with a 10-c/m background. If 1 h is allocated for counting the background and the source, calculate the optimum counting times for each.

13-12. A 10-min sample count yields 1000 counts on a G-M detector system that registers 2340 background counts in 1 h. If 2 h is available for a recount, what is the optimum division of counting time?

13-13. Four aliquots of a sample were measured with count rates and respective standard deviations of 95 ± 3, 105 ± 10, 94 ± 6, and 118 ± 12. Determine (**a**) the weighted sample mean and (**b**) its 1σ uncertainty.

13-14. An environmental sample yielded 530 counts in 10 min and a 30-min background for the detector registered 50 c/m. (**a**) Determine the probability that the sample count is different from background. (**b**) Determine whether this difference is significant at the 95% confidence level.

14

NEUTRONS

The possible existence of an atom of mass 1 (and zero) charge...[with] novel properties...to move freely through matter...unite with the nucleus or be disintegrated by its intense field... . How (else) on earth (are you) to build up a big nucleus with a large positive charge.

—Lord Ernest Rutherford, 1920

The neutron, which was discovered by Chadwick in 1932 after about 3 weeks of tireless experimentation and calculation (following which he wished "to be chloroformed and put to bed for a fortnight"), had actually been predicted by Rutherford some 10 years earlier (see Chapter 3). Neutrons interact with matter principally by elastic scattering in light materials and are slowed to thermal energies. They can also be depleted from a beam by absorption reactions that yield new products, some of which may be radioactive. The effects and products of such interactions need to be considered in neutron dosimetry, neutron shielding, and detection of neutrons.

NEUTRON SOURCES

Nuclear reactors and accelerators produce neutrons with a wide range of energies; these are immediate sources, and the production of neutrons terminates when they are shut down. Various neutron generators depend on irradiation of a target material, and neutrons from these sources are also terminated when the source of irradiation is removed. Alpha/neutron sources contain Be or ^{2}H mixed with an alpha-emitting material, which of course persists and changes according to the radioactivity of the alpha source used. Similarly, Be and ^{2}H can also be used with a high-energy gamma emitter as a photoneutron, or (γ, n) source of neutrons. Neutrons from accelerators or nuclear reactors typically emanate as a beam, and these are readily characterized in terms of a fluence of neutrons per unit area (n/cm^2) or a fluence rate or flux (n/cm$^2 \cdot$ s); point sources, disc sources, and volume sources can be similarly characterized. Table 14-1 lists common sources of neutrons, their energy range, and the average energy obtained.

The *deuterium–tritium* (D-T) *generator* is a straightforward source of high-energy neutrons in which a tritium target is bombarded by deuterons that have been accel-

TABLE 14-1. Neutron Sources By Production Mode and Energy

Source	Reaction	Energy Range (MeV)	Average Energy (MeV)
^{124}Sb–Be	(γ,n)	[a]	0.024
^{88}Y–Be	(γ,n)	[a]	0.16
^{24}Na–D_2O	(γ,n)	[a]	0.22
^{88}Y–Be	(γ,n)	[a]	0.31
^{24}Na–Be	(γ,n)	[a]	0.83
Fission	(n,n)	0–8 MeV	2
^{2}H–^{2}H (D-D)	(d,n)	[a]	3.27
^{226}Ra–Be	(α,n)	0–8 MeV	5
^{239}Pu–Be	(α,n)	0–8 MeV	4.5
^{252}Cf	SF	0–10 MeV	2.3
^{2}H–^{3}H (D-T)	(d,n)	[a]	14.1

[a]Essentially monoenergetic, depending on self-absorption in the source.

erated to about 200 keV. The reaction is exoergic and releases 17.6 MeV, of which 14.1 MeV is given to the ejected neutron. The neutrons so produced are monoenergetic and are ejected isotropically, which causes the flux to fall off as the square of the distance.

Cyclotron-produced neutrons are of high energy and are produced by accelerating high-energy deuterons onto a Be target. Neutrons released in the ^{9}Be(d,n)^{10}B reaction are peaked in the direction of the deuteron beam and, depending on the number of deuterons in the incident beam, can produce intense focused beams of neutrons. The neutrons are not, however, monoenergetic but are distributed around a peak energy.

Photoneutron sources yield monoenergetic neutrons, which can be quite useful for many research purposes, calibration of instruments, and neutron dosimetry.

Alpha–neutron sources consist of radium, polonium, and plutonium intermixed with beryllium which, because of the Q-value of the reaction, yields specific-energy neutrons; however, self-absorption of the alpha particles in the source causes the neutrons to be distributed about an average value that is usually several MeV (see Table 14-1). Such sources are always a compromise between a desired source lifetime and the amount of alpha-emitting material that must be mixed with Be to obtain the desired source strength and energy range.

Spontaneous fission neutron sources are also available, with ^{252}Cf being the most common and practical source because of its yield of 2.3×10^{12} neutrons/s per gram (4.3×10^{9} per curie). The neutrons are emitted with a spectrum of energies from thermal up to several MeV, with an average energy of about 2.3 MeV. The half-life of ^{252}Cf is 2.638 y, and transformation occurs by fission about 3% of the time and by alpha-particle emission about 97% of the time (see Appendix D).

Most neutron sources produce energetic neutrons (see Table 14-1) that are termed *fast* neutrons because of their velocities at these energies. Although neutrons are born fast, they quickly undergo various interactions with media and their energy

TABLE 14-2. Neutron Groups by Energy

Cold ($T < 20°C$)	< 0.0253 eV
Thermal	0.0253 eV
Epithermal[a]	0.0253–ca. 1 eV
Epicadmium[b]	> 1 eV
Slow[c]	0.0253–100 eV
Intermediate[c]	0.5–10^4 eV
Fast	0.01–10 MeV
High energy	> 10 MeV

[a]Energies corresponding to room temperature and extending up to the sharp absorption cutoff by a cadmium absorber.
[b]Neutrons that are transparent to a cadmium absorber (strong resonance capture of about 20,000 b at about 0.4 eV).
[c]Slow and intermediate classes are often used interchangeably and/or without distinction.

is degraded. Neutrons are best classified in terms of their energies into the broad groups designated in Table 14-2.

NEUTRON PARAMETERS

The two most important properties of neutrons relative to radiation protection are the probability of interaction in a medium, denoted by the cross section (see Chapter 9), and the energy transferred to or deposited in the medium. Cross sections are related directly to neutron energy and the absorbing medium, as illustrated in Figure 14-1 for boron and cadmium. A distinguishing characteristic of the cross section for some absorbers (e.g., cadmium in Figure 14-1) is the presence of resonance peaks, while others (e.g., boron in Figure 14-1) have large regions in which the cross section decreases uniformly with neutron energy even though resonances may exist at higher energy ranges. Such decreases are governed by the $1/v$ law, where v is the neutron velocity, which, in turn, is proportional to $\sqrt{E}$. Because of the $1/v$ dependence of many materials, it has been standard practice to specify all neutron absorption cross sections, whether absorbers are $1/v$ or not, at 0.0253 eV, which corresponds to the velocity a neutron has at standard room temperature, or 2200 m/s. For $1/v$ absorbers, it is straightforward to determine the absorption cross section $\sigma_a(E)$ at any other energy, as long as it is in the $1/v$ region, by

$$\sigma_a(E) = \sigma_a(E_0)\sqrt{\frac{E_0}{E}}$$

where E and E_0 are any two known energies and $\sigma_a(E_0)$ corresponds to the lower one; $\sigma_a(E_0)$ is typically at thermal energy because precise values have been determined for many elements.

Example 14-1: Determine σ_a for neutrons of 10 eV in hydrogen (σ_a at thermal energy = 0.333 b).

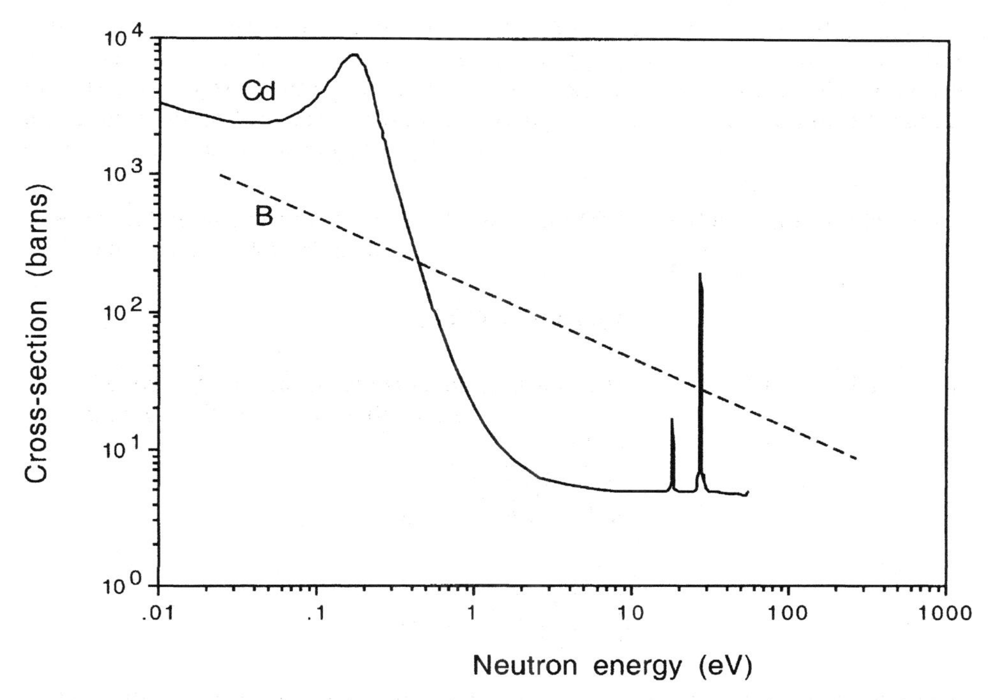

FIGURE 14-1. Neutron absorption cross section versus neutron energy for boron and cadmium.

SOLUTION: The absorption cross section for H is $1/v$; therefore,

$$\sigma_a(10 \text{ eV}) = \sigma_{a,\text{th}}\sqrt{\frac{E_0}{E}}$$

$$= 0.333 \text{ b}\sqrt{\frac{0.0253 \text{ eV}}{10 \text{ eV}}} = 0.0167 \text{ b}$$

NEUTRON INTERACTIONS

Radiation protection for neutrons involves three types of interactions: elastic scattering, inelastic scattering, and capture. Each of these occurs between the neutron and the nuclei of target atoms; electrons of atoms are rarely involved in the primary interaction. *Elastic scattering interactions* are quite effective in slowing down neutrons in light materials, especially hydrogen. The energy spectrum will, of course, change significantly, as these scattering reactions moderate the original neutrons in the beam. A free neutron will eventually be captured; however, since free neutrons are unstable, some may undergo radioactive transformation ($T_{1/2} = 10.25$ m) to create a hydrogen ion and a β^- particle before capture can occur. Both will be absorbed nearby. *Inelastic scattering interactions* are also effective in slowing neutrons. These produce excitation of nuclei in the absorbing medium, and this energy is released almost immediately by the emission of a photon. Inelastic scattering is most prominent for fast neutrons and heavy nuclei; consequently, fast-neutron interactions often produce an associated source of gamma rays, especially if the neutron energy is above 1 or 2 MeV and the absorbing medium contains high-Z atoms. Neutron shields will usually absorb the photons released in inelastic scattering reactions because the thickness required to absorb most of the neutrons is also sufficient to attenuate any gamma rays released by the excited nuclei.

Capture reactions are likely to occur after elastic and inelastic interactions slow neutrons down (or moderate them) to resonance or thermal energies so they can be readily absorbed in target nuclei of the absorbing medium. Capture gamma rays can be quite energetic at several MeV and, depending on the shield material, may require additional consideration because of their very high energy. The amount of energy transferred to a target atom during elastic and inelasic interactions is most useful for radiation dose determinations because the recoiling target atoms will deposit all the energy transferred to them within a few micrometers of the sites of interactions. For elastic collisions the average energy, $\overline{E}$, transferred to recoiling target atoms by neutrons of initial energy E is

$$E - \overline{E}' = \tfrac{1}{2}(1 - \alpha)E$$

where

$$\alpha = \left(\frac{A-1}{A+1}\right)^2$$

and A is the mass number of the target nucleus.

Scattering interactions in hydrogen, a large component of tissue, are unique. Inelastic scattering cannot occur in either hydrogen or deuterium since these nuclei have no excited states. Resonance scattering or absorption also does not occur and σ_s is constant up to above 10^4 eV (see Figure 14-3) and σ_a follows the $1/v$ law. For elastic scattering interactions, E'_{min} is zero, and the average energy transferred is

$$E - \overline{E}' = \tfrac{1}{2}(1 - \alpha)E = \tfrac{1}{2}E$$

since α for ^{1}H is zero. Hydrogen is thus important for neutron dosimetry because one-half of the energy of intermediate and fast neutrons is transferred to the recoiling hydrogen atoms. This energy transfer relationship also explains why hydrogenous materials are so effective in slowing down (or moderating) high-energy neutrons.

For heavier media, the amount of energy transferred by elastic scattering interations decreases appreciably and is less than 1% for lead and uranium absorbers. Consequently, the principal mechanism of energy loss by high-energy neutrons in heavy materials is by inelastic scattering interactions. Under these conditions, σ_{is} is generally quite a bit less (on the order of 2 to 3 b) than elastic scattering cross sections for fast neutrons ($E > 1$ MeV), but it is still the most effective mechanism for slowing neutrons in dense materials because of the large energy decrease that occurs with these interactions. Neutrons of energy E will have an average energy $\overline{E}'_{is}$ after inelastic scattering in a medium of atomic mass A of approximately

$$\overline{E}'_{is} \simeq 6.4\sqrt{\frac{E}{A}}$$

For example, if iron is used to shield 14.1 MeV neutrons from a D-T generator, the average energy after an inelastic collision with an iron atom ($A = 56$) will be about 3.2 MeV. Elastic scattering of 14.1 MeV neutrons in water or paraffin only reduces their energy by about half, or to 7 MeV. On the other hand, neutron energy deposition in tissue, made up of light elements, is due almost solely to elastic scattering since inelastic scattering interactions are very small in light elements.

Cross-section data for inelastic scattering are sparse. Most of the available data are for uranium, plutonium, and so on, and mainly because of the need to describe neutron energy spectra related to fission. Inelastic cross sections for neutron interactions in these elements are also relatively small (on the order of 1 to 3 b).

Neutron Attenuation and Absorption

It is practical, useful, and convenient to represent neutron intensity in terms of the number per unit area, either as a fluence (n/cm^2) or as a fluence rate or flux (n/cm$^2 \cdot$ s). The interactions that slow neutrons down and cause their eventual removal from a beam are probabilistic; they either occur or they don't. Consequently, a flux of neutrons of intensity, I, will be diminished in a thickness x of absorber proportional to the intensity of the neutron source and the neutron removal coefficient, Σ_{nr}, of the absorbing material, or

$$-\frac{dI}{dx} = \Sigma_{nr} I$$

which, like photon attenuation, has the solution

$$I(x) = I_0 e^{-\Sigma_{nr} x}$$

where I_0 is the initial intensity and $I(x)$ refers to those neutrons that penetrate a distance x in an absorber without a collision; therefore, $e^{-\Sigma_{nr} x}$ represents the probability that a given neutron travels a distance x without an interaction. Conceptually, Σ_{nr} can be thought of as the probability per unit path length that a neutron will undergo an interaction as it moves through an absorber and be removed from the beam either by absorption or scattering. In this context, then, it very much resembles the attenuation coefficient for photons in good (or narrow-beam) geometry, and can be similarity developed and used for neutron shielding and dosimetry.

The features of neutron beams, including the concept of narrow-beam effects, are shown schematically in Figure 14-2. The various interactions serve to remove a neutron from the beam such that it does not reach the receptor of interest (e.g., a detector or a person). In this respect, elastic and inelastic scattering interactions deflect neutrons out of the beam, and Σ_{nr} accounts for all the processes that do so. However, neutrons scattered from the narrow beam are likely to undergo other scattering interactions and be deflected back into the beam and reach the receptor. These more realistic, or poor geometry conditions, are accounted for with a neutron buildup factor.

When no hydrogenous materials are present, the neutron removal coefficient, Σ_{nr}, is determined by the macroscopic cross section, $\Sigma = N\sigma_t$, where N is the number of target atoms/cm^3 in an absorber and σ_t is the total cross section in barns

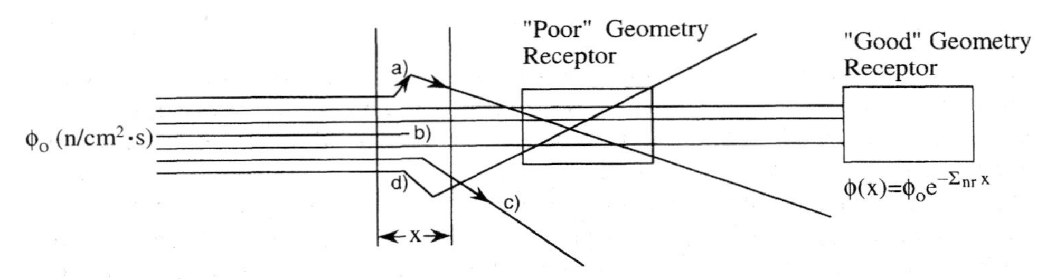

FIGURE 14-2. Schematic of interactions in an absorber of thickness x that deplete a beam of neutrons by (a) inelastic scattering followed by photon emission and another scattering reaction back into the beam, (b) absorption or capture, (c) elastic scattering out of the beam, or (d) elastic scattering with additional scattering back into the beam. However, in poor geometry conditions some of the scattered neutrons are scattered back into the beam, which must be accounted for with a neutron buildup factor.

(10^{-24} cm²/atom) for each atom in a unit volume of absorber. Therefore, Σ_{nr} has units of cm^{-1} and is closely related physically to the attenuation coefficient for photons (and the beta absorption coefficient for electrons), and is in fact used in a similar way; it can be converted to a neutron mass coefficient (cm²/g) by dividing by the density of the absorber, or

$$\text{neutron mass coefficient (cm}^2\text{/g)} = \frac{\Sigma_{nr}}{\rho}$$

A related concept is the *mean free path*, which is the average distance a neutron of a given energy will travel before it undergoes an interaction. The mean free path can also be thought of as the average thickness of a medium in which an interaction is likely to occur and is similar to the mean life of a radioactive atom. It has the value

$$\text{mean free path} = \frac{1}{\Sigma_{nr}}$$

Example 14-2: Determine the neutron removal coefficient and the mean free path of 100-keV neutrons in liquid sodium that has a total cross section of 3.4 b at 100 keV.

SOLUTION: Sodium at STP contains 0.0254×10^{24} atoms/cm³; therefore, the neutron removal coefficient in this case is the macroscopic cross section

$$\Sigma_{nr} = 0.0254 \times 10^{24} \text{ atoms/cm}^3 \times 3.4 \times 10^{-24} \text{ cm}^2\text{/atom} = 0.0864 \text{ cm}^{-1}$$

The mean free path (i.e., the average travel distance without an interaction) is

$$\text{mean free path} = \frac{1}{\Sigma} = \frac{1}{0.0864} = 11.6 \text{ cm}$$

Note that in these types of calculations it is somewhat convenient to express the atom density in units of 10^{24} since it cancels with the 10^{-24} unit for the barn.

TABLE 14-3. Atom Density in Tissue

Element	Atoms/cm^3 or g
Hydrogen	5.98×10^{22}
Oxygen	2.45×10^{22}
Carbon	9.03×10^{21}
Nitrogen	1.29×10^{21}
Sodium	3.93×10^{19}
Chlorine	1.70×10^{19}

Source: Adapted from ICRP, 1975.

NEUTRON DOSIMETRY

The composition of tissue, which is the primary medium of interest in neutron dosimetry, is shown in Table 14-3. The macroscopic cross section, $\Sigma = N_i\sigma_i$, is used to determine energy deposition in tissue due to interactions with its various components. It is important to recognize that it is different from the total neutron removal coefficient used for shielding calculations (see below); that is, the approach for energy deposition in tissue uses specific values for the interaction processes involved, whereas shielding calculations use values of Σ_{nr} specifically measured for shielding materials.

Fast neutrons lose energy in tissue primarily by elastic scattering, and slow and thermal neutrons undergo capture reactions primarily by $^1H(n,\gamma)^2H$ and $^{14}N(n,p)^{14}C$ reactions. The (n,γ) reaction in hydrogen releases a capture photon of 2.225 MeV, which may also deposit some of its energy in the body. The (n, p) reaction with nitrogen produces a 0.626-MeV proton, and all of this energy will be deposited near the site of the interaction. Neutron interactions may also occur with carbon and oxygen in tissue. Since H, N, C, and O are important to neutron energy deposition in tissue, their cross sections versus neutron energy are plotted in Figures 14-3, 14-4, 14-5, and 14-6, respectively.

Dosimetry for Fast Neutrons

For fast neutrons up to about 20 MeV, the main mechanism of energy transfer in tissue is by elastic collisions with light elements. These interactions occur mostly in hydrogen for two reasons: It represents the largest number of tissue atoms, and since one-half the energy is transferred, it represents the largest fractional transfer of energy. The concept of *first collision dose* is used to describe this energy deposition, which is due to the recoiling hydrogen atom; the scattered neutron is likely to leave the body of a person before it undergoes another scattering interaction, and even if it does, the energy deposited will be only 25% of the initial energy. The first collision dose concept for fast neutrons does not, however, account for energy deposition due to absorption reactions or those neutrons that undergo additional scattering before exiting the body. These processes will obviously occur, but since a person is not very thick, the energy deposited in humans will be relatively smaller compared to the first collision interactions. A 5 MeV neutron, for example, has a macroscopic cross section in soft tissue of 0.051 cm^{-1}, so its mean free path is $1/0.051 = 20$ cm,

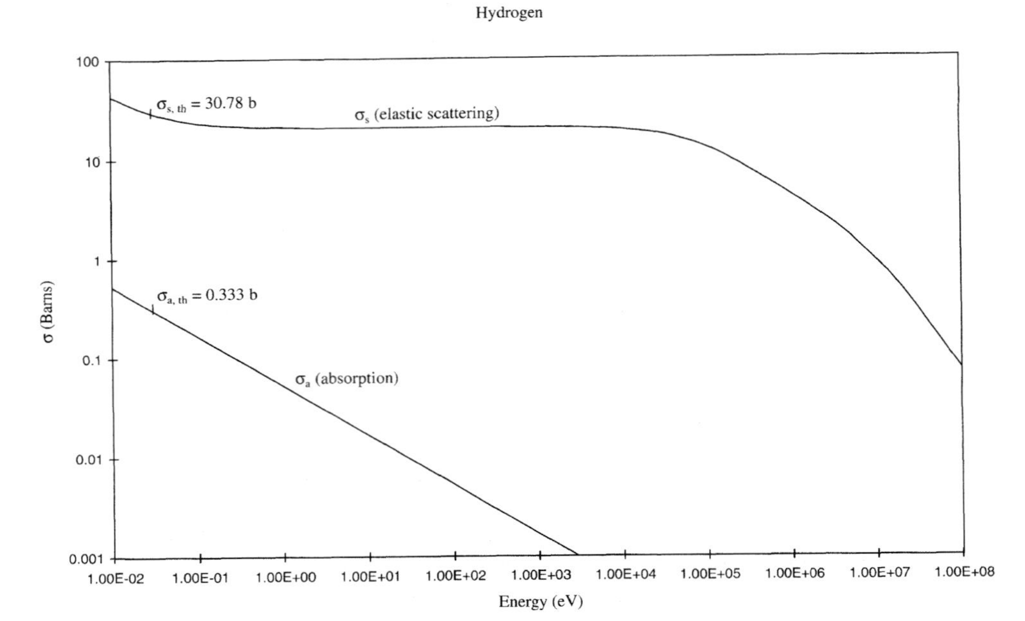

FIGURE 14-3. Neutron absorption (capture) and elastic scattering cross sections (in barns) in hydrogen. The curve for elastic scattering is effectively the total neutron interaction in hydrogen since σ_a is small.

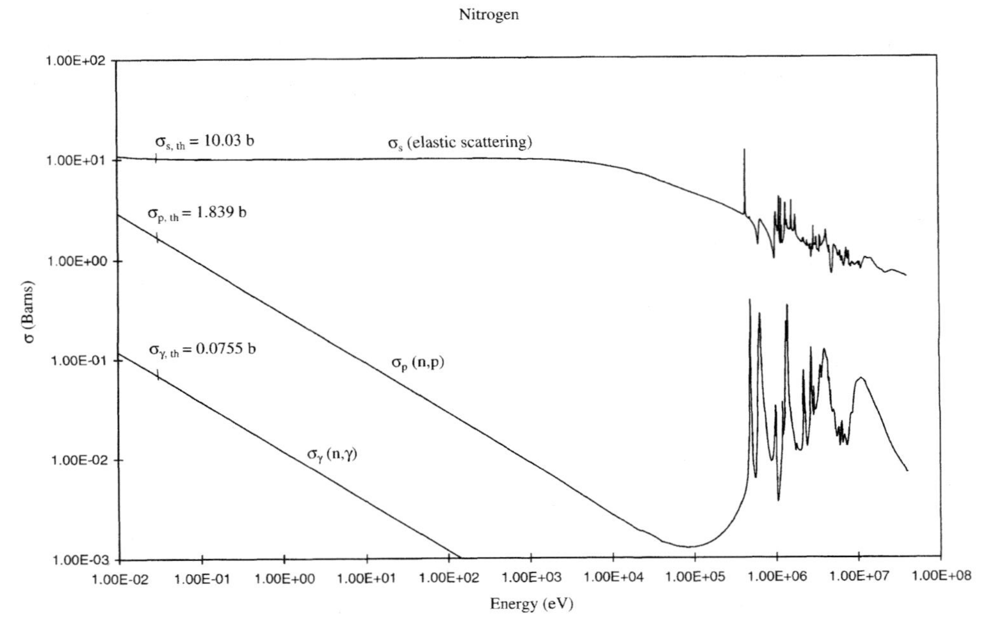

FIGURE 14-4. Neutron absorption (capture) and elastic scattering cross sections (in barns) for nitrogen.

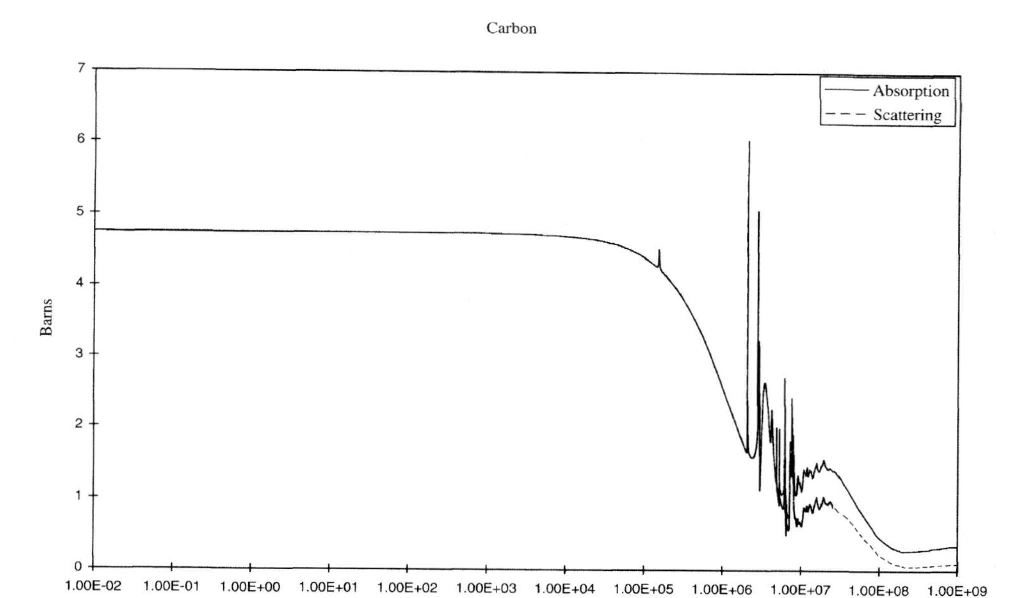

FIGURE 14-5. Total neutron (absorption + scattering) and elastic scattering cross sections (in barns) for carbon.

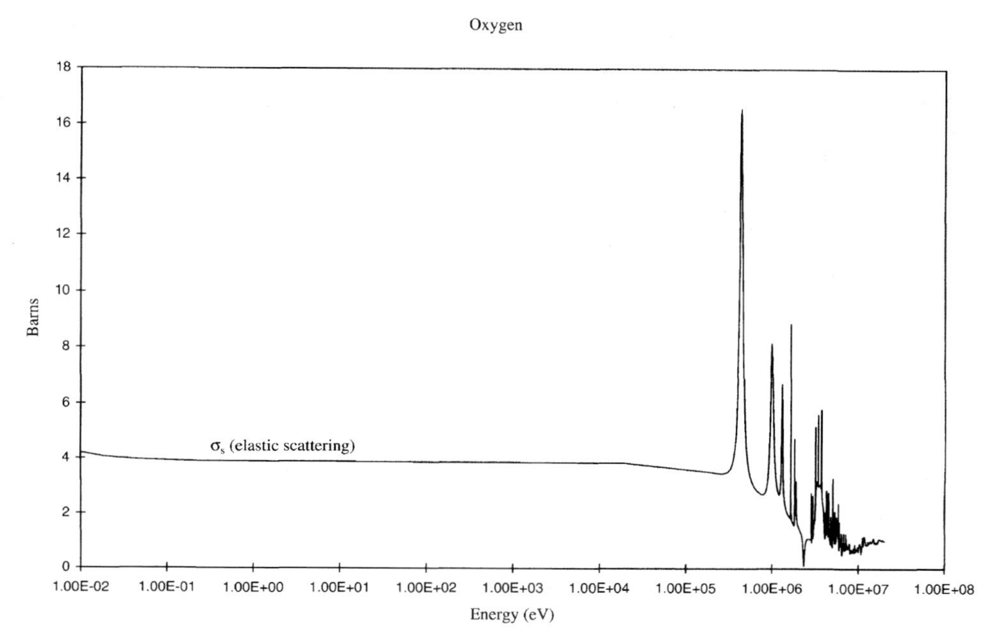

FIGURE 14-6. Elastic scattering cross section (in barns) for oxygen; the total cross section is also equal to the elastic scattering cross section.

which suggests that a 5 MeV neutron will have only one interaction (a first-collision dose) before it penetrates through the total body.

Example 14-3: Calculate the first-collision dose in tissue due to scattering interactions in hydrogen for a fluence of 5 MeV neutrons.

SOLUTION: The density of H atoms in tissue is 5.98×10^{22}/per gram and the cross section for elastic scattering of 5 MeV neutrons is about 1.5 b (Figure 14-3) The mean energy loss per collision is about one-half the incident neutron energy, or $E = 2.5$ MeV per collision; thus the dose for each neutron per square centimeter due to hydrogen collisions is

$$D = (\text{n/cm}^2) \times \frac{5.98 \times 10^{22}\ \text{atoms/g} \times 1.5 \times 10^{-24}\ \text{cm}^2/\text{atom} \times 2.5\ \text{MeV} \times 1.6022 \times 10^{-6}\ \text{erg/MeV}}{100\ \text{ergs/g} \cdot \text{rad}}$$

$$= 3.6 \times 10^{-9}\ \text{rad/(neutron/cm}^2)$$

The first collision dose is due entirely to the fraction of the energy transferred to the recoiling nucleus; the scattered neutron is not considered after the primary interaction. Scattering interactions are obviously possible with all the constituents of tissue; therefore, a more accurate procedure, as illustrated in Example 14-4, is to consider them in determining the absorbed dose.

Example 14-4: What is the absorbed dose rate to soft tissue for a fluence of 5 MeV neutrons?

SOLUTION: It is convenient in calculating the first collision dose rate from neutrons of energy E to first determine a neutron energy mass absorption coefficient for all the constituents, or

$$\sum f_i N_i \sigma_i$$

where f_i is the fraction of the neutron energy transferred in an elastic collision with a nucleus. It is determined from the mass of the neutron (mass = 1) and the mass M_i of the recoiling atom as

$$f_i = \frac{2M_i}{(M_i + 1)^2}$$

which can be combined, as shown in Table 14-4, with the neutron scattering cross section σ_s for each of the elements in tissue to yield a mass energy absorption coefficient in tissue for 5 MeV neutrons of 0.0512 cm^2/g.

The neutron dose to tissue for a unit flux of 5 MeV neutrons is

$$D_n = \frac{1\ \text{n/cm}^2 \times 5\ \text{MeV/n} \times 1.6022 \times 10^{-6}\ \text{erg/MeV} \times 0.0512\ \text{cm}^2/\text{g}}{100\ \text{ergs/g} \cdot \text{rad}}$$

$$= 4.1 \times 10^{-9}\ \text{rad/(neutron/cm}^2)$$

TABLE 14-4. Neutron Absorption Factors for 5 MeV Neutrons in Tissue

Element	N_i (atoms/g)	f_i	σ_s (cm^2)	$N_i\sigma_s f_i$
Hydrogen	5.98×10^{22}	0.500	1.50×10^{-24}	4.485×10^{-2}
Oxygen	2.69×10^{22}	0.111	1.55×10^{-24}	4.628×10^{-3}
Carbon	6.41×10^{21}	0.142	1.65×10^{-24}	1.502×10^{-3}
Nitrogen	1.49×10^{21}	0.124	1.00×10^{-24}	1.848×10^{-4}
Sodium	3.93×10^{19}	0.080	2.3×10^{-24}	7.231×10^{-6}
Chlorine	1.70×10^{19}	0.053	2.8×10^{-24}	2.523×10^{-6}
				$\sum_i N_i\sigma_S f_i = 0.0512$ cm^2/g

which is about 12% higher than the dose calculated in Example 14-3, due to interactions with hydrogen only (i.e., 88% of the dose due to elastic interactions is due to hydrogen interactions).

Example 14-4 considered only monoenergetic neutrons of 5 MeV. If a beam contains neutrons of several energies, a similar calculation must be carried out separately for each energy group.

Dose from Thermal Neutrons

Thermal neutrons incident on tissue transfer energy only if they are absorbed because they are already at thermal energies (i.e., all elastic and inelastic reactions that could occur have already done so). Only two reactions need be considered: the (n,p) reaction with nitrogen, which produces recoil protons of 0.626 MeV [the Q-value of the ^{14}N(n,p)^{14}C reaction] and the (n,γ) reaction with hydrogen, which yields 2.225 MeV gamma rays [the Q-value of the ^{1}H(n,γ)^{2}H reaction]. All the recoil protons from the (n,p) reaction will almost assuredly be absorbed; however, only part of the capture gamma rays from the (n,γ) reactions can be assumed to be absorbed to deposit energy.

The radiation dose rate for a thermal neutron fluence rate (or flux) of ϕ due to 0.626 MeV protons from (n,p) interactions with nitrogen in tissue is

$$\dot{D}_{np} = \frac{\phi N_N \sigma_N E_p \times 1.6022 \times 10^{-6} \text{ erg/MeV} \times 3600 \text{ s/h}}{100 \text{ ergs/g} \cdot \text{rad}}$$

and since there are 1.49×10^{21} nitrogen atoms per gram of tissue and the thermal neutron absorption cross section for nitrogen is 1.83×10^{-24} cm^2,

$$\dot{D}_{np} = 9.845 \times 10^{-8}\phi \text{ rad/h}$$

Radiation dose from the gamma rays produced by the ^{1}H(n,γ)^{2}H reaction can be significant if the whole body is exposed. In these reactions the neutron dose is actually delivered by the 2.225 MeV gamma rays released in the reaction. Since hydrogen is distributed uniformly in the body, the total body is both the source organ that produces the dose and the target organ that receives the dose. The gamma emission source term, in γ/s per gram, is determined by the thermal neutron flux ϕ that produces (n,γ) reactions with 5.8×10^{22} hydrogen atoms per gram of tissue with a cross section of 0.333 b, and is

$$\begin{aligned}\gamma/\text{s} \cdot \text{g} &= \phi N_\text{H} \sigma_\text{H} \\ &= 1.93 \times 10^{-2} \phi\end{aligned}$$

where ϕ is the thermal neutron flux (neutrons/cm$^2 \cdot$ s).

The radiation dose due to thermal neutrons is thus produced by a combination of proton recoils and gamma radiation from neutron absorption reactions, as illustrated in Example 14-5.

Example 14-5: Determine the dose rate to the total body uniformly exposed to a thermal flux of 10,000 neutrons/cm^2 per second from (a) the (n,p) reactions in nitrogen, (b) the gamma rays released by (n,γ) reactions in hydrogen, and (c) the total absorbed dose and dose equivalent.

SOLUTION: (a) The energy deposition from recoil protons due to (n,p) reactions with nitrogen atoms in tissue ($\sigma = 1.83$ b) is due to the protons ejected by (n,p) reactions. These protons are charged particles and will deposit their energy within a few micrometers of the sites of interaction; therefore,

$$\begin{aligned}\dot{D}_{np} &= 10^4 \text{ neutrons/cm}^2 \cdot \text{s} \times 1.49 \times 10^{21} \text{ atoms/g} \times 1.83 \times 10^{-24} \text{ cm}^2\text{/atom} \\ &\quad \times 0.626 \text{ MeV} \times 1.6022 \times 10^{-6} \text{ erg/MeV} \times 3600 \text{ s/h} \\ &= 9.845 \times 10^{-2} \text{ erg/g} \cdot \text{h} = 0.98 \text{ mrad/h}\end{aligned}$$

(b) The flux of 2.225 MeV prompt gamma rays released by (n,γ) reactions in hydrogen is determined by the number of (n,γ) reactions, or

$$\begin{aligned}\gamma/\text{s} \cdot \text{g} &= 10^4 \text{ neutrons/cm}^2 \cdot \text{s} \times 5.98 \times 10^{22} \text{ atoms/g} \times 0.333 \times 10^{-24} \text{ cm}^2\text{/atom} \\ &= 198.5\ \gamma/\text{s} \cdot \text{g}\end{aligned}$$

These gamma rays occur uniformly throughout the body so the total body is both the source of the energy flux and the target organ in which the energy is absorbed. For such high-energy gamma rays, only a fraction of the energy emitted will be absorbed since many of the photons will escape the body without interacting. This "absorbed fraction," AF (Total body $\leftarrow$ Total body), for 2.225 MeV gamma rays is 0.278, therefore, the dose rate for the total body as a source organ irradiating the total body as a target organ is

$$\begin{aligned}\dot{D}_\gamma &= \frac{198.5\ \gamma/\text{s} \cdot \text{g} \times 2.225 \text{ MeV} \times 1.6022 \times 10^{-6} \text{ erg/MeV} \times 0.278 \times 3600 \text{ s/h}}{100 \text{ ergs/g} \cdot \text{rad}} \\ &= 7.08 \times 10^{-3} \text{ rad/h} \quad \text{or} \quad 7.08 \text{ mrad/h}\end{aligned}$$

(c) The proton recoil dose and the gamma-ray dose can be added together if, and only if, each is muliplied by the appropriate quality factor to convert absorbed dose to an effective dose equivalent in millirems. The quality factor is 1.0 for photons and 2.0 for thermal neutrons, therefore,

$$\dot{D}_n = \dot{D}_{np} + \dot{D}_\gamma = (7.08 \text{ mrad/h} \times 1.0) + (0.98 \text{ mrad/h} \times 2.0) = 9.04 \text{ mrem/h}$$

As demonstrated in Example 14-5, about 87% of the absorbed dose from thermal neutrons is due to gamma rays produced by neutron capture in hydrogen. Auxier et al. (in Attix and Roesch, 1968) have shown that the gamma component is about 86% for neutrons up to about 0.01 MeV (10^4 eV) and then declines rapidly as elastic scattering with hydrogen in tissue and associated energy deposition by recoil protons becomes more dominant than the (n, γ) reaction. For neutrons above about 1 MeV, the absorbed dose due to gamma rays becomes negligible.

Monte Carlo Calculations of Neutron Dose

The most accurate representation of dose equivalent for neutrons of different energies is based on Monte Carlo calculations of multiple neutron scattering and energy deposition as neutrons are transmitted through tissue. In these calculations neutron transport through a target is based on a statistical distribution of neutron events, and the dose equivalent is determined not only by the energy deposited by each interacting event but also by the quality factor for each neutron energy; therefore, the dose equivalent constantly changes as a fast neutron undergoes repeated collisions and produces recoiling nuclei of various energies. Monte Carlo techniques, which are possible with modern computers, allows the deposited energy from each neutron to be followed through multiple neutron histories and the appropriate quality factor to be assigned for each energy sequence to obtain the distribution of dose equivalent in the absorbing medium. When done for a large number of neutron histories, the variance in the calculated quantities can be reduced considerably.

The results of Monte Carlo calculations for thermal neutrons (0.025 MeV) and for 5 MeV neutrons incident on a slab of tissue 30 cm thick (about the thickness of the body) are shown in Figures 14-7*a* and *b*. The total dose for 5 MeV neutrons (Figure 14-7*b*) builds up somewhat in the first few centimeters of tissue depth and then decreases by a factor of about 10 at 26 cm as the neutrons in the beam become degraded in energy and are eventually absorbed, primarily by (n, γ) reactions in hydrogen.

When 5 MeV neutrons penetrate into tissue, they are moderated to thermal energies which increases their absorption by (n, γ) reactions in hydrogen as reflected in the rise of the gamma-dose curve between 6 and 14 cm. The first-collision dose of 3.6×10^{-9} rad/(neutron/cm^2) in Example 14-3 is less than the entrance dose [5.2×10^{-9} rad/(neutron/cm^2)] shown in Figure 14-7*b* because the Monte Carlo calculation accounts for increases in the first-collision dose due to backscatter within the tissue slab. Most of the dose from 5-MeV neutrons is due to recoiling hydrogen nuclei caused by elastic scattering collisions with hydrogen atoms, and very little is due to capture gammas.

The dose due to thermal neutrons in Figure 14-7*a* is quite different from 5 MeV neutrons in that the dose due to hydrogen recoil nuclei only exceeds the gamma

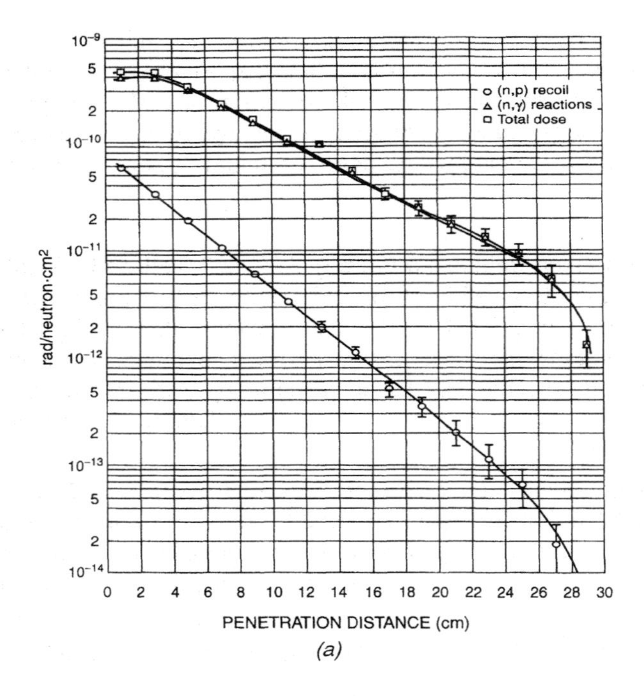

(a)

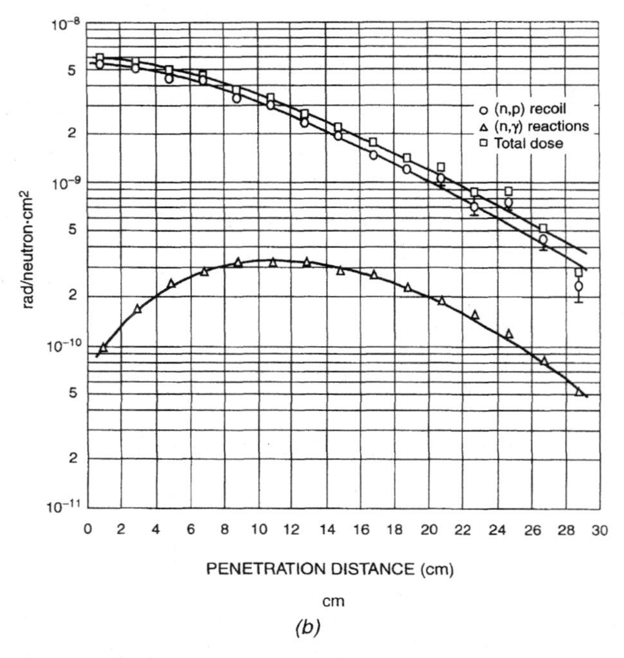

(b)

FIGURE 14-7. Radiation dose (rad) per unit neutron fluence (neutrons/cm^2) versus depth in tissue due to protons, photons, and recoil nuclei for (*a*) thermal neutrons and (*b*) 5 MeV neutrons. (From NCRP, 1971.)

dose over the first 3 cm. The thermal neutron density reaches a maximum at about 10 cm and then decreases due to absorption. Thermal neutrons are not energetic enough to produce recoiling hydrogen nuclei protons in collisions with hydrogen; however, protons are produced by absorption of thermal neutrons in nitrogen by $^{14}N(n,p)^{14}C$ reactions ($\sigma = 1.83$ b and $Q = 0.626$ MeV).

Kerma for Neutrons

Kerma and absorbed dose are essentially equal for (n, p) interactions with nitrogen in tissue because the energies of the recoil products are deposited within a few micrometers of the sites of interaction. Charged particle equilibrium is thus quickly established, and bremsstrahlung production, which could carry energy away, is highly unlikely for these heavier recoil nuclei. Since elastic scattering is the dominant interaction for intermediate and fast neutrons, kerma is equal to deposited dose; the same is true for (n, p) interactions with tissue nitrogen at thermal energies. On the other hand, kerma and absorbed dose are significantly different for (n,γ) absorption in hydrogen because a significant distance is required for equilibrium to be established for each 2.225 MeV gamma ray emitted. Kerma from the indirect (n,γ) reaction with hydrogen will dominate in a tissue mass the size of an adult human because of its uniform disribution in tissue and its larger proportion of tissue (about 40 times more H than N). For thermal neutrons, kerma from the $^{1}H(n,\gamma)^{2}H$ reaction is about 25 times larger than the (n, p) reaction in nitrogen, which is also true for intermediate neutrons because they become thermalized in passing through body tissue.

Dose Equivalent versus Neutron Flux

As shown in Example 14-5, it is straightforward to calculate a neutron dose rate once the incident flux is known. This relationship between dose rate and neutron flux is presented in Table 14-5 and Figure 14-8 in terms of the neutron flux at various energies required to produce a dose equivalent rate of 1 mrem/h (the total dose equivalent for a given neutron fluence can be obtained by multiplying the dose rate by the exposure time). The two sets of data contained in Table 14-5 and Figure 14-8 are different because of differing interpretations in the value of the quality factor for neutrons of different energies. The upper curve in Figure 14-8 reflects current regulations; however, the NCRP has recently recommended increasing the neutron quality factor by a factor of about 2.0 to 2.5, and the effects of these changes are reflected in the lower curve. Whether and/or when these recommendations will be incorporated into regulations is not known, but as shown in Figure 14-8, the dose equivalent for a given flux will be higher if the NCRP recommendations are considered. The curves in Figure 14-8 also demonstrate that a unit flux of high-energy neutrons will yield a higher dose than the same flux of neutrons at lower energy primarily because of the increase in the quality factor at higher energies.

Example 14-6: Estimate the dose equivalent rate 1 m from an unshielded ^{239}Pu–Be source that emits 3×10^7 neutrons/s with an average energy of 4.5 MeV.

TABLE 14-5. Fluence Rates for Monoenergetic Neutrons That Correspond to a Dose-Equivalent Rate of 1 mrem/h

Neutron Energy (eV)	Fluence Rate (neutrons/cm$^2 \cdot$ s) for 1 mrem/h	
	10 CFR 20[a]	NCRP-112[b]
0.025 (thermal)	272	112
0.1	272	112
1	224	112
10	224	112
10^2	232	116
10^3	272	112
10^4	280	120
10^5	46.0	16.0
5×10^5	10.8	6.40
10^6	7.6	3.88
5×10^6	6.4	3.88
10^7	6.8	3.20
1.4×10^7	4.8	2.72
6×10^7	4.4	—
10^8	5.6	—

[a]Adapted from Title 10, Code of Federal Regulations, Part 20 (1993) by the U.S. Nuclear Regulatory Commission (the rates incorporate neutron quality factors of 2.0 for thermal and low-energy neutrons and 2.5 to 11 for higher-energy neutrons).
[b]These fluence rates are based on NCRP Report 112 (1987), which recommends increasing Q for neutrons by a factor of 2.5 for thermal neutrons and 2.0 for all other energies.

SOLUTION: Point sources that emit S neutrons/s can be specified at a distance r (cm) in terms of a neutron flux ϕ,

$$\phi(\text{neutrons/cm}^2 \cdot \text{s}) = \frac{S(\text{neutrons/s})}{4\pi r^2}$$

Table 14-5 indicates that the flux corresponding to a dose-equivalent rate of 1.0 mrem/h due to 4.5 MeV neutrons is about 6.4 neutrons/cm$^2 \cdot$ s or 0.156 (mrem/h)/(neutrons/cm$^2 \cdot$ s). The dose equivalent rate at 100 cm from the unshielded neutron source is

$$\dot{D}_0 = \frac{3 \times 10^7 \times 0.156}{4\pi(100)^2} = 37.2 \text{ mrem/h}$$

Boron Neutron Capture Therapy

Boron neutron capture therapy (BNCT) is an experimental radiation therapy that takes advantage of the neutron absoption properties of boron for the treatment of

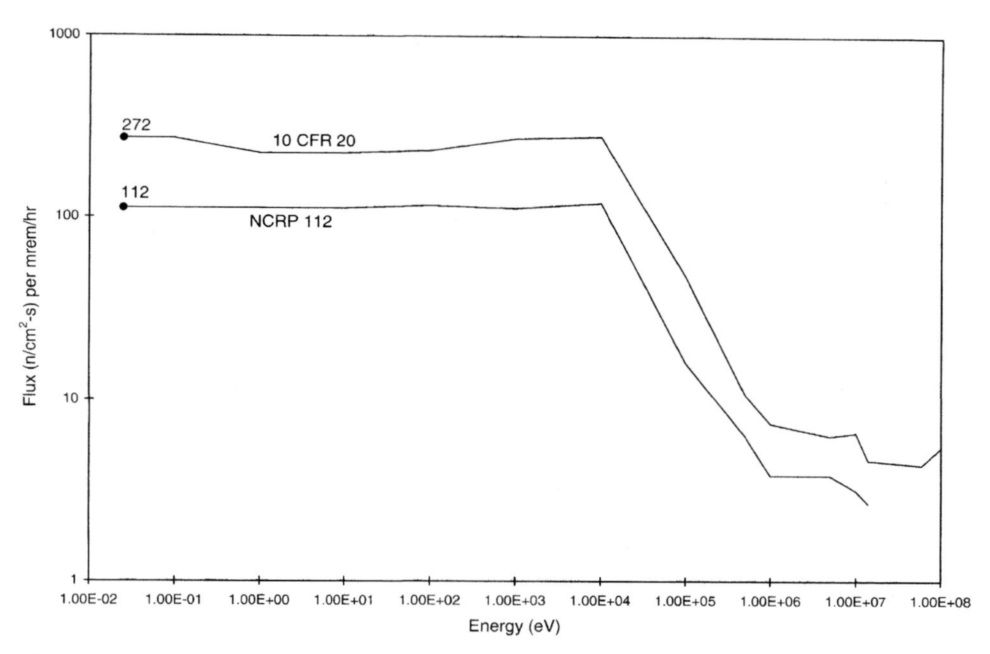

FIGURE 14-8. Fluence rates for neutrons of energy E that produce 1 mrem/h as stated in Federal Regulations (10CFR20) and as recommended by NCRP (1987), based on a reevaluation of neutron quality factors.

malignant tumors, especially those in the brain. Boron-10 (^{10}B), a constituent of natural boron, has a large cross section for thermal neutron absorption (3840 b), and the $^{10}B(n, \alpha)^7Li$ reaction yields

$$^{10}_{5}B + ^{1}_{0}n \begin{cases} \xrightarrow{4\%} \ ^{7}_{3}Li + ^{4}_{2}He + 2.79 \text{ MeV} \qquad (\sigma_{th} = 3840 \text{ b}) \\ \xrightarrow{96\%} \ ^{7}_{3}Li^{*} + ^{4}_{2}He + 2.31 \text{ MeV} \end{cases}$$

(*Excitation energy emitted as a 0.48 MeV photon.)

The ^{7}Li nucleus and the alpha particle have energies of 0.83 and 1.47 MeV, respectively, which is deposited within a distance of about 10 μm in tissue. This energy is deposited in a single cell and/or its immediate neighbor to cause cell death. The success of the BCNT technique depends on selectively concentrating ^{10}B in the tumor mass so that irradiation of the area with thermal neutrons will produce high-energy deposition in tumor cells with minimal damage to normal tissue.

NEUTRON SHIELDING

Neutron shields incorporate materials that enhance interactions that deplete neutrons from the primary beam. For fast neutrons, materials that slow neutrons down through elastic and inelastic scattering are used in combination with materials that enhance capture of neutrons when they reach thermal energies. Hydrogen is very effective in removing neutrons, especially energetic ones, from a beam because collisions with hydrogen, the dominant interaction, diminish the energy by about one-half. The cross section for the scattered neutrons is considerably greater because of the lower energy and the mean free path length is reduced considerably; therefore, a collision with hydrogen effectively removes a fast neutron from the beam.

Neutron Shielding Materials

Many shielding materials produce capture gamma rays or photons from inelastic scattering interactions. These can be minimized by adding lithium or boron to the shield. Lithium-6, which has a large $(n_{\text{thermal}}, \alpha)$ cross section of 941 b does not yield capture gamma rays by neutron absorption, and the helium and tritium atoms that are formed are easily absorbed. Natural boron contains 20% ^{10}B and is a very effective shielding material for thermal neutrons through $^{10}B(n,\alpha)^7Li$ reactions. About 96% of the ^7Li atoms created in these reactions are in an excited state which is relieved promptly by emission of a 0.48 MeV gamma ray, but these are easier to shield than the 2.225 MeV gamma rays from hydrogen capture and boron is commonly used in neutron shields.

Hydrogenous materials such as paraffin and water make efficient neutron shields because of the effectiveness of elastic scattering with hydrogen atoms but not without potential problems. Paraffin is flammable, water can leak and evaporate, and thermal neutron capture produces 2.225 MeV gamma rays. When H_2O is used as a neutron shield, it is necessary to prevent leakage, minimize corrosion, and keep contaminants out, which is generally done by demineralization.

Metals such as lead, iron, tungsten, and depleted uranium are relatively poor shield materials for neutrons; however, they are often used as a gamma shield, especially around nuclear reactors, and their neutron shielding properties are important because of such uses. Lead and iron can produce capture gamma rays of 7.4 and 7.6 MeV, respectively, although with low probability and (n,γ) interactions in ^{58}Fe produce radioactive ^{59}Fe ($T_{1/2} = 44.51$ d), which emits 1.1 and 1.29 MeV gamma rays. *Tungsten* is dense and is almost as effective as lead as a gamma shield. It is much better than lead for neutron attenuation, although secondary gamma radiation is produced due to capture reactions. *Depleted uranium*, which is readily available from nuclear fuel enrichment processes, is very dense ($\rho = 19$) and is the best attenuator available on a volume basis for gamma rays. Neutron attenuation in U is about the same as in lead, and even though it doesn't produce significant capture gamma rays, fast fission reactions may yield gamma-emitting fission products.

Boron is often incorporated into neutron shields because of its absorption cross section (760 b) and the large (n,γ) cross section of ^{10}B (3840 b). The alpha particles from ^{10}B reactions are easily absorbed and the 0.48 MeV gamma rays from the excited ^7Li product, which occurs in 96% of the interactions, is not too difficult to shield. Borax (sodium borate) is a crystalline powdery material that is easily shaped into various shield configurations, is not subject to leakage, and is cheap and effective. Borated water and borated polyethylene are useful, as are boron oxide, boric acid, and boron carbide (B_4C). A sandwich material called *boral* is available that consists of an Al–B_4C mixture clad in aluminum. Boron has also been added to various steels to preferentially absorb thermal neutrons and reduce activation products and associated gamma rays.

Concrete and earth are many times the shield materials of choice for neutron sources, especially around nuclear reactors and accelerators since they are dense, contain hydrogen and other light materials that promote neutron capture, and can be shaped easily. Barytes, iron–portland, and Colemanite–aggregate are some of the many varieties of concrete that are used in neutron shields. Barytes concrete contains boron for neutron absorption and hydrogen for moderation of fast neutrons; it also is a good gamma shield. Colemanite is used as an aggregate and contains hydrated calcium borate, which enhances neutron capture because of the boron it contains. Additives that promote neutron capture can be incorporated into concrete; however, care must be taken to assure that cracks, access ports, ducts, or other penetrations do not permit the escape of neutrons.

Polyethylene is a pure hydrocarbon that contains 18% more hydrogen per unit volume than water (about 8×10^{22} atoms of H/cm^3 versus 5.98×10^{22} atoms of H/cm^3 for H_2O). Unfortunately, it softens at about 110°C and will burn; a more dense ($\rho = 0.96$) variety is available that softens at about 200°C but with slightly diminished neutron removal properties. Boron can also be added to polyethylene to absorb thermal neutrons from hydrogen interactions. *Water extended polyethylene* (or WEP) is a special formulation of polyethylene that is an especially effective neutron shield.

Lithium hydride contains about 12.6% H by weight and is a very effective material for neutron attenuation. It is, however, difficult to fabricate into solid shields. It also actively combines with water, and because of this property needs to be protected from water by encapsulation or other means. Lithium hydroxide is often mixed with water to absorb thermal neutrons and has been used as a burnable reactivity shim in nuclear reactors or added to water shields to absorb thermal neutrons after they are slowed down. Capture reactions in lithium do not produce gamma rays, but absorption of neutrons by ^{6}Li in natural lithium produces tritium, which can be minimized by using lithium depleted in ^{6}Li.

Cadmium has a high (n, γ) neutron capture cross section (2450 b) and is frequently used as a neutron absorber, but like hydrogen has the disadvantage of emitting energetic 9.05 MeV capture gamma rays, which themselves require shielding.

Neutron Shielding Calculations

In concept, neutron attenuation and absorption in a shielding material are similar to good geometry photon attenuation and absorption and can be represented by an exponential function based on absorber thickness and the neutron removal cross section Σ_{nr}, or

$$I(x) = I_0 e^{-\Sigma_{nr} x}$$

where for radiation protection considerations, neutron intensity is expressed as a flux (or fluence rate) of neutrons/cm$^2 \cdot$ s. The flux at a distance r from a point source of neutrons is straightforward but may be more difficult to establish for other sources.

The data in Table 14-5 and Figure 14-8 can be used as a starting point to determine the dose equivalent rate or $\dot{D}(x)$ outside a shield of thickness x if the neutron flux is known. The shielded value $\dot{D}(x)$ can then be obtained by

$$\dot{D}(x) = \dot{D}_0 B e^{-\Sigma_{nr} x}$$

where Σ_{nr} is the neutron removal cross section and $\dot{D}_0$ is the dose-equivalent rate per unit neutron fluence rate [i.e., (mrem/h)/(neutron/cm$^2 \cdot$ s) for neutrons of the source energy]. The buildup factor B is necessary to account for neutrons that are scattered back into the beam after they undergo an elastic or inelastic collision in the absorber. Scattered neutrons can contribute significantly to the dose outside a shield, and since Σ_{nr} is basically an attenuation coefficient (somewhat similar to photons), it is necessary to adjust calculated neutron intensities by a buildup factor. Buildup factors for neutrons are sparse; however, a neutron buildup factor of about 5.0 is appropriate for most calculations in which water or paraffin shields of 20 cm or more are used.

Neutron Removal Coefficients

Whereas photon attenuation and absorption coefficients vary smoothly with atomic number and energy, neutron removal coefficients can change irregularly from element to element because neutron cross sections have complicated resonance structures as their energies change. Neutron removal coefficients are further complicated by the amount of hydrogenous material incorporated in the shield to take advantage of its slowing down properties; therefore, Σ_{nr} is determined experimentally for each shield material as a combined removal coefficient. These are included in Table 14-6

TABLE 14-6. Neutron Removal Coefficients, Σ_{nr}, for Fission Neutrons in Several Common Materials

Medium	Σ_{nr} (cm^{-1})[a]
Sodium	0.032
Graphite	0.078
Carbon	0.084
Concrete (6% H_2O)	0.089
D_2O	0.092
Zirconium	0.101
H_2O	0.103
Paraffin	0.106
Polyethylene	0.111
Lead	0.118
Beryllium	0.132
Iron	0.156
Copper	0.167
Uranium	0.182
Tungsten	0.212

Source: USAEC.

[a]Materials are surrounded by sufficient hydrogenous material to absorb neutrons that are degraded in energy due to scattering interactions.

for shields with a thickness of less than five mean free paths and at least 6 g/cm^2 of hydrogenous material; also included are values for the substance alone (i.e., no hydrogen absorber).

The *macroscopic cross section*, Σ, is the neutron removal coefficient for materials or elements used without a layer of hydrogenous material. It is calculated as

$$\Sigma = N_i \sigma_i$$

where the microscopic cross section σ_i is obtained from the chart of the nuclides for thermal neutrons or plots (e.g., Figure 14-1 or Figures 14-3 to 14-6) for other energies. If a material such as lead, iron, or concrete is used by itself, the macroscopic cross section Σ should be used for determining the change in neutron intensity; if, however, the material is used with a sufficient amount of hydrogenous material, the neutron removal coefficients Σ_{nr} listed in Table 14-6 should be used.

The effectiveness of a neutron shield is dependent on Σ_{nr}; therefore, considerable care must be exercised in determining Σ_{nr} for shielding calculations. When shield materials meet the conditions for the values in Table 14-6, these should be used in shielding calculations. If, however, the shield material is used alone, the macroscopic cross section, calculated from the total microscopic cross section for the particular neutron energy, should be used.

Empirical relationship for mass neutron removal coefficients, Σ_{nr}/ρ (cm^2/g), have been derived (USAEC, 1973) as:

$$\Sigma_{nr}/\rho \ (\text{cm}^2/\text{g}) = \begin{cases} 0.190Z^{-0.43} & \text{for} \quad Z \le 8 \\ 0.125Z^{-0.565} & \text{for} \quad Z > 8 \end{cases}$$

or more generically,

$$\Sigma_{nr}/\rho \ (\text{cm}^2/\text{g}) = 0.206A^{-1/3}Z^{-0.294}$$

where A is the atomic mass number and Z is the atomic number of the element. The neutron removal coefficient, Σ_{nr}, is obtained by multiplying the neutron mass removal coefficient by the density of the absorber.

Lead plus polyethylene is a convenient shield that can be consructed in layers to use the mass attenuation of lead followed by the neutron removal properties of hydrogen-rich polyethylene. Neutron removal coefficients for lead plus a thickness t of polyethylene have been determined experimentally and can be represented by the fit equation,

$$\Sigma_{nr}(t) = 0.1106(1 - 0.9836e^{-0.109t})$$

where t is in centimeters of polyethylene. This relationship is good for up to 40 cm of polyethylene, and it is noted that Σ_{nr} is stated for lead without reference to its thickness.

Example 14-7: Determine the neutron removal cross section for a lead shield with a 20-cm-thick layer of polyethylene enclosing it.

SOLUTION:

$$\Sigma_{nr}(t = 20\ \text{cm}) = 0.1106(1 - 0.9836e^{(-0.109)(20)})$$
$$= 0.098\ \text{cm}^{-1}$$

Mixtures of nuclei may include natural or enriched abundances of isotopes of an element, or chemical mixtures such as B in H_2O. Neutron removal cross sections for such mixtures and compounds can also be obtained by first determining the effective microscopic cross section σ for the compound as shown in Examples 14-8 and 14-9.

Example 14-8: Determine Σ_{nr} for paraffin ($C_{22}H_{46}$; $\rho = 0.87$) if Σ_{nr} for carbon (A = 12.0) is 0.6 cm^{-1} and 0.61 cm^{-1} for hydrogen (A = 1.0).

SOLUTION: The molecular weight for paraffin is $(46 \times 1) + (22 \times 12) = 310$. The neutron removal coeffiecient is

$$\Sigma_{nr} = \left(\frac{46 \times 1}{310} \times 0.87 \times 0.61\right) + \left(\frac{22 \times 12}{310} \times 0.87 \times 0.6\right)$$
$$= 0.12\ \text{cm}^{-1}$$

Example 14-9: The scattering cross section for 1 MeV neutrons in H is 3 b and 8 b for O. Determine the scattering cross section for H_2O and the macroscopic scattering cross section.

SOLUTION: Since each water molecule contains two atoms of H and one of O,

$$\sigma_{s,\text{H}_2\text{O}} = (2)3\ \text{b} + (1)8\ \text{b} = 14\ \text{b}$$

and the macroscopic scattering cross section is

$$\Sigma_{s,\text{H}_2\text{O}} = N\sigma_t$$
$$= \frac{6.022 \times 10^{24}\ \text{mol/mol}}{18\ \text{g/mol}} \times 14 \times 10^{-24}\ \text{cm}^2$$
$$= 0.468\ \text{cm}^2/\text{g}$$

and since the density of H_2O is 1.0:

$$\Sigma_{s,\text{H}_2\text{O}} = 0.468\ \text{cm}^2/\text{g} \times 1\ \text{g/cm}^3$$
$$= 0.468\ \text{cm}^{-1}$$

These approaches assume that the different nuclei in a mixture act independently of each other such that the atoms of each represent independent targets for interactions. This is a valid assumption for all neutron interactions except elastic scattering of low-energy neutrons by molecules and solids; these must be obtained by experiment. Consequently, the results of Example 14-9 for H_2O would not be applicable for thermal neutrons since it yields a value of 46 b; the experimental, and correct, value is 103 b.

Example 14-10: Determine the dose rate for the Pu–Be source in Example 14-6 if shielded by 25 cm of water.

SOLUTION: A 25-cm water shield obviously contains sufficient hydrogen to assure absorption of scattered neutrons. The dose equivalent, including a buildup factor of 5, is determined with Σ_{nr} from Table 14-6 of 0.103 cm^{-1}, or

$$\dot{D}(25 \text{ cm H}_2\text{O}) = \dot{D}_0 B e^{-\Sigma_{nr}x} = 5 \times 37.2 e^{(-0.103)(25 \text{ cm})}$$

$$= 14.2 \text{ mrem/h}$$

Neutron Attenuation in Concrete

Since concrete is a widely used neutron shield, Clark et al. at ORNL performed Monte Carlo calculations of the dose rate outside concrete slabs of various thicknesses. Figure 14-9 shows the dose rate versus slab thickness for 2 MeV (typical

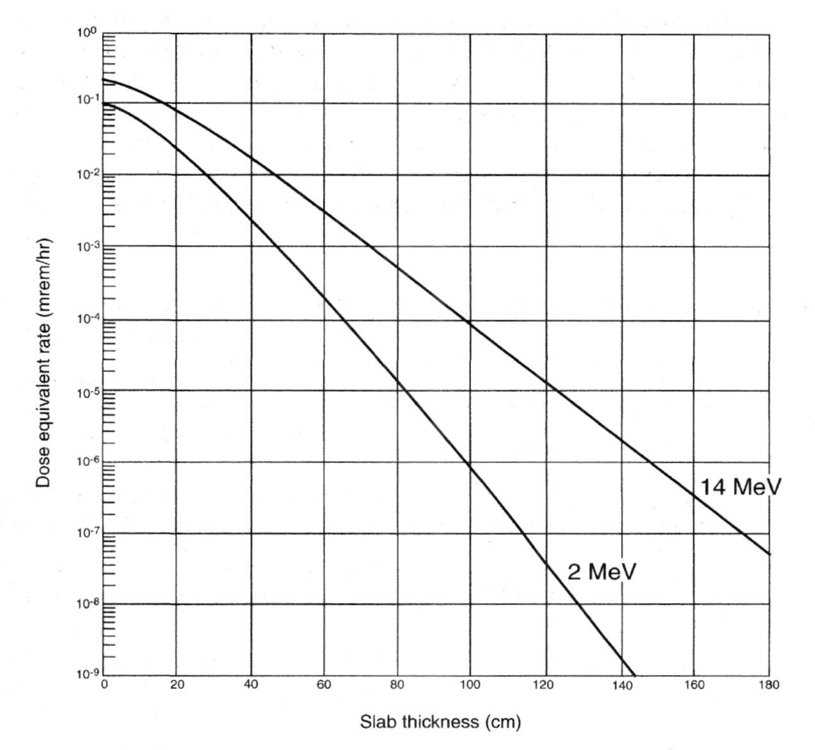

FIGURE 14-9. Dose equivalent rate (mrem/h) for unit intensity (n/cm$^2\cdot$s) for 2 MeV neutrons (lower curve) and 14 MeV neutrons (upper curve) versus thickness of concrete shielding. (Adapted from Clark et al., 1966, based on a quality factor of 10.)

of fission reactors) and 14 MeV neutrons (as produced in D-T interactions). These may be used for the design of neutron shields for these common sources; dose rates for other energies can be approximated by interpolation of these values or from the original reference.

Example 14-11: A flux of 10^3 fission neutrons/cm$^2 \cdot$ s with $\overline{E} \simeq 2$ MeV impinges on a concrete slab 60 cm thick. What is the dose equivalent rate on the exit side of the slab?

SOLUTION: From Figure 14-9, the dose rate after penetration of a 60-cm slab of concrete is 2×10^{-4} mrem/h per unit flux; therefore,

$$\dot{D} = 10^3 \text{ neutrons/cm}^2 \cdot \text{s} \times 2 \times 10^{-4} \text{ (mrem/h)/(neutron/cm}^2 \cdot \text{s)} = 0.2 \text{ mrem/h}$$

NEUTRON DETECTION

Since neutrons are not ionizing particles, their detection depends on the ionizing properties of the products of neutron capture reactions, the three most important of which are:

$$^{10}_{5}\text{B} + ^{1}_{0}\text{n} \begin{cases} \xrightarrow{4\%} \; ^{7}_{3}\text{Li} + ^{4}_{2}\text{He} + 2.79 \text{ MeV} & (\sigma_{\text{th}} = 3840 \text{ b}) \\ \xrightarrow{96\%} \; ^{7}_{3}\text{Li}^{*} + ^{4}_{2}\text{He} + 2.31 \text{ MeV} & \end{cases}$$

(* Excitation energy emitted as 0.48 MeV photon)

$$^{6}_{3}\text{Li} + ^{1}_{0}\text{n} \longrightarrow \; ^{3}_{1}\text{H} + ^{4}_{2}\text{He} + 4.78 \text{ MeV} \qquad (\sigma_{\text{th}} = 941 \text{ b})$$

$$^{3}_{2}\text{He} + ^{1}_{0}\text{n} \longrightarrow \; ^{3}_{1}\text{H} + ^{1}_{1}\text{H} + 0.76 \text{ MeV} \qquad (\sigma_{\text{th}} = 5330 \text{ b})$$

The alpha particles, protons, and recoil atoms produced by these reactions are energetic (moderate to high Q-values) and highly ionizing. These properties as well as the large capture cross sections for thermal neutrons in ^{10}B, ^{6}Li, and ^{3}He are used in the design of neutron detectors.

Measurement of Thermal Neutrons

Various measurement techniques can be used for thermal neutrons. Most of these depend on the thermal neutron cross section of selected materials, from which one can deduce the total fluence rate of thermal neutrons striking an absorber.

Boron-type detectors are used to detect thermal neutrons because of the high cross section for (n, α) reactions in ^{10}B. Boron trifluoride (BF_3) gas, often enriched in ^{10}B, can be used as a detector gas in a proportional counter to detect thermal neutrons. The pulse heights produced in the BF_3 counter are large due to the Q-value of 2.31 MeV (and 2.79 MeV in 4% of reactions that produce ground-state ^{7}Li). These large pulses allow discrimination against gamma rays, which are usually present with neutrons, that also may ionize the BF_3 gas but with much smaller pulses (multiple pulses from photon interactions in the BF_3 counter can become a problem if the gamma fields are intense).

Other neutron detectors are designed by lining the interior walls of the detector wall with a boron compound and filling it with a gas that may be more suitable for proportional counting than BF_3. Boron can also be impregnated in scintillators (e.g., ZnS) for slow-neutron detection. Each of these methods also relies on the high cross section of boron and the ionization produced by the recoil products.

Lithium detectors incorporate lithium ($\sigma_a = 941$ b for ^{6}Li) to detect slow neutrons. The $^6\text{Li}(n,\alpha)^3\text{H}$ reaction has a higher Q-value than the (n,α) reaction in boron, which provides potentially better gamma-ray discrimination, but at reduced sensitivity due to the lower cross section. The isotope ^{6}Li is only 7.5% abundant in nature, but lithium enriched in ^{6}Li is available. Lithium-6 fabricated into crystals of LiI(Eu) is similar to NaI(Tl) for photon detection, but gamma-ray discrimination in such crystal detectors is poor compared to BF_3 gas. Lithium compounds can also be mixed with ZnS to make small detectors which have good gamma-ray discrimination because secondary electrons produced by gamma rays easily escape the crystal without interacting to produce a detector pulse.

Helium-filled detectors take advantage of the high cross section for (n, p) reactions in ^{3}He at 5330 b. A ^{3}He proportional counter will lose pulses because of the low energies supplied to the products by the low Q-value; however, ^{3}He is a better counter gas and can be operated at high pressures with good detection efficiency. Gamma-ray discrimination is poor because the lower-energy protons produce smaller pulses due to the low Q-value.

Fission counters use fissile material coated onto the inner surface of an ionization chamber. Isotopes of ^{233}U, ^{235}U, or ^{239}Pu fission readily when exposed to thermal neutrons and produce very energetic fission fragments since they share about 165 MeV of kinetic energy. The fission explosion literally tears the fragments away from their orbital electron fields, and each fragment has a charge of about +20, which produces extremely large pulses; thus slow-neutron counting can be done at low levels, even in a high background. Since fissile materials are also alpha emitters, alpha-particle pulses will also be produced, but these are considerably smaller and can easily be discriminated against. Since the fission fragments are highly charged and very energetic, sufficient ionization occurs for fission counters to be operated as ion chambers containing standard air.

Measurement of Intermediate and Fast Neutrons

Two general methods of measuring intermediate and fast neutrons are first to moderate them to thermal energies so they can be detected with a thermal neutron detector and to use an assortment of activation foils to detect them directly. The two principal measurement systems that moderate fast neutrons are the long counter and Bonner spheres.

The *long counter* (Figure 14-10) contains a BF_3 tube surrounded by an inner paraffin moderator; it responds uniformly to neutron energies from about 10 keV to 5 MeV. Neutrons incident on the front cause a direct response after being thermalized in the inner paraffin layer. Those from other directions are either reflected

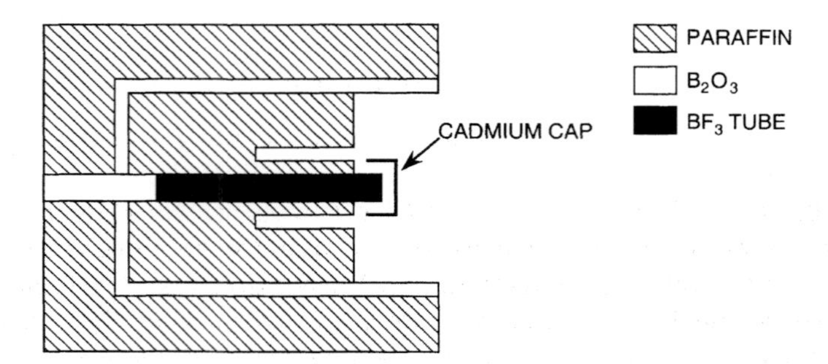

FIGURE 14-10. Schematic of the long counter, in which a BF_3 tube is surrounded by layers of paraffin and metal to reflect and thermalize fast neutrons so they will be absorbed by (n, α) reactions in the boron trifluoride (BF_3) gas.

or thermalized by the outer paraffin jacket and then absorbed in a layer of boron oxide. With this arrangement, the probability that a moderated neutron will enter the BF_3 tube and be counted is not dependent on the initial energy; therefore, no information on the spectral distribution of neutron energies is obtained.

Bonner-sphere detectors consist of a series of polyethylene moderating spheres of different diameters that enclose small lithium iodide scintillators at their centers. Bonner spheres can be used to obtain neutron spectral information. The spheres range in diameter from 2 to 12 in., and each provides varying degrees of moderation for neutrons of different energies. Each sphere is calibrated with monoenergetic neutrons ranging in energy from thermal neutrons to 10 MeV or so. An unfolding procedure is used to infer information about the neutron spectrum based on the calibration curves and the measured count rates, but the unfolding procedure is not very precise.

A Bonner-sphere detector is very useful for neutron dose measurements because, as shown in Figure 14-11, the detector response inside a 10 to 12 in. polyethylene sphere is very close to the dose equivalent delivered by neutrons at any given energy. The detector is also totally insensitive to gamma radiation, which is a distinct advantage. Since the energy response is similar to the dose equivalent per neutron, the instrument is often called a *rem ball* and is often used as a neutron rem meter. The correspondence between dose equivalent and neutron energy for the rem ball appears to be a coincidence rather than a distinct physical correlation, but a fortunate one, nevertheless, for neutron fields of unknown energy. As shown in Figure 14-11, the response of the 10-in. Bonner sphere with a LiI scintillation detector is high for high-energy neutrons (with a high quality factor) and lower for the lower-energy neutrons (with corresponding lower quality factor), which causes the counts to be weighted automatically as neutron energy changes. The detector overresponds up to a factor of about 2 (greatest for 10 keV neutrons) for neutron energies below about 100 keV, which is conservative for radiation protection decisions. On the other hand, for neutrons above 8 MeV or so it underresponds sometimes by a considerable amount, and caution must be exercised when using Bonner spheres for high-energy neutrons. Larger spheres and unfolding procedures to determine neutron fluence and energy spectra are needed if accurate dose determinations are to be made for neutrons above 5 MeV or so.

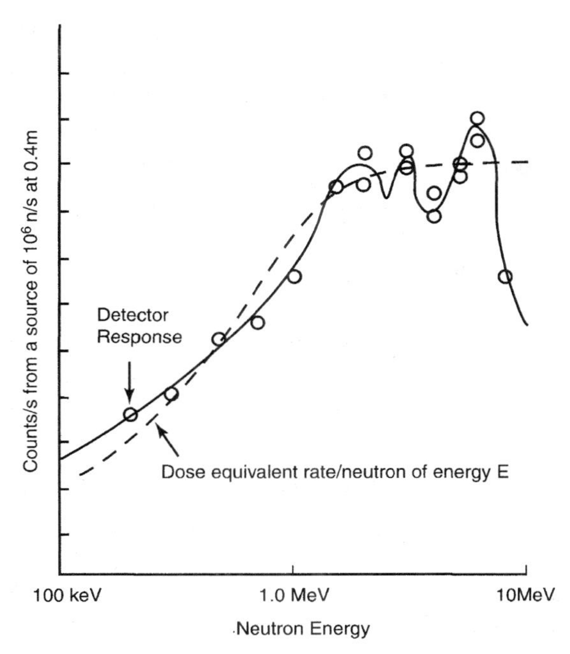

FIGURE 14-11. Response (arbitrary units of c/s at 40 cm) of a 10-cm Bonner-sphere rem-ball detector to fast neutrons and the corresponding dose equivalent per neutron versus neutron energy, showing a relatively close relationship for neutron energies up to about 5 MeV and substantial underresponse ≥ 8 MeV.

Neutron Foils

Neutron foil activation is one of the best methods for determining neutron spectral information because different foils have differing threshold energies for activation, as shown in Table 14-7. For example, if an exposed indium foil shows induced activity from ^{115m}In and a ^{58}Ni foil exposed simultaneously shows no ^{58}Co activity, the neutron energies are likely to be between 0.5 and 1.9 MeV. They can also be used to capture neutrons and induce radioactivity for analysis after an actual exposure or if one is suspected. These form the basis of personal neutron dosimeters or stationary monitors for areas where neutron exposure or nuclear criticality could occur. The amount of induced activity will depend on a number of factors—the element chosen, the mass of the foil, the neutron energy spectrum, the capture cross section, and the time of irradiation. Mn, Co, Cu, Ag, In, Dy, and Au are examples of foils used to detect thermal neutrons.

An effective and somewhat clever method of determining the fraction of thermal neutrons in a beam is to take advantage of the high neutron absorption property of cadmium. As shown in Figure 12-1, Cd effectively removes all neutrons of energy below about 0.4 eV and is completely transparent to neutrons above about 1 eV. If a foil such as indium or gold is irradiated bare and an identical one is wrapped in a Cd foil 0.1 to 0.2 in. thick and irradiated under the same conditions, the activity of each is a measure of the number of thermal and fast neutrons in the beam. The activity of the bare foil is due to thermal plus fast neutrons while the activity of the

TABLE 14-7. Threshold Activation Energies for Neutron Foils

Reaction	$T_{1/2}$	Threshold Energy (MeV)
$^{115}In(n,n,\gamma)^{115m}In$	4.49 h	0.5
$^{54}Fe(n,p)^{56}Mn$	312 d	2.2
$^{58}Ni(n,p)^{58}Co$	70.9 d	2.9
$^{27}Al(n,p)^{27}Mg$	9.45 m	3.8
$^{64}Zn(n,p)^{64}Cu$	12.8 h	4.0
$^{56}Fe(n,p)^{56}Mn$	2.58 d	4.9
$^{59}Co(n,\alpha)^{56}Mn$	2.58 d	5.2
$^{24}Mg(n,p)^{24}Na$	15 h	6.0
$^{27}Al(n,\alpha)^{27}Mg$	15 h	8.1
$^{197}Au(n,2n)^{196}Au$	6.18 d	8.6
$^{19}F(n,2n)^{18}F$	110 m	11.6
$^{58}Ni(n,2n)^{57}Ni$	36 h	13.0

cadmium-covered foil is due only to fast neutrons, since essentially all the thermal neutrons will be absorbed by the cadmium. The ratio of thermal to fast neutrons is given by the cadmium ratio

$$\text{Cd ratio} = \frac{\text{Activity(bare)}}{\text{Activity(with Cd)}}$$

This technique provides a rough division between thermal and fast neutrons since σ_a falls off as $1/v$ for most foils (i.e., the induced activity from fast neutrons, which is a function of σ_a, will decrease as the neutron energy increases). For gold, $\sigma_a = 98$ b for thermal neutrons and the $1/v$ relation holds up to about 0.1 eV, but due to strong resonance absorption of about 40,000 b at 5 eV, a non-$1/v$ correction factor needs to be applied when gold foils are used.

Fission foils can also be used for fission neutron fields such as may be encountered around nuclear reactors. The basic techniques for use of these foils was developed by Hurst, Mills, et al. in the early 1950s. Foils of ^{237}Np, ^{238}U, and ^{239}Pu are generally used. ^{237}Np has a fission threshold at about 0.4 MeV and ^{238}U at about 1.4 MeV; a ^{239}Pu foil encased in a spherical shell of boron with density thickness of 1 to 2 g/cm^2 has a pseudothreshold of about 0.01 MeV even though it is fissionable at all energies and especially so at thermal energies ($\sigma_f \sim 740$ b). Selective counting of a gamma-emitting fission product such as ^{140}Ba–La or ^{65}Zn for each foil provides measurements that can be used to determine the relative distribution of neutron fluences between 0.01 and 0.4 MeV, 0.4 and 1.4 MeV, and above 1.4 MeV. Unfortunately, gram quantities of strictly regulated fissionable material are required, and if used as a personal dosimeter to measure low neutron exposures, fission foil dosimeters can be somewhat heavy.

Carbon activation can also be used as a threshold detector for neutrons with energies above 20 MeV. The $^{12}C(n,2n)^{11}C$ reaction has a threshold at about 20 MeV, and the cross section for the reaction is essentially constant from 20 to 40 MeV at 20 mb. The carbon detector can be shaped around a NaI(Tl) detector to provide good geometry for counting (e.g., in the form of a cylinder) or constructed in a disc shape that can be placed on the detector to measure the ^{11}C activation produced. Since ^{11}C transforms by positron emission, the 0.511 MeV annihilation photons are easily measured. The geometry and efficiency can be determined by calibrating the system with a standard of a positron emitter such as ^{22}Na, or if an accelerator is available, the positron emitter can be produced locally (e.g., ^{18}F or ^{11}C itself).

Albedo Dosimeters

Thermoluminescent crystals made of lithium fluoride (LiF) can be used to measure fast neutrons indirectly by using the body to moderate and reflect them to the LiF crystal, which is sensitive to thermal neutrons. Fast neutrons that strike the human body (or a phantom can be used) undergo a series of elastic collisions until they reach thermal energies. A fraction of these are in turn diffused out of the body and are captured by ^{6}Li in the LiF dosimeter crystal with a (n, α) cross section of 941 b. It is known as an *albedo* (Greek; meaning "reflection") dosimeter because it depends on the body to "reflect" neutrons back to the LiF crystal; therefore, the dosimeter must be worn on the surface of the body and it must be calibrated on a slab of tissue-equivalent material. The albedo factor is the fraction of incident neutrons reflected back to the dosimeter; it decreases gradually from about 0.8 for thermal neutrons to about 0.1 for 5 MeV neutrons. Intermediate values of the albedo factor are 0.5 at 1 eV, 0.4 at 10 eV, 0.3 at 1 keV, and 0.2 at 100 keV.

Flux Depression of Neutrons

It is often important to determine if a neutron flux is significantly altered as it passes through an absorber, be it a sample, a detector, or perhaps a person. Such alteration constitutes flux depression. It is calculated as a ratio of the attenuated neutron flux to that before attenuation occurs, ϕ_0, via the relationship

$$\phi(x) = \phi_0 e^{-\Sigma x}$$

where Σ is the macroscopic cross section for a medium and x is its thickness.

Example 14-12: Determine the flux depression for thermal neutrons in a BF_3 long counter that is 30 cm long and 5 cm in diameter if the boron content is essentially all ^{10}B.

SOLUTION: The macroscopic cross section Σ for the BF_3 gas is, neglecting the minor contribution of F,

$$\Sigma = \frac{0.6022 \times 10^{24} \text{ atoms/mol}}{22{,}400 \text{ cm}^3\text{/mol}} \times 3840 \times 10^{-24} \text{ cm}^2\text{/atom}$$

$$= 0.1032 \text{ cm}^{-1}$$

The fractional flux depression for the diameter of the counter is

$$\frac{\phi(5\ \text{cm})}{\phi_0} = e^{(-0.1032\ \text{cm}^{-1})(5\ \text{cm})} = 0.597$$

and for the length of the counter

$$\frac{\phi(30\ \text{cm})}{\phi_0} = e^{(-0.1032\ \text{cm}^{-1})(30\ \text{cm})} = 0.045$$

Therefore, only 59.7% of the flux penetrates across the detector and only 4.5% the length of the detector, and flux depression is significant.

SUMMARY

The two most important properties of neutrons relative to radiation protection are the probability of interaction in a medium (denoted by the cross section) and the energy transferred to or deposited in the medium. Most neutron sources produce energetic neutrons that are classified as fast because of their velocities at these energies; however, they quickly undergo various interactions with media and their energy is degraded to intermediate and thermal energies, the other two major classifications. Such sources are readily characterized in terms of a fluence of neutrons per unit area (neutrons/cm^2) or a fluence rate or flux (neutrons/cm$^2 \cdot$ s).

The intensity of a neutron beam will be diminished when it strikes a receptor of interest, which may be an absorber to change or attenuate the beam, a shielding material, a detector, or a person that may receive a proportionate dose. Three types of interactions are important: elastic scattering, inelastic scattering, and capture. Scattering interactions, whether elastic or inelastic, reduce the neutron energy and transfer the energy lost to the absorbing medium, where it is deposited by recoiling nuclei. Capture interactions yield new products that may emit capture gamma rays, eject a charged particle from the compound nucleus formed, or produce radioactive atoms that emit various forms of radiation as they undergo radioactive transformation after formation. All of these processes require consideration in determinations of neutron dose, design of shields, and in methods of detection.

REFERENCES AND ADDITIONAL RESOURCES

Attix, F. H., *Introduction to Radiological Physics and Radiation Dosimetry*, Wiley, New York, 1986, Chapter 16.

Attix, F. H., and W. C. Roesch, eds., *Radiation Dosimetry*, Vol. 1, Academic Press, San Diego, CA, 1968, Chapter 6.

Clark, F. H., N. A. Betz, and J. Brown, *Monte Carlo Calculations of the Penetration of Normally Incident Neutron Beams Through Concrete*, Report ORNL-3926, Oak Ridge National Laboratory, Oak Ridge, TN, 1966.

Hurst, G. S., J. A. Harter, P. N. Hemsley, W. A. Mills, M. Slater, and P. W. Reinhardt, The neutron long counter, *Rev. Sci. Instr.* **27**, (1956), 153.

International Commission on Radiological Protection, ICRP Publication 23, Report of the Task Group on Reference Man, Pergamon, Oxford, 1975.

Lamarsh, J. R., *Introduction to Nuclear Engineering*, 2nd ed., Addison-Wesley, Reading, MA, 1983.

NCRP, Protection Against Neutron Radiation, Report 38, National Council on Radiation Protection, Washington, DC, 1971.

NCRP, Calibration of Survey Instruments Used in Radiation Protection for Assessment of Radiation Fields and Radioactive Surface Contamination, Report 112, National Council on Radiation Protection, Washington, DC, 1987.

ACKNOWLEDGMENTS

This chapter came about because of the patient and skillful work of Chul Lee, a graduate of the University of Michigan Radiological Health Program.

PROBLEMS

14-1. Estimate the dose equivalent rate at a distance of 120 cm from a ^{210}Po–B (average energy of 2.5 MeV) point source that emits 2×10^7 neutrons/s and is shielded by 30 cm of water.

14-2. What is the maximum number of neutrons/s that a ^{210}Po–Be (average energy of 4 MeV) point source can emit if it is to be stored behind 65 cm of paraffin and the dose equivalent rate is not to exceed 2 mrem/h at a distance of 1 m?

14-3. The cross section for the ^{6}Li(n, α)^{3}H reaction is 941 b for thermal neutrons and ^{6}Li is a $1/v$ absorber. If the average neutron energy in a zone of the K reactor is 5 eV, what is the absorption cross section for ^{6}Li targets placed in the zone?

14-4. Calculate (**a**) the amount of energy transferred to carbon atoms in a graphite moderator due to elastic scattering of neutrons with an average energy of about 2 MeV, and (**b**) the fraction of the initial energy transferred.

14-5. If 5 cm of lead is added to the inside of a concrete shield, what fractional increase in removal of fast-fission neutrons will occur?

14-6. Determine the tissue dose to a person exposed to a fluence of 10^7 thermal neutrons due to (n, p) absorption reactions in nitrogen.

14-7. Determine the dose equivalent rate for neutrons emitted from a point source of ^{252}Cf if the source stength is 10 mCi, the fraction of transformations that occur by spontaneous fission is 0.029, 2.45 neutrons are emitted per fission, and the receptor is positioned at 80 cm from the source.

14-8. A neutron source of intermediate energy produces an intensity of 10^7 neutrons/cm$^2 \cdot$ s. If a lead shield 2 cm thick is placed in the beam followed by 30 cm of water, what would the intensity be just outside the water barrier?

14-9. Calculate the dose equivalent rate for the unshielded and shielded conditions in Problem 14-8.

14-10. The fluence rate of 14.1 MeV neutrons several feet from a D-T generator is 10^3 neutrons/cm$^2 \cdot$ s. If a 60-cm slab of concrete is used to shield the beam, what would be the dose equivalent rate just outside the slab?

14-11. A carbon disk is exposed to a flux of neutrons around a high-energy accelerator for 20 min. It is counted immediately after exposure and 1000 0.511 MeV gammas are recorded. If the system geometry/efficiency is 4%, what is the flux of neutrons with energies above 20 MeV?

15

X RAYS

I have discovered something very interesting.
—W. C. Roentgen, November 8, 1895

The production, use, and control of x rays is an important aspect of radiation protection. When used in medicine, the medical goal as well as the protection goal is to optimize the parameters of purposeful exposures to obtain the greatest benefit (i.e., diagnostic information or treatment) while avoiding unnecessary radiation exposure of patients and workers. Fortunately, these are compatible goals: the sharpest and clearest x-ray pictures are obtained when those features that optimize exposure are also used. It is, unfortunately, also possible to obtain acceptable x-ray images with unnecessary exposures of patients and, even worse, to obtain x-ray images that do not contain the requisite information. Both circumstances impose unnecessary radiation and can be especially harmful when a patient does not receive essential medical information.

As shown in Figure 15-1, there are four major interconnected categories of physical principles in an x-ray system designed to optimize image quality and patient exposure:

1. The x-ray tube and associated filters and collimators which are designed and operated to direct a beam of x-ray photons of sufficient quality and quantity at the patient to image the body part under study
2. Interactions of the x rays that strike the patient to yield a pattern of absorption and attenuation that will produce information once captured and made visible
3. The image capture system consisting of an air gap or grid to reduce scattered photons, intensifying phosphors which capture x-ray photons and produce visible light, and a recording system (film or electronic) to capture the pattern of photons transmitted through the patient to produce a diagnostic image
4. Processing of the interactions produced by the captured photons to make the information they contain visible either by electronic means or by film development

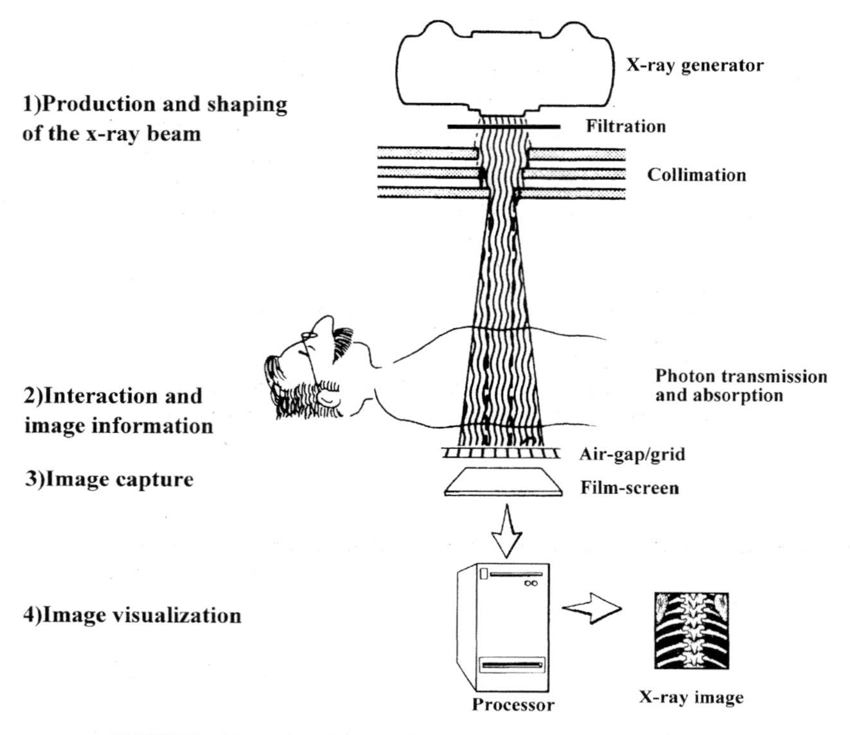

FIGURE 15-1. Four major components of an x-ray system.

The purpose of this chapter is to describe these four components of the x-ray system and how they are interrelated to optimize image quality and patient exposure. In so doing, we build upon the information on atomic physics and photon interactions presented in Chapters 4 and 7, respectively. As noted in earlier chapters, these considerations were set in motion soon after Roentgen's discovery of x rays on November 8, 1895. It is perhaps worth noting that Goodspeed at the University of Pennsylvania in 1890, some five years before Roentgen's discovery, accidentally recorded an x-ray image of two coins on a photographic plate lying near a Crooke's tube, but he failed to deduce the cause and missed one of the greatest discoveries in human history. Roengten followed up on his interesting observation, whereas Goodspeed (and probably various others) did not. Such is the nature of discovery!

PRODUCING AND SHAPING THE X-RAY BEAM

The production of x rays makes use of three properties of the atoms in the target of the x-ray tube: the electric field of the nucleus; the binding energy of orbital electrons; and the propensity for atoms to exist in the lowest energy states possible. X rays are produced when high-velocity electrons undergo accelerations as they are stopped in the target anode of an x-ray tube. These electrons are boiled off a tungsten filament centered in the cathode (negative electrode). By placing a high

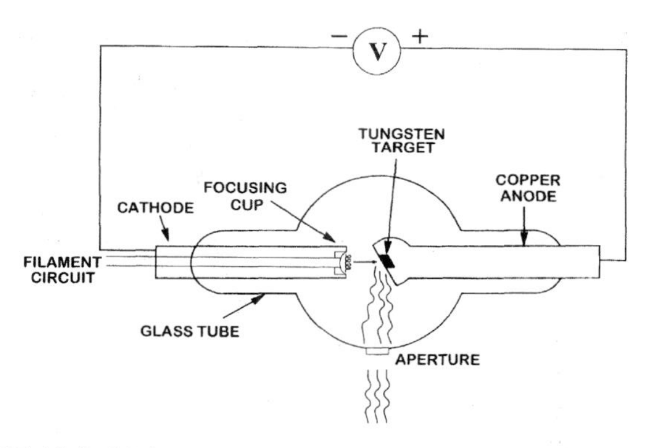

FIGURE 15-2. Major components of a stationary anode x-ray tube.

voltage across the tube, the electrons are accelerated toward the anode, which is commonly made of tungsten. Figure 15-2 illustrates this process in an x-ray tube that has a stationary anode. Stationary anodes consist of a small plate of tungsten that is 2 or 3 mm thick and 1 cm or more in length and width. This plate is angled at 15 to 20° to focus the x rays produced in it and is embedded in a large mass of copper to dissipate the intense heat produced by stopping accelerated electrons, most of which do not produce x rays. The high atomic number of tungsten (74) is efficient for the production of x rays and its high melting point (3370°C) is resistant to heat damage. A single x-ray exposure may raise the temperature of the bombarded area of the tungsten target by 1000°C or more, and since copper melts at 1070°C, the tungsten target needs to be large enough to allow for some cooling around the edges.

An x-ray generator supplies electrical energy to the x-ray tube and regulates the length of the radiographic exposure. The x-ray tube requires two sources of energy, one to heat the filament and the other to accelerate electrons between the cathode and the anode. The filament circuit contains a variable resistance and a step-down transformer; this circuit serves as the current selector (mA) and when set for a given exposure time(s) is the mA · s setting that determines the number of x-ray photons produced. The cathode–anode circuit, called the high-voltage circuit, contains an autotransformer and a step-up transformer. The autotransformer serves as the kVp selector, which determines the energy of the x-ray photons produced. X-ray generators are run by alternating current (ac) which is rectified (usually by silicon diodes) to transmit current in only one direction.

X rays are generated by the production of bremsstrahlung and characteristic x-radiation. Bremsstrahlung, braking radiation, occurs from interactions of the electrons with the nucleus of the target atoms. Most electrons hitting a target will undergo several interactions with target atoms before they are stopped; thus, only part of the electron energy appears in the form of radiation each time it is braked. This creates a wide spectrum of emitted x-ray energies. Occasionally, an electron will collide head-on with a target atom, converting all its energy into a single x-ray photon. If all the energy of the electron is given to the resulting x-ray photon, it will have a maximum energy equal to that of the electron or $h\nu = eV$. This photon will have a minimum wavelength of

$$E = eV = h\nu_{\max} = \frac{hc}{\lambda_{\min}}$$

or

$$\lambda_{\min} = \frac{1.24 \times 10^{-6}}{eV(\text{volts})}\ \text{m} = \frac{1.24 \times 10^{-9}}{\text{keV}}\ \text{m}$$

Example 15-1: What is the minimum wavelength of x rays produced in an x-ray tube operated at a peak voltage of 100 kV?

SOLUTION: The maximum energy (eV) that an electron can acquire is 100 keV, and the minimum wavelength will be

$$\lambda_{\min} = \frac{1.24 \times 10^{-9}}{100\ \text{keV}} = 1.24 \times 10^{-11}\ \text{m} = 0.0124\ \text{nm}$$

Most of the x rays produced in Example 15-1 will have wavelengths longer than 0.0124 nm, and in fact, will produce the spectrum shown in Figure 15-3. Over 99% of the energy in the incident electrons will appear as heat energy since x-ray production is an inefficient process, which may account for them not being discovered earlier. Filtration by the glass wall of the tube removes some of the low-energy x rays from the beam; other filters may be inserted in the beam to further modify the energy spectrum.

Characteristic x rays are introduced into the spectrum of emitted x rays when the high-speed electrons have sufficient energy to dislodge orbital electrons from the

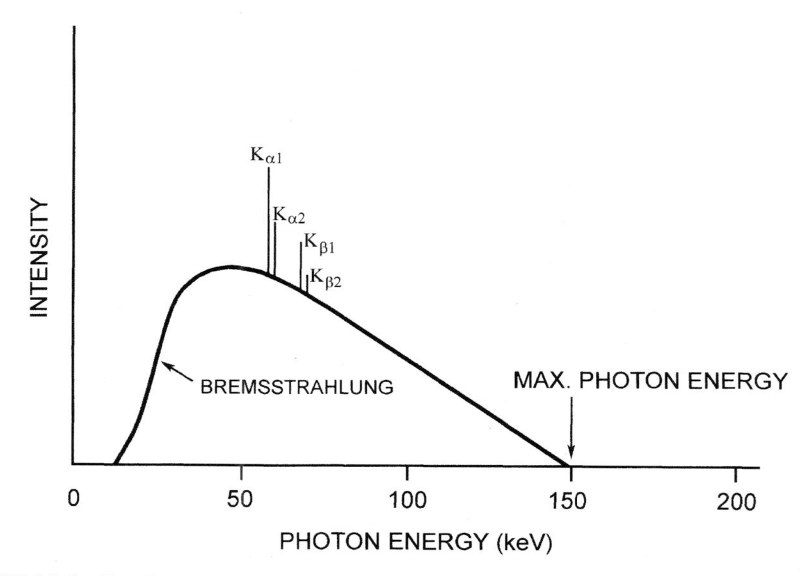

FIGURE 15-3. Continuous spectrum of x-ray energies with $E_{\max}$ corresponding to a maximum x-ray energy of 150 keV across the x-ray tube.

target material. Such collisions create shell vacancies that are filled by less tightly bound electrons, and in order to reach the lower-energy state of the vacated shell, these electrons must emit electromagnetic radiation. This radiation is characteristic of the target atoms, hence the term *characteristic x rays*. The spikes in the spectrum shown in Figure 15-3 are the K-characteristic x rays of tungsten and are superimposed on the continuous spectrum of bremsstrahlung. The $K_{\alpha 1}$ (59.3 keV) and $K_{\alpha 2}$ (57.9 keV) characteristic x rays occur when L-shell electrons from the L_{III} and L_{II} subshells, respectively, fill K-shell vacancies. The $K_{\beta 1}$ (keV) represents an M shell-to-K shell transition, and the $K_{\beta 2}$ (69 keV) x rays in tungsten occur from N-shell transitions. K-characteristic x rays are produced when the tube voltage is above 69 kVp in tungsten targets; at 80 kVp characteristic x rays represent about 10% of the useful x-ray beam and about 28% at 150 kVp. The relative contribution of characteristic radiation in tungsten decreases significantly above 150 kVp and is a negligible fraction above about 300 kVp.

X-Ray Yield: Bremsstrahlung Production Fraction

The production and yield of bremsstrahlung is complex. Whether a given electron converts part (or all) of its energy to radiation emission depends on the path of the electron toward a target nucleus and the degree of deflection that occurs. The deflecting force is also dependent on the nuclear charge of the target material. The incoming path of the electron establishes the separation distance and hence the magnitude of the coulombic force of attraction between the positively charged target nucleus and the negatively charged electron. Such deflection, if it occurs, produces radial acceleration, which in turn results in the emission of bremsstrahlung.

The yield or fraction of bremsstrahlung produced is proportional to the atomic number of the target (or absorbing) material and the energy of the electrons striking the target, which of course decreases rapidly as the electrons traverse the target material. Describing the process analytically is difficult; consequently, the best data are those obtained by experiment even though various models have been constructed to fit empirical data. One such expression for the fraction of the incident electron energy that is converted to radiative energy is

$$Y \simeq \frac{6 \times 10^{-4} ZE}{1 + 6 \times 10^{-4} ZE}$$

where Y is the fraction of the electron energy E in MeV that is converted to bremsstrahlung by a medium of atomic number Z. In the absence of more specific data, this expression provides a useful relationship for radiation protection of the fraction of the electron energy that is converted to photons that must be accounted for in dosimetry or radiation shielding.

Berger and Seltzer (1983) have determined the fraction of the initial electron energy that is converted to radiation, results of which are listed in Table 15-1 for monoenergetic electrons on several absorbers, some of which are commonly used in x-ray generators. These data, which are plotted in Figure 15-4, illustrate that the radiation yield increases with electron energy and the Z of the absorber. Berger and Seltzer also demonstrated that the photons produced by a beam of monoenergetic

TABLE 15-1. Percent Radiation Yield for Electrons of Initial Energy *E* on Different Absorbers

	Absorber, *Z*						
E (keV)	Air	Tissue	Cu (29)	Mo (42)[a]	Sn (50)	W (74)	Pb (82)
20	0.019	0.013	0.090	0.128	0.152	0.220	0.243
30	0.026	0.018	0.130	0.188	0.224	0.330	0.366
40	0.033	0.023	0.167	0.245	0.293	0.438	0.487
50	0.039	0.027	0.203	0.300	0.360	0.543	0.606
55	0.042	0.029	0.219	0.326	0.392	0.594	0.664
60	0.045	0.031	0.236	0.352	0.423	0.645	0.721
70	0.050	0.035	0.267	0.402	0.485	0.745	0.835
80	0.056	0.039	0.298	0.451	0.545	0.843	0.946
90	0.061	0.043	0.327	0.497	0.602	0.939	1.055
100	0.066	0.046	0.355	0.543	0.658	1.032	1.162
125	0.078	0.055	0.421	0.651	0.792	1.257	1.410
150	0.090	0.063	0.482	0.751	0.917	1.470	1.664
200	0.111	0.078	0.595	0.937	1.148	1.865	2.118
300	0.150	0.106	0.795	1.261	1.548	2.558	2.917

[a]By interpolation.

electrons on a target is peaked in the direction of the beam; therefore, determining the fluence or flux of photons produced by an electron source (e.g., an x-ray machine or accelerator) can be problematic. If, however, the photon flux or fluence can be estimated, it can be used in the usual way in radiation-protection determinations.

Example 15-2: What is the energy fluence at 1 m from 10^8 electrons of 100 keV striking a tungsten target, assuming that the bremsstahlung produced is radiated isotropically?

SOLUTION: From Figure 15-4 and Table 15-1, the fractional photon yield from 100 keV electrons on a tungsten target is determined to be 0.01032. (By comparison, the photon yield calculated from the empirical fit formula is 0.044.) The maximum total energy from 10^8 photons is $10^8 \times 0.1$ MeV = 10^7 MeV. Multiplication by the yield gives an energy conversion to bremsstrahlung of 1.032×10^5 MeV. The energy fluence at 1 m for 7×10^5 MeV of photon energy is

$$\text{photon fluence} = \frac{1.032 \times 10^5 \text{ MeV}}{4\pi(100 \text{ cm})^2} = 0.82 \text{ MeV/cm}^2$$

For assessing the radiation hazard, one would conservatively assume that each photon has the maximum energy of 100 keV; however, it is probably more reasonable from the shape of the bremsstrahlung spectrum to use an average energy of about 33 keV. A further adjustment of the photon fluence could also be made to account for forward peaking of the bremsstrahlung.

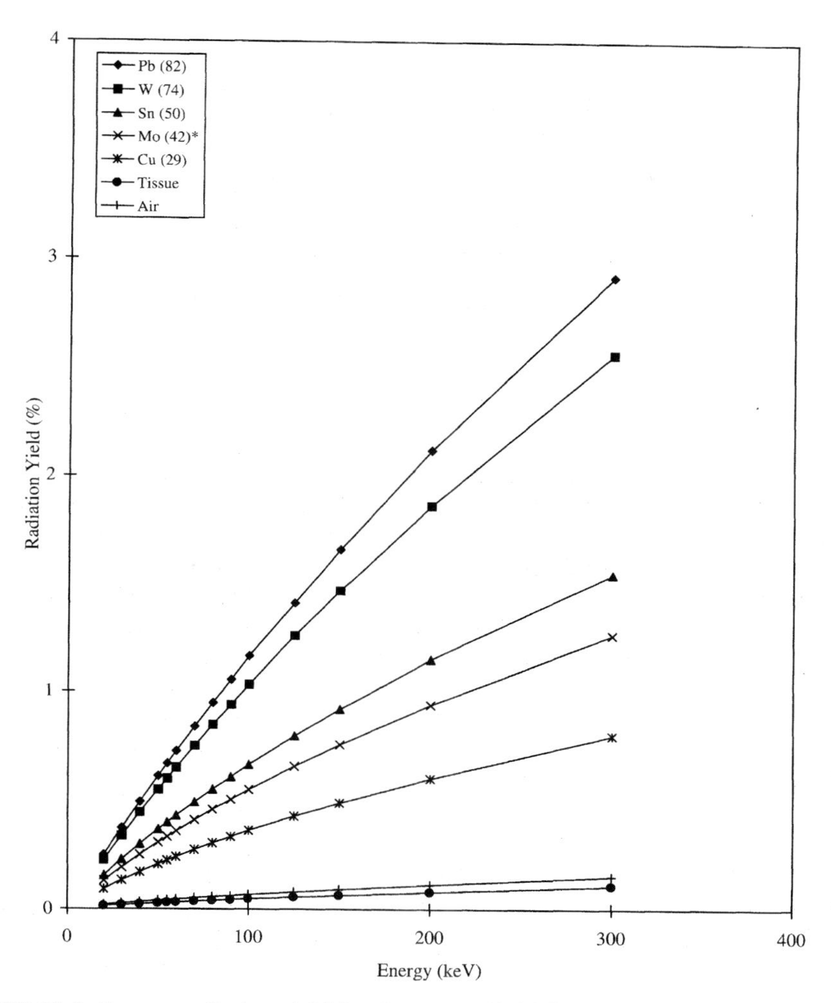

FIGURE 15-4. Percent radiation yield for electrons of initial energy E on different absorbers.

X-Ray Tubes

X-ray tubes contain an anode and a cathode sealed in a vacuum to preclude interactions that would reduce the number or speed of accelerated electrons striking the anode target. Interactions with air or other gas molecules would yield secondary electrons, which would not only reduce the speed of the accelerated electrons but would result in wide variations in the number striking the target and the energy of the x rays produced. Vacuums are not easy to maintain; consequently, any connecting wires must be sealed into the wall of the glass x-ray tube and special care must be taken to prevent breaking the vacuum seal by differential expansion of tube components due to high temperatures. The latter is usually achieved by use of special alloys in tube components that provide the same degree of expansion during heating and cooling of the tube.

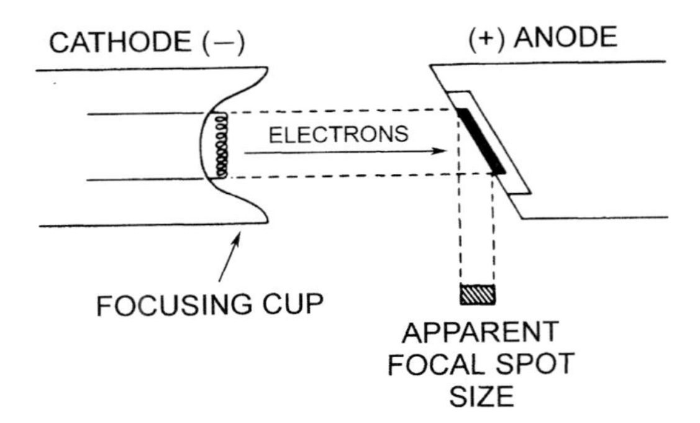

FIGURE 15-5. Use of a focusing cup to achieve a thin electron beam and an angled anode to achieve a small apparent focal spot.

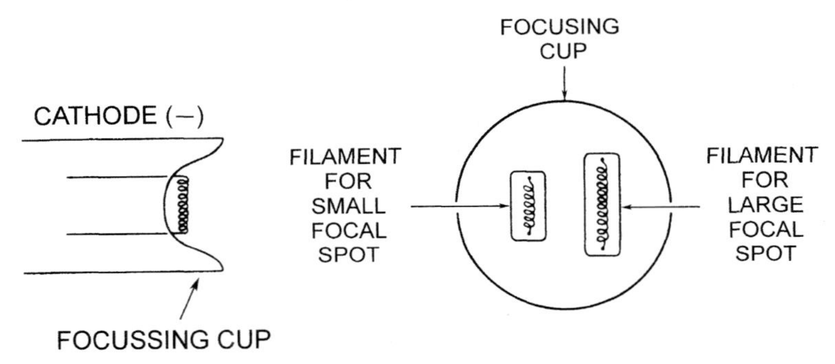

FIGURE 15-6. Typical focusing cup arrangement with a double filament in the cathode of an x-ray tube.

The cathode, or negative terminal of the x-ray tube, is constructed as a metallic focusing cup. It also contains the filament, which is heated by about 10 V and 3 to 5 A to produce electrons. The cathode focusing cup is usually made of nickel and is shaped around the filament (Figures 15-2 and 15-5) in such a way as to produce electrical forces which direct the electrons from the filament toward the target in the required size and shape. Double filaments mounted side by side in the focusing cup are common (Figure 15-6); the shorter one produces a small focal spot (see below) and the longer one is generally used for larger exposures, which require a larger target area (i.e., focal spot) to dissipate the heat. The filament is made of tungsten wire that is about 0.2 mm in diameter and coiled to form a spiral about 0.2 cm in diameter and 1 cm or less in length. It is heated to about 2200°C, and when so heated emits a useful number of electrons through a process known as the *Edison effect*. Although more efficient emitters are available, tungsten is used because it can be extruded into thin wires, has a high melting point (3370°C), and lasts a long time with minimal vaporization.

Overheating the filament is not good because tungsten vaporized from the filament can produce a thin coating on the inner surface of the x-ray tube, where it acts to attenuate the x rays emitted, thus changing the quality of the beam. This effect is reduced by use of an automatic filament-boosting circuit to reduce overheating.

This circuit quickly raises the filament current from a standby value (about 5 mA) to the required value before the exposure is made, and then immediately returns it to the standby value after the exposure. A bronze-colored tint that deepens with age is produced by the tungsten vaporized from the filament (and occasionally from the anode), and this buildup may eventually cause arcing between the tube wall and the electrodes.

Electrons emitted from the filament form a small cloud next to the filament, called the *space charge*. For x-ray tube voltages below about 40 kVp, the residual space charge limits the number of electrons pulled away from the filament, and thus it limits the current flowing in the x-ray tube. Above 40 kVp, however, the tube becomes saturated, and further increases in kilovoltage across the tube produce very little change in tube current. Different x-ray tubes have different saturation voltages, and different amounts of space-charge compensation are required if operated below the saturation voltage.

The rate of flow of electrons in the x-ray tube determines the number (or quantity) of x rays produced. This is characterized as the x-ray tube current, which is usually measured in milliamperes. The combination of tube current and the time it exists determines the total number of electrons that strike the target and the number of x-ray photons produced for a given exposure, measured in milliampere · seconds (mA · s). For example, in a given unit of time, say 1 s, a tube current of 200 mA will produce twice as many electrons as will a current of 100 mA.

A coulomb of electricity is produced by 6.24×10^{18} electrons each carrying a charge of 1.6022×10^{-19} C. An ampere is 1 C/s; thus an x-ray tube with a current of 100 mA (0.1 A) produces a "flow" of 6.24×10^{17} electrons/s from the cathode to the anode.

Focal Spot

The *focal spot* is the area of the tungsten target that is bombarded by electrons to produce x rays. The focal spot size is determined by the dimensions and position of the filament and the construction of the focusing cup. Since less than 1% of the energy carried by the electrons is converted into x rays, the anode can get very hot and can be damaged unless this bombarding source of heat energy is distributed uniformly over a sizable area. This need to distribute heat energy conflicts with the fact that small focal spots are best for producing good radiographic detail because x rays stream from the target as though from a point source. Two approaches, both clever, are used both to provide a small focal spot and to take advantage of the need for a large impact area to allow greater heat loading: the line focus principle and the rotating anode.

X rays emerging from the tube can be made to appear to emanate from a small area if the anode is constructed with an angle perpendicular to the incident beam. This angle, which is also shown in Figure 15-5, typically ranges between 6 and 20°. This construction, called the *line focus principle*, causes the emitted x rays to project an apparent focal spot that is considerably smaller than that of the actual area being bombarded. The size of the projected focal spot is directly related to the sine of the angle of the anode (i.e., smaller angles produce smaller focal spots). Focal spot size is expressed in terms of the apparent or projected focal spot, and typical sizes are 0.3, 0.6, 1.0, and 1.2 mm.

Example 15-3: If a beam of electrons 2 mm wide and 7 mm high is incident on a tungsten target angled at 16°, what are the dimensions of the apparent focal spot?

SOLUTION: The horizontal width projection will be the same as the width of the electron beam striking the target, or 2 mm. The other dimension will be $h \sin 16°$ or 7 mm $\times$ 0.27 = 1.93 mm; therefore, the x rays will be projected from an area 2 mm $\times$ 1.93 mm and the area of the focal spot will "appear" to be about 4 mm^2.

A phenomenon called the *heel effect* creates a practical limit to how much the anode angle can be decreased to obtain small focal spots. The lower side of the anode is shaped somewhat like the heel of the human foot, and a significant fraction of x rays emitted from the target in this direction will be absorbed. This absorption reduces the number of x rays emitted on the anode side, as illustrated in Figure 15-7, and the reduction in beam intensity toward the anode side of the tube can be quite noticeable for large films. A 14- by 17-in. film exposed at 40 in. away will receive a relative exposure of 73% on the anode side and 105% on the cathode side. At 72 in. the variation in exposure is roughly 87 to 104%, and the variation in relative exposure due to the heel effect is considerably less. Diagnostic x-ray tubes usually have an anode angle of 15° or more to reduce the heel effect.

The heel effect can be used to advantage for radiographing body parts of different thicknesses by placing the thicker parts toward the cathode. For example, an

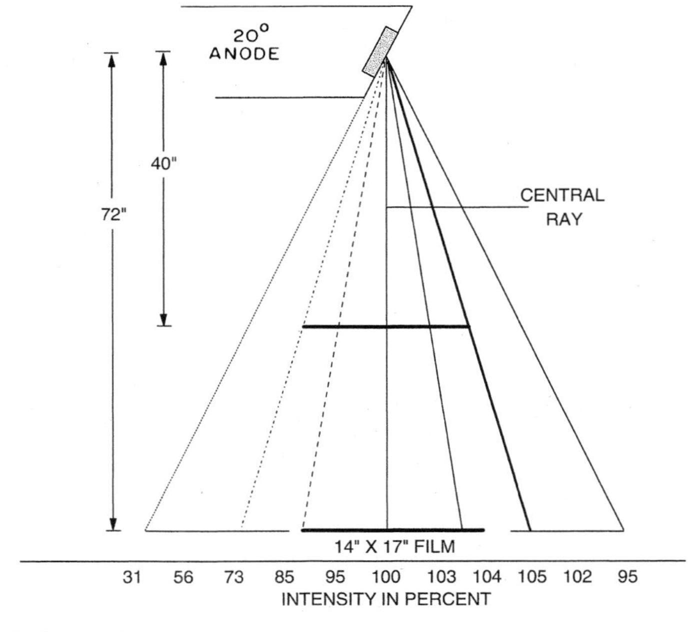

FIGURE 15-7. Reduction in x-ray intensity due to the heel effect.

anteroposterior (AP) film of the thoracic spine should be made such that the upper thoracic spine (i.e., thin side) is below the anode, where the relative intensity is lower, so that the thicker portion will receive the greater intensity.

Rotating Anodes

Although modern x-ray generators can deliver large amounts of power, the output of an x-ray tube is limited by its ability to dissipate heat generated at the anode. Rotating anodes made of tungsten or an alloy of tungsten are commonly used to dissipate this heat, which allows larger exposures and more frequent use. Rotating anodes are shaped into a large disc that is beveled at an angle of 6 to 20° to achieve a small apparent focal spot and rotated at about 3600 revolutions per minute (rpm) while an exposure is made to distribute the induced heat over a larger area. For example, a focused electron beam impinging on a stationary target beveled at 16.5° (focal spot of about 2 mm × 2 mm) will strike an area that is 7 mm high by 2 mm wide, an area that is 14 mm^2 (see Figure 15-8). In a stationary anode, the entire heat load would be delivered to this one small 14-mm^2 area of the target. If, however, the target is rotated, the electrons will bombard a constantly changing area and the heat will be distributed over an area of about 1760 mm^2. The effective focal spot will, of course, appear to remain stationary.

A typical rotating anode x-ray tube is shown in Figure 15-9. The beveled circular anode assembly is usually constructed of an alloy of 90% tungsten and 10% rhenium which has a higher thermal capacity than pure tungsten, to resist surface roughening. Heating the rotating anode assembly can cause the bearings to expand and bind. This problem can be reduced by a molybdenum stem (2600°C melting point and poor heat conductor) to provide a partial heat barrier between the tungsten disc and the bearings. Increasing the speed of rotation also increases heat dissipation

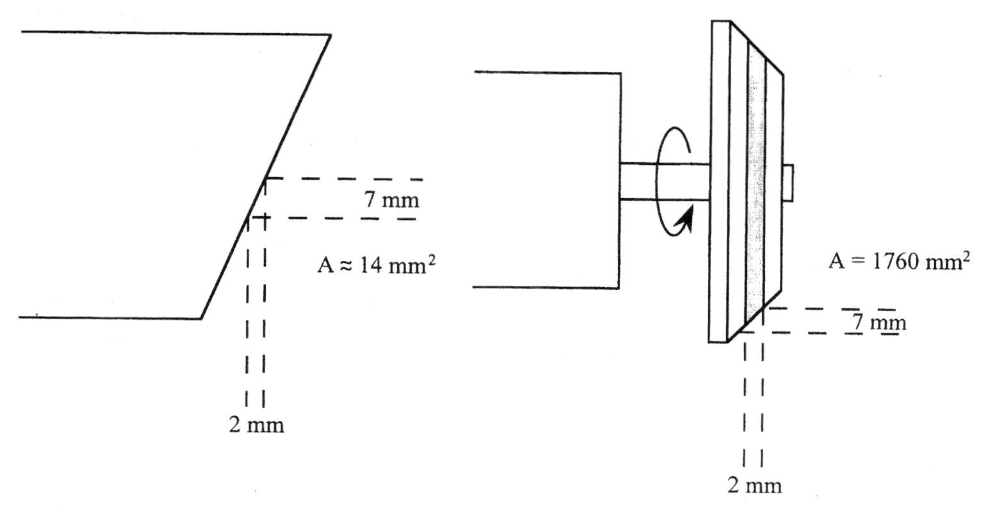

FIGURE 15-8. Heat dissipation area of an x-ray tube with a stationary anode versus one with a rotating anode.

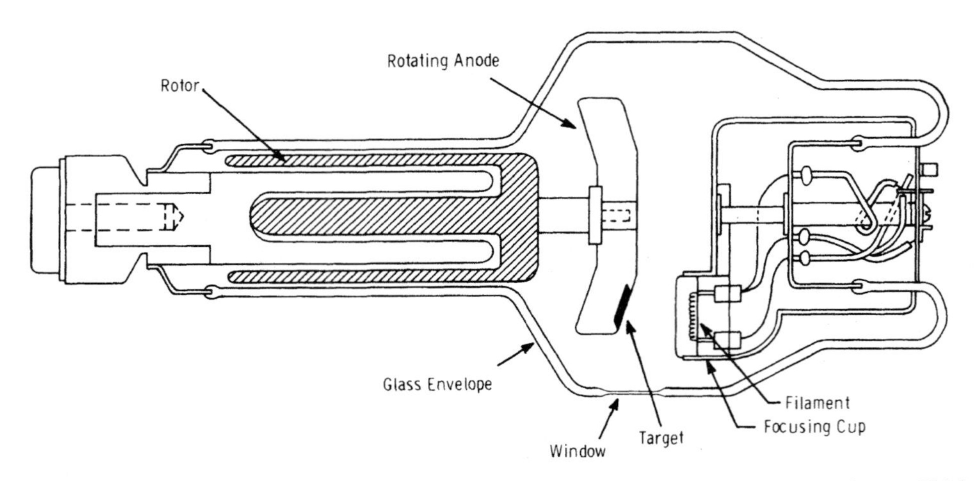

FIGURE 15-9. Principal parts of a modern rotating-anode x-ray tube. (From Bushong, 1984, with permission.)

by shortening the time that any given area of the target is exposed to the electron beam. Speeds of 10,000 rpm are not uncommon.

Equipment Techniques

The *falling load principle* and *automatic exposure controls* can be used to optimize exposure time for requisite film exposures. In the falling load principle, a falling load generator operates the x-ray tube at its maximum kilowatt rating during the course of the exposure. This yields a shorter exposure time than would be obtained with a fixed mA · s technique. For example, Figure 15-10 shows the theoretical tube rating for an x-ray tube operated at 70 kVp. If an exposure of 200 mA · s is desired for this tube using fixed settings, it would be necessary to operate it at 200 mA for 1.0 s. Shorter exposure times are not possible because the tube is limited [i.e., it can only be operated for 0.5 s when set to a tube current of 300 mA (yielding 150 mA · s), 0.3 s at 400 mA (for 120 mA · s), etc.]. If, however, a falling load is used, 200 mA · s can be obtained in 0.5 s as follows:

600 mA	for	0.05 s = 30 mA · s
500 mA	for	0.15 s = 75 mA · s
400 mA	for	0.10 s = 40 mA · s
300 mA	for	0.20 s = 60 mA · s

or a total of 205 mA · s in 0.5 s. Use of a shorter exposure time may mean the difference between a sharp radiograph and a blurred one, due to patient movement.

Automatic exposure controls, or phototimers, are used to reduce operator error associated with setting mechanical or electronic timers and/or estimating patient thickness, both of which can yield improperly exposed radiographs. The amount

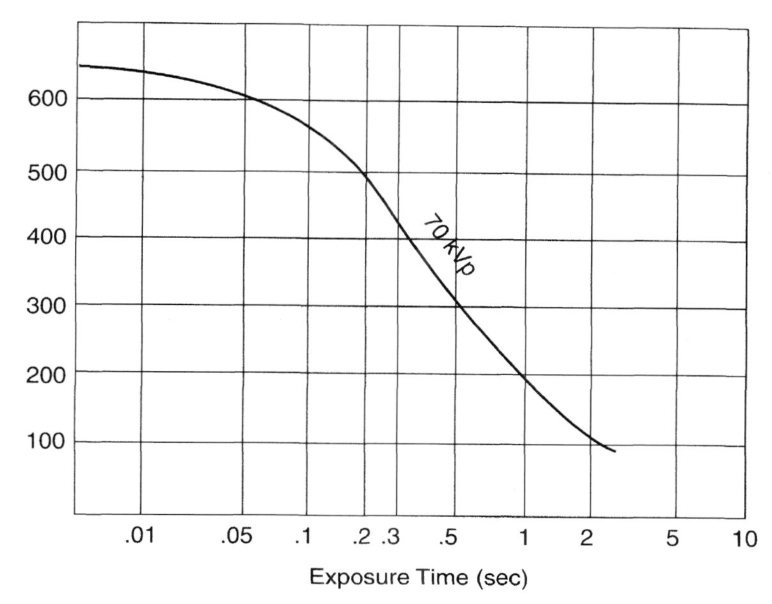

FIGURE 15-10. Theoretical tube rating for an x-ray tube operated at 70 kVp.

of radiation corresponding to a correct exposure can be measured by a phototimer, which can be set to shut off the x-ray tube when the amount of x-radiation required for the exposure has accumulated at the film.

Photomultiplier detectors, ion chambers, and solid-state detectors are used as phototimers, and they can be located in front of or behind the x-ray film cassette. Photomultiplier detectors are typically made of Lucite coated with a light-emitting phosphor. The light emitted when the phosphor is struck by x rays is amplified by a photomultiplier tube to produce a current that controls a timing circuit. Ionization chambers perform a similar function, typically on the entrance side of the patient, by using ionization current produced when x rays ionize the gas, typically air, inside the chamber. Solid-state detectors are now being used to provide a reliable and durable detector that produces current that can be used to control a timing circuit.

Intensity of X-Ray Beams

The kVp determines the maximum energy of the x rays produced and also influences the quantity at a given energy, as shown in Figure 15-11. The energy of the x rays generated depends almost entirely on the x-ray tube potential, and the number produced is proportional to the Z of the target, the square of the kilovoltage [$(\text{kVp})^2$], and the product of x-ray tube current (mA) and the exposure time (in seconds), or mA · s. Increasing the mA · s if kVp is held constant, as shown in Figure 15-12, increases the number of x rays emitted; however, the distribution of x-ray energies remains essentially the same.

The intensity of an x-ray beam is the number of photons in the beam multiplied by the energy of each photon. Intensity is commonly measured in roentgens per minute

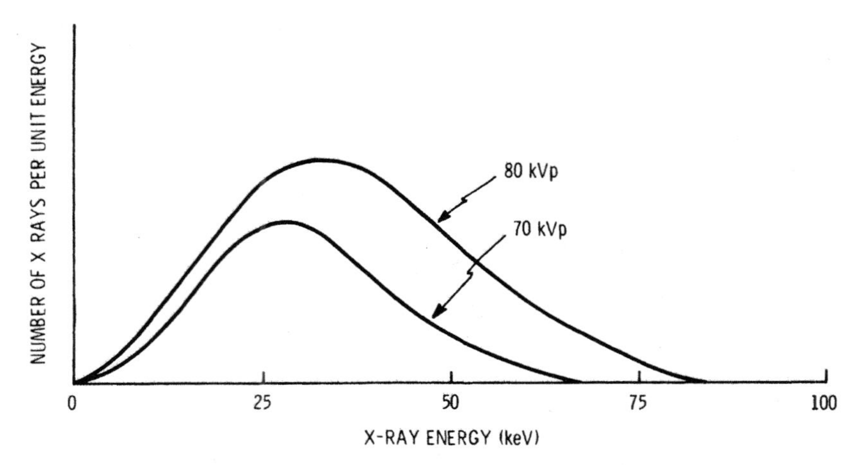

FIGURE 15-11. Influence of x-ray tube voltage on the energy distribution of emitted x rays in which an increase in kVp increases both the average and maximum energy of emitted x rays.

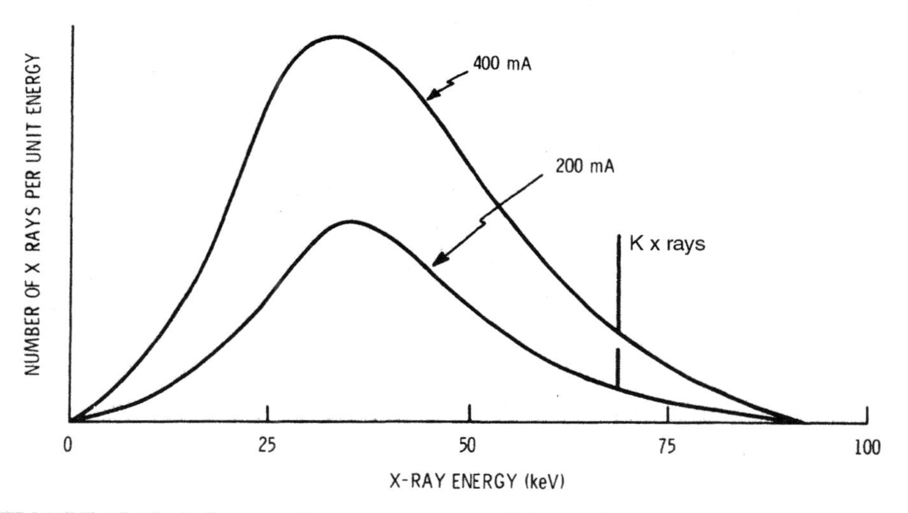

FIGURE 15-12. Influence of mA · s at constant kVp on the number of x rays emitted.

(R/min) or, in SI units, as coulombs per kilogram of air (C/kg). The intensity of an x-ray beam is determined by the kilovoltage across the tube, the x-ray tube current, target material, and filtration. Tungsten, a high-Z material ($Z = 74$), is most often used as the x-ray target because of its relatively high bremsstrahlung yield at a given tube potential and current and its high melting point for heat dissipation. Higher-Z materials such as platinum ($Z = 78$) and gold ($Z = 79$) would produce more x rays per mA · s, but their melting points (1770 and 1063°C, respectively) limit their use.

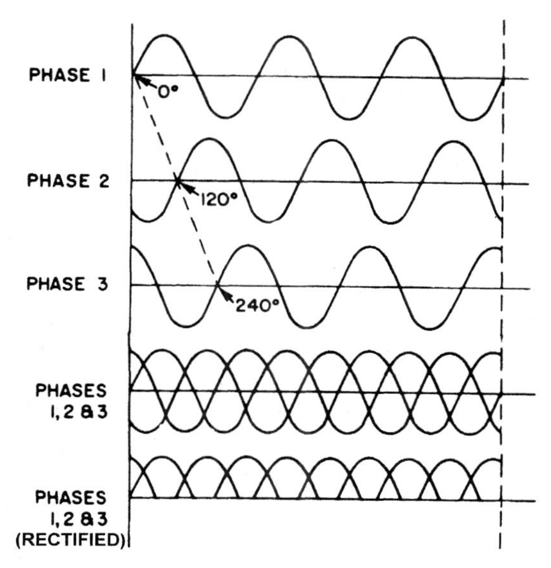

FIGURE 15-13. Three-phase ac waveforms that are composited and rectified (the ripple effect in the rectified waveform can be reduced to an almost steady output by increased phasing of voltage pulses).

The yield of characteristic radiation is also influenced by Z and tube voltage and can be a major contributor to the beam when low-Z targets are used with a low tube voltage. Both of these effects can be used to advantage for x-ray procedures such as mammography, which requires low-energy uniform fields of x rays. For example, a molybdenum ($Z = 42$) anode tube operated at approximately 40 kVp yields a significant number of 17.5-keV K_α and 19.6 keV K_β x rays, which can, in turn, be filtered with a molybdenum filter to further increase the fraction of characteristic radiation. In tungsten, K x rays range from 57 to 69 keV (by comparison, those in tin range from 25 to 29 keV) and their contribution to the x-ray beam is much less than bremsstrahlung production.

X-ray machines do not operate at fixed voltages because power is supplied by an alternating current (ac). It is necessary to convert the ac voltage to direct current (dc) and to smooth it out as much as possible to achieve an approximately constant voltage in one direction across the x-ray tube. Alternating current is changed into dc by passing the ac through a rectifier circuit which allows current to flow in only one direction. Because only half of the ac voltage (from cathode to anode) is used to produce x rays, the waveform is called *half-wave rectification*. Figure 15-13 shows the incoming waveform, the waveform after half-wave rectification, and the effect of combining three-phase circuitry and rectification. Single-phase generators at 60 pulses per second have choppy half-wave rectification. This ripple can be smoothed by three-phase six-pulse generators operated at 360 pulses per second to produce a theoretical ripple factor of 13.5%; also available are 12-pulse generators at 720 pulses per second, which have a ripple factor of about 3.5%. Modern rectifiers use solid-state devices made of silicon, and these are smaller, more reliable, and have a longer life than that of vacuum-tube rectifiers, which are rarely used anymore.

TABLE 15-2. Percent Attenuation of Monochromatic Radiation by Aluminum Filters

Photon Energy (keV)	Percent of Photons Attenuated in Aluminum Filters of:			
	1 mm	2 mm	3 mm	10 mm
10	100	100	100	100
20	58	82	92	100
30	24	42	56	93
40	12	23	32	73
50	8	16	22	57
60	6	12	18	48
80	5	10	14	48
100	4	8	12	35

X-Ray Filters

X rays that reach a patient need to have certain characteristics to produce a good-quality image while minimizing patient exposure. Filters are most useful for shaping the x-ray beam to increase the number of photons that are best for imaging certain tissues and to reduce those that increase patient dose or decrease image contrast. X rays leaving the target of an x-ray tube have a wide energy spectrum. Since the mean energy is about one-third of the peak energy, the beam contains a large number of low-energy photons. These photons will be absorbed in the first few centimeters of a patient to produce dose but not image; therefore, it is desirable to eliminate them and "harden" the beam so that it contains a larger proportion of higher-energy photons. This can be done effectively by filtration. Some filtration has already occurred in the glass of the x-ray tube, which has an inherent filtration equivalent to that of 0.5 to 1.0 mm of aluminum. When low-energy photons are needed (to provide tissue contrast) a beryllium ($Z = 4$) window, which is quite transparent for low-energy photons, is inserted in the tube wall.

Aluminum ($Z = 13$) is an excellent filter material for low-energy radiation and for practical reasons is the filter material most often used (see Table 15-2). Copper filters require less thickness for equivalent attenuation; however, photoelectric attenuation in copper yields characteristic radiation with an energy of about 8 keV. When copper is used, it is most often used with a layer of aluminum facing the patient to absorb K x rays from copper to decrease skin dose; characteristic radiation from aluminum (1.5 keV) is absorbed in the air gap before reaching the patient.

The NCRP recommends that the total inherent and added filtration be 0.5 mm aluminum equivalent for x-ray tubes operated below 50 kVp. Between 50 and 70 kVp the filtration should be 1.5 mm Al, and above 70 kVp it should be equivalent to 2.5 mm Al. As shown in Figure 15-14, a filter reduces the total number of photons in the x-ray beam, but more important, those that are eliminated are of low energy. The peak of the spectrum is shifted from 25 up to 35 keV, thus increasing the mean energy of the x-ray beam. Although filtration reduces the number of x rays in the beam, the patient receives less radiation dose than from an unfiltered beam because the higher-energy photons that remain are more likely to contribute to the image, whereas the lower-energy photons would only deposit energy in tissue. It may be

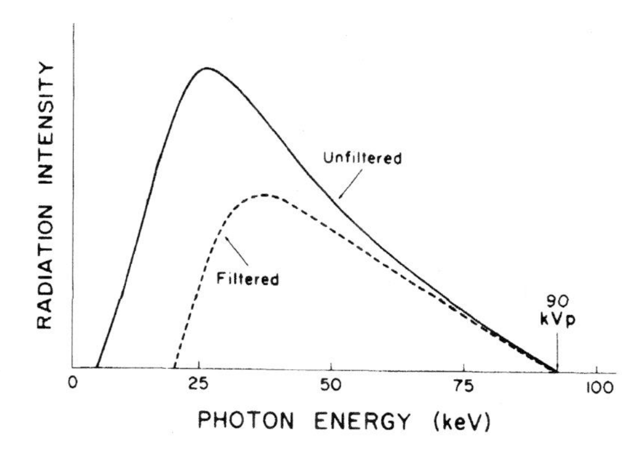

FIGURE 15-14. Effect of filtration on average energy and intensity of a continuous x-ray spectrum produced by 90 kVp electrons on a tungsten target.

TABLE 15-3. Effect of Aluminum Filtration on Entrance Skin Exposure for a Typical 60 kVp X-Ray Beam

mm Al	ESE (mR)
None	2380
0.5	1850
1.0	1270
3.0	465

necessary to increase the current of the x-ray tube (mA · s) to produce the required number of photons, but because of filtration more of these will be of high energy, and the total number that interact in the patient for a given radiographic exposure actually decreases. The effect of filters on patient exposure protection is shown in Table 15-3 for a 60-kVp beam on a pelvic phantom. Entrance skin exposure (ESE) can be reduced by as much as 80% with 3 mm of aluminum filtration for this technique (or a factor of about 5).

Wedge filters are used occasionally in diagnostic radiology to adjust for variations in patient thickness, thus achieving a beam that promotes more uniform film density after it exits the patient. The effect of a wedge filter to compensate for differences in patient thickness is shown in Figure 15-15. Less radiation is absorbed by the thinner part of the filter, so more is available to penetrate the thicker part of the patient.

K-edge filters use high-Z materials to take advantage of K-absorption edge effects, especially when used with high-speed intensifying screens and high-capacity x-ray tubes. Table 15-4 lists K-edge absorption energies for several elements that can be used to take advantage of the K-edge effect. Examples of K-edge filters are shown in Figures 15-16 and 15-17.

Contrast agents of barium and iodine also use the K-edge effect to advantage in imaging various organ systems. Both iodine and barium provide excellent contrast for x-ray energies slightly above their respective K-absorption edges. The mass absorption coefficient of iodine is 6.6 cm^2/g just below the K-absorption edge of

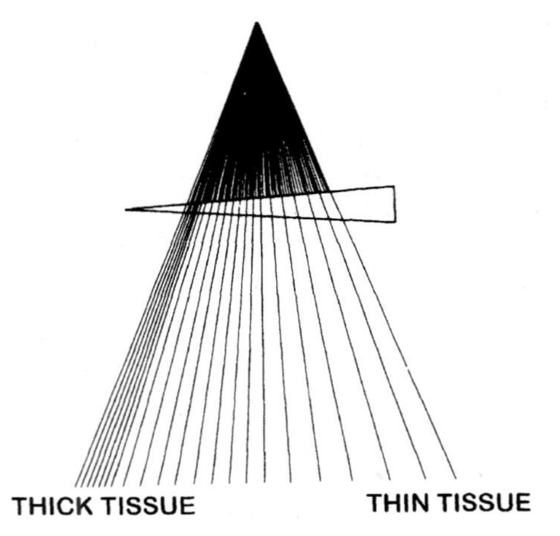

FIGURE 15-15. Use of a wedge filter to shape an x-ray beam for radiography of nonuniform anatomy as may be encountered in a thorasic spine examination.

TABLE 15-4. K-Edge Absorption Energy of Various Elements

Element *Z*	K-Edge Energy (keV)
Aluminum 13	1.6
Molybdenum 42	20.0
Gadolinium 64	50.2
Holmium 67	55.6
Erbium 68	57.5
Ytterbium 70	61.3
Tungsten 74	69.5
Iodine 53	33.2
Barium 56	37.5

33.2 keV, but jumps to 36 cm^2/g just above it; and, a similar effect occurs for barium just below and just above 37.5 keV.

Proper selection and use of a K-edge filter can produce a significantly narrower spectrum of photon energies with decreased numbers of both low- and high-energy photons. Figure 15-16 shows how a gadolinium filter enhances the number of medium energy photons in an x-ray beam. Gadolinium has a sharp cutoff at 50.2 keV, and when used with 60 kVp x rays, a high-quality beam is obtained for imaging thin body parts while eliminating low-energy photons that contribute only patient dose. Film contrast is increased by reducing the number of higher-energy photons because at these energies more photoelectric interactions, will occur than Compton interactions, which produce scatter radiation. It is necessary, however, to increase the tube current (more mA · s) to compensate for this reduction in photons; however, overall patient dose is reduced considerably (it is about half even when mA · s is doubled).

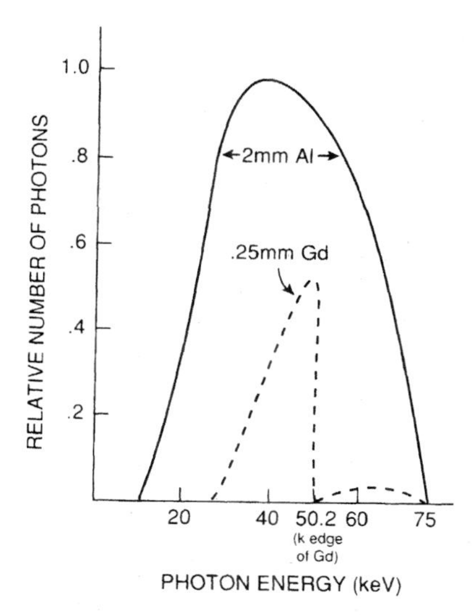

FIGURE 15-16. Effect of a gadolinium filter on the bremsstrahlung spectrum of a 75-kVp x-ray beam with 2-mm Al filtration. (From Curry et al., 1990, with permission.)

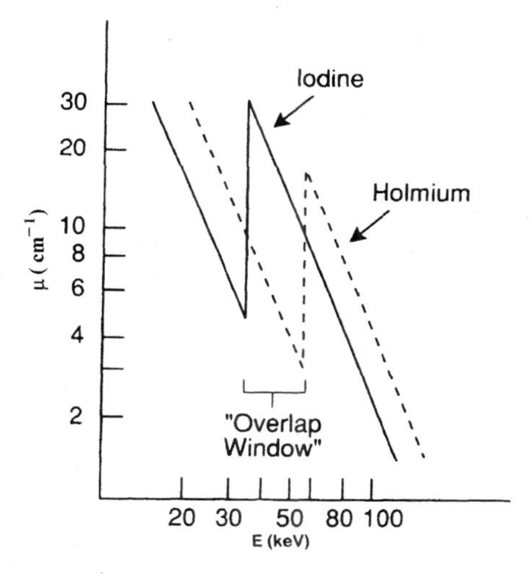

FIGURE 15-17. Attenuation of photons by iodine and holmium in which the transmission of holmium enhances transmission of x-ray photons below 55.6 keV and absorbs those above it, while attenuation in iodine produces an overlap window of high contrast. (Adapted from Curry et al., 1990, with permission.)

K-edge absorption effects of two or more materials can be used in conjunction to achieve special conditions, as shown in Figure 15-17 for a holmium filter and iodine contrast. The attenuation coefficient of iodine increases dramatically at the 33.2-keV K edge of iodine, and then decreases. The attenuation coefficient of holmium decreases steadily from about 33 keV to 55.6 keV (the K edge of holmium), at which point attenuation by holmium increases dramatically to provide efficient removal of photons above this energy. This creates a "window" from 33 to 55 keV in which high transmission of photons by the holmium filter overlaps the region of high attenuation by iodine. The effect in barium would be almost identical for improving film contrast except that its K edge is at 37.5 keV.

Film-screen mammography is another example of the use of the K-edge effect to optimize the x-ray beam, and a very important one. A typical mammographic unit uses a molybdenum target to take advantage of the 17.5-keV K_α and 19.6 keV K_β characteristic radiation produced in molybdenum. These energies are excellent for imaging the soft tissue of the breast. The molybdenum target tube is typically operated at 30 to 40 kVp for mammographic imaging, and at these potentials a considerable amount of bremsstrahlung with energies above 20 keV is produced; unfortunately, these energies reduce contrast in breast tissue. If, however, a filter of molybdenum (the same material as the target) is inserted in the beam, x rays just above its 20 keV K edge will be attenuated very strongly by photoelectric interactions, but 57% of the 17.5 keV K_α rays and 67% of the 19.6 keV K_β x rays will be transmitted. This yields a narrowly focused beam of relatively low-energy x rays that is very useful for imaging breast tissue.

In summary, the K-edge effect is pronounced, but when selected wisely, filters that take advantage of the effect can be very effective in optimizing beam quality for select procedures. When such filters are used in x-ray beams, low-energy x rays are absorbed to eliminate useless radiation dose and the "shaped" beam will produce higher-quality images by reducing scatter radiation. Their function is governed by the physics of photon interactions in various elements, in particular attenuation by photoelectric and Compton interactions.

X-RAY IMAGE FORMATION

Once the x-ray beam from the x-ray tube has been filtered and collimated to produce the required number and quality (energy) of the x rays striking a patient, a number of factors become important in producing the x-ray image: the number of x rays and the distribution of energies; the pattern of penetration, absorption, and scatter of x rays in the patient (i.e., the image information); physical interactions in the film screen to encode the image; and processing of the film to make it visible to the diagnostician.

Interactions of x-ray photons in a patient to yield diagnostic information occur primarily through the photoelectric effect and Compton scattering. Coherent scattering is numerically unimportant, and pair production and photodisintegration do not occur at most diagnostic x-ray energies. The photoelectric effect, which is dominant for low-energy photons and high-*Z* absorbers, generates no significant scatter radiation and produces high contrast in the x-ray image, but due to total absorption of photon energy, can contribute significantly to patient exposure. This can be offset by using higher-energy x-ray techniques, but Compton interactions increase at these higher energies with a reduction in radiographic image contrast due to an increase in scatter radiation. Photons that are absorbed are removed from the x-ray beam, and this absence will appear as information in the x-ray film; those that are scattered are deflected into a random course and no longer carry useful information because their direction is random. Scattered photons darken the film, creating film fog that may completely obscure the image (it is still there but cannot be seen). Consequently, any chosen x-ray technique becomes a trade-off between image quality and radiation exposure, and proper balancing of the physics of these systems to meet both goals is desirable.

Coherent scattering consists of Thomson scattering, which involves a single electron, and Rayleigh scattering; however, only Rayleigh scattering occurs to any extent at diagnostic x-ray energies. In Rayleigh scattering, a low-energy photon encounters the electrons of an atom and causes them to vibrate at the frequency of the radiation, and since it is a charged particle, it emits radiation. The process may be envisioned as the absorption of radiation, vibration of the atom, and emission of radiation as the atom returns to its undisturbed state. The only effect of coherent scattering is to change the direction of the incident radiation somewhat; no energy is transferred and no absorption occurs. Its effect is less than 5% over the diagnostic energy range, and although it may contribute to film fog, the amount is too small to be important.

The *photoelectric effect* produces radiographic images of excellent quality because no scatter radiation is involved. Tissue contrast is enhanced because photoelectric interactions are a function of Z^5, which magnifies the difference in tissues composed of different elements, especially bone and soft tissue. Photoelectric interactions are dominant for low-energy x rays, but since all the photon energy is absorbed, it may make a significant contribution to patient dose. This can be offset somewhat by using high-kVp techniques as long as appropriate consideration is given to scatter radiation produced by Compton interactions.

Compton interactions yield scattered photons and recoil electrons. Table 15-5 shows that x-ray photons up to 150 keV will retain most of their original energy even after being scattered through large angles. This is an important consideration for producing radiographic images.

Photons scattered at 30° or less retain almost all of their original energy; thus they cannot be filtered from the x-ray beam (because they are too energetic) or removed by grids (because their angle of deflection is too small); therefore, their effect is to lessen the sharpness of tissue images. Even when deflected by 90°, Compton-scattered photons will retain most of their original energy, which can create an exposure field for medical personnel in a room with a patient, as in fluoroscopy. Although the likelihood of Compton interactions changes linearly with the atomic number of the absorber, it is much more dependent on the energy of the radiation

TABLE 15-5. Energy of Compton-Scattered Photons for Various Angles of Deflection

Incident	Energy (keV) at Scatter Angle of:		
Energy (keV)	30°	60°	90°
25	25	24	24
50	50	48	46
75	74	70	66
100	98	91	84
150	146	131	116

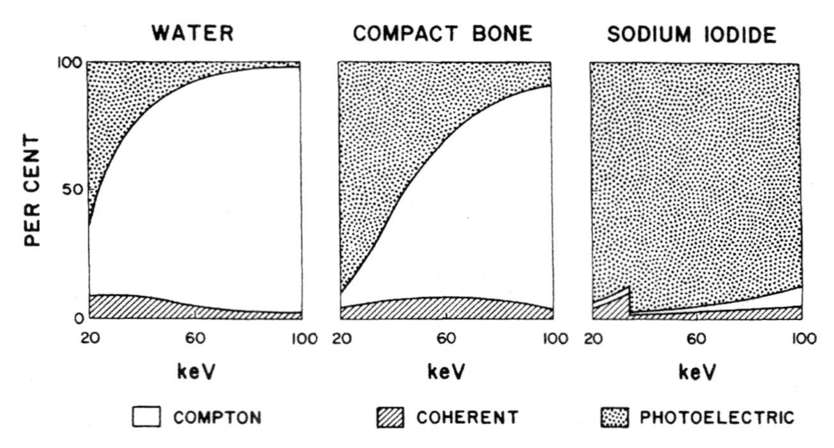

FIGURE 15-18. Percent of coherent, photoelectric, and Compton reactions in water, compact bone, and sodium iodide. (From Curry et al., 1990, with permission.)

and the electron density of the absorber. The number of Compton-scattering interactions gradually diminishes as photon energy increases, so that high-kVp techniques produce a higher percentage of photons that pass undeviated through the patient.

Figure 15-18 shows the relative contribution of coherent, photoelectric, and Compton interactions for water, compact bone, and sodium iodide for photon energies of 20 to 100 keV. Water is illustrative of low-*Z* tissues such as fat and muscle. Compact bone contains a large amount of calcium and is representative of intermediate-*Z* elements. Iodine is a high-*Z* material and because of this is a good contrast agent for diagnostic procedures; barium, another contrast agent, has essentially the same pattern of interactions as iodine. Compton scattering reactions dominate above 20 to 30 keV in water and tissue. At lower energies, photoelectric interactions are of most significance in bone and contrast agents, both high-*Z* materials. Coherent scattering is obviously a minor contributor at all energies.

X-Ray Imaging Dosimetry

An important quantity in describing the radiation dose to a patient from x rays is the mass energy-absorption coefficient, μ_{en}/ρ. Table 15-6 contains values of this parameter for muscle and compact bone. The mass energy-absorption coefficient gives the fraction of the photon energy absorbed per unit mass of the medium due to photon interactions. When a beam of photons is incident on a medium, the product of the mass energy-absorption coefficient in units of cm^2/g and the photon energy carried in 1 cm^2 of the beam gives the energy absorbed per gram. This can be converted readily into radiation absorbed dose (rad), which is the absorption of 100 ergs of energy in 1 g of tissue (or other medium), as shown in Example 15-4.

The mass energy absorption coefficient should not be confused with the mass attenuation coefficient, which determines the number of photons interacting per unit mass of medium rather than the energy absorbed. The mass attenuation coefficient is used in the calculation of the number of photons that reach a point, while the mass

TABLE 15-6. Mass-Energy Absorption, μ_{en}/ρ, Coefficients (cm^2/g) in Good Geometry for Photons in Muscle and Compact Bone

	μ_{en}/ρ (cm^2/g)	
Photon Energy (keV)	Muscle ($\rho = 1.05$)	Cortical Bone ($\rho = 1.92$)
10	4.964	26.800
20	0.564	3.601
30	0.161	1.070
40	0.072	0.451
50	0.043	0.234
60	0.033	0.140
80	0.026	0.069
100	0.025	0.046
150	0.027	0.032
200	0.029	0.030
300	0.032	0.030

Source: Data from Hubbell and Seltzer, 1995.

energy absorption coefficient is used to calculate the amount of energy absorbed from the interaction of photons once they get there.

Example 15-4: A fluence of 10^9 photons/cm^2, all with an energy of 80 keV, is incident on a person. How much energy is absorbed in muscle tissue at the point of incidence?

SOLUTION: The photon energy passing through 1 cm^2 = 0.08 MeV/photon $\times$ 10^9 photons/cm^2 = 8×10^7 MeV/cm^2. From Table 15-6 the mass energy absorption coefficient in muscle for 80 keV photons is 0.026 cm^2/g, and the energy absorbed per gram is

$$E_{abs} = 8 \times 10^7 \times 0.026 = 2.08 \times 10^6 \text{ MeV/g}$$

This corresponds to an absorbed energy of 3.33 ergs/g, which is 3.33×10^{-2} rad, or 33.3 mrad.

When a beam of x-ray photons is used to produce an x-ray image optimal patterns of lightness and darkening can be established to produce a latent image of tissue differences which can be brought out by film development. The paths of individual photons determine the image as shown in Figure 15-19, in which various darkness patterns on the film depict the amount of matter through which they had to penetrate to reach the film. Dense matter reduces the intensity of the emergent beam, yielding light areas on the developed film; interactions with low-density materials have minimal effect on the intensity of the emergent beam and film darkening will increase. The amount of darkening is characterized as film density, which is shown in Figure 15-19 as low density when the incoming x rays are intercepted by dense objects such as bone, arrowheads, and shrapnel and as high density when they encounter an air cavity. A density of 1.0 means that the light transmission is $\frac{1}{10}$ that for unexposed film; for a density of 2.0 the transmission is $\frac{1}{100}$; and so on.

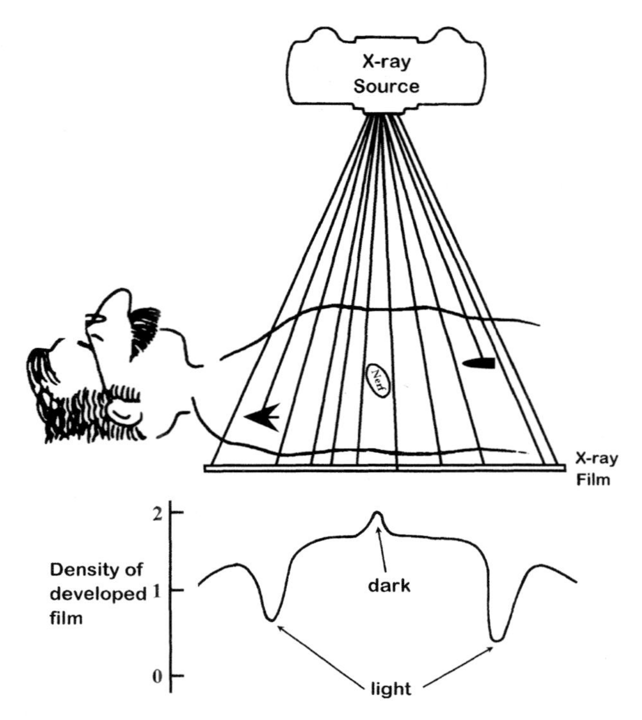

FIGURE 15-19. Effects of various types of matter in a noted physicist on film density due to differential absorption of x-ray photons.

The absorbed dose at the entrance surface of the patient is higher than the exit dose by a factor depending on the number of photons absorbed as the radiation penetrates the patient. The intensity of 30 keV photons after passing through 10 cm of tissue is reduced through mass absorption (see Table 15-6); thus the dose to tissue in the first centimeter on the entrance side is 4.3 times higher than the dose on the exit side. For 60 keV photons the entrance dose is about 1.3 times greater. Since it is the number of photons that exit the patient that determines whether an acceptable radiographic image is produced, higher-energy photons yield lower entrance doses for a given procedure. Heavier patients will also require more dose than thinner ones because more photons have to strike the patient to have a sufficient number, after absorption in tissue, on the exit side for producing an image on film.

A very important factor to consider in selecting the machine kVp (photon energy) is how effective a given photon energy will be in revealing tissue differences. An abnormal mass of tissue that is 0.5 cm thick has an attenuation coefficient of $0.34\ \text{cm}^{-1}$ for 80 keV photons and the fraction of photons removed is approximately $0.5 \times 0.34 = 0.17$, or about 17%. However, at 100 keV, the attenuation coefficient is $0.161\ \text{cm}^{-1}$, and only 8% of the photons would be attenuated by the mass; therefore, the lower-energy x rays would be better for producing an image contrast in such tissue. Such results are even better for higher-Z materials such as bone, and sharper contrast and image detail can be obtained if lower-kVp x rays are used to image bone versus surrounding tissues. On the other hand, these advantages need to be considered in light of the lower entrance doses provided by higher-energy x rays for a given film exposure due to their greater penetrating power.

The matter is even more complicated because x rays are never produced at a single energy but rather, are produced over a wide range of photon energies. The low-energy photons in x-ray spectra are greatly attenuated, and these photons contribute very little, if anything, to the image. All they do is produce very high doses at the surface of the body and for a short distance along their path; thus their number should be minimized by filtration and other techniques.

CAPTURE OF THE X-RAY IMAGE

The pattern of x rays exiting the patient (or any other object being radiographed) contains the information for producing the radiographic image. This image pattern is determined by the differential degrees of attenuation of the x-ray beam as it passes through the region of interest. Intuitively, dense objects such as metal, bone, or injected contrast material will absorb more of the x rays than will less dense matter such as air cavities, fat, or tissue. The desired image capture system is one that most sharply delineates these density differences so that the location and pattern of abnormalities can be brought forth into a visual image and thus diagnosed. The image pattern is a combination of x-ray photons that can be "seen" because they have been removed or absorbed to varying degrees. The most usual image capture system is film; however, recent sensors combined with computerized storage, processing, and enhancement techniques can be used instead of x-ray film as an image capture system.

Regardless of which image capture system is used, it is essential that the pattern of x-ray photons that intereact with the sensitive elements of the receptor to encode visual information be such that the necessary information is contained therein so that the processing system (whether film development or an electronic image) can produce a visual image. The discussion that follows is for image capture on film; however, many of the same adjustments to enhance image information and optimize patient exposure apply equally well to electronic imaging. In fact, one of the advantages of electronic image processing is the ability to amplify such enhancements with a significant decrease in patient exposure because of the wide flexibility in data capture and processing.

A large amount of diagnostic radiography uses high-kVp electrons, beam filtration, and collimation to yield an x-ray beam of relatively high energy photons incident on a patient to reduce radiation exposure. This approach represents a tradeoff between low-energy photons in which photoelectric interactions provide good contrast but result in more exposure and more penetrating higher-energy photons in which Compton interactions yield scattered photons that distort the image.

Air Gaps and Grids

Scattered photons exiting a patient can distort the film image; therefore, it is desirable to remove them before they reach the x-ray film. This can be accomplished by either an air-gap technique or a grid. Both require other adjustments in the image capture system to optimize image quality. The *air-gap technique* involves placing

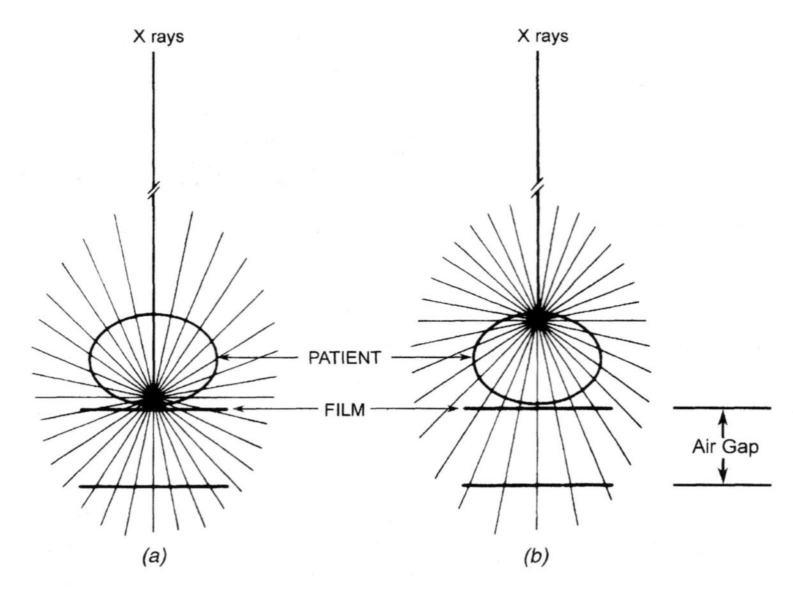

FIGURE 15-20. Use of an air gap between patient and film causes much of the scattered radiation to miss the film, an effect that is greatest for objects closest to the film (*a*), and somewhat lower for tissue elements farther from the film (*b*).

the film a distance from the patient to create an air gap between the two such that photons scattered at large angles simply miss the film (see Figure 15-20); those with minimal scatter will still strike the film and produce useful information even though the image will be less sharp. The air-gap technique is most effective if the object being imaged is close to the exit side of the patient: for example, imaging the spine with the film near the patient's back (Figure 15-20*a*). Figure 15-20 illustrates the use of an air gap for such a circumstance, and Figure 15-21 illustrates the effect of air-gap distance on reducing the ratio of scattered to primary photons for various patient thicknesses. The air-gap technique has its greatest practical application for a specific procedure such as chest radiography, where an optimal arrangement of x-ray tube, patient distance, and air-gap thickness can be established empirically and then left in place.

Grids can also be used to reduce scatter radiation and are generally required for most general-purpose radiography units because these units are used for many types of procedures and it is difficult to establish optimal air gaps for each. Grids are made of a series of lead strips separated by spacers that are transparent to x rays. Primary photons will pass through to the film, whereas most of the photons scattered by the patient strike the grid at an angle and are absorbed in the lead strips before they can reach the film. Grids are manufactured with spacers made of organic material or aluminum, and patient exposure is generally higher because of the need to overcome the absorption of primary photons by the lead strips in the grid.

An air-gap technique has a major advantage over grids in that the number of scattered photons is reduced without having to increase exposure of a patient, as illustrated in Figure 15-22 for chest radiography. Figure 15-22 compares the relevant

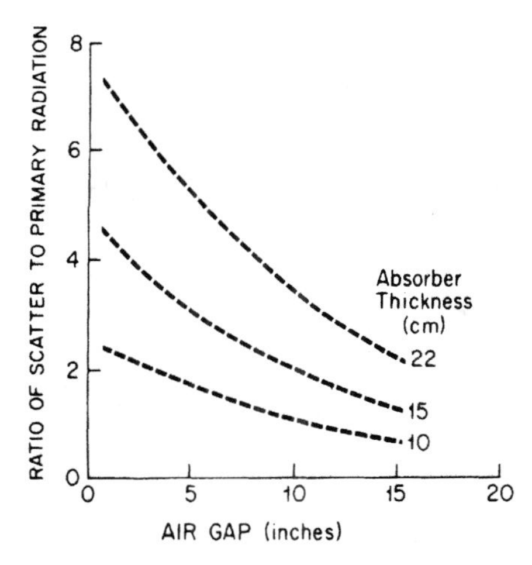

FIGURE 15-21. Ratio of scatter to primary radiation reaching the x-ray film for various air gaps and patient thickness.

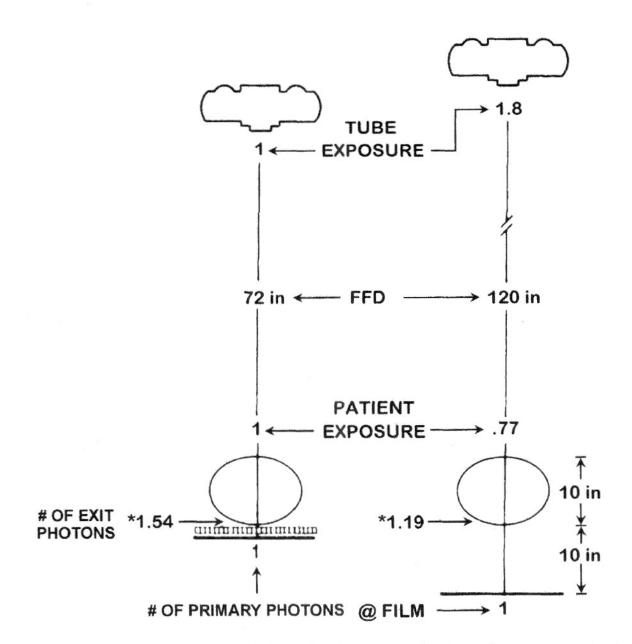

FIGURE 15-22. Comparison of the grid technique and the air-gap technique for chest radiography.

factors for a 72-in. focal spot–film technique using a grid with a 120-in. focal spot–film distance and a 10-in. air gap behind the patient (the air-gap technique requires a larger distance between the focal spot of the x-ray table and the film, and because of the inverse square law, it is necessary to increase the output of the x-ray tube). Whereas the air-gap technique requires 1.19 photons to overcome

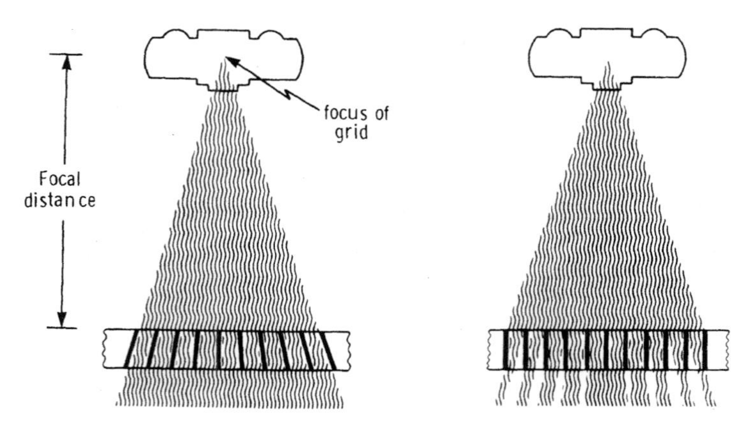

FIGURE 15-23. Focused grid with grid strips angled to the path of x-ray photons and a linear grid with parallel strips, resulting in some cutoff of the x-ray beam at the outer edge. (Adapted from Bushong, 1984, with permission.)

inverse-square losses and produce a unit of exposure, the grid technique requires 1.54 photons because of the absorption of primary photons by the lead strips in the grid. Consequently, patient exposure is about 23% less for the air-gap technique which is a significant but not a major savings in dose because chest radiography is generally a relatively low-dose procedure anyway.

Grids are of two types: If the lead strips are evenly spaced, the grid is a linear grid; if angled to be parallel to the x rays emanating from the focal spot of the x-ray tube, it is a focused grid. These two types are shown in Figure 15-23. Depending on the spacing and height of the lead strips and the focal line, angles will be such that primary photons will also be absorbed in the grid strips, especially toward the edge of the grid, a phenomenon called *grid cutoff*. This effect can be quite noticeable for large films and the wrong grid. If a focused grid is accidentally used upside down (a not uncommon occurrence), cutoff of much of the primary beam will occur, which is easily recognized because the processed film will be dark down the center and bright (underexposed) toward the edges.

Grids are characterized by a number of factors, most of which are interrelated. Some of the more important ones are grid ratio, lines per inch, stationary versus moving, and the Bucky factor. The *grid ratio* is defined as the ratio of the height of the lead strips to the spacing between them:

$$\text{grid ratio} = \frac{\text{height of strips (mm)}}{\text{spacing distance (mm)}}$$

For example, a grid constructed of lead strips 2.0 mm high and spaced 0.25 mm apart will have a grid ratio of 2.0/0.25, or 8 : 1. Grid ratios usually range from 4 : 1 up to 16 : 1. Generally, a higher-ratio grid removes more scattered photons than one of lower ratio. The grid ratio only tells the ratio of the height of lead strips to the spacing between them. A grid ratio of 10 : 1 is the same whether the height/spacing ratio is 3 mm/0.3 mm or 2 mm/0.2 mm, even though the two grids are quite different. Most manufacturers of grids use strips of sufficient thickness to absorb scattered photons and of sufficient height to capture the majority of those that strike the grid at an angle significantly different from the path of the primary beam.

The grid ratio provides no information on the thickness of the lead strips themselves, a value that is addressed to some degree by stating the number of lines per inch. This quantity is calculated by adding the thickness of the lead strips (in millimeters) and that of the spacers (in millimeters) together and dividing this sum into 25.4 mm/in. A grid with many lines per inch (over 100) is generally thinner and has a lower lead content than that of a grid of comparable ratio with fewer lines per inch.

The lead in a grid requires exposure to be increased to overcome the absorption of primary photons. The degree to which exposure must be increased is known as the *Bucky factor*, which is the ratio of the amount of exposure required to produce an equivalent exposure of the x-ray film with a grid to that without. An 8 : 1 grid requires a Bucky factor of about 3.5 for a 70 kVp technique and 4.0 for 120 kVp; for a 16 : 1 grid these factors are 4.5 and 6.0, respectively. Generally, an 8 : 1 grid ratio is satisfactory for x ray energies less than 90 kVp; for higher-energy radiation, grids with a ratio of 12 : 1 or more are better. High-ratio grids yield maximum contrast, but the required increase in exposure may not justify it; this is a trade-off that needs to made carefully but many times is not. Such adjustments require full understanding of the interrelated physical factors involved and how a change in one affects the others. Diagnosticians who insist on films with very sharp contrast may use higher ratio grids than necessary, but the sharpness obtained will be at the expense of the patient who may receive more exposure than is necessary to provide an accurate diagnosis. Use of a grid thus becomes a compromise between improved contrast by absorbing scattered photons and an increase in patient exposure.

Since grids impose absorbing material into the x-ray beam, it is often desirable to use a moving grid to avoid imaging the strips on the processed x-ray film. Moving grids have to be moved quickly and uniformly to accomplish this effect, and often are made to oscillate several times during the exposure. Optimizing all of these parameters to prevent random density variations on processed film is difficult, and many diagnosticians simply use stationary grids and interpret around the lines.

X-Ray Film and Screen Systems

X-ray film provides a medium in which invisible information contained in the x-ray beam transmitted through an object is captured and made visible. X-ray films are not particularly sensitive to x rays directly; therefore, an intensifying screen is used in contact with the film to convert information in the transmitted x-ray beam into light photons, which then expose the film with considerably increased amplification. When so exposed, the film emulsion contains a latent image in a spatial pattern of silver bromide crystals sensitized by exposure but not yet developed, and therefore, invisible to the eye.

Due to its thin construction and relatively low-Z components, radiographic film by itself responds poorly to x rays; thus intensifying screens (Figure 15-24*b*) are used to amplify the effect of the x rays striking the film, to minimize patient exposures and preserve x-ray tubes. The intensifying screen is coated with a material that readily absorbs x rays and emits light photons, which then expose the film.

Silver halide is quite sensitive to ultraviolet, violet, and blue light below 500 nm, and screens coated with calcium tungstate are good for producing these wavelengths. Rare-earth phosphors such as gadolinium and lanthanum are better absorbers of x-ray photons and their response is faster; however, these compounds emit green light. Films with relatively long wavelength spectral sensitivity have been developed to accommodate these factors by adding green-absorbing dyes into the emulsion layer to reduce the low-frequency cutoff to about 575 nm. The performance of intensifying screens can be further enhanced by use of a reflecting layer to direct light photons, which are emitted in all directions by screen phosphors, back into the film emulsion.

The fundamental process for the formation of a latent image in the radiographic emulsion is the decomposition of ionic silver bromide crystals by radiolysis, which occurs in the following sequence:

1. An x-ray photon produces visible-light photons, which in turn produce electron–hole pairs in the crystal when they are absorbed.
2. The electrons liberated by the visible-light photons are trapped at impurity sites or defects in the crystal, causing silver ions to migrate to them and form atomic silver.
3. Migration and trapping continues until sufficient clusters of atomic silver atoms exist to be made visible through chemical development.

The primary points of absorption of incident visible light (and to a lesser extent x-rays directly) are bromine atoms in the silver bromide crystals. Absorption of light photons by a bromine atom liberates a photoelectron, which then migrates through the crystal lattice until it is trapped in a defect to create a sensitivity speck. The electric field surrounding the trapped electron is capable of attracting a mobile silver cation, which then combines with the electron to form an atom of atomic silver. The residual bromine (or the hole) diffuses out of the crystal and into the surrounding gelatin matrix. The atomic silver atom also acts as an electron trap, which in turn attracts more silver cations to form a cluster of atomic silver atoms. These clusters are made visible when the film is developed. Clusters of less than three atoms will not catalyze development, and single silver atoms are likely to dissociate. The stability of the development center is also reduced with respect to the passage of time, which results in the phenomenon of latent image fading. The radiographic image itself consists of numerous black specks of atomic silver in a spatial distribution corresponding to the relative intensities of the x-ray beam that impinged on the film. The invisible latent image is then converted into a visible image through chemical development.

Although radiographic film products are manufactured with a variety of speed and contrast characteristics, they have many similar properties. All x-ray film has a flexible base onto which a thin coat of adhesive is applied to firmly attach a photographic emulsion layer. Both sides of the base can be coated to form a double-emulsion film, and each emulsion layer is sealed with a protective topcoat. These features are shown in Figure 15-24*b*. The film base is usually blue-tinted polyester 0.178 mm thick. It is dimensionally and chemically stable, waterproof, and flame resistant; however, film emulsion will not stick to it directly, hence the need for an adhesive layer.

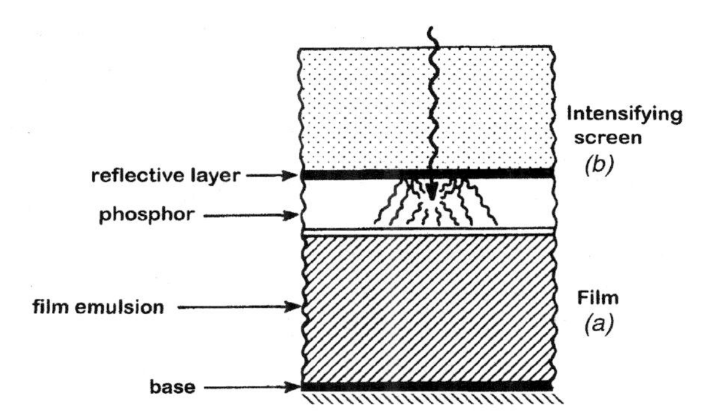

FIGURE 15-24. Cross section of a typical double-emulsion radiographic film (*a*) overlain with an intensifying screen (*b*) with a reflective layer and phosphor to increase the number of light photons reaching the film.

The film emulsion is a composite of colloidal gelatin that contains photosensitive silver halide crystals and other materials. The gelatin is optically transparent, liquefies when heated (to allow uniform mixing of silver halide crystals throughout), but remains solid at film processing temperatures. It also absorbs water and swells, permitting water-based processor chemicals to reach the silver halide during film development. The typical film emulsion layer is 0.0127 mm thick.

Silver halide in the film emulsion is light sensitive and is a mixture of 90 to 99% silver bromide and 1 to 10% silver iodide. Silver halide crystals contain silver iodobromate in a cubic lattice composed of ionic silver (Ag^+), ionic bromine (Br^-), and ionic iodide (I^-), which significantly increases the sensitivity of the emulsion compared to pure silver bromide. Silver sulfide is also added to the film emulsion; it attaches to the surface of the silver halide crystal to form a sensitivity speck and functions as an electron trap in the process of latent image formation and conversion of the silver halide crystal into a visible silver globule. Each crystal in medical x-ray film is 1.0 to 1.5 μm in diameter and contains about 10 billion atoms.

IMAGE PROCESSING

The latent image created by x rays and light from intensifying screen phosphors does not become visible until it is processed, either electronically or by film developing. When exposed film is developed, the sensitized silver halide crystals are chemically reduced, which has the effect of amplifying the radiation exposure with a tremendous gain in image density, probably by a factor of at least 10^8. The developer solution acts to convert all of the silver ions in the crystal into an irregular globule.

Film process solutions typically consist of developing agents, accelerators, preservatives, hardeners, and restrainers, mixed together in a water-based solvent. Table 15-7 lists these and their function in film development. The composition and relative concentrations of the developer solution components produce different degrees of photographic effect; however, the formulation of commercially available developer

TABLE 15-7. Components of Developer Chemicals and Their Functional Characteristics

Phenidone	Reducing agent; produces low contrast and midtones over a wide range of temperature and pH
Hydroquinone	Reducing agent; produces high contrast and dark tones, but only for high processor temperatures and pH > 7.0
Potassium hydroxide, sodium carbonate, and sodium metaborate	Accelerating agent; swells gelatin so that Ag crystals are exposed to the reducing agent; buffers the developer solution and maintains pH > 7.0
Sodium sulfite	Preservative; scavenges oxygen from developer solution to prevent oxidation of reducing agents and moderates the activity of the reducing agents
Glutaraldehyde	Hardener; prevents excessive swelling and softening of the film emulsion
Potassium bromide	Limits chemical reduction to developing sensitized silver halide crystals only

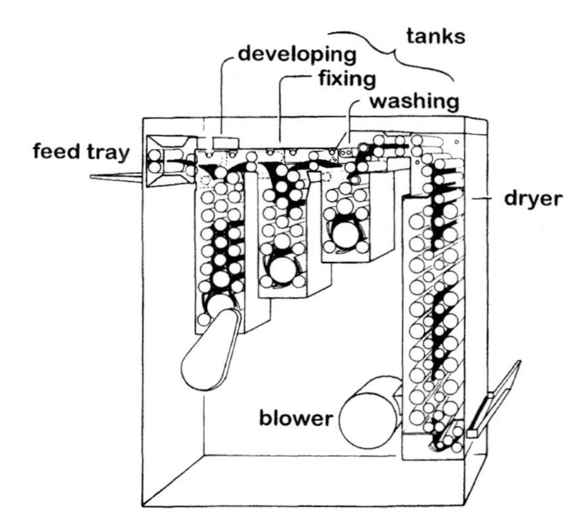

FIGURE 15-25. Major components of an automatic film processor. (Adapted from Bushong, 1984, with permission.)

solutions is relatively generic. The solutions are used in a definite sequence either by manual processing or in automatic film processors. When film is processed automatically, the film is moved by a series of rollers through the various stages, as shown in Figure 15-25.

Processing of radiographic film produces images with different degrees of film blackness or optical density. Optical density (OD) is defined as

$$\mathrm{OD} = \log_{10} \frac{1}{T} = \log_{10} \frac{I_0}{I}$$

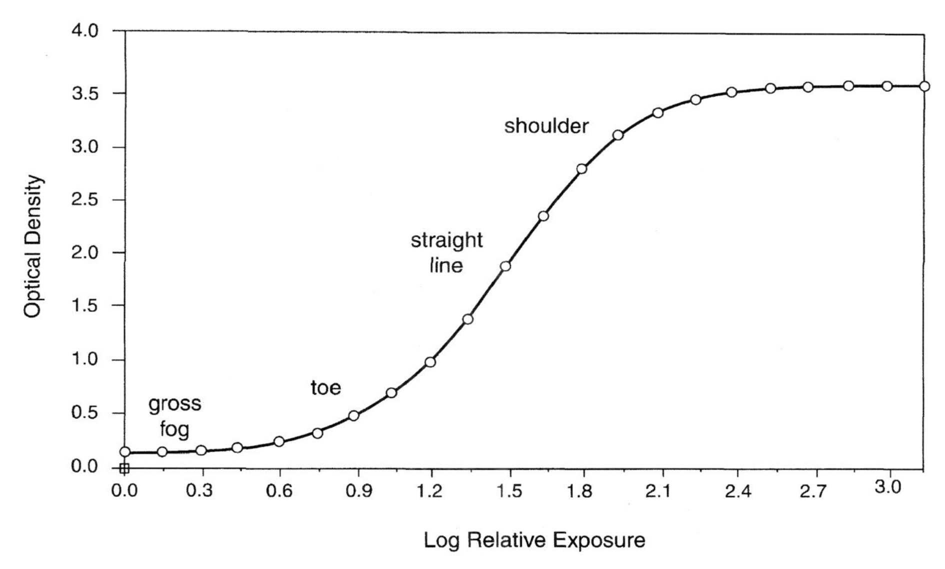

FIGURE 15-26. Typical H and D (or characteristic) curve of optical density produced in x ray film by varying amounts of x-ray exposure and its four principal regions.

where T is the fractional transmission of light through the film. T is equal to the intensity of the light transmitted through the film, I, divided by the intensity of the light before attenuation by the film, I_0. The OD of film base material is about 0.07 OD, and residual emulsion adds another 0.05 OD per emulsion layer. The combination of base plus emulsion density, called the *gross fog*, results in a baseline density of 0.12 to 0.20 OD. Useful image densities in diagnostic radiology range from about 0.30 OD, which corresponds to transmission of 50% of the incident light, to about 2.0 OD, or 1% transmission of the incident light.

The relationship between optical density (OD), and exposure is plotted as a curve, as shown in Figure 15-26, which is known as the *characteristic curve* for the film. This curve is called the *H and D curve* after Hurter and Driffield, who first published it. The labels H and D can be used as memonics: D to represent density (optical density), and H, a symbol used in radiation protection for dose equivalent, can be related to the intensity of photons that exposed the film to produce the density (optical density) observed. A typical H and D curve is produced by exposing film to a series of increasing exposures, processing the film, measuring the OD that results, and then plotting the optical density against the amount of exposure. By convention, the horizontal scale of the characteristic curve is presented in terms of log relative exposure in order to display a wide range of exposures, and more important, to simplify analysis of the characteristic curve. The H and D curve in Figure 15-26 is plotted in increments of 0.3 units of log relative exposure, because doubling the film exposure increases the log relative exposure by 0.3, and vice versa.

The characteristic curve is roughly divided into four regions: gross fog, the toe, the straight line, and the shoulder. The gross fog region occurs at optical densities of 0.12 to 0.20, due to insufficient sensitization of the emulsion and dissociation

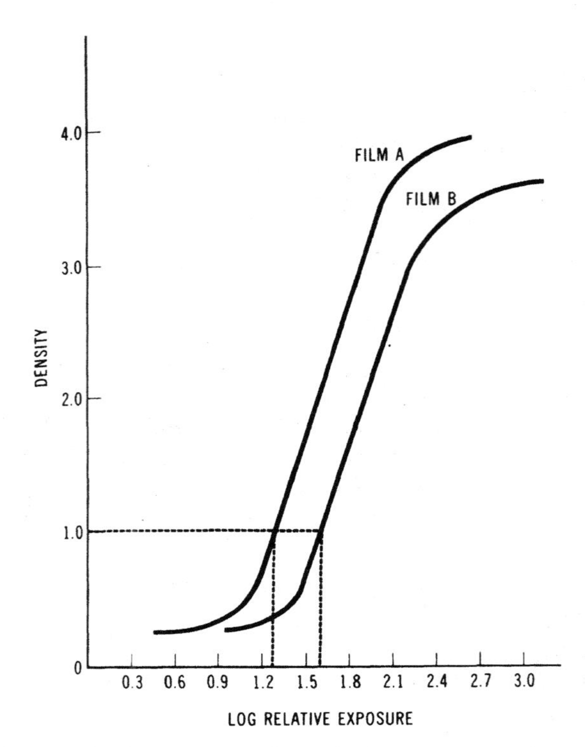

FIGURE 15-27. Speed of a film is the reciprocal of the exposure, in roentgens, needed to produce a density of 1.0. Film A is faster than film B.

of development centers. With increased film exposure, a toe region of development centers occurs which is a transition from insufficient film exposure (gross fog) to a region of useful image density (the straight-line region) which includes densities above the minimum film density, or *D-min*, for producing a useful image. D-min values of 0.25 to 0.30 OD are typical. In the straight-line region the optical density is approximately proportional to the log relative exposure to the film. In this region, incremental increases in relative exposure produce corresponding increases in optical density and include densities between 0.30 and 2.0 OD; this is the most useful range for radiographic imaging. The shoulder region is in essence a practical limit of optical density for radiographic film. This maximum film density, or *D-max*, occurs at an OD of about 3.0 to 3.5.

Two important photographic properties described by the characteristic curve are relative film speed and film contrast. *Relative film speed* is defined as the log relative exposure that results in an OD of 1.0 over gross fog; it may also be referred to as the *film-process speed* since increasing film-process speed moves the characteristic curve to the left on the log relative exposure axis (see Figure 15-27). The quantitative difference in speed between two films is equal to the antilog of the difference in the log of relative exposure required by each film to result in an OD of 1.0 over gross fog. For example, in Figure 15-27 the log relative exposure required to produce an OD of 1.0 over gross fog is 1.25 for film A and 1.65 for film B. The antilog of the difference in log relative exposure is 2.5; therefore, film A is 2.5 times as fast as film B.

The slope of the straight-line region of the characteristic curve describes the rate of change in OD with the rate of change in log relative exposure, and the value of this slope is a numeric indication of the contrast level of the film. *Film contrast* is best described by the average gradient, G, which is determined by first plotting the characteristic curve of a given system and using the following equation to calculate G:

$$G = \frac{(2.0 + \text{gross fog}) - (0.25 + \text{gross fog})}{E_2 - E_1}$$

where G is the average gradient, E_2 the log relative exposure required to produce an OD of (2.0 + gross fog), and E_1 the log relative exposure required to produce an OD of (0.25 + gross fog). Most radiographic films have an average gradient between 1.5 and 3.5, and at these values the contrast between tissue types will be exaggerated. Films with G values in the upper part of this range are called contrast films because small differences in relative exposure produce large differences in optical density, yielding radiographic images that appear black and white. Films with low G values are called *latitude films*. Small changes in relative exposure of the film produce small differences in optical density, and the resulting radiographic images appear as shades of gray.

Film gamma is defined as the slope of a tangent to the point of inflection in the straight-line region of the characteristic curve (the point of inflection occurs where the curve changes from concave-upward to concave-downward). The gamma value describes the contrast only for the narrow range of OD that occurs on the steepest part of the straight-line region and excludes other radiographically important densities. For this reason, the average gradient is most frequently used to describe the average contrast over the range of clinically useful OD.

Optimization of Film Processing

Chemical processing, radiographic exposure factors, radiation dose, and radiographic image quality are all interrelated. Chemical processing affects the position and shape of the characteristic curve, which is the key determinant in radiographic imaging. Radiographic exposure factors determine whether a desired range of optical densities will occur in the processed radiographic image after being attenuated differentially through the patient. The overall optical density of the radiographic image is controlled by the total quantity of x-ray photons used for the exposure (i.e., the mA · s setting), as are the skin entrance exposure and radiation absorbed dose. The mA · s required for sufficient image density is inversely related to the relative film speed achieved through processing. The greater the relative film speed, the less mA · s (and radiation dose) necessary to produce the required film density. The mA · s values are defined clinically to assure that diagnostic-quality radiographs are produced, and the appropriate kVp is chosen to achieve the desired differential attenuation of the x-ray beam through the patient, a nonhomogeneous absorber. Altering the penetrating ability of the x-ray beam is done to assure that the x-ray beam, after differential exposures through a patient, will produce optical densities in the straight-line region of the characteristic curve.

TABLE 15-8. Comparison of Fractional Change in Contrast and Relative Speed for Indicated Factors Relative to Recommended Factors (Shown in Bold Type) and Normalized to 1.00 for Speed and Contrast

Changing Temperature			Changing pH			Changing Specific Gravity		
T (°C)	*G*	Speed	pH	*G*	Speed	Sp. Gr.	*G*	Speed
33	0.91	0.78	7.0	1.02	0.91	1.04	1.00	0.55
34	0.98	0.89	8.0	1.01	0.96	1.05	1.00	0.85
35	1.00	1.00	**9.0**	1.00	1.00	**1.06**	1.00	1.00
36	1.01	1.11	10.0	0.96	1.06	1.07	1.00	1.14
37	0.99	1.24	11.0	0.86	1.11	1.08	0.98	1.28
38	0.97	1.35	12.0	0.76	1.17	1.09	0.99	1.42

Developer activity and photographic response are governed by several variables: the developer tank dimensions and volume, the speed and path length through the developer tank, the length of time that the film is in the developer solution, the developer solution recirculation and film agitation, and the ability of the thermometer to accurately indicate the temperature throughout the developer solution tank. The mechanical process factors are fixed by manufacturers; thus adjustments in clinical photographic response must be achieved by changing temperature, pH, and specific gravity of developer chemicals. Kasza has studied means to optimize the performance of film, developer chemistry, and the automatic film processor by varying one of these factors while holding the other two constant. The measured photographic response of each parameter is shown in Table 15-8 and Figure 15-28.

The average gradient and relative film speed values in Table 15-8 have been normalized to 1.00 for a developer solution of 35°C, a pH of 9.0, and a specific gravity of 1.06, as recommended by the manufacturer. The data in Table 15-8 show that both average gradient and relative film speed can be improved by increasing the developer temperature from 35 to 36°C; these simultaneous increases occur only at the recommended pH of 9. Increasing the recommended developer solution specific gravity from 1.06 to 1.07 results in no change in average gradient but does increase the relative film speed. For this film–process combination, optimum photographic results are obtained when the developer temperature is 36°C, pH is 9.0, and specific gravity is 1.07; this combination increases relative film speed by about 25% and the average gradient about 2%.

Film/processor performance can be optimized by first increasing the developer temperature. After the developer temperature is increased, solution pH should be measured to determine if the higher temperature has reduced the pH to unacceptably low levels, necessitating use of a buffer. Finally, the specific gravity is measured. If the specific gravity is less than the product specification, a higher-density solution should be obtained.

The most common deficiency related to film/processor systems is reduced relative film speed, which is often compensated for by increasing the mA · s, which unfortunately and generally unnecessarily increases radiation doses to patients. In extreme situations, the image receptor may be so slow that the kVp must be increased, which also results in low contrast and poor diagnostic quality.

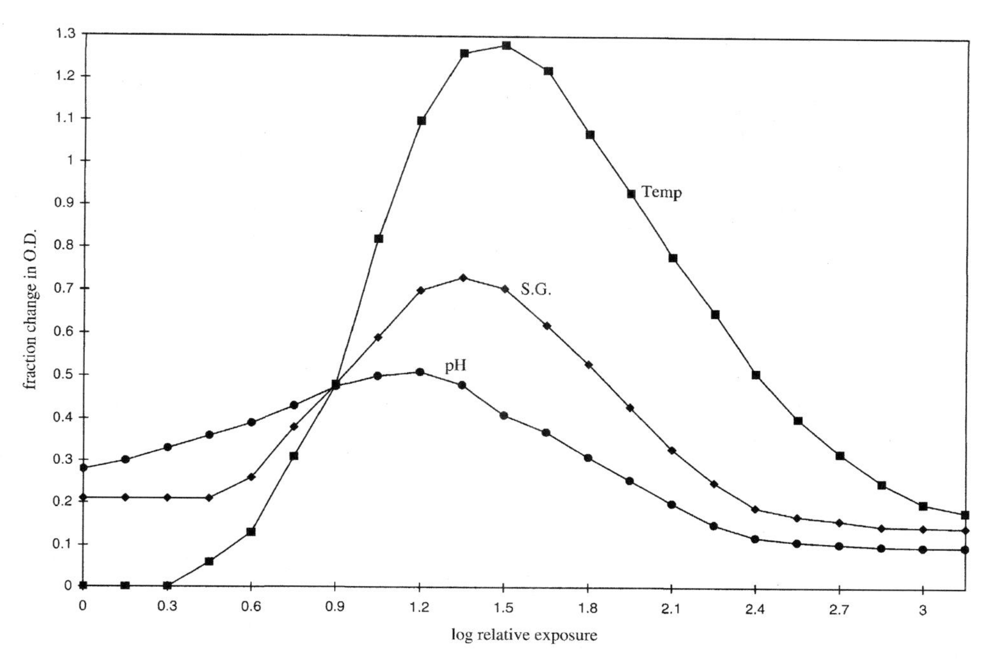

FIGURE 15-28. Influences of changes in developer temperature, pH, and specific gravity (S.G.) on the characteristic curve for radiographic film. (From Kasza, 1989.)

To visualize slight tissue density differences, relatively low kVp values must be used to create a broad exposure differential to accentuate radiographic contrast. Use of film/processing systems with a high average gradient allows use of a high-kVp technique to achieve the same radiographic contrast. High-kVp techniques are desirable because although dose increases with kVp per unit exposure, the mA · s required to produce image density decreases rapidly with concommitant reductions in net exposure and dose (for every 15% increase in kVp there is a 50% decrease in the mA · s required to obtain equivalent film density). Therefore, it is desirable to produce high average gradients in film processing so that high-kVp techniques can be used both to reduce patient dose and to achieve the appropriate radiographic contrast.

Dedicated film-screen mammography with a molybdenum target x-ray tube is one example of what happens when film processing considerations are not assessed properly. Mammographic technique factors are often modified to compensate for inadequacies traceable to the performance of the film processor rather than adjusting the processor to meet some objective level of performance, a circumstance that Kasza (1989) has observed frequently. For imaging facilities where such compensatory x-ray technique factors are used, there is found, at best, an increase in patient radiation exposure and dose over what is necessary, and at worst, poor image quality due to less-than-optimal technique factors. Film-screen mammography places extraordinary demands on radiological imaging; therefore, it is necessary to predetermine processing factors based on radiation dose and image-quality considerations and to use the most efficient, highest-quality processing techniques available.

RADIATION PROTECTION FOR X RAYS

Three groups of people should be considered for protection in the use of x rays: the patient, workers who perform examinations, and members of the public, most of whom can be presumed to be exposed only occasionally when attending to or visiting patients. Two axioms, applicable to all three groups, govern the degree of protection achieved: minimizing the number of x rays actually taken and assuring that those that are performed are done by well-trained personnel using optimal techniques.

Protection from appropriately prescribed x rays (or others for that matter) begins with assuring that the x-ray-generator/image-capture and processing system has optimized the interrelated physics parameters to provide a high-quality x-ray beam for the examination to be performed. Such a determination requires considerable professional judgment by persons skilled in radiological physics. Such judgments have been generalized into recommended factors by professional groups such as the National Council on Radiation Protection and Measurements (NCRP), the American College of Radiology (ACR), and others to guide good practice. Many of these findings have been encoded in regulations by the U.S. Food and Drug Administration (FDA) to assure that optimal systems are used.

Potential risks of the medical uses of ionizing radiation are difficult to assess, especially in light of the tremendous medical benefit they provide. It is, however, generally recognized by many professional medical organizations that more examinations than necessary are performed and that many of these are done with less than optimal technique. Many members of the medical profession do not necessarily think in terms of radiological protection; however, they will generally take steps to benefit their patients when made aware of improvements that can be made in diagnostic quality of x-ray procedures when optimal exposure is achieved. Fortunately, both medical goals and radiation protection goals are achieved because the sharpest and clearest radiographs with the maximum amount of diagnostic information is obtained when those features that optimize radiation exposure are used. The challenge to radiation protection personnel is to understand the interrelated factors and present information and services to eliminate unproductive exposures.

Assuring Patient Protection in X-Ray Diagnosis

The first element in protecting patient interests is to assure that the examination is in fact necessary (i.e., don't prescribe the study unless it is medically justified). Considerable pressures exist for physicians to prescribe x-ray examinations even if their professional judgment would suffice: They are vulnerable to a malpractice suit if a diagnosis is missed; patients expect and sometimes demand definitive proof; and criticism by medical colleagues for not meeting the standards of the neighborhood. It currently appears much safer to perform many x-ray examinations routinely even if a clinical diagnosis can be made without them; however, this may change under cost containment of medicine or if litigious elements of society associate cancer cases with unnecessary radiation exposure due to overexposure from less-than-optimal techniques or retakes caused by operators with minimal or improper qualifications.

The second element in radiation protection of patients is to assure that good-quality radiographs that contain the requisite diagnostic information are obtained by using properly calibrated and maintained x-ray equipment that is kept in optimal operating condition by quality control procedures. Mammography examinations indicate a trend toward expected performance since they are now required to meet American College of Radiology guides for imaging objects in a breast phantom while keeping exposures within recommended guidelines.

The remaining elements for patient protection from medical x rays are based on good practices, some of which are:

1. Confine the field size to the regions being examined through proper collimation and shielding, especially for the reproductive regions.
2. Use the maximum distance practicable between the x-ray source and the patient.
3. Use the highest x-ray tube voltage practicable, and use proper filtration of the x-ray beam to give the minimum absorbed dose consistent with producing a satisfactory radiograph.
4. Pay particular attention to the film processor, especially assurance of accurate processing temperature and the quality and strengths of process chemicals.
5. Use fast film/screen combinations and short exposure times.
6. Plan all exposures carefully and minimize retakes.

Radiation Exposure from X Rays

Patient exposure is determined by the particular tube current and beam quality (i.e., tube voltage and beam filtration) used in an x-ray radiographic procedure. From the foregoing discussion, it is apparent that any given system integrates many factors, all of which can influence exposure. To assure that x-ray beams are of sufficient quality, the FDA requires a minimum half-value layer (HVL) for x-ray tube voltages (kVp) as listed in Table 15-9; similar values have been recommended by NCRP and ACR. In this context, the HVL is a measure (and it must be measured) of the "hardness" (or overall quality) of the x rays in the beam. It is *not* the actual amount of filtration in the beam but is measured after technique factors, including inherent and added filtration, have been established. The HVL is measured by placing layers of absorber, usually sheets of type 1100 aluminum or copper, in the beam and determining the amount required to reduce the measured intensity or exposure of the existing beam to one-half its original value (the absorbers are, of course, removed before actual examinations are performed).

Although an x-ray machine that meets the HVL_{min} requirement does not completely assure that an x-ray beam will be optimal for a given examination, it does generally assure that most of the soft x rays that contribute only patient exposure are eliminated. As such, it is a convenient and good measure of minimum beam quality that can readily be determined. If the HVL_{min} requirement is not met, it can be reasonably assumed that patients will be exposed to unproductive soft x rays for the given kVp setting and will receive more radiation dose than is necessary. If the x-ray beam does not meet the minimum HVL requirements, it is necessary to add

TABLE 15-9. Minimum Half-Value Layers (HVL_{min}) for Diagnostic X-Ray Equipment and Dental X-Ray Equipment Above 70 kVp

X-Ray Tube Voltage (kVp)[a]	Single-Phase Unit HVL_{min} (mm Al)	Three-Phase Unit HVL_{min} (mm Al)
30	0.3	0.4
40	0.4	—
50	1.2	1.5
60	1.3	—
70	1.5	2
80	2.3	—
90	2.5	3.1
100	2.7	—
110	3.0	3.6
120	3.2	—
130	3.5	4.2
140	3.8	—
150	4.1	4.8

Source: Data from NCRP, 1977.
[a]Dental x-ray units operated below 70 kVp are required to have a HVL $\geq$ 1.5 mm of Al.

filtration to the exit port of the tube to increase the number of energetic photons in the beam.

Increasing the HVL above the HVL_{min} value can often decrease patient exposure without significantly increasing tube loading. For example, increasing the HVL from 2.5 mm of aluminum to 3.0 mm (at 80 kVp) reduces adult patient exposure by some 25%, with minimal effect on tube loading, image contrast, or radiographic density. Similar benefits occur for pediatric patients, but reductions in exposure are smaller because of thinner body parts, and a slight decrease in film contrast may be noted.

Entrance skin exposure (ESE) is generally used to designate patient exposure since the absorbed dose is a complex function of many interrelated parameters, most of which are affected by the spectral energy distribution associated with a particular x-ray examination. The relationships between exposure rate in air per mA · s for diagnostic x-ray equipment operated at different tube potentials (kVp) and amounts of Al filtration are shown in Figure 15-29. The data in Figure 15-29 are for a three-phase fully rectified diagnostic x-ray machine and a source-to-film distance of 100 cm; exposure rates for a fully rectified single-phase x-ray machine would be a factor of about 1.7 lower. The exposure rate in air is estimated by first determining the product of the tube current (mA) and the exposure time usually measured in seconds (or mA · s). Next, the exposure per mA · s is determined from Figure 15-29 for the tube voltage (kVp) and the filtration (expressed as mm Al or equivalent) used in the technique, which in turn is multiplied by the total mA · s used in the examination to obtain the exposure. If the filtration is not known, a nominal value of 2.5 mm Al can be used, and if the distance is different from the 100 cm given in Figure 15-29, it will be necessary to correct the exposure by the inverse-square law to account for it. Each of these factors is considered in Example 15-5.

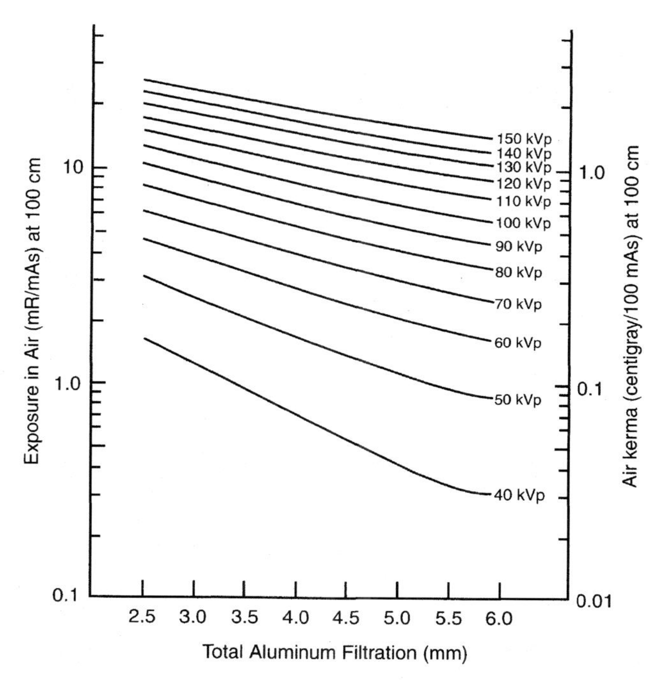

FIGURE 15-29. Exposure rates in air, without backscatter, from typical diagnostic three-phase x-ray equipment for a target–skin distance of 100 cm; these values would be a factor of 1.7 lower for a single-phase high-voltage supply.

TABLE 15-10. Backscatter Factors to Convert X-Ray Exposure in Air to Entrance Skin Exposure by Accounting for Photon Scattering Interactions in Tissue

X-Ray Tube Voltage (kVp)	Backscatter Factor	
	8 × 10 in. Field	14 × 17 in. Field
40	1.16	1.16
60	1.27	1.27
80	1.34	1.35
100	1.38	1.40
130	1.41	1.45
150	1.42	1.46

Source: Data from Trout et al., 1962.

Since radiation will scatter from within the body of the patient, it is necessary to account for the backscatter contribution to patient dose. This is done by an appropriate backscatter factor (BSF), which is a function of the size of the field, the x-ray energy, and the medium being irradiated; values for tissue are given in Table 15-10.

Example 15-5: An x-ray unit is operated at 90 kVp with 2.5-mm Al filtration and a target distance of 166 cm. The film size is 14 × 17 in. and the usual exposure is at 300 mA for $\frac{1}{120}$ s. What is the estimated exposure to the patient?

SOLUTION: From Figure 15-29, the exposure at 100 cm is 8.5 mR/mA · s and the total mA · s for an exposure is: mA · s = $\frac{1}{120} \times 300 = 2.5$ mA · s. The estimated exposure is

$$\text{exposure} = 8.5\ \text{mR/mA}\cdot\text{s} \times 2.5\ \text{mA}\cdot\text{s} \times \frac{(100\ \text{cm})^2}{(166\ \text{cm})^2} = 7.7\ \text{mR}$$

which is in air. The exposure at the surface of the patient must include consideration of backscatter, which by interpolation in Table 15-10 is 1.38 for 90 keV x rays; therefore, the entrance skin exposure is

$$\text{ESE} = 7.7\ \text{mR} \times 1.38 = 10.6\ \text{mR}$$

Film/screen combinations and beam quality affect entrance exposure because the amount of incident radiation varies with the amount absorbed in penetrating a given patient thickness and the sensitivity of various combinations of film and intensifying screens used to capture the image. Average exposures to produce a radiographic image on slow, medium, and fast blue- and green-sensitive films are listed in Table 15-11 when used with medium-speed calcium tungstate ($CaWO_4$) screens (5 to 6 line pairs/mm). The values in the table exclude the effects of grids and film processing on the amount of radiation delivered.

The exposure values listed in Table 15-11 are for double screens, one on each side of a double emulsion film, and are reasonably constant for x-ray tube potentials $\geq$ 70 kVp; the data in Table 15-11 indicate that the exposure required for slow-speed film is twice that for medium-speed films and one-half for fast films. Furthermore, if a slow-speed screen is used for a given film speed, the exposure required increases by a factor of 2.0, and if a fast screen is used, it decreases by a factor of 2.0. Certain rare-earth screens are rated as extrafast, and these require only one-third of the exposure needed to produce an image on medium-speed rare-earth screens (i.e., the exposure values listed in Table 15-11 for rare-earth screens would be divided by 3.0 to obtain the correct exposure for extrafast rare-earth screens).

The exposure values in Table 15-11 establish only the average amount of exposure incident on a particular film/screen unit to yield a radiographic image after

TABLE 15-11. Exposure to Produce an Image on Slow-, Medium-, and Fast-Speed Films Used with Medium-Speed Calcium Tungstate ($CaWO_4$) and Rare-Earth Screens

	Exposure Versus Film Speed		
Film/screen combination[a]	Slow	Medium	Fast
Blue film/$CaWO_4$	2 mR	1 mR	0.5 mR
Green film/rare earth[b]	1 mR	0.5 mR	0.25 mR

[a]For slow screens (7 to 8.5 line pairs/mm), these values increase by a factor of 2.0; for fast screens (3.5 to 4.3 line pairs/mm, they decrease by a factor of 2.0.

[b]For extrafast rare-earth screens, exposures decrease by a factor of 3.0.

TABLE 15-12. Absorption Factors (Ratio of Entry to Exit Dose) for General Radiographic Procedures

Tissue Thickness (cm)	Absorption Factor Versus HVL (mm Al)				
	1.5	2.0	3.0	4.0	5.0
0	1.00	1.00	1.00	1.00	1.00
3	2.10	1.81	1.54	1.35	1.17
6	4.35	3.13	2.50	2.12	1.80
8	6.41	4.42	3.42	2.90	2.40
10	9.35	5.99	4.61	3.82	3.20
12	13.7	8.20	6.25	5.18	4.30
14	20.0	11.2	8.55	7.04	5.70
16	28.6	15.4	11.8	9.52	7.00
20	63.0	29.5	22.0	17.3	13.7
25	165	65.5	49.0	36.5	27.8
30	440	146	108	78.0	57.0
35	1150	330	240	166	118

Source: NCRP Report 102, 1989.

penetration through the patient; it is a function of beam quality (represented by the HVL of the beam), the thickness of the tissue layer in the primary beam, and the entrance exposure. The amount of entrance exposure for a given film/screen combination will obviously be quite a bit higher, to allow for absorption in the tissue thickness being imaged, and is a function of the same parameters, which can be represented as an absorption factor (AF). Values of AF are listed in Table 15-12 for combinations of tissue thickness and HVLs of x-ray beams and are accurate to 10% for source to skin distances of 45 to 90 cm. All other factors related to x-ray beam quality (e.g., kVp) are incorporated in the HVL determination; thus the estimate of entrance skin exposure for the patient depends on an accurate determination of the HVL of the incident beam.

Example 15-6: A general radiographic x-ray system was determined to have a nominal beam quality represented by a HVL of 2.5 mm Al and to use a medium-speed $CaWO_4$ screen with medium-speed double-emulsion blue film: (*a*) What is the entrance skin exposure for imaging a tissue area that is 15 cm thick? (*b*) How would the exposure change if green-sensitive fast film were used with a fast-speed rare-earth screen? (*c*) How would it change with medium-speed green-sensitive film and an extrafast rare-earth screen?

SOLUTION: (a) From Table 15-11, the required exposure for a medium-speed $CaWO_4$ screen with medium-speed, double-emulsion blue film is 1 mR, and by interpolation in Table 15-12, the absorption factor for 15 cm of tissue and a HVL of 2.5 mm Al is 11.7; thus the ESE is

$$\text{ESE} = 1 \text{ mR} \times 11.7 = 11.7 \text{ mR}$$

(b) For a fast-speed green-sensitive film and a fast-speed rare-earth screen, the required exposure is one-half the value of 0.25 mR for medium-speed green film.

Since AF remains the same, the ESE is

$$\text{ESE} = 0.5 \times 0.25 \text{ mR} \times 11.7 = 1.46 \text{ mR}$$

(c) For green-sensitive medium-speed film with an extrafast rare-earth screen, the exposure is one-third of that for the medium-speed film–screen combination, which is given in Table 15-11 as 0.5 mR. Since AF remains the same, the ESE is

$$\text{ESE} = 0.333 \times 0.5 \text{ mR} \times 11.7 = 1.95 \text{ mR}$$

Good Practice

In January 1978, President Jimmy Carter, acting under the legislative authority given to the president to issue federal radiation guidance, issued "Radiation Protection Guidance to Federal Agencies for Diagnostic X Rays." The guidance contains 12 recommendations to reduce exposure to diagnostic uses of x rays by (1) minimizing the number of x rays taken by avoiding the prescription of clinically unproductive examinations and minimizing the number of radiographic views required, and (2) using optimal technique for x-ray procedures.

The federal radiation guidance addressed overprescription of x rays by recommending that only persons with specified qualifications be authorized to prescribe x-rays. Routine examinations were discouraged in general, and the guidance recommended that the following examinations not be performed:

1. Chest and lower-back x-ray examinations in routine physical examinations or as a routine requirement for employment
2. Tuberculosis screening by chest radiography
3. Chest x rays for routine hospital admission of patients under age 20 or lateral chest x rays for patients under age 40 unless a clinical indication of chest disease exists
4. Chest radiography in routine prenatal care
5. Mammography examinations of women under age 50 who neither exhibit symptoms nor have a personal or strong family history of breast cancer
6. Full-mouth series and bitewing radiographs in routine dental care in the absence of a clinical evaluation

Most professional medical groups support these prescription guides for routine/screening procedures, and even though controversy exists, a study by the National Cancer Institute has affirmed that risk–benefit considerations do not support frequent mammographic screening of asymtomatic women under age 50 for breast cancer.

To assure that equipment, technique, and operator performance were optimal, the guidance recommended that persons operating x-ray equipment should be qualified to perform examinations and should use optimal technique factors to reduce retakes and unnecessary exposure. To guide good technique, the guidance provided a set of entrance skin exposure guides (ESEGs) for certain routine nonspecialty

TABLE 15-13. Entrance Skin Exposure Guides for Medical X Rays

Examination (Projection)	ESE-1976 (mR)	ESE-1988 (mR)	
		200 speed	400 speed
Abdomen (A/P), grid	750	490	300
Lumbar spine (A/P)	1000	450	350
Full spine (A/P)	300	260	145
Cervical spine (A/P)	250	135	95
Skull (lateral)	300	145	70
Chest (P/A)			
No grid	25	15	5
Mammography			
No grid	—	345	—
Grid	—	690	—
Dental (bitewing)	700		
At 75 kVp	—	100–140 (E-speed film)	

Source: Data from EPA, 1976; CRCPD, 1988.

examinations (see Table 15-13). These exposure guides were an attempt to provide a framework which suggested that procedures which exceeded the exposure guides used less than optimal technique and that improvements were warranted. With reasonably good equipment and technique factors, they were easy to meet, but unfortunately, many facilities did not meet them even if exposure was determined, which often it was not. In 1988, the National Conference of Radiation Control Program Directors (CRCPD) of the various states provided an additional and improved set of ESEGs (also shown in Table 15-13) and added guides for mammography. Since states have regulatory authority for uses of medical radiation, the CRCPD has provided significant leadership in assessing and improving patient exposures for routine radiographic procedures.

Few states regulate patient doses from diagnostic x rays. The state of Illinois has, however, specified maximum patient doses for routine radiographic examinations which have greatly improved patient protection and improved radiographic quality. Medical radiographs are limited to a maximum of 1400 mR per radiograph, and it is recommended that the measured value at the tabletop for each radiographic exposure not exceed 100 mR. Illinois has also limited routine intraoral radiography to an incident exposure of 100 mR per radiograph and a limit of 500 mR per examination.

A national conference on referral criteria for x-ray examinations was held in 1978 by the Bureau of Radiological Health (now the National Center for Devices and Radiological Health) of the U.S. Public Health Service. The conference affirmed President Carter's federal radiation guidance by identifying major factors causing excessive radiological exposure: excessive radiation per film, excessive films per examination, excessive examinations per patient, lack of radiological screening of requests, and defensive medicine. Retakes because of faulty films were found to require about 27 million additional films per year. It was suggested that routine skull examinations for patients with headache or vertigo without other physical signs could be limited to a single film for screening. A periodic upper gastrointestinal series for patients with healed duodenal ulcers was found to be generally unnecessary because the patient's response could usually be determined by the clinical course observed; and similar findings were also made for routine "screening" x rays, except for high-risk groups. Defensive medicine, a consequence of existing malpractice mores, was believed to comprise a significant percentage of the millions of x-ray examinations given each year.

X-RAY SHIELDING

Shielding is placed around x-ray units and x-ray rooms to maintain exposures of workers and members of the public below prescribed limits and at levels as low as reasonable achievable (ALARA) within these limits. Since x-ray beams have a spectrum of energies, the usual shielding calculations with photon attenuation coefficients and buildup factors do not produce accurate results since these are based on monoenergetic photons (see Chapter 8). Instead, measurements of the reduction in transmission of x rays generated at different values of kVp have been performed, and the data have been developed into empirical relationships between exposure, kVp, and barrier thickness.

Shielding Calculations

The primary determinants of the x-ray exposure rate outside a protection shield are the operating kVp of the x-ray tube (a measure of the degree of penetration of the x-rays) and the total mA consumed, which determines the number of photons emitted. The kVp of the beam dominates the fractional transmission of x rays through a shield rather than the amount of filtration in the primary beam, which though very important for patient protection and diagnostic film quality by removing soft x rays, has little effect on the overall transmission of x-ray photons.

The workload of the x-ray unit is usually given in mA · min per week of tube operating time and is the primary determinant of the total exposure delivered. The workload of an x-ray unit is used to design the primary barrier that surrounds the unit to assure that worker exposure is within a specified value, usually ≤ 0.1 mSv/week. Secondary barriers are also needed to maintain exposures of the public in hallways, waiting rooms, and so on, below guidelines for exposure of the general population. Exposure rates are usually estimated at 1 m and then adjusted for the location of interest by the inverse-square law.

The design of x-ray shields assumes that x rays from a pacticular x-ray unit are generated at a single kVp and that this beam strikes the primary barrier unaltered. Curves for such beams have been developed for different barrier materials, four sets of which are plotted in Figure 15-30 for lead, Figure 15-31 for steel, Figure 15-32 for plate glass, and Figure 15-33 for gypsum. These curves are based on recent data by Archer et al. (1994) for three-phase x-ray generators to update data provided in NCRP Report 49 (1988); they also provide data for single-phase generators but recommend that shield designs be based on three-phase units since the shielding will be overestimated. The data in the curves relate exposure rate in dose units per mA · min at 1 m for different kVp x-ray beams and for variations in shield thickness.

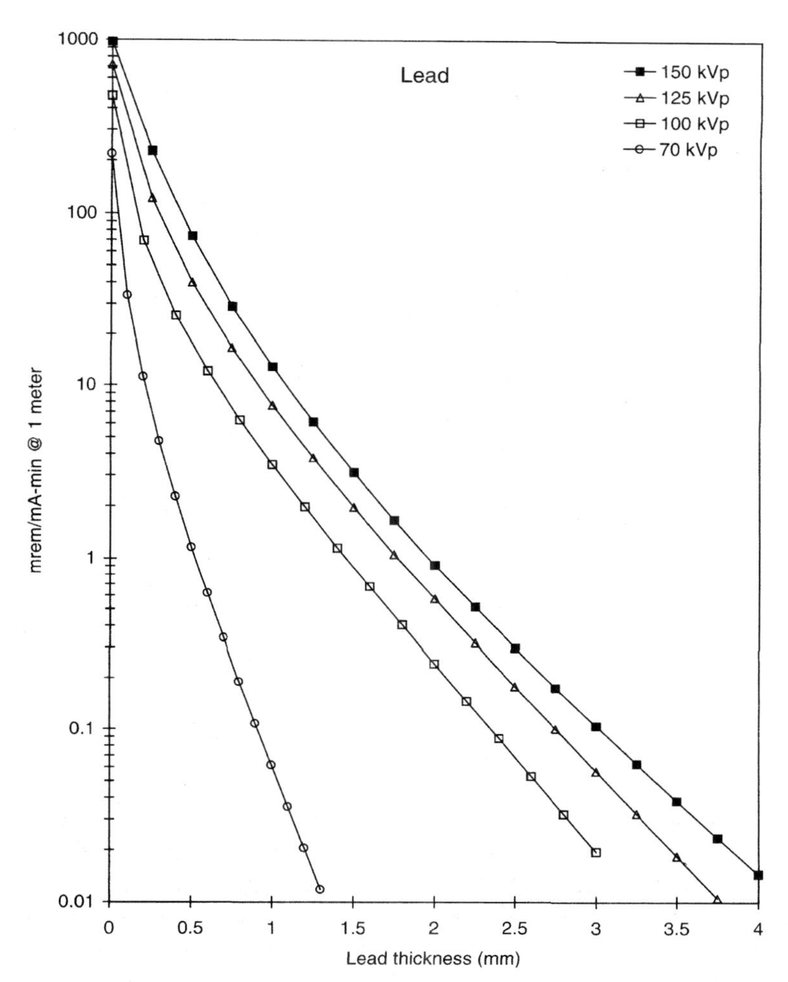

FIGURE 15-30. Dose equivalent rate per mA · min at 1 m for three-phase x-ray generators at various kVp values after transmission through thickness of lead. (Adapted from Archer et al., 1994.)

Similar curves exist for other media and for higher-energy special-purpose x-ray units such as those used in cancer therapy and testing of materials (see NCRP, 1988). These curves can be used to determine exposure versus shield thickness as shown in Example 15-7.

Example 15-7: An x-ray unit is normally operated at 125 kVp and 200 mA such that the average time the tube is energized is 90 s per week. If the unit is used in a configuration where it is pointed toward a hallway 4 m away one-third of the time (commonly called the *use factor*), what is the exposure in the hallway if it is occupied 8 h everyday and the beam must first penetrate a wall lined with 1 mm of lead sheet?

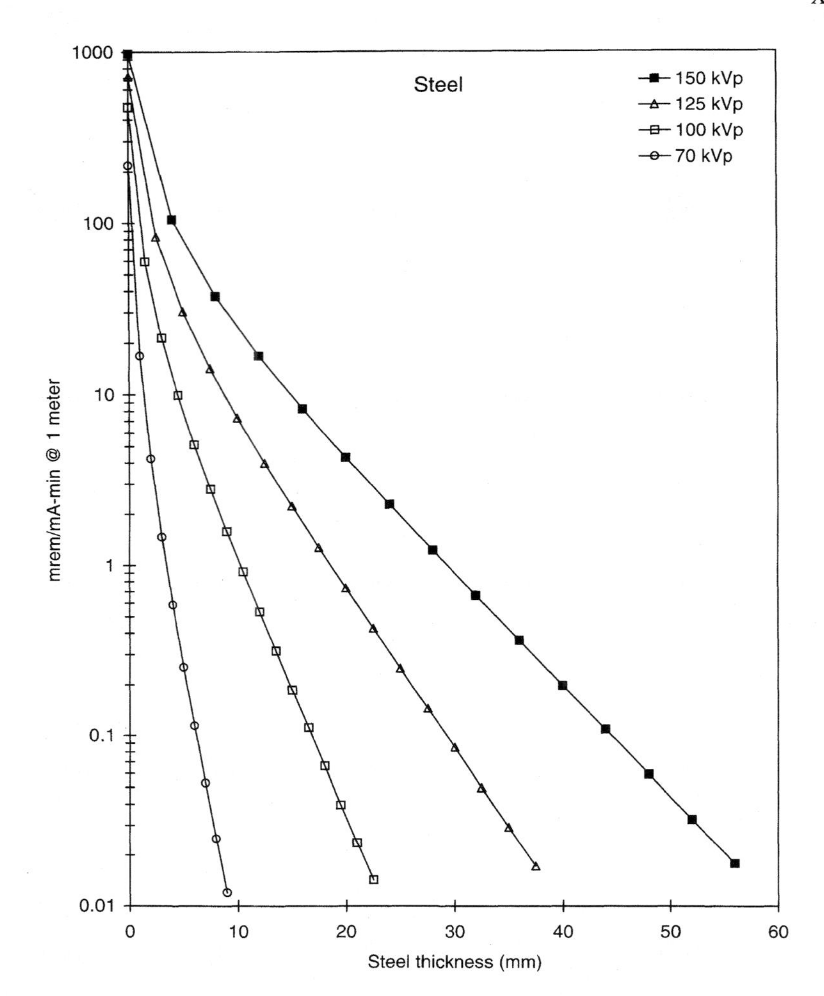

FIGURE 15-31. Dose equivalent rate per mA · min at 1 m for three-phase x-ray generators at various kVp values after transmission through thickness of steel. (Adapted from Archer et al., 1994.)

SOLUTION: The workload is 200 mA × 1.5 min/wk = 300 mA · min/wk, and since it is directed toward the hallway only one-third of the time, the workload contributing to exposure in the hallway is 100 mA · min/wk. From Figure 15-30, the exposure rate at 1 m for a 125 kVp unit and a 1-mm thickness of lead is 9 mrem/mA · min; therefore, the exposure in the hallway at 4 m distance, ignoring other materials in the wall, is

$$\text{exposure} = 9\ \text{mrem/mA} \cdot \text{min} \times \frac{100\ \text{mA} \cdot \text{min/wk}}{(4\ \text{m})^2} = 56\ \text{mrem/wk}$$

which presumes full occupancy of the hallway for the week during each 8-h period the x-ray unit is operated. Adjustments for real occupancy could reduce this estimate considerably.

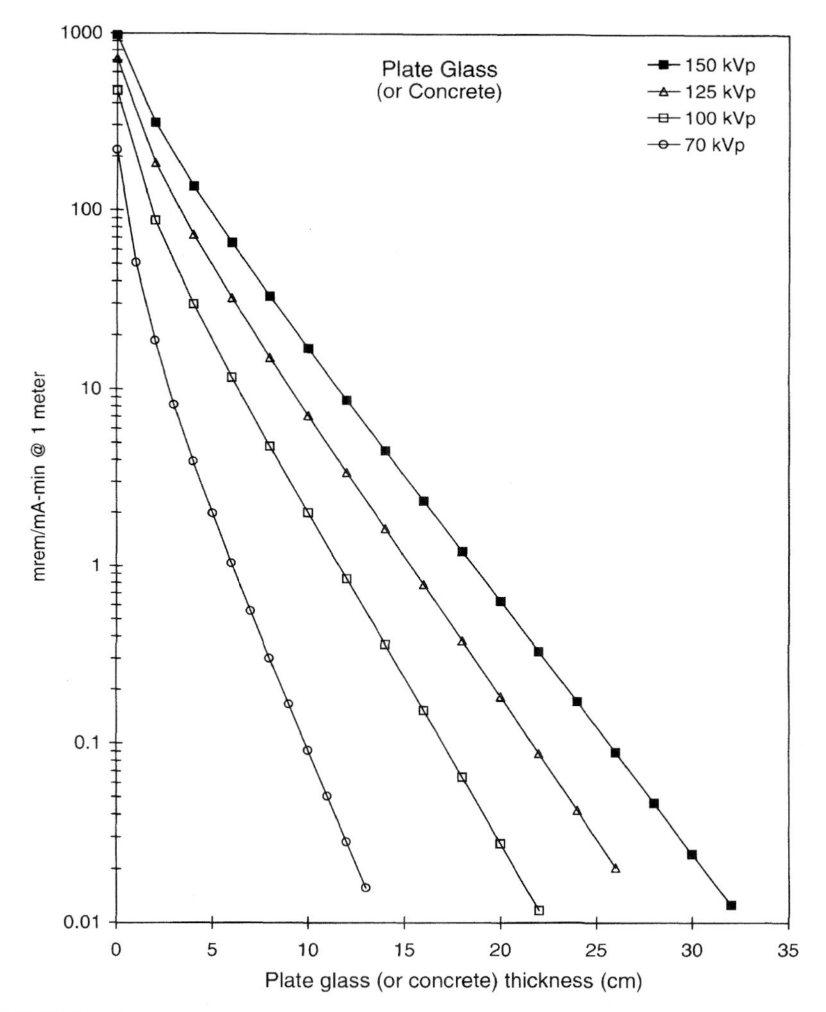

FIGURE 15-32. Dose equivalent rate per mA · min at 1 m for three-phase x-ray generators at various kVp values after transmission through thicknesses (in cm) of plate glass or concrete (p = 2.45). (Adapted from Archer et al., 1994.)

Two other considerations yield significant reductions in determinations of worker/public exposures: typical rather than conservative use patterns and attenuation provided by the x-ray capture system. Use factors for a general-purpose radiographic unit can reasonably be predicted because patterns of medical practice can be fairly standard. Dixon and Simpkin have established that a general-purpose radiographic room performs procedures with the beam toward the floor 89% of the time, toward the wall bucky 9% of the time, and toward other walls only 2% of the time, as shown in Figure 15-34. These use patterns indicate that chest radiography with a high-kVp technique is generally of low use even though barrier penetration is relatively higher because of the higher-energy x rays. On the other hand, most general radiography is done at 70 to 90 kVp but with higher use factors. The workload

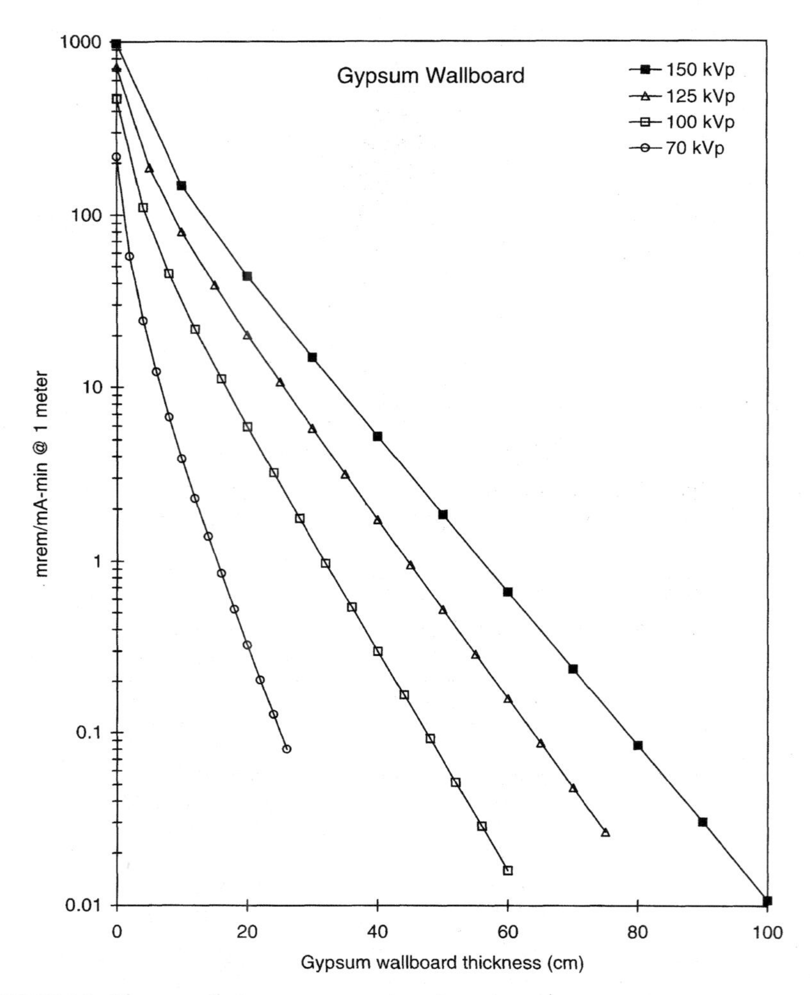

FIGURE 15-33. Dose equivalent rate per mA · min at 1 m for three-phase x-ray generators at various kVp values after transmission through thickness of gypsum. (Adapted from Archer et al., 1994.)

patterns in Figure 15-34 can be used to derive weighted values of total workload per patient, which is just the area under the curve.

Attenuation by the grid, the film cassette and holder, and the radiographic table reduce the x-ray beam intensity significantly before it strikes the walls or the floor of the x-ray room, thus reducing the amount of barrier material needed for shielding. The effectiveness of the x-ray capture system is provided in Table 15-14 and illustrated in Example 15-8.

Example 15-8: What additional thickness of lead would need to be included to reduce the weekly exposure rate calculated in Example 15-7 to 1 mrem/wk if credit is given for attenuation in the x-ray capture system?

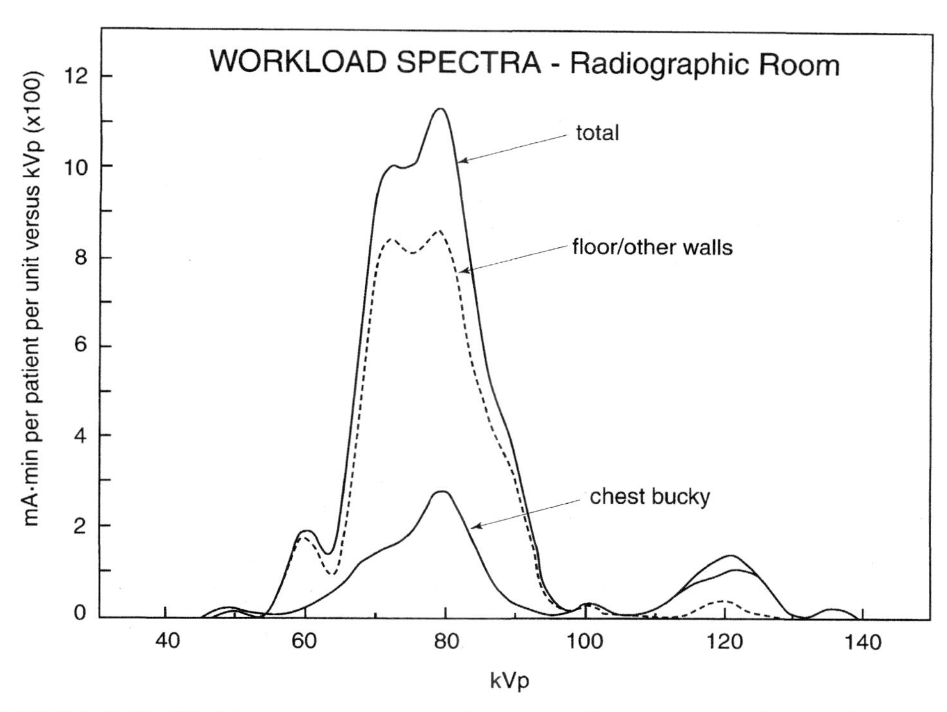

FIGURE 15-34. Workload spectra, smoothed for ease of interpretation, for a radiographic room determined by a survey of clinical sites. The total spectrum has been separated into one component in which the primary beam is directed against the chest bucky, and a second one in which it is directed toward the floor and other walls. (From Simpkin, 1996.)

TABLE 15-14. Thicknesses of X-Ray Shielding Materials That Provide Attenuation Equivalent to That of the Grid, Cassette Assembly, and Radiographic Table of a Typical Three-Phase X-Ray Unit

	Equivalent Thickness (mm) for X-Ray Tube Voltage (kVp) of:						
Material	50	70	80	90	100	120	140
Lead	0.5	0.75	0.09	1	1.05	0.88	0.7
Iron	3	5	6.5	7	7.2	7.5	7.2
Concrete	52	70	76	75	75	72	70
Glass	65	85	90	90	90	85	76
Gypsum	170	210	225	230	235	230	210

Source: Adapted from Dixon and Simpkins, 1998.

SOLUTION: The required reduction in beam intensity to achieve 1 mrem/wk at 4 m is a factor of 56 due to an exposure rate with 1 mm Pb of 9 mrem/mA · min. With the same use factor 1 mrem/wk would require a rate of 0.16 mrem/mA · min, which from Figure 15-30 corresponds to about 2.5 mm Pb, or about 1.5 mm additional. As shown in Table 15-14, the x-ray capture unit represents an equivalent thickness of about 0.8 mm of Pb for 125 kVp x rays; therefore, only about 0.7 mm of Pb (1.5 – 0.8 = 0.7 mm) is required.

The more realistic conditions of Example 15-8 are also conservative because attenuation of the x-ray beam by the patient, who is almost always in the primary beam, is not accounted for. Dimunition of the x-ray beam is even greater for fluoroscopy units, including spot filming during these procedures, because the primary beam is intercepted by a fairly massive image receptor system and attenuated below federal standards for scatter radiation.

SUMMARY

The discovery and subsequent uses of x rays have been a boon for humankind, especially since W. C. Roentgen considered his discovery a gift to science and did not seek to patent the device. Although many improvements in equipment and supporting facilities have occurred since, the essential features of x rays remain as he described them in 1895.

Two compatible goals in medical uses of x rays are to optimize the parameters of purposeful exposures to maximize diagnostic information while avoiding unnecessary radiation exposure of patients and workers. Four major interconnected categories of physical principles are fundamental to these two goals: (1) production of x-ray photons of sufficient quality and quantity at the patient to image the body part under study, (2) the pattern of absorption and attenuation interactions that produce information that can be captured and made visible, (3) capture and recording by film or electronic means of the image produced by the pattern of photons transmitted through the patient, and (4) processing of the diagnostic image captured. Optimization of these interconnected features is complicated because x rays are produced with a wide range of photon energies and with a large number of low-energy x rays that contribute very little, if anything, to the image but if not removed, can deliver very high doses to patients.

Production and shaping of the x-ray beam are directed at producing a pattern of photons on the exit side of the patient that sharply delineates differences in tissue density such that they yield a visual image in an image capture system. The photons can be "seen" because they have been removed or absorbed to varying degrees. The features of the image capture system also determine the number and quality of photons generated and directed at the patient, especially the various contributions of film speed, intensifying screens, and grids that are selected. The information contained in the photons that reach the image capture system does not become visible until it is processed, either electronically or by film developing. If a proper balance is not achieved between temperature, pH, and specific gravity of film-developing solutions, the relative film speed will be reduced, which unfortunately can be, and often is, compensated for by increasing the mA · s with a corresponding (and often unnecessary) increase in patient exposure.

Radiation protection for patients, workers, and members of the public requires minimizing the number of x rays prescribed and assuring optimal performance of those that are actually taken. Although each x-ray image receives a quality evaluation as it is interpreted, the patient exposure, which can vary considerably for established procedures, is usually not evaluated; therefore, it is useful to measure exposures periodically and to compare them to recommended exposure guides. Exposure guides can indicate less-than-optimal technique, and perhaps more important, unproductive exposure of the patient.

Shielding of x-ray units and x-ray rooms is an important feature for radiation protection of workers and members of the public. Since x-ray beams have a spectrum of energies, x-ray shielding calculations are based on empirical relationships between exposure, kVp (generally assumed to be constant), and barrier thickness. Attenuation of the x-ray beam by the grid, the film cassette and holder, and the radiographic table can in combination reduce the x-ray beam intensity considerably and the barrier thickness required to meet exposure objectives. Designs that account for these more realistic conditions are also somewhat conservative because attenuation of the x-ray beam by the patient, who is almost always in the primary beam, is usually not considered.

ACKNOWLEDGMENTS

This chapter was compiled with the help of Chul Lee and Thomas Kasza, with special thanks to S. C. Bushong, the authors of Christensen's *Physics of Radiology*, and Jacob Shapiro, whose writings are a tribute to understanding and optimizing the use of x rays.

REFERENCES AND ADDITIONAL RESOURCES

Archer, B. R., T. R. Fewell, B. J. Conway, and P. W. Quinn, Attenuation properties of diagnostic x-ray shielding materials, *Medical Physics* **21**, 9 (1994). (Part of Report of Task Group No. 9, Am. Assoc. of Physicists in Medicine.)

Berger, M. J., and S. M. Selzer, Stopping powers and ranges of electrons and positions, NBSIR 82-2550-A, National Bureau of Standards, Washington, DC, 1983.

Bushong, S. C., *Radiologic Science for Technologists*, 3rd ed., C. V. Mosby, St. Louis, MO, 1984.

CRCPD, Average Patient Exposure Guides 1988. Conference of Radiation Control Program Directors, Inc., Publ. 88-5, Frankfort, KY, 1988.

Curry, T. S., III, J. E. Dowdey, and R. C. Murray, Jr., *Christensen's Physics of Radiology*, 4th ed., Lea & Febiger, Philadelphia, PA, 1990.

Dixon, R. L., and D. J. Simpkin, Primary shielding barriers for diagnostic X-ray facilities: A new model, *Health Physics* **74**, 2 (1998).

Hubbell, J. H., and S. M. Seltzer, *Tables of X-ray Mass Attenuation Coefficients and Mass Energy-Absorption Coefficients* 1 keV *to* 20 MeV *for Elements* Z = 1 *to* 92 *and* 48 *Additional Substances of Dosimetric Interest*, NISTIR 5632, National Institute of Standards and Technology, Gaithersburg, MD, 1995.

Kasza, T., Effect of film processing techniques on radiation exposure, Master's thesis, University of Michigan, School of Public Health, Ann Arbor, MI, 1989.

National Council on Radiation Protection and Measurements, Structure Shielding Design and Evaluation for Medical Use of X rays and Gamma Rays of Energies up to 10 MeV, NCRP Report 49, Bethesda, MD, 1976.

Shapiro, J., *Radiation Protection: A Guide for Scientists and Physicians*, 3rd ed., Harvard University Press, Cambridge, MA, 1990.

Simpkin, D. J., Evaluation of NCRP Report 49 Assumptions on workloads and use factors in dianogistic radiology facilities, *Medical Physics* **23** (1996), 577–584.

Trout, E. D., J. P. Kelly, and A. C. Lucas, The effect of kilovoltage and filtration on depth dose. In *Technological Needs for Reduction of Patient Dosage from Diagnostic Radiology*, M. L. Janowen and Charles C. Thomas, eds., Springfield, MO, 1962.

U.S. Environmental Protection Agency, Radiation Protection Guidance for Diagnostic X-rays, Federal Guidance Report No. 9, EPA 520/4-76-019, 1976.

PROBLEMS

15-1. X-ray machines produce leakage radiation and scattered radiation. Which is the most penetrating, and why?

15-2. A 150 kVp diagnostic x-ray machine is installed in a room with a 4-in. plaster wall along a 4-ft hallway and a 6-in. concrete wall that adjoins a city street. The average weekly workload is 25 mA · min, during which the useful beam is pointed horizontally in the direction of the hall and 100 mA · min when it is pointed horizontally in the direction of the outside wall next to the street. (**a**) What are the exposure rates in the hall and outside the concrete wall? (**b**) Calculate any additional thickness of concrete that is needed for the primary protective barrier for each.

15-3. For Problem 15-2, what additional thickness of lead added to the plaster wall will provide an adequate primary barrier for (**a**) the hall and (**b**) the public street?

15-4. What thickness is required for a plaster wall between an x-ray machine and an adjoining laboratory to keep exposure below 10 mR/week if the x-ray machine operates at 150 kVp for 25 min each day?

15-5. What is the weekly exposure rate in the center of an office directly below the 4-in. concrete floor of a dentist's office if a 100 kVp x-ray machine is operated at an average weekly workload of 100 mA · min?

15-6. A dental x-ray machine that will operate at 70 kVp with a weekly workload of 50 mA · min is to be placed 4 ft from the outside of a (soft) brick wall where a public sidewalk is located. How thick must the brick wall be to provide an adequate primary protective barrier?

APPENDIX A

CONSTANTS OF NATURE AND SELECTED PARTICLE MASSES

Speed of light	c	2.99792458×10^{8} m/s		
Charge of electron	e	$1.60217733 \times 10^{-19}$ C		
Boltzmann constant	k	1.380658×10^{-23} J/K		
		8.617385×10^{-5} eV/K		
Faraday's constant		96485.309 C/mol		
Planck's constant	h	$6.6260755 \times 10^{-34}$ J · s		
		$4.1356692 \times 10^{-15}$ eV · s		
Gravitational constant	G	6.67259×10^{-11} $m^3/kg \cdot s^2$		
Avogadro's number	N_A	6.0221367×10^{23} mol^{-1}		
Universal gas constant	R	8.314510 J/mol · K		
Stefan–Boltzmann constant	σ	5.6705×10^{-8} $W/m^2 \cdot K^4$		
Rydberg constant	R	10973731.534 m^{-1}		
Bohr radius	a_0	$0.529177249 \times 10^{-10}$ m		
Fine structure constant	a	1/137.0359895		
Electron volt	eV	$1.60217733 \times 10^{-19}$ J		
Joule	J	6.2415×10^{-18} eV		
		10^{7} ergs		
Erg		10^{-7} J		
Unified mass unit		1.66054×10^{-27} kg	1.00000 u	931.502 MeV
Electron	e	$9.1093897 \times 10^{-31}$ kg	5.4857990×10^{-4} u	0.5109990 MeV
Proton	p	$1.6726231 \times 10^{-27}$ kg	1.00727647 u	938.27231 MeV
Neutron	n	$1.6749286 \times 10^{-27}$ kg	1.00866490 u	939.56563 MeV
Deuteron	d	$3.3435856 \times 10^{-27}$ kg	2.01355320 u	1875.62883 MeV
Tritium (^3H)	T	$5.0073595 \times 10^{-27}$ kg	3.015500688 u	2808.94492 MeV
Alpha	α	$6.6446618 \times 10^{-27}$ kg	4.001506178 u	3727.38025 MeV
Helium	He	$6.6464828 \times 10^{-27}$ kg	4.002603250 u	3728.43293 MeV

Source: *CODATA Recommended Values of the Fundamental Physical Constants*, National Institue of Science and Technology, Gaithersburg, MD, 1986. Internet address: *physics.nist.gov/PhysRefData/codata86/codata86.html*.

APPENDIX B

ATOMIC MASSES AND BINDING ENERGIES FOR SELECTED ISOTOPES OF THE ELEMENTS

Z		A	Binding Energy (MeV)	Atomic Mass (amu)
0	n	1	0	1.008664923
1	H	1	0	1.007825032
		2	2.225	2.014101778
		3	8.482	3.016049268
2	He	3	7.718	3.016029310
		4	28.296	4.002603250
3	Li	6	31.995	6.015122281
		7	39.245	7.016004049
		8	41.277	8.022486670
4	Be	7	37.6	7.016929246
		8	56.5	8.005305094
		9	58.165	9.012182135
		10	64.977	10.013533720
		11	65.481	11.021657653
5	B	8	37.738	8.024606713
		9	56.314	9.013328806
		10	64.751	10.012937027
		11	76.205	11.009305466
		12	79.575	12.014352109
6	C	10	60.321	10.016853110
		11	73.44	11.011433818
		12	92.162	12.000000000
		13	97.108	13.003354838
		14	105.285	14.003241988
		15	106.503	15.010599258
7	N	13	94.105	13.005738584
		14	104.659	14.003074005
		15	115.492	15.000108898
		16	117.981	16.006101417
		17	123.865	17.008449673

Z		A	Binding Energy (MeV)	Atomic Mass (amu)
8	O	14	98.733	14.008595285
		15	111.956	15.003065386
		16	127.619	15.994914622
		17	131.763	16.999131501
		18	139.807	17.999160419
		19	143.763	19.003578730
		20	151.371	20.004076150
9	F	17	128.22	17.002095238
		18	137.369	18.000937667
		19	147.801	18.998403205
		20	154.403	19.999981324
		21	162.504	20.999948921
10	Ne	18	132.153	18.005697066
		19	143.781	19.001879839
		20	160.645	19.992440176
		21	167.406	20.993846744
		22	177.77	21.991385510
		23	182.971	22.994467337
		24	191.836	23.993615074
11	Na	21	163.076	20.997655099
		22	174.145	21.994436782
		23	186.564	22.989769675
		24	193.523	23.990963332
		25	202.535	24.989954352
		26	208.151	25.992589898
12	Mg	22	168.578	21.999574055
		23	181.725	22.994124850
		24	198.257	23.985041898
		25	205.588	24.985837023
		26	216.681	25.982593040
		27	223.124	26.984340742
		28	231.628	27.983876703

Z		A	Binding Energy (MeV)	Atomic Mass (amu)
13	Al	25	200.528	24.990428555
		26	211.894	25.986891659
		27	224.952	26.981538441
		28	232.677	27.981910184
		29	242.113	28.980444848
14	Si	26	206.046	25.992329935
		27	219.357	26.986704764
		28	236.537	27.976926533
		29	245.01	28.976494719
		30	255.62	29.973770218
		31	262.207	30.975363275
		32	271.41	31.974148129
15	P	29	239.285	28.981801376
		30	250.605	29.978313807
		31	262.917	30.973761512
		32	270.852	31.973907163
		33	280.956	32.971725281
16	S	30	243.685	29.984902954
		31	256.738	30.979554421
		32	271.781	31.972070690
		33	280.422	32.971458497
		34	291.839	33.967866831
		35	298.825	34.969032140
		36	308.714	35.967080880
		37	313.018	36.971125716
17	Cl	33	274.057	32.977451798
		34	285.566	33.973761967
		35	298.21	34.968852707
		36	306.789	35.968306945
		37	317.1	36.965902600
		38	323.208	37.968010550
		39	331.282	38.968007677

Z	El	A		
18	Ar	34	278.721	33.980270118
		35	291.462	34.975256726
		36	306.716	35.967546282
		37	315.505	36.966775912
		38	327.343	37.962732161
		39	333.941	38.964313413
		40	343.81	39.962383123
		41	349.909	40.964500828
		42	359.335	41.963046386
19	K	37	308.573	36.973376915
		38	320.647	37.969080107
		39	333.724	38.963706861
		40	341.523	39.963998672
		41	351.618	40.961825972
		42	359.152	41.962403059
		43	368.795	42.960715746
20	Ca	38	313.122	37.976318637
		39	326.411	38.970717729
		40	342.052	39.962591155
		41	350.415	40.962278349
		42	361.895	41.958618337
		43	369.828	42.958766833
		44	380.96	43.955481094
		45	388.375	44.956185938
		46	398.769	45.953692759
		47	406.045	46.954546459
		48	415.991	47.952533512
		49	421.138	48.955673302
21	Sc	43	366.825	42.961150980
		44	376.525	43.959403048
		45	387.849	44.955910243
		46	396.61	45.955170250
		47	407.254	46.952408027
		48	415.487	47.952234991
22	Ti	44	375.475	43.959690235
		45	385.005	44.958124349
		46	398.194	45.952629491
		47	407.072	46.951763792
		48	418.699	47.947947053
		49	426.841	48.947870789
		50	437.78	49.944792069
		51	444.153	50.946616017
		52	451.961	51.946898175
23	V	48	413.904	47.952254480
		49	425.457	48.948516914
		50	434.79	49.947162792
		51	445.841	50.943963675
		52	453.152	51.944779658
		53	461.631	52.944342517
24	Cr	48	411.462	47.954035861
		49	422.044	48.951341135
		50	435.044	49.946049607
		51	444.306	50.944771767
		52	456.345	51.940511904
		53	464.284	52.940653781
		54	474.003	53.938884921
		55	480.25	54.940844164
		56	488.506	55.940645238
25	Mn	52	450.851	51.945570079
		53	462.905	52.941294702
		54	471.844	53.940363247
		55	482.07	54.938049636
		56	489.341	55.938909366
		57	497.991	56.938287458
26	Fe	52	447.697	51.948116526
		53	458.38	52.945312282
		54	471.759	53.939614836
		55	481.057	54.938298029
		56	492.254	55.934942133
		57	499.9	56.935398707
		58	509.944	57.933280458
		59	516.525	58.934880493
		60	525.345	59.934076943
		61	530.927	60.936749461
27	Co	57	498.282	56.936296235
		58	506.855	57.935757571
		59	517.308	58.933200194
		60	524.8	59.933822196
		61	534.122	60.932479381
28	Ni	56	483.988	55.942136339
		57	494.235	56.939800489
		58	506.454	57.935347922
		59	515.453	58.934351553
		60	526.842	59.930790633
		61	534.662	60.931060442
		62	545.259	61.928348763
		63	552.097	62.929672948
		64	561.755	63.927969574
		65	567.853	64.930088013
29	Cu	61	531.642	60.933462181
		62	540.528	61.932587299
		63	551.381	62.929601079
		64	559.297	63.929767865
		65	569.207	64.927793707
		66	576.273	65.928873041
		67	585.391	66.927750294

Z		A	Binding Energy (MeV)	Atomic Mass (amu)
30	Zn	62	538.119	61.934334132
		63	547.232	62.933215563
		64	559.094	63.929146578
		65	567.073	64.929245079
		66	578.133	65.926036763
		67	585.185	66.927130859
		68	595.383	67.924847566
		69	601.866	68.926553538
		70	611.081	69.925324870
		71	616.915	70.927727195
31	Ga	67	583.402	66.928204915
		68	591.68	67.927983497
		69	601.989	68.925580912
		70	609.644	69.926027741
		71	618.948	70.924705010
		72	625.469	71.926369350
		73	634.657	72.925169832
32	Ge	68	590.792	67.928097266
		69	598.98	68.927972002
		70	610.518	69.924250365
		71	617.934	70.924953991
		72	628.685	71.922076184
		73	635.468	72.923459361
		74	645.665	73.921178213
		75	652.17	74.922859494
		76	661.598	75.921402716
		77	667.671	76.923548462
33	As	73	634.345	72.923825288
		74	642.32	73.923929076
		75	652.564	74.921596417
		76	659.892	75.922393933
		77	669.59	76.920647703
34	Se	72	622.43	71.927112313
		73	630.823	72.926766800
		74	642.89	73.922476561
		75	650.918	74.922523571
		76	662.072	75.919214107
		77	669.491	76.919914610
		78	679.989	77.917309522
		79	686.951	78.918499802
		80	696.865	79.916521828
		81	703.566	80.917992931
		82	712.842	81.916700000
		83	718.66	82.919119072
35	Br	77	667.343	76.921380123
		78	675.633	77.921146130
		79	686.32	78.918337647
		80	694.212	79.918529952
		81	704.369	80.916291060
		82	711.962	81.916804666
		83	721.547	82.915180219
36	Kr	76	654.235	75.925948304
		77	663.499	76.924667880
		78	675.558	77.920386271
		79	683.912	78.920082992
		80	695.434	79.916378040
		81	703.306	80.916592419
		82	714.272	81.913484601
		83	721.737	82.914135952
		84	732.257	83.911506627
		85	739.378	84.912526954
		86	749.235	85.910610313
		87	754.75	86.913354251
37	Rb	83	720.045	82.915111951
		84	728.794	83.914384676
		85	739.283	84.911789341
		86	747.934	85.911167080
		87	757.853	86.909183465
		88	763.936	87.911318556
38	Sr	82	708.128	81.918401258
		83	716.987	82.917555029
		84	728.906	83.913424778
		85	737.436	84.912932689
		86	748.926	85.909262351
		87	757.354	86.908879316
		88	768.467	87.905614339
		89	774.825	88.907452906
		90	782.631	89.907737596
39	Y	87	754.71	86.910877833
		88	764.062	87.909503361
		89	775.538	88.905847902
		90	782.395	89.907151443
		91	790.325	90.907303415
40	Zr	88	762.606	87.910226179
		89	771.923	88.908888916
		90	783.893	89.904703679
		91	791.087	90.905644968
		92	799.722	91.905040106
		93	806.456	92.906475627
		94	814.676	93.906315765
		95	821.139	94.908042739
		96	828.993	95.908275675
		97	834.573	96.910950716
41	Nb	91	789.052	90.906990538
		92	796.934	91.907193214
		93	805.765	92.906377543
		94	812.993	93.907283457
		95	821.482	94.906835178

42	Mo	90	773.728	89.913936161
		91	783.835	90.911750754
		92	796.508	91.906810480
		93	804.578	92.906812213
		94	814.256	93.905087578
		95	821.625	94.905841487
		96	830.779	95.904678904
		97	837.6	96.906021033
		98	846.243	97.905407846
		99	852.168	98.907711598
		100	860.458	99.907477149
		101	865.856	100.910346543
43	Tc	95	819.152	94.907656454
		96	827.024	95.907870803
		97	836.498	96.906364843
		98	843.776	97.907215692
		99	852.743	98.906254554
		100	859.507	99.907657594
44	Ru	94	806.849	93.911359569
		95	815.802	94.910412729
		96	826.496	95.907597681
		97	834.607	96.907554546
		98	844.791	97.905287111
		99	852.254	98.905939307
		100	861.928	99.904219664
		101	868.73	100.905582219
		102	877.949	101.904349503
		103	884.182	102.906323677
		104	893.085	103.905430145
		105	898.995	104.907750341
45	Rh	101	867.406	100.906163526
		102	874.844	101.906842845
		103	884.163	102.905504182
		104	891.162	103.906655315
		105	900.13	104.905692444
46	Pd	100	856.371	99.908504596
		101	864.643	100.908289144
		102	875.213	101.905607716
		103	882.837	102.906087204
		104	892.82	103.904034912
		105	899.914	104.905084046
		106	909.477	105.903483087
		107	916.016	106.905128453
		108	925.236	107.903894451
		109	931.39	108.905953535
		110	940.207	109.905152385
		111	945.958	110.907643952
47	Ag	105	897.787	104.906528234
		106	905.729	105.906666431
		107	915.266	106.905093020
		108	922.536	107.905953705
		109	931.723	108.904755514
		110	938.532	109.906110460
48	Cd	104	885.841	103.909848091
		105	894.266	104.909467818
		106	905.141	105.906458007
		107	913.067	106.906614232
		108	923.402	107.904183403
		109	930.727	108.904985569
		110	940.642	109.903005578
		111	947.618	110.904181628
		112	957.016	111.902757226
		113	963.556	112.904400947
		114	972.599	113.903358121
		115	978.74	114.905430553
		116	987.44	115.904755434
		117	993.217	116.907218242
49	In	111	945.97	110.905110677
		112	953.648	111.905533338
		113	963.091	112.904061223
		114	970.365	113.904916758
		115	979.404	114.903878328
		116	986.188	115.905259995
50	Sn	110	934.562	109.907852688
		111	942.743	110.907735404
		112	953.529	111.904820810
		113	961.272	112.905173373
		114	971.571	113.902781816
		115	979.117	114.903345973
		116	988.68	115.901744149
		117	995.625	116.902953765
		118	1,004.952	117.901606328
		119	1,011.437	118.903308880
		120	1,020.544	119.902196571
		121	1,026.715	120.904236867
		122	1,035.529	121.903440138
		123	1,041.475	122.905721901
		124	1,049.962	123.905274630
		125	1,055.695	124.907784924
		126	1,063.889	125.907653953
51	Sb	119	1,010.061	118.903946460
		120	1,017.081	119.905074315
		121	1,026.323	120.903818044
		122	1,033.13	121.905175415
		123	1,042.095	122.904215696
		124	1,048.563	123.905937525
		125	1,057.276	124.905247804
52	Te	118	999.457	117.905825187
		119	1,006.985	118.906408110
		120	1,017.281	119.904019891
		121	1,024.505	120.904929815
		122	1,034.33	121.903047064
		123	1,041.259	122.904272951
		124	1,050.685	123.902819466
		125	1,057.261	124.904424718
		126	1,066.375	125.903305543
		127	1,072.665	126.905217290
		128	1,081.441	127.904461383
		129	1,087.524	128.906595593
		130	1,095.943	129.906222753
		131	1,101.872	130.908521880

Z	A	Binding Energy (MeV)	Atomic Mass (amu)
53 I	125	1,056.293	124.904624150
	126	1,063.437	125.905619387
	127	1,072.58	126.904468420
	128	1,079.406	127.905805254
	129	1,088.239	128.904987487
	130	1,094.74	129.906674018
54 Xe	122	1,027.641	121.908548396
	123	1,035.785	122.908470748
	124	1,046.254	123.905895774
	125	1,053.858	124.906398236
	126	1,063.913	125.904268868
	127	1,071.136	126.905179581
	128	1,080.743	127.903530436
	129	1,087.651	128.904779458
	130	1,096.907	129.903507903
	131	1,103.512	130.905081920
	132	1,112.447	131.904154457
	133	1,118.887	132.905905660
	134	1,127.435	133.905394504
	135	1,133.817	134.907207499
	136	1,141.877	135.907219526
	137	1,145.903	136.911562939
55 Cs	131	1,102.377	130.905460232
	132	1,109.545	131.906429799
	133	1,118.532	132.905446870
	134	1,125.424	133.906713419
	135	1,134.186	134.905971903
	136	1,141.015	135.907305741
	137	1,149.293	136.907083505
56 Ba	128	1,074.727	127.908308870
	129	1,082.459	128.908673749
	130	1,092.731	129.906310478
	131	1,100.225	130.906930798
	132	1,110.042	131.905056152
	133	1,117.232	132.906002368
	134	1,126.7	133.904503347
	135	1,133.673	134.905682749
	136	1,142.78	135.904570109
	137	1,149.686	136.905821414
	138	1,158.298	137.905241273
	139	1,163.021	138.908835384
57 La	136	1,139.128	135.907651181
	137	1,148.304	136.906465656
	138	1,155.778	137.907106826
	139	1,164.556	138.906348160
	140	1,169.717	139.909472552
	141	1,176.405	140.910957016
58 Ce	134	1,120.922	133.909026379
	135	1,128.882	134.909145555
	136	1,138.819	135.907143574
	137	1,146.299	136.907777634
	138	1,156.04	137.905985574
	139	1,163.495	138.906646605
	140	1,172.696	139.905434035
	141	1,178.125	140.908271103
	142	1,185.294	141.909239733
	143	1,190.439	142.912381158
	144	1,197.335	143.913642686
59 Pr	139	1,160.584	138.908932181
	140	1,168.526	139.909071204
	141	1,177.923	140.907647726
	142	1,183.766	141.910039865
	143	1,191.118	142.910812233
60 Nd	140	1,167.521	139.909309824
	141	1,175.318	140.909604800
	142	1,185.146	141.907718643
	143	1,191.27	142.909809626
	144	1,199.087	143.910082629
	145	1,204.842	144.912568847
	146	1,212.407	145.913112139
	147	1,217.7	146.916095794
	148	1,225.032	147.916888516
	149	1,230.071	148.920144190
	150	1,237.451	149.920886563
	151	1,242.785	150.923824739
61 Pm	143	1,189.446	142.910927571
	144	1,195.973	143.912585768
	145	1,203.897	144.912743879
	146	1,210.153	145.914692165
	147	1,217.813	146.915133898
	148	1,223.71	147.917467786
	149	1,230.979	148.918329195
62 Sm	142	1,176.619	141.915193274
	143	1,185.221	142.914623555
	144	1,195.741	143.911994730
	145	1,202.498	144.913405611
	146	1,210.913	145.913036760
	147	1,217.255	146.914893275
	148	1,225.396	147.914817914
	149	1,231.268	148.917179521
	150	1,239.254	149.917271454
	151	1,244.85	150.919928351
	152	1,253.108	151.919728244
	153	1,258.976	152.922093907
	154	1,266.943	153.922205303
	155	1,272.75	154.924635940

Z	Element	A		
63	Eu	149	1,229.79	148.917925922
		150	1,236.211	149.919698294
		151	1,244.144	150.919846022
		152	1,250.451	151.921740399
		153	1,259.001	152.921226219
		154	1,265.443	153.922975386
		155	1,273.595	154.922889429
64	Gd	150	1,236.4	149.918655455
		151	1,242.898	150.920344273
		152	1,251.488	151.919787882
		153	1,257.735	152.921746283
		154	1,266.629	153.920862271
		155	1,273.065	154.922618801
		156	1,281.601	155.922119552
		157	1,287.961	156.923956686
		158	1,295.898	157.924100533
		159	1,301.842	158.926385075
		160	1,309.293	159.927050616
		161	1,314.928	160.929665688
65	Tb	157	1,287.119	156.924021155
		158	1,293.896	157.925410260
		159	1,302.03	158.925343135
		160	1,308.405	159.927164021
		161	1,316.102	160.927566289
66	Dy	154	1,261.749	153.924422046
		155	1,268.584	154.925748950
		156	1,278.025	155.924278273
		157	1,284.995	156.925461256
		158	1,294.05	157.924404637
		159	1,300.882	158.925735660
		160	1,309.458	159.925193718
		161	1,315.912	160.926929595
		162	1,324.109	161.926794731
		163	1,330.38	162.928727532
		164	1,338.038	163.929171165
		165	1,343.754	164.931699828
67	Ho	163	1,329.595	162.928730286
		164	1,336.269	163.930230577
		165	1,344.258	164.930319169
		166	1,350.501	165.932281267
		167	1,357.786	166.933126195
68	Er	161	1,311.486	160.930001348
		162	1,320.7	161.928774923
		163	1,327.603	162.930029273
		164	1,336.449	163.929196996
		165	1,343.1	164.930722800
		166	1,351.574	165.930289970
		167	1,358.01	166.932045448
		168	1,365.781	167.932367781
		169	1,371.784	168.934588082
		170	1,379.043	169.935460334
		171	1,384.725	170.938025885
69	Tm	167	1,356.479	166.932848844
		168	1,363.32	167.934170375
		169	1,371.353	168.934211117
		170	1,377.946	169.935797877
		171	1,385.433	170.936425817
70	Yb	166	1,346.666	165.933879623
		167	1,353.743	166.934946862
		168	1,362.794	167.933894465
		169	1,369.662	168.935187120
		170	1,378.132	169.934758652
		171	1,384.747	170.936322297
		172	1,392.767	171.936377696
		173	1,399.134	172.938206756
		174	1,406.599	173.938858101
		175	1,412.421	174.941272494
		176	1,419.285	175.942568409
		177	1,424.852	176.945257126
71	Lu	173	1,397.681	172.938926901
		174	1,404.442	173.940333522
		175	1,412.109	174.940767904
		176	1,418.397	175.942682399
		177	1,425.469	176.943754987
72	Hf	172	1,388.333	171.939457980
		173	1,395.294	181.406500000
		174	1,403.933	173.940040159
		175	1,410.642	174.941502991
		176	1,418.807	175.941401828
		177	1,425.185	176.943220013
		178	1,432.811	177.943697732
		179	1,438.91	178.945815073
		180	1,446.298	179.946548760
		181	1,451.994	180.949099124
73	Ta	179	1,438.017	178.945934113
		180	1,444.662	179.947465655
		181	1,452.239	180.947996346
		182	1,458.302	181.950152414
74	W	178	1,429.243	177.945848364
		179	1,436.175	178.947071733
		180	1,444.587	179.946705734
		181	1,451.268	180.948198054
		182	1,459.333	181.948205519
		183	1,465.524	182.950224458
		184	1,472.935	183.950932553
		185	1,478.689	184.953420586
		186	1,485.883	185.954362204
		187	1,491.35	186.957158365
75	Re	183	1,464.185	182.950821349
		184	1,470.67	183.952524289
		185	1,478.34	184.952955747
		186	1,484.519	185.954986529
		187	1,491.879	186.955750787
		188	1,497.75	187.958112287

Z		A	Binding Energy (MeV)	Atomic Mass (amu)
76	Os	182	1,454.06	181.952186222
		183	1,461.271	191.531100000
		184	1,469.919	183.952490808
		185	1,476.545	184.954043023
		186	1,484.806	185.953838355
		187	1,491.099	186.955747928
		188	1,499.088	187.955835993
		189	1,505.009	188.958144866
		190	1,512.801	189.958445210
		191	1,518.559	190.960927951
		192	1,526.117	191.961479047
		193	1,531.702	192.964148083
77	Ir	189	1,503.694	188.958716473
		190	1,510.018	189.960592299
		191	1,518.091	190.960591191
		192	1,524.289	191.962602198
		193	1,532.06	192.962923700
		194	1,538.127	193.965075610
78	Pt	188	1,494.208	187.959395697
		189	1,500.941	188.960831900
		190	1,509.853	189.959930073
		191	1,516.29	190.961684653
		192	1,524.966	191.961035158
		193	1,531.221	192.962984504
		194	1,539.592	193.962663581
		195	1,545.697	194.964774449
		196	1,553.619	195.964934884
		197	1,559.465	196.967323401
		198	1,567.022	197.967876009
		199	1,572.578	198.970576213
79	Au	195	1,544.688	194.965017928
		196	1,551.331	195.966551315
		197	1,559.402	196.966551609
		198	1,565.914	197.968225244
		199	1,573.498	198.968748016
80	Hg	194	1,535.495	193.965381832
		195	1,542.395	194.966638981
		196	1,551.234	195.965814846
		197	1,558.02	196.967195333
		198	1,566.504	197.966751830
		199	1,573.168	198.968262489
		200	1,581.197	199.968308726
		201	1,587.427	200.970285275
		202	1,595.181	201.970625604
		203	1,601.174	202.972857096
		204	1,608.669	203.973475640
		205	1,614.336	204.976056104
81	Tl	201	1,586.161	200.970803770
		202	1,593.034	201.972090569
		203	1,600.883	202.972329088
		204	1,607.539	203.973848646
		205	1,615.085	204.974412270
		206	1,621.589	205.976095321
82	Pb	202	1,592.202	201.972143786
		203	1,599.126	202.973375491
		204	1,607.52	203.973028761
		205	1,614.252	204.974467112
		206	1,622.34	205.974449002
		207	1,629.078	206.975880605
		208	1,636.446	207.976635850
		209	1,640.382	208.981074801
		210	1,645.567	209.984173129
83	Bi	207	1,625.897	206.978455217
		208	1,632.784	207.979726699
		209	1,640.244	208.980383241
		210	1,644.849	209.984104944
		211	1,649.983	210.987258139
84	Po	207	1,622.206	206.981578228
		208	1,630.601	207.981231059
		209	1,637.568	208.982415788
		210	1,645.228	209.982857396
85	At	209	1,633.3	208.986158678
		210	1,640.465	209.987131308
		211	1,648.211	210.987480806
86	Rn	211	1,644.536	210.990585410
		222	1,708.184	222.017570472
87	Fr	212	1,646.6	211.996194988
		223	1,713.461	223.019730712
88	Ra	223	1,713.828	223.018497140
		224	1,720.311	224.020202004
		225	1,725.213	225.023604463
		226	1,731.61	226.025402555
89	Ac	225	1,724.788	225.023220576
		226	1,730.187	226.026089848
		227	1,736.715	227.027746979
90	Th	229	1,748.341	229.031755340
		230	1,755.135	230.033126574
		231	1,760.253	231.036297060
		232	1,766.691	232.038050360
		233	1,771.478	233.041576923
91	Pa	230	1,753.043	230.034532562
		231	1,759.86	231.035878898
		232	1,765.414	232.038581720
92	U	233	1,771.728	233.039628196
		234	1,778.572	234.040945606
		235	1,783.87	235.043923062
		236	1,790.415	236.045561897
		237	1,795.541	237.048723955
		238	1,801.695	238.050782583
		239	1,806.501	239.054287777
93	Np	236	1,788.703	236.046559724
		237	1,795.277	237.048167253
		238	1,800.765	238.050940464

94 Pu	238	1,801.275	238.049553400
	239	1,806.921	239.052156519
	240	1,813.455	240.053807460
	241	1,818.697	241.056845291
	242	1,825.006	242.058736847
95 Am	241	1,817.935	241.056822944
	242	1,823.473	242.059543039
	243	1,829.84	243.061372686
96 Cm	246	1,847.827	246.067217551
	247	1,852.983	247.070346811
	248	1,859.196	248.072342247
97 Bk	247	1,852.246	247.070298533
98 Cf	251	1,875.103	251.079580056
99 Es	252	1,879.232	252.082972247
100 Fm	257	1,907.511	257.095098635
101 Md	258	1,911.701	258.098425321
102 No	259	1,916.569	260.010240000
103 Lr	260	1,919.621	261.055720000
104 Db	261	1,923.949	262.087520000
105 Jl	262	1,926.206	263.141530000
106 Rf	258	1,894.073	259.131510000
	259	1,900.745	260.146520000
	260	1,909.019	260.114435447
	261	1,915.447	262.161990000
	262	1,923.259	263.164770000
	263	1,929.62	264.183130000
	264	1,937.123	265.189240000
	265	1,943.199	266.210660000
	266	1,950.468	267.219280000
107 Bh	260	1,901.373	261.218030000
	261	1,909.447	262.218000000
	262	1,916.393	263.230090000
	263	1,924.336	264.231460000
	264	1,930.932	265.247300000
	265	1,938.568	266.251980000
	266	1,944.952	267.270090000
	267	1,952.342	268.277400000
108 Hn	263	1,918.371	264.287100000
	264	1,926.724	264.128408258
	265	1,933.311	266.300010000
	266	1,941.345	267.300420000
	267	1,947.803	268.317740000
	268	1,955.518	269.321560000
	269	1,961.765	270.341140000
109 Mt	265	1,926.413	266.365670000
	266	1,933.205	267.379400000
	267	1,941.662	268.375260000
	268	1,948.532	269.388160000
	269	1,956.333	270.391060000
	270	1,962.898	271.407230000
	271	1,970.498	272.412290000
110 Xa	267	1,934.89	268.439560000
	268	1,943.359	269.435290000
	269	1,949.927	270.451440000
	270	1,958.48	271.446260000
	271	1,965.199	272.460780000
	272	1,973.054	273.463100000
	273	1,978.392	274.492450000
111 Xb	272	1,965.596	273.534770000

Source: Adapted from G. Audi and A. H. Wapstra, *Nuclear Physics A* **595**, (4), 409–480, December 25, 1995.

APPENDIX C

ELECTRON BINDING ENERGIES (eV)

This appendix lists the binding energy in eV of an orbital electron in each shell and subshell of each of the elements by increasing Z (e.g., 1 H with Z = 1, 2 He with Z = 2, etc.).

Subshell	1 H	2 He	3 Li	4 Be	5 B	6 C	7 N	8 O	9 F	10 Ne
K	13.6	24.6	54.7	111.5	188	284.2	409.9	543.1	696.7	870.2
L_I							37.3	41.6		48.5
L_{II}										21.7
L_{III}										21.6

Subshell	11 Na	12 Mg	13 Al	14 Si	15 P	16 S	17 Cl	18 Ar	19 K	20 Ca
K	1,070.8	1,303	1,559	1,839	2,145.5	2,472	2,822	3,205.9	3,608.4	4,038.5
L_I	63.5	88.6	117.8	149.7	189	230.9	270	326.3	378.6	438.4
L_{II}	30.4	49.6	72.9	99.8	136	163.6	202	250.6	297.3	349.7
L_{III}	30.5	49.2	72.5	99.2	135	162.5	200	248.4	294.6	346.2
M_I								29.3	34.8	44.3
M_{II}								15.9	18.3	25.4
M_{III}								15.7	18.3	25.4

Subshell	21 Sc	22 Ti	23 V	24 Cr	25 Mn	26 Fe	27 Co	28 Ni	29 Cu	30 Zn
K	4,492	4,966	5,465	5,989	6,539	7,112	7,709	8,333	8,979	9,659
L_I	498	560.9	626.7	696	769.1	844.6	925.1	1,008.6	1,096.7	1,196.2
L_{II}	403.6	460.2	519.8	583.8	649.9	719.9	793.2	870	952.3	1,044.9
L_{III}	398.7	453.8	512.1	574.1	638.7	706.8	778.1	852.7	932.5	1,021.8
M_I	51.1	58.7	66.3	74.1	82.3	91.3	101	110.8	122.5	139.8
M_{II}	28.3	32.6	37.2	42.2	47.2	52.7	58.9	68	77.3	91.4
M_{III}	28.3	32.6	37.2	42.2	47.2	52.7	58.9	66.2	75.1	88.6
M_{IV}										10.2
M_V										10.1

Subshell	31 Ga	32 Ge	33 As	34 Se	35 Br	36 Kr	37 Rb	38 Sr	39 Y	40 Zr
K	10,367	11,103	11,867	12,658	13,474	14,326	15,200	16,105	17,038	17,998
L_I	1,299	1,414.6	1,527	1,652	1,782	1,921	2,065	2,216	2,373	2,532
L_{II}	1,143.2	1,248.1	1,359.1	1,474.3	1,596	1,730.9	1,864	2,007	2,156	2,307
L_{III}	1,116.4	1,217	1,323.6	1,433.9	1,550	1,678.4	1,804	1,940	2,080	2,223
M_I	159.5	180.1	204.7	229.6	257	292.8	326.7	358.7	392	430.3
M_{II}	103.5	124.9	146.2	166.5	189	222.2	248.7	280.3	310.6	343.5
M_{III}	100	120.8	141.2	160.7	182	214.4	239.1	270	298.8	329.8
M_{IV}	18.7	29.8	41.7	55.5	70	95	113	136	157.7	181.1
M_V	18.7	29.2	41.7	54.6	69	93.8	112	134.2	155.8	178.8
N_I						27.5	30.5	38.9	43.8	50.6
N_{II}						14.1	16.3	21.6	24.4	28.5
N_{III}						14.1	15.3	20.1	23.1	27.1

Subshell	41 Nb	42 Mo	43 Tc	44 Ru	45 Rh	46 Pd	47 Ag	48 Cd	49 In	50 Sn
K	18,986	20,000	21,044	22,117	23,220	24,350	25,514	26,711	27,940	29,200
L_I	2,698	2,866	3,043	3,224	3,412	3,604	3,806	4,018	4,238	4,465
L_{II}	2,465	2,625	2,793	2,967	3,146	3,330	3,524	3,727	3,938	4,156
L_{III}	2,371	2,520	2,677	2,838	3,004	3,173	3,351	3,538	3,730	3,929
M_I	466.6	506.3	544	586.1	628.1	671.6	719	772	827.2	884.7
M_{II}	376.1	411.6	447.6	483.7	521.3	559.9	603.8	652.6	703.2	756.5
M_{III}	360.6	394	417.7	461.5	496.5	532.3	573	618.4	665.3	714.6
M_{IV}	205	231.1	257.6	284.2	311.9	340.5	374	411.9	451.4	493.2
M_V	202.3	227.9	253.9	280	307.2	335.2	368	405.2	443.9	484.9
N_I	56.4	63.2	69.5	75	81.4	87.1	97	109.8	122.9	137.1
N_{II}	32.6	37.6	42.3	46.5	50.5	55.7	63.7	63.9	73.5	83.6
N_{III}	30.8	35.5	39.9	43.2	47.3	50.9	58.3	63.9	73.5	83.6
N_{IV}							11.7	17.7	24.9	
N_V							10.7	16.9	23.9	

Subshell	51 Sb	52 Te	53 I	54 Xe	55 Cs	56 Ba	57 La	58 Ce	59 Pr	60 Nd
K	30,491	31,814	33,169	34,561	35,985	37,441	38,925	40,443	41,991	43,569
L_I	4,698	4,939	5,188	5,453	5,714	5,989	6,266	6,548	6,835	7,126
L_{II}	4,380	4,612	4,852	5,107	5,359	5,624	5,891	6,164	6,440	6,722
L_{III}	4,132	4,341	4,557	4,786	5,012	5,247	5,483	5,723	5,964	6,208
M_I	946	1,006	1,072	1,148.7	1,211	1,293	1,362	1,436	1,511	1,575
M_{II}	812.7	870.8	931	1,002.1	1,071	1,137	1,209	1,274	1,337	1,403
M_{III}	766.4	820	875	940.6	1,003	1,063	1,128	1,187	1,242	1,297
M_{IV}	537.5	583.4	631	689	740.5	795.7	853	902.4	948.3	1,003.3
M_V	528.2	573	620	676.4	726.6	780.5	836	883.8	928.8	980.4
N_I	153.2	169.4	186	213.2	232.3	253.5	274.7	291	304.5	319.2
N_{II}	95.6	103.3	123	146.7	172.4	192	63.9	223.3	236.3	243.3
N_{III}	95.6	103.3	123	145.5	161.3	178.6	196	206.5	217.6	224.6
N_{IV}	33.3	41.9	50	69.5	79.8	92.6	105.3	109	115.1	120.5
N_V	32.1	40.4	50	67.5	77.5	89.9	102.5	—	115.1	120.5
N_{VI}				—	—	—	—	0.1	2	1.5
N_{VII}				—	—	—	—	0.1	2	1.5
O_I				23.3	22.7	30.3	34.3	37.8	37.4	37.5
O_{II}				13.4	14.2	17	19.3	19.8	22.3	21.1
O_{III}				12.1	12.1	14.8	16.8	17	22.3	21.1

Subshell	61 Pm	62 Sm	63 Eu	64 Gd	65 Tb	66 Dy	67 Ho	68 Er	69 Tm	70 Yb
K	45,184	46,834	48,519	50,239	51,996	53,789	55,618	57,486	59,390	61,332
L_I	7,428	7,737	8,052	8,376	8,708	9,046	9,394	9,751	10,116	10,486
L_{II}	7,013	7,312	7,617	7,930	8,252	8,581	8,918	9,264	9,617	9,978
L_{III}	6,459	6,716	6,977	7,243	7,514	7,790	8,071	8,358	8,648	8,944
M_I	—	1,723	1,800	1,881	1,968	2,047	2,128	2,206	2,307	2,398
M_{II}	1,471.4	1,541	1,614	1,688	1,768	1,842	1,923	2,006	2,090	2,173
M_{III}	1,357	1,419.8	1,481	1,544	1,611	1,676	1,741	1,812	1,885	1,950
M_{IV}	1,052	1,110.9	1,158.6	1,221.9	1,276.9	1,333	1,392	1,453	1,515	1,576
M_V	1,027	1,083.4	1,127.5	1,189.6	1,241.1	1,292	1,351	1,409	1,468	1,528
N_I	—	347.2	360	378.6	396	414.2	432.4	449.8	470.9	480.5
N_{II}	242	265.6	284	286	322.4	333.5	343.5	366.2	385.9	388.7
N_{III}	242	247.4	257	271	284.1	293.2	308.2	320.2	332.6	339.7
N_{IV}	120	129	133	—	150.5	153.6	160	167.6	175.5	191.2
N_V	120	129	127.7	142.6	150.5	153.6	160	167.6	175.5	182.4
N_{VI}		5.2	0	8.6	7.7	8	8.6	—	—	2.5
N_{VII}		5.2	0	8.6	2.4	4.3	5.2	4.7	4.6	1.3
O_I		37.4	32	36	45.6	49.9	49.3	50.6	54.7	52
O_{II}		21.3	22	20	28.7	26.3	30.8	31.4	31.8	30.3
O_{III}		21.3	22	20	22.6	26.3	24.1	24.7	25	24.1

Subshell	71 Lu	72 Hf	73 Ta	74 W	75 Re	76 Os	77 Ir	78 Pt	79 Au	80 Hg
K	63,314	65,351	67,416	69,525	71,676	73,871	76,111	78,395	80,725	83,102
L_I	10,870	11,271	11,682	12,100	12,527	12,968	13,419	13,880	14,353	14,839
L_{II}	10,349	10,739	11,136	11,544	11,959	12,385	12,824	13,273	13,734	14,209
L_{III}	9,244	9,561	9,881	10,207	10,535	10,871	11,215	11,564	11,919	12,284
M_I	2,491	2,601	2,708	2,820	2,932	3,049	3,174	3,296	3,425	3,562
M_{II}	2,264	2,365	2,469	2,575	2,682	2,792	2,909	3,027	3,148	3,279
M_{III}	2,024	2,107	2,194	2,281	2,367	2,457	2,551	2,645	2,743	2,847
M_{IV}	1,639	1,716	1,793	1,949	1,949	2,031	2,116	2,202	2,291	2,385
M_V	1,589	1,662	1,735	1,809	1,883	1,960	2,040	2,122	2,206	2,295
N_I	506.8	538	563.4	594.1	625.4	658.2	691.1	725.4	762.1	802.2
N_{II}	412.4	438.2	463.4	490.4	518.7	549.1	577.8	609.1	642.7	680.2
N_{III}	359.2	380.7	400.9	423.6	446.8	470.7	495.8	519.4	546.3	576.6
N_{IV}	206.1	220	237.9	255.9	273.9	293.1	311.9	331.6	353.2	378.2
N_V	196.3	211.5	226.4	243.5	260.5	278.5	296.3	314.6	335.1	358.8
N_{VI}	8.9	15.9	23.5	33.6	42.9	53.4	63.8	74.5	87.6	104
N_{VII}	7.5	14.2	21.6	31.4	40.5	50.7	60.8	71.2	83.9	99.9
O_I	57.3	64.2	69.7	75.6	83	84	95.2	101.7	107.2	127
O_{II}	33.6	38	42.2	453	45.6	58	63	65.3	74.2	83.1
O_{III}	26.7	29.9	32.7	36.8	34.6	44.5	48	51.7	57.2	64.5
O_{IV}										9.6
O_V										7.8

Subshell	81 Tl	82 Pb	83 Bi	84 Po	85 At	86 Rn	87 Fr	88 Ra	89 Ac	90 Th
K	85,530	88,005	90,526	93,105	95,730	98,404	101,137	103,922	106,755	109,651
L_{I}	15,347	15,861	16,388	16,939	17,493	18,049	18,639	19,237	19,840	20,472
L_{II}	14,698	15,200	15,711	16,244	16,785	17,337	17,907	18,484	19,083	19,693
L_{III}	12,658	13,055	13,419	13,814	14,214	14,619	15,031	15,444	15,871	16,300
M_{I}	3,704	3,851	3,999	4,149	4,317	4,482	4,652	4,822	5,002	5,182
M_{II}	3,416	3,554	3,696	3,854	4,008	4,159	4,327	4,490	4,656	4,830
M_{III}	2,957	3,066	3,177	3,302	3,426	3,538	3,663	3,792	3,909	4,046
M_{IV}	2,485	2,586	2,688	2,798	2,909	3,022	3,136	3,248	3,370	3,491
M_{V}	2,389	2,484	2,580	2,683	2,787	2,892	3,000	3,105	3,219	3,332
N_{I}	846.2	891.8	939	995	1,042	1,097	1,153	1,208	1,269	1,330
N_{II}	720.5	761.9	805.2	851	886	929	980	1,058	1,080	1,168
N_{III}	609.5	643.5	678.8	705	740	768	810	879	890	966.4
N_{IV}	405.7	434.3	464	500	533	567	603	636	675	712.1
N_{V}	385	412.2	440.1	473	507	541	577	603	639	675.2
N_{VI}	122.2	141.7	162.3	184	210	238	268	299	319	342.4
N_{VII}	117.8	136.9	157	184	210	238	268	299	319	333.1
O_{I}	136	147	159.3	177	195	214	234	254	272	290
O_{II}	94.6	106.4	119	132	148	164	182	200	215	229
O_{III}	73.5	83.3	92.6	104	115	127	140	153	167	182
O_{IV}	14.7	20.7	26.9	31	40	48	58	68	80	92.5
O_{V}	12.5	18.1	23.8	31	40	48	58	68	80	85.4
P_{I}						26	34	44	—	41.4
P_{II}							15	19	—	24.5
P_{III}							15	19	—	16.6

Subshell	91 Pa	92 U
K	11,2601	11,5606
L_{I}	21,105	21,757
L_{II}	20,314	20,948
L_{III}	16,733	17,166
M_{I}	5,367	5,548
M_{II}	5,001	5,182
M_{III}	4,174	4,303
M_{IV}	3,611	3,728
M_{V}	3,442	3,552
N_{I}	1,387	1,439
N_{II}	1,224	1,271
N_{III}	1,007	1,043
N_{IV}	743	778.3
N_{V}	708	736.2
N_{VI}	371	388.2
N_{VII}	360	377.4
O_{I}	310	321
O_{II}	232	257
O_{III}	232	192
O_{IV}	94	102.8
O_{V}	94	94.2
P_{I}	—	43.9
P_{II}	—	26.8
P_{III}	—	16.8

Source: David R. Lide, ed.-in-chief, *CRC Handbook of Chemistry and Physics*, 74th ed., CRC Press, Boca Raton, FL, 1993.

APPENDIX D

RADIOACTIVE TRANSFORMATION DATA

Transformation diagrams and lists are provided in this appendix for about 100 radionuclides of interest to radiation protection. These were compiled from resources at the National Nuclear Data Center (NNDC), which is operated for the U.S. Department of Energy by Brookhaven National Laboratory; it is an invaluable national resource for such data and can be accessed electronically at *www.nndc.bnl.gov*. The NNDC database also contains lists prepared for the Medical Internal Radiation Dosimetry (MIRD) Committee of the Society of Nuclear Medicine and the International Commission on Radiation Protection (ICRP); these lists were compiled for internal dosimetry calculations and contain all the radiation information necessary for such calculations. The MIRD compilation (published in 1989) is more recent but lists only those radionuclides of interest to nuclear medicine.

The decay-scheme drawings in this appendix show the radionuclide, its mass number and atomic number in conventional notation, and the half-life in parentheses to the right [e.g., ${}^{14}_{6}C$ (5730 y)]; the predominant transformation mode(s) and the energy change associated with each; the energy of the emitted particle or radiation; the percentage of transformations of the parent, unless noted otherwise, that produce the emission; and the product nuclide and whether it is stable or radioactive (if a half-life is shown). Since radioactive transformation is principally an energy change as one radionuclide transforms to another, the various energy levels associated with particle emissions and subsequent emissions of gamma rays and/or conversion electrons (ce) of the transforming nuclide above the ground state of its product are also shown.

As helpful as it is to have a pictorial decay scheme, it is difficult to obtain complete information on the emissions from a given radionuclide; therefore, lists of the yield and energy of each significant particle emission and/or radiation are provided with each diagram to amplify the pictorial information significantly. The mode of production of the radioisotope is shown for each table of radiations together with the date the data were last reviewed by the NNDC. For example, for ^{32}P there are three primary modes of production: an (n,γ) reaction with ^{31}P, irradiation of stable ^{34}S with deuterons to produce ^{32}P by a (d,α) reaction, and an (n,p) reaction with stable ^{32}S. The data were last reviewed by the National Nuclear Data Center on February 3, 1993. The listing shows one beta particle, β_1^-, emitted with a yield, Y_i, of 100% with an average energy per transformation of 0.6949 MeV; the maximum beta energy is provided in the pictorial decay scheme as 1.7104 MeV.

Many of the decay schemes and associated data lists are quite complex. These have been simplified for most radiation protection purposes by eliminating radiations that contribute less than 1% (and in some noted cases, 2%) of the energy emitted per transformation of the parent unless they are necessary to understand the modes of transformation. Radiations are listed by the same designators used by NNDC, MIRD, and ICRP, and even though the listings are somewhat more simplified, the format is compatible with the more extensive lists should they need to be consulted. The abbreviations used in the data tables are described in Table D-1; the designations for various characteristic x rays ($K_{\alpha 1}$, $K_{\alpha 2}$, L_{α}, etc.) and the energy transition levels, established by historical precedence, are detailed in Figure D-1.

The radionuclide data listings in this appendix are listed in Table D-2. The data were obtained from the NNDC in May 1999 and are current as of that date. The date listed on the decay scheme is usually different because it is the last date the experimental data for the radionuclide were evaluated by scientists at NNDC. For those circumstances where the most recent authoritative data are desirable, it is prudent to access the NNDC database to determine if a recent evaluation has changed the listed information.

ACKNOWLEDGMENTS

The transformation schemes and data lists in this appendix are due to the patient and dedicated efforts of my research associates Chul Lee and Patricia Ellis, and J. I. Tuli and the staff and resources of the National Nuclear Data Center.

TABLE D-1. Designations of Characteristic X Rays and Associated Shell Changes According to Historical Precedent

Radiation	Description
$K_{\alpha 1}$, $K_{\alpha 2}$, $K_{\beta 1}$, $K_{\beta 2}$, $L_{\alpha 1}$, $L_{\alpha 2}$, $L_{\beta 1}$, $L_{\beta 2}$, M, N, etc.	Characteristic x rays according to shell being filled; α, β, γ, δ designate in order the outer shell from which the electron comes; subscripts designate subshells. For small energy separations such as occur in outer shells, they are sometimes not broken down into separate components but are designated simply for the total shell (e.g., K, L, M x rays). The specific shell (subshell) transformations are shown in Figure D-1 in accordance with their historical designation—for example K_{β_1} denotes translocation from the MIII subshell, whereas $K\beta_2$ represents, as originally formulated, translocations from the N shell.
Auger-K, Auger-L, Auger-M	Auger-K, L, M, etc. denotes Auger electron emission due to an initial vacancy in the K, L, M shell, etc. An Auger-K electron is due to a vacancy in the K shell which is usually filled by an L-shell electron and subsequent ejection of an L-shell electron (the Auger), yielding vacancies in the L-shell (actually, a composite of several Auger transitions of similar energy because there are three subshells in the L shell). Similar definitions apply to the other Auger transitions listed. The MIRD listing uses KLL for this transition, and KLX, KXY, LMM, LMX, etc., where X and Y refer to any shell higher than the L shell for K-series Auger transitions and to any shell for L- and M-series Auger transitions.
γ_i	Gamma rays by i, ordered according to increasing gamma-ray energy.
$\gamma^{\pm}$	Photons of 0.511 MeV from the annihilation of positrons with electrons.
β_i^+ and β_i^-	β^+ and β^- particles, ordered by i according to increasing endpoint energy.
ce-K, γ_i, ce-L_1, γ_i, ce-N+, γ_i	Internal conversion electrons associated with the ith gamma ray, γ_i; K, L, and N designate the shell from which the internally converted electron is ejected; shell subscripts denote the respective subshell; + designates all subshells.
α_i	Alpha particles ordered according to increasing alpha-particle energy.
α recoil	Recoil atom associated with α_i.
SFn, SFf, SF$p\gamma$, SF$d\gamma$, and SFβ	Spontaneous fission neutrons, fragments, prompt gamma rays, delayed gamma rays, and beta particles (averaged over a continuous spectrum of energies).

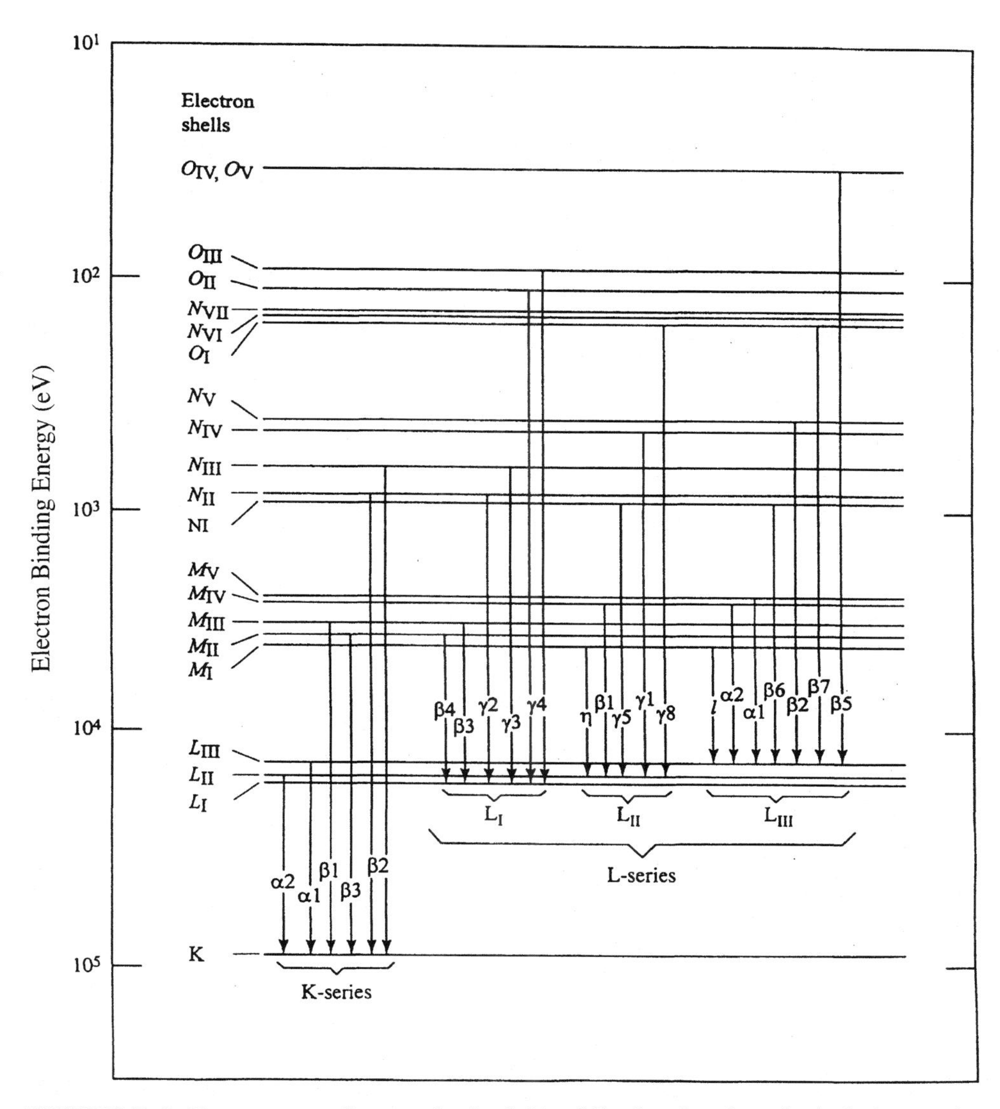

FIGURE D-1. X-ray energy diagram for lead ($Z = 82$), showing the principal characteristic x-ray transitions and their standard nomenclature. (From Arthur B. Chilton, J. Kenneth Shultis, and Richard E. Faw, *Principles of Radiation Shielding*, Prentice Hall, Upper Saddle River, NJ, 1984, p. 107.)

TABLE D-2. Radionuclides by Atomic Number (*Z*), Nuclide Symbol (Nuc.), and Mass Number (*A*) Included in Appendix D, Ordered Principally by *Z*

Z	Nuc.	*A*	*Z*	Nuc.	*A*	*Z*	Nuc.	*A*	*Z*	Nuc.	*A*
1	H	3	28	Ni	59	48	Cd	109	80	Hg	197
4	Be	7		Ni	63	49	In	113m		Hg	203
6	C	11		Ni	65	50	Sn	113	81	Tl	201
	C	14	29	Cu	64	51	Sb	122		Tl	204
7	N	13	30	Zn	65		Sb	124	84	Po	210
	N	16	31	Ga	67		Sb	125	86	Rn	222
8	O	15		Ga	68	52	Te	125m	88	Ra	226
9	F	18	32	Ge	68	53	I	123		Ra	228
11	Na	22	33	As	76		I	125	90	Th	230
	Na	24	34	Se	72		I	128	92	U	233
12	Mg	27		Se	75		I	129		U	238
15	P	32	35	Br	82		I	130	94	Pu	236
	P	33	36	Kr	85m		I	131		Pu	238
16	S	35			85	54	Xe	133m		Pu	239
17	Cl	36	37	Rb	86			133		Pu	240
	Cl	38	38	Sr	85	55	Cs	134m		Pu	241
18	Ar	41		Sr	89			134		Pu	242
19	K	40		Sr	90		Cs	137	95	Am	241
20	Ca	41	39	Y	88	56	Ba	133m	96	Cm	242
	Ca	45		Y	89m			133	98	Cf	252
21	Sc	46		Y	90		Ba	137m			
	Sc	47	40	Zr	95		Ba	140			
24	Cr	51	41	Nb	95m	57	La	140			
25	Mn	54		Nb	95	58	Ce	141			
	Mn	56	42	Mo	99		Ce	144			
26	Fe	55	43	Tc	99m	59	Pr	144			
	Fe	59		Tc	99	63	Eu	154			
27	Co	57	44	Ru	103	64	Gd	148			
	Co	58m		Ru	106	74	W	187			
	Co	58	45	Rh	103m	77	Ir	192			
	Co	60m		Rh	106	79	Au	195			
	Co	60	47	Ag	110m		Au	198			

RADIONUCLIDE DECAY SCHEMES

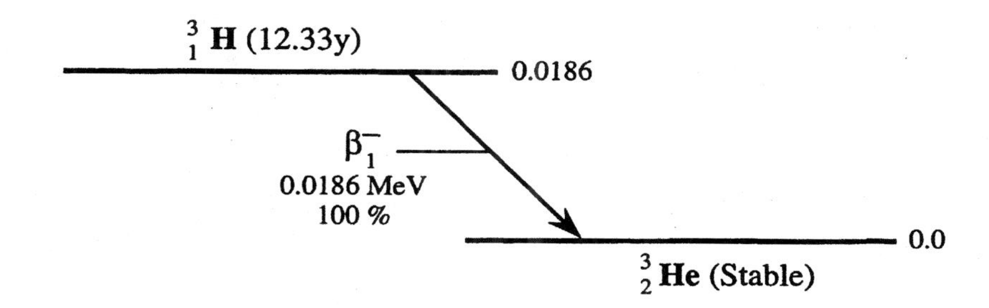

^{6}Li(n, α); Natural [^{14}N(n, t); ^{16}O(n, t)] (July 13, 1998)

Radiation	Y_i (%)	E_i (MeV)
β_1^-	100.0	0.0057[a]

[a]Average energy.

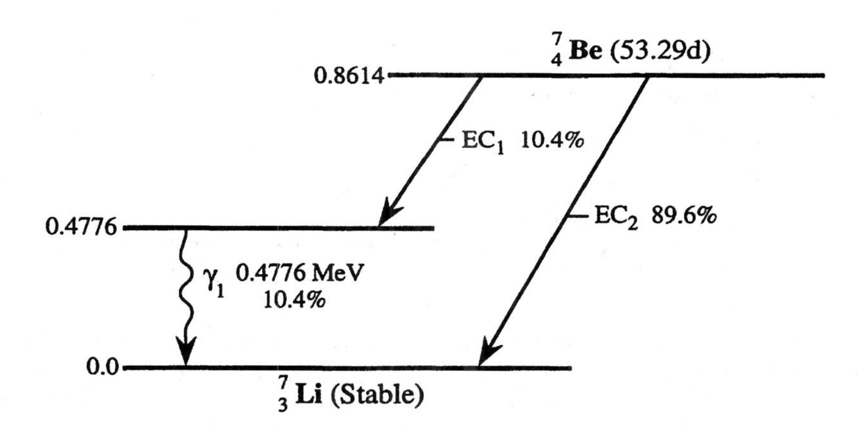

^{6}Li(d, n); ^{10}B(p, α); ^{12}C(^{3}He, 2α); Natural (Jan. 28, 1987)

Radiation	Y_i (%)	E_i (MeV)
γ_1	10.4	0.4776
ce-K, γ_1	8.04×10^{-6}	0.4776
ce-L_1, γ_1	1.18×10^{-7}	0.4776[a]
$K_{\alpha 1}$ x ray	0.0163	0.0001

[a]Maximum energy for subshell.

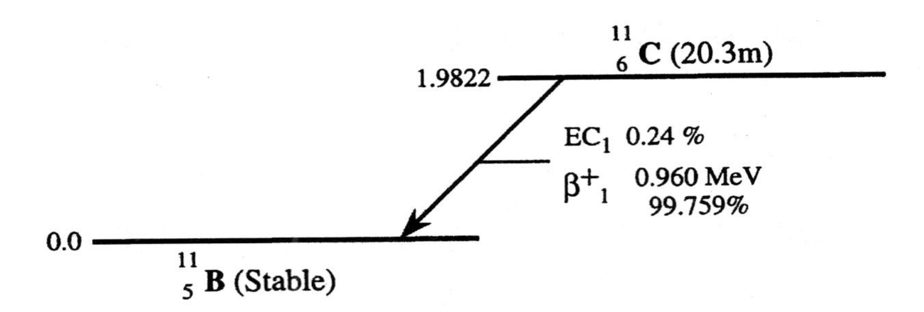

^{11}B(p,n); ^{10}B(p,γ); ^{10}B(d,n); ^{14}N(p,α) (Aug. 27, 1993)

Radiation	Y_i (%)	E_i (MeV)
β_1^+	99.759	0.3856[a]
$\gamma^\pm$	199.518	0.5110
K x ray	2.22×10^{-4}	0.0002[a]
Auger-K	0.222	0.0002[a]

[a]Average energy.

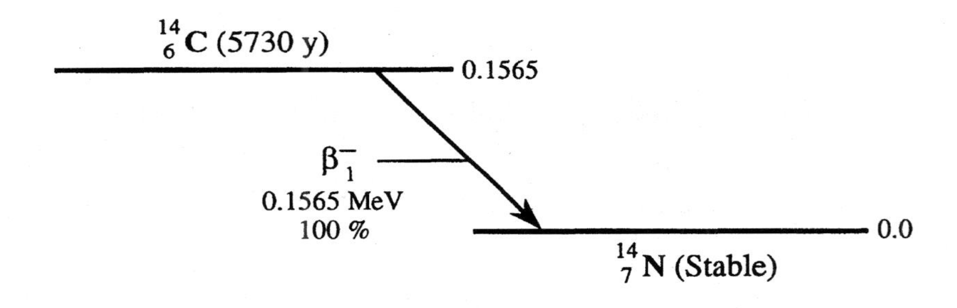

^{14}N(n,p); Natural [^{14}N(n,p)] (Dec. 1, 1993)

Radiation	Y_i (%)	E_i (MeV)
β_1^-	100.0	0.0495[a]

[a]Average energy.

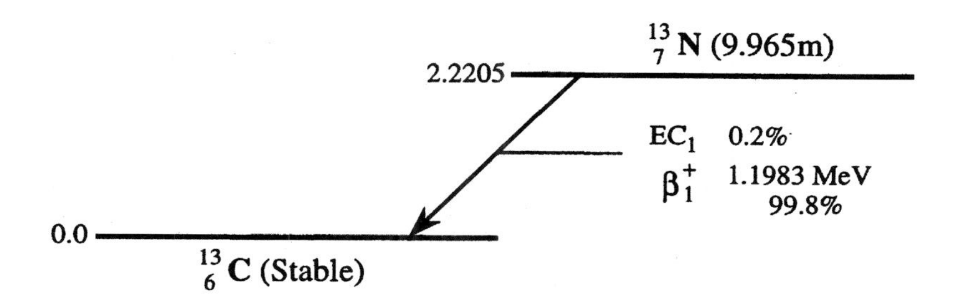

^{10}B(α,n); ^{12}C(d,n); ^{13}C(p,n); ^{12}C(p,γ) (Dec. 1, 1993)

Radiation	Y_i (%)	E_i (MeV)
β_1^+	99.8	0.4918[a]
$\gamma^{\pm}$	199.6	0.5110
K x ray	3.72×10^{-6}	0.0003[a]
Auger-K	0.186	0.0003[a]

[a]Average energy.

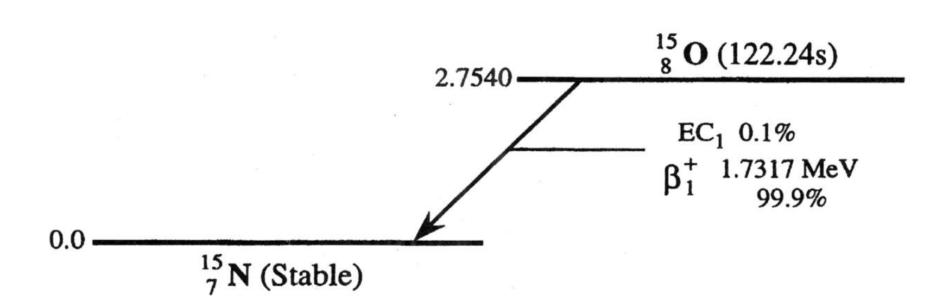

^{14}N(d,n); ^{14}N(p,γ); ^{16}O(^{3}He,α); ^{12}C(α,n) (Dec. 1, 1993)

Radiation	Y_i (%)	E_i (MeV)
β_1^+	99.9	0.7353[a]
$\gamma^{\pm}$	199.8	0.5110
K x ray	3.30×10^{-4}	0.0004[a]
Auger-K	0.094	0.0004[a]

[a]Average energy.

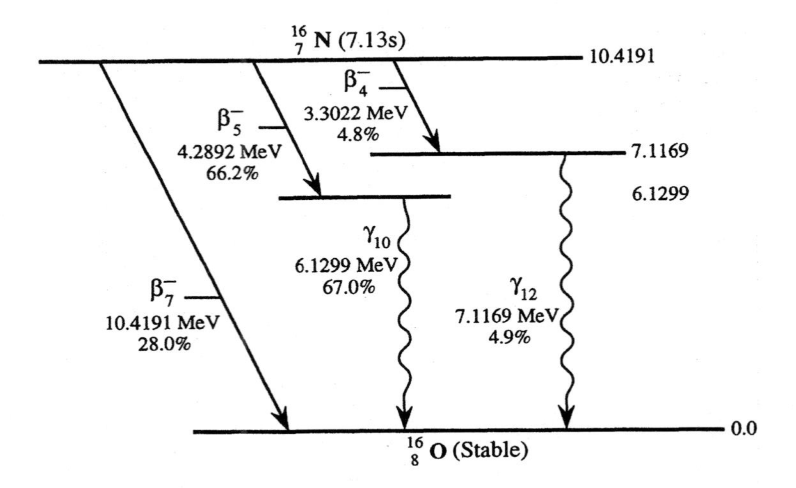

$^{15}N(d,p)$; $^{16}O(n,p)$; $^{19}F(n,\alpha)$; $^{15}N(n,\gamma)$ (Feb. 2, 1999)

Radiation	Y_i (%)	E_i (MeV)
β_3^-	1.06	0.6306[a]
β_4^-	4.80	1.4620[a]
β_5^-	66.2	1.9420[a]
β_7^-	28.0	4.9800[a]
γ_{10}	67.0	6.1299
γ_{12}	4.90	7.1169

[a]Average energy.

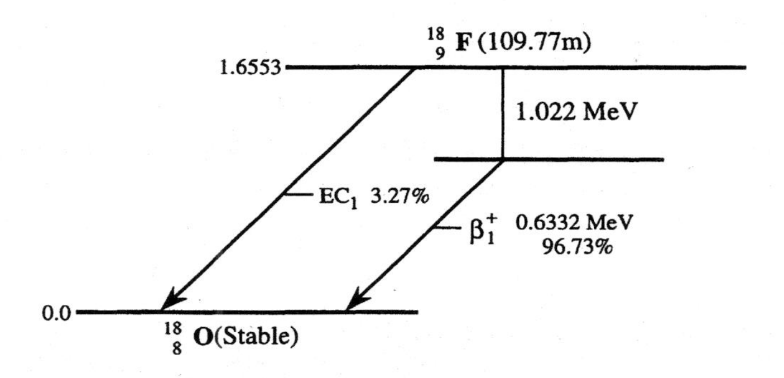

$^{18}O(p,n)$; $^{16}O(t,n)$; $^{16}O(^3He,p)$; $^{19}F(n,2n)$; $^{19}F(d,t)$; Ne(d,α) (Nov. 19, 1996)

Radiation	Y_i (%)	E_i (MeV)
β_1^+	96.73	0.2498[a]
$\gamma^{\pm}$	193.46	0.5110
K x ray	0.018	0.0005[a]
Auger-K	3.07	0.0005[a]

[a]Average energy.

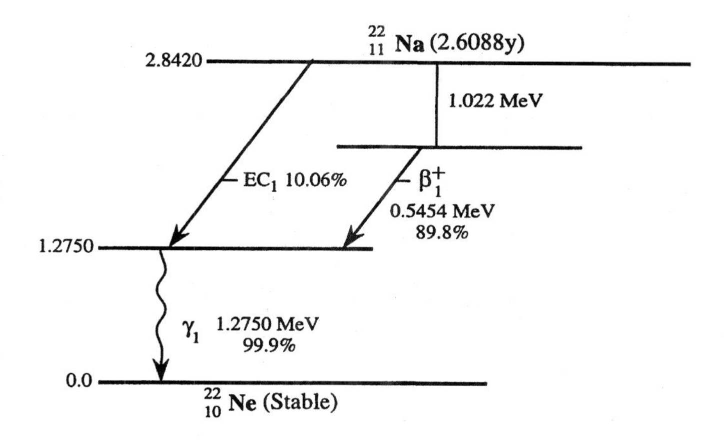

$^{19}F(\alpha,n)$; $^{24}Mg(d,\alpha)$		(Oct. 6, 1994)
Radiation	Y_i (%)	E_i (MeV)
β_1^+	89.8	0.2155[a]
$\gamma^{\pm}$	179.6	0.5110
γ_1	99.9	1.2750
K x ray	0.125	0.0008[a]
Auger-K	9.20	0.0008[a]

[a] Average energy.

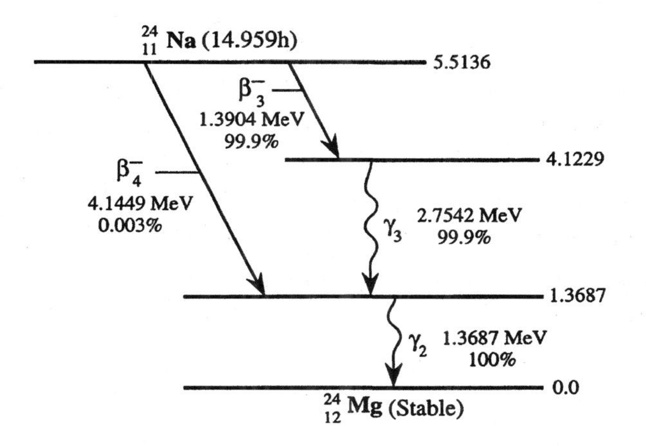

$^{23}Na(n,\gamma)$		(June 24, 1993)
Radiation	Y_i (%)	E_i (MeV)
β_3^-	99.9	0.5541[a]
β_4^-	0.003	1.8650[a]
γ_2	100.0	1.3687[a]
γ_3	99.9	2.7542

[a] Average energy.

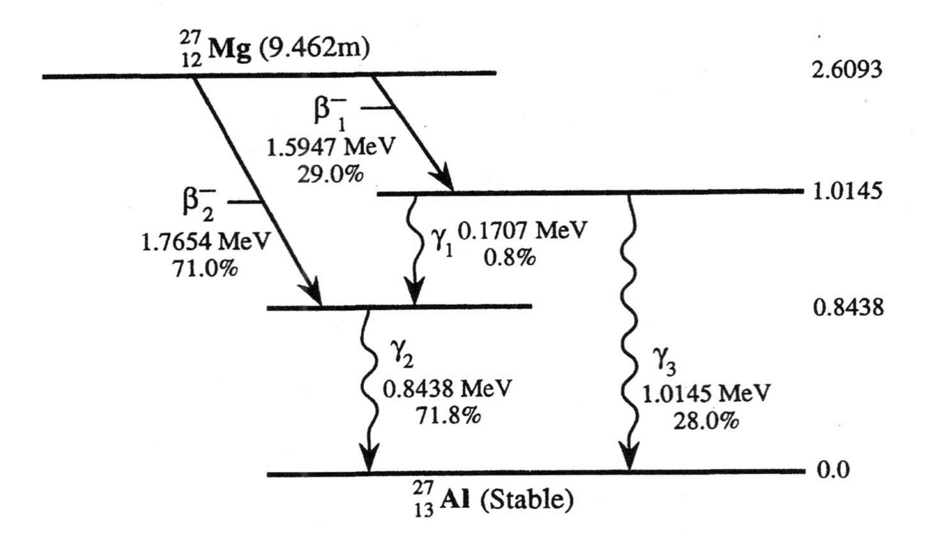

^{26}Mg(n, γ)		(Oct. 6, 1994)
Radiation	Y_i (%)	E_i (MeV)
β_1^-	29.0	0.6457[a]
β_2^-	71.0	0.7242[a]
γ_1	0.8	0.1707
γ_2	71.8	0.8438
γ_3	28.0	1.0145

[a] Average energy.

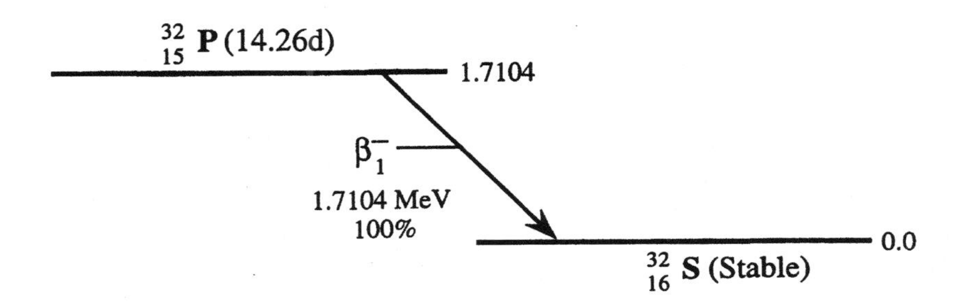

^{31}P(n, γ); ^{34}S(d, α); ^{32}S(n, p)		(Feb. 3, 1993)
Radiation	Y_i (%)	E_i (MeV)
β_1^-	100.0	0.6949[a]

[a] Average energy.

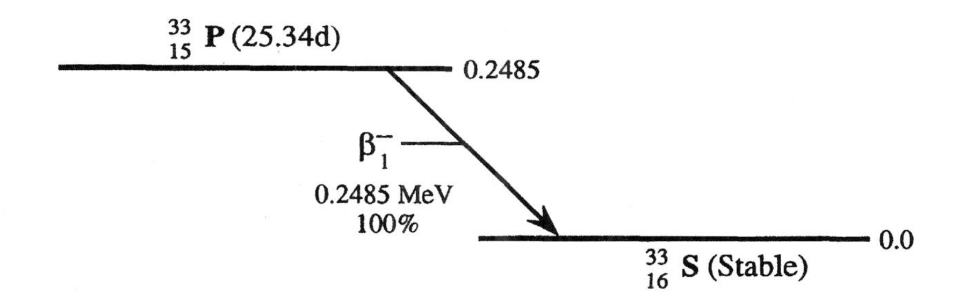

^{33}S(n,p); ^{37}Cl(γ,α) (June 24, 1993)

Radiation	Y_i (%)	E_i (MeV)
β_1^-	100.0	0.0764[a]

[a] Average energy.

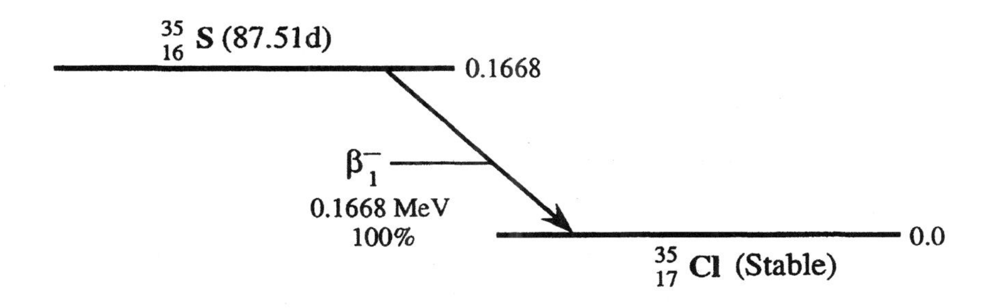

^{34}S(n,γ); ^{37}Cl(d,α) (June 24, 1993)

Radiation	Y_i (%)	E_i (MeV)
β_1^-	100.0	0.0486[a]

[a] Average energy.

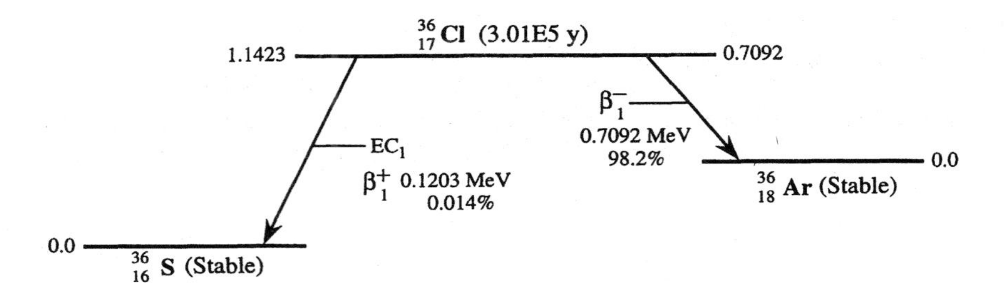

^{35}Cl(n, γ) (June 24, 1993)

Radiation	Y_i (%)	E_i (MeV)
β_1^+	0.014	0.0502[a]
$\gamma^{\pm}$	0.028	0.5110
β_1^-	98.2	0.2512[a]
K x ray	0.130	0.0023[a]
Auger-K	1.57	0.0021[a]

[a]Average energy.

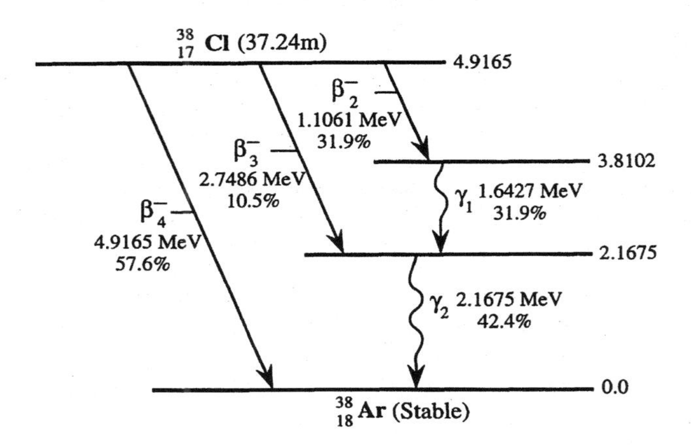

^{37}Cl(n, γ) (June 24, 1993)

Radiation	Y_i (%)	E_i (MeV)
β_2^-	31.9	0.4200[a]
β_3^-	10.5	1.1810[a]
β_4^-	57.6	2.2440[a]
γ_1	31.9	1.6427
γ_2	42.4	2.1675

[a]Average energy.

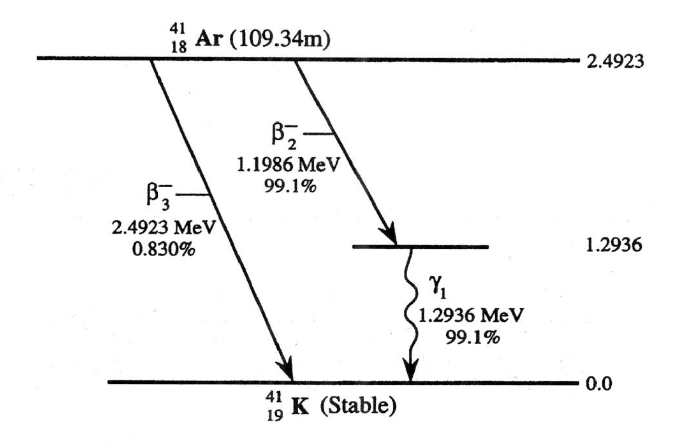

^{40}Ar(n, γ) (Oct. 6, 1994)

Radiation	Y_i (%)	E_i (MeV)
β_2^-	99.1	0.4595[a]
β_3^-	0.83	1.0770[a]
γ_1	99.1	1.2940

[a]Average energy.

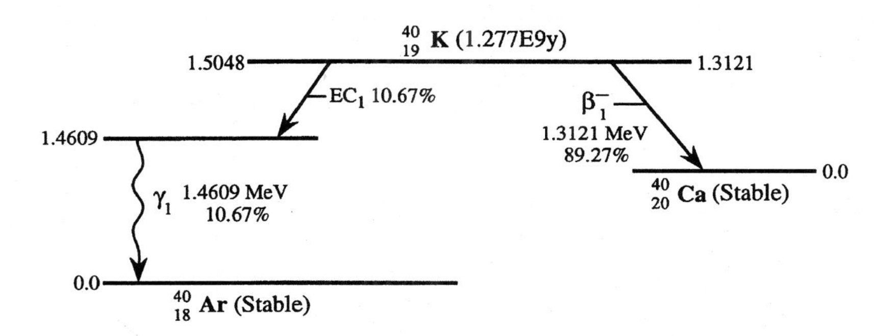

Natural (June 24, 1993)

Radiation	Y_i (%)	E_i (MeV)
γ_1	10.67	1.4609
β_1^-	89.27	0.5606[a]
K x ray	0.938	0.0030[a]
Auger-K	7.22	0.0027[a]

[a]Average energy.

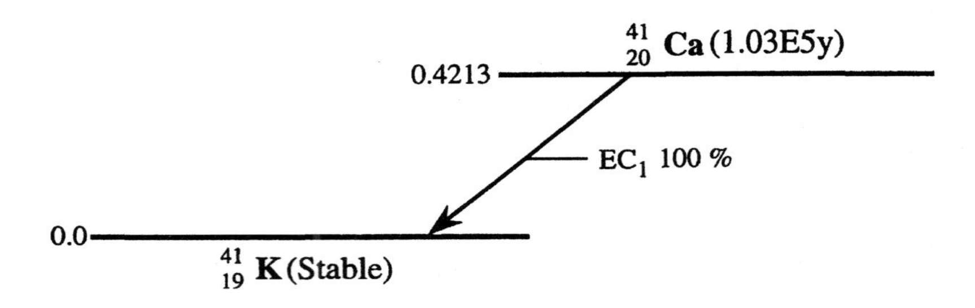

^{40}Ca(n, γ) (June 24, 1993)

Radiation	Y_i (%)	E_i (MeV)
$K_{\alpha 1}$ x ray	7.59	0.0033
$K_{\alpha 2}$ x ray	3.83	0.0033
K_{β} x ray	1.40	0.0036[a]
Auger-K	76.6	0.0030[a]

[a]Average energy.

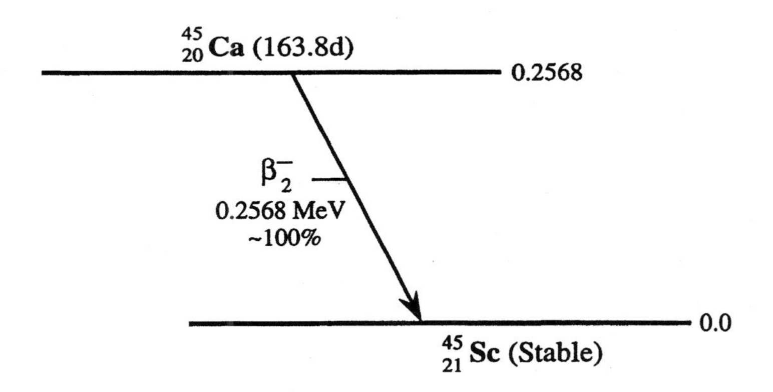

^{44}Ca(n, γ) (June 22, 1995)

Radiation	Y_i (%)	E_i (MeV)
β^-_2	100.0	0.0772[a]
Auger-K	0.0012	0.0036[a]

[a]Average energy.

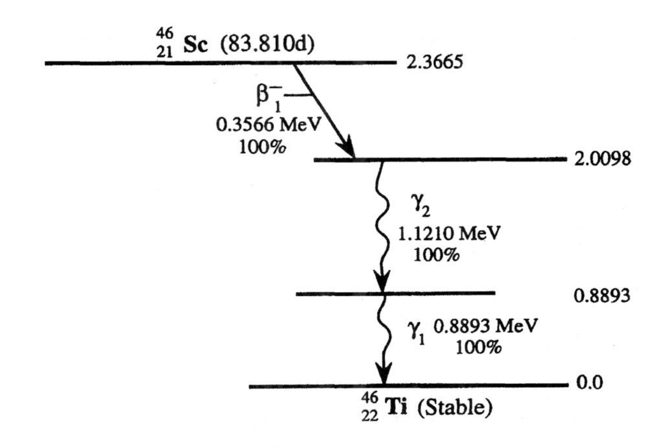

^{45}Sc(n, γ)		(Apr. 28, 1993)
Radiation	Y_i (%)	E_i (MeV)
β_1^-	100.0	0.1118[a]
γ_1	100.0	0.8893
γ_2	100.0	1.1210

[a]Average energy.

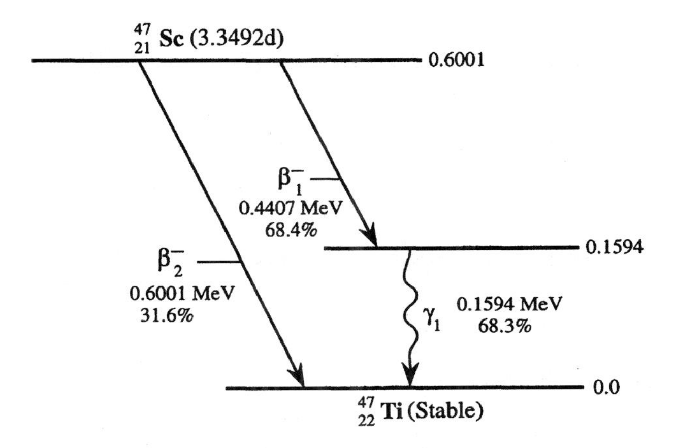

Transformation Product of ^{47}Ca		(Mar. 15, 1995)
Radiation	Y_i (%)	E_i (MeV)
β_1^-	68.4	0.1426[a]
β_2^-	31.6	0.2039[a]
γ_1	68.3	0.1594
ce-K, γ_1	0.277	0.1544
Auger-K	0.214	0.0040[a]
Auger-L	0.434	0.0004[a]

[a]Average energy.

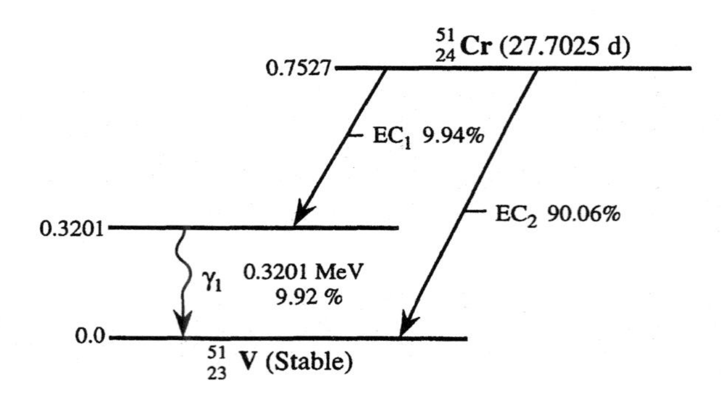

^{50}Cr(n,γ) (June 17, 1997)

Radiation	Y_i (%)	E_i (MeV)
γ_1	9.92	0.3201
ce-K, γ_1	0.0167	0.3146
ce-L, γ_1	0.0016	0.3195[a]
ce-M, γ_1	2.58×10^{-4}	0.3200[a]
$K_{\alpha 1}$ x ray	13.1	0.0050
$K_{\alpha 2}$ x ray	6.6	0.0049
K_β x ray	2.62	0.0054[b]
L x ray	0.334	0.0005[b]
Auger-K	66.9	0.0044[b]

[a]Maximum energy for subshell.
[b]Average energy.

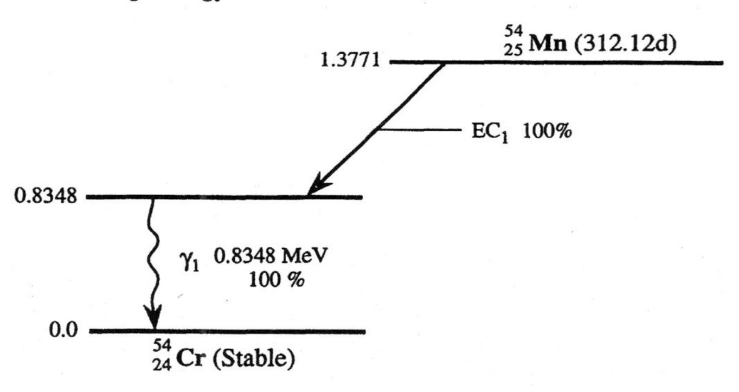

^{56}Fe(d,α); ^{51}V(α,n); ^{53}Cr(d,n); ^{54}Cr(p,n) (Nov. 9, 1995)

Radiation	Y_i (%)	E_i (MeV)
γ_1	100.0	0.8348
ce-K, γ_1	0.022	0.8289
$K_{\alpha 1}$ x ray	14.7	0.0054
$K_{\alpha 2}$ x ray	7.43	0.0054
K_β x ray	2.95	0.0060[a]
L x ray	0.37	0.0006[a]
Auger-K	63.9	0.0048[a]

[a]Average energy.

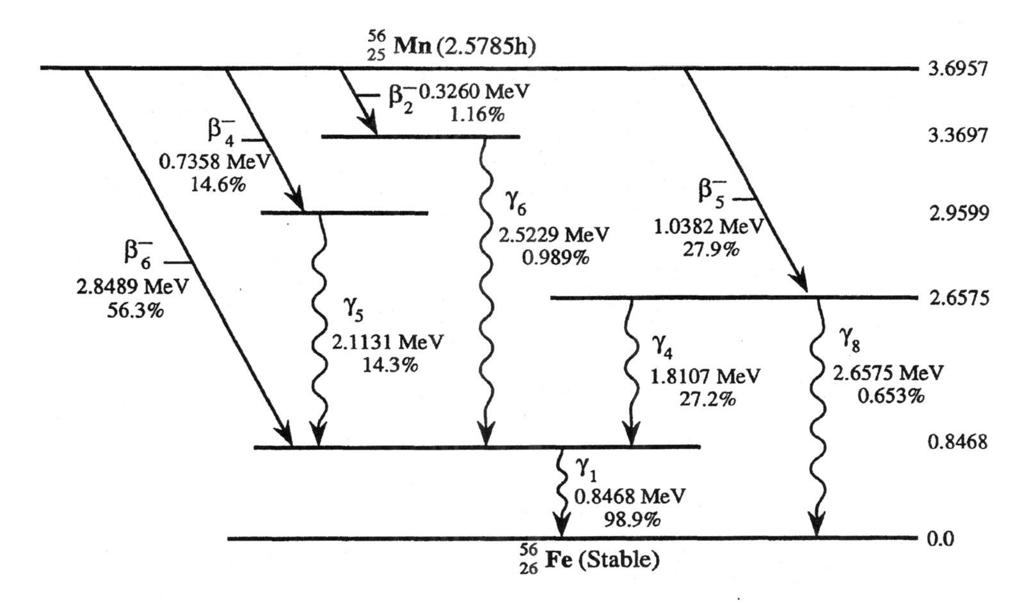

^{55}Mn(n,γ)		(Jan. 14, 1993)
Radiation	Y_i (%)	E_i (MeV)
β_2^-	1.16	0.0992[a]
β_4^-	14.6	0.2553[a]
β_5^-	27.9	0.3820[a]
β_7^-	56.3	1.2170[a]
γ_1	98.9	0.8468
γ_4	27.2	1.8107
γ_5	14.3	2.1131
γ_6	0.989	2.5229
γ_8	0.653	2.6575

[a] Average energy.

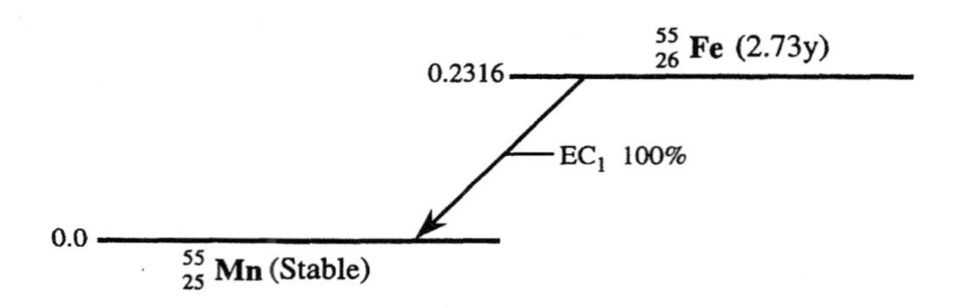

^{54}Fe(n,γ) (Nov. 9, 1995)

Radiation	Y_i (%)	E_i (MeV)
$K_{\alpha 1}$ x ray	16.3	0.0059[a]
$K_{\alpha 2}$ x ray	8.24	0.0059[a]
K_{β} x ray	3.29	0.0065[b]
L x ray	0.421	0.0006[b]
Auger-K	60.7	0.0052[b]

[a]$K_{\alpha 1}$ energy is 5.899 keV, $K_{\alpha 2}$ is 5.888 keV.
[b]Average energy.

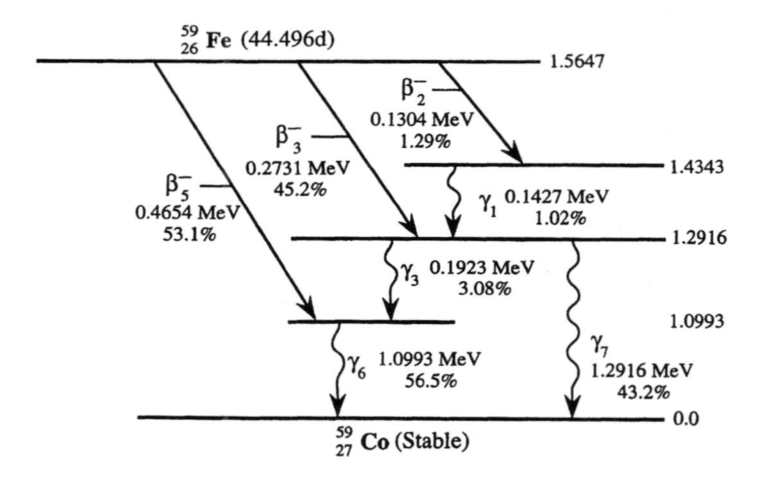

^{58}Fe(n,γ) (Nov. 5, 1993)

Radiation	Y_i (%)	E_i (MeV)
β_2^-	1.31	0.0356[a]
β_3^-	45.3	0.0808[a]
β_5^-	53.1	0.1491[a]
γ_1	1.02	0.1427
γ_3	3.08	0.1923
γ_6	56.5	1.0993
γ_7	43.2	1.2916

[a]Average energy.

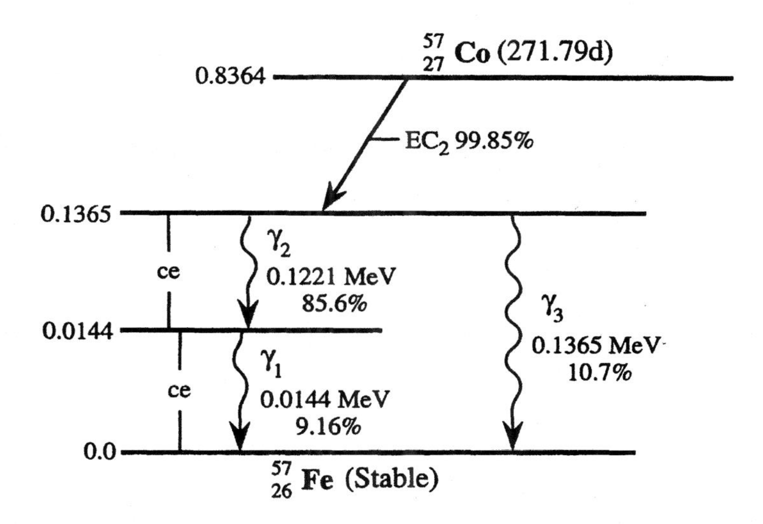

^{58}Ni(γ,p); ^{56}Fe(d,n); ^{56}Fe(p,γ); ^{55}Mn(α,2n) (Dec. 3, 1998)

Radiation	Y_i (%)	E_i (MeV)
γ_1	9.16	0.0144
ce-K, γ_1	71.1	0.0073
ce-L, γ_1	7.36	0.0136[a]
γ_2	85.6	0.1221
ce-K, γ_2	1.83	0.1149
ce-L, γ_2	0.192	0.1212[a]
ce-M$^+$, γ_2	0.0317	0.1220[a]
γ_3	10.7	0.1365
ce-K, γ_3	1.3	0.1294
$K_{\alpha 1}$ x ray	33.7	0.0064
$K_{\alpha 2}$ x ray	17.2	0.0064
K_β x ray	6.97	0.0071[b]
L x ray	1.52	0.0007[b]
Auger-K	105.0	0.0056[b]

[a]Maximum energy for subshell.
[b]Average energy.

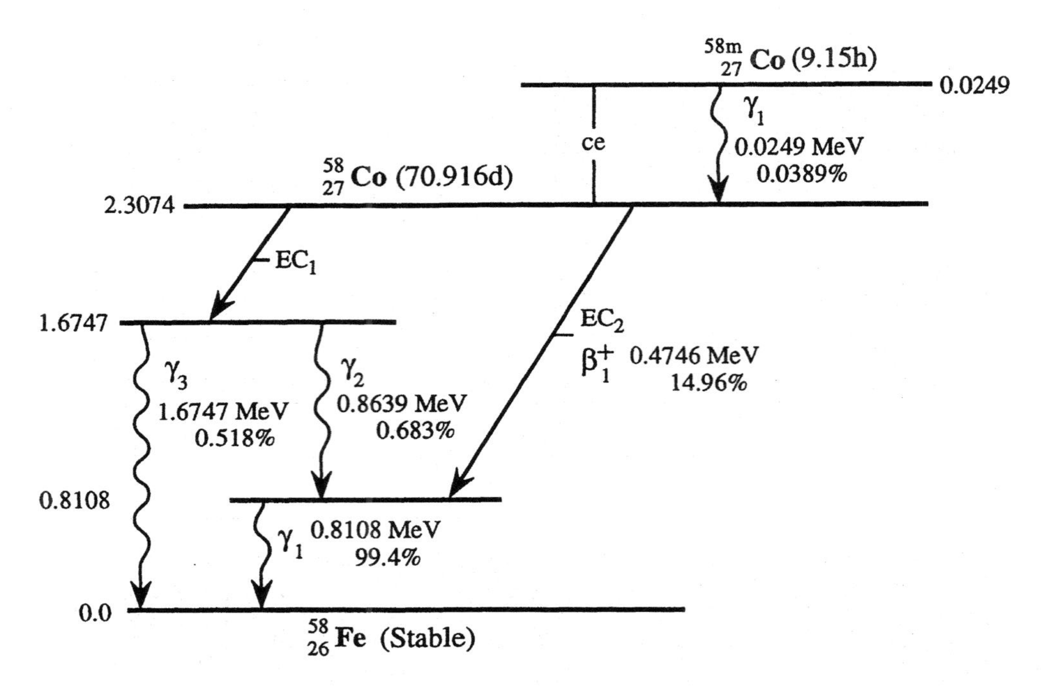

^{55}Mn(α,n); ^{58}Ni(n,p) (May 8, 1997)

Radiation	Y_i (%)	E_i (MeV)
^{58m}Co		
γ_1	0.0389	0.0249
ce-K, γ_1	71.6	0.0172
ce-L, γ_1	23.6	0.0240[a]
$K_{\alpha1}$ x ray	16.0	0.0069
$K_{\alpha2}$ x ray	8.09	0.0069
K_{β} x ray	3.25	0.0077[b]
L x ray	0.437	0.0008[b]
Auger-K	44.3	0.0061[b]
^{58}Co		
β_1^+	14.96	0.2011[b]
$\gamma^{\pm}$	29.92	0.5110
γ_1	99.4	0.8108
γ_2	0.683	0.8640
γ_3	0.518	1.6750
$K_{\alpha1}$ x ray	15.4	0.0064
$K_{\alpha2}$ x ray	7.79	0.0064
K_{β} x ray	3.1	0.0071[b]
Auger-K	49.4	0.0056[b]

[a] Maximum energy for subshell.
[b] Average energy.

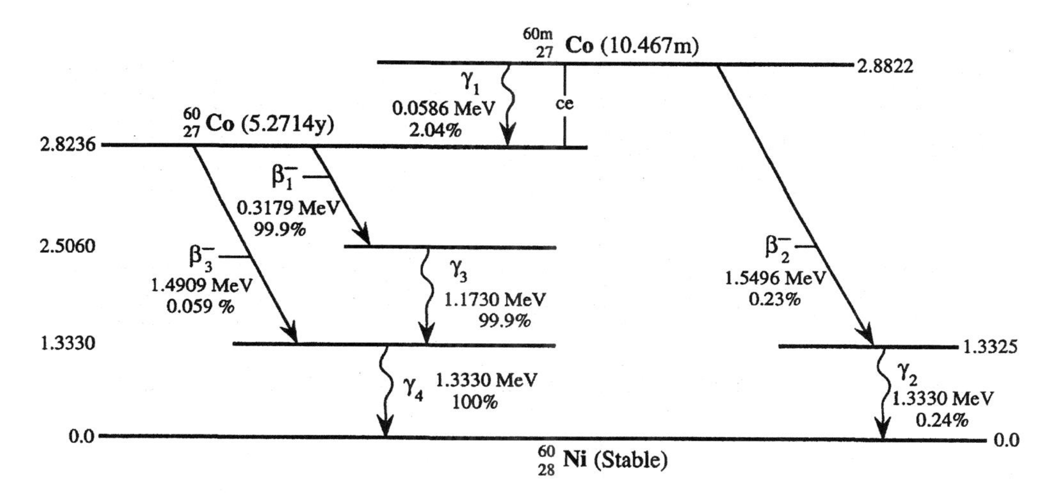

^{59}Co(n, γ)		(July 8, 1993)
Radiation	Y_i (%)	E_i (MeV)
^{60m}Co		
γ_1	2.04	0.0586
β_2^-	0.23	0.6064[a]
γ_2	0.24	1.3330
^{60}Co[b]		
β_1^-	99.9	0.0958[a]
β_3^-	0.059	0.6259[a]
γ_3	99.9	1.1730
γ_4	100.0	1.3330

[a] Average energy.
[b] Yields are per transformation of ^{60}Co.

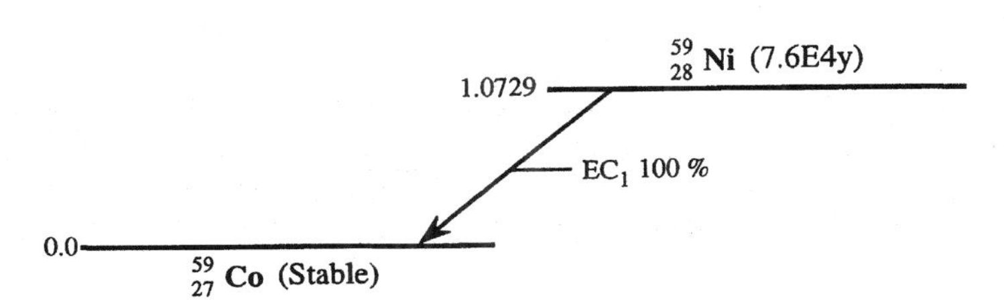

^{58}Ni(n, γ); ^{59}Co(d, 2n)		(Nov. 5, 1993)
Radiation	Y_i (%)	E_i (MeV)
$K_{\alpha 1}$ x ray	19.8	0.0069
$K_{\alpha 2}$ x ray	10.0	0.0069
K_{β} x ray	4.02	0.0077[a]
L x ray	0.473	0.0008[a]
Auger-K	54.9	0.0061[a]

[a] Average energy.

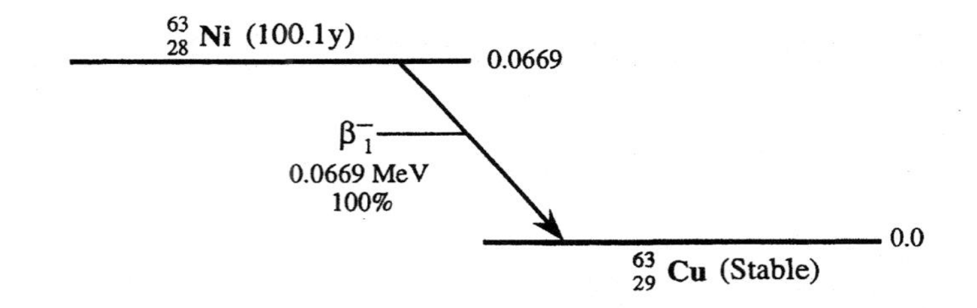

^{62}Ni(n, γ) (Dec. 10, 1991)

Radiation	Y_i (%)	E_i (MeV)
β_1^-	100.0	0.0174[a]

[a]Average energy.

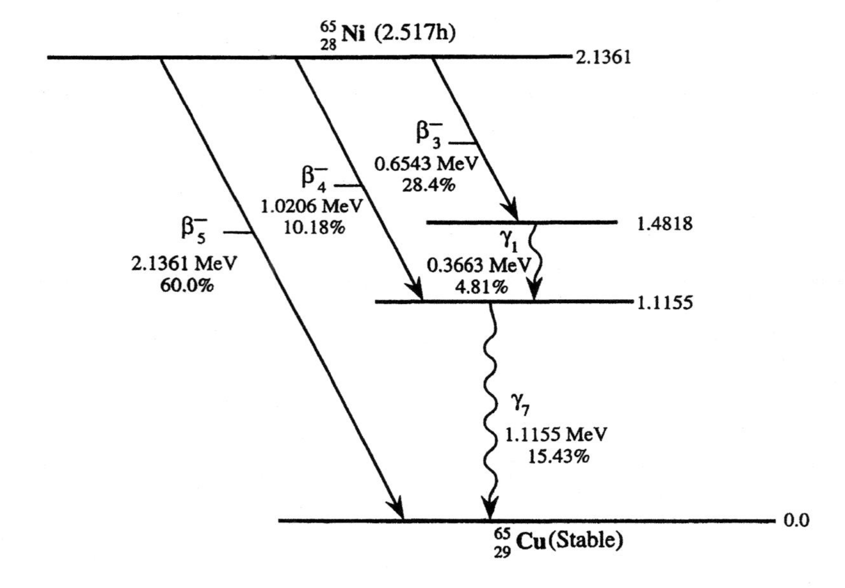

^{64}Ni(n, γ) (Aug. 31, 1993)

Radiation	Y_i (%)	E_i (MeV)
β_3^-	28.4	0.2205[a]
β_4^-	10.18	0.3717[a]
β_5^-	60.0	0.8754[a]
γ_1	4.81	0.3663
γ_7	15.4	1.1160
γ_8	23.59	1.4820

[a]Average energy.

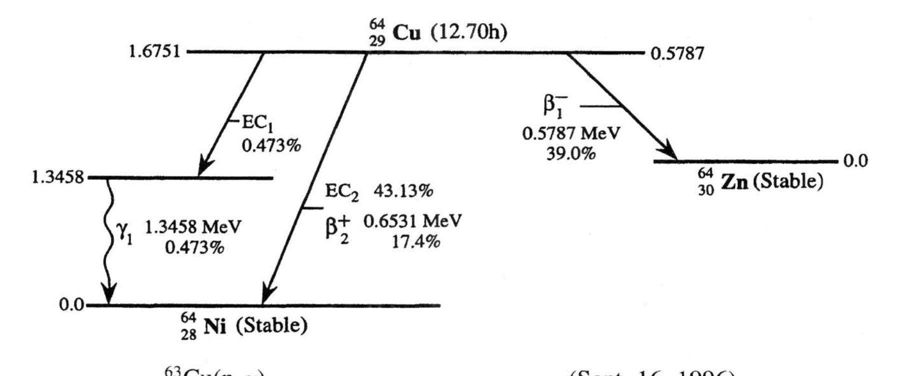

^{63}Cu(n, γ) (Sept. 16, 1996)

Radiation	Y_i (%)	E_i (MeV)
β_1^-	39.0	0.1902[a]
β_1^+	17.4	0.2782[a]
$\gamma^\pm$	34.8	0.5110
γ_1	0.473	1.3460
$K_{\alpha 1}$ x ray	9.36	0.0075
$K_{\alpha 2}$ x ray	4.75	0.0075
K_β x ray	1.91	0.0083[a]
L x ray	0.219	0.0009[a]
Auger-K	22.7	0.0065[a]

[a] Average energy.

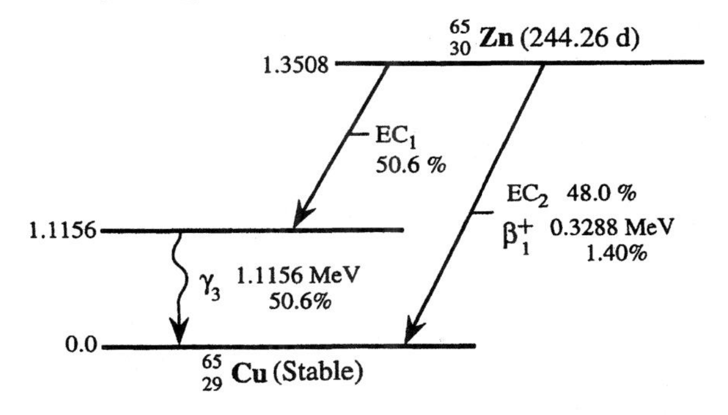

^{64}Zn(n, γ) (Aug. 31, 1993)

Radiation	Y_i (%)	E_i (MeV)
β_1^+	1.40	0.1430[a]
$\gamma^\pm$	2.81	0.5110
γ_3	50.6	1.1160
$K_{\alpha 1}$ x ray	22.6	0.0080
$K_{\alpha 2}$ x ray	11.5	0.0080
K_β x ray	4.61	0.0089[a]
L x ray	0.574	0.0009[a]
Auger-K	48.3	0.0070[a]

[a] Average energy.

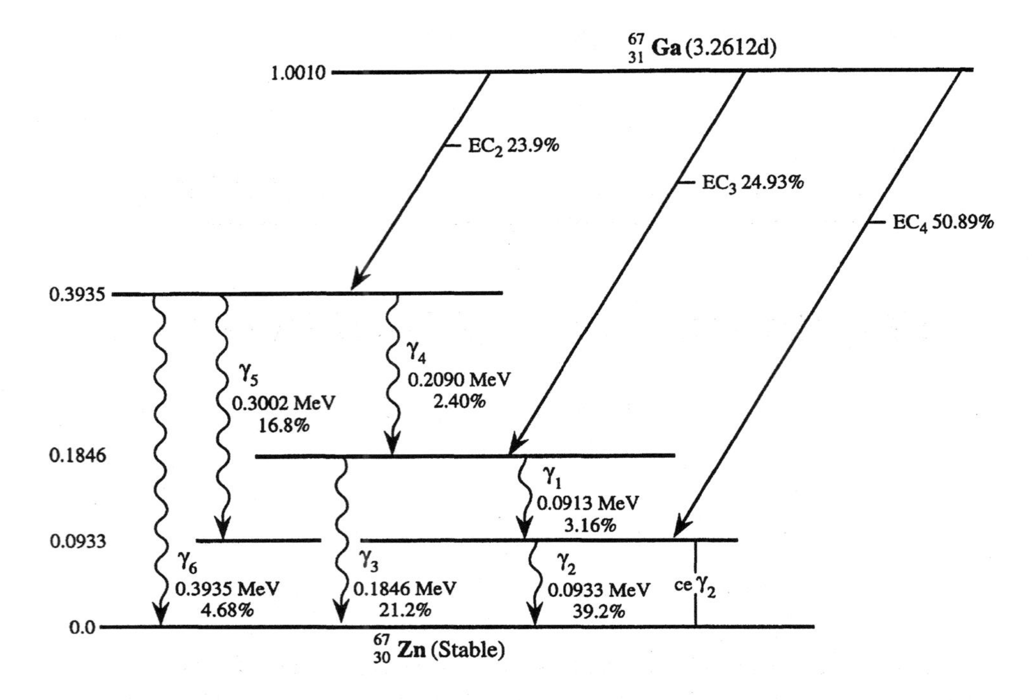

^{67}Zn(d, n); ^{65}Cu(α, 2n) (Dec. 10, 1991)

Radiation	Y_i (%)	E_i (MeV)
γ_1	3.16	0.0913
γ_2	39.2	0.0933
ce-K, γ_2	30.2	0.0837
ce-L, γ_2	3.61	0.0921[a]
γ_3	21.2	0.1846
γ_4	2.4	0.2090
γ_5	16.8	0.3002
γ_6	4.68	0.3935
$K_{\alpha 1}$ x ray	33.5	0.0086
$K_{\alpha 2}$ x ray	17.2	0.0086
K_{β} x ray	7.18	0.0096[b]
L x ray	1.81	0.0010[b]
Auger-K	61.1	0.0075[b]
Auger-L	166.0	0.0010[b]

[a] Average energy for subshell.
[b] Average energy.

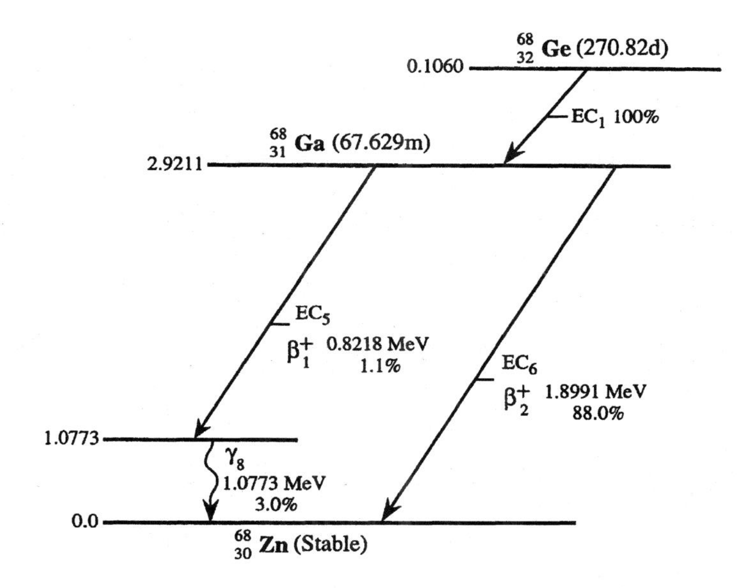

^{68}Ge: ^{66}Zn(α,2n)
^{68}Ga: From ^{68}Ge; ^{65}Cu(α,n); ^{68}Zn(p,n); ^{57}Zn(d,n) (Jan. 4, 1996)

Radiation	Y_i (%)	E_i (MeV)
^{68}Ge		
$K_{\alpha 1}$ x ray	25.6	0.0093
$K_{\alpha 2}$ x ray	13.1	0.0092
K_{β} x ray	5.45	0.0103[a]
L x ray	0.673	0.0011[a]
Auger-K	42.4	0.0080[a]
^{68}Ga		
β_1^+	1.1	0.3526[a]
β_2^+	88.0	0.8360[a]
$\gamma_{\pm}$	178.2	0.5110
γ_8	3.0	1.0770
$K_{\alpha 1}$ x ray	2.71	0.0086
$K_{\alpha 2}$ x ray	1.38	0.0086
K_{β} x ray	0.553	0.0096[a]
Auger-K	5.05	0.0075[a]

[a]Average energy.

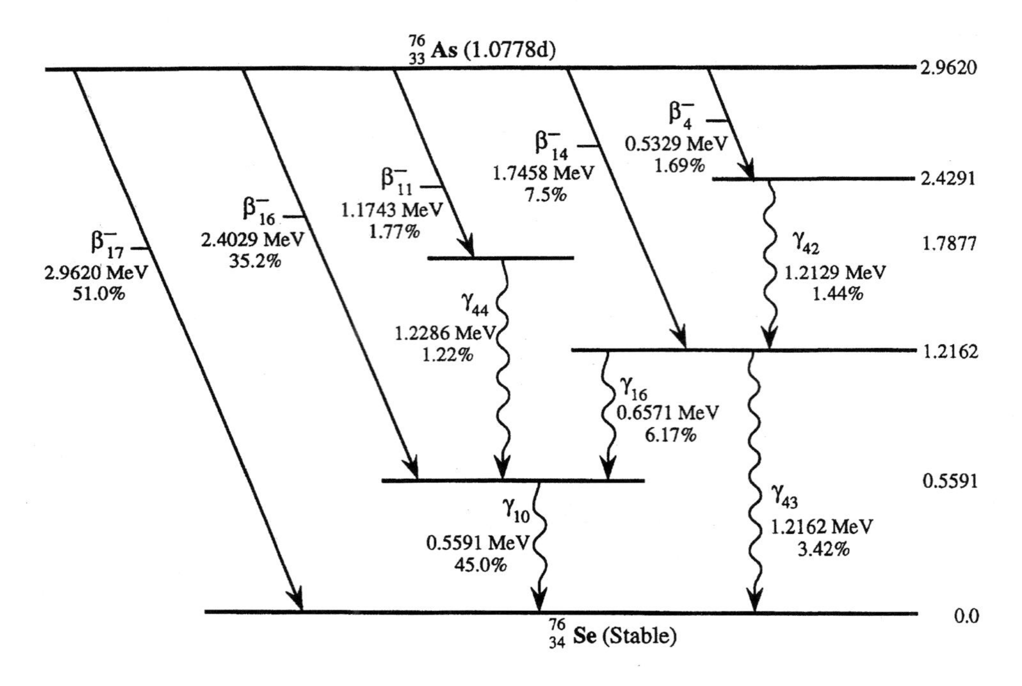

^{75}As(n, γ) (Mar. 15, 1995)

Radiation	Y_i (%)	E_i (MeV)
β^-_4	1.69	0.1736[a]
β^-_{11}	1.77	0.4363[a]
β^-_{14}	7.50	0.6915[a]
β^-_{16}	35.2	0.9963[a]
β^-_{17}	51.0	1.2670[a]
γ_{10}	45.0	0.5591
γ_{11}	1.20	0.5632
γ_{16}	6.17	0.6571
γ_{42}	1.44	1.2129
γ_{43}	3.42	1.2162
γ_{44}	1.22	1.2286

[a] Average energy.

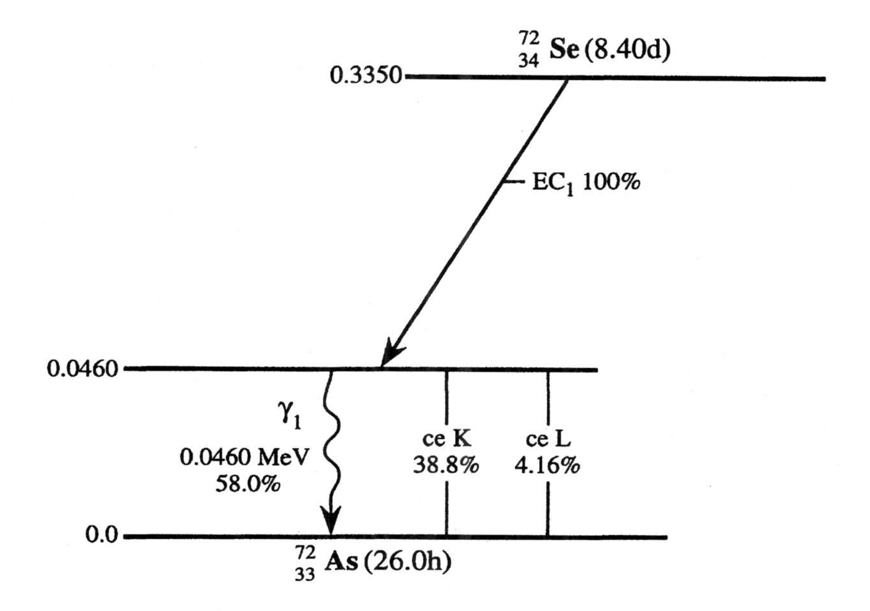

^{75}As(d, 5n); ^{70}Ge(α, 2n) (Jan. 26, 1995)

Radiation	Y_i (%)	E_i (MeV)
γ_1	58.0	0.0460
ce-K, γ_1	38.8	0.0341
ce-L, γ_1	4.16	0.0445[a]
$K_{\alpha 1}$ x ray	41.4	0.0105
$K_{\alpha 2}$ x ray	21.4	0.0105
K_{β} x ray	9.83	0.0117[b]
L x ray	2.64	0.0013[b]
Auger-K	53.6	0.0091[b]
Auger-L	168.0	0.0012[b]

[a] Average energy for subshell.
[b] Average energy.

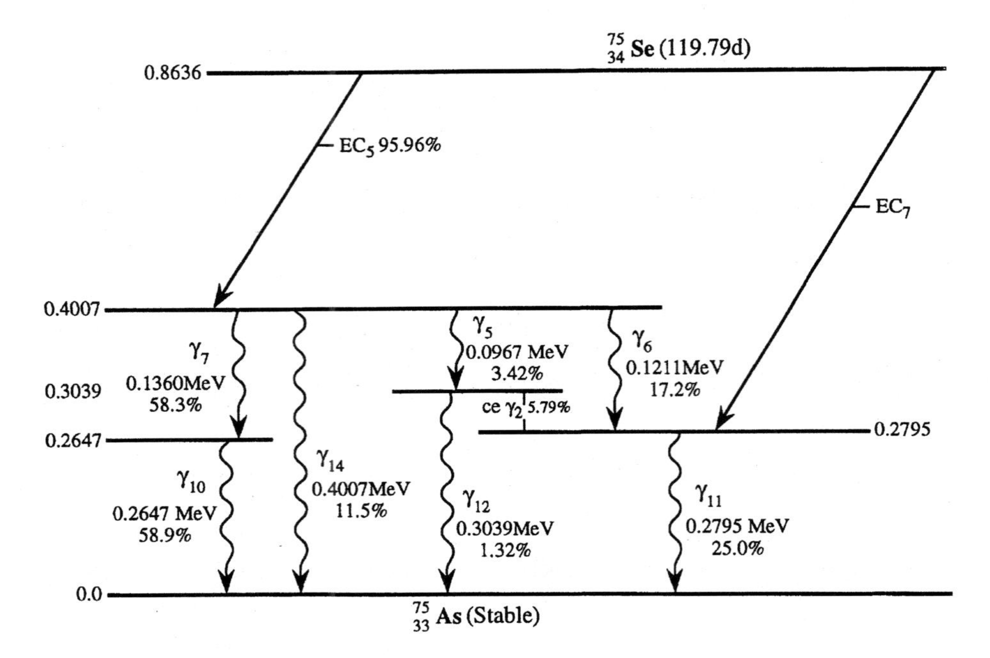

^{74}Se(n, γ); ^{75}As(d, n); ^{75}As(p, n) (Aug. 15, 1996)

Radiation	Y_i (%)	E_i (MeV)
ce-K, γ_2	4.57	0.0125
ce-L, γ_2	1.22	0.0229[a]
γ_3	1.11	0.0661
γ_5	3.42	0.0967
γ_6	17.2	0.1211
γ_7	58.3	0.1360
ce-K, γ_7	1.54	0.1241
ce-L, γ_7	0.163	0.1345[a]
ce-M, γ_7	0.0255	0.1358[a]
γ_8	1.48	0.1986
γ_{10}	58.9	0.2647
γ_{11}	25.0	0.2795
γ_{12}	1.32	0.3039
γ_{14}	11.5	0.4007
$K_{\alpha 1}$ x ray	46.2	0.0105
$K_{\alpha 2}$ x ray	23.8	0.0105
K_{β} x ray	11.0	0.0117[b]
Auger-K	59.8	0.0091[b]

[a]Maximum energy for subshell.
[b]Average energy.

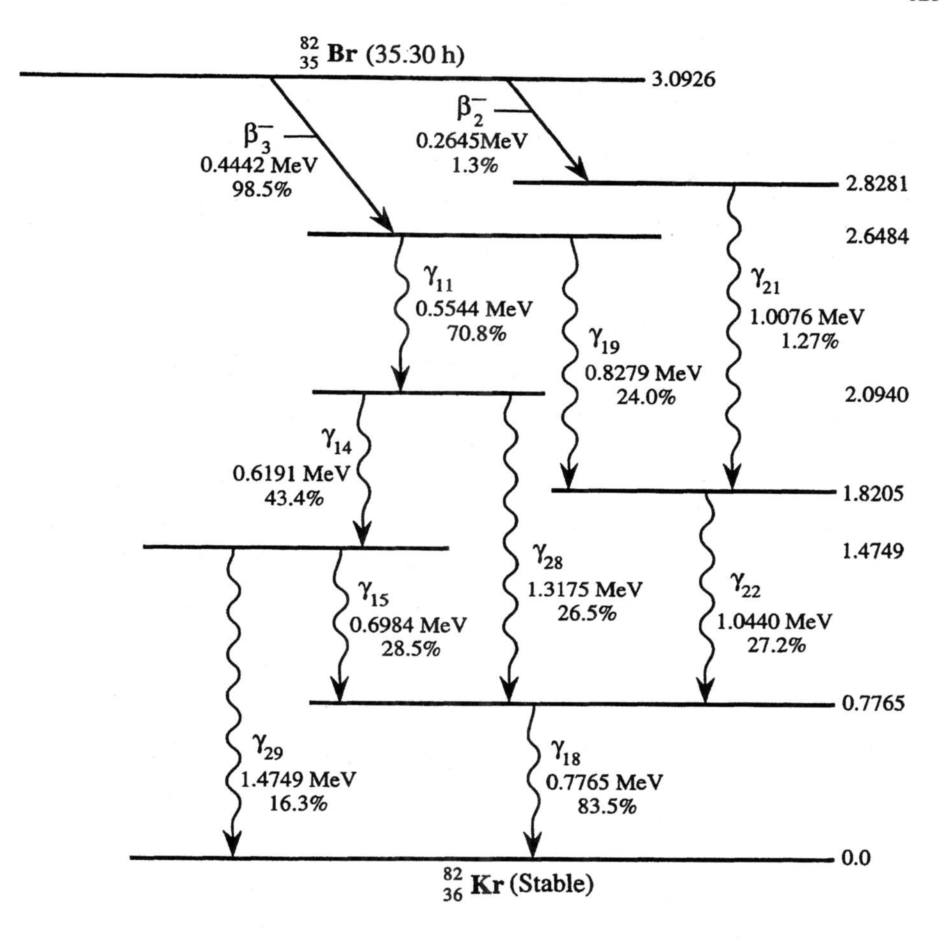

^{81}Br(n, γ); Independent Yield FP Shielded from β^- Decay (Dec. 12, 1995)

Radiation	Y_i (%)	E_i (MeV)
β_2^-	1.3	0.0762[a]
β_3^-	98.5	0.1377[a]
γ_{11}	70.8	0.5544
γ_{14}	43.4	0.6191
γ_{15}	28.5	0.6984
γ_{18}	83.5	0.7765
ce-K, γ_{18}	0.0685	0.7622
γ_{19}	24.0	0.8279
γ_{21}	1.27	1.0076
γ_{22}	27.2	1.0440
γ_{28}	26.5	1.3175
γ_{29}	16.3	1.4749

[a]Average energy; 5 other β^- with yields $\leq 0.4\%$.

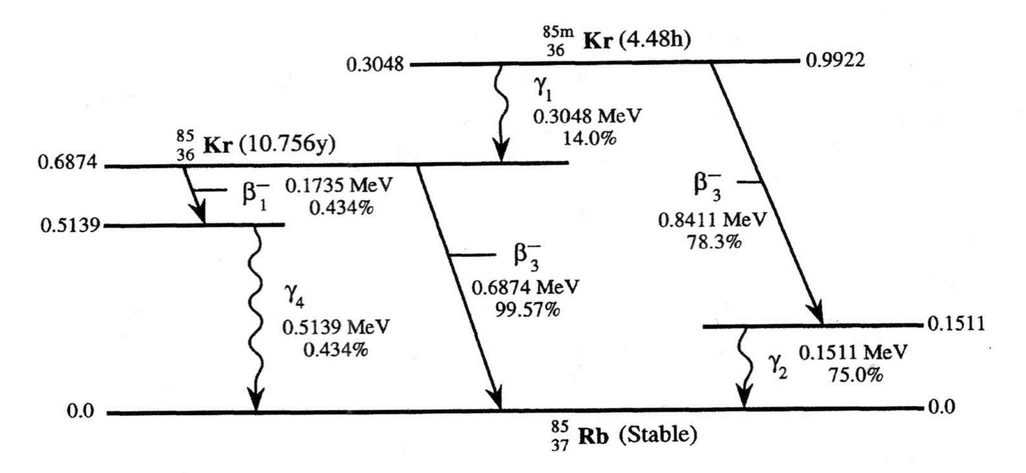

FP; ^{84}Kr(n,γ)^{85m}Kr; ^{84}Kr(n,γ)^{85}Kr		(Apr. 26, 1991)
Radiation	Y_i (%)	E_i (MeV)
^{85m}Kr (IT)		
γ_1	14.0	0.3049
ce-K, γ_1	6.15	0.2905
ce-L, γ_1	0.932	0.3029[a]
ce-M, γ_1	0.182	0.3046[a]
$K_{\alpha1}$ x ray	2.25	0.0127
$K_{\alpha2}$ x ray	1.17	0.0126
K_{β} x ray	0.594	0.0141[b]
Auger-K	2.14	0.0108[b]
^{85m}Kr (β^-)		
β_3^-	78.3	0.2905[b]
γ_1	0.3	0.1299
γ_2	75.0	0.1511
ce-K, γ_2	3.13	0.1360
$K_{\alpha1}$ x ray	1.18	0.0134
Auger-K	1.02	0.0114[b]
^{85}Kr (β^-)		
β_1^-	0.434	0.0477[b]
β_3^-	99.57	0.2516[b]
γ_4	0.434	0.5139

[a]Maximum energy for subshell.
[b]Average energy.

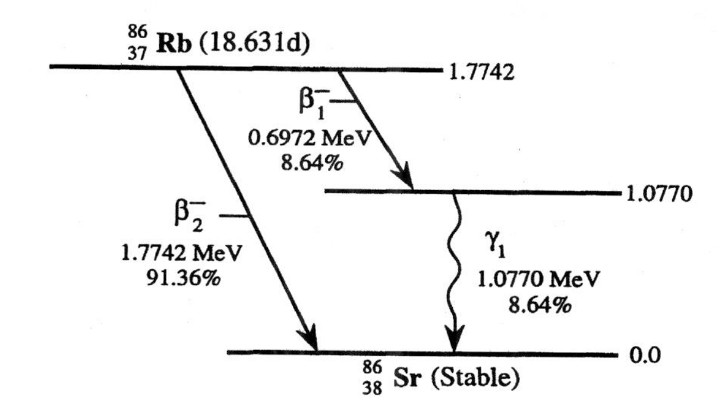

^{85}Rb(n, γ); FP[a] (Apr. 24, 1997)

Radiation	Y_i (%)	E_i (MeV)
β_1^-	8.64	0.2325[b]
β_2^-	91.4	0.7094[b]
γ_1	8.64	1.0770

[a]EC to ^{86}Kr (stable) also occurs < 0.01%.
[b]Average energy.

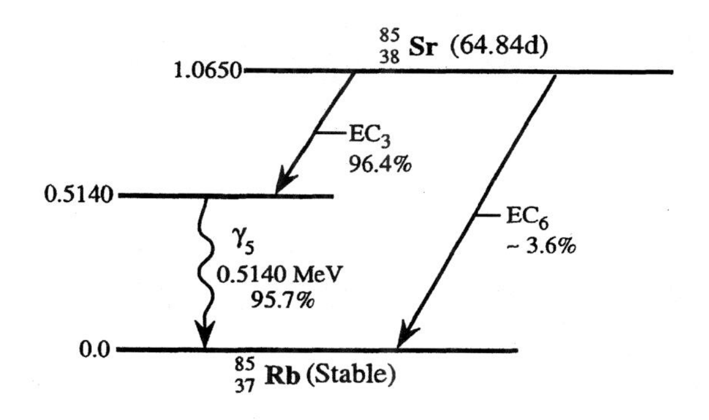

^{84}Sr(n, γ); ^{85}Rb(p, n); ^{85}Rb(d, 2n) (Apr. 26, 1991)

Radiation	Y_i (%)	E_i (MeV)
γ_5	95.70	0.5140
ce-K, γ_5	0.603	0.4988
ce-L, γ_5	0.068	0.5119[a]
$K_{\alpha 1}$ x ray	33.1	0.0134
$K_{\alpha 2}$ x ray	17.2	0.0133
K_{β} x ray	8.98	0.0150[b]
L x ray	2.59	0.0017[b]
Auger-K	28.6	0.0114[b]

[a]Maximum energy for subshell.
[b]Average energy.

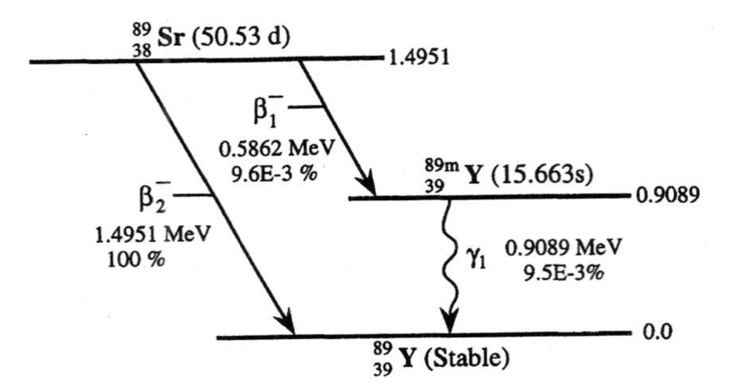

FP; ^{88}Sr(d,p); ^{88}Sr(n,γ) (Nov. 6, 1998)

Radiation	Y_i (%)	E_i (MeV)
β_1^-	0.0096	0.1890[a]
β_2^-	100.0	0.58346[a]
γ_1[b]	0.0095	0.9089

[a]Average energy.
[b]Yields of ^{89m}Y radiations are per transformation of ^{89}Sr.

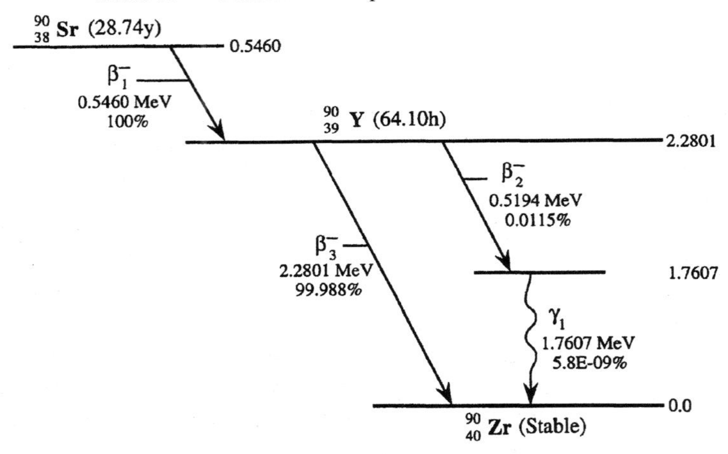

FP; ^{89}Sr(n,γ)^{90}Sr; ^{89}Sr(n,γ)^{90}Y (Jan. 7, 1998)

Radiation	Y_i (%)	E_i (MeV)
^{90}Sr		
β_1^-	100.0	0.1958[a]
^{90}Y		
β_1^-	1.40×10^{-6}	0.0250[a]
β_2^-	0.0115	0.1856[a]
β_3^-	99.988	0.9337[a]
γ_1	5.80×10^{-9}	1.7607
ce-K, γ_1	0.0102	1.7430
ce-L, γ_1	0.0013	1.7580[b]
$K_{\alpha 1}$ x ray	0.0042	0.0158
$K_{\alpha 2}$ x ray	0.0022	0.0157
K_β x ray	0.0012	0.0177[a]
Auger-K	0.0028	0.0134[a]

[a]Average energy.
[b]Maximum energy for subshell.

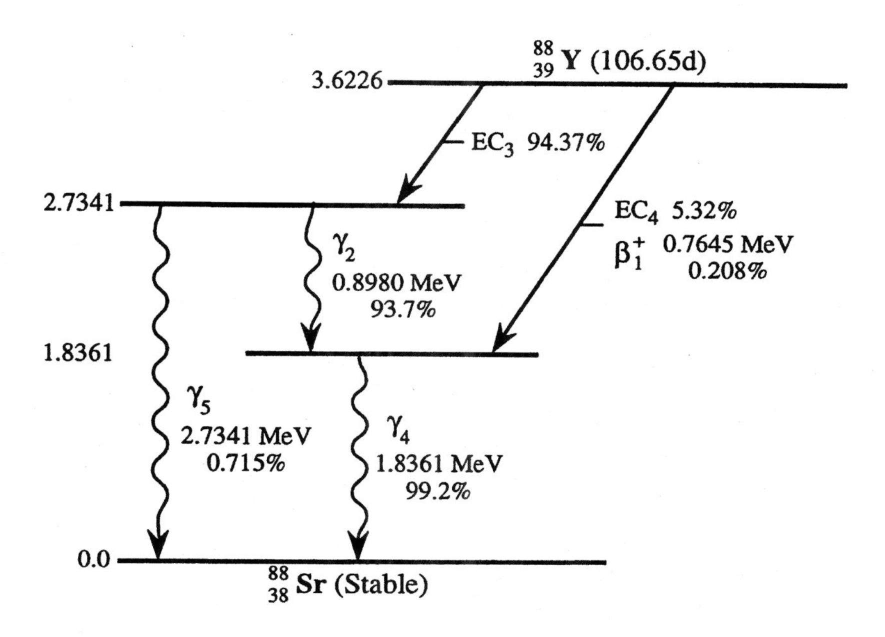

^{88}Sr(p, n); ^{88}Sr(d, 2n) (Aug. 25, 1988)

Radiation	Y_i (%)	E_i (MeV)
β_1^+	0.208	0.3595[a]
$\gamma^{\pm}$	0.416	0.5110
γ_2	93.7	0.8980
ce-K, γ_2	0.0253	0.8819
γ_4	99.2	1.8361
ce-K, γ_4	0.0129	1.8200
γ_5	0.715	2.7341
$K_{\alpha 1}$ x ray	33.7	0.0142
$K_{\alpha 2}$ x ray	17.5	0.0141
K_{β} x ray	9.40	0.0158[a]
L x ray	2.79	0.0018[a]
Auger-K	26.4	0.0121[a]

[a] Average energy.

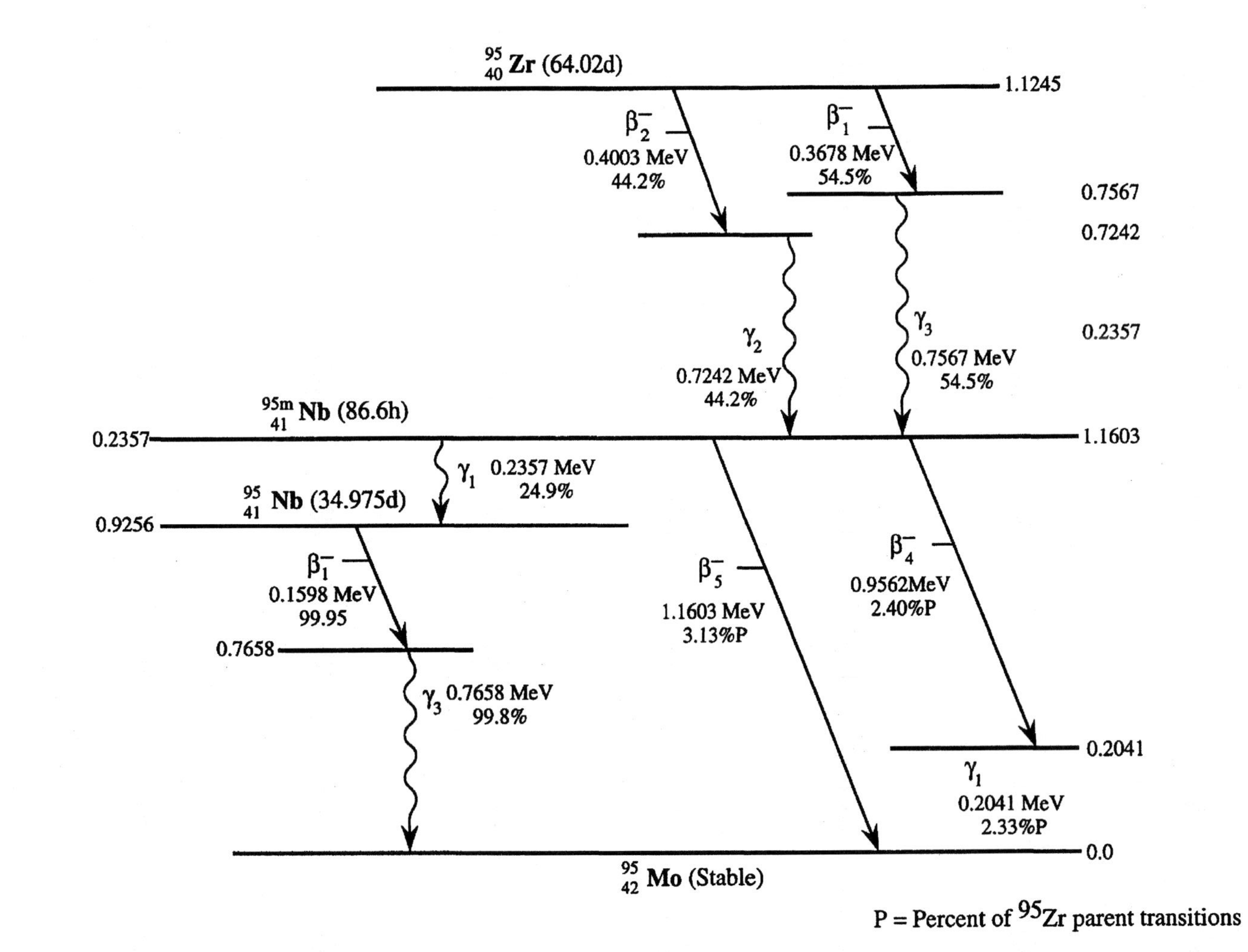

P = Percent of ^{95}Zr parent transitions

FP (Apr. 5, 1994)

Radiation	Y_i (%)	E_i (MeV)
^{95}Zr		
β_1^-	54.5	0.1097[a]
β_2^-	44.2	0.1209[a]
β_3^-	1.13	0.3276[a]
γ_2	44.2	0.7242
γ_3	54.5	0.7567
^{95m}Nb-IT		
γ_1	24.9	0.2357
^{95m}Nb(β^-)		
β_4^-	2.4	0.3346[a]
β_5^-	3.13	0.4374[a]
γ_1	2.33	0.2041
$K_{\alpha 1}$ x ray	23.1	0.0165
$K_{\alpha 2}$ x ray	12.1	0.0165
K_{β} x ray	6.87	0.0186[a]
L x ray	2.42	0.0022[a]
Auger-K	13.9	0.0140[a]
^{95}Nb		
β_1^-	99.95	0.0434[a]
γ_3	99.8	0.7658

[a] Average energy.

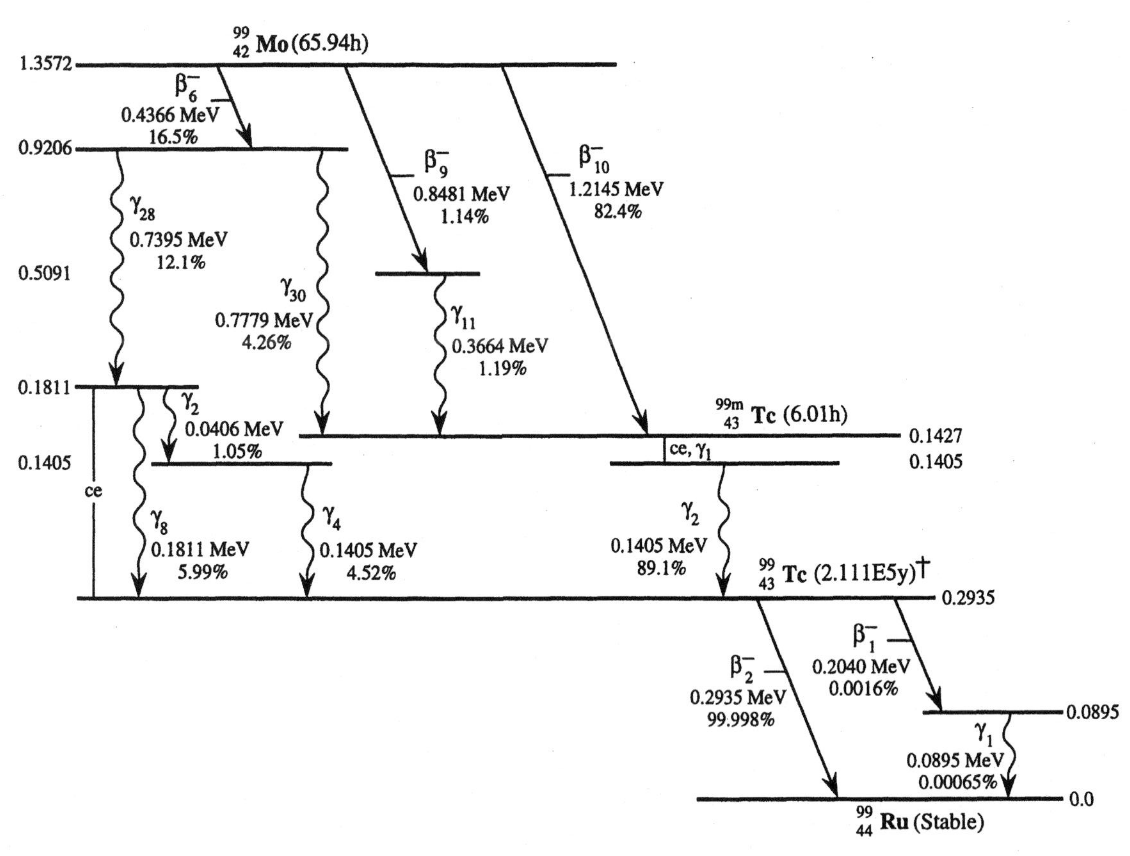

† NOTE: 87.907% of transitions of ^{99}Mo produce the 0.1427 level of ^{99m}Tc. Metastable ^{99m}Tc is relieved in 89.9% of its transitions by a 0.1405 MeV gamma ray; therefore, the 0.1405 MeV γ is produced in 78.32% of the transformation of ^{99}Mo. However, if the ^{99}Mo parent and the ^{99m}Tc product are together and in equilibrium, the 0.1405 MeV gamma is produced in 82.84 % of transformations of ^{99}Mo (this accounts for the 4.52% contribution through β^-_6 and γ_4).

^{98}Mo(n,γ); FP (Jan. 9, 1995)

Radiation	Y_i (%)	E_i (MeV)
^{99}Mo		
β_6^-	16.4	0.1331[a]
β_9^-	1.14	0.2897[a]
β_{10}^-	82.4	0.4428[a]
γ_2	1.05	0.0406
ce-K, γ_2	3.44	0.0195
γ_4	4.52	0.1405
ce-K, γ_4	0.449	0.1195
γ_8	5.99	0.1811
ce-K, γ_8	0.753	0.1600
γ_{11}	1.19	0.3664
γ_{28}	12.1	0.7395
γ_{30}	4.26	0.7779
$K_{\alpha 1}$ x ray	2.00	0.0184
$K_{\alpha 2}$ x ray	1.04	0.0183
^{99m}Tc		
ce-M, γ_1	74.4	0.0016[b]
γ_2	89.1	0.1405
ce-K, γ_2	8.84	0.1195
ce-L, γ_2	1.07	0.1375[b]
ce-M, γ_2	0.194	0.1400[b]
ce-N$^+$, γ_2	0.0374	0.1404[b]
$K_{\alpha 1}$ x ray	4.02	0.0184
$K_{\alpha 2}$ x ray	2.1	0.0183
Auger-K	2.07	0.0155[a]
^{99}Tc		
β_1^-	0.0016	0.0817[a]
β_2^-	99.998	0.0846[a]
γ_1	6.50×10^{-4}	0.0895
ce-K, γ_1	7.86×10^{-4}	0.0674
ce-L, γ_1	1.73×10^{-4}	0.0863[b]
K_α x ray	5.08×10^{-4}	0.0193[a]
K_β x ray	1.02×10^{-4}	0.0217[a]

[a]Maximum energy for subshell.
[b]Average energy.

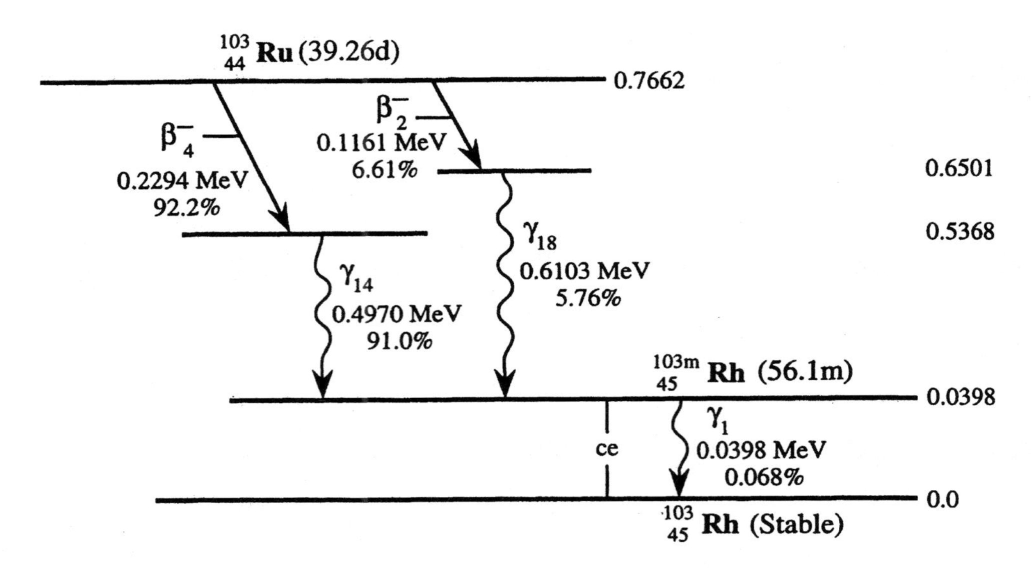

^{102}Ru(n,γ); FP (Apr. 28, 1993)

Radiation	Y_i (%)	E_i (MeV)
β_2^-	6.61	0.0307[a]
β_4^-	92.2	0.0641[a]
β_7^-	0.87	0.2559[a]
γ_1[b]	0.0684	0.0398
ce-K, γ_1	9.89	0.0165
ce-L, γ_1	73.3	0.0364[c]
ce-M, γ_1	14.5	0.0391[c]
γ_{14}	91.0	0.4971
γ_{18}	5.76	0.6103
$K_{\alpha 1}$ x ray	4.89	0.0202
$K_{\alpha 2}$ x ray	2.57	0.0201
Auger-K	2.14	0.0170[a]

[a]Average energy.
[b]Gamma from ^{103m}Rh.
[c]Maximum energy for subshell.

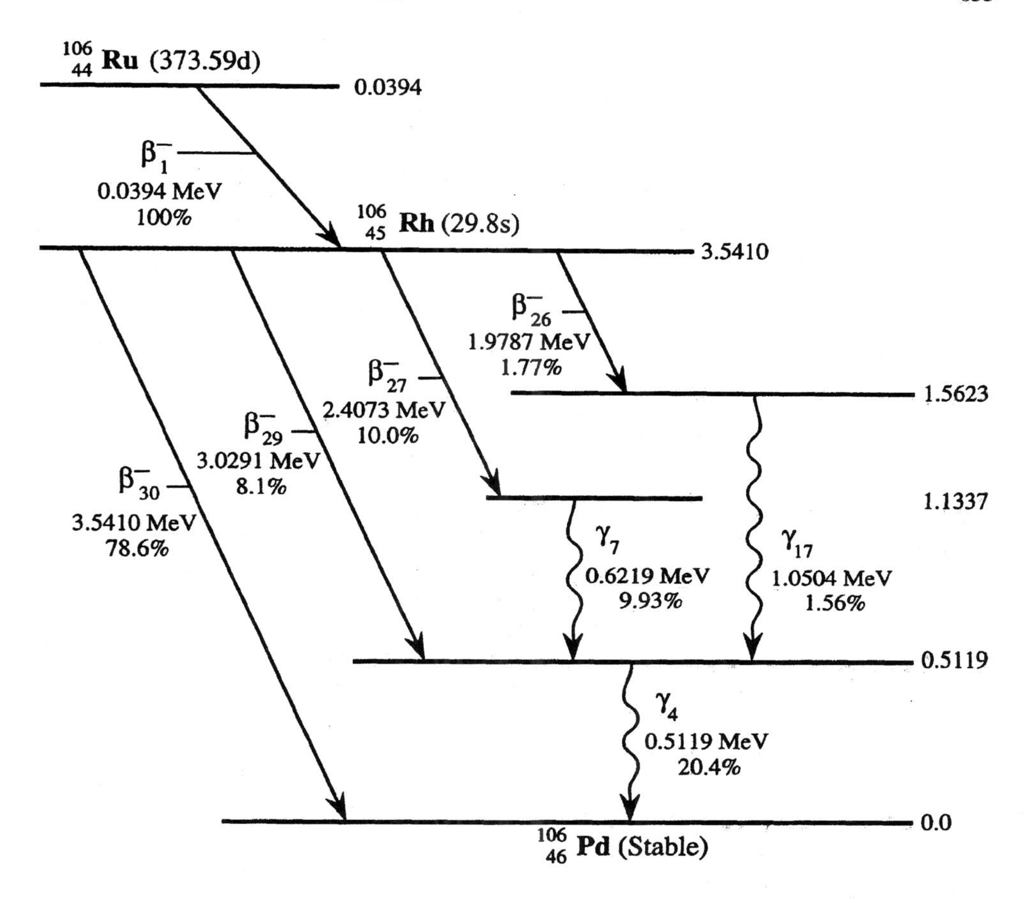

FP (Aug. 16, 1994)

Radiation	Y_i (%)	E_i (MeV)
^{106}Ru		
β_1^-	100.0	0.0100[a]
^{106}Rh		
β_{26}^-	1.77	0.7790[a]
β_{27}^-	10.0	0.9760[a]
β_{29}^-	8.1	1.2670[a]
β_{30}^-	78.6	1.5080[a]
γ_4	20.4	0.5119
γ_7	9.93	0.6219
γ_{17}	1.56	1.0504

[a]Average energy.

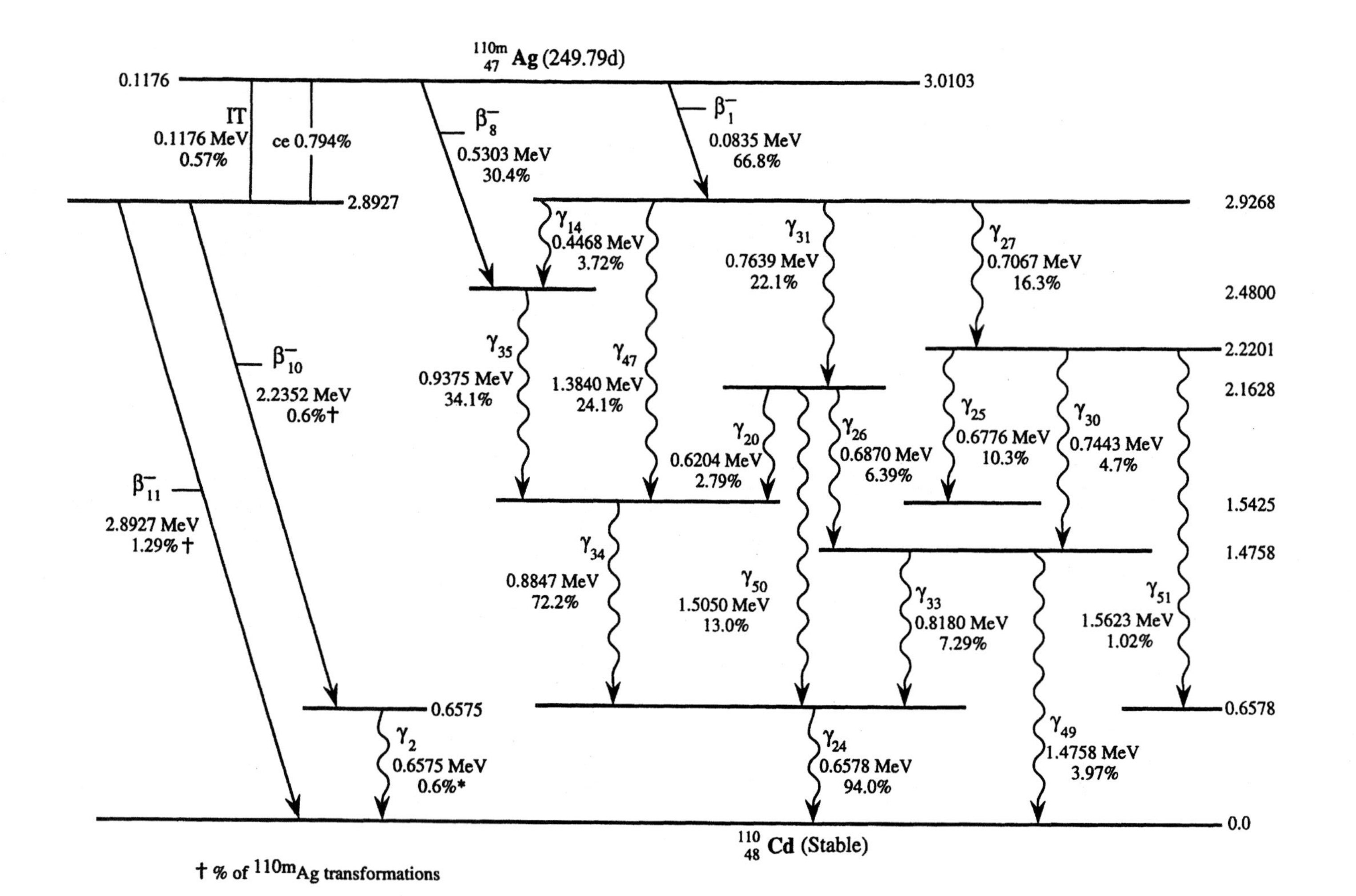

110m 47 Ag (249.79d)
0.1176
3.0103
IT
0.1176 MeV
0.57%
ce 0.794%
β⁻8
0.5303 MeV
30.4%
β⁻1
0.0835 MeV
66.8%
2.8927
2.9268
γ14
0.4468 MeV
3.72%
γ31
0.7639 MeV
22.1%
γ27
0.7067 MeV
16.3%
2.4800
2.2201
2.1628
β⁻10
2.2352 MeV
0.6%†
γ35
0.9375 MeV
34.1%
γ47
1.3840 MeV
24.1%
γ20
0.6204 MeV
2.79%
γ26
0.6870 MeV
6.39%
γ25
0.6776 MeV
10.3%
γ30
0.7443 MeV
4.7%
1.5425
1.4758
β⁻11
2.8927 MeV
1.29%†
γ34
0.8847 MeV
72.2%
γ50
1.5050 MeV
13.0%
γ33
0.8180 MeV
7.29%
γ51
1.5623 MeV
1.02%
0.6575
0.6578
γ2
0.6575 MeV
0.6%*
γ24
0.6578 MeV
94.0%
γ49
1.4758 MeV
3.97%
0.0
110 48 Cd (Stable)
† % of 110mAg transformations

$^{109}Ag(n,\gamma)$ (Jan. 14, 1993)

Radiation[a]	Y_i (%)	E_i (MeV)
ce-K, γ_2	0.794	0.0910
β_1^-	66.8	0.0218[b]
β_8^-	30.4	0.1655[b]
γ_{14}	3.72	0.4468
γ_{20}	2.79	0.6204
γ_{24}	94.0	0.6578
γ_{25}	10.3	0.6776
γ_{26}	6.39	0.6870
γ_{27}	16.3	0.7067
γ_{30}	4.7	0.7443
γ_{31}	22.1	0.7639
γ_{33}	7.29	0.8180
γ_{34}	72.2	0.8847
γ_{35}	34.1	0.9375
ce-K, γ_{35}	0.0403	0.9108
γ_{47}	24.1	1.3840
γ_{49}	3.97	1.4758
γ_{50}	13.0	1.5050
γ_{51}	1.02	1.5623

[a]Radiations from ^{110m}Ag only; transformations through ^{110}Ag ($T_{1/2}$ = 24.6 s are negligible.

[b]Average energy.

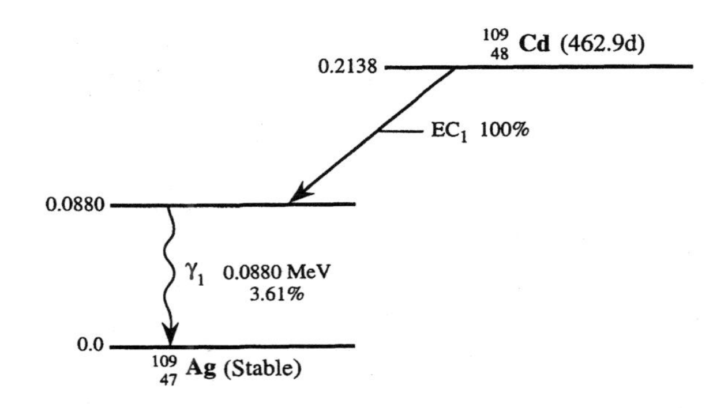

^{108}Cd(n,γ); ^{109}Ag(d,2n) (Feb. 14, 1994)

Radiation	Y_i (%)	E_i (MeV)
γ_1	3.61	0.0880
ce-K, γ_1	41.7	0.0625
ce-L, γ_1	44.0	0.0842[a]
ce-M, γ_1	8.95	0.0873[a]
ce-N$^+$, γ_1	1.59	0.0879[a]
$K_{\alpha 1}$ x ray	55.2	0.0222
$K_{\alpha 2}$ x ray	29.1	0.0220
K_{β} x ray	17.8	0.0249[b]
L x ray	11.2	0.0030[b]
Auger-K	20.9	0.0185[b]

[a]Maximum energy for subshell.
[b]Average energy.

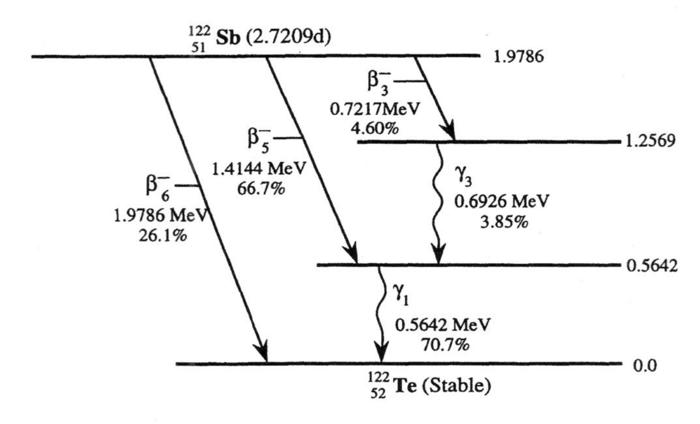

^{121}Sb(n,γ) (July 8, 1994)

Radiation	Y_i (%)	E_i (MeV)
β_3^-	4.6	0.2357[a]
β_4^-	66.7	0.5212[a]
β_5^-	26.1	0.7710[a]
γ_1	70.7	0.5642
γ_3	3.85	0.6926
Auger-L	2.06	0.0030[a]

[a]Average energy.

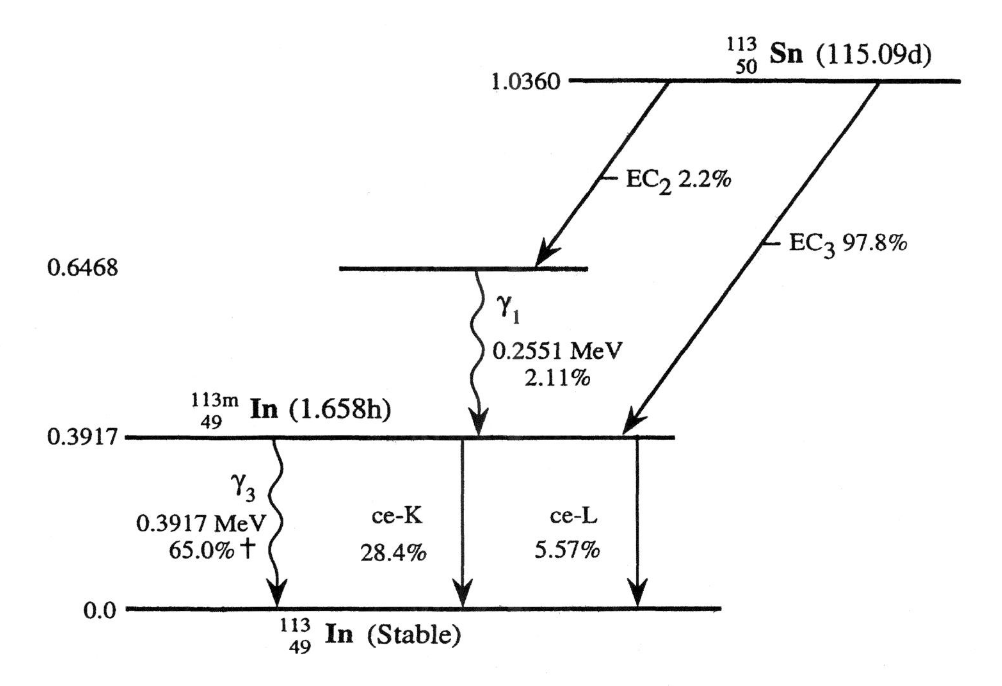

† Gamma yield per transformation of ^{113}Sn is 65%; the independent gamma yield for ^{113m}In alone is 64.2%.

^{113}Sn: ^{112}Sn(n, γ); ^{113}In(p, n)
^{113m}In: Product of ^{113}Sn (Jan. 28, 1999)

Radiation	Y_i (%)	E_i (MeV)
^{113}Sn		
γ_1	2.11	0.2551
γ_3	65.0	0.3917
ce-K, γ_3	28.4	0.3637
ce-L, γ_3	5.57	0.3875[a]
ce-M, γ_3	1.11	0.3909[a]
$K_{\alpha1}$ x ray	52.0	0.0242
$K_{\alpha2}$ x ray	27.7	0.0240
K_{β} x ray	17.4	0.0273[b]
L x ray	8.54	0.0033[b]
Auger-K	16.9	0.0201[b]
Auger-L	116.0	0.0028[b]
^{113m}In		
γ_1	64.2	0.3917[b]

[a]Maximum energy for subshell.
[b]Average energy.

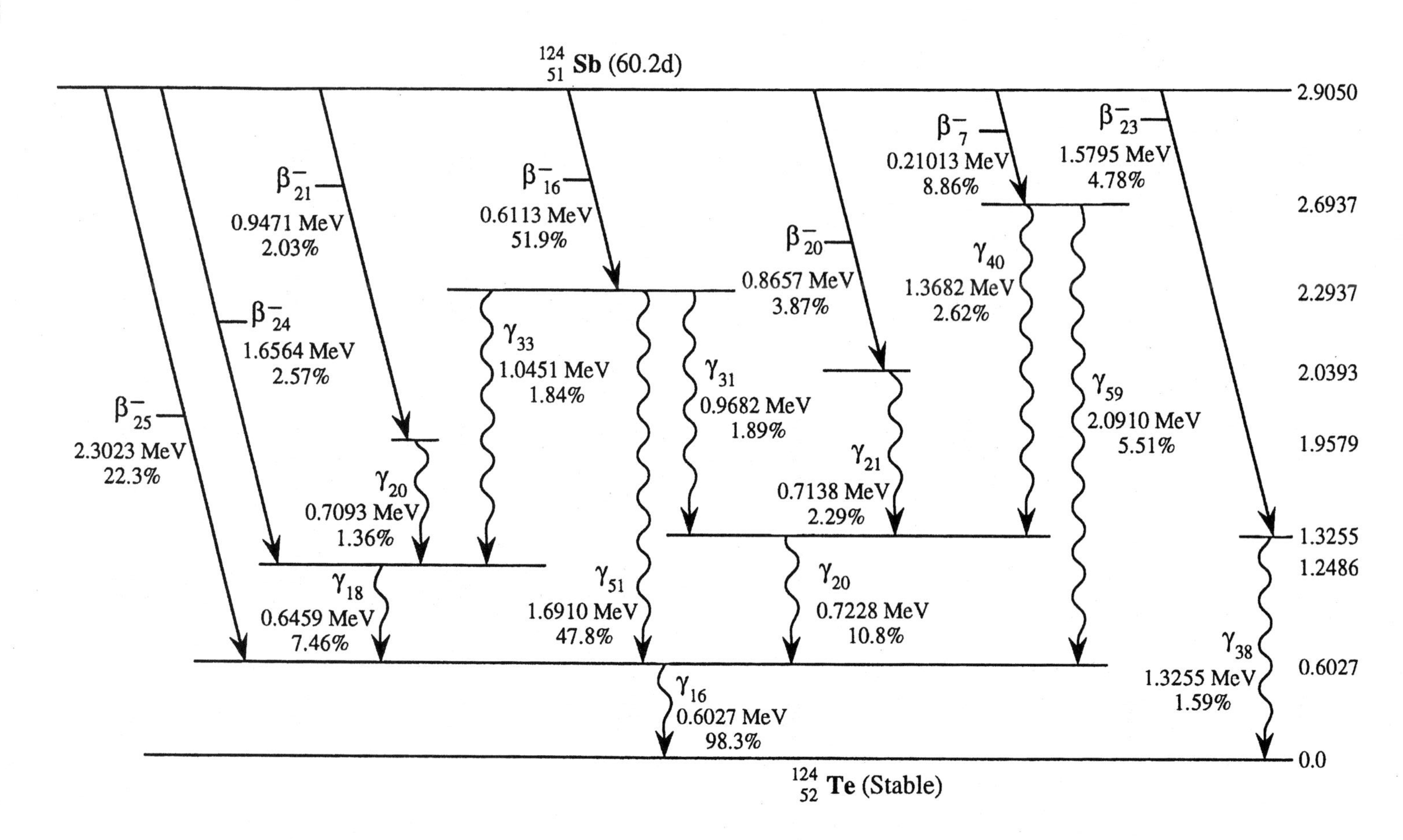

$^{124}_{51}$Sb (60.2d)
2.9050
2.6937
2.2937
2.0393
1.9579
1.3255
1.2486
0.6027
0.0
β^-_{25} 2.3023 MeV 22.3%
β^-_{24} 1.6564 MeV 2.57%
β^-_{21} 0.9471 MeV 2.03%
β^-_{16} 0.6113 MeV 51.9%
β^-_{20} 0.8657 MeV 3.87%
β^-_{7} 0.21013 MeV 8.86%
β^-_{23} 1.5795 MeV 4.78%
γ_{20} 0.7093 MeV 1.36%
γ_{33} 1.0451 MeV 1.84%
γ_{18} 0.6459 MeV 7.46%
γ_{51} 1.6910 MeV 47.8%
γ_{31} 0.9682 MeV 1.89%
γ_{21} 0.7138 MeV 2.29%
γ_{20} 0.7228 MeV 10.8%
γ_{40} 1.3682 MeV 2.62%
γ_{59} 2.0910 MeV 5.51%
γ_{16} 0.6027 MeV 98.3%
γ_{38} 1.3255 MeV 1.59%
$^{124}_{52}$Te (Stable)

^{123}Sb(n,γ) (May 8, 1997)

Radiation[a]	Y_i (%)	E_i (MeV)
β_7^-	8.86	0.0581[b]
β_{16}^-	51.9	0.1938[b]
β_{20}^-	3.87	0.2918[b]
β_{21}^-	2.03	0.3245[b]
β_{23}^-	4.78	0.5932[b]
β_{24}^-	2.57	0.6271[b]
β_{25}^-	22.3	0.9184[b]
γ_{16}	98.3	0.6027
ce-K, γ_{16}	0.413	0.5709
γ_{18}	7.46	0.6459
γ_{20}	1.36	0.7093
γ_{21}	2.29	0.7138
γ_{22}	10.8	0.7228
γ_{31}	1.89	0.9682
γ_{33}	1.84	1.0450
γ_{38}	1.59	1.3255
γ_{40}	2.62	1.3682
γ_{51}	47.8	1.6910
γ_{59}	5.51	2.0910

[a]Several other βs and γs with yields less than 1%. ^{124}Sb is also formed with metastable ^{124m}Sb, which undergoes isomeric transition 75% of the time with a half-life of 93 s and is dominated by internal conversion; 25% of ^{124m}Sb transforms to ^{124}Te (stable) by β^- emission.

[b]Average energy.

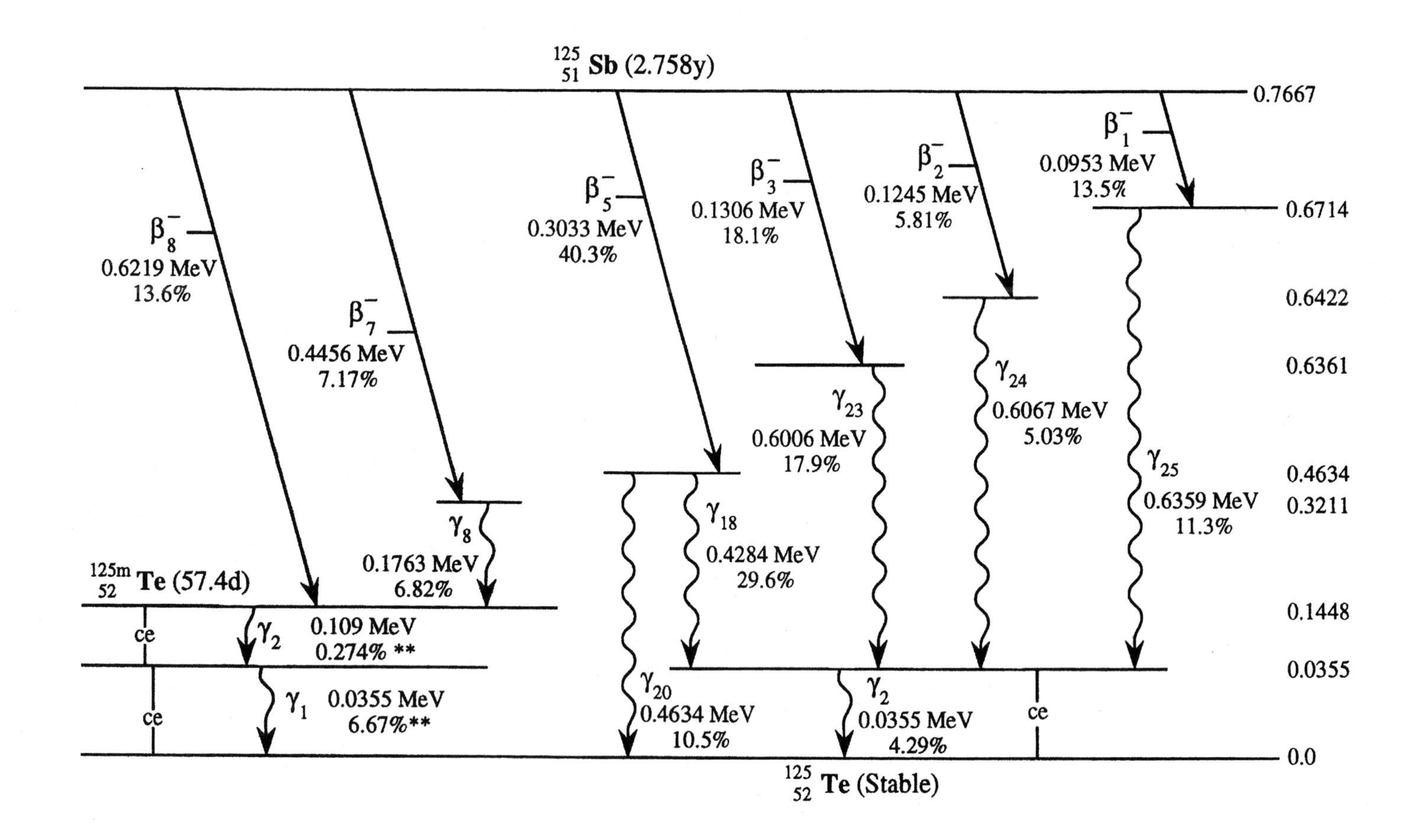

$^{125}_{51}$Sb (2.758y)
0.7667
0.6714
0.6422
0.6361
0.4634
0.3211
0.1448
0.0355
0.0
β^-_8 0.6219 MeV 13.6%
β^-_7 0.4456 MeV 7.17%
β^-_5 0.3033 MeV 40.3%
β^-_3 0.1306 MeV 18.1%
β^-_2 0.1245 MeV 5.81%
β^-_1 0.0953 MeV 13.5%
γ_8 0.1763 MeV 6.82%
γ_{18} 0.4284 MeV 29.6%
γ_{20} 0.4634 MeV 10.5%
γ_{23} 0.6006 MeV 17.9%
γ_{24} 0.6067 MeV 5.03%
γ_{25} 0.6359 MeV 11.3%
γ_2 0.0355 MeV 4.29%
ce
$^{125m}_{52}$Te (57.4d)
γ_2 0.109 MeV 0.274% **
γ_1 0.0355 MeV 6.67%**
ce
ce
$^{125}_{52}$Te (Stable)

$^{124}Sn(n,\gamma)$; $^{125}Sn(\beta^-)$ (Feb. 10, 1994)

Radiation	Y_i (%)	E_i (MeV)
^{125}Sb		
β_1^-	13.5	0.0249[a]
β_2^-	5.8	0.0330[a]
β_3^-	18.1	0.0348[a]
β_4^-	1.59	0.0675[a]
β_5^-	40.3	0.0870[a]
β_7^-	7.17	0.1345[a]
β_8^-	13.6	0.2155[a]
γ_2	4.29	0.0355
ce-K, γ_2	51.5	0.0037
ce-L, γ_2	6.91	0.0306[b]
ce-M, γ_2	1.38	0.0345[b]
γ_7	6.82	0.1763
γ_{18}	29.6	0.4279
γ_{20}	10.4	0.4634
γ_{23}	17.9	0.6006
γ_{24}	5.03	0.6067
γ_{25}	11.3	0.6359
γ_{27}	1.79	0.6714
$K_{\alpha1}$ x ray	24.7	0.0275
$K_{\alpha2}$ x ray	13.3	0.0272
K_β x ray	8.58	0.0310[a]
Auger-K	6.64	0.0227[a]
^{125m}Te		
γ_1	6.67[c]	0.0355
ce-K, γ_1	80.0	0.0037
ce-L, γ_1	10.7	0.0306[b]
ce-M, γ_1	2.15	0.0345[b]
γ_2	0.274[c]	0.1093
ce-K, γ_2	51.8	0.0775
ce-L, γ_2	37.3	0.1043[b]
ce-M, γ_2	8.55	0.1083[b]
ce-N^+, γ_2	2.24	0.1091[b]
$K_{\alpha1}$ x ray	61.3	0.0275
$K_{\alpha2}$ x ray	32.8	0.0272
K_β x ray	21.3	0.0310[a]
L x ray	15.2	0.0038[a]
Auger-K	16.5	0.0227[a]

[a] Average energy.
[b] Maximum energy for subshell.
[c] Yield from ^{125m}Te only.

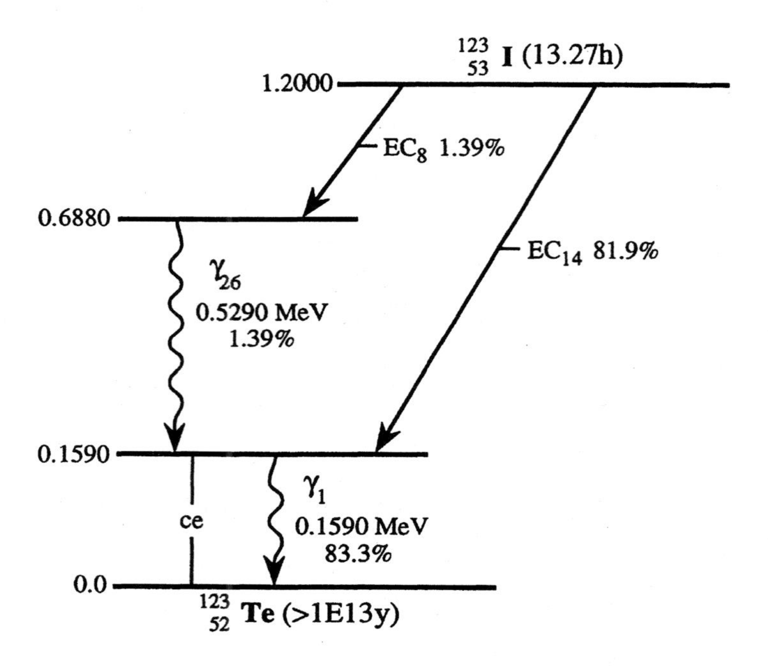

^{121}Sb(α,n) (Feb. 15, 1994)

Radiation	Y_i (%)	E_i (MeV)
γ_1	83.3	0.1590
ce-K, γ_1	13.6	0.1272
ce-L, γ_1	1.77	0.1540[a]
γ_{26}	1.39	0.5290
$K_{\alpha 1}$ x ray	46.0	0.0275
$K_{\alpha 2}$ x ray	24.6	0.0272
K_{β} x ray	16.0	0.0310[b]
L x ray	9.34	0.0038[b]
Auger-K	12.3	0.0227[b]

[a]Maximum energy for subshell.
[b]Average energy.

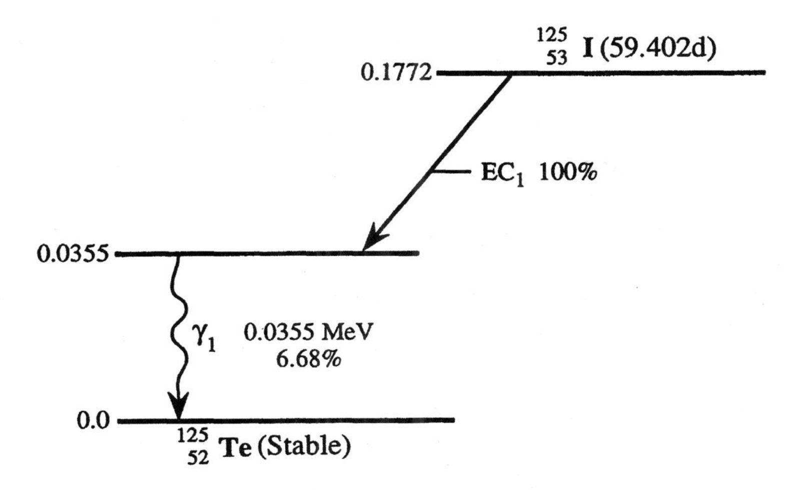

$^{123}Sb(\alpha,2n)$; Daughter of ^{125}Xe; Deuterons on Te (Feb. 10, 1994)

Radiation	Y_i (%)	E_i (MeV)
γ_1	6.68	0.0355
ce-K, γ_1	80.2	0.0037
ce-L, γ_1	10.8	0.0306[a]
ce-M, γ_1	2.14	0.0345[a]
$K_{\alpha 1}$ x ray	74.3	0.0275
$K_{\alpha 2}$ x ray	39.8	0.0272
K_{β} x ray	25.8	0.0310[b]
L x ray	15.5	0.0038[b]
Auger-K	20.0	0.0227[b]

[a]Maximum energy for subshell.
[b]Average energy.

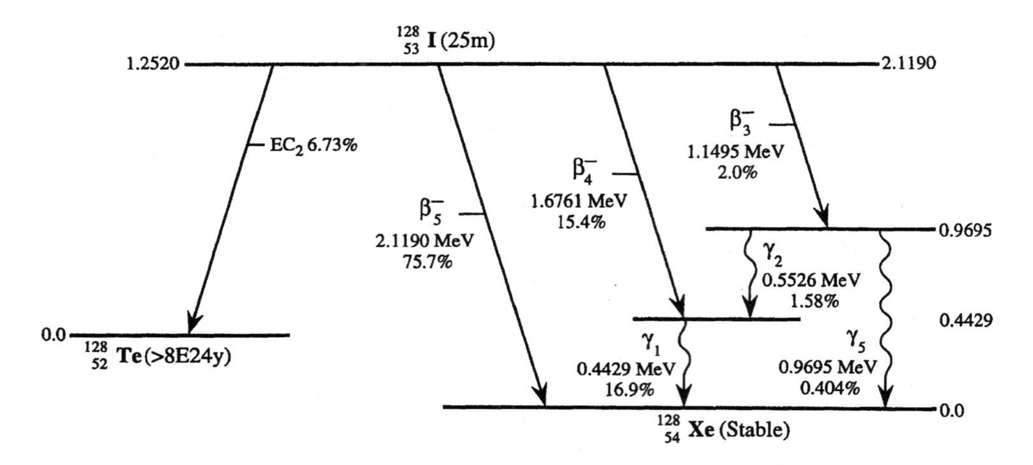

^{127}I(n, γ)		(Mar. 31, 1998)
Radiation	Y_i (%)	E_i (MeV)
EC transition		
γ_1	0.165	0.7435
$K_{\alpha1}$ x ray	2.77	0.0275
$K_{\alpha2}$ x ray	1.49	0.0272
K_{β} x ray	0.963	0.0310[a]
$\beta-$ transition		
β_3^-	2.0	0.4084[a]
β_4^-	15.4	0.6355[a]
β_5^-	75.7	0.8335[a]
γ_1	16.9	0.4429
γ_2	1.58	0.5266
γ_5	0.404	0.9695

[a]Average energy.

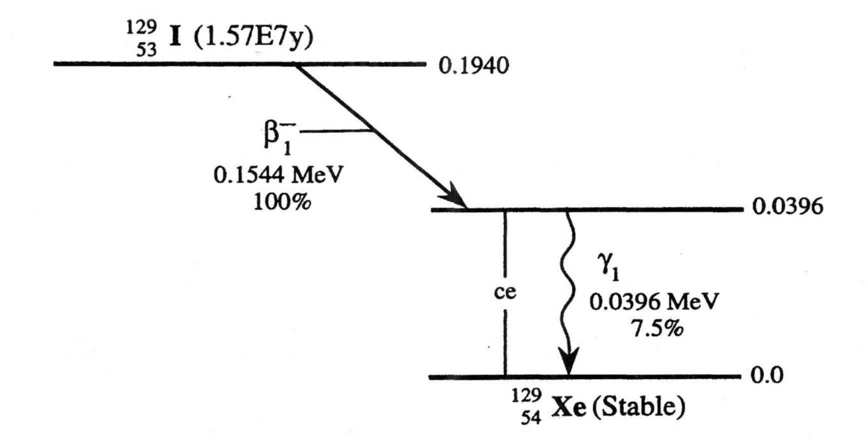

FP; ^{128}Te(n,γ); ^{129}Te(β^-) (Feb. 29, 1996)

Radiation	Y_i (%)	E_i (MeV)
β_1^-	100.0	0.0409[a]
γ_1	7.5	0.0396
ce-K, γ_1	79.9	0.0030
ce-L, γ_1	10.7	0.0321
ce-M, γ_1	2.16	0.0365
$K_{\alpha 1}$ x ray	37.5	0.0298
$K_{\alpha 2}$ x ray	20.2	0.0295
K_{β} x ray	13.33	0.0336
L x ray	8.3	0.0041
Auger-K	8.85	0.0246
Auger-L	74.0	0.0034

[a] Average energy.

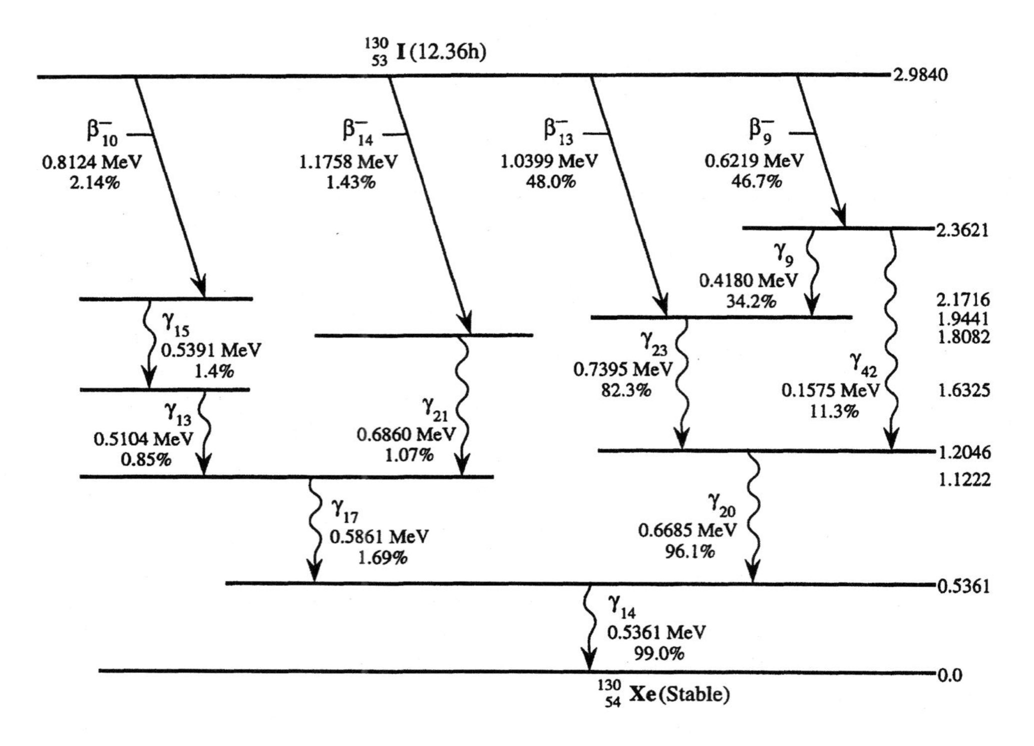

^{129}I(n, γ); ^{130}Te(d, 2n); ^{130}Te(p, n); ^{133}Cs(n, α) (Feb. 20, 1990)

Radiation	Y_i (%)	E_i (MeV)
β_9^-	46.7	0.1970[a]
β_{10}^-	2.14	0.2700[a]
β_{13}^-	48.0	0.3610[a]
β_{14}^-	1.43	0.4180[a]
γ_9	34.2	0.4180
γ_{13}	0.852	0.5104
γ_{14}	99.0	0.5361
ce-K, γ_{14}	0.623	0.5015
ce-L, γ_{14}	0.0891	0.5306[b]
γ_{15}	1.4	0.5391
γ_{17}	1.69	0.5861
γ_{20}	96.1	0.6685
γ_{21}	1.07	0.6860
γ_{23}	82.3	0.7395
γ_{42}	11.3	1.1570

[a]Average energy.
[b]Maximum energy for subshell.

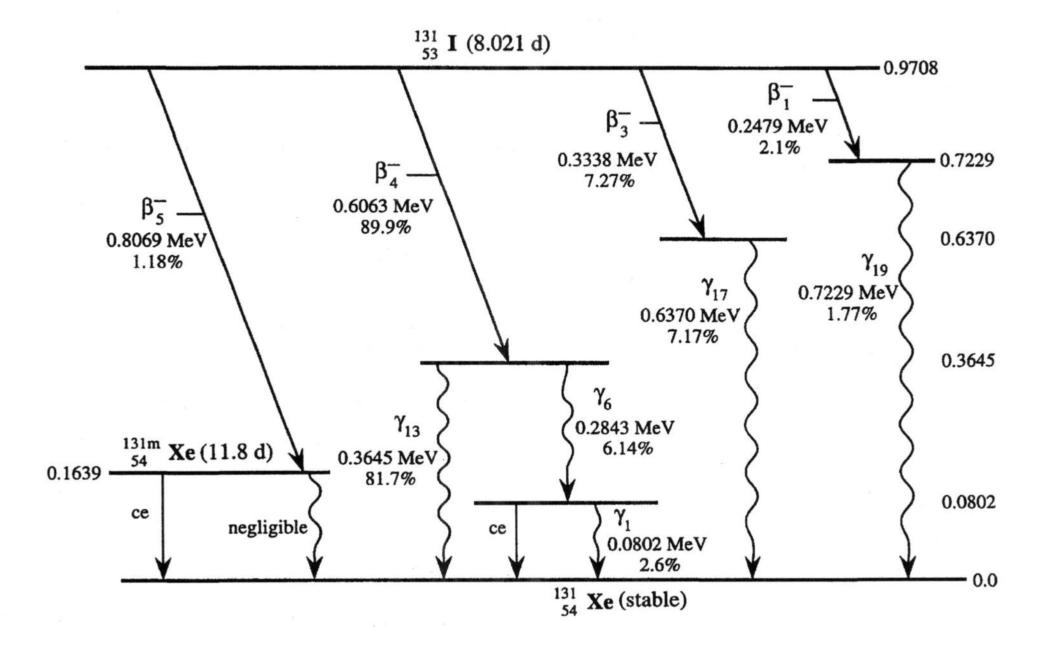

FP[a] (Dec. 14, 1994)

Radiation	Y_i (%)	E_i (MeV)
β_1^-	2.10	0.0694[b]
β_3^-	7.27	0.0966[b]
β_4^-	89.9	0.1916[b]
β_5^-	1.18	0.8069[b]
γ_1	2.62	0.0802
ce-K, γ_1	3.54	0.0456
γ_6	6.14	0.2843
γ_{13}	81.7	0.3645
ce-K, γ_{13}	1.55	0.3299
ce-L, γ_{13}	0.246	0.3590[c]
γ_{17}	7.17	0.6370
γ_{19}	1.77	0.7229
$K_{\alpha 1}$ x ray	2.56	0.0298
$K_{\alpha 2}$ x ray	1.38	0.0295

[a] ^{131m}Xe product, yield 1.18%; ^{131}Xe yield 98.8%.
[b] Average energy.
[c] Maximum energy for subshell.

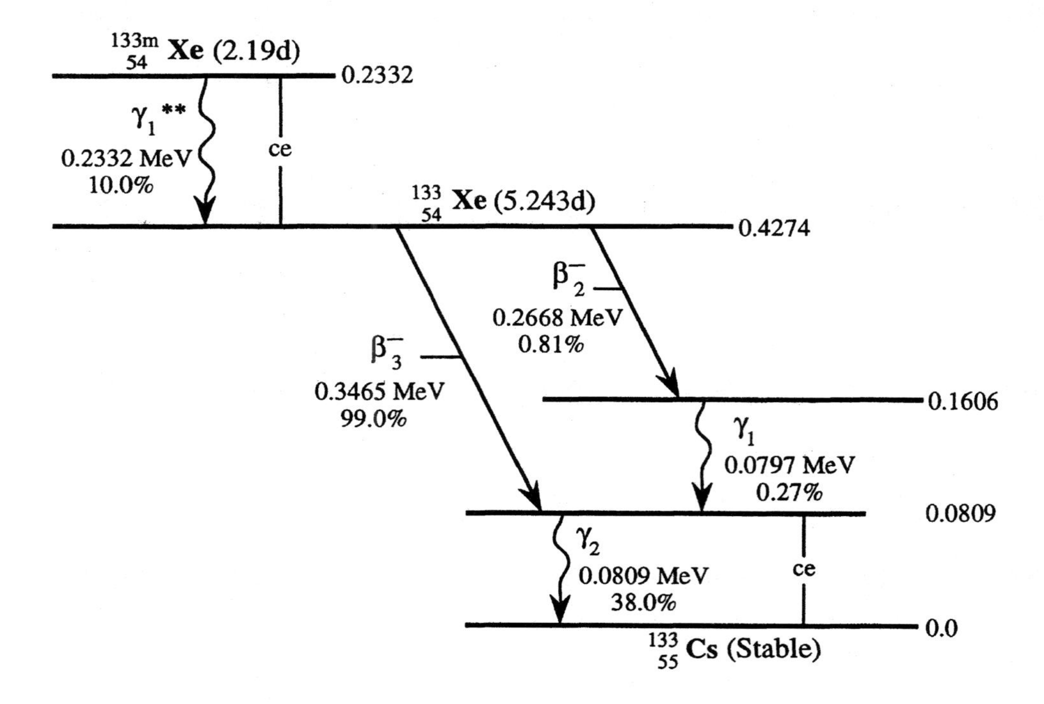

133m 54 Xe (2.19d)
0.2332
γ1 **
0.2332 MeV
10.0%
ce
133 54 Xe (5.243d)
0.4274
β2−
0.2668 MeV
0.81%
β3−
0.3465 MeV
99.0%
0.1606
γ1
0.0797 MeV
0.27%
0.0809
γ2
0.0809 MeV
38.0%
ce
0.0
133 55 Cs (Stable)

FP; ^{132}Xe(n,γ) (Oct. 6, 1995)

Radiation	Y_i (%)	E_i (MeV)
^{133m}Xe		
γ_1 [a]	10.0	0.2332
ce-K, γ_1 [a]	63.5	0.1987
ce-L, γ_1 [a]	20.7	0.2278[b]
ce-M, γ_1 [a]	4.57	0.2321[b]
ce-N$^+$, γ_1 [a]	1.22	0.2330[b]
$K_{\alpha 1}$ x ray	29.7	0.0298
$K_{\alpha 2}$ x ray	16.0	0.0295
K_β x ray	10.7	0.0336[c]
L x ray	7.60	0.0041[c]
Auger-K	7.08	0.0246[c]
^{133}Xe		
β_2^-	0.81	0.0750[c]
β_3^-	99.0	0.1005[c]
γ_1	0.27	0.0796
ce-K, γ_1	0.41	0.0436
γ_2	38.0	0.0809
ce-K, γ_2	55.1	0.0450
ce-L, γ_2	8.21	0.0753[b]
ce-M, γ_2	1.69	0.0798[b]
$K_{\alpha 1}$ x ray	26.2	0.0310
$K_{\alpha 2}$ x ray	14.1	0.0306
K_β x ray	9.47	0.0350[c]
L x ray	6.06	0.0043[c]
Auger-K	5.86	0.0255[c]

[a]Transformation yield of ^{133m}Xe alone; ^{133}Xe also has a listed γ_1 yield and energy, but it is associated with ^{133}Xe transformation only.
[b]Maximum energy for subshell.
[c]Average energy.

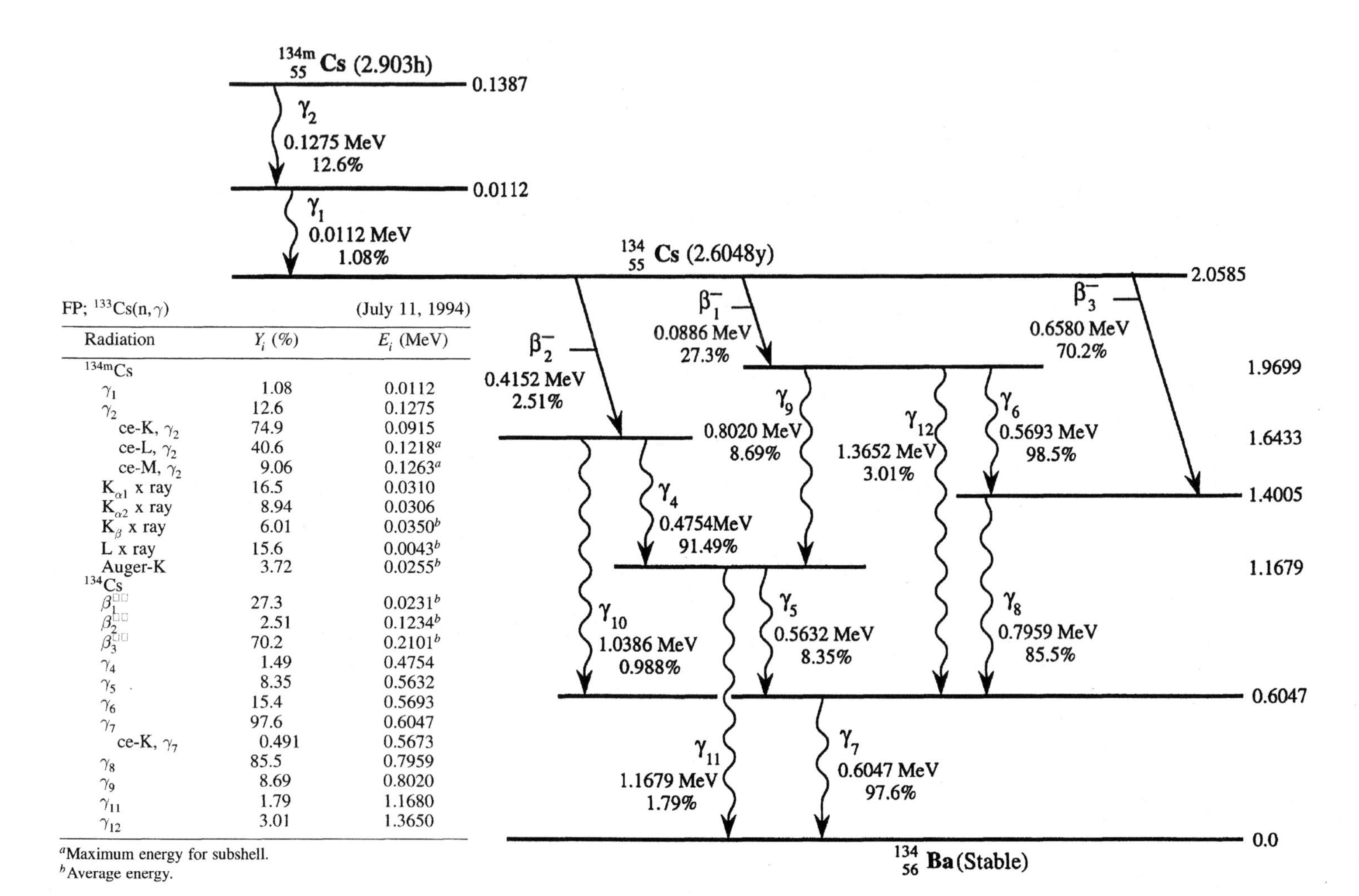

FP; ^{133}Cs(n,γ) (July 11, 1994)

Radiation	Y_i (%)	E_i (MeV)
^{134m}Cs		
γ_1	1.08	0.0112
γ_2	12.6	0.1275
ce-K, γ_2	74.9	0.0915
ce-L, γ_2	40.6	0.1218[a]
ce-M, γ_2	9.06	0.1263[a]
$K_{\alpha 1}$ x ray	16.5	0.0310
$K_{\alpha 2}$ x ray	8.94	0.0306
K_{β} x ray	6.01	0.0350[b]
L x ray	15.6	0.0043[b]
Auger-K	3.72	0.0255[b]
^{134}Cs		
β_1	27.3	0.0231[b]
β_2	2.51	0.1234[b]
β_3	70.2	0.2101[b]
γ_4	1.49	0.4754
γ_5	8.35	0.5632
γ_6	15.4	0.5693
γ_7	97.6	0.6047
ce-K, γ_7	0.491	0.5673
γ_8	85.5	0.7959
γ_9	8.69	0.8020
γ_{11}	1.79	1.1680
γ_{12}	3.01	1.3650

[a]Maximum energy for subshell.
[b]Average energy.

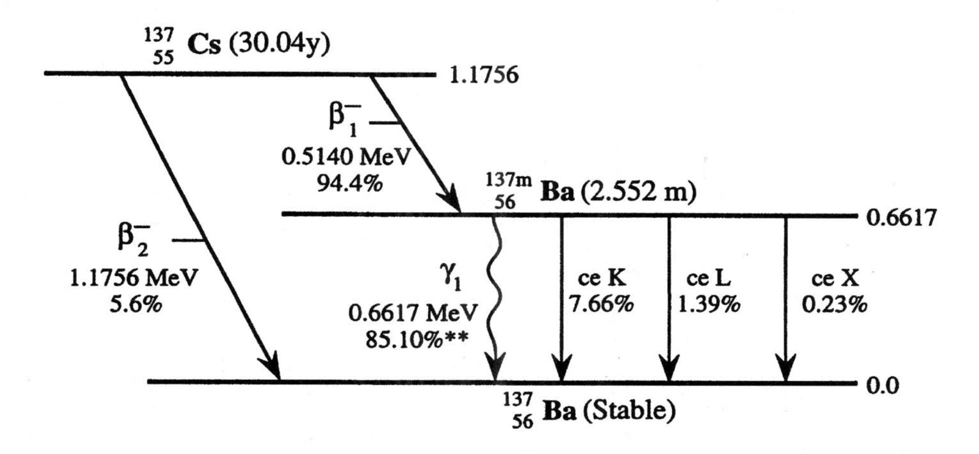

FP (Nov. 4, 1994)

Radiation	Y_i (%)	E_i (MeV)
β_1^-	94.40	0.1734[a]
β_2^-	5.60	0.4163[a]
γ_1	85.10[b]	0.6617
ce-K, γ_1	7.66	0.6242
ce-L_1, γ_1	1.12	0.6557
ce-L_2, γ_1	0.16	0.6560
ce-L_3, γ_1	0.13	0.6564
$K_{\alpha 1}$ x ray	3.62	0.0322
$K_{\alpha 2}$ x ray	1.96	0.0318
K_β x ray	1.32	0.0364[a]
L x ray	1.00	0.0045[a]
Auger-K	0.76	0.0264[a]
Auger-L	7.20	0.0367[a]

[a] Average energy.
[b] Photon yield per transformation of ^{137}Cs; photon yield from each transformation of ^{137m}Ba is 90.11%.

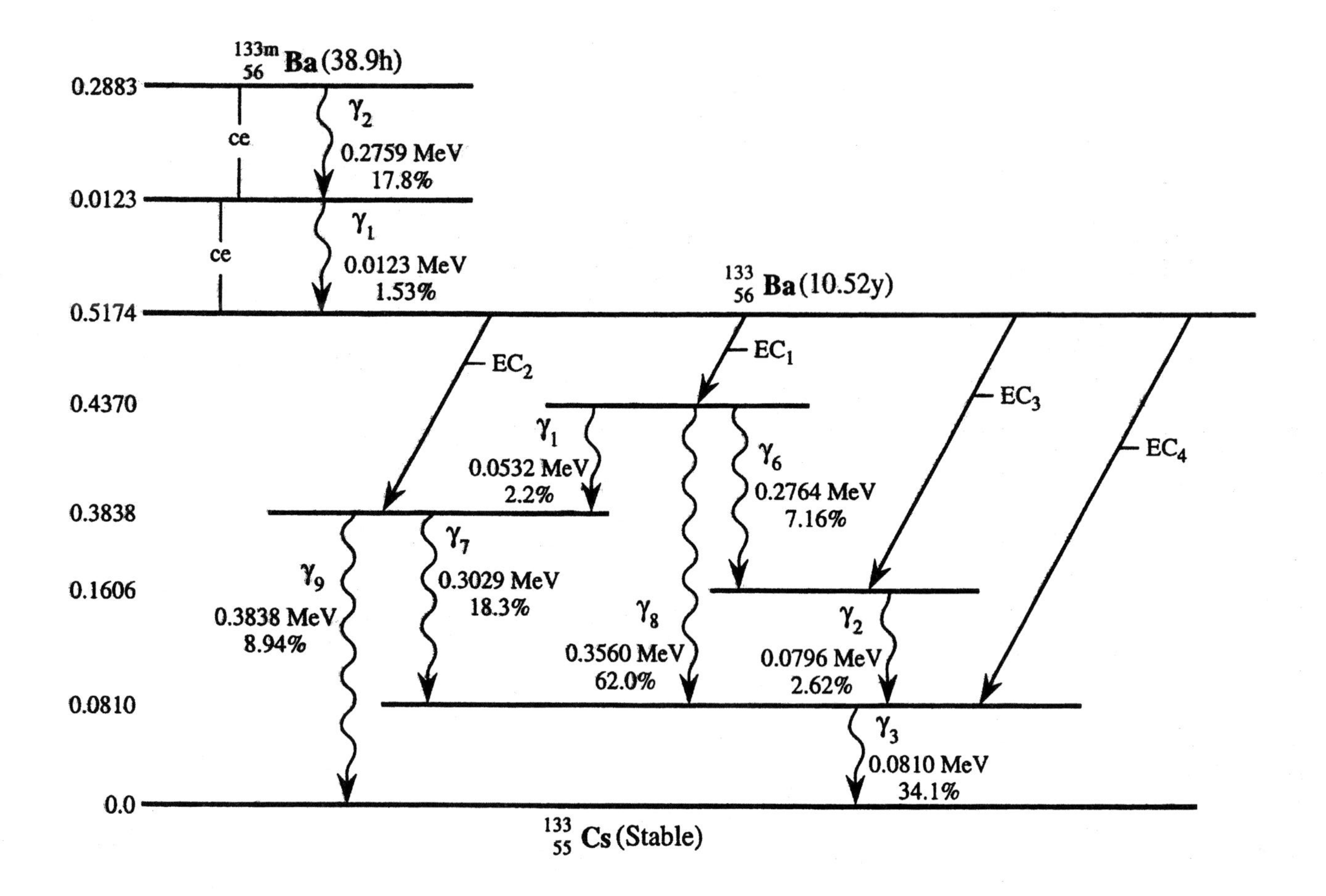

$^{133m}_{56}$Ba (38.9h)
0.2883
ce
γ_2
0.2759 MeV
17.8%
0.0123
ce
γ_1
0.0123 MeV
1.53%
0.5174
$^{133}_{56}$Ba (10.52y)
EC$_2$
EC$_1$
EC$_3$
EC$_4$
0.4370
γ_1
0.0532 MeV
2.2%
γ_6
0.2764 MeV
7.16%
0.3838
γ_7
0.3029 MeV
18.3%
γ_9
0.3838 MeV
8.94%
0.1606
γ_8
0.3560 MeV
62.0%
γ_2
0.0796 MeV
2.62%
0.0810
γ_3
0.0810 MeV
34.1%
0.0
$^{133}_{55}$Cs (Stable)

$^{132}Ba(n,\gamma)$; $^{133}Cs(p,n)$		(Oct. 6, 1995)
Radiation	Y_i (%)	E_i (MeV)
^{133m}Ba		
γ_1	1.53	0.0123
ce-L, γ_1	84.5	0.0063[a]
ce-M, γ_1	17.3	0.0110[a]
γ_2	17.8	0.2759
ce-K, γ_2	60.3	0.2385
ce-L, γ_2	18.4	0.2699[a]
ce-M, γ_2	4.12	0.2746[a]
ce-N^+, γ_2	1.17	0.2757[a]
$K_{\alpha 1}$ x ray	28.0	0.0322
$K_{\alpha 2}$ x ray	15.2	0.0318
K_β x ray	10.2	0.0364[b]
L x ray	18.5	0.0045[b]
^{133}Ba		
γ_1	2.2	0.0532
ce-K, γ_1	11.0	0.0172
ce-L, γ_1	1.87	0.0475[a]
γ_2	2.62	0.0796
ce-K, γ_2	3.97	0.0436
ce-L, γ_2	0.574	0.0739[a]
γ_3	34.1	0.0810
ce-K, γ_3	43.9	0.0450
ce-L, γ_3	7.36	0.0753[a]
ce-M, γ_3	1.51	0.0798[a]
ce-N^+, γ_3	0.392	0.0808[a]
γ_6	7.16	0.2764
γ_7	18.3	0.3029
ce-K, γ_7	0.69	0.2669
γ_8	62.0	0.3560
ce-K, γ_8	1.31	0.3200
ce-L, γ_8	0.216	0.3503[a]
γ_9	8.94	0.3838
$K_{\alpha 1}$ x ray	62.5	0.0310
$K_{\alpha 2}$ x ray	33.7	0.0306
K_β x ray	22.4	0.0350[b]
L x ray	15.7	0.0043[b]
Auger-K	13.9	0.0255[b]

[a]Maximum energy for subshell.

[b]Average energy.

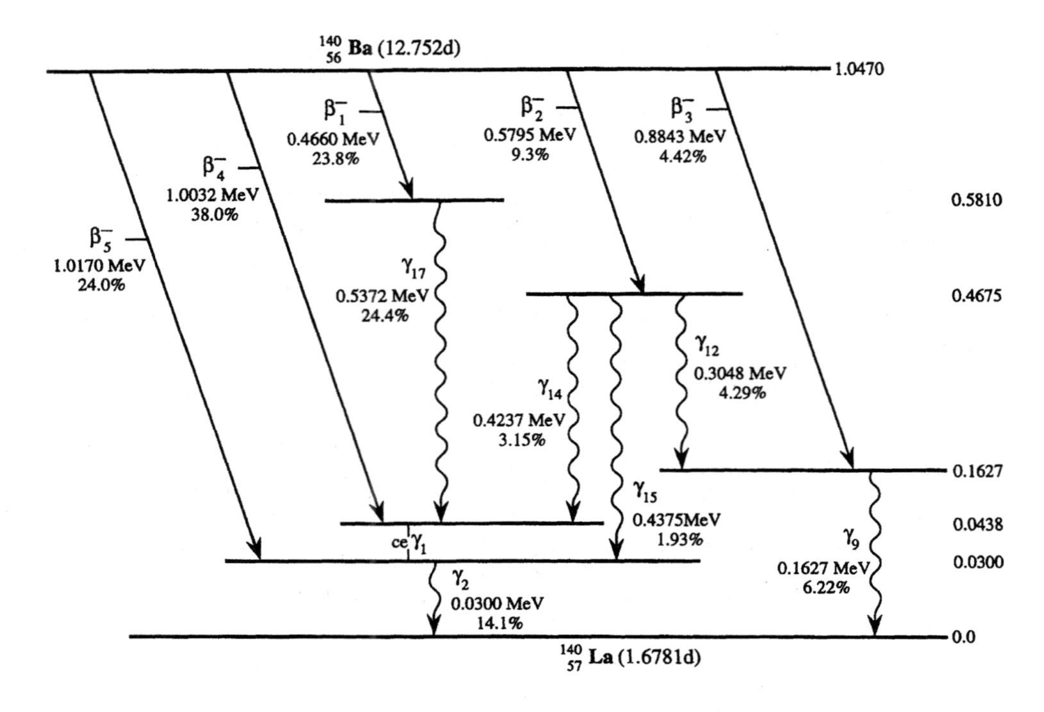

FP (Jan. 26, 1995)

Radiation	Y_i (%)	E_i (MeV)
β_1^-	23.8	0.1360[a]
β_2^-	9.3	0.1760[a]
β_3^-	4.42	0.3050[a]
β_4^-	38.0	0.3390[a]
β_5^-	24.0	0.3570[a]
ce-L, γ_1	53.7	0.0076[b]
ce-M, γ_1	11.1	0.0125[b]
γ_2	14.1	0.0300
ce-L, γ_2	61.2	0.0237[b]
ce-M, γ_2	12.6	0.0286[b]
γ_9	6.22	0.1627
ce-K, γ_9	1.48	0.1237
γ_{12}	4.29	0.3048
γ_{14}	3.15	0.4237
γ_{15}	1.93	0.4375
γ_{17}	24.4	0.5372
$K_{\alpha 1}$ x ray	0.813	0.0334
L x ray	15.2	0.0047[a]

[a] Average energy.

[b] Maximum energy for subshell.

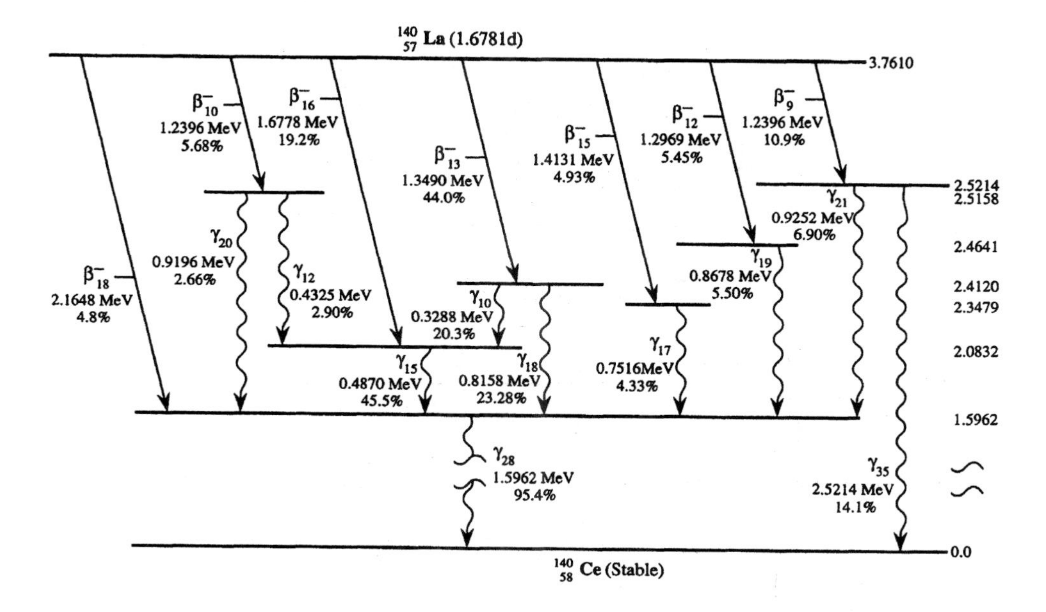

^{139}La(n, γ); FP (Nov. 13, 1991)

Radiation	Y_i (%)	E_i (MeV)
β_9^-	10.9	0.4415[a]
β_{10}^-	5.68	0.4439[a]
β_{12}^-	5.45	0.4656[a]
β_{13}^-	44.0	0.4877[a]
β_{15}^-	4.93	0.5151[a]
β_{16}^-	19.2	0.6298[a]
β_{18}^-	4.8	0.8465[a]
γ_{10}	20.3	0.3288
ce-K, γ_{10}	0.804	0.2883
γ_{12}	2.90	0.4325
γ_{15}	45.5	0.4870
γ_{17}	4.33	0.7516
γ_{18}	23.28	0.8158
γ_{19}	5.5	0.8678
γ_{20}	2.66	0.9196
γ_{21}	6.90	0.9252
γ_{28}	95.4	1.5962
γ_{35}	3.46	2.5214

[a] Average energy.

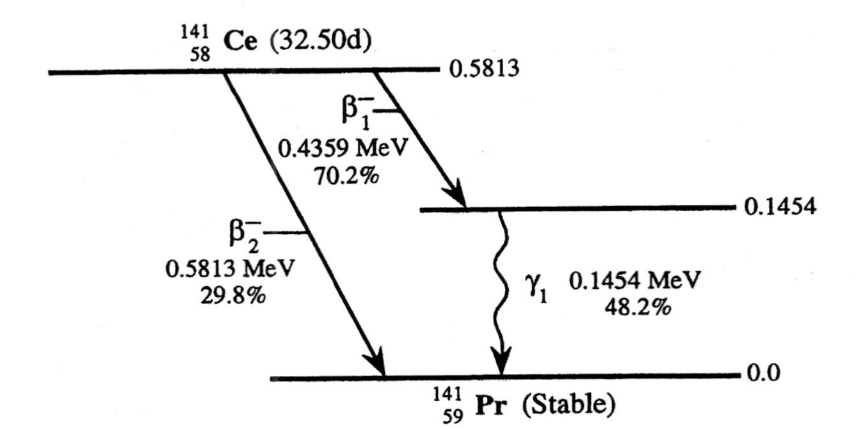

^{140}Ce(n,γ); Decay Product of ^{141}La; ^{141}Pr(n,p); FP (Nov. 13, 1991)

Radiation	Y_i (%)	E_i (MeV)
β_1^-	70.2	0.1300[a]
β_2^-	29.8	0.1811[a]
γ_1	48.2	0.1454
ce-K, γ_1	18.7	0.1034
ce-L, γ_1	2.58	0.1386[b]
ce-M, γ_1	0.542	0.1439[b]
ce-N$^+$, γ_1	0.148	0.1451[b]
$K_{\alpha 1}$ x ray	8.91	0.0360
$K_{\alpha 2}$ x ray	4.86	0.0356
K_β x ray	3.35	0.0407[a]
L x ray	2.64	0.0050[a]
Auger-K	1.59	0.0294[a]

[a] Average energy.
[b] Maximum energy for subshell.

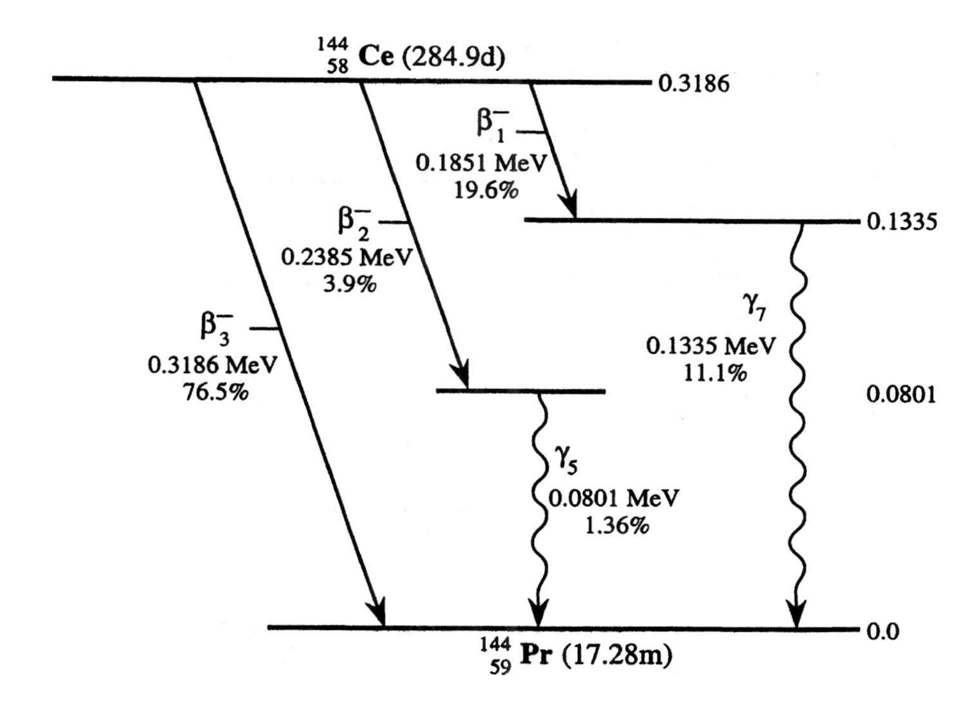

FP (Aug. 22, 1989)

Radiation	Y_i (%)	E_i (MeV)
β_1^-	19.6	0.0502[a]
β_2^-	3.9	0.0661[a]
β_3^-	76.5	0.0911[a]
γ_1	0.2	0.0336
ce-L, γ_1	0.747	0.0267[b]
γ_2	0.257	0.0410
ce-L, γ_2	0.561	0.0342[b]
ce-M, γ_4	0.153	0.0575[b]
γ_5	1.36	0.0801
ce-K, γ_5	2.89	0.0381
ce-L, γ_5	0.4	0.0733[b]
γ_7	11.1	0.1335
ce-K, γ_7	5.47	0.0915
ce-L, γ_7	0.751	0.1267[b]
ce-M, γ_7	0.157	0.1320[b]
$K_{\alpha 1}$ x ray	4.53	0.0360
$K_{\alpha 2}$ x ray	2.47	0.0356
K_β x ray	1.7	0.0407[a]
L x ray	1.61	0.0050[a]

[a] Average energy.
[b] Maximum energy for subshell.

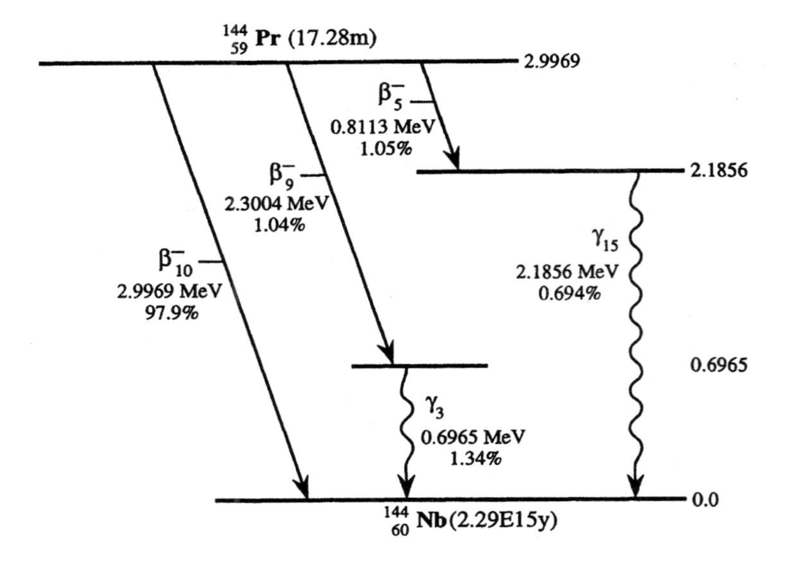

FP (Aug. 22, 1989)

Radiation	Y_i (%)	E_i (MeV)
β_5^-	1.05	0.2666[a]
β_9^-	1.04	0.9040[a]
β_{10}^-	97.9	1.2210[a]
γ_3	1.34	0.6965
γ_{15}	0.694	2.1860

[a] Average energy.

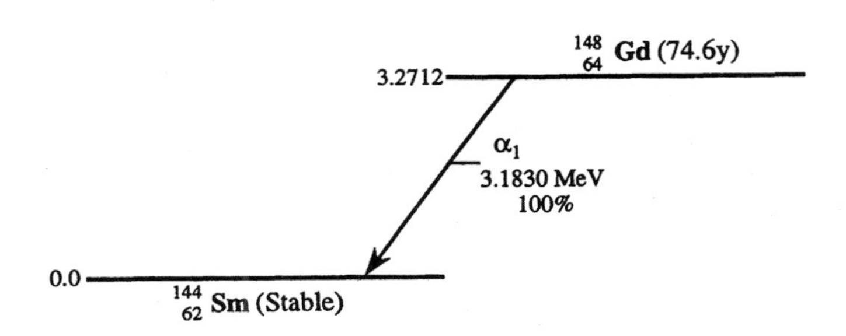

^{147}Sm(α,3n); ^{151}Eu(p,4n) (May 7, 1998)

Radiation	Y_i (%)	E_i (MeV)
α_1	100.0	3.1830
α recoil	100.0	0.0862

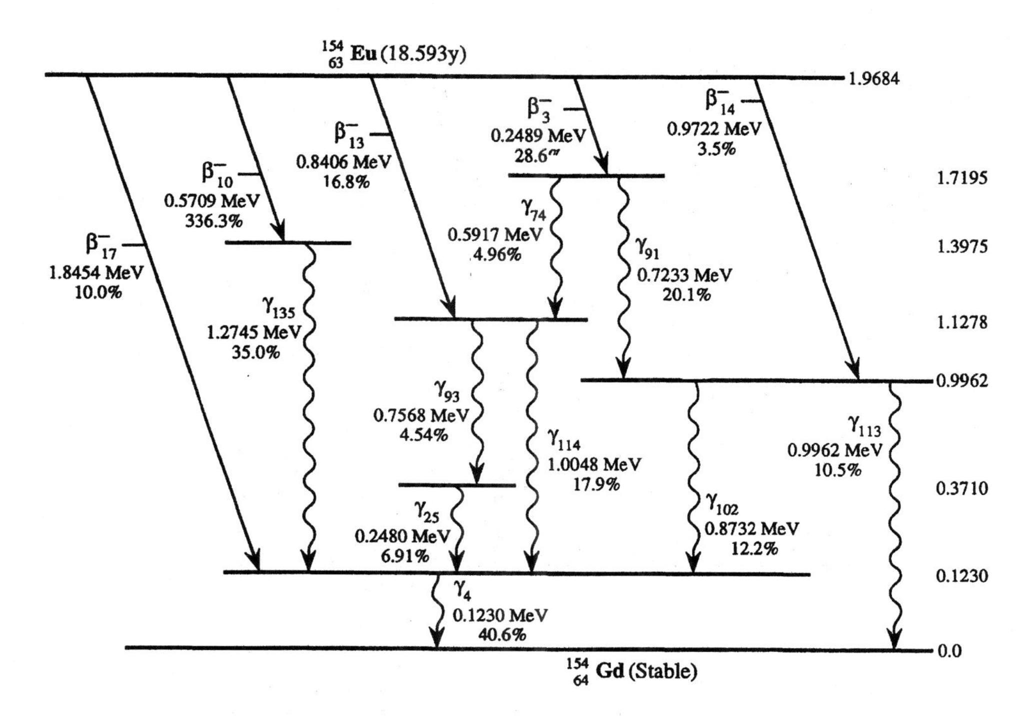

^{153}Eu(n,γ); FP (Independent Yield) (Dec. 3, 1998)

Radiation	Y_i (%)	E_i (MeV)
β_3^-	28.6	0.0693[a]
β_6^-	1.64	0.1014[a]
β_{10}^-	36.3	0.1762[a]
β_{13}^-	16.8	0.2766[a]
β_{14}^-	3.5	0.3281[a]
β_{17}^-	10.0	0.6956[a]
γ_4	40.6	0.1230
ce-K, γ_4	26.8	0.0728
ce-L, γ_4	16.8	0.1147[b]
ce-M, γ_4	3.91	0.1212[b]
ce-N$^+$, γ_4	1.1	0.1227[b]
γ_{25}	6.91	0.2480
γ_{74}	4.96	0.5917
γ_{89}	1.79	0.6924
γ_{91}	20.1	0.7233
γ_{93}	4.54	0.7568
γ_{102}	12.2	0.8732
γ_{113}	10.5	0.9962
γ_{114}	17.9	1.0048
γ_{135}	35.0	1.2745
γ_{154}	1.79	1.5965

[a]Average energy.

[b]Maximum energy for subshell.

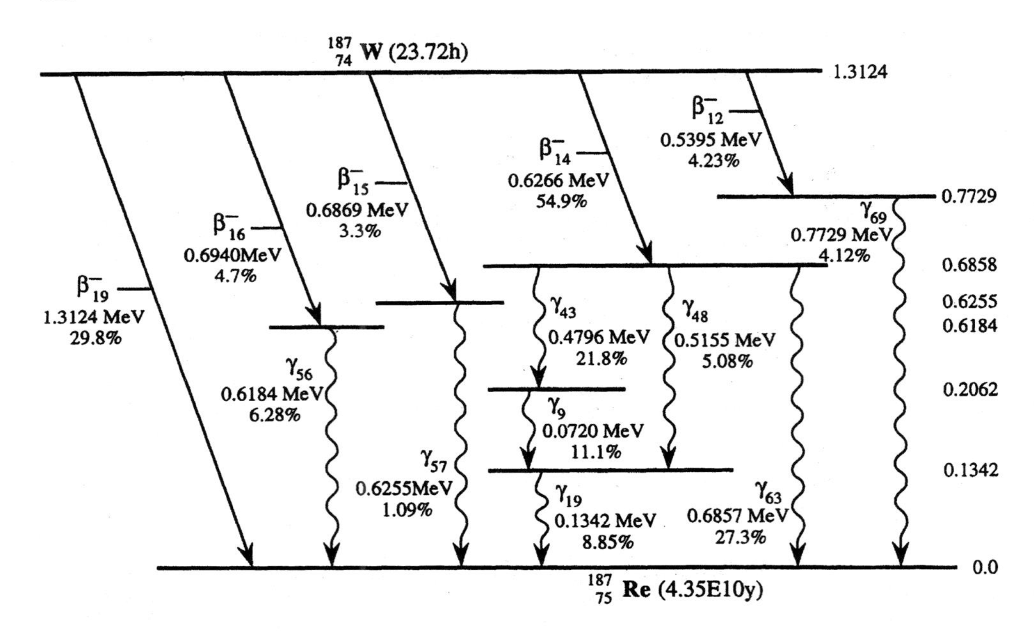

^{186}W(n, γ) (Feb. 20, 1991)

Radiation	Y_i (%)	E_i (MeV)
β^-_{12}	4.23	0.1629[a]
β^-_{14}	54.9	0.1934[a]
β^-_{15}	3.30	0.2150[a]
β^-_{16}	4.70	0.2176[a]
β^-_{19}	29.8	0.4570[a]
γ_9	11.1	0.0720
ce-L, γ_9	1.57	0.0595[b]
γ_{19}	8.85	0.1342
ce-K, γ_{19}	16.5	0.0626
ce-L, γ_{19}	2.79	0.1217[b]
ce-M, γ_{19}	0.640	0.1313[b]
γ_{43}	21.8	0.4796
γ_{48}	5.08	0.5515
γ_{56}	6.28	0.6183
γ_{57}	1.09	0.6255
γ_{63}	27.3	0.6857
γ_{67}	0.298	0.7452
γ_{69}	4.12	0.7729
$K_{\alpha1}$ x ray	12.5	0.0611
$K_{\alpha2}$ x ray	7.23	0.0597
K_β x ray	5.36	0.0693[a]
L x ray	8.25	0.0087[a]
Auger-K	1.07	0.0470[a]

[a]Average energy.
[b]Maximum energy for subshell.

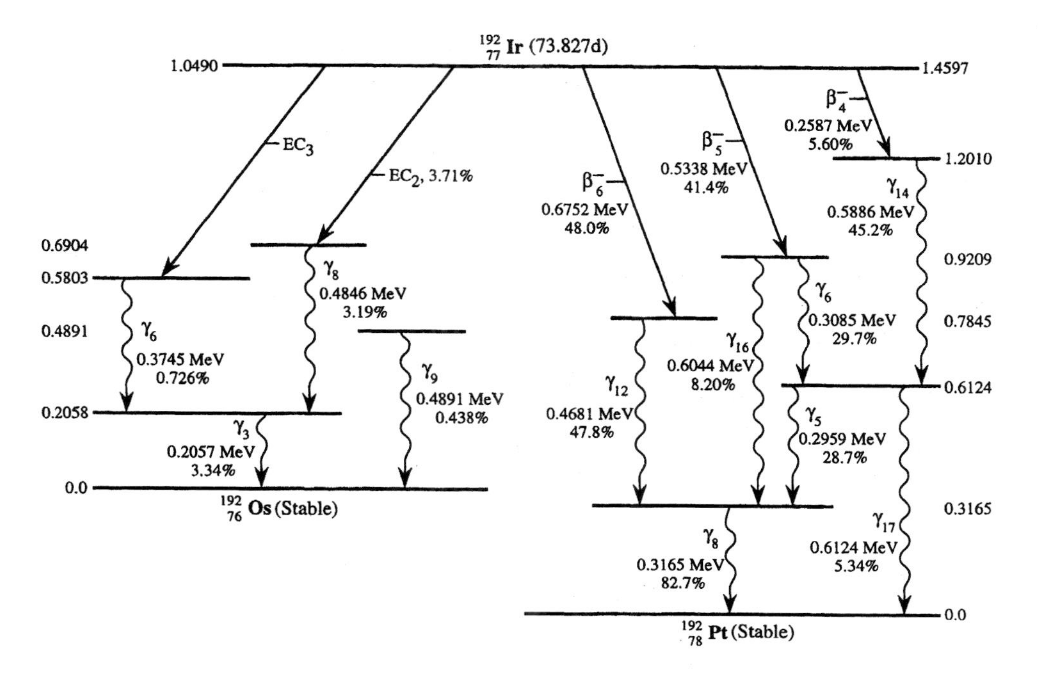

^{191}Ir(n,γ); ^{192}Os(d,2n) (Sept. 4, 1996)

Radiation	Y_i (%)	E_i (MeV)
Via EC		
γ_2	0.473	0.2013
γ_3	3.3	0.2057
γ_8	3.18	0.4846
$K_{\alpha 1}$ x ray	2.09	0.0630
Via beta minus		
β_4^-	5.61	0.0717[a]
β_5^-	41.4	0.1622[a]
β_6^-	48.0	0.2099[a]
γ_5	28.7	0.2960
ce-K, γ_5	1.88	0.2176
γ_6	29.7	0.3085
ce-K, γ_6	1.80	0.2301
γ_8	82.7	0.3165
ce-K, γ_8	4.44	0.2381
ce-L, γ_8	1.95	0.3026[b]
γ_{12}	47.8	0.4681
ce-K, γ_{12}	1.02	0.3897
γ_{14}	4.52	0.5886
γ_{17}	8.20	0.6044
γ_{18}	5.34	0.6124
$K_{\alpha 1}$ x ray	4.55	0.0668
$K_{\alpha 2}$ x ray	2.66	0.0651
$K\beta$ x ray	1.97	0.0757[a]

[a]Average energy.
[b]Maximum energy for subshell.

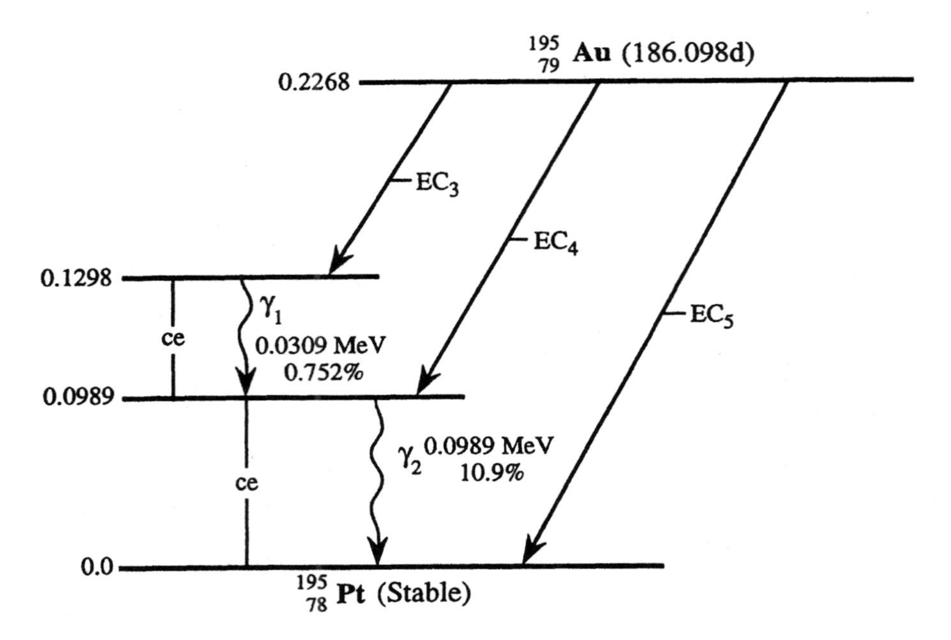

Deuterons on Pt; ^{193}Ir(α,2n); ^{195}Pt(p,n) (July 8, 1994)

Radiation	Y_i (%)	E_i (MeV)
γ_1	0.752	0.0309
ce-L, γ_1	22.7	0.0170[a]
ce-M, γ_1	5.23	0.0276[a]
γ_2	10.9	0.0989
ce-K, γ_2	63.1	0.0205
ce-L, γ_2	11.1	0.0850[a]
ce-M, γ_2	2.57	0.0956[a]
ce-N$^+$, γ_2	0.814	0.0982[a]
$K_{\alpha 1}$ x ray	49.2	0.0668
$K_{\alpha 2}$ x ray	28.7	0.0651
K_β x ray	21.4	0.0757[b]
L x ray	56.6	0.0094[b]
Auger-K	3.89	0.0510[b]

[a]Maximum energy for subshell.
[b]Average energy.

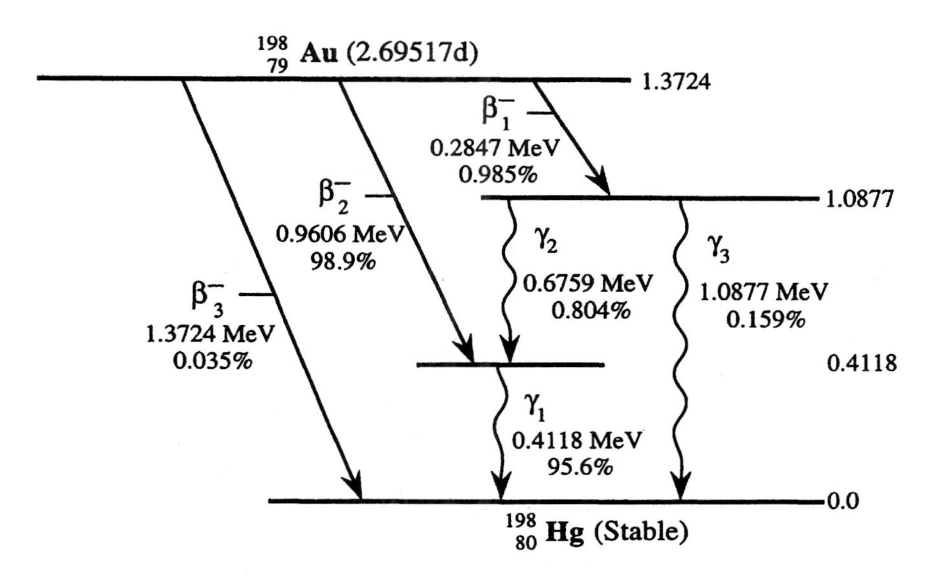

^{197}Au(n, γ); ^{198}Pt(p, n) (Mar. 22, 1995)

Radiation	Y_i (%)	E_i (MeV)
β_1^-	0.985	0.0794[a]
β_2^-	99.0	0.3147[a]
β_3^-	0.025	0.4673[a]
γ_1	95.6	0.4118
ce-K, γ_1	2.88	0.3287
ce-L, γ_1	1.02	0.3970[b]
ce-M, γ_1	0.254	0.4082[b]
ce-N$^+$, γ_1	0.0793	0.4110[b]
γ_2	0.804	0.6759
γ_3	0.159	1.0880
$K_{\alpha1}$ x ray	1.38	0.0708
$K_{\alpha2}$ x ray	0.811	0.0689
K_β x ray	0.607	0.0803[a]
L x ray	1.28	0.0100[a]
Auger-K	0.0979	0.0538[a]

[a]Average energy.
[b]Maximum energy for subshell.

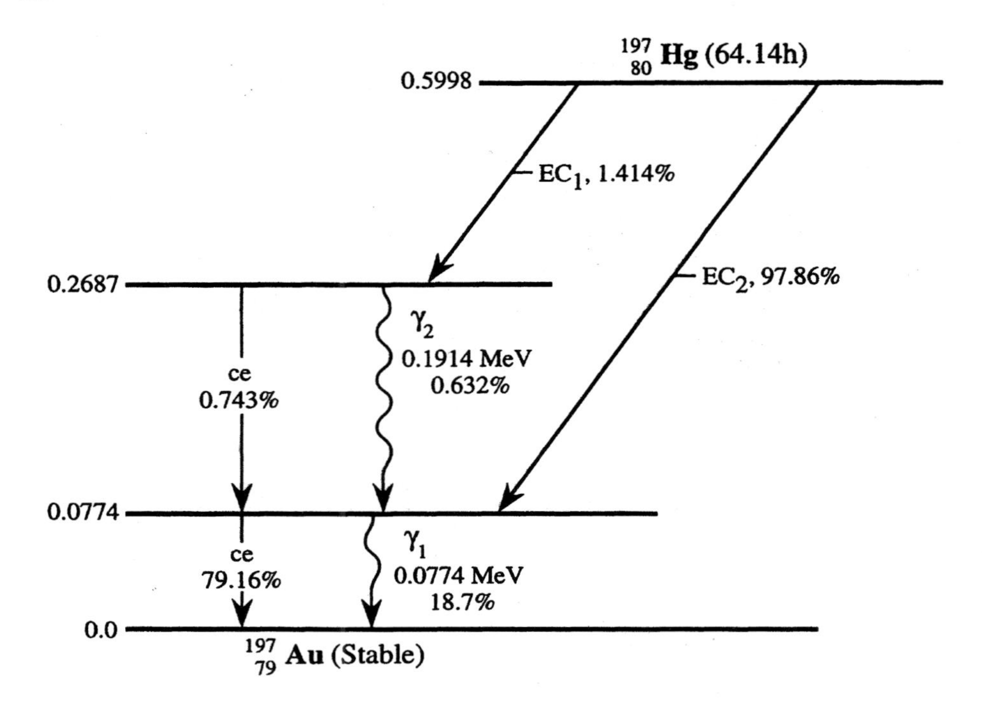

^{196}Hg(n,γ); ^{197m}Hg[a] (Jan. 4, 1996)

Radiation	Y_i (%)	E_i (MeV)
γ_1	18.7	0.0774
ce-L, γ_1	60.0	0.0630[b]
ce-M, γ_1	14.6	0.0739[b]
ce-N$^+$, γ_1	4.56	0.0766[b]
γ_2	0.632	0.1914
$K_{\alpha 1}$ x ray	54.6	0.0688
$K_{\alpha 2}$ x ray	31.9	0.0670
K_β x ray	23.9	0.0780[c]
L x ray	64.8	0.0097[c]
Auger-K	4.11	0.0524[c]

[a]Contribution of ^{197m}Hg ($T_{1/2}$ = 23.8 h) to the activity of a ^{197}Hg source may be significant.
[b]Maximum energy for subshell.
[c]Average energy.

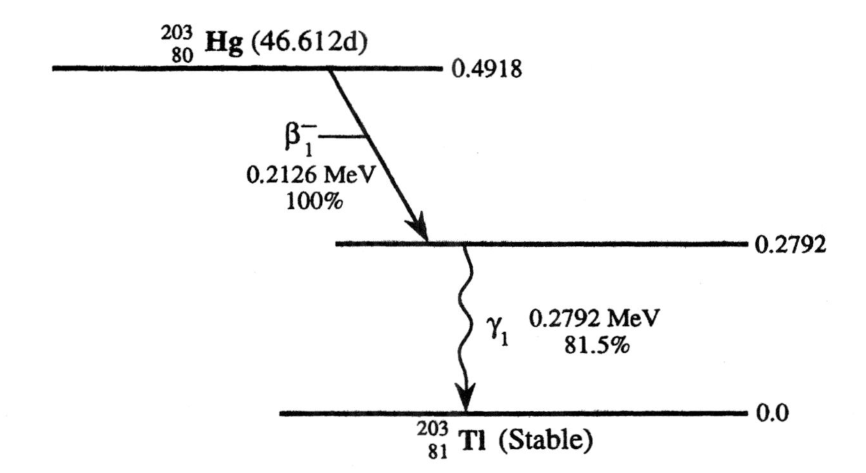

^{202}Hg(n,γ) (Nov. 5, 1993)

Radiation	Y_i (%)	E_i (MeV)
β_1^-	100.0	0.0577[a]
γ_1	81.5	0.2792
ce-K, γ_1	13.4	0.1937
ce-L, γ_1	3.91	0.2638[b]
ce-M$^+$, γ_1	1.3	0.2769[b]
$K_{\alpha 1}$ x ray	6.36	0.0729
$K_{\alpha 2}$ x ray	3.75	0.0708
K_β x ray	2.81	0.0826[a]
L x ray	5.89	0.0103[a]
Auger-K	0.44	0.0552[a]

[a]Average energy.
[b]Maximum energy for subshell.

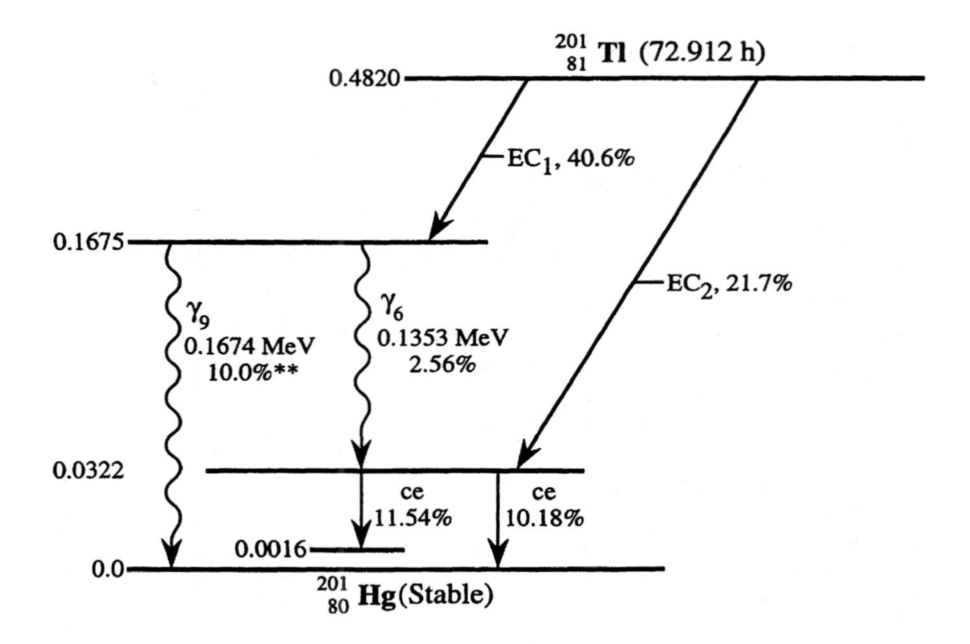

^{201}Pb(EC); ^{201m}Tl; ^{199}Hg(d,γ); ^{200}Hg(d,n) (July 8, 1994)

Radiation	Y_i (%)	E_i (MeV)
ce-L, γ_4	9.36	0.0158[a]
ce-M, γ_4	2.18	0.0270[a]
ce-L, γ_5	8.26	0.0174[a]
ce-M, γ_5	1.92	0.0286[a]
γ_6	2.56	0.1353
ce-K, γ_6	7.23	0.0522
ce-L, γ_6	1.23	0.1205[a]
γ_9	10.0	0.1674
ce-K, γ_9	15.4	0.0843
ce-L, γ_9	2.62	0.1526[a]
$K_{\alpha1}$ x ray	46.4	0.0708
$K_{\alpha2}$ x ray	27.3	0.0689
K_{β} x ray	20.4	0.0803[b]
L x ray	45.6	0.0100[b]
Auger-K	3.29	0.0538[b]

[a]Maximum energy for subshell.
[b]Average energy.

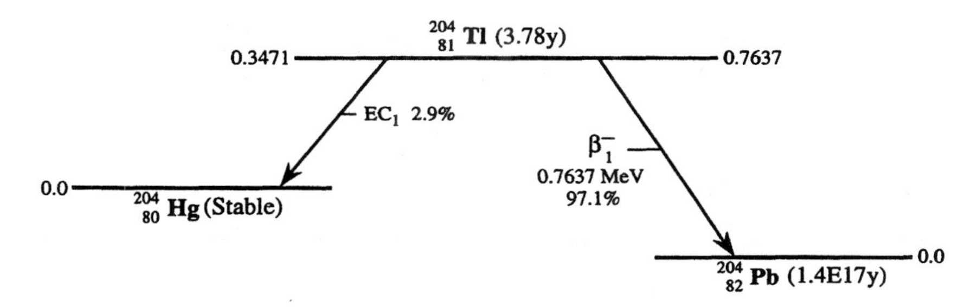

^{203}Tl(n,γ) (Nov. 4, 1994)

Radiation	Y_i (%)	E_i (MeV)
Via EC		
$K_{\alpha1}$ x ray	0.811	0.0708
$K_{\alpha2}$ x ray	0.477	0.0689
K_{β} x ray	0.357	0.0803[a]
L x ray	0.855	0.0100[a]
Via beta minus		
β_1^-	97.1	0.2440[a]
Auger-L	1.4	0.0076[a]

[a] Average energy.

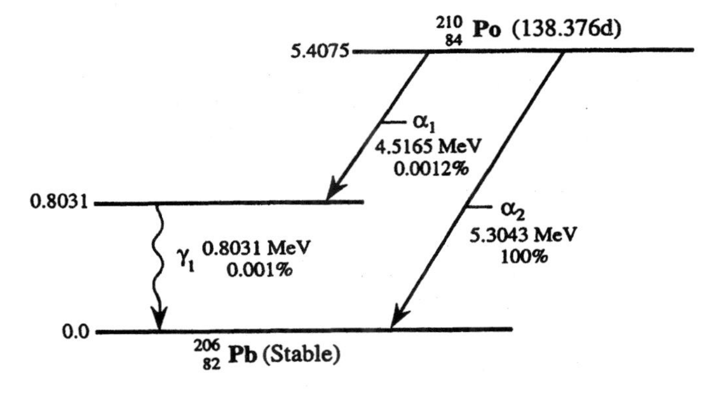

^{238}U Series (May 7, 1998)

Radiation	Y_i (%)	E_i (MeV)
α_1	0.0012	4.5165
α recoil	0.0012	0.0862
α_2	100.0	5.3043
α recoil	100.0	0.1013
γ_1	0.0012	0.8031
ce-K, γ_1	9.79×10^{-6}	0.7151
ce-L, γ_1	2.13×10^{-6}	0.7872[a]
$K_{\alpha1}$ x ray	4.65×10^{-6}	0.0750
$K_{\alpha2}$ x ray	2.76×10^{-6}	0.0728
K_{β} x ray	2.07×10^{-6}	0.0849[b]
L x ray	4.12×10^{-6}	0.0106[b]
Auger-K	3.12×10^{-7}	0.0567[b]

[a] Maximum energy for subshell.
[b] Average energy.

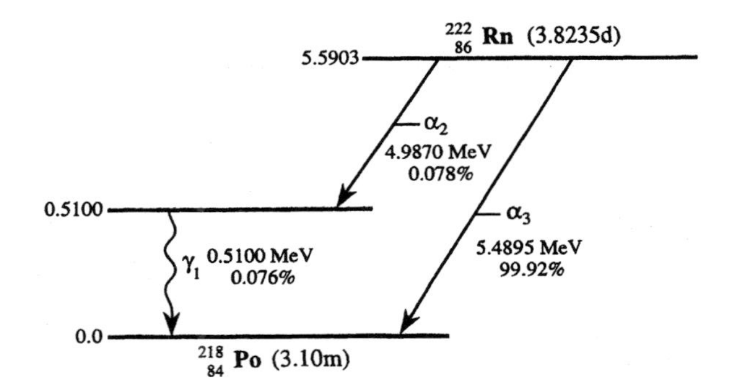

^{238}U Series (Jan. 4, 1996)

Radiation	Y_i (%)	E_i (MeV)
α_2	0.078	4.9870
α recoil	0.078	0.0900
α_3	99.9	5.4895
α recoil	99.9	0.0991
γ_1	0.076	0.5100

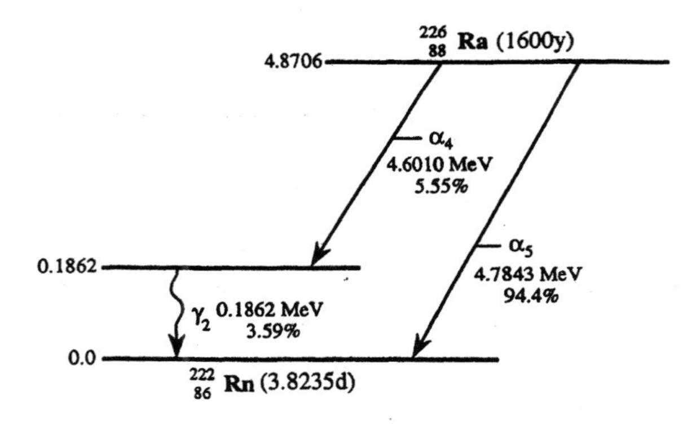

^{238}U Series (Jan. 29, 1996)

Radiation	Y_i (%)	E_i (MeV)
α_4	5.55	4.6010
α recoil	5.55	0.0816
α_5	94.4	4.7843
α recoil	94.4	0.0848
γ_2	3.59	0.1862
ce-K, γ_2	0.693	0.0878
ce-L, γ_2	1.32	0.1682[a]
ce-M, γ_2	0.351	0.1817[a]
ce-N$^+$, γ_2	0.112	0.1851[a]
$K_{\alpha 1}$ x ray	0.328	0.0838
$K_{\alpha 2}$ x ray	0.197	0.0811
K_β x ray	0.149	0.0949[b]
L x ray	0.880	0.0117[b]
Auger-K	0.0193	0.0627[b]

[a]Maximum energy for subshell.
[b]Average energy.

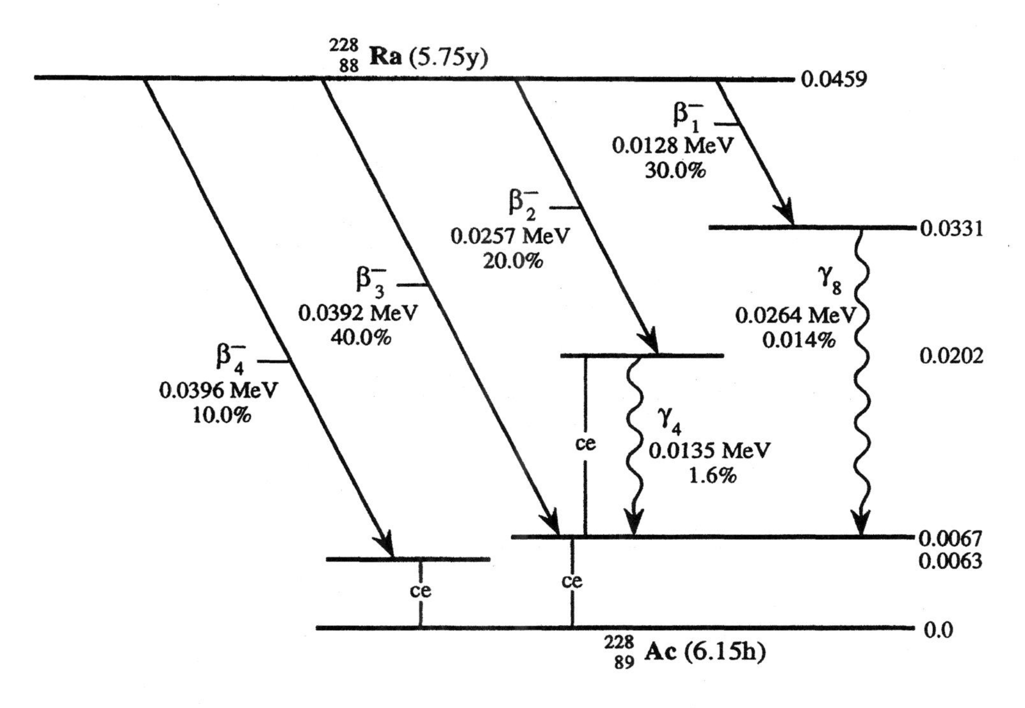

^{232}Th Series (Apr. 24, 1997)

Radiation	Y_i (%)	E_i (MeV)
β_1^-	30.0	0.0032[a]
β_2^-	20.0	0.0065[a]
β_3^-	40.0	0.0099[a]
β_4^-	10.0	0.0100[a]
ce-M, γ_1	7.5	0.0013[b]
ce-M, γ_2	37.5	0.0017[b]
ce-M, γ_3	2.25	0.0074[b]
γ_4	1.60	0.0135
ce-M, γ_4	7.31	0.0085[b]
ce-L, γ_8	2.21	0.0066[b]
L x ray	1.13	0.0127[a]
Auger-K	1.08	0.0093[a]

[a] Average energy.
[b] Maximum energy for subshell.

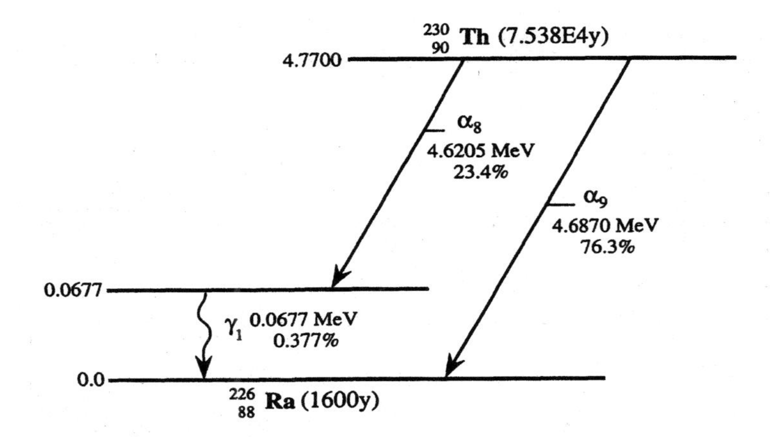

^{238}U Series (Apr. 5, 1996)

Radiation	Y_i (%)	E_i (MeV)
α_8	23.4	4.6205
α recoil	23.4	0.0805
α_9	76.3	4.6870
α recoil	76.3	0.0817
γ_1	0.377	0.0677
ce-L, γ_1	17.0	0.0484[a]
ce-M, γ_1	4.6	0.0629[a]
ce-N$^+$, γ_1	1.66	0.0665[a]
L x ray	8.55	0.0123[b]

[a]Maximum energy for subshell.
[b]Average energy.

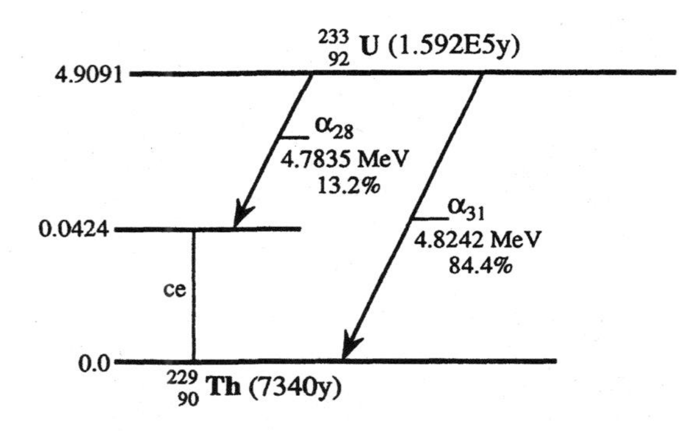

^{232}Th(n,γ); ^{233}Th(β^-); ^{233}Pa(β^-) (Feb. 16, 1990)

Radiation[a]	Y_i (%)	E_i (MeV)
α_{28}	13.2	4.7835
α recoil	13.2	0.0823
α_{31}	84.4	4.8242
α recoil	84.4	0.0830
L x ray	5.28	0.0130[b]

[a]Numerous conversion electrons with yields $< 1\%$ omitted; yields of gamma rays insignificant.
[b]Average energy.

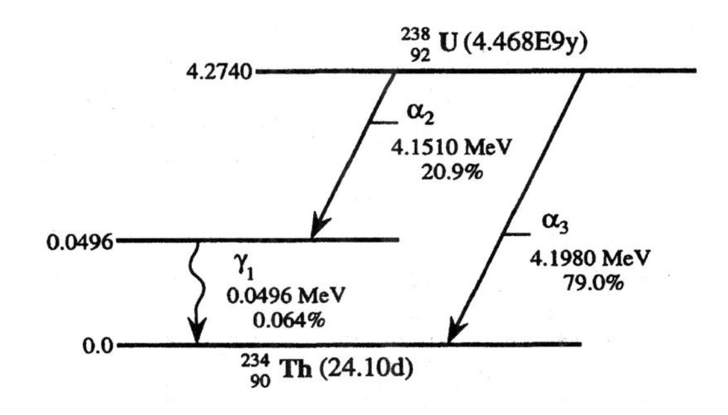

Natural (Apr. 8, 1994)

Radiation	Y_i (%)	E_i (MeV)
α_2	20.9	4.1510
α recoil	20.9	0.0699
α_3	79.0	4.1980
α recoil	79.0	0.0707
γ_1	0.064	0.0496
ce-L, γ_1	15.3	0.0291[a]
ce-M, γ_1	4.18	0.0444[a]
$K_{\alpha 1}$ x ray	0.0011	0.0934
L x ray	8.01	0.0130[b]

[a]Maximum energy for subshell.
[b]Average energy.

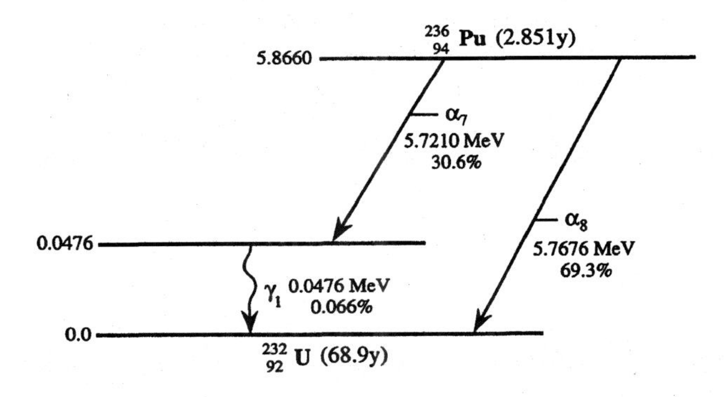

β^- Transformation of ^{236}Np; ^{235}U(α,3n) (Aug. 7, 1991)

Radiation	Y_i (%)	E_i (MeV)
α_7	30.6	5.7210
α_7 recoil	30.6	0.0972
α_8	69.3	5.7676
α_8 recoil	69.3	0.0980
γ_1	0.066	0.0476
ce-L, γ_1	22.5	0.0258[a]
ce-M, γ_1	6.21	0.0420[a]
ce-N$^+$, γ_1	2.05	0.0461[a]
L x ray	12.6	0.0136[b]

[a]Maximum energy for subshell.
[b]Average energy.

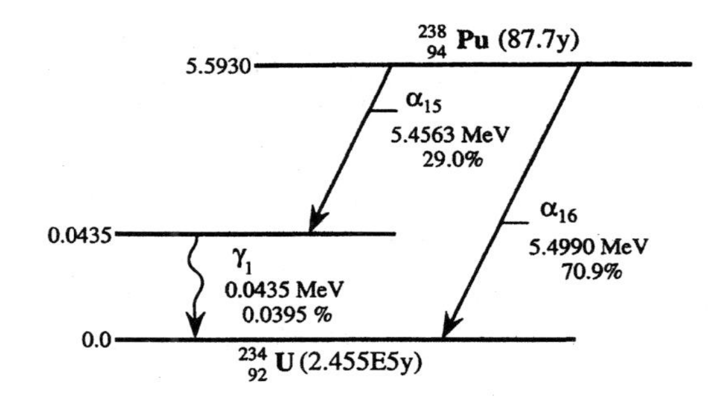

β^- Transformation from ^{237}Np(n, γ); ^{238}Np; Product of ^{242}Cm (Apr. 8, 1994)

Radiation	Y_i (%)	E_i (MeV)
α_{15}	29.0	5.4563
α recoil	29.0	0.0919
α_{16}	70.9	5.4990
α recoil	70.9	0.0926
γ_1	0.0395	0.0435
ce-L, γ_1	20.9	0.0217[a]
ce-M, γ_1	5.76	0.0380[a]
L x ray	11.7	0.0136[b]

[a]Maximum energy for subshell.
[b]Average energy.

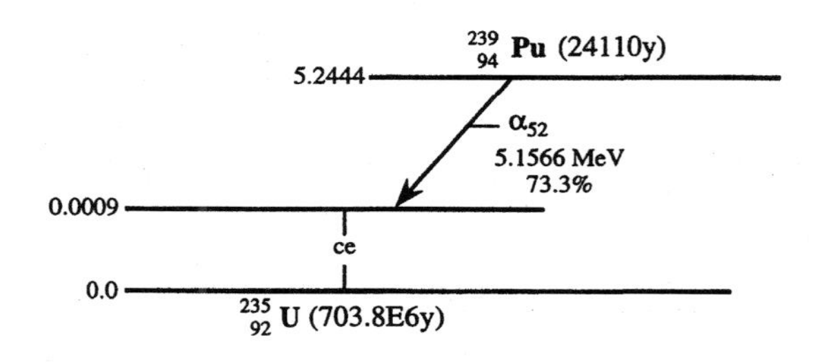

β^- Transformation from ^{238}U(n, γ); ^{239}U[a] (Aug. 31, 1993)

Radiation	Y_i (%)	E_i (MeV)
α_{52}	73.3	5.1566
α recoil	73.3	0.0865
ce-M, γ_2	12.9	0.0074[b]
ce-N$^+$, γ_2	5.52	0.0115[b]
ce-L, γ_4	2.36	0.0169[b]
ce-M, γ_4	0.63	0.0331[b]
ce-L, γ_{10}	6.24	0.0299[b]
ce-M, γ_{10}	1.72	0.0461[b]
ce-N$^+$, γ_{10}	0.642	0.0502[b]
L x ray	4.91	0.0136[c]
Auger-K	2.86×10^{-4}	0.0726[c]

[a]Greatly simplified; 52 separate α energies, very closely spread; numerous conversion electrons from 168 γ-energy states.
[b]Maximum energy for subshell.
[c]Average energy.

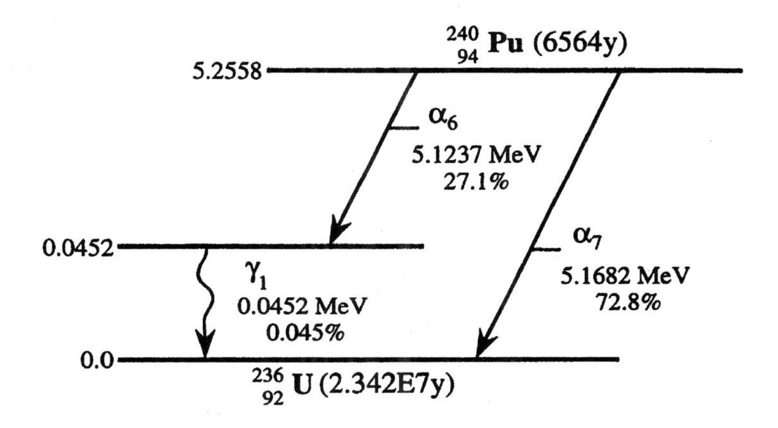

Multiple Neutron Capture from ^{238}U, ^{239}Pu (Aug. 7, 1991)

Radiation	Y_i (%)	E_i (MeV)
α_6	27.1	5.1237
α recoil	27.1	0.0856
α_7	72.8	5.1682
α recoil	72.8	0.0863
γ_1	0.045	0.0452
ce-L, γ_1	19.7	0.0235[a]
ce-M, γ_1	5.41	0.0397[a]
L x ray	11.0	0.0136[b]
Auger-K	2.00×10^{-6}	0.0726[b]

[a]Maximum energy for subshell.
[b]Average energy.

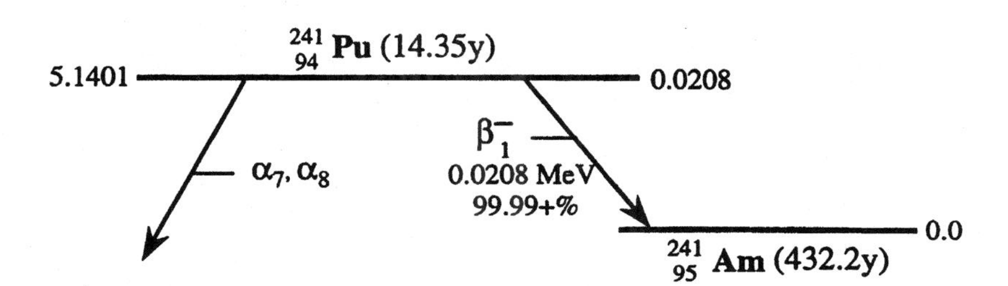

Multiple Neutron Capture from ^{238}U, ^{239}Pu, etc. (May 26, 1995)

Radiation	Y_i (%)	E_i (MeV)
α_7	0.0021	4.8963
β_1^-	100.0	0.0052[a]

[a]Average energy.

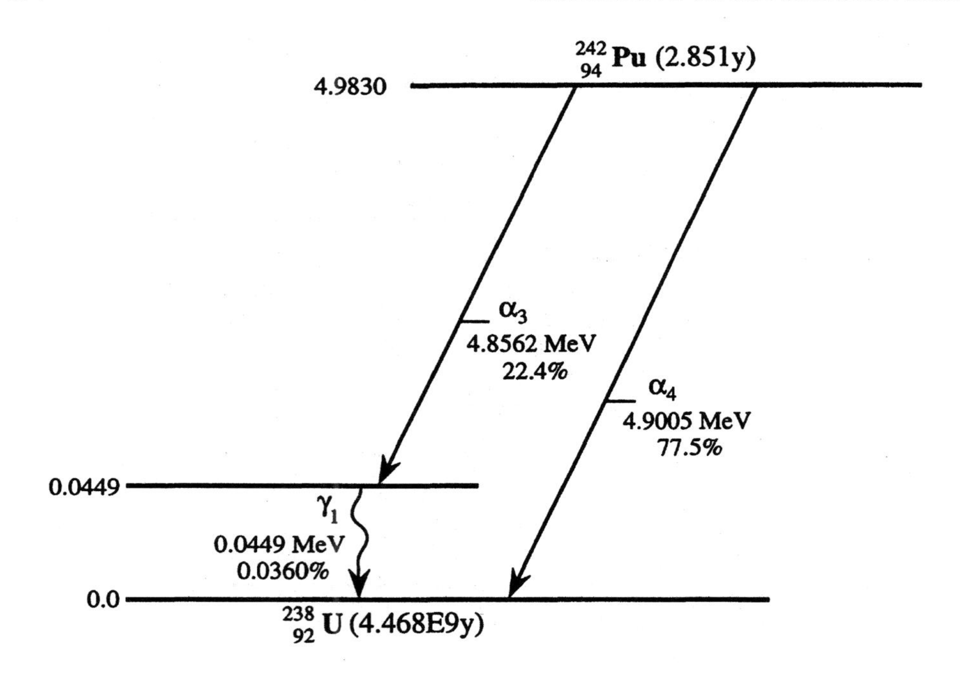

Multiple Neutron Capture from ^{238}U (Aug. 25, 1988)

Radiation	Y_i (%)	E_i (MeV)
α_3	22.4	4.8562
α recoil	22.4	0.0804
α_4	77.5	4.9005
α recoil	77.5	0.0811
γ_1	0.036	0.0449
ce-L, γ_1	16.3	0.0232[a]
ce-M, γ_1	4.5	0.0394[a]
ce-N$^+$, γ_1	1.48	0.0435[a]
L x ray	9.13	0.0136[b]

[a]Average energy.
[b]Maximum energy for subshell.

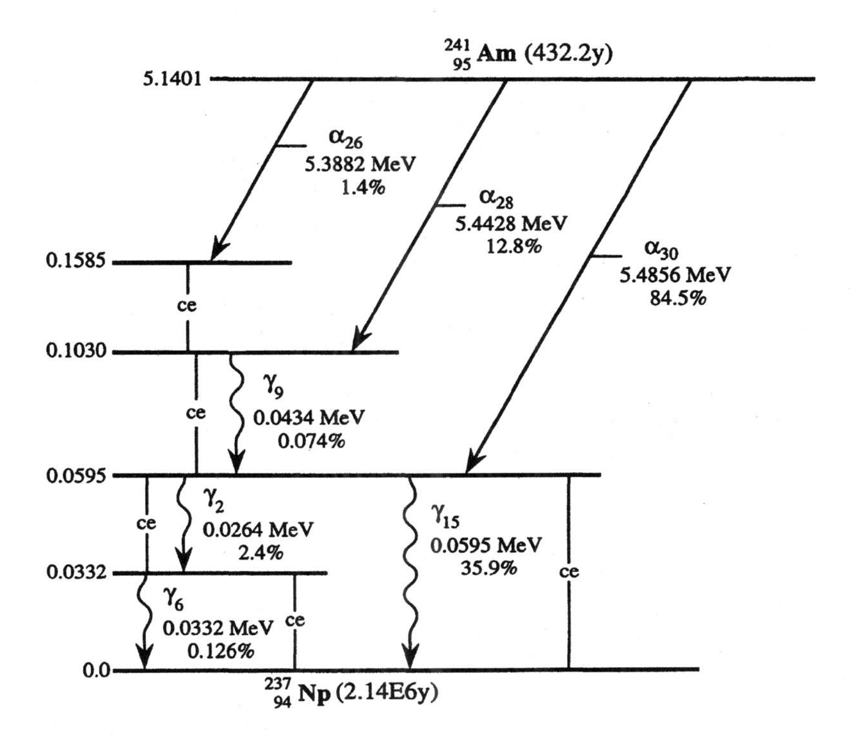

Transformation Product of ^{241}Pu (May 26, 1995)

Radiation	Y_i (%)	E_i (MeV)
α_{26}	1.6	5.3882
α recoil	1.6	0.0896
α_{28}	13.0	5.4428
α recoil	13.0	0.0905
α_{30}	0.34	5.5440
α recoil	0.34	0.0922
γ_2	2.40	0.0264
ce-L, γ_2	14.4	0.0039[a]
ce-M, γ_2	3.84	0.0206[a]
γ_6	0.126	0.0332
ce-L, γ_6	17.4	0.0108[a]
ce-M, γ_6	4.41	0.0275[a]
γ_9	0.074	0.0434
ce-L, γ_9	9.05	0.0210[a]
ce-M, γ_9	2.39	0.0377[a]
γ_{15}	35.9	0.0595
ce-L, γ_{15}	30.2	0.0371[a]
ce-M, γ_{15}	8.11	0.0538[a]
ce-N$^+$, γ_{15}	3.37	0.0580[a]
Auger-L	30.3	0.0101[b]

[a] Maximum energy for subshell.
[b] Average energy.

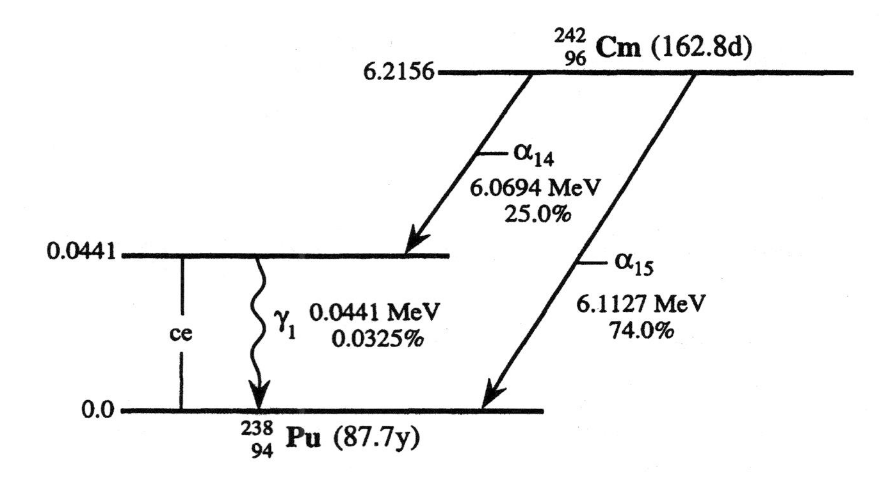

Product of ^{241}Am(n, γ) ^{242m}Am; Multiple Neutron Capture in ^{238}U (Dec. 14, 1995)

Radiation	Y_i (%)	E_i (MeV)
α_{14}	25.0	6.0694
α recoil	25.0	0.1005
α_{15}	74.0	6.1127
α recoil	74.0	0.1012
γ_1	0.0325	0.0441
ce-L, γ_1	19.0	0.0210[a]
ce-M, γ_1	5.25	0.0382[a]
ce-N$^+$, γ_1	1.74	0.0425[a]
L x ray	11.4	0.0143[b]

[a]Maximum energy for subshell.
[b]Average energy.

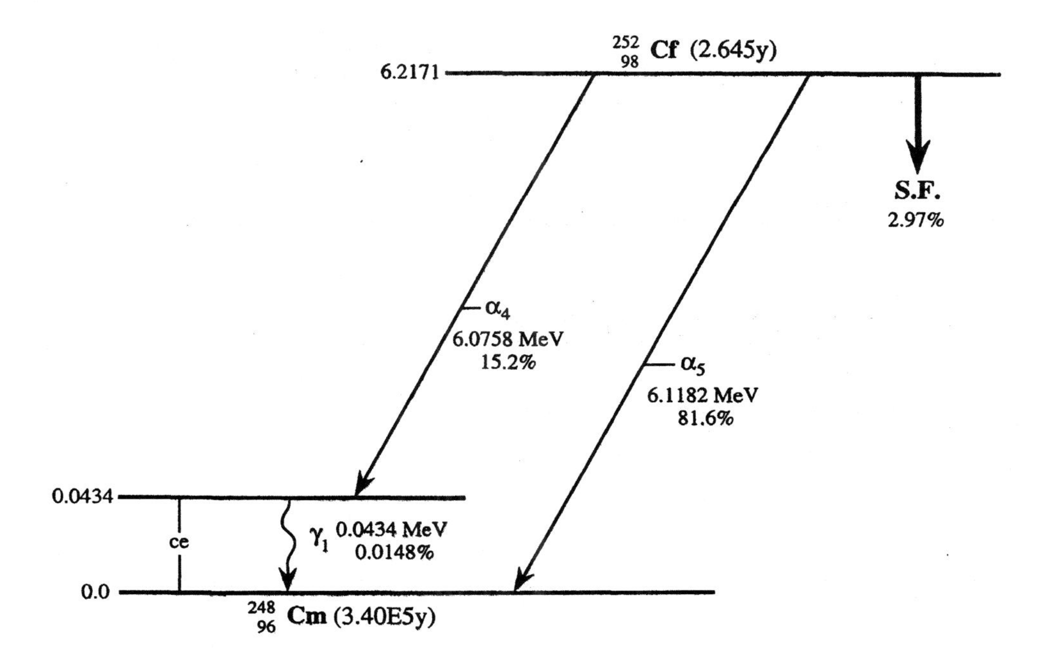

Multiple Neutron Capture from ^{238}U (Feb. 7, 1990)

Radiation	Y_i (%)	E_i (MeV)
α_3[a]	0.233	5.9765
α_4[a]	15.2	6.0758
α_5[a]	81.6	6.1182
γ_1	0.0148	0.0434
ce-L, γ_1	11.0	0.0189[b]
ce-M, γ_1	3.07	0.0371[b]
L x ray	7.12	0.0150[c]
Neutrons	6–12	Fast
FP	5.9	~ 170

[a]Each α emission also has recoil energy = [transition E − $E_{\alpha i}$].
[b]Maximum energy for subshell.
[c]Average energy.

APPENDIX E

FISSION PRODUCT CHAINS

Fission product chains are presented here for mass numbers 72 through 165 showing the cumulative percent thermal neutron fission product yield for ^{235}U and (^{239}Pu) fission just below each nuclide in the mass chain, and for each radioactive nuclide the half-life and branching ratios, if applicable. The number of short-lived precursors that begin each ^{235}U and (^{239}Pu) fission mass chain is also shown.

Mass Number	Number of Prec.							
72	1 (1)	Co (0.12s) → 5.81×10^{-8} (4.14×10^{-8})	Ni (2.1s) → 7.66×10^{-6} (8.54×10^{-6})	Cu (6.6s) → 2.04×10^{-5} (6.09×10^{-5})	Zn (46.5h) → 2.65×10^{-5} (0.0001)	Ga (14.1h) → 2.65×10^{-5} (0.0001)	**Ge (Stable)** 2.65×10^{-5} (0.0001)	
						↗	mGe (0.53s) 0.0001 (0.0003) ↓	
73	1 (0)	Co (0.13s) → 2.21×10^{-8} (6.81×10^{-9})	Ni (0.9s) → 7.16×10^{-6} (3.50×10^{-6})	Cu (3.9s) → 5.42×10^{-5} (7.46×10^{-5})	Zn (24s) → 0.0001 (0.0002)	Ga (4.87h) 0.0001 (0.0003)	**Ge (Stable)** 0.0001 (0.0003)	
74	0 (0)	Co (0.09s) → 3.00×10^{-9} (7.26×10^{-10})	Ni (1.1s) → 4.58×10^{-6} (1.35×10^{-6})	Cu (1.6s) → 7.28×10^{-5} (6.44×10^{-5})	Zn (1.6m) → 0.0003 (0.0005)	Ga (8.1m) → 0.0003 (0.0006)	**Ge (Stable)** 0.0003 (0.0006)	
						4% ↗	mGe (48.5s) 4.70×10^{-5} (0.0001) ↓	
75	0 (0)	Co (0.08s) → 3.81×10^{-10} (5.50×10^{-11})	Ni (0.23s) → 1.53×10^{-6} (2.64×10^{-7})	Cu (1.3s) → 0.0001 (4.45×10^{-5})	Zn (10.2s) → 0.0009 (0.0007)	Ga (2.1m) 96% → 0.0011 (0.0013)	Ge (1.38h) → 0.0011 (0.0014)	**As (Stable)** 0.0011 (0.0014)
76	1 (1)	Ni (0.31s) → 5.90×10^{-7} (3.94×10^{-8})	Cu (0.64s) → 0.0001 (1.66×10^{-5})	Zn (5.7s) → 0.0019 (0.0009)	Ga (29s) → 0.0030 (0.0024)	**Ge (Stable)** 0.0031 (0.0029)		

Mass Number	Number of Prec.							
						mGe (53s) 0.0065 (0.0043) (from Ga: 88%; to Ge: 21%; to As: 79%)		mSe (17.5s) 2.39×10^{-5} (2.20×10^{-5}) (from As: 0.3%)
77	0 (0)	Ni (0.1s) → 5.64×10^{-8} (3.45×10^{-9})	Cu (0.47s) → 4.46×10^{-5} (5.44×10^{-6})	Zn (2.1s) → 0.0032 (0.0007)	Ga (13s) → (12%) 0.0073 (0.0045)	Ge (11.3h) → 0.0028 (0.0038)	As (38.8h) → (99.7%) 0.0080 (0.0072)	**Se (Stable)** 0.0080 (0.0072)
78	0 (0)	Ni (0.13s) → 5.03×10^{-9} (2.38×10^{-10})	Cu (0.34s) → 1.09×10^{-5} (9.79×10^{-7})	Zn (1.5s) → 0.0036 (0.0004)	Ga (5.09s) → 0.0139 (0.0060)	Ge (1.45h) → 0.0208 (0.0178)	As (1.51h) → 0.0210 (0.0188)	**Se (Stable)** 0.0210 (0.0188)
							mSe (3.92m) ↓ 0.0447 (0.0435) (from As)	mBr (4.86s) ↓ 1.04×10^{-9} (1.91×10^{-7})
79	1 (1)		Zn (1s) → 0.0016 (0.0001)	Ga (2.85s) → 0.0187 (0.0056)	Ge (19.0s) → 0.0421 (0.0354)	As (9m) 0.0447 (0.0434)	Se (6.5E4y) → 0.0447 (0.0437)	**Br (Stable)** 0.0447 (0.0437)
80	1 (0)	Cu (0.09s) → 5.24×10^{-8} (7.23×10^{-9})	Zn (0.54s) → 0.0002 (3.30×10^{-5})	Ga (1.70s) → 0.0119 (0.0037)	Ge (29.5) → 0.1146 (0.0587)	As (16s) → 0.1283 (0.0906)	**Se (Stable)** 0.1288 (0.0939)	
							mSe (57.3m) ↓ 0.0069 (0.0211)	
81	2 (2)			Ga (1.22s) → 0.0082 (0.0015)	Ge (7.6s) → 0.1348 (0.0565)	As (33s) → 0.1957 (0.1591)	Se (18.5m) → 0.2037 (0.1834)	**Br (Stable)** 0.2037 (0.1836)

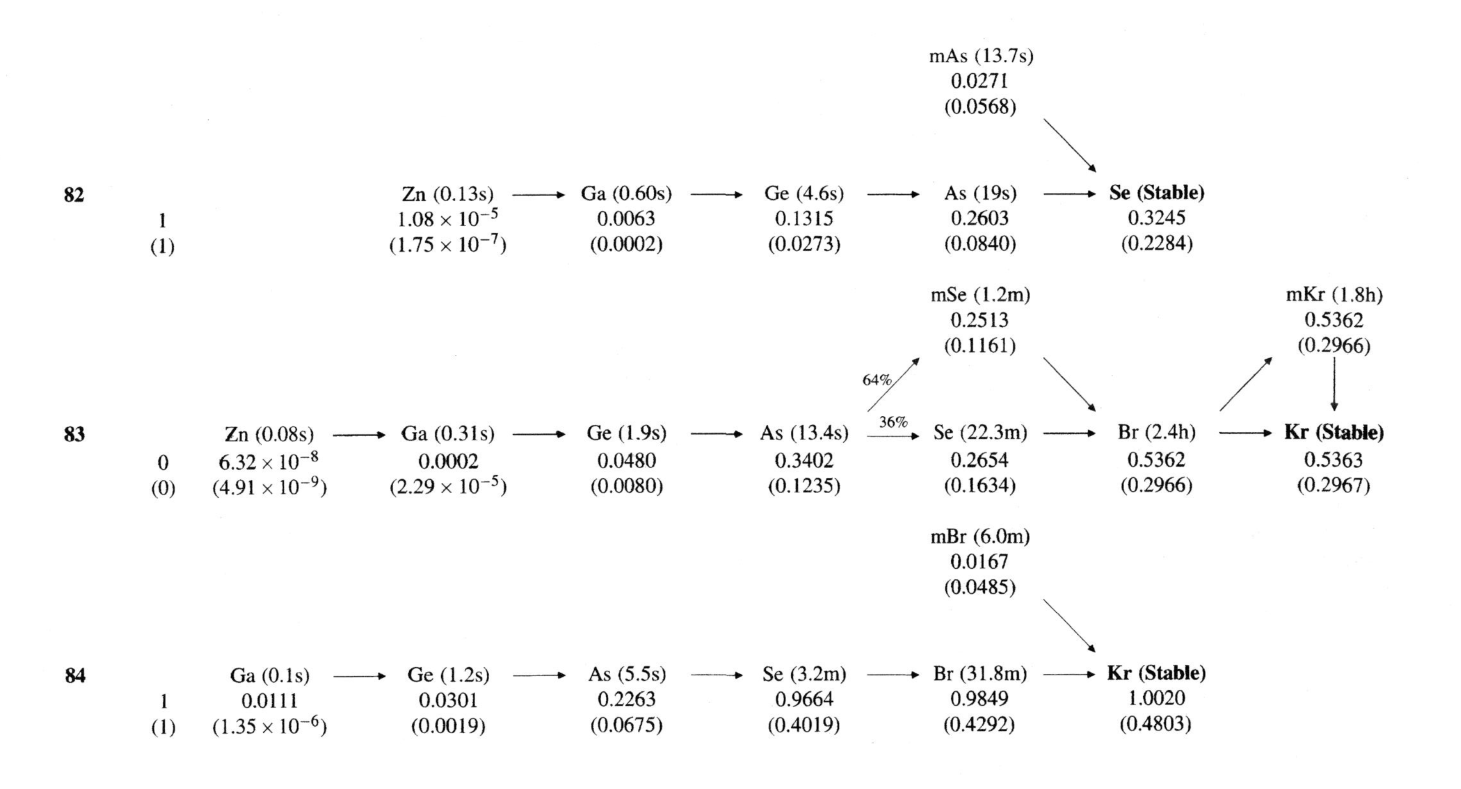

82
1
(1)
Zn (0.13s)
1.08×10^{-5}
(1.75×10^{-7})
Ga (0.60s)
0.0063
(0.0002)
Ge (4.6s)
0.1315
(0.0273)
mAs (13.7s)
0.0271
(0.0568)
As (19s)
0.2603
(0.0840)
Se (Stable)
0.3245
(0.2284)
83
0
(0)
Zn (0.08s)
6.32×10^{-8}
(4.91×10^{-9})
Ga (0.31s)
0.0002
(2.29×10^{-5})
Ge (1.9s)
0.0480
(0.0080)
As (13.4s)
0.3402
(0.1235)
64%
36%
mSe (1.2m)
0.2513
(0.1161)
Se (22.3m)
0.2654
(0.1634)
Br (2.4h)
0.5362
(0.2966)
mKr (1.8h)
0.5362
(0.2966)
Kr (Stable)
0.5363
(0.2967)
84
1
(1)
Ga (0.1s)
0.0111
(1.35×10^{-6})
Ge (1.2s)
0.0301
(0.0019)
As (5.5s)
0.2263
(0.0675)
Se (3.2m)
0.9664
(0.4019)
mBr (6.0m)
0.0167
(0.0485)
Br (31.8m)
0.9849
(0.4292)
Kr (Stable)
1.0020
(0.4803)

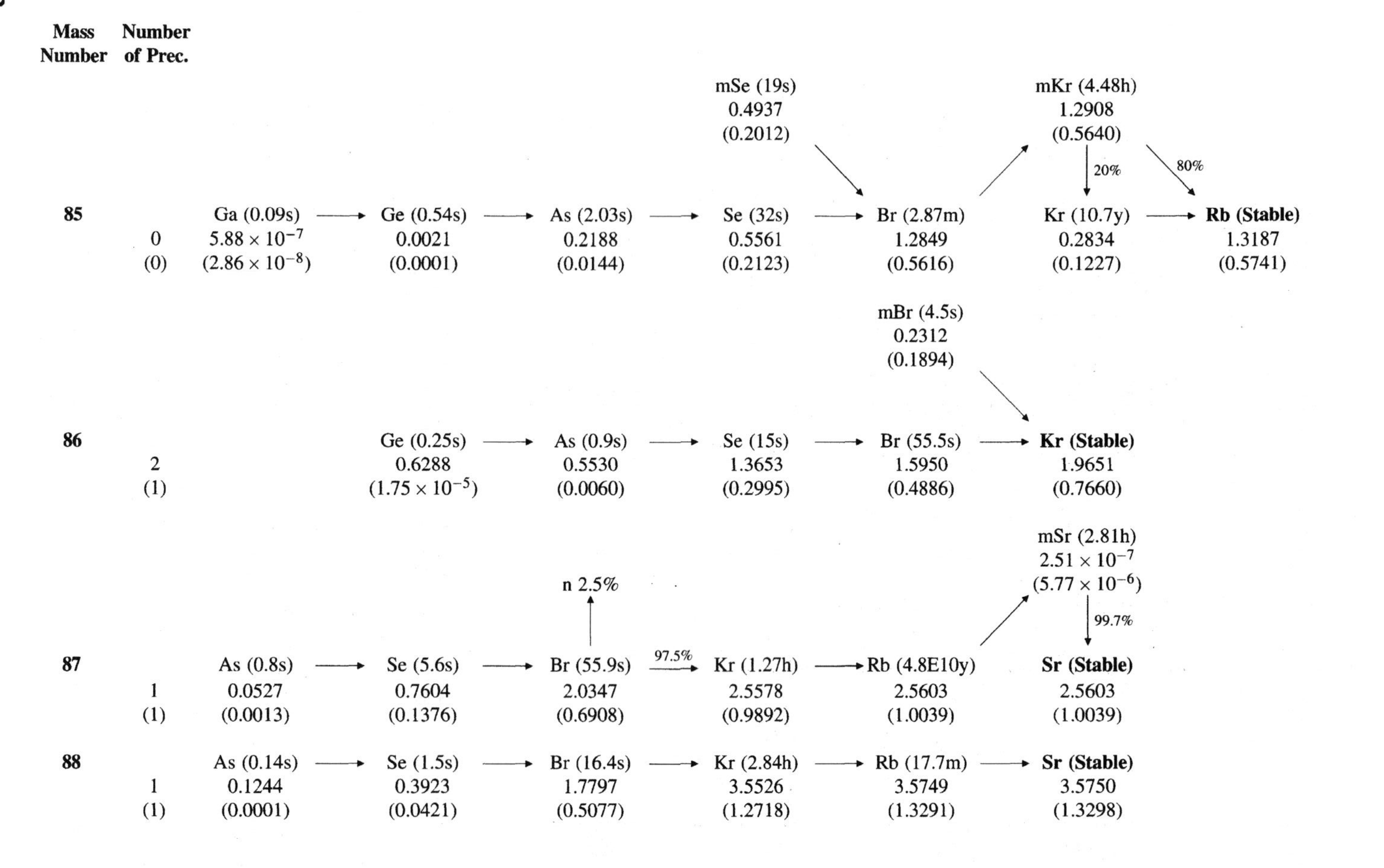
Mass Number
Number of Prec.
85
0
(0)
Ga (0.09s)
5.88 × 10^−7
(2.86 × 10^−8)
Ge (0.54s)
0.0021
(0.0001)
As (2.03s)
0.2188
(0.0144)
mSe (19s)
0.4937
(0.2012)
Se (32s)
0.5561
(0.2123)
Br (2.87m)
1.2849
(0.5616)
mKr (4.48h)
1.2908
(0.5640)
20%
80%
Kr (10.7y)
0.2834
(0.1227)
Rb (Stable)
1.3187
(0.5741)
86
2
(1)
Ge (0.25s)
0.6288
(1.75 × 10^−5)
As (0.9s)
0.5530
(0.0060)
Se (15s)
1.3653
(0.2995)
mBr (4.5s)
0.2312
(0.1894)
Br (55.5s)
1.5950
(0.4886)
Kr (Stable)
1.9651
(0.7660)
87
1
(1)
As (0.8s)
0.0527
(0.0013)
Se (5.6s)
0.7604
(0.1376)
n 2.5%
Br (55.9s)
2.0347
(0.6908)
97.5%
Kr (1.27h)
2.5578
(0.9892)
Rb (4.8E10y)
2.5603
(1.0039)
mSr (2.81h)
2.51 × 10^−7
(5.77 × 10^−6)
99.7%
Sr (Stable)
2.5603
(1.0039)
88
1
(1)
As (0.14s)
0.1244
(0.0001)
Se (1.5s)
0.3923
(0.0421)
Br (16.4s)
1.7797
(0.5077)
Kr (2.84h)
3.5526
(1.2718)
Rb (17.7m)
3.5749
(1.3291)
Sr (Stable)
3.5750
(1.3298)

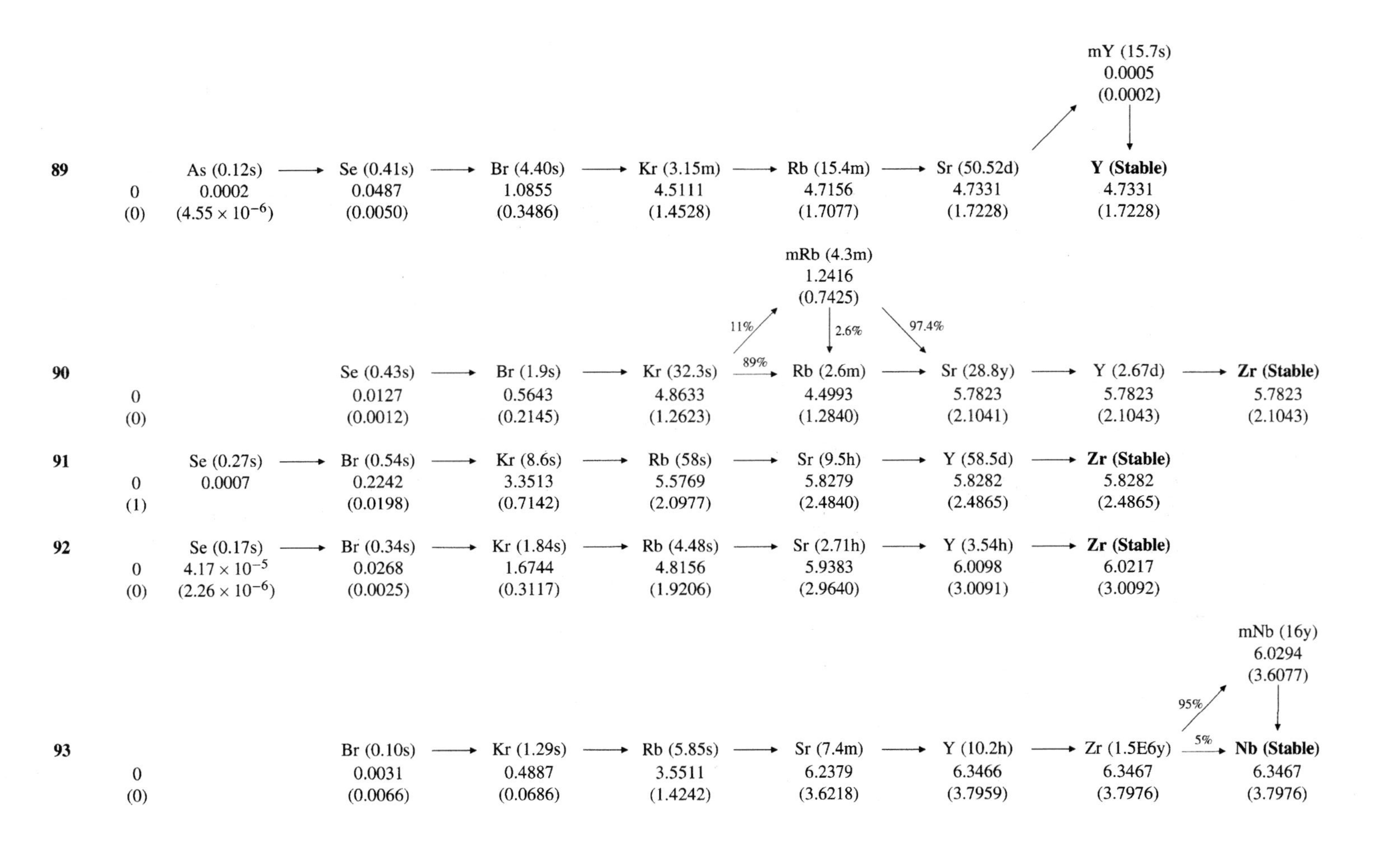
89
0
(0)
As (0.12s)
0.0002
(4.55 × 10−6)
Se (0.41s)
0.0487
(0.0050)
Br (4.40s)
1.0855
(0.3486)
Kr (3.15m)
4.5111
(1.4528)
Rb (15.4m)
4.7156
(1.7077)
Sr (50.52d)
4.7331
(1.7228)
mY (15.7s)
0.0005
(0.0002)
Y (Stable)
4.7331
(1.7228)
90
0
(0)
Se (0.43s)
0.0127
(0.0012)
Br (1.9s)
0.5643
(0.2145)
Kr (32.3s)
4.8633
(1.2623)
11%
89%
mRb (4.3m)
1.2416
(0.7425)
2.6%
97.4%
Rb (2.6m)
4.4993
(1.2840)
Sr (28.8y)
5.7823
(2.1041)
Y (2.67d)
5.7823
(2.1043)
Zr (Stable)
5.7823
(2.1043)
91
0
(1)
Se (0.27s)
0.0007
Br (0.54s)
0.2242
(0.0198)
Kr (8.6s)
3.3513
(0.7142)
Rb (58s)
5.5769
(2.0977)
Sr (9.5h)
5.8279
(2.4840)
Y (58.5d)
5.8282
(2.4865)
Zr (Stable)
5.8282
(2.4865)
92
0
(0)
Se (0.17s)
4.17 × 10−5
(2.26 × 10−6)
Br (0.34s)
0.0268
(0.0025)
Kr (1.84s)
1.6744
(0.3117)
Rb (4.48s)
4.8156
(1.9206)
Sr (2.71h)
5.9383
(2.9640)
Y (3.54h)
6.0098
(3.0091)
Zr (Stable)
6.0217
(3.0092)
93
0
(0)
Br (0.10s)
0.0031
(0.0066)
Kr (1.29s)
0.4887
(0.0686)
Rb (5.85s)
3.5511
(1.4242)
Sr (7.4m)
6.2379
(3.6218)
Y (10.2h)
6.3466
(3.7959)
Zr (1.5E6y)
6.3467
(3.7976)
95%
5%
mNb (16y)
6.0294
(3.6077)
Nb (Stable)
6.3467
(3.7976)

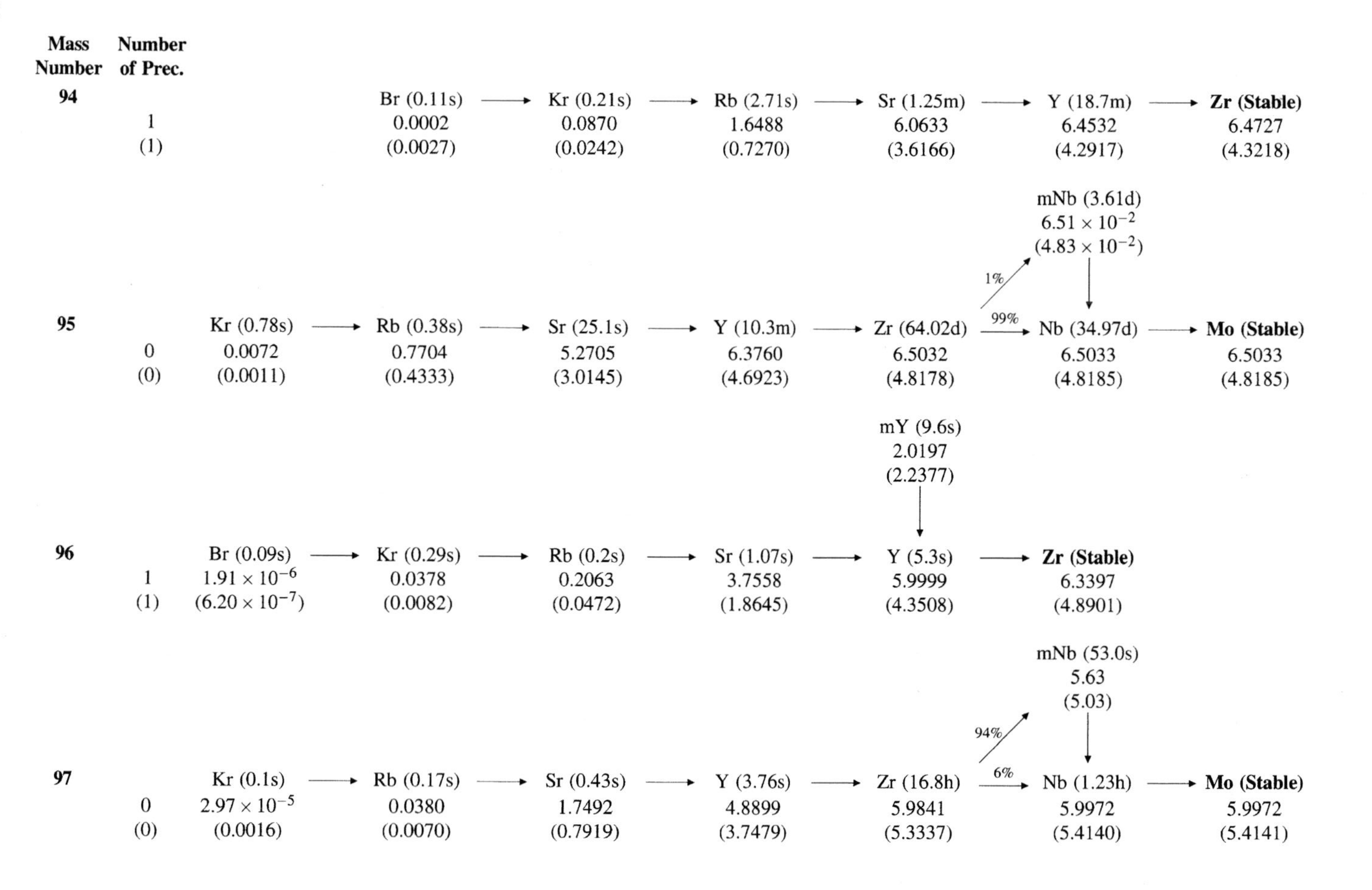

Mass Number
Number of Prec.
94
1
(1)
Br (0.11s)
0.0002
(0.0027)
Kr (0.21s)
0.0870
(0.0242)
Rb (2.71s)
1.6488
(0.7270)
Sr (1.25m)
6.0633
(3.6166)
Y (18.7m)
6.4532
(4.2917)
Zr (Stable)
6.4727
(4.3218)
95
0
(0)
Kr (0.78s)
0.0072
(0.0011)
Rb (0.38s)
0.7704
(0.4333)
Sr (25.1s)
5.2705
(3.0145)
Y (10.3m)
6.3760
(4.6923)
Zr (64.02d)
6.5032
(4.8178)
1%
99%
mNb (3.61d)
6.51 × 10^−2
(4.83 × 10^−2)
Nb (34.97d)
6.5033
(4.8185)
Mo (Stable)
6.5033
(4.8185)
96
1
(1)
Br (0.09s)
1.91 × 10^−6
(6.20 × 10^−7)
Kr (0.29s)
0.0378
(0.0082)
Rb (0.2s)
0.2063
(0.0472)
Sr (1.07s)
3.7558
(1.8645)
mY (9.6s)
2.0197
(2.2377)
Y (5.3s)
5.9999
(4.3508)
Zr (Stable)
6.3397
(4.8901)
97
0
(0)
Kr (0.1s)
2.97 × 10^−5
(0.0016)
Rb (0.17s)
0.0380
(0.0070)
Sr (0.43s)
1.7492
(0.7919)
Y (3.76s)
4.8899
(3.7479)
Zr (16.8h)
5.9841
(5.3337)
94%
6%
mNb (53.0s)
5.63
(5.03)
Nb (1.23h)
5.9972
(5.4140)
Mo (Stable)
5.9972
(5.4141)

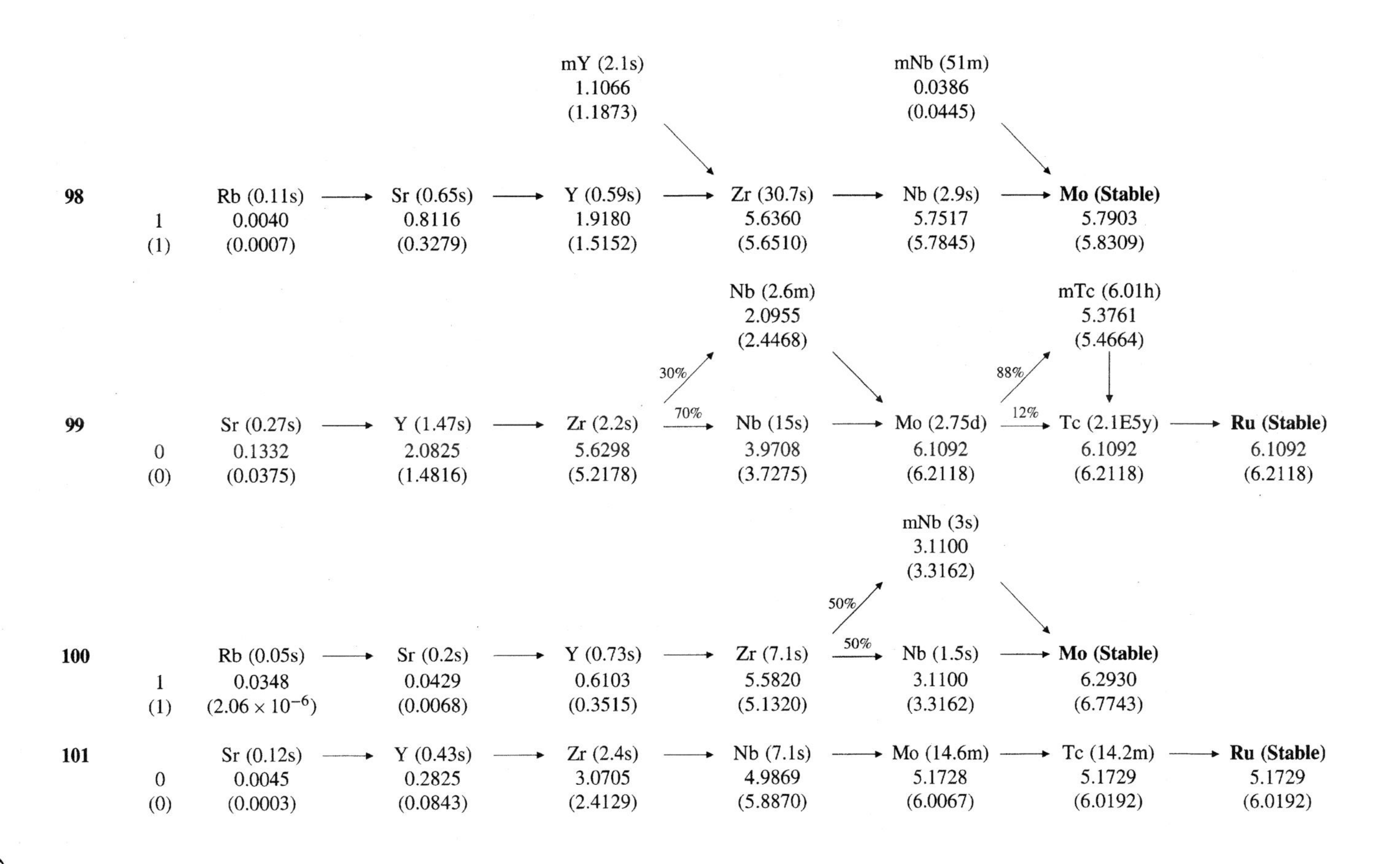
mY (2.1s)
1.1066
(1.1873)
mNb (51m)
0.0386
(0.0445)
98
1
(1)
Rb (0.11s)
0.0040
(0.0007)
Sr (0.65s)
0.8116
(0.3279)
Y (0.59s)
1.9180
(1.5152)
Zr (30.7s)
5.6360
(5.6510)
Nb (2.9s)
5.7517
(5.7845)
Mo (Stable)
5.7903
(5.8309)
Nb (2.6m)
2.0955
(2.4468)
mTc (6.01h)
5.3761
(5.4664)
30%
70%
88%
12%
99
0
(0)
Sr (0.27s)
0.1332
(0.0375)
Y (1.47s)
2.0825
(1.4816)
Zr (2.2s)
5.6298
(5.2178)
Nb (15s)
3.9708
(3.7275)
Mo (2.75d)
6.1092
(6.2118)
Tc (2.1E5y)
6.1092
(6.2118)
Ru (Stable)
6.1092
(6.2118)
mNb (3s)
3.1100
(3.3162)
50%
50%
100
1
(1)
Rb (0.05s)
0.0348
(2.06 × 10⁻⁶)
Sr (0.2s)
0.0429
(0.0068)
Y (0.73s)
0.6103
(0.3515)
Zr (7.1s)
5.5820
(5.1320)
Nb (1.5s)
3.1100
(3.3162)
Mo (Stable)
6.2930
(6.7743)
101
0
(0)
Sr (0.12s)
0.0045
(0.0003)
Y (0.43s)
0.2825
(0.0843)
Zr (2.4s)
3.0705
(2.4129)
Nb (7.1s)
4.9869
(5.8870)
Mo (14.6m)
5.1728
(6.0067)
Tc (14.2m)
5.1729
(6.0192)
Ru (Stable)
5.1729
(6.0192)

Mass Number	Number of Prec.							
						mTc (4.4m) 4.2892 (6.0926)		
					↗	↓ 5%	↘ 95%	
102	0 (0)	Sr (0.07s) → 0.0002 (8.03×10^{-6})	Y (0.30s) → 0.2682 (0.0061)	Zr (2.9s) → 2.0496 (1.2240)	Nb (1.3s) → 3.6286 (4.3022)	Mo (11.3m) 4.2797 (6.0589)	Tc (5.3s) → 0.2240 (0.3383)	**Ru (Stable)** 4.2988 (6.1263)
								mRh (56.1m) 3.0008 (6.9250)
							↗ 99%	↓
103	0 (0)	Y (0.26s) → 0.0026 (0.0010)	Zr (1.3s) → 0.5015 (0.2245)	Nb (1.5s) → 1.9120 (2.9272)	Mo (1.13m) → 2.9489 (6.7403)	Tc (54s) → 3.0311 (6.9936)	Ru (39.3d) → 1% 3.0311 (6.9949)	**Rh (Stable)** 3.0311 (6.9949)
104	0 (0)	Sr (0.16s) → 1.31×10^{-7} (9.44×10^{-9})	Y (0.13s) → 0.0006 (7.34×10^{-5})	Zr (1.2s) → 0.0834 (0.0611)	Nb (4.8s) → 0.6549 (1.2244)	Mo (60s) → 1.7879 (5.5283)	Tc (18.2m) → 1.8805 (6.0704)	**Ru (Stable)** 1.8808 (6.0948)
							mRh (40s) 0.2603 (1.5238)	
						↗ 27%	↓ 99.7%	
105	1 (1)	Zr (1.0s) → 0.1159 (0.0064)	Nb (3s) → 0.2532 (0.5006)	Mo (36s) → 0.9158 (4.0040)	Tc (7.6m) → 0.9642 (5.5982)	Ru (4.44h) → 73% 0.9642 (5.6436)	Rh (35.4h) → 0.9642 (5.6438)	**Pd (Stable)** 0.9642 (5.6438)

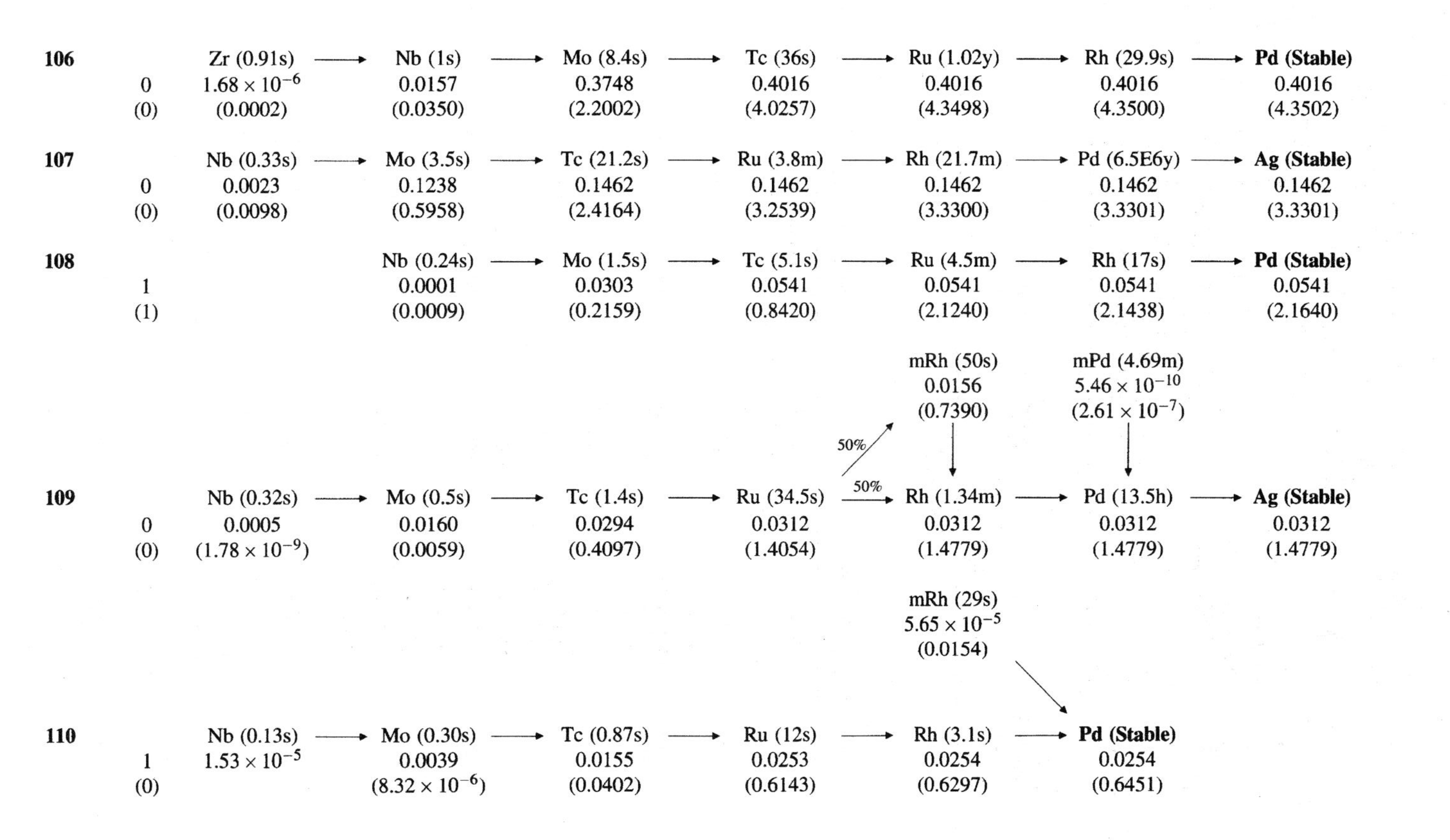
106
0
(0)
Zr (0.91s)
1.68 × 10^−6
(0.0002)
Nb (1s)
0.0157
(0.0350)
Mo (8.4s)
0.3748
(2.2002)
Tc (36s)
0.4016
(4.0257)
Ru (1.02y)
0.4016
(4.3498)
Rh (29.9s)
0.4016
(4.3500)
Pd (Stable)
0.4016
(4.3502)
107
0
(0)
Nb (0.33s)
0.0023
(0.0098)
Mo (3.5s)
0.1238
(0.5958)
Tc (21.2s)
0.1462
(2.4164)
Ru (3.8m)
0.1462
(3.2539)
Rh (21.7m)
0.1462
(3.3300)
Pd (6.5E6y)
0.1462
(3.3301)
Ag (Stable)
0.1462
(3.3301)
108
1
(1)
Nb (0.24s)
0.0001
(0.0009)
Mo (1.5s)
0.0303
(0.2159)
Tc (5.1s)
0.0541
(0.8420)
Ru (4.5m)
0.0541
(2.1240)
Rh (17s)
0.0541
(2.1438)
Pd (Stable)
0.0541
(2.1640)
mRh (50s)
0.0156
(0.7390)
mPd (4.69m)
5.46 × 10^−10
(2.61 × 10^−7)
50%
50%
109
0
(0)
Nb (0.32s)
0.0005
(1.78 × 10^−9)
Mo (0.5s)
0.0160
(0.0059)
Tc (1.4s)
0.0294
(0.4097)
Ru (34.5s)
0.0312
(1.4054)
Rh (1.34m)
0.0312
(1.4779)
Pd (13.5h)
0.0312
(1.4779)
Ag (Stable)
0.0312
(1.4779)
mRh (29s)
5.65 × 10^−5
(0.0154)
110
1
(0)
Nb (0.13s)
1.53 × 10^−5
Mo (0.30s)
0.0039
(8.32 × 10^−6)
Tc (0.87s)
0.0155
(0.0402)
Ru (12s)
0.0253
(0.6143)
Rh (3.1s)
0.0254
(0.6297)
Pd (Stable)
0.0254
(0.6451)

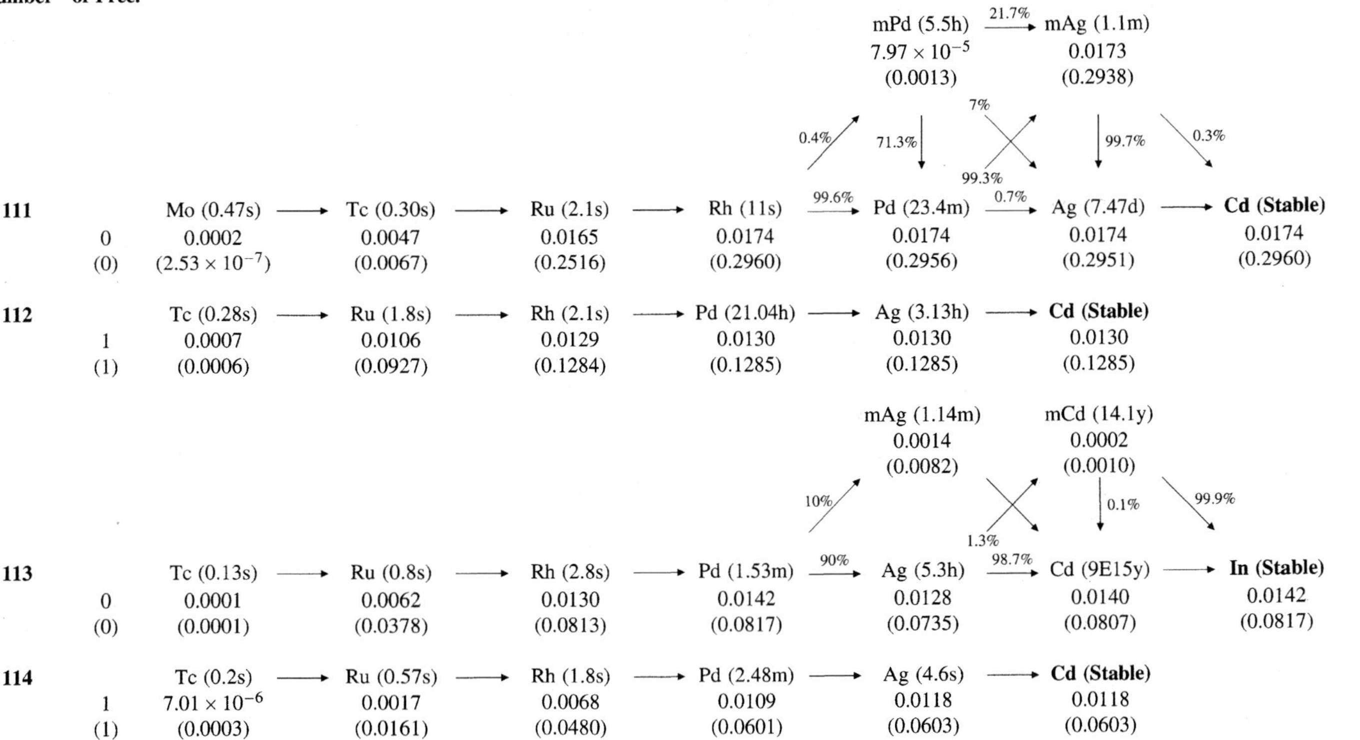

Mass Number
Number of Prec.
111
0
(0)
Mo (0.47s)
0.0002
(2.53 × 10^−7)
Tc (0.30s)
0.0047
(0.0067)
Ru (2.1s)
0.0165
(0.2516)
Rh (11s)
0.0174
(0.2960)
0.4%
99.6%
mPd (5.5h)
7.97 × 10^−5
(0.0013)
71.3%
7%
99.3%
21.7%
mAg (1.1m)
0.0173
(0.2938)
99.7%
0.3%
Pd (23.4m)
0.0174
(0.2956)
0.7%
Ag (7.47d)
0.0174
(0.2951)
Cd (Stable)
0.0174
(0.2960)
112
1
(1)
Tc (0.28s)
0.0007
(0.0006)
Ru (1.8s)
0.0106
(0.0927)
Rh (2.1s)
0.0129
(0.1284)
Pd (21.04h)
0.0130
(0.1285)
Ag (3.13h)
0.0130
(0.1285)
Cd (Stable)
0.0130
(0.1285)
mAg (1.14m)
0.0014
(0.0082)
mCd (14.1y)
0.0002
(0.0010)
10%
0.1%
99.9%
1.3%
113
0
(0)
Tc (0.13s)
0.0001
(0.0001)
Ru (0.8s)
0.0062
(0.0378)
Rh (2.8s)
0.0130
(0.0813)
Pd (1.53m)
0.0142
(0.0817)
90%
Ag (5.3h)
0.0128
(0.0735)
98.7%
Cd (9E15y)
0.0140
(0.0807)
In (Stable)
0.0142
(0.0817)
114
1
(1)
Tc (0.2s)
7.01 × 10^−6
(0.0003)
Ru (0.57s)
0.0017
(0.0161)
Rh (1.8s)
0.0068
(0.0480)
Pd (2.48m)
0.0109
(0.0601)
Ag (4.6s)
0.0118
(0.0603)
Cd (Stable)
0.0118
(0.0603)

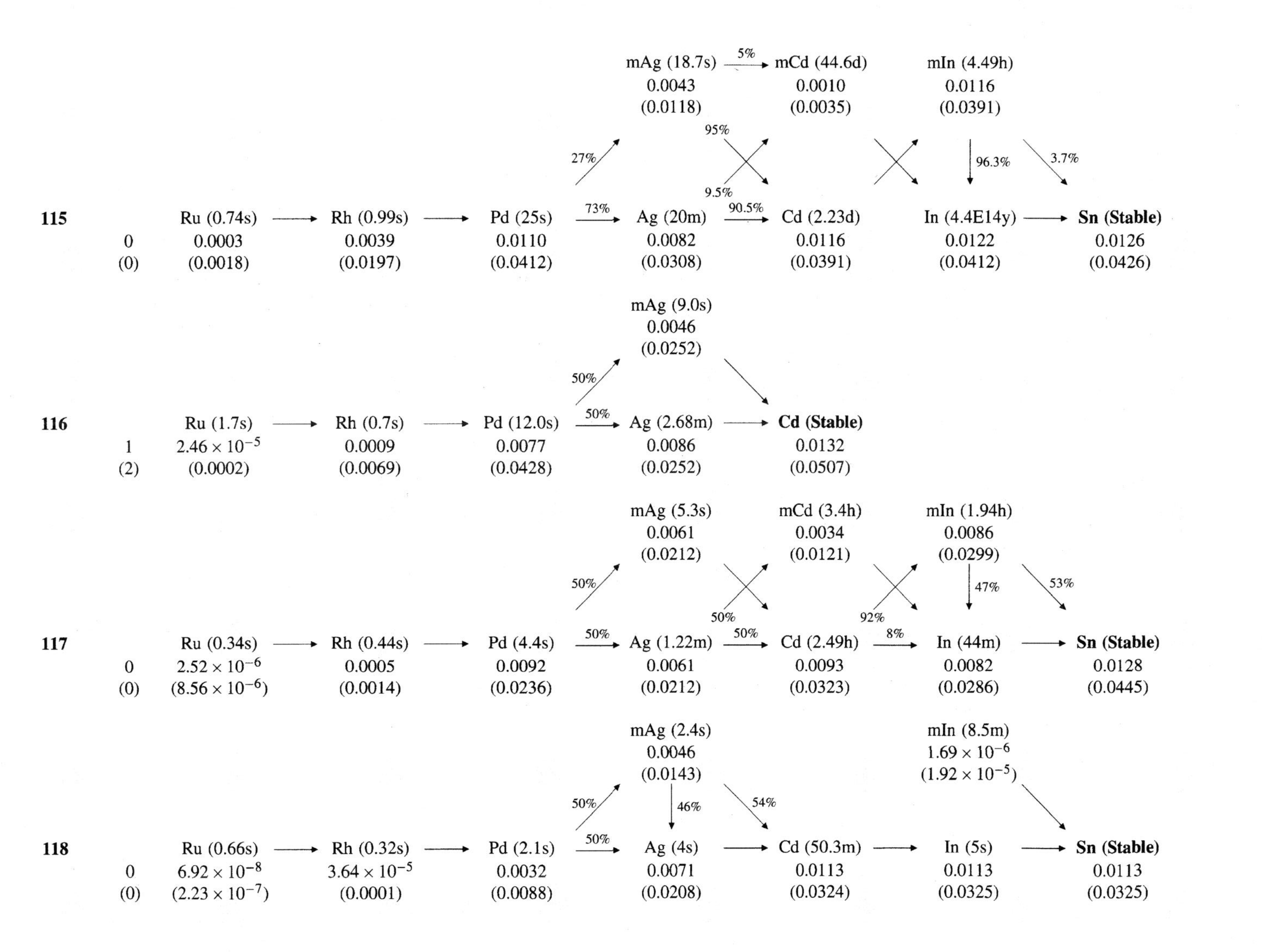

115
0
(0)
Ru (0.74s)
0.0003
(0.0018)
Rh (0.99s)
0.0039
(0.0197)
Pd (25s)
0.0110
(0.0412)
27%
73%
mAg (18.7s)
0.0043
(0.0118)
Ag (20m)
0.0082
(0.0308)
5%
95%
9.5%
90.5%
mCd (44.6d)
0.0010
(0.0035)
Cd (2.23d)
0.0116
(0.0391)
mIn (4.49h)
0.0116
(0.0391)
96.3%
3.7%
In (4.4E14y)
0.0122
(0.0412)
Sn (Stable)
0.0126
(0.0426)
116
1
(2)
Ru (1.7s)
2.46×10^{-5}
(0.0002)
Rh (0.7s)
0.0009
(0.0069)
Pd (12.0s)
0.0077
(0.0428)
50%
50%
mAg (9.0s)
0.0046
(0.0252)
Ag (2.68m)
0.0086
(0.0252)
Cd (Stable)
0.0132
(0.0507)
117
0
(0)
Ru (0.34s)
2.52×10^{-6}
(8.56×10^{-6})
Rh (0.44s)
0.0005
(0.0014)
Pd (4.4s)
0.0092
(0.0236)
50%
50%
mAg (5.3s)
0.0061
(0.0212)
Ag (1.22m)
0.0061
(0.0212)
50%
50%
mCd (3.4h)
0.0034
(0.0121)
Cd (2.49h)
0.0093
(0.0323)
92%
8%
mIn (1.94h)
0.0086
(0.0299)
47%
53%
In (44m)
0.0082
(0.0286)
Sn (Stable)
0.0128
(0.0445)
118
0
(0)
Ru (0.66s)
6.92×10^{-8}
(2.23×10^{-7})
Rh (0.32s)
3.64×10^{-5}
(0.0001)
Pd (2.1s)
0.0032
(0.0088)
50%
50%
mAg (2.4s)
0.0046
(0.0143)
46%
54%
Ag (4s)
0.0071
(0.0208)
Cd (50.3m)
0.0113
(0.0324)
mIn (8.5m)
1.69×10^{-6}
(1.92×10^{-5})
In (5s)
0.0113
(0.0325)
Sn (Stable)
0.0113
(0.0325)

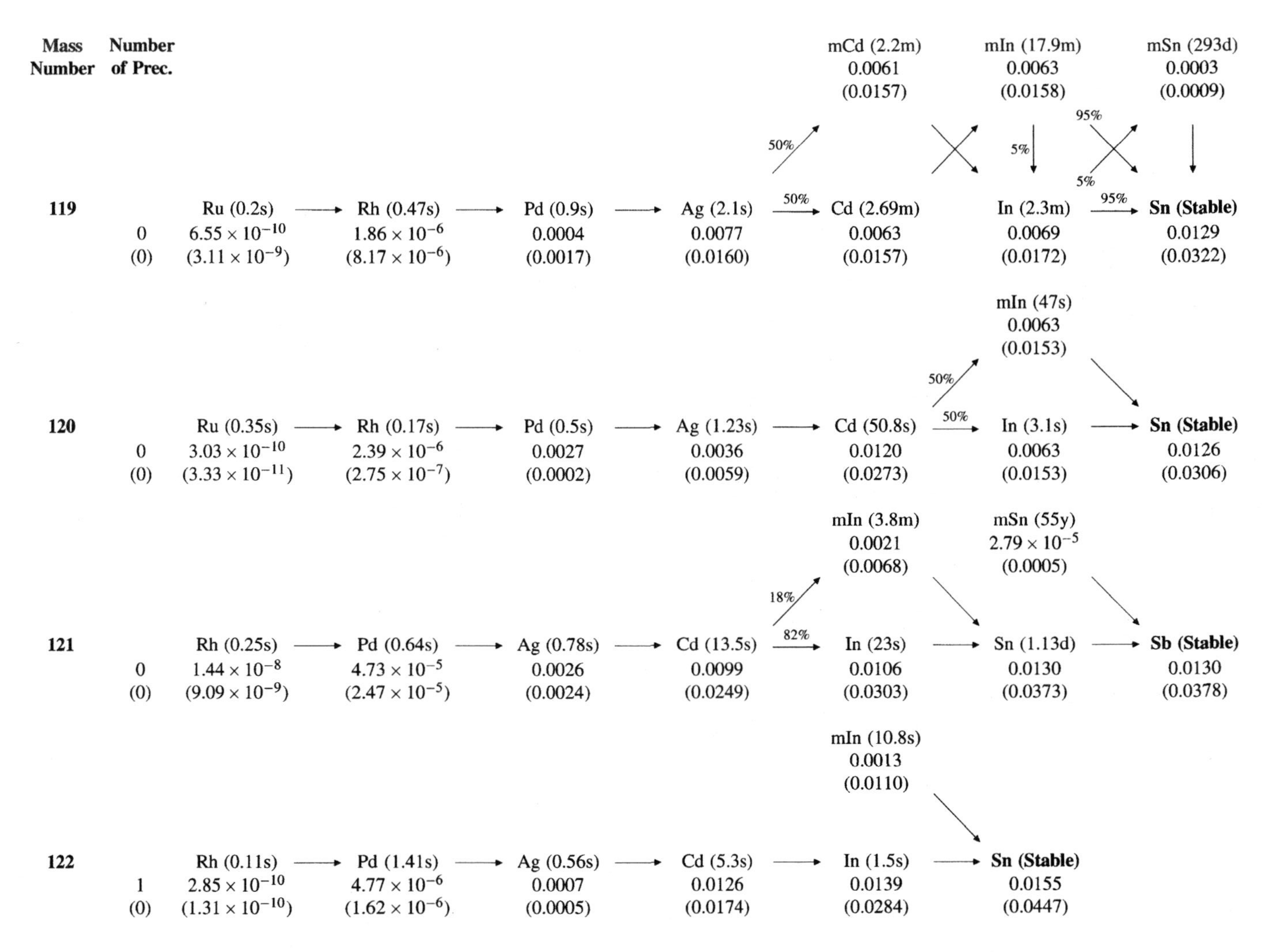
Mass Number
Number of Prec.
119
0
(0)
Ru (0.2s)
6.55 × 10^-10
(3.11 × 10^-9)
Rh (0.47s)
1.86 × 10^-6
(8.17 × 10^-6)
Pd (0.9s)
0.0004
(0.0017)
Ag (2.1s)
0.0077
(0.0160)
50%
50%
mCd (2.2m)
0.0061
(0.0157)
Cd (2.69m)
0.0063
(0.0157)
mIn (17.9m)
0.0063
(0.0158)
5%
In (2.3m)
0.0069
(0.0172)
95%
5%
95%
mSn (293d)
0.0003
(0.0009)
Sn (Stable)
0.0129
(0.0322)
120
0
(0)
Ru (0.35s)
3.03 × 10^-10
(3.33 × 10^-11)
Rh (0.17s)
2.39 × 10^-6
(2.75 × 10^-7)
Pd (0.5s)
0.0027
(0.0002)
Ag (1.23s)
0.0036
(0.0059)
Cd (50.8s)
0.0120
(0.0273)
50%
50%
mIn (47s)
0.0063
(0.0153)
In (3.1s)
0.0063
(0.0153)
Sn (Stable)
0.0126
(0.0306)
121
0
(0)
Rh (0.25s)
1.44 × 10^-8
(9.09 × 10^-9)
Pd (0.64s)
4.73 × 10^-5
(2.47 × 10^-5)
Ag (0.78s)
0.0026
(0.0024)
Cd (13.5s)
0.0099
(0.0249)
18%
82%
mIn (3.8m)
0.0021
(0.0068)
In (23s)
0.0106
(0.0303)
mSn (55y)
2.79 × 10^-5
(0.0005)
Sn (1.13d)
0.0130
(0.0373)
Sb (Stable)
0.0130
(0.0378)
122
1
(0)
Rh (0.11s)
2.85 × 10^-10
(1.31 × 10^-10)
Pd (1.41s)
4.77 × 10^-6
(1.62 × 10^-6)
Ag (0.56s)
0.0007
(0.0005)
Cd (5.3s)
0.0126
(0.0174)
mIn (10.8s)
0.0013
(0.0110)
In (1.5s)
0.0139
(0.0284)
Sn (Stable)
0.0155
(0.0447)

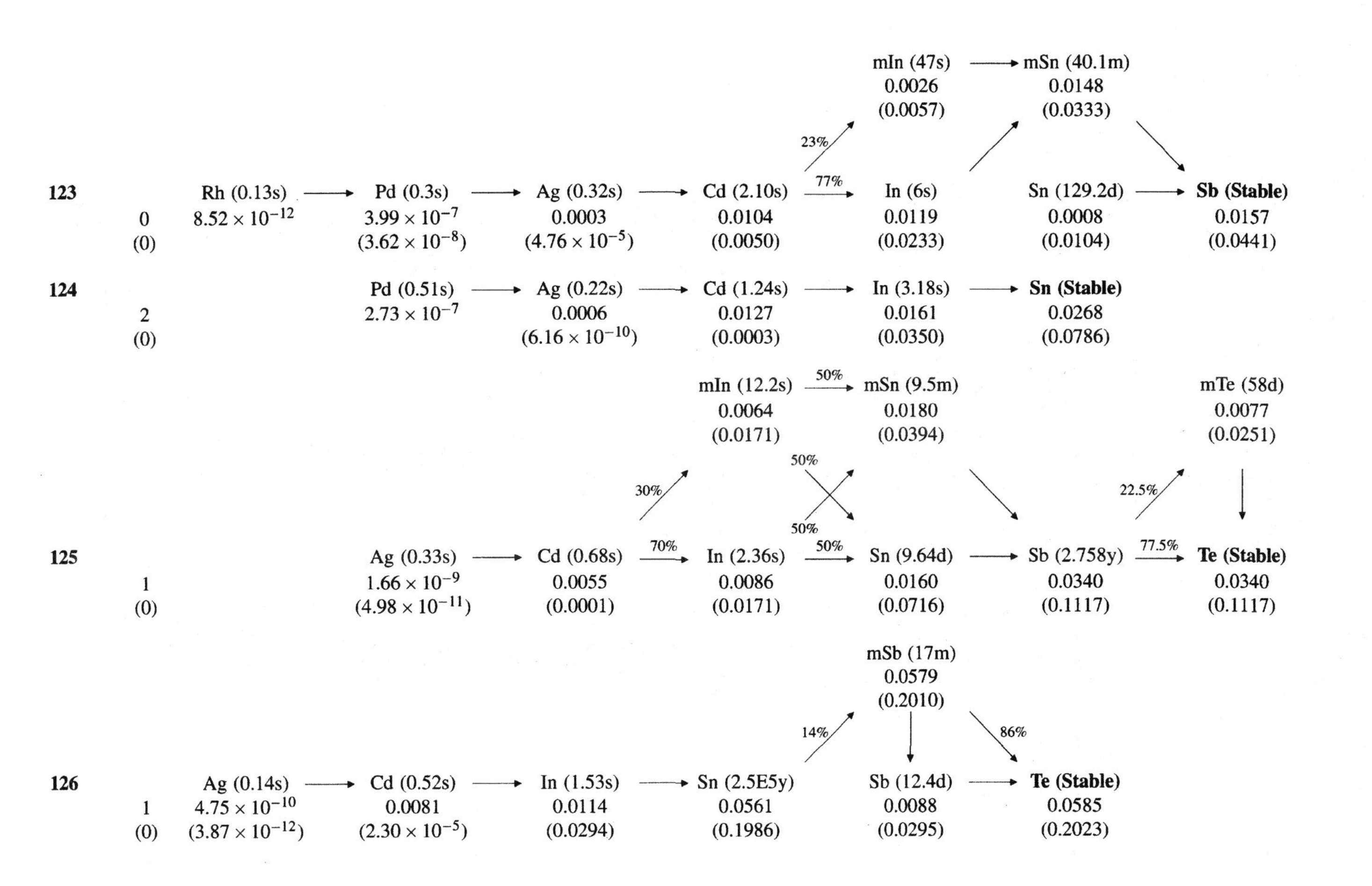

123
0
(0)
Rh (0.13s)
8.52 × 10−12
Pd (0.3s)
3.99 × 10−7
(3.62 × 10−8)
Ag (0.32s)
0.0003
(4.76 × 10−5)
Cd (2.10s)
0.0104
(0.0050)
23%
77%
mIn (47s)
0.0026
(0.0057)
In (6s)
0.0119
(0.0233)
mSn (40.1m)
0.0148
(0.0333)
Sn (129.2d)
0.0008
(0.0104)
Sb (Stable)
0.0157
(0.0441)
124
2
(0)
Pd (0.51s)
2.73 × 10−7
Ag (0.22s)
0.0006
(6.16 × 10−10)
Cd (1.24s)
0.0127
(0.0003)
In (3.18s)
0.0161
(0.0350)
Sn (Stable)
0.0268
(0.0786)
125
1
(0)
Ag (0.33s)
1.66 × 10−9
(4.98 × 10−11)
Cd (0.68s)
0.0055
(0.0001)
30%
70%
mIn (12.2s)
0.0064
(0.0171)
In (2.36s)
0.0086
(0.0171)
50%
50%
50%
50%
mSn (9.5m)
0.0180
(0.0394)
Sn (9.64d)
0.0160
(0.0716)
Sb (2.758y)
0.0340
(0.1117)
22.5%
77.5%
mTe (58d)
0.0077
(0.0251)
Te (Stable)
0.0340
(0.1117)
126
1
(0)
Ag (0.14s)
4.75 × 10−10
(3.87 × 10−12)
Cd (0.52s)
0.0081
(2.30 × 10−5)
In (1.53s)
0.0114
(0.0294)
Sn (2.5E5y)
0.0561
(0.1986)
14%
mSb (17m)
0.0579
(0.2010)
86%
Sb (12.4d)
0.0088
(0.0295)
Te (Stable)
0.0585
(0.2023)

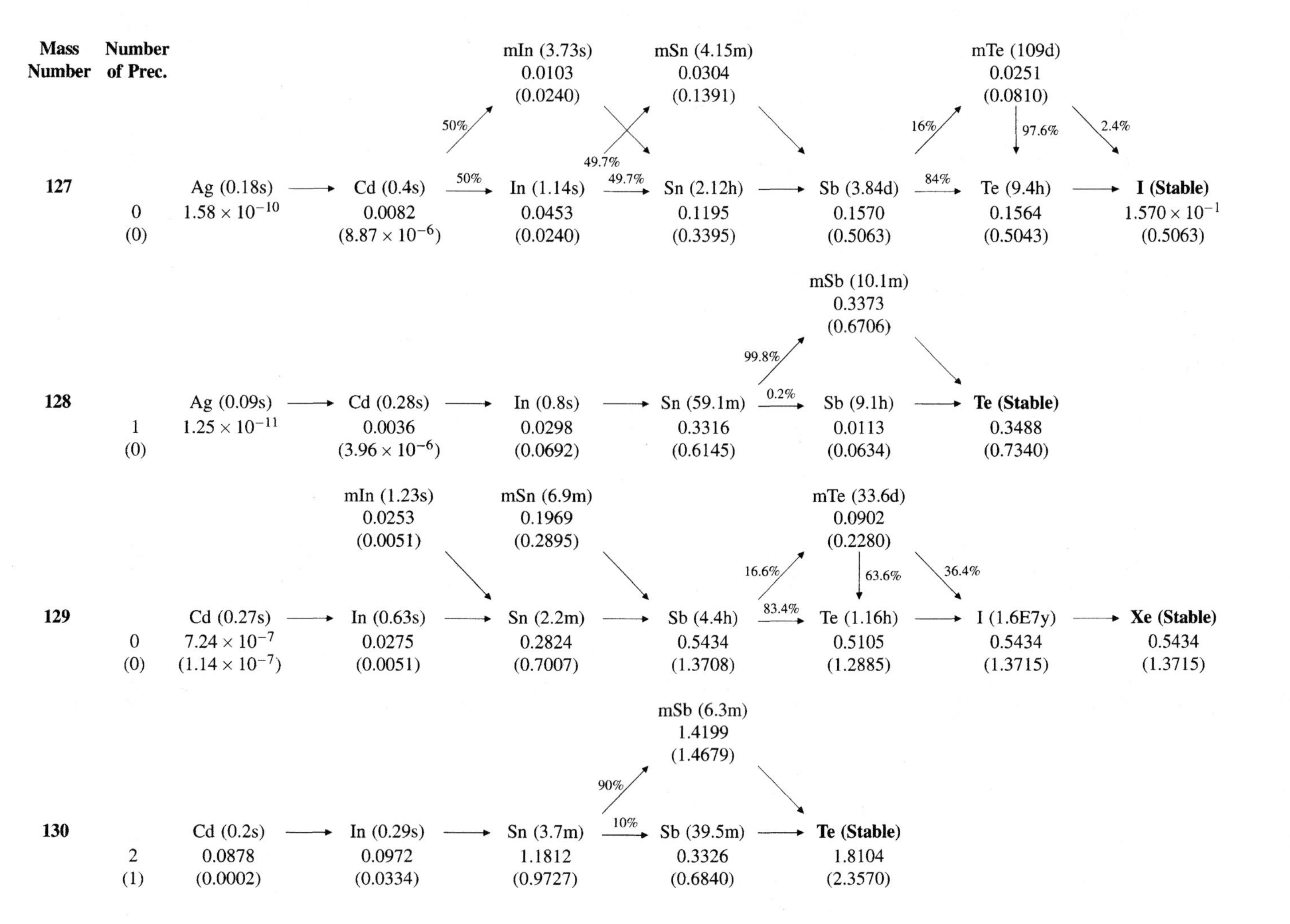

Mass Number
Number of Prec.
127
0
(0)
Ag (0.18s)
1.58 × 10^-10
Cd (0.4s)
0.0082
(8.87 × 10^-6)
50%
50%
mIn (3.73s)
0.0103
(0.0240)
In (1.14s)
0.0453
(0.0240)
49.7%
49.7%
mSn (4.15m)
0.0304
(0.1391)
Sn (2.12h)
0.1195
(0.3395)
Sb (3.84d)
0.1570
(0.5063)
16%
84%
mTe (109d)
0.0251
(0.0810)
97.6%
2.4%
Te (9.4h)
0.1564
(0.5043)
I (Stable)
1.570 × 10^-1
(0.5063)
128
1
(0)
Ag (0.09s)
1.25 × 10^-11
Cd (0.28s)
0.0036
(3.96 × 10^-6)
In (0.8s)
0.0298
(0.0692)
Sn (59.1m)
0.3316
(0.6145)
99.8%
0.2%
mSb (10.1m)
0.3373
(0.6706)
Sb (9.1h)
0.0113
(0.0634)
Te (Stable)
0.3488
(0.7340)
129
0
(0)
Cd (0.27s)
7.24 × 10^-7
(1.14 × 10^-7)
mIn (1.23s)
0.0253
(0.0051)
In (0.63s)
0.0275
(0.0051)
mSn (6.9m)
0.1969
(0.2895)
Sn (2.2m)
0.2824
(0.7007)
Sb (4.4h)
0.5434
(1.3708)
16.6%
83.4%
mTe (33.6d)
0.0902
(0.2280)
63.6%
36.4%
Te (1.16h)
0.5105
(1.2885)
I (1.6E7y)
0.5434
(1.3715)
Xe (Stable)
0.5434
(1.3715)
130
2
(1)
Cd (0.2s)
0.0878
(0.0002)
In (0.29s)
0.0972
(0.0334)
Sn (3.7m)
1.1812
(0.9727)
90%
10%
mSb (6.3m)
1.4199
(1.4679)
Sb (39.5m)
0.3326
(0.6840)
Te (Stable)
1.8104
(2.3570)

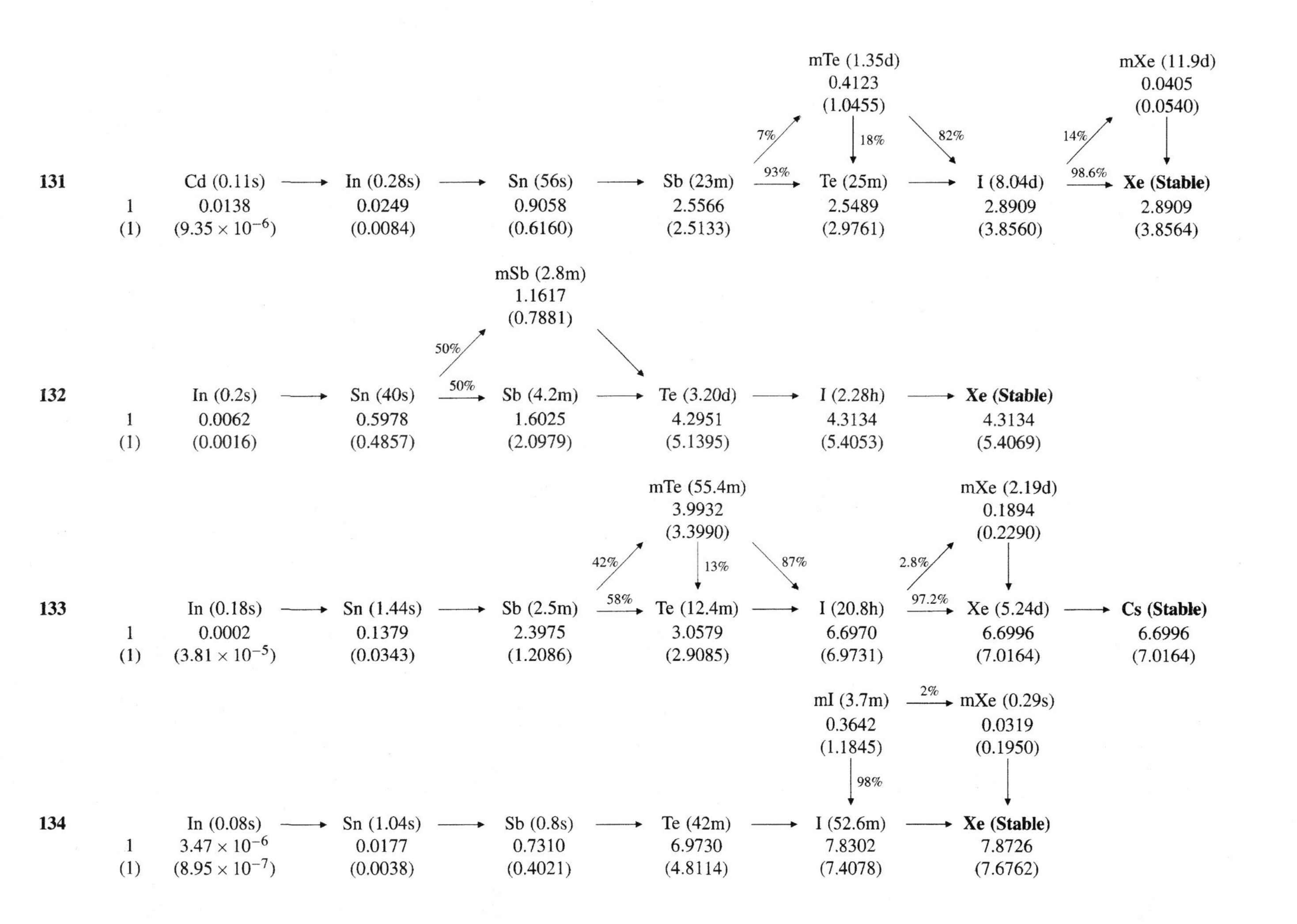

131
1
(1)
Cd (0.11s)
0.0138
(9.35 × 10⁻⁶)
In (0.28s)
0.0249
(0.0084)
Sn (56s)
0.9058
(0.6160)
Sb (23m)
2.5566
(2.5133)
7%
93%
mTe (1.35d)
0.4123
(1.0455)
18%
Te (25m)
2.5489
(2.9761)
82%
I (8.04d)
2.8909
(3.8560)
14%
98.6%
mXe (11.9d)
0.0405
(0.0540)
Xe (Stable)
2.8909
(3.8564)
132
1
(1)
In (0.2s)
0.0062
(0.0016)
Sn (40s)
0.5978
(0.4857)
50%
50%
mSb (2.8m)
1.1617
(0.7881)
Sb (4.2m)
1.6025
(2.0979)
Te (3.20d)
4.2951
(5.1395)
I (2.28h)
4.3134
(5.4053)
Xe (Stable)
4.3134
(5.4069)
133
1
(1)
In (0.18s)
0.0002
(3.81 × 10⁻⁵)
Sn (1.44s)
0.1379
(0.0343)
Sb (2.5m)
2.3975
(1.2086)
42%
58%
mTe (55.4m)
3.9932
(3.3990)
13%
Te (12.4m)
3.0579
(2.9085)
87%
I (20.8h)
6.6970
(6.9731)
2.8%
97.2%
mXe (2.19d)
0.1894
(0.2290)
Xe (5.24d)
6.6996
(7.0164)
Cs (Stable)
6.6996
(7.0164)
mI (3.7m)
0.3642
(1.1845)
2%
mXe (0.29s)
0.0319
(0.1950)
98%
134
1
(1)
In (0.08s)
3.47 × 10⁻⁶
(8.95 × 10⁻⁷)
Sn (1.04s)
0.0177
(0.0038)
Sb (0.8s)
0.7310
(0.4021)
Te (42m)
6.9730
(4.8114)
I (52.6m)
7.8302
(7.4078)
Xe (Stable)
7.8726
(7.6762)

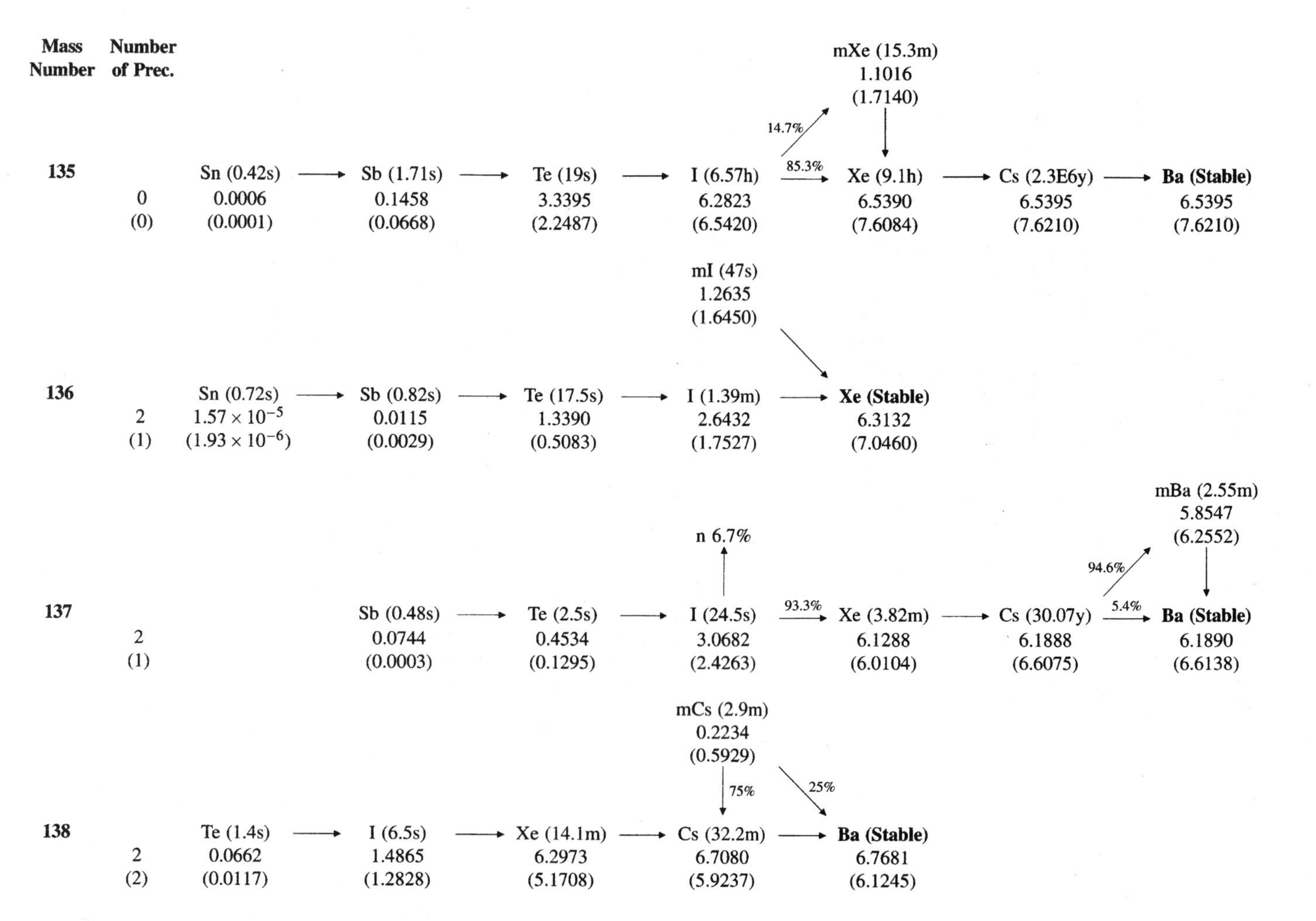

Mass Number
Number of Prec.
135
0
(0)
Sn (0.42s)
0.0006
(0.0001)
Sb (1.71s)
0.1458
(0.0668)
Te (19s)
3.3395
(2.2487)
I (6.57h)
6.2823
(6.5420)
14.7%
85.3%
mXe (15.3m)
1.1016
(1.7140)
Xe (9.1h)
6.5390
(7.6084)
Cs (2.3E6y)
6.5395
(7.6210)
Ba (Stable)
6.5395
(7.6210)
136
2
(1)
Sn (0.72s)
1.57 × 10^−5
(1.93 × 10^−6)
Sb (0.82s)
0.0115
(0.0029)
Te (17.5s)
1.3390
(0.5083)
I (1.39m)
2.6432
(1.7527)
mI (47s)
1.2635
(1.6450)
Xe (Stable)
6.3132
(7.0460)
137
2
(1)
Sb (0.48s)
0.0744
(0.0003)
Te (2.5s)
0.4534
(0.1295)
I (24.5s)
3.0682
(2.4263)
n 6.7%
93.3%
Xe (3.82m)
6.1288
(6.0104)
Cs (30.07y)
6.1888
(6.6075)
94.6%
5.4%
mBa (2.55m)
5.8547
(6.2552)
Ba (Stable)
6.1890
(6.6138)
138
2
(2)
Te (1.4s)
0.0662
(0.0117)
I (6.5s)
1.4865
(1.2828)
Xe (14.1m)
6.2973
(5.1708)
mCs (2.9m)
0.2234
(0.5929)
75%
25%
Cs (32.2m)
6.7080
(5.9237)
Ba (Stable)
6.7681
(6.1245)

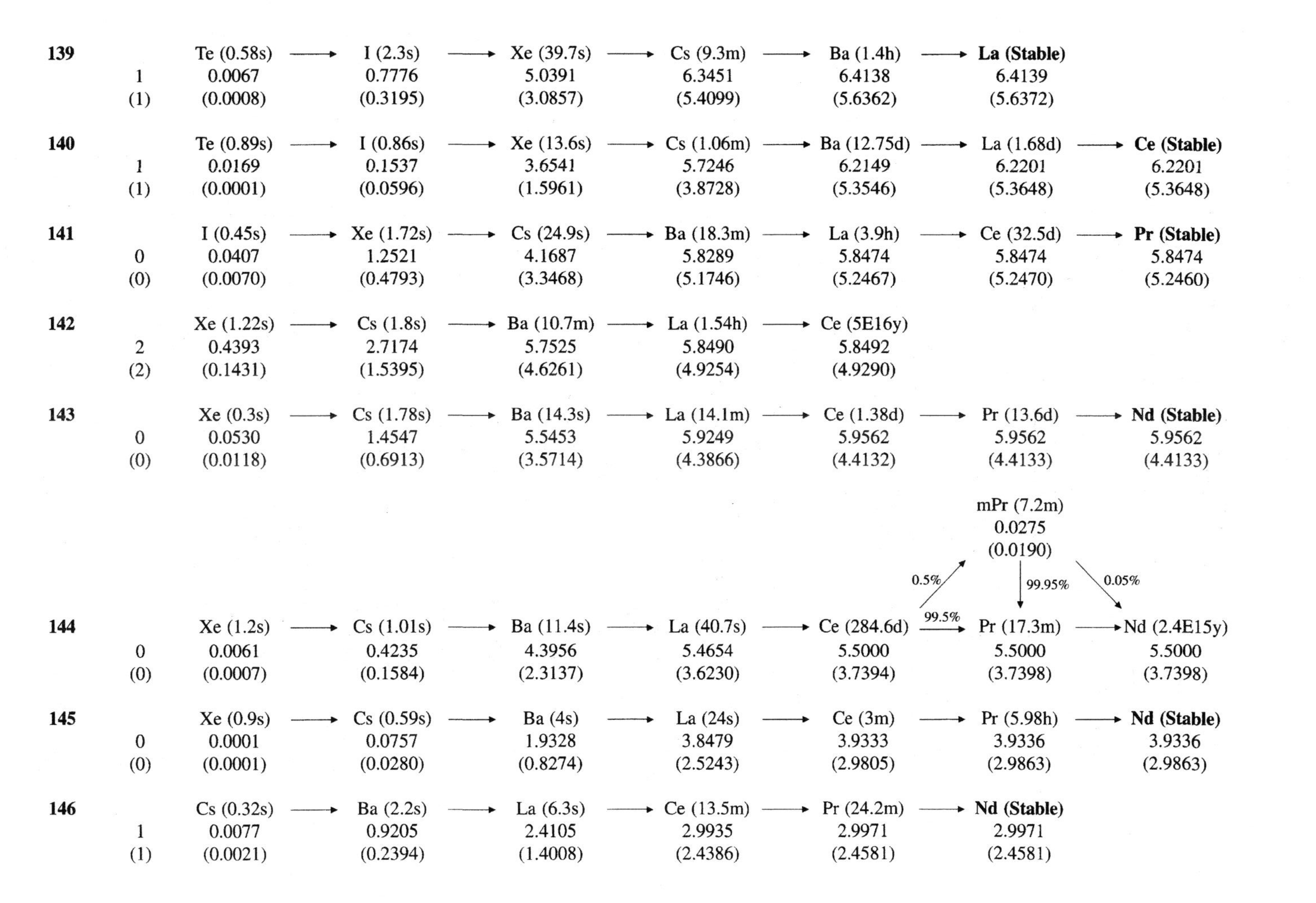
139
Te (0.58s) → I (2.3s) → Xe (39.7s) → Cs (9.3m) → Ba (1.4h) → La (Stable)
1 0.0067 0.7776 5.0391 6.3451 6.4138 6.4139
(1) (0.0008) (0.3195) (3.0857) (5.4099) (5.6362) (5.6372)
140
Te (0.89s) → I (0.86s) → Xe (13.6s) → Cs (1.06m) → Ba (12.75d) → La (1.68d) → Ce (Stable)
1 0.0169 0.1537 3.6541 5.7246 6.2149 6.2201 6.2201
(1) (0.0001) (0.0596) (1.5961) (3.8728) (5.3546) (5.3648) (5.3648)
141
I (0.45s) → Xe (1.72s) → Cs (24.9s) → Ba (18.3m) → La (3.9h) → Ce (32.5d) → Pr (Stable)
0 0.0407 1.2521 4.1687 5.8289 5.8474 5.8474 5.8474
(0) (0.0070) (0.4793) (3.3468) (5.1746) (5.2467) (5.2470) (5.2460)
142
Xe (1.22s) → Cs (1.8s) → Ba (10.7m) → La (1.54h) → Ce (5E16y)
2 0.4393 2.7174 5.7525 5.8490 5.8492
(2) (0.1431) (1.5395) (4.6261) (4.9254) (4.9290)
143
Xe (0.3s) → Cs (1.78s) → Ba (14.3s) → La (14.1m) → Ce (1.38d) → Pr (13.6d) → Nd (Stable)
0 0.0530 1.4547 5.5453 5.9249 5.9562 5.9562 5.9562
(0) (0.0118) (0.6913) (3.5714) (4.3866) (4.4132) (4.4133) (4.4133)
mPr (7.2m)
0.0275
(0.0190)
0.5%
99.95%
0.05%
99.5%
144
Xe (1.2s) → Cs (1.01s) → Ba (11.4s) → La (40.7s) → Ce (284.6d) → Pr (17.3m) → Nd (2.4E15y)
0 0.0061 0.4235 4.3956 5.4654 5.5000 5.5000 5.5000
(0) (0.0007) (0.1584) (2.3137) (3.6230) (3.7394) (3.7398) (3.7398)
145
Xe (0.9s) → Cs (0.59s) → Ba (4s) → La (24s) → Ce (3m) → Pr (5.98h) → Nd (Stable)
0 0.0001 0.0757 1.9328 3.8479 3.9333 3.9336 3.9336
(0) (0.0001) (0.0280) (0.8274) (2.5243) (2.9805) (2.9863) (2.9863)
146
Cs (0.32s) → Ba (2.2s) → La (6.3s) → Ce (13.5m) → Pr (24.2m) → Nd (Stable)
1 0.0077 0.9205 2.4105 2.9935 2.9971 2.9971
(1) (0.0021) (0.2394) (1.4008) (2.4386) (2.4581) (2.4581)

Mass Number	Number of Prec.													
147		Ba (0.89s)	→	La (4.02s)	→	Ce (56s)	→	Pr (13.4m)	→	Nd (11d)	→	Pm (2.62y)	→	Sm (1E11y)
	0	0.2459		0.8892		1.8872		2.2468		2.2469		2.2469		2.2469
	(0)	(0.0353)		(0.6459)		(1.8622)		(2.0014)		(2.0030)		(2.0030)		(2.0030)
148		Cs (0.15s)	→	Ba (0.64s)	→	La (1.1s)	→	Ce (56s)	→	Pr (2.27m)	→	**Nd (Stable)**		
	1	1.31×10^{-5}		0.0222		0.3585		1.5949		1.6727		1.6737		
	(1)	(1.51×10^{-6})		(0.0026)		(0.1197)		(1.0105)		(1.6284)		(1.6421)		
149		Ba (0.34s)	→	La (1.1s)	→	Ce (5.2s)	→	Pr (2.3m)	→	Nd (1.73h)	→	Pm (2.21d)	→	**Sm (Stable)**
	0	0.0010		0.0809		0.7781		1.0749		1.0817		1.0817		1.0817
	(0)	(0.0002)		(0.0366)		(0.5982)		(1.1665)		(1.2163)		(1.2166)		(1.2166)
150		Ba (0.96s)	89.1% →	La (0.61s)	99.6% →	Ce (4.4s)	→	Pr (6.2s)	→	**Nd (Stable)**				
	1	0.0001		0.0105		0.4022		0.6206		0.6533				
	(1)	(1.20×10^{-5})		(0.0052)		(0.2832)		(0.7926)		(0.9663)				
151		La (0.72s)	→	Ce (1s)	→	Pr (18.9s)	→	Nd (12.4m)	→	Pm (1.18d)	→	Sm (90y)	→	**Eu (Stable)**
	0	0.0010		0.1001		0.3386		0.4182		0.4188		0.4188		0.4188
	(0)	(0.0005)		(0.0636)		(0.4357)		(0.7276)		(0.7384)		(0.7384)		(0.7384)
												mPm (13.8m)	↘	
												0.0014		
												(0.0167)		
152		Ba (0.42s)	→	La (0.29s)	→	Ce (3.1s)	→	Pr (3.2s)	→	Nd (11.4m)	→	Pm (4.1m)	→	**Sm (Stable)**
	0	1.49×10^{-8}		4.53×10^{-5}		0.0206		0.1231		0.2641		0.2655		0.2669
	(0)	(5.36×10^{-9})		(2.41×10^{-5})		(0.0118)		(0.1723)		(0.5424)		(0.5592)		0.5763
153		La (0.33s)	→	Ce (1.47s)	→	Pr (4.3s)	→	Nd (28.9s)	→	Pm (5.4m)	→	Sm (1.93d)	→	**Eu (Stable)**
	0	1.42×10^{-6}		0.0017		0.0385		0.1494		0.1582		0.1583		0.1583
	(0)	(8.51×10^{-7})		(0.0011)		(0.0504)		(0.2908)		(0.3591)		(0.3613)		(0.3613)

						mPm (2.7m) ↘		
						0.0054		
						(0.0449)		
154		Ce (2.02s) →	Pr (2.3s) →	Nd (25.9s) →	Pm (1.7m) →	**Sm (Stable)**		
	1	0.0001	0.0051	0.0631	0.0686	0.0744		
	(1)	(0.0001)	(0.0098)	(0.1605)	(0.2054)	(0.2598)		
155		Ce (0.53s) →	Pr (1.12s) →	Nd (8.9s) →	Pm (42s) →	Sm (22.2m) →	Eu (4.75y) →	**Gd (Stable)**
	0	2.55×10^{-6}	0.0007	0.0182	0.0308	0.0321	0.0321	0.0321
	(0)	(3.46×10^{-6})	(0.0014)	(0.0522)	(0.1443)	(0.1655)	(0.1657)	(0.1657)
156		Ce (0.6s) →	Pr (0.38s) →	Nd (5.5s) →	Pm (26.7s) →	Sm (9.4h) →	Eu (15.2d) →	**Gd (Stable)**
	0	5.73×10^{-8}	4.07×10^{-5}	0.0047	0.0118	0.0148	0.0149	0.0149
	(0)	(1.23×10^{-7})	(0.0001)	(0.0161)	(0.0783)	(0.1228)	(0.1240)	(0.1240)
157		Ce (0.21s) →	Pr (0.38s) →	Nd (2.48s) →	Pm (10.9s) →	Sm (8m) →	Eu (15.2h) →	**Gd (Stable)**
	0	4.59×10^{-10}	1.65×10^{-6}	0.0005	0.0034	0.0061	0.0062	0.0062
	(0)	(1.79×10^{-9})	(7.19×10^{-6})	(0.0022)	(0.0296)	(0.0707)	(0.0741)	(0.0742)
158		Nd (2.69s) →	Pm (5.0s) →	Sm (5.3m) →	Eu (45.9m) →	**Gd (Stable)**		
	2	0.0001	0.0008	0.0032	0.0033	0.0033		
	(2)	(0.0002)	(0.0066)	(0.0359)	(0.0413)	(0.0414)		
159		Nd (0.64s) →	Pm (3s) →	Sm (11.4s) →	Eu (18.1m) →	Gd (18.5h) →	**Tb (Stable)**	
	1	1.77×10^{-6}	0.0001	0.0009	0.0010	0.0010	0.0010	
	(1)	(1.19×10^{-5})	(0.0012)	(0.0134)	(0.0202)	(0.0207)	(0.0207)	
160		Nd (0.79s) →	Pm (0.73s) →	Sm (9.6s) →	Eu (38s) →	**Gd (Stable)**		
	1	5.04×10^{-8}	7.60×10^{-6}	0.0002	0.0003	0.0003		
	(1)	(5.06×10^{-7})	(0.0001)	(0.0041)	(0.0087)	(0.0097)		
161		Nd (0.31s) →	Pm (0.79s) →	Sm (4.78s) →	Eu (26s) →	Gd (3.66m) →	Tb (6.91d) →	**Dy (Stable)**
	0	6.16×10^{-10}	4.41×10^{-7}	2.88×10^{-5}	0.0001	0.0001	0.0001	0.0001
	(0)	(1.13×10^{-8})	(1.04×10^{-5})	(0.0008)	(0.0036)	(0.0048)	(0.0048)	(0.0048)

Mass Number	Number of Prec.							
						mTb (2.23h) 3.61×10^{-7} (0.0001) (2% from Gd; → Dy)		
162	1 (1)	Pm (0.32s) → 6.80×10^{-9} (4.16×10^{-7})	Sm (5.26s) → 2.03×10^{-6} (0.0001)	Eu (11s) → 9.02×10^{-6} (0.0010)	Gd (8.4m) → 98% / 2% ↗ 1.58×10^{-5} (0.0022)	Tb (7.6m) → 1.55×10^{-5} (0.0022)	**Dy (Stable)** 1.59×10^{-5} (0.0022)	
163	1 (2)		Sm (1.27s) → 1.38×10^{-7} (7.96×10^{-6})	Eu (7.6s) → 2.08×10^{-6} (0.0002)	Gd (1.13m) → 5.89×10^{-6} (0.0008)	Tb (19.5m) → 6.11×10^{-6} (0.0009)	**Dy (Stable)** 6.11×10^{-6} (0.0009)	
164	1 (1)		Sm (1.39s) → 6.04×10^{-9} (4.17×10^{-7})	Eu (1.58s) → 2.07×10^{-7} (2.56×10^{-5})	Gd (45s) → 1.70×10^{-6} (0.0002)	Tb (3m) → 1.88×10^{-6} (0.0003)	**Dy (Stable)** 1.88×10^{-6} (0.0003)	
							mDy (1.26m) 4.69×10^{-7} (0.0001) (50% from Tb; 97.5% → Dy, 2.5% → Ho)	
165	0 (1)		Sm (0.45s) → 2.55×10^{-10} (1.40×10^{-8})	Eu (1.35s) → 3.90×10^{-8} (3.05×10^{-6})	Gd (0.71m) → 6.53×10^{-7} (0.0001)	Tb (2.1m) → 50% / 50% ↗ 9.34×10^{-7} (0.0001)	Dy (2.33h) → 9.40×10^{-7} (0.0001)	**Ho (Stable)** 9.51×10^{-7} (0.0001)

Sources: Data for fission product yields are from T. R. England, and B. F. Rider, *Evaluation and Compilation of Fission Product Yields*, LA-UR-94-3106, ENDF-349, Los Alamos National Laboratory, Los Alamos, NM, October 1994. Data for half-lives are from the chart of the nuclides, *Nuclides and Isotopes*, 15th ed., General Electric Company and KAPL, Inc., San Jose, CA, 1996. Data for branching ratios are calculated from fission yields.

APPENDIX F

NEUTRON-PARTICLE EMISSION CROSS-SECTION DATA

Cross Section versus Neutron Energy for Charged Particle Emission (CPE), (n, 2n), (n, n′), and (n, fission) Reactions for Selected Target Elements (10^{-24} cm^2 = 1 barn).

Source: ICRU, Neutron Dosimetry for Biology and Medicine, ICRU Report 26, International Commission on Radiation Units and Measurements, 7910 Woodmont Avenue, Bethesda, MD 20814.

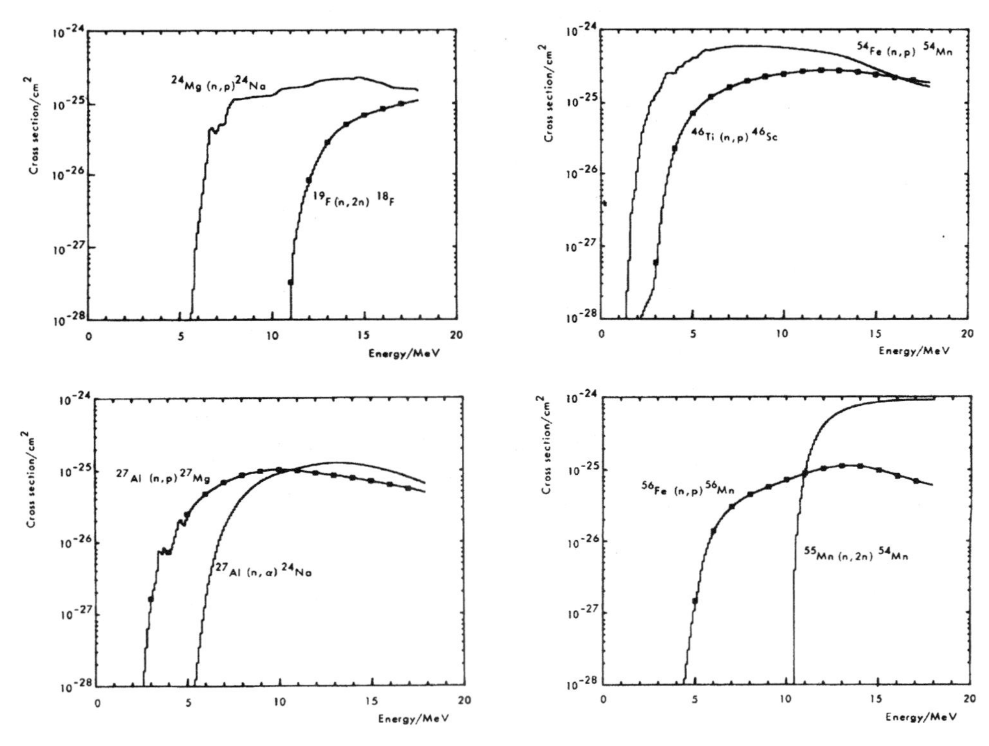

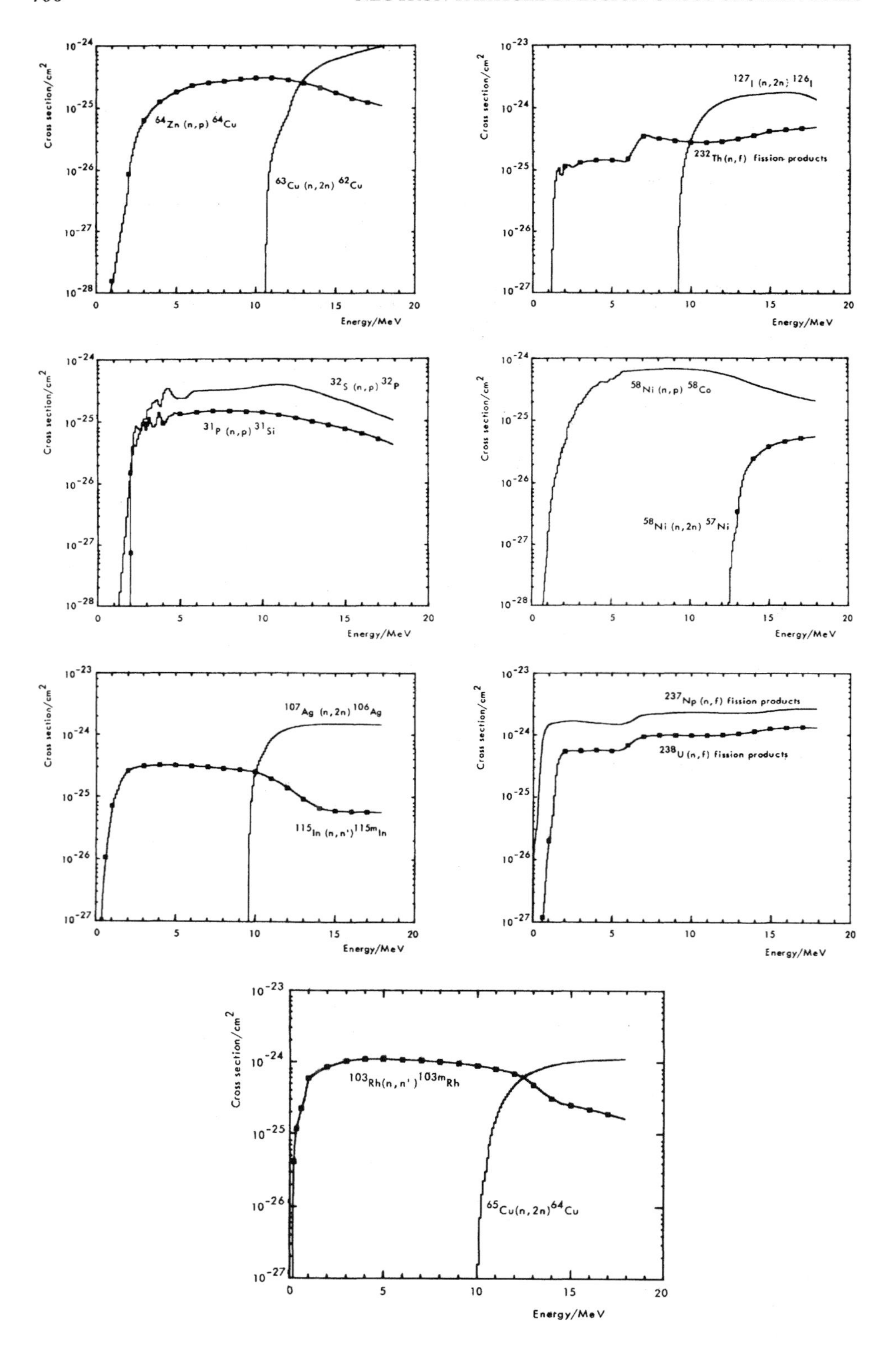

Cross section/cm2
Energy/MeV
64Zn (n,p) 64Cu
63Cu (n,2n) 62Cu
127I (n,2n) 126I
232Th (n,f) fission products
32S (n,p) 32P
31P (n,p) 31Si
58Ni (n,p) 58Co
58Ni (n,2n) 57Ni
107Ag (n,2n) 106Ag
115In (n,n')115mIn
237Np (n,f) fission products
238U (n,f) fission products
103Rh(n,n')103mRh
65Cu(n,2n)64Cu

ANSWERS TO SELECTED PROBLEMS

CHAPTER 1

1-1. **(a)** 6p, 8n; **(c)** 54p, 79n

1-3. 5.9737×10^{23} atoms

1-4. 3.34707×10^{-24} g

1-5. **(b)** 3.9 fermi

1-6. 6.507×10^{21} atoms

1-7. **(c)** 700 keV or 1.215×10^{-13} J

CHAPTER 2

2-2. **(b)** 6.45757×10^{-30} kg or $7.088m_0$

2-3. 625.52 MeV

2-4. 21.05 MeV

2-5. $0.566c$

2-6. $v = 0.866c$

2-7. $m = 30.354m_0$ of electron; $1.016m_0$ of proton

2-8. $v = 0.9957c$

2-9. **(a)** 37.6 MeV or 5.37 MeV/neutron; **(c)** 1,801.72 MeV or 7.7 MeV/neutron

CHAPTER 3

3-1. 2.254 eV

3-2. **(a)** 1.98 eV; **(b)** 1.98 eV; **(c)** 4.784×10^{14}/s

3-3. **(a)** 1.2264×10^{-9} m

3-4. **(a)** 2.0×10^{-7} m; **(b)** 5.32×10^{14}/s

3-5. 6.04962×10^{-34} J · s

3-6. **(b)** 2.18 eV; 5.27×10^{14}/s

3-7. 4.133×10^{-10} m

3-9. **(a)** 1.226×10^{-9} m; **(b)** 1.4×10^{-12} m (relativistic); **(c)** 1.798×10^{-12} m; **(d)** 1.59×10^{-38} m

3-11. 15,000 V

3-12. 6.077×10^{23} atoms/mole

CHAPTER 4

4-2. **(a)** 17.35 MeV; **(c)** 2.1257 MeV; **(e)** −0.2813 MeV; **(g)** 4.9658 MeV

4-3. **(a)** −1.853 MeV; **(b)** −3.005 MeV; **(c)** −2.454 MeV

4-4. **(b)** 2.575 MeV

4-5. ^{12}C + n; E = 10 + 4.9458 MeV

4-6. **(a)** 13.125, 18.7213, and 4.9463 MeV

4-7. **(a)** 8.691 MeV; **(b)** 15.9566 MeV; **(c)** 17.5337 MeV

4-9. 9.049 MeV

4-10. 8.665 MeV

4-11. 205 MeV

4-12. 14.1 MeV

4-13. **(b)** 34.7 MeV

4-14. 2.73 MeV

4-15. 0.1565 MeV

CHAPTER 5

5-1. 2403 Ci

5-3. 1.81 h

5-4. 209 y

5-5. **(a)** 2 half lives or 276 d; **(b)** 8.88×10^{-4} g

5-6. **(a)** 1.23×10^{6} Ci; **(b)** 7.3×10^{-11} g

5-9. 11.5 μCi ^{40}K

5-10. 304 mCi

5-11. **(a)** 532 mCi; **(b)** 14.93 d

5-13. **(b)** 0.637 of original

5-16. **(a)** 8.3 μg; **(b)** 1.33 μg of ^{3}H

5-17. 5.28 min

5-19. **(a)** 1.94 mCi; **(b)** 9.97 mCi

5-20. 9.98 mCi at 746 h

5-21. **(b)** 146.9 Ci at 5.6533 d

5-23. 1.28 mg

CHAPTER 6

6-1. $\sim$97%

6-2. 36.5 d

6-5. 1.794×10^{6} ergs

6-7. 3520 y

6-9. 2.6×10^{6} y

6-10. 16.3 y

6-12. 43 pCi/g

6-17. 0.096 WL

CHAPTER 7

7-1. **(a)** 6.5 MeV

7-3. 1.728 MeV, 1.479 MeV for e^{+}, 0.249 MeV e^{-}

7-5. 37.7 rads/hr (3 βs plus conversion electrons in 0.2 cm of tissue)

7-7. **(a)** 12 keV

7-9. **(a)** $\Gamma = 0.463$ R/hr Ci @ 1 m; **(b)** 72.7 mrads/hr in tissue

7-10. **(a)** $\Gamma = 0.602 + 0.649$ R/hr Ci @ 1 cm

7-11. **(a)** $\phi = 2.62\beta/\text{cm}^2$ s; $\phi_{\beta,E} = 183$ MeV/cm^2 s; $D = 97$ mrad/hr

7-13. **(a)** 1.77 rads/hr; **(b)** 0.996 rads/hr

CHAPTER 8

8-1. 42 μm (M = 12)

8-3. **(a)** $x = 0.42$ cm

8-4. 8.42 cm

8-7. 0.7176

8-9. 5.4 cm

8-10. 2.1 mR/hr

8-11. 11.6 cm including buildup

8-12. 0.367 mR/hr

8-13. $\sim$ 20 cm including buildup

8-16. 0.04 mR/hr

8-19. 32.3 R/hr

8-22. 12.5 μR/hr

CHAPTER 9

9-1. 3.58×10^9 t/s

9-2. 60.37 mCi

9-4. **(a)** 11,600 Ci ^{60m}Co; **(b)** 11,154 Ci ^{60}Co

9-5. 7,668 Ci

9-6. 1.74×10^{-9} g of Mn

9-8. 1.9×10^8 t/s

9-9. ^{203}Hg = 0.152 Ci

9-10. **(b)** 3.16×10^{10} t/s of ^{192}Ir per mg of iridium

9-11. 5.94×10^{10} t/s of ^{204}Tl minus target depletion

9-12. 298 MBq

CHAPTER 10

10-1. **(a)** 3.57%

10-2. **(a)** 4.34×10^5 Ci ^{133m}Xe; 6.74×10^6 Ci ^{133}Xe

10-3. 1.4×10^6 Ci

10-4. **(a)** 6.96×10^7 Ci; **(b)** $\sim$ 130 d, $A = 1.66 \times 10^8$ Ci; **(c)** 5.5×10^7 Ci

10-5. **(a)** 9.6 μCi

10-6. 1.4×10^6 Ci

10-7. **(a)** 1.09×10^8 Ci; **(b)** 1.05×10^8 Ci

10-8. **(a)** 1.54×10^4 Ci; **(b)** 1.54×10^4 Ci or the same

CHAPTER 11

11-4. 9.3×10^6 Ci (thermal fission)

11-6. 6.58×10^7 Ci

11-8. **(a)** 1.06×10^5 Ci; **(b)** 5574 + 1723 Ci

11-9. 800–2000 rads

11-10. 1.222 mCi (thermal fission)

CHAPTER 12

12-1. ce-K = 12 keV; ce-L = 35 keV

12-3. 0.796 MeV

12-4. 166.8 keV

12-6. **(a)** extrapolated range $\approx$ 760 mg/cm^2; **(b)** $\sim$1.7 MeV; **(c)** 1.68 MeV

12-10. 26 c/m

CHAPTER 13

13-1. **(a)** 100 ± 10 c or 50 ± 5 c/m; **(b)** 100 ± 19.6 c or 50 ± 9.8 c/m

13-3. **(a)** 200 ± 10 c/m; **(b)** 5.41 min (1.645σ at 90% confidence)

13-5. **(a)** 1955 – 2045 cts; **(b)** 1912 – 2088 cts.

13-7. **(a)** 342,860 c/m; **(b)** $2.143 \times 10^6 \pm 472$ d/m

13-9. 25 ± 3.75 c/m

13-11. $t_{s+b} = 43$ min; $t_b = 17$ min

13-19. 74.3% probability that sample contains activity above background; at 95% confidence it does not.

CHAPTER 14

14-1. 0.72 mrem/hr

14-3. 66.94 b

14-6. 0.273 mrad or 0.684 mrem (for $Q = 2.5$)

14-8. 3.6×10^5 n/cm$^2 \cdot$ s

14-10. 3.5 mrem/hr

14-11. 5×10^7 n/cm$^2 \cdot$ s (based on 1 gram target)

CHAPTER 15

15-2. Hall = 1250 mrem/wk @ 2 m; outside = 100 mrem/wk @ 2 m; **(b)** For hall $\approx$ 15 cm concrete; outside $\approx$ 3 cm additional (based on plate glass and 10 mrem/wk)

15-5. 500 mrem/wk @ 1 m

15-6. 9 cm (assumes plate glass attenuation = concrete and 10 mrem/wk)

Acknowledgments Richard Harvey, Arthur Ray Morton III, David Nestle, Daniel Hamilton, and Matthew Peck, all of whom have been graduate students in the University of Michigan Radiological Health Program, contributed answers to problems.

INDEX